Rock Mechanics & Strata Control

Rock Mechanics & Strata Control: Theory, Practice, and Application serves as a handbook that examines many of the fundamental and practical aspects of rock mechanics and strata control needed to help ensure safe and effective surface and underground mining. Clearly written and comprehensive in scope, this book includes numerous worked examples to elaborate on how to interpret and use the rock mechanics and support principles presented. It also includes fundamental coverage of major aspects of the topic that students and practitioners would find useful. This book is aimed primarily as a teaching and reference book for students in mining engineering and other associated disciplines, such as civil engineering, geotechnical and geological engineering, and geology. It will also be useful for practitioners working in the industry as a reference, showing numerous practical application examples.

This book:

- Focuses on rock mechanics and strata control from a green and sustainable point of view.
- Includes numerous examples and case studies showing how to apply concepts and formula.

Rock Mechanics & Strata Control

Theory, Practice, and Application

A.J.S. (Sam) Spearing and Todd R. Kostecki

CRC Press
Taylor & Francis Group
Boca Raton London New York

CRC Press is an imprint of the
Taylor & Francis Group, an **informa** business

Designed cover image: Shutterstock

First edition published 2026
by CRC Press
6000 Broken Sound Parkway NW, Suite 300, Boca Raton, FL 33487-2742

and by CRC Press
4 Park Square, Milton Park, Abingdon, Oxon, OX14 4RN

CRC Press is an imprint of Taylor & Francis Group, LLC

Library of Congress Cataloging-in-Publication Data
Names: Spearing, A. J. S. (Sam) author | Kostecki, T. R. author
Title: Rock mechanics & strata control : theory, practice, and application /
A.J.S. (Sam) Spearing and T.R. Kostecki.
Description: First edition. | Boca Raton, FL : CRC Press, 2026. |
Includes bibliographical references and index. |
Identifiers: LCCN 2025014619 | ISBN 9781032703787 hbk |
ISBN 9781032704890 pbk | ISBN 9781032704845 ebk
Subjects: LCSH: Rock mechanics—Handbooks, manuals, etc. | Ground control
(Mining)—Handbooks, manuals, etc. | Mine safety—Handbooks, manuals, etc.
Classification: LCC TN292 .S625 2026
LC record available at https://lccn.loc.gov/2025014619

ISBN: 978-1-032-70378-7 (hbk)
ISBN: 978-1-032-70489-0 (pbk)
ISBN: 978-1-032-70484-5 (ebk)

DOI: 10.1201/9781032704845

Typeset in Times
by codeMantra

Contents

About the Authors

Sam Spearing has held various senior mining and academic positions in Australia, China, and the United Sates related to the industry. He has been involved directly in the mining industry for many years and has undertaken mining and underground project work in over 40 countries. He therefore has a combination of direct mining, mining supply, and academic experience that gives him a broad understanding and perspective of the industry. He has written over 70 peer-reviewed journal papers, this is his fourth book that he has authored or co-authored, and he has written several chapters in other books. He has co-invented safety- and support-related products that have been successfully commercialized and most continue to be used internationally. He is involved in mine design, rock mechanics, safety and support with specialized experience in backfilling and shotcrete, and training. He has lived and worked on five continents.

Todd R. Kostecki has served in engineering roles in mine safety and ground control. Academically he has taught courses within the mining engineering discipline and holds degrees in civil and mining engineering including a PhD from Southern Illinois University Carbondale. He has co-authored about 20 publications on various topics related to mining, rock mechanics, and strata control.

1 The Basic Sciences

1.1 INTRODUCTION

Why is mining important?

To put it simply, "If it can't be grown, it must be mined." Note, original source unknown.

For example, mining provides:

- Metals (such as iron, nickel, lead, silicon, gold, copper, and platinum).
- Nonmetals (such as limestone, uranium, stone, and graphite).
- Phosphates and nitrates, which are used mainly for fertilizers.
- Coal, which was vital historically for power generation, is still commonly used for this purpose, and is still important to make steel.
- Rare Earths.

Mining is therefore essential to our life and will continue to be critical along with farming.

Most building materials, for example, are made from minerals and metals:

- Components such as insulation, roofing, nails, glass, pipes, electrical wires, concrete, and brick are made from minerals and metals.
- Modern businesses are very dependent on metals and minerals.
- Computers contain over 35 metals and minerals.

Civilization has been defined until recently by mining:

- Stone Age – pre-4000 BC.
- Bronze Age – 4000 BC to 1500 BC.
- Iron Age – 1500 BC to about 1800 AD.
- Steel Age – 1800 AD to about 1945.
- Nuclear Age – After 1945.

Computers and electronics have become an important part of our daily life in the digital and information age and are heavily dependent on metals and minerals for their production. Beyond necessities, the demand for electric vehicles and batteries represents a modern thrust of the mining industry in first-world countries. Globally, mining contributes heavily to the economy, either directly or indirectly, in regions where it is located, and it is a vital source of employment.

1.2 UNITS OF MEASURE

The two systems of measure are used globally:

- The Standard International (SI), or metric system.
- The United States Customary, or British, Imperial System. This system is officially only used in the US, Liberia, and Myanmar, but the UK and some Commonwealth countries still refer to it occasionally.

DOI: 10.1201/9781032704845-1

In some cases, tolerances are necessary. For further information, the *Journal of Petroleum Technology* (1977) is an example of an expanded reference for SI systems with some reference to Imperial units.

1.2.1 Basic Units

Table 1.1 shows the basic units for both systems.

1.2.2 Derived Units

Table 1.2 shows the derived units for both systems.

1.2.3 Powers and Prefixes

The power and prefixes used are given in Table 1.3.

TABLE 1.1
Basic Units for Standard International (SI) and Imperial Systems

Basic Units	SI Units		Imperial Units	
	Unit	Symbol	Unit	Symbol
Length, L	meter	m	foot	ft
Mass, m	kilogram	kg	pound	lb
Time, t	second	s	second	s
Temperature, T	Kelvin	K	Fahrenheit	°F

Note: In practice, degrees Centigrade (°C) is almost universally used in SI Units, rather than K.

TABLE 1.2
Derived Units for SI and Imperial Systems

Quantity	Standard International (SI)			Imperial		
	Unit	Symbol	Derivation	Unit	Symbol	Derivation
Force, F	Newton	N	kg m/s^2	pound force	lb_f	lb_m ft/s^2
Pressure, p	Pascal	Pa	N/m^2	pound force/L^2	psi	lb_f/in^2
Energy or Work, W	Joule	J	N m	foot pound force	ft lb_f	lb_m ft^2/s^2
Power, P	Watt	W	(N m)/s	horsepower	hp	ft lb_f/s
Frequency, f	Hertz	Hz	1 cycle/s	revolution/min	rpm	1 cycle/min

TABLE 1.3
Common Power and Prefixes for Basic Science and Engineering

Power	10^{12}	10^9	10^6	10^3	10^2	10^1	1	10^{-1}	10^{-2}	10^{-3}	10^{-6}	10^{-9}	10^{-12}
Prefix	Tera	Giga	Mega	kilo	hecto	deca	unity	deci	centi	milli	micro	nano	pico
Symbol	T	G	M	k	h	da		d	c	m	μ	n	p

1.3 BASIC CONCEPTS

1.3.1 Acceleration, Velocity, Distance, and Time

Acceleration (a) is the rate of change (Δ) of velocity (v) as a function of time (t). It is a vector, having magnitude and direction, in contrast to scalars that possess only magnitude, or tensors that have magnitude, direction, and act on a plane. If the acceleration is in the same direction as the velocity, then we take it as positive. Acceleration is expressed as follows:

$$a = \Delta v/t \tag{1.1}$$

According to Equation 1.1, if the rate change of velocity is zero, so is the acceleration. For most basic problems, this equation is applied when approximating the translation of a body along a straight line at constant, non-zero acceleration. Time is always positive.

Expression of the rate change velocity in terms of initial velocity (u), in some cases expressed as v_i, and final velocity (v), in some cases expressed as v_f, with a rearrangement of the terms in Equation 1.1, produces the following form:

$$v = u + at \tag{1.2}$$

Displacement (s), or distance, can be resolved with a known initial velocity, acceleration, and time:

$$s = ut + \tfrac{1}{2}\left(at^2\right) \tag{1.3}$$

Similarly, if the final velocity is known instead of the acceleration:

$$s = [(u + v)/2]t \tag{1.4}$$

Finally, in the cases where time is unknown:

$$v^2 = u^2 + 2as \tag{1.5}$$

Velocities are expressed in distance (s) per time (t) such as m/s or ft/s. Acceleration has the only unique units in distance per time squared, for example, m/s^2.

1.3.2 Density

Density (ρ) is a measure of mass (m) and volume (V) and is fundamentally an important measure to understand in basic rock mechanics. It is expressed as follows:

$$\rho = m/V \tag{1.6}$$

A common mistake is the use of weight, in place of mass, in Equation 1.6. In the SI system, density is typically expressed as g/cm^3, kg/m^3, or t/m^3. Mass and volume represent the numerator and denominator, respectively.

Specific gravity is the ratio of the density of a material to the density of water at a specified temperature and pressure, and it is dimensionless. The density of water is typically taken as $1{,}000\,kg/m^3$, which is density at 4°C. This makes the calculations of specific gravity very simple and convenient.

In some cases, the use of slugs, or slug-inch, known as a slinch, may be used to express density of the form $slugs/ft^3$ or $slinch/in^3$. The slug is a measure of mass accelerated at $1\,ft/s^2$ by 1 lb_f. Using Newton's Second Law of Motion, we can show that the slug is equivalent to 1 $lb_f\ s^2/ft$.

The slinch being the inch-associated measure of the slug deduces toward the conversion of 12 slugs equivalent to 1 slinch.

1.3.3 Mass

Mass is the quantity of matter, or material, in a body, or volume. Alternatively, it is a materials weight divided by the acceleration due to gravity. The SI unit of mass is the kilogram where 1,000 kg is equivalent to 1 tonne, or metric ton. The Imperial system uses the long ton, and short ton, equivalent to 2,240 lb_m and 2,000 lb_m, respectively. The variations of ton can also be expressed as a force.

A fundamental property of mass is the mutual attraction it experiences toward another mass in the vicinity. The attraction increases directly with the increasing size of the mass and decreases rapidly with increasing distance. This is covered by Newton's Law of Universal Gravitation.

Mass (m) and weight (mg) on earth differentiate by the acceleration due to gravity (g), which is 9.81 m/s² or 32.2 ft/s². Mass is a scalar having only a magnitude, whereas weight is a vector. Mass is therefore always a constant, whereas weight varies depending on the value of gravity.

1.3.4 Gravity

Gravity is the universal force of attraction acting between all bodies of matter. Gravity is therefore the mass attraction exerted by the earth and all other large bodies as defined in 1.3.3.

1.3.5 Force

A force (F) moves or tries to move an object from a stationary position or to change the rate or direction of movement of a steadily moving object. A unit of force is the Newton, which, acting on a mass of 1 kg, will give it an acceleration of 1 m/s². The force exerted by gravity on a mass of 1 kg is 9.81 N. Force has direction and magnitude, and is therefore a vector. The force of gravity is always directed precisely downward:

$$F = ma \tag{1.7}$$

The force acting on a ventilation door, for example, acts along the direction of the tunnel (the air flow direction). In this example, because the force (from the air flow) is perpendicular to the surface of the vent door, it is known as a normal force. A force acting parallel to a surface is known as a shear force. Any force can be a combination of a normal and a shear force, acting at 90° from each other and are referred to as the two components of the resultant force. A force acting at a distance from a point, or axis, of reference on a body is considered a moment. Much like forces, multiple moments can be resolved to a resultant. Forces can occur internally or externally on a body.

1.3.6 Stress

Stress is defined as force, or load, divided by the area (A) on which it acts. Stress is a tensor. For a system of forces to be in equilibrium, i.e., stationary, each action is balanced by an equal-and-opposite reaction. Normal stress is denoted by the small Greek letter sigma (σ). Shear stress is expressed by the Greek letter tau (τ). The expression for normal stress is as follows:

$$\sigma = F/A \tag{1.8}$$

Units are the Pascal, abbreviated as Pa, or N/m². The Imperial system normally expresses stress in lb/in², psi, or lb/ft², psf. In rock mechanics practice, stresses can be large, so it is usually appropriate to utilize the prefix Mega, i.e., Mega Pascal (MPa), which is equivalent to one million Pa.

The basic types of stress are as follows:

- Compressive where a normal force will squash, shorten, or compress the body.
- Tensile where a normal force will extend or stretch the body.
- Shear where a non-normal force will distort the body.

It is a common mistake in mining to assume compressive stresses as negative since most basic engineering courses treat it as such; however, if compressive stresses are universally understood to be the stress of interest, for example, in pillar design, then a sign may not accompany the stress magnitude, or it may be taken as positive.

1.3.7 Displacement

Displacement is the distance a body moves when subjected to a sufficient force. The unit of displacement is a meter or foot. In mining, generally limiting the displacement of the rock around an excavation is important for safety and it is usually more appropriate to use millimeters or inches.

1.3.8 Deformation

If a body is prevented from moving, a sufficient force will introduce deformation, which is most often a change in shape since no material is infinitely rigid. Deformation may not always be detectable by the human eye, and the nature of the deformation depends on the material properties of the body, type of force, force magnitude, and its boundary conditions. Translation, rotation, distortion (change of shape), or combination thereof is possible under the action of forces.

1.3.9 Strain

The deformation of a stressed body is strain. It is commonly identified by the Greek letter epsilon (ε) and is expressed as the change in length (ΔL) divided by the original length (L_o):

$$\varepsilon = \Delta L / L_o \tag{1.9}$$

For simple problems, compressive stress produces compressive strain (shortening) in the same direction, which, in rock mechanics, is typically accompanied with a positive (+), or no, sign. It is emphasized that strain is a unique measure having no units or units of m/m, mm/mm, in/in, and so on. Strain may also be shown as a percentage (%).

1.3.10 Elasticity

Elasticity explains the tendency of strain or deformation to disappear, or recover, when the stress or force which caused it is removed. In theory, elastic strain is recoverable, and the original shape and dimension of the body are restored when the stresses are removed. On the contrary, unrecoverable strain remains permanently when the stresses are removed and is termed non-elastic strain or, sometimes, plastic strain. Elasticity describes the nature of the strain not its amount. Simple problems are associated with theoretical elasticity, as the loads are minimal, and the behavior of materials is thoroughly tested, understood, and predictable. However, practical problems envelope the elastic range of materials and are almost always associated with some degree of plasticity, especially in mining.

1.3.11 Young's Modulus

Young's Modulus (E) is a material property that describes the relationship between stress and strain in elastic theory. The more an elastic material is stressed, or loaded, the greater will be its

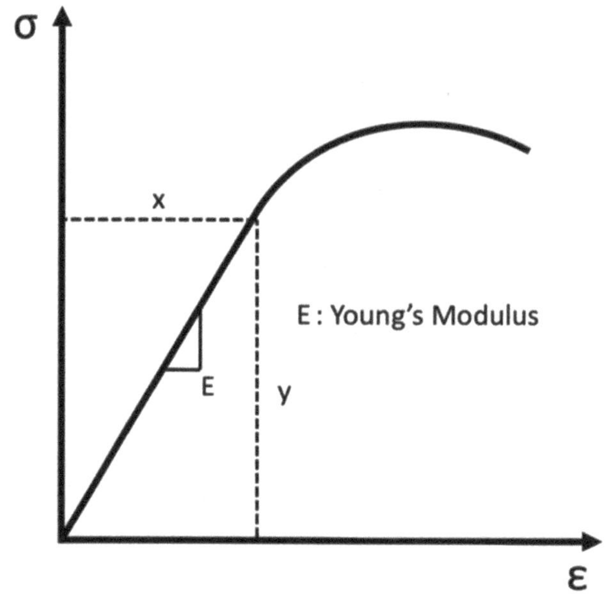

FIGURE 1.1 Young's Modulus from a stress–strain graph in the elastic portion of the test.

strain, or deformation. For the so-called linearly elastic material, E is a constant value. In practical circumstances, E has a range for natural materials such as rock.

If a body is tested under increasing measured stress, or load, the corresponding strain, or deformation, can also be measured and recorded. The change in these two quantities under load can be plotted on a graph that is known as a stress–strain plot as shown in Figure 1.1.

If the same body is stress unloaded and the corresponding strain recovers exactly along the loading path, then the material is considered perfectly elastic. For small strains, this is true of rock such as quartzite. If the stress–strain plot forms a straight line, the material is said to be linear elastic.

The slope of the line on the graph in Figure 1.1 is Young's Modulus. It is true up to the proportional limit:

$$E = \sigma / \varepsilon \quad (1.10)$$

For rock testing, the linear-elastic portion is generally small or ill-defined, and even more so for crude tests, therefore, the tangent, or the secant Young's Modulus is defined as shown in Figure 1.2. The secant modulus is the gradient of the line from the origin to a fixed percentage of the ultimate strength of the rock specimen. The tangent modulus is the tangent to the stress–strain curve, taken at a fixed percentage of the ultimate strength. The moduli can be significantly different, so it should be specified when they are quoted. The units mimic stress units and are typically MPa or, sometimes, GPa (1 Giga Pascal = 1,000 Mega Pascal). Note that the term stiffness may be used in the place of Young's Modulus, especially for stress versus strain plots.

1.3.12 Modulus of Rigidity

The Modulus of Rigidity, or Shear Modulus (G), is the coefficient of elasticity for a shearing force. It is defined as the ratio of shear stress (τ) to the displacement per unit sample length, i.e., shear strain (γ):

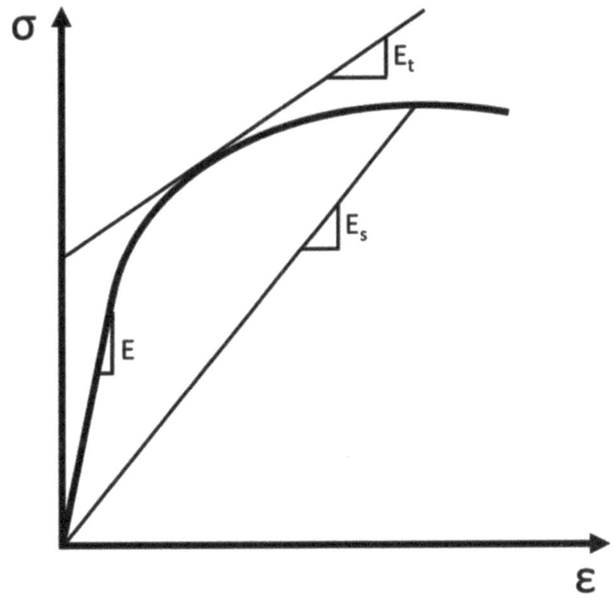

FIGURE 1.2 Tangent and secant moduli.

$$G = \tau/\gamma \tag{1.11}$$

In two dimensions, it is a similar concept of Young's Modulus but under pure shear loading as shown in Figure 1.3. Shear strain is expressed in radians as it is defined as the change in the right angle of an element under shear load. Clearly, if the stress is three dimensional, then the distortion of the object would result in an angular change in three dimensions. Units are similarly expressed in MPa or GPa. It is not unusual to confuse the Shear Modulus (G) with the Bulk Modulus (K) when first learning these coefficients. It should be understood that these are two different parameters.

If Young's Modulus (E) and Poisson's Ratio (ν) are known, then the Shear Modulus is resolved as follows:

$$G = E/2(1 + v) \tag{1.12}$$

1.3.13 Poisson's Ratio

In a conventional load-deformation test on an elastic material, the main deformation, or strain, is observed in the direction of the applied load, or stress. This is known as the axial, or longitudinal, strain. It is expressed with a Greek letter epsilon (ε) but with an associated subscript to indicate the direction; e.g., ε_z would relate to the strain in the z-direction.

In the perpendicular direction of load, a transverse strain will be observed in the opposite sense to that of the axial strain. Similarly, the transverse strain is also expressed with a Greek letter epsilon (ε) but with a different subscript to indicate its direction; e.g., ε_y would relate to the strain in the y-direction.

In a practical sense, if the material shortens in the axial direction (compression), it will expand in the transverse direction, and vice versa. In solid mechanics, this is considered the Poisson effect, and its measure can be written as Poisson's Ratio (ν):

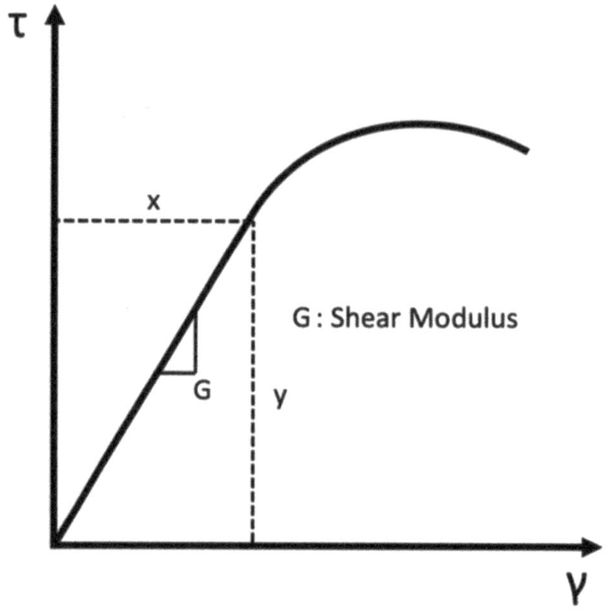

FIGURE 1.3 Modulus of Rigidity.

$$v = -\varepsilon_y / \varepsilon_z \tag{1.13}$$

Of course, the subscript can relate to any two directions, and it is shown in the form of Equation 1.13, as an example.

1.3.14 Energy or Work

When a force (F) is applied to move a body through a distance (d), work (W) is done which is equal to the product of the force and the distance (as shown in Figure 1.4):

$$W = Fd \tag{1.14}$$

Similarly, when a body is strained by an applied stress, work is also done to cause the deformation. If the strained body is elastic, energy is stored which can be recovered when the strain is allowed to relax by removal of the stresses. The amount of stored strain energy is proportional to the stress times the strain (similar to Figure 1.4), the area under the load-deformation curve. Thus, energy and work are two similar terms for the physical quantity. Associated material properties with these concepts are toughness and resilience. Though essential for understanding material behavior, they are often not used in engineering mining practice.

1.3.15 Friction

Friction is the resistance against the sliding of one body over another. The shear (sliding) force (F_s) necessary to overcome frictional resistance is proportional to the normal force (F_n). The normal force can be thought of as clamping surfaces together. The relationship is given by the coefficient of friction, represented as the Greek letter mu (μ). The magnitude of μ depends on the nature or smoothness of the surfaces. A subscript is sometimes associated with μ to denote static (s) or kinetic (k) friction. If the two bodies are ice, for example, then μ will be closer to zero, whereas, if the two bodies are rock and asphalt, μ will be greater.

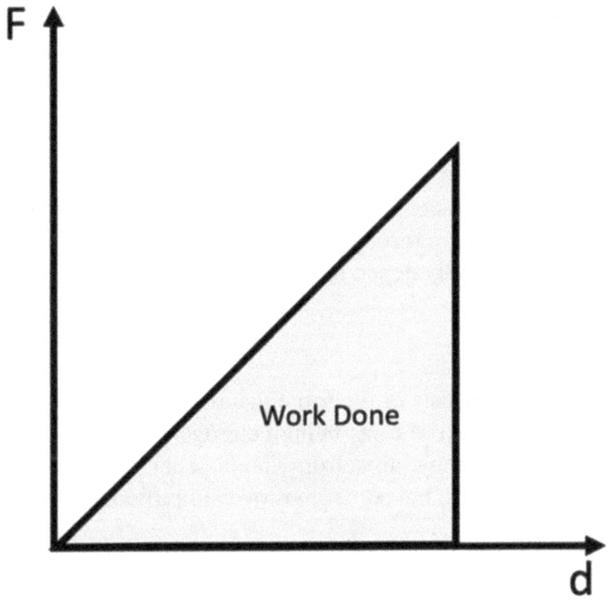

FIGURE 1.4 Example of work done.

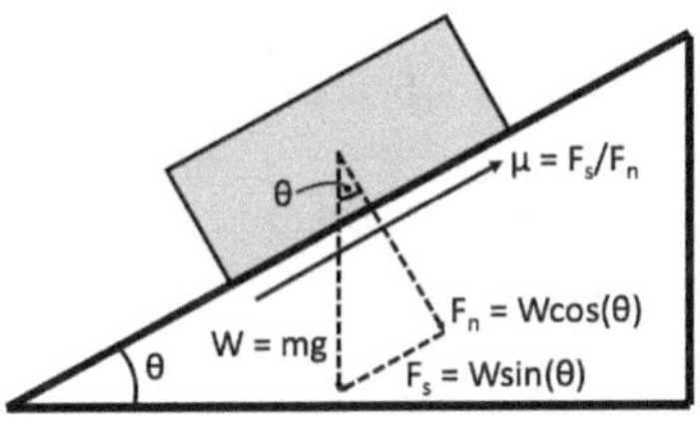

FIGURE 1.5 Block on an inclined plane showing the effect of friction to resist sliding.

A simple form of this concept is in Figure 1.5 for a rectangular body on an inclined plane at some angle theta (θ) to the horizontal. The weight of the body (W) can be determined by multiplying the mass of the body (m) times its acceleration due to gravity (g). In the scenario in this figure, the weight of the body is always taken in the direction in which gravity acts, i.e., toward the level plane, and forms the hypotenuse of a right-angled triangle. The force is directed through the centroid of the body. The normal force, which acts normal to the plane on which the body rests, occurs at the same angle theta (θ) as the inclined plane. The shear force acts parallel to the inclined plane. Once the weight of the block and the angle of the plane are known, the normal and shear forces can be easily resolved as well as the coefficient of friction:

$$\text{Cosine}(\theta) = \text{adjacent/hypotenuse} = F_n / W \tag{1.15}$$

$$\text{Sine}(\theta) = \text{opposite/hypotenuse} = F_s / W \tag{1.16}$$

$$\mu = \text{Tangent}(\theta) = \text{Sine}(\theta)/\text{Cosine}(\theta) = F_s / F_n \tag{1.17}$$

In practical problems, the angle of internal friction, or friction angle, represented by the Greek letter phi (ϕ), is the measure of a material, such as rock, to withstand shear stress. Another frictional property of practical importance is the angle of repose (θ_R) which is best found by field measurement. In the case of loose rock in a dump, it physically represents the angle the material will take before sliding failure (the maximum angle of repose).

1.3.16 Isotropy

A material is isotropic when its physical properties, such as Young's Modulus or Poisson's Ratio, velocity of sound, or light propagation through it, are constant in all directions within the material. If material properties are directionally dependent, then it is referred to as anisotropic.

1.3.17 Homogeneity

A body is homogenous when it consists of the same material throughout. If it does not contain the same material throughout, it is referred to as being heterogenous. Figure 1.6 shows diagrammatically the differences between isotropy, anisotropy, homogeneity, and heterogeneity. Problems in practice, stated or not, are simplified, but sometimes oversimplified, by assuming homogenous and isotropic conditions.

1.3.18 Hooke's Law

Hooke's Law states that the amount of deformation of an elastic object is proportional to the force (F) in Newtons, applied to deform it (x):

$$F = kx \tag{1.18}$$

For Equation 1.18, k is a force constant, or stiffness, in N/m.

1.3.19 Plasticity

A body undergoes plastic deformation when there is permanent unrecoverable strain when stressed. A certain minimum stress, known as the yield stress (σ_y), must be reached before unrecoverable extension, or contraction, occurs. Basic elastic, or linear, compared to plastic, or non-linear, behavior is best understood graphically as in Figure 1.7.

An ideal viscous material has a specific relationship between stress, strain, and time. For this case, the cause from the applied stress and the strain effect can be linear or non-linear.

For a linear viscous material,

$$\sigma = \acute{\eta}\, d\varepsilon/dt \tag{1.19}$$

where the Greek letter eta ($\acute{\eta}$) is a constant for the linear viscous material under consideration. The latter portion of Equation 1.19 represents change in strain (ε) with respect to time (t).

For a non-linear viscous material,

$$\sigma = \mu (d\varepsilon/dt)^n \tag{1.20}$$

where the Greek letter mu (μ) and n are material characteristic constants. Note that the exponent (n) included with the change in strain with respect to time ($d\varepsilon/dt$) mathematically signals non-linearity. Unlike elastic behavior, viscous flow is non-recoverable, and therefore a permanent deformation

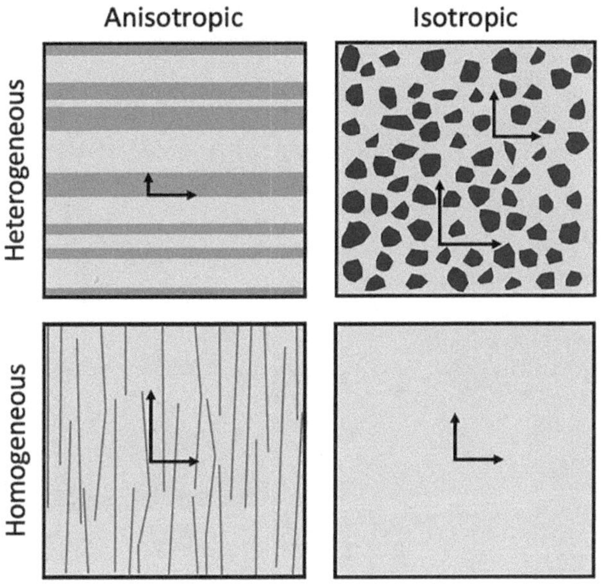

FIGURE 1.6 Diagram showing different materials (Hall, 2015).

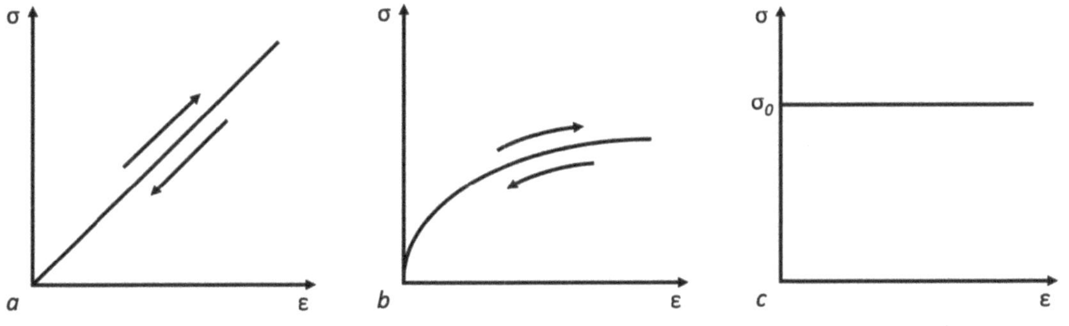

FIGURE 1.7 Material idealized behavior: (a) linear, (b) non-linear, and (c) perfectly plastic.

takes place. Idealized plastic behavior – often more theoretical and simplified than actual – is shown in Figure 1.7.

1.4 WORKED EXAMPLES

In any calculation, it is essential to always use consistent units, an appropriate number of significant figures (usually dictated by the parameters given) and to include the units.

Example 1.4.1

If an object is initially not moving but after 2 seconds its velocity is 8 m/s, what is its acceleration?

Solution:

$$v = u + at$$

$$8\ \text{m/s} = 0 + a(2\ \text{s})$$

$$a = 4\ \text{m/s}^2$$

Example 1.4.2

If an object is moving at 12 m/s and 4 seconds later it is stationary, what is the deceleration?

Solution:

$$v = u + at$$

$$0 = 12\ \text{m/s} + (a)(4\ \text{s})$$

$$(a)(4\ \text{s}) = -12\ \text{m/s}$$

$$a = -3\ \text{m/s}^2$$

Note that the negative (–) sign implies deceleration.

Example 1.4.3

If a rock falls 6 m from the mine roof, what is its velocity?

Solution:

$$v^2 = u^2 + 2as$$

$$v^2 = 0^2 + (2)(9.81\ \text{m/s}^2)(6\ \text{m})$$

$$v^2 = 117.7\ \text{m}^2/\text{s}^2$$

$$v = 10.8\ \text{m/s}$$

Example 1.4.4

Determine the time it takes a mine cart to travel 200 ft at a constant velocity of 5 ft/s?

Solution:

$$s = [(u + v)/2]t$$

$$200\ \text{ft} = [(5 + 5\ \text{ft/s})/2]t$$

$$t = 40\ \text{s}$$

Example 1.4.5

A mine skip moves for approximately 10 seconds between two mine shaft levels. The velocity as it reaches the deeper (second) level is 50 ft/s. The skip was accelerating at 0.5 ft/s^2. Show the distance traveled using Equations 1.3–1.5.

Solution:

$$v = u + at$$

$$50\ \text{ft/s} = u + \left(0.5\ \text{ft/s}^2\right)\left(10\ \text{s}\right)$$

$$u = 45\ \text{ft/s}$$

Equation 1.3:

$$s = ut + ½\left(at^2\right)$$

$$s = \left(45\ \text{ft/s}\right)\left(10\ \text{s}\right) + ½\left[\left(0.5\ \text{ft/s}^2\right)\left(10\ \text{s}\right)^2\right]$$

$$s = 450 + 25\ \text{ft}$$

$$s = 475\ \text{ft}$$

Equation 1.4:

$$s = \left[(u + v)/2\right]t$$

$$s = \left[\left(45 + 50\ \text{ft/s}\right)/2\right]\left(10\ \text{s}\right)$$

$$s = 475\ \text{ft}$$

Equation 1.5:

$$v^2 = u^2 + 2as$$

$$\left(50\ \text{ft/s}\right)^2 = \left(45\ \text{ft/s}\right)^2 + 2\left(0.5\ \text{ft/s}^2\right)s$$

$$2{,}500\ \text{ft}^2/\text{s}^2 = 2{,}025\ \text{ft}^2/\text{s}^2 + \left(1\ \text{ft/s}^2\right)s$$

$$s = 475\ \text{ft}$$

Example 1.4.6

If the density of aluminum is 2.75 g/cm^3, what volume does 3.0 kg occupy?

Solution:

$$\rho = m/V$$

$$2.75\ \text{g/cm}^3 = 3.0\ \text{kg}/V$$

$$V = (3{,}000\ g)/(2.75\ g/cm^3)$$

$$V = 1{,}091\ cm^3$$

Example 1.4.7

A block of wood with dimensions of 20 cm by 10 cm by 15 cm weighs 1.8 kg. What is its density?

Solution:

$$V = (20\ cm)(10\ cm)(15\ cm) = 3{,}000\ cm^3$$

$$\rho = m/V$$

$$\rho = 1{,}800\ g/3{,}000\ cm^3$$

$$\rho = 0.60\ g/cm^3$$

Example 1.4.8

A rock has a density of 2.75 t/m^3 with dimensions of 2.0 m by 1.0 m by 2.5 m. What is its mass?

Solution:

$$V = (2.0\ m)(1.0\ m)(2.5\ m)$$

$$V = 5.0\ m^3$$

$$\rho = m/V$$

$$2.75\ t/m^3 = m/5.0\ m^3$$

$$m = 13.75\ t$$

Example 1.4.9

The mass and volume for three blocks are provided in Figure 1.8. Determine the density of each block, and then find the total density (ρ_T) of the three blocks.

Solution:

Block 1:

$$\rho = m/V$$

$$\rho = 200\ kg/0.2\ m^3$$

$$\rho = 1{,}000\ kg/m^3$$

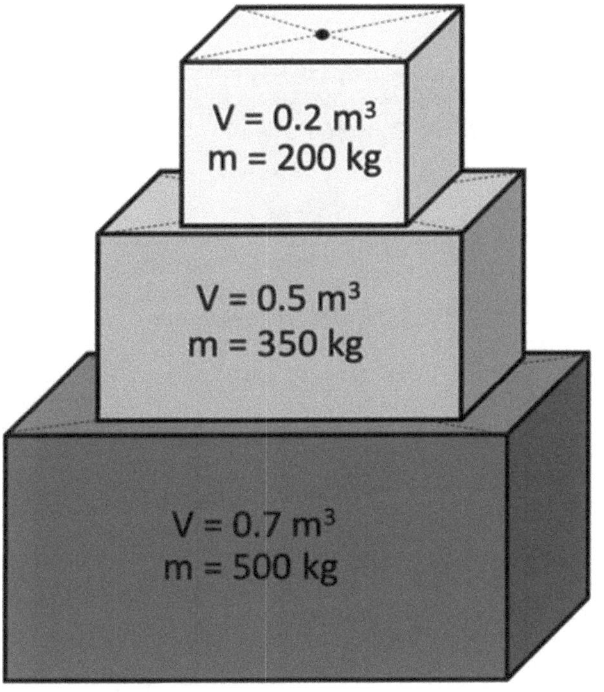

FIGURE 1.8 Mass-volume example with three blocks.

Block 2:

$$\rho = m/V$$

$$\rho = 350\ \mathrm{kg}/0.5\ \mathrm{m}^3$$

$$\rho = 700\ \mathrm{kg/m}^3$$

Block 3:

$$\rho = m/V$$

$$\rho = 500\ \mathrm{kg}/0.7\ \mathrm{m}^3$$

$$\rho = 714\ \mathrm{kg/m}^3$$

Total

$$\rho_T = m_T/V_T$$

$$\rho_T = (200 + 350 + 500\ \mathrm{kg})/(0.2 + 0.5 + 0.7)$$

$$\rho_T = 750\ \mathrm{kg/m}^3$$

Example 1.4.10

Relative density (RD), or the specific gravity (SG), is the ratio of a material density ($\rho_{material}$) to a reference density ($\rho_{reference}$). What is the specific gravity of a mineral that has a density of 9.0 g/cm^3 relative to water (use ρ_{water}=1 g/cm^3). Convert the density of the mineral to lb/ft^3.

Solution:

$$RD_{mineral} = \rho_{mineral} / \rho_{water}$$

$$RD_{mineral} = \left(9.0 \text{ g/cm}^3\right) / \left(1 \text{ g/cm}^3\right)$$

$$RD_{mineral} = 9.0$$

$$\rho_{mineral} = \left(RD_{mineral}\right)\left(\rho_{water}\right)$$

$$\rho_{mineral} = (9.0)\left(62.4 \text{ lb/ft}^3\right)$$

$$\rho_{mineral} = 562 \text{ lb/ft}^3$$

Example 1.4.11

A force of 90 N gives an object an acceleration of 6.5 m/s^2, what is the mass of the object?

Solution:

$$F = ma$$

$$m = \left(90 \text{ N}\right) / (6.5 \text{ m/s}^2)$$

$$m = 13.8 \text{ kg}$$

Example 1.4.12

Find the weight of a 300 g object on earth. Also, find the weight of a 3 lb object on earth.

Solution:

$$W = mg$$

$$W = \left(300 \text{ g}\right)\left(1 \text{ kg}/1{,}000 \text{ g}\right)\left(9.81 \text{ m/s}^2\right)$$

$$W = 2.94 \text{ N}$$

For the latter part of the question, no calculation is needed to determine the weight of a 3 lb object on earth, and it is 3 lb.

Example 1.4.13

A 250 kg mine cart is pulled by a cable with a horizontal force of 600 N. Ignoring friction, how long does it take for the cart to reach a speed of 8 m/s? How far will it have traveled?

Solution:

$$F = ma$$

$$a = (600\ \mathrm{N})/(250\ \mathrm{kg})$$

$$a = 2.4\ \mathrm{m/s^2}$$

$$v = u + at$$

$$8\ \mathrm{m/s} = 0\ \mathrm{m/s} + (2.4\ \mathrm{m/s^2})(t)$$

$$t = 3.3\ \mathrm{s}$$

$$s = ut + \tfrac{1}{2}\, at^2$$

$$s = (0\ \mathrm{m/s})(t) + (\tfrac{1}{2})(2.4\ \mathrm{m/s^2})\left[(3.3\ \mathrm{s})^2\right]$$

$$s = 13.3\ \mathrm{m}$$

Example 1.4.14

A 6 kg mass hangs at the end of a cable. Find the tension (T) in the cable if the acceleration on the mass is 1.8 m/s² down. What is T if the acceleration is 1.8 m/s² up? What is T if acceleration is 9.81 m/s² up?

Solution:

$$W = mg$$

$$W = (6\ \mathrm{kg})(9.81\ \mathrm{m/s^2})$$

$$W = 58.9\ \mathrm{N}$$

Account for the tension (T) in the cable by substituting for F (since they are both forces) in F = ma, and including the appropriate sign and weight (W) of the object.

$$T = W \pm F$$

$$T = W \pm ma$$

$$T = 58.9\ \mathrm{N} + (6\ \mathrm{kg})(1.8\ \mathrm{m/s^2})$$

$$T = 58.9\ \mathrm{N} + 10.8\ \mathrm{N}$$

$$T = 69.7\ \text{N}$$

$$T = W \pm ma$$

$$T = 58.9\ \text{N} - 10.8\ \text{N}$$

$$T = 48.1\ \text{N}$$

$$T = W \pm ma$$

$$T = 58.9\ \text{N} - 58.9\ \text{N}$$

$$T = 0\ \text{N}$$

Example 1.4.15

Show that 1 slug is equal to 1 $lb_f\, s^2/ft$ on earth using the conversion of 1 slug equal to 32.2 lb_m.

Given:

$$1\ \text{slug} = 32.2\ \text{lb}_\text{m}$$

$$a = 32.2\ \text{ft/s}^2$$

Solution:

$$F = ma$$

$$1\ \text{lb}_\text{f} = \left(1\ \text{lb}_\text{m}\right)\left(32.2\ \text{ft/s}^2\right)$$

$$1\ \text{lb}_\text{f} = 32.2\ \text{lb}_\text{m}\text{ft/s}^2$$

$$1\ \text{lb}_\text{f} = \left(1\ \text{slug}/32.2\ \text{lb}_\text{m}\right)\left(32.2\ \text{lb}_\text{m}\text{ft/s}^2\right)$$

Rearrangement of terms:

$$1\ \text{lb}_\text{f}\text{s}^2/\text{ft} = 1\ \text{slug}$$

Example 1.4.16

A tank contains a pool of mercury and water. The mercury is 0.3 m deep, and the water is 1.2 m deep. Find the pressure (stress) exerted on the bottom of the tank. Note: $RD_{mercury} = 13.6$ and $RD_{water} = 1.0$.

Solution:

Find the density of the water and mercury:

$$RD_{mercury} = \rho_{mercury}/\rho_{water}$$

$$\rho_{mercury} = (13.6)\left(1{,}000\ \text{kg/m}^3\right)$$

$$\rho_{mercury} = 13{,}600\ kg/m^3$$

$$RD_{water} = \rho_{water}/\rho_{water}$$

$$\rho_{water} = (1.0)(1{,}000\ kg/m^3)$$

$$\rho_{water} = 1{,}000\ kg/m^3$$

To find the stress, the units must be of the form force/area:

$$\sigma_{mercury} = \rho_{mercury} g h_{mercury}$$

$$\sigma_{mercury} = (13{,}600\ kg/m^3)(9.81\ m/s^2)(0.3\ m)$$

$$\sigma_{mercury} = 40{,}024\ N/m^2 = 40.0\ kPa$$

$$\sigma_{water} = \rho_{water} g h_{water}$$

$$\sigma_{mercury} = (1{,}000\ kg/m^3)(9.81\ m/s^2)(1.2\ m)$$

$$\sigma_{mercury} = 11{,}772\ N/m^2 = 11.8\ kPa$$

$$\sigma_{total} = \sigma_{water} + \sigma_{mercury}$$

$$\sigma_{total} = 11.8\ kPa + 40.0\ kPa$$

$$\sigma_{total} = 51.8\ kPa$$

Example 1.4.17

A metal wire is 2.5 mm in diameter and 2.0 m long. A force of 12 N acts at the end of the wire along the axis of its length. The wire expands 0.3 mm due to the force. Assuming elastic behavior, find the stress in MPa, the strain in m/m, mm/mm, με, and as a percent, and finally, calculate Young's Modulus, E.

Solution:

Stress (σ)

$$\sigma = F/A$$

$$Area_{circle} = \pi d^2/4$$

$$\sigma = 12\ N/\left\{\left[3.14(2.5\ mm)^2\right]/4\right\}$$

$$\sigma = (12\ N)/(4.91\ mm^2)$$

$$\sigma = (2.44\ N/mm^2)(1{,}000\ mm/1\ m)^2$$

$$\sigma = 2.44\ MPa$$

Strain (ε)

$$\varepsilon = \Delta L / L_o$$

$$\varepsilon = (2.0003\ \text{m} - 2.0000\ \text{m})/(2.0000\ \text{m})$$

$$\varepsilon = 0.00015\ \text{m/m}$$

$$\varepsilon = 0.00015\ \text{mm/mm}$$

$$\varepsilon = 150\mu\varepsilon$$

$$\varepsilon = 0.015\%$$

Young's Modulus (E)

$$E = \sigma/\varepsilon$$

$$E = (2.44\ \text{MPa})/(0.00015\ \text{mm/mm})$$

$$E = 16{,}267\ \text{MPa}$$

Example 1.4.18

Find the force needed to yield a 6 mm diameter rod if its tensile yield strength is 70 MPa.

Solution:

$$\sigma = F/A$$

$$\sigma_{yield} = F_{yield}/A$$

$$F_{yield} = \sigma_{yield}\left(\pi d^2/4\right)$$

$$F_{yield} = (70\ \text{MPa})\left\{\left[3.14(6\ \text{mm})^2\right]/4\right\}$$

$$F_{yield} = \left(70\ \text{N/mm}^2\right)\left(28.3\ \text{mm}^2\right)$$

$$F_{yield} = 1{,}978\ \text{N}$$

Example 1.4.19

A steel wire is 0.95 m long and stretched by 0.03 m. Determine the tensile strain (ε_t). The steel wire is then compressed by 0.03 m. Determine the compressive strain (ε_c).

Solution:

$$\varepsilon_t = \Delta L / L_o$$

$$\varepsilon_t = (L_f - L_o)/L_o$$

$$\varepsilon_t = (0.98\ \text{m} - 0.95\ \text{mm})/(0.95\ \text{m})$$

$$\varepsilon_t = 0.032\ \text{m/m}$$

$$\varepsilon_c = \Delta L / L_o$$

$$\varepsilon_c = (L_f - L_o)/L_o$$

$$\varepsilon_c = (0.92\ \text{m} - 0.95\ \text{m})/(0.95\ \text{m})$$

$$\varepsilon_c = -0.032\ \text{m/m}$$

Example 1.4.20

Calculate the shear strain at the origin for the square element shown in Figure 1.9.

Solution:

$$\text{Tan}(\alpha_{dx}) = 5\ \text{mm}/10\ \text{mm}$$

$$\alpha_{dx} = \text{Tan}^{-1}(5\ \text{mm}/10\ \text{mm})$$

$$\alpha_{dx} = 26.6°(\pi/180°)$$

$$\alpha_{dx} = 0.463\ \text{radians}$$

$$\text{Tan}(\alpha_{dy}) = 4\ \text{mm}/10\ \text{mm}$$

$$\alpha_{dy} = \text{Tan}^{-1}(4\ \text{mm}/10\ \text{mm})$$

$$\alpha_{dy} = 21.8°(\pi/180°)$$

$$\alpha_{dy} = 0.380\ \text{radians}$$

$$\gamma_{xy} = \alpha_{dx} + \alpha_{dy}$$

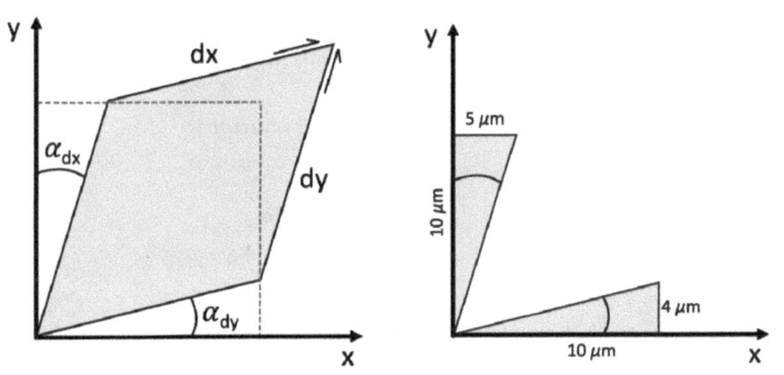

FIGURE 1.9 Simple shear-strain example.

$$\gamma_{xy} = 0.463 \text{ radians} + 0.380 \text{ radians}$$

$$\gamma_{xy} = 0.843 \text{ radians}$$

Example 1.4.21

A copper wire is 2.0 m long with a diameter of 2 mm. If Young's Modulus (E) is 117.6 GPa, what tension will cause an elongation of 1 mm?

Solution:

$$E = \sigma/\varepsilon$$

$$E = (F/A)/(\Delta L/L_o)$$

$$E = (F/\pi r^2)/[(L_f - L_o)/L_o]$$

$$117.6\times 10^9 \text{Pa} = \{F/[\pi(0.001 \text{ m})^2]\}/[(2.001 - 2.000)/(2.000)]$$

$$117.6\times 10^9 \text{Pa} = \{F/[\pi(10^{-6} \text{m}^2)]\}/(0.0005)$$

$$F = 184.6 \text{ N}$$

Example 1.4.22

A steel bar has a cross-sectional area of 100 mm² and a gauge length of 50 mm. The stress gradient of the elastic section is 410 kN/mm. Find Young's Modulus.

Solution:

$$E = \sigma/\varepsilon$$

The equation becomes

$$E = (F/A)(L_o/\Delta L)$$

$$E = (F/\Delta L)(L_o)/A$$

$$E = [(410{,}000 \text{ N/mm})(50 \text{ mm})]/(100 \text{ mm}^2)$$

$$E = (205{,}000 \text{ N/mm}^2)(1{,}000 \text{ mm/1 m})^2$$

$$E = 205 \text{ GPa}$$

Example 1.4.23

If a metal rod of length 1 m stretches to 1,020 mm under load and its diameter decreases from 20 to 19.9 mm, find Poisson's Ratio.

Solution:

$$\nu = -\varepsilon_{lat}/\varepsilon_{long}$$

$$\varepsilon = \Delta L/L_o$$

$$\varepsilon_{lat} = (19.9\ \text{mm} - 20.0\ \text{mm})/(20\ \text{mm}) = -0.005\ \text{mm/mm}$$

$$\varepsilon_{long} = (1{,}020\ \text{mm} - 1{,}000\ \text{mm})/(1{,}000\ \text{mm}) = 0.02\ \text{mm/mm}$$

$$\nu = -\varepsilon_{lat}/\varepsilon_{long}$$

$$\nu = -(-0.005)/(0.02) = 0.25$$

Example 1.4.24

A wood timber supports an opening as shown in Figure 1.10. It is braced at its sides with additional wood to strengthen the main timber. If the timber is Western Larch with a Poisson's Ratio of 0.49 and the wood bracing the timber is Douglas Fir with a Poisson's Ratio of 0.29, determine the transverse displacement of the bracing if the vertical timber undergoes a displacement of 5 mm due to the applied force. Assume that the wood has a width of 0.15 m.

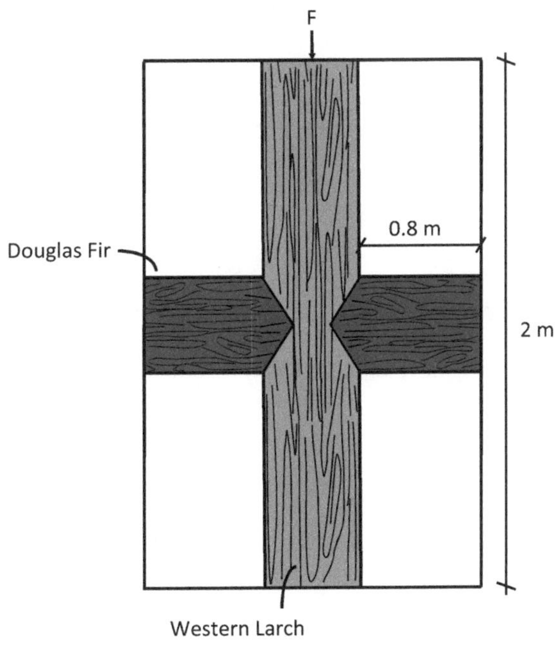

FIGURE 1.10 The wood timber support.

Solution:

Western Larch:

$$\varepsilon = \Delta L / L_o$$

$$\varepsilon_{long} = \left[\left(1{,}995\ \text{mm} - 2{,}000\ \text{mm}\right)/2{,}000\ \text{mm}\right] = -0.0025\ \text{mm/mm}$$

$$\nu = -\varepsilon_{lat} / \varepsilon_{long}$$

$$0.49 = -\varepsilon_{lat} / -0.0025\ \text{mm/mm}$$

$$\varepsilon_{lat} = 0.001225\ \text{mm/mm}$$

$$\varepsilon_{lat} = \Delta L / L_o$$

$$0.001225\ \text{mm/mm} = \Delta L / 150\ \text{mm}$$

$$\Delta L = 0.18375\ \text{mm, or } 0.2\ \text{mm}$$

The lateral displacement of the main timber due to the Poisson effect is 0.2 mm. Each side of the main timber would therefore expand a magnitude of 0.2 mm/2 = 0.1 mm.

Douglas Fir:

$$\varepsilon = \Delta L / L_o$$

$$\varepsilon_{long} = \left[\left(799.9\ \text{mm} - 800\ \text{mm}\right)/800\ \text{mm}\right] = -0.000125\ \text{mm/mm}$$

$$\nu = -\varepsilon_{lat} / \varepsilon_{long}$$

$$0.29 = -\varepsilon_{lat} / -0.000125\ \text{mm/mm}$$

$$\varepsilon_{lat} = 0.00003625\ \text{mm/mm}$$

$$\varepsilon_{lat} = \Delta L / L_o$$

$$0.00003625\ \text{mm/mm} = \Delta L / (150\ \text{mm})$$

$$\Delta L = 0.00544\ \text{mm, or effectively zero displacement}$$

Example 1.4.25

A box is pulled by two forces as shown in Figure 1.11. Find the work done by each of the forces and the total work done if the box is moved 70 cm along the y-axis.

Resolve the forces along the dotted line. For this example, using the Cosine is appropriate and the forces in the Figure are treated as the hypotenuses.

Solution:

$$\cos(\theta) = \text{adjacent/hypotenuse}$$

$$\cos(30°) = \text{adjacent}/85\ \text{N}$$

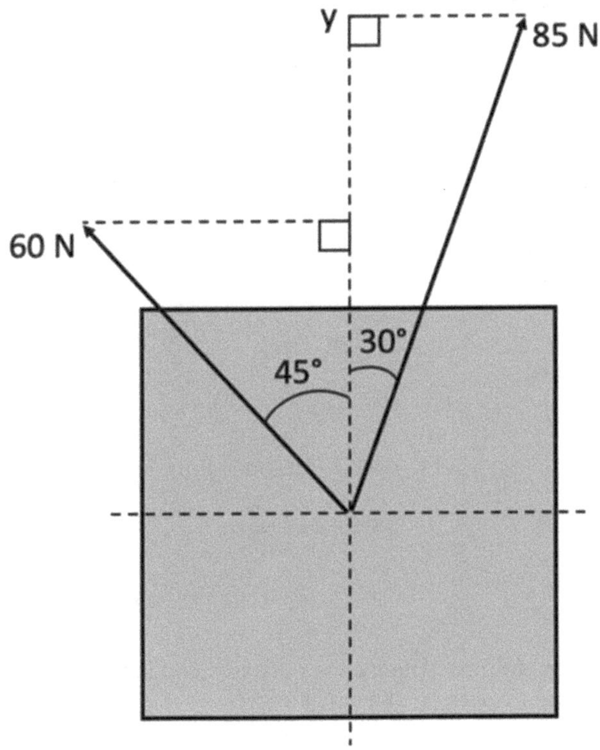

FIGURE 1.11 Work/energy problem for block pulled by two forces.

$$\text{Adjacent} = 73.6 \text{ N}$$

The units of energy are the Joule, N-m, and force-distance, and so the conversion can be completed as follows:

$$(73.6 \text{ N})(0.70 \text{ m}) = 51.5 \text{ J}$$

Similarly:

$$\cos(\theta) = \text{adjacent/hypotenuse}$$

$$\cos(45°) = \text{adjacent}/60 \text{ N}$$

$$\text{Adjacent} = 42.4 \text{ N}$$

$$(42.4 \text{ N})(0.70 \text{ m}) = 29.7 \text{ J}$$

Finally, the summation:

$$\text{Energy} = 51.5 \text{ J} + 29.7 \text{ J} = 81.2 \text{ J}$$

Example 1.4.26

Provide a short definition of Potential Energy, the equation, and the metric units. If an object of mass is 10 kg and the height of the object is 5 m, what is the potential energy?

Solution:

Potential Energy (PE) is the stored energy in an object. The potential energy is expressed as follows:

$$PE = mgh$$

where
m: mass of the object.
g: acceleration due to gravity.
h: height of the object.
Units are kg m^2/s^2, N-m, and J:

$$PE = mgh$$

$$PE = (10\ kg)(9.81\ m/s^2)(5\ m)$$

$$PE = 490.5\ (kg - m/s^2)(m)$$

$$PE = 490.5\ N - m$$

Example 1.4.27

Provide a short definition of Kinetic Energy, the equation, and the metric units. If a mass is 10 kg and the velocity is 9.81 m/s, what is the kinetic energy?

Solution:

Kinetic Energy (KE) is the energy of an object in motion. The kinetic energy is expressed as follows:

$$KE = \frac{1}{2}(mv^2)$$

where
m: mass of the object.
v: velocity of the object.
Units are kg m^2/s^2, N-m, and J:

$$KE = \frac{1}{2}(mv^2)$$

$$KE = \frac{1}{2}(10\ kg)(9.81\ m/s)^2$$

$$KE = 482\ kg - m^2/s^2$$

$$KE = 482\ N - m$$

Example 1.4.28

Provide a short definition of the law of conservation of energy, the equation, and the metric units. Determine the mechanical energy (E) for a 5 kg mass, dropped from 10 m, with a velocity of 14 m/s?

Law of conservation of energy states that for a closed system, energy remains constant and can neither be created nor destroyed:

$$E = PE + KE$$

where
E: mechanical energy or work.
PE: potential energy.
KE: kinetic energy.
Units are kg m^2/s^2, N-m, and J:

Initially the ball has only potential energy:

$$E = PE + KE$$
$$PE = mgh$$
$$PE = (5\ kg)(9.81\ m/s^2)(10\ m)$$
$$PE = 490\ (kg\text{-}m/s^2)(m)$$
$$PE = 490\ N\text{-}m$$
$$E = PE + KE$$
$$E = 490\ N\text{-}m + 0\ N\text{-}m$$
$$E = 490\ N\text{-}m$$

Free of external forces the potential energy converts to kinetic energy on impact:

$$KE = \tfrac{1}{2}\ (mv^2)$$
$$KE = \tfrac{1}{2}\ (5\ kg)(14\ m/s)^2$$
$$KE = 490\ kg\text{-}m^2/s^2$$
$$KE = 490\ N\text{-}m$$
$$E = PE + KE$$
$$E = 0\ N\text{-}m + 490\ N\text{-}m$$
$$E = 490\ N\text{-}m$$

Example 1.4.29

A baseball player slides to a stop at home base on level ground. Using energy considerations, calculate the distance the 75 kg baseball player slides, given that his initial speed is 5.5 m/s and the force of friction against his slide is a constant 425 N.

Solution:

Using conservation of energy:

$$E = PE + KE$$
$$W = PE + KE$$
$$Fd = mgh + \tfrac{1}{2}(mv^2)$$

Potential energy is zero, and so it can be ignored:

$$d = ½(mv^2)/F$$

$$d = \{½(75\ kg)(5.5\ m/s)^2\}/425\ N$$

$$d = 2.7\ m$$

Example 1.4.30

Determine the necessary force to give a 25 Mg locomotive an acceleration of 1.5 m/s² on a level track with a coefficient of rolling friction (μ) of 0.04 as shown in Figure 1.12?

The friction that works against the force attempting to propel the locomotive must be accounted for:

$$F = ma$$

$$F - \mu N = ma$$

$$N = mg$$

$$F = (0.04)(25{,}000\ kg)(9.81\ m/s^2) + (25{,}000\ kg)(1.5\ m/s^2)$$

$$F = 9{,}810\ N + 37{,}500\ N$$

$$F = 47{,}310\ N$$

$$F = 47.3\ kN$$

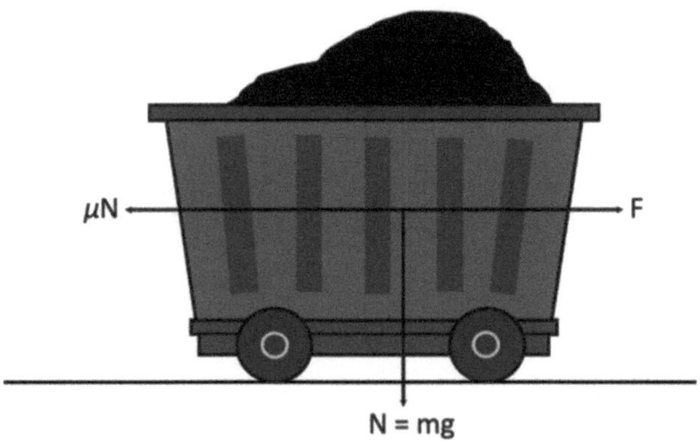

FIGURE 1.12 Locomotive example for rolling friction.

Example 1.4.31

Assume that the locomotive in the previous example is rolling and has overcome the initial coefficient of friction. How much power is needed to reach a velocity of 25 m/s? If there are 746 Watts per horsepower, convert the final answer to the latter units.

Solution:

Power takes the unit of N-m/s:

$$P = Fv$$

$$P = (47{,}310\ N)(25\ m/s)$$

$$P = 1{,}182{,}750\ N - m/s$$

$$P = 1{,}182{,}750\ Watts$$

$$P = 1{,}182.8\ kW$$

$$P = 1{,}585\ HP$$

Example 1.4.32

Determine the coefficient of friction (μ) from Figure 1.13 if W is 1 tonne-force and θ is 30° to the horizontal.

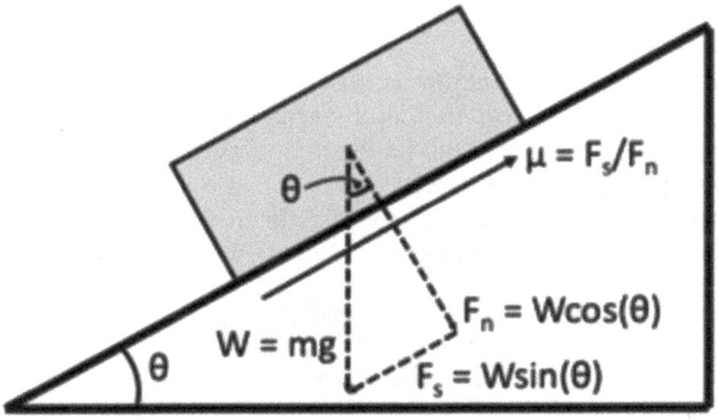

FIGURE 1.13 Coefficient of sliding friction for an inclined block.

Solution:

$$\mu = F_s/F_n$$

$$F_n = W * \cos(\theta)$$

$$F_n = (1{,}000\ kgf) * \cos(30^\circ)$$

$$F_n = 866\ kgf$$

$$F_s = W * \sin(\theta)$$

$$F_s = (1{,}000\ kgf) * \sin(30^\circ)$$

$$F_s = 500\ kgf$$

$$\mu = F_s/F_n$$

$$\mu = 500/866 \text{ kgf}$$

$$\mu = 0.58$$

Alternatively,

$$\mu = F_s/F_n$$

Here, F_s is the opposite, F_n is the adjacent, for the internal triangle, and then μ can be rewritten as:

$$\mu = \text{opposite/adjacent}$$

$$\text{Tan}(\theta) = \text{opposite/adjacent}$$

Therefore,

$$\mu = \tan(30°)$$

$$\mu = 0.58$$

Example 1.4.33

As an exercise assume an exposed fracture is observed at the top of a compacted rock pile as shown in Figure 1.14. Upon inspection, the depth of the fracture is estimated as shown.

If the mass of the rock is 150 kg and the limiting coefficient of friction (μ) for the rock is 0.50, determine if the limiting μ is being exceeded.

If the angle of repose of the rock is roughly 35°, determine if the angle at the piles steepest slope, along r–r′ exceeds the angle of repose.

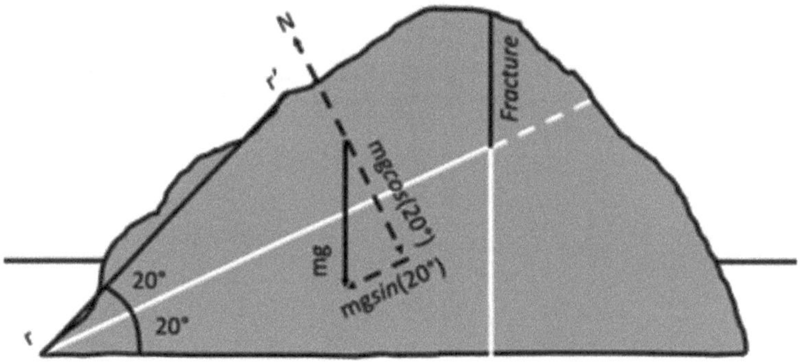

FIGURE 1.14 Coefficient of friction and angle of repose for rock pile.

Solution:

$$F_n = (150 \text{ kg})(9.81 \text{ m/s}^2)\cos(20°)$$

$$F_n = 1{,}382 \text{ N}$$

$$F_s = (150 \text{ kg})(9.81 \text{ m/s}^2)\sin(20°)$$

$$F_s = 503 \text{ N}$$

$$\mu = F_s/F_n$$

$$\mu = 503\ \text{N}/1{,}382\ \text{N}$$

$$\mu = 0.36 < 0.50$$

$$\text{The angle of repose} = 20° + 20° = 40° > 35°$$

Example 1.4.34

Determine the distance a feather and bowling ball will fall in 4 seconds on earth if there is no air resistance. Use Imperial units. Concept for this example was found at NASA (2022).

Solution:

Galileo proved that without drag, objects fall at the same acceleration. On earth, the object is subject to the gravitational constant of 32.2 ft/s^2. If the objects do not encounter drag, they fall the same distance after the same time. The solution is given in Table 1.4. The velocity (v) is equal to acceleration multiplied by time (at), and distance (d) is equal to ½ at^2.

TABLE 1.4
The Solution to the Problem

Acceleration (ft/s^2)	Time (s)	Velocity (ft/s)	Distance (ft)
32.2	1	32.2	16.1
	2	64.4	64.4
	3	96.6	144.9
	4	128.8	257.6

Example 1.4.35

Determine the height of an impoundment if an angle of 10°45′0″ is measured from an Abney Level. The horizontal distance on level ground is 125 ft (as shown in Figure 1.15). Abney Levels are simple survey tools for measuring heights of geotechnical structures. In this example, the angle is presented in degrees, minutes, and seconds, and so it should be converted to decimal degrees.

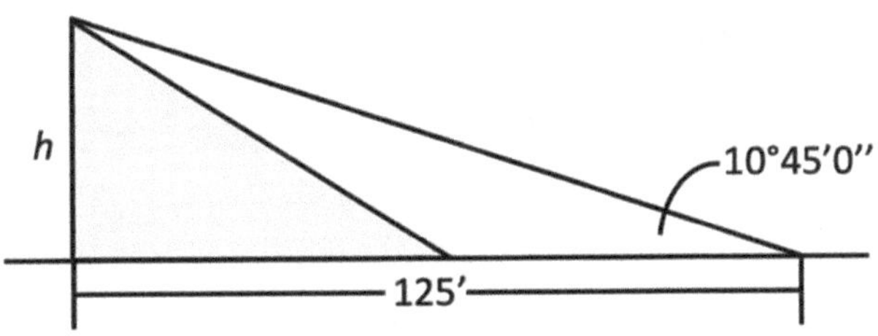

FIGURE 1.15 A section of the problem.

Solution:

$$10°45'0'' = 10° + (45'/60') + (0''/3{,}600'') = 10.75°$$

$$\tan(\theta) = \text{opposite/adjacent} = \text{vertical/horizontal}$$

$$\tan(10.75°) = \text{vertical/125 ft}$$

$$\text{Vertical height} = 24\text{ ft}$$

Example 1.4.36

Determine the height of a structure if the upper portion is measured at an angle of 10°45′0″ with an Abney Level, as shown in Figure 1.16. The lower portion is at an angle of –8°15′0″. The horizontal distance is 125 ft.

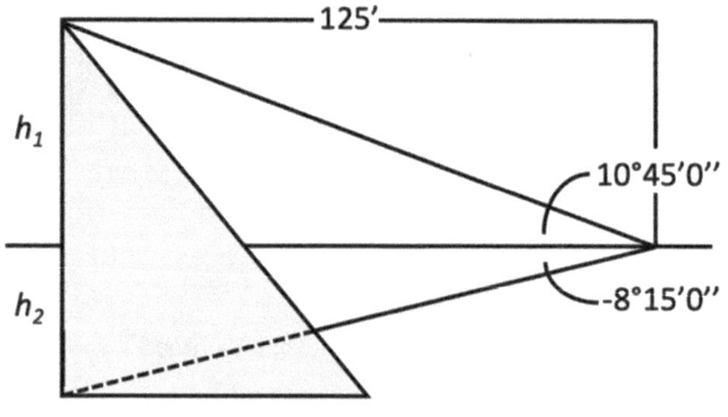

FIGURE 1.16 Outline of the problem.

Solution:

$$10°45'0'' = 10° + (45'/60') + (0''/3{,}600'') = 10.75°$$

$$-8°15'0'' = -8° + (-15'/60') + (0''/3{,}600'') = -8.25°$$

$$\tan(\theta) = \text{opposite/adjacent} = \text{vertical/horizontal}$$

$$\tan(10.75°) = \text{vertical/125 ft}$$

$$\text{Vertical height} = 24\text{ ft}$$

$$\tan(\theta) = \text{opposite/adjacent} = \text{vertical/horizontal}$$

$$\tan(8.25°) = \text{vertical/125 ft}$$

$$\text{Vertical height} = 18\ \text{ft}$$

$$\text{Total height} = 24 + 18 = 42\ \text{ft}$$

Example 1.4.37

A small section of rock is ejected under pressure at a sudden velocity of 100 mph. The unit weight (γ) of the rock is 85 lbs/ft^3. Determine the acceleration of the ejected rock if the travel time was nearly instantaneous at 0.25 s.

Solution:

In this example, the unit weight has no bearing on the final answer. The problem can be approached using an equation of motion:

$$v = u + at$$

$$(100\ \text{mph})(5{,}280\ \text{ft/1 mile})(1\ \text{hr/3,600 s}) = 0\ \text{ft/s} + (a)(0.25\ \text{s})$$

$$a = 587\ \text{ft/s}^2$$

Example 1.4.38

The weight of a canary in a cage is approximately 2 lbs. Determine the moment produced when the cage is hung from the spad (wedge) in a mine post which has an exposed effective length of 2 inches as shown in Figure 1.17.

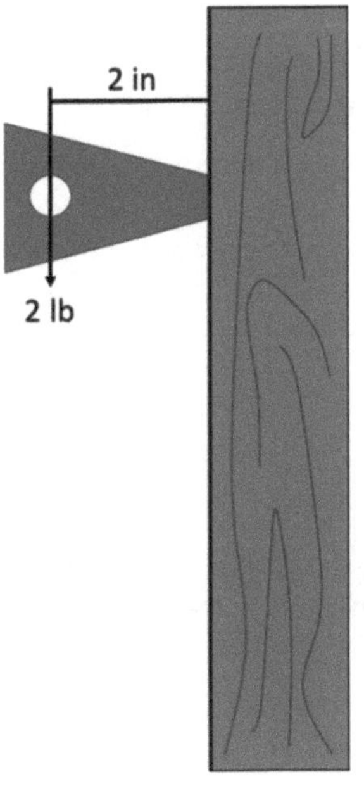

FIGURE 1.17 Section showing the post and wedge (spad).

Solution:

Moments act through a body and are equal to a force times the distance to the point of interest:

$$M = Fd$$

$$M = (2\ \text{lbs})(2\ \text{in})$$

$$M = 4\ \text{in} - \text{lbs}$$

Example 1.4.39

An ideal spring is compressed under a force of 0.5 N for a distance of 0.1 m. Determine the stiffness of a spring using Hooke's Law. Compare the differences between stiffness (k) and Elastic Modulus (E).

Solution:

$$F = kx$$

$$0.5\ \text{N} = k(0.1\ \text{m})$$

$$k = 5\ \text{N/m}$$

Table 1.5 shows the differences between stiffness and the Elastic Modulus.

TABLE 1.5
Stiffness and Elastic Modulus

	Stiffness	Modulus of Elasticity
Symbol	K	E
Units	Force/Displacement	(Force/Area)/Strain
Function	= F/x	$= \sigma/\varepsilon$

Example 1.4.40

Two springs support a mass in series and in parallel as shown below. If the stiffness of spring one (k_1) is 10 lb/in and spring two (k_2) is 20 lb/in, determine the equivalent stiffness (k_{eq}) of the springs (as shown in Figure 1.18). Comment on how the mass would displace.

Conceptually, the stiffness can be thought of as rock stiffness instead of a spring.

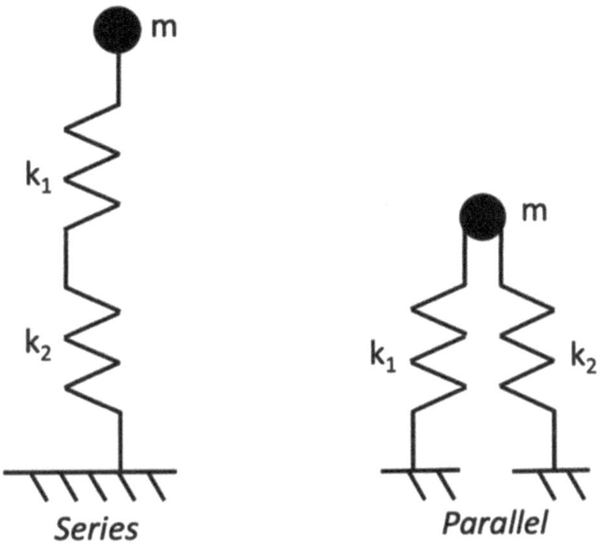

FIGURE 1.18 Springs in series and parallel.

Solution:

$$\text{Springs in series}: 1/k_{eq} = 1/k_1 + 1/k_2$$

$$1/k_{eq} = 1/k_1 + 1/k_2$$

$$1/k_{eq} = \left[1/(10\ \text{lb/in})\right] + \left[1/(20\ \text{lb/in})\right]$$

$$1/k_{eq} = 0.1\ \text{in/lb} + 0.05\ \text{in/lb}$$

$$k_{eq} = 6.7\ \text{lb/in}$$

$$\text{Spring in parallel}: k_{eq} = k_1 + k_2$$

$$k_{eq} = k_1 + k_2$$

$$k_{eq} = 10\ \text{lb/in} + 20\ \text{lb/in}$$

$$k_{eq} = 30\ \text{lb/in}$$

When subjected to the same mass, the springs would displace more in series and less in parallel.

Example 1.4.41

The Archimedes Principle states that the buoyant force on a body is a function of the products of the unit weight of the fluid displaced and the volume of the object. If a specimen of pumice with dimensions 2 in by 10 in by 8 in is forced under water, calculate the buoyant force. Use a density of water of 1.94 slugs/ft^3.

Solution:

$$F_b = \rho g V$$

$$\rho_{water} = 1.94 \text{ slugs/ft}^3$$

$$g = 32.2 \text{ ft/s}^2$$

$$\gamma = \left[1.94 \left(\text{lb}_f\text{s}^2/\text{ft}\right)/\text{ft}^3\right]\left(32.2 \text{ ft/s}^2\right)$$

$$\gamma = \left(1.94 \text{ lb}_f\text{s}^2/\text{ft}^4\right)\left(32.2 \text{ ft/s}^2\right)$$

$$\gamma = 62.4 \text{ lb}_f/\text{ft}^3$$

$$F_b = \left(62.4 \text{ lb}_f/\text{ft}^3\right)\left(2 \text{ in}\right)\left(10 \text{ in}\right)\left(8 \text{ in}\right)\left(1 \text{ ft}^3/1{,}728 \text{ in}^3\right)$$

$$F_b = 5.8 \text{ lb}_f$$

Example 1.4.42

A drain is necessary to convey water away from an area. Given a vertical distance of 7.85 ft and a horizontal distance of 133 ft, determine the slope of the drain in percent and provide the Secant of the angle.

Percent of a slope is most easily remembered as the rise (height) over the run (horizontal length) of a right triangle:

Solution:

$$\text{Slope} = \text{rise/run}$$

$$\text{Slope} = 7.85 \text{ ft}/133 \text{ ft}$$

$$\text{Slope} = 0.059$$

$$\text{Slope}(\%) = (0.059)(100)$$

$$\text{Slope}(\%) = 5.9\%$$

$$\theta = \tan^{-1}\left(\text{opposite/adjacent}\right)$$

$$\theta = \tan^{-1}\left(7.85 \text{ ft}/133 \text{ ft}\right)$$

$$\theta = 3.38°$$

$$\sec(\theta) = 1/\cos(\theta)$$

$$\sec(\theta) = 1/\cos(3.38°)$$

$$\text{Sec}(3.38°) = 1.002$$

Example 1.4.43

Estimate the force at the bottom of a square column of rock which has an area of 1,600 ft^2 if the column is 200 ft tall and the weight of the rock increases with height at a rate of 0.8 psi/ft.

Solution:

$$\text{The stress at the bottom of the column is } (0.8 \text{ psi / ft})(200 \text{ ft}) = 160 \text{ psi}$$

$$\text{Area}(\text{in}^2) = (1{,}600 \text{ ft}^2)(12 \text{ in/1 ft})(12 \text{ in/1 ft})$$

$$\text{Area}(\text{in}^2) = 230{,}400 \text{ in}^2$$

$$\text{Force at the bottom of the column is therefore}: (160 \text{ lb/in}^2)(230{,}400 \text{ in}^2) = 36{,}864 \text{ kips}$$

Example 1.4.44

Define and illustrate, in a sketch, the difference between the elastic, tangent, and secant modulus. If the following strains (ε), and stresses (σ) correspond to points of intersection on the stress–strain curve, use them to determine the elastic, tangent, and secant moduli.

Given:

Elastic Modulus:

$$\varepsilon = 0.5\ \mu\varepsilon$$

$$\sigma = 4 \text{ MPa}$$

Solution:

Tangent Modulus:

Assume this line crosses the y-axis at $\sigma = 4$ MPa:

$$\varepsilon = 1.25\ \mu\varepsilon$$

$\sigma = 6$ MPa (touches the stress–strain curve)

Secant Modulus:

$$\varepsilon = 2.5\ \mu\varepsilon$$

$$\sigma = 7.5 \text{ MPa}$$

The Elastic Modulus (E) is the slope of the stress–strain curve from the origin to the proportional limit. The Secant Modulus (E_s) is the gradient of the stress–strain curve from the origin to a fixed percentage of the peak stress of the rock specimen. The Tangent Modulus (E_t) is the tangent to the stress–strain curve typically at a fixed percentage of the ultimate strength. The moduli are shown in Figure 1.19:

$$E = \sigma/\varepsilon = 4 \text{ MPa}/0.5 \times 10^{-6} = 8{,}000{,}000 \text{ MPa} = 8{,}000 \text{ GPa}$$

$$E_t = \sigma/\varepsilon = (6 - 4 \text{ MPa})/1.25 \times 10^{-6} = 1{,}600{,}000 \text{ MPa} = 1{,}600 \text{ GPa}$$

$$E_S = \sigma/\varepsilon = 7.5 \text{ MPa}/2.5 \times 10^{-6} = 3{,}000{,}000 \text{ MPa} = 3{,}000 \text{ GPa}$$

Example 1.4.45

A circle has a radius of 10 ft. What are the dimensions of a square of equal area?

Solution:

$$\text{Area of circle} = \pi * r^2$$

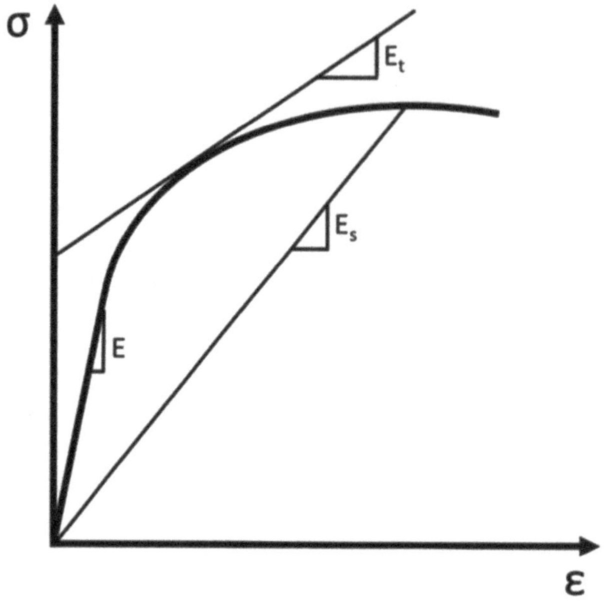

FIGURE 1.19 The tangent and secant Young's Moduli.

$$\text{Area of circle} = 3.142 * 10^2$$

$$\text{Area of circle} = 314.2\ \text{ft}^2$$

$$\text{The dimensions of an equivalent square}(s * s):$$

$$\text{Area of square} = s * s$$

$$314.2\ \text{ft}^2 = s^2$$

$$s = 17.7\ \text{ft}$$

Example 1.4.46

If cylindrical rock specimen has failed in compression at a stress of 130.52 MPa and a strain of 1.4 millistrain, what is Young's Modulus of the rock?

Note: The sign convention for stresses and strains is discussed in Chapter 2.

Solution:

$$E = \sigma/\varepsilon$$

$$E = 130.52\ \text{MPa}/0.0014$$

$$E = 93{,}229\ \text{MPa}$$

$$E = 93.2\ \text{GPa}$$

Example 1.4.47

A cylindrical rock specimen has a 44 mm diameter and failed under a uniaxial load of 550 kN at a strain of 2 millistrain. What was the stress at failure and its Young's Modulus?

Solution:

$$\sigma = F/A$$

$$\sigma = 550,000\ \mathrm{N}/[\pi(0.022\ \mathrm{m})^2]$$

$$\sigma = 550,000\ \mathrm{N}/[0.00152]$$

$$\sigma = 361,842,105\ \mathrm{Pa}$$

$$\sigma = 361.8\ \mathrm{MPa}$$

$$E = \sigma/\varepsilon$$

$$E = 361.8\ \mathrm{MPa}/0.002\ \varepsilon$$

$$E = 181\ \mathrm{GPa}$$

Example 1.4.48

A metal bolt of 2-m length stretches to 2,020 mm under load. Its diameter decreases from 25 to 24.8 mm. What is the value of Poisson's Ratio?

Solution:

$$\varepsilon_{\mathrm{lat}} = (25.0 - 24.8)/25$$

$$\varepsilon_{\mathrm{lat}} = 0.008$$

$$\varepsilon_{\mathrm{long}} = (2,020 - 2,000)/2,000 = 0.01$$

$$\varepsilon_{\mathrm{long}} = 0.01$$

$$\nu = \varepsilon_{\mathrm{lat}}/\varepsilon_{\mathrm{long}}$$

$$\nu = 0.008/0.01$$

$$\nu = 0.8$$

Example 1.4.49

A point-anchored rockbolt is 4 m long with a cross-sectional area of 0.5 cm^2. If the bolt carries a mass of 225 kg and stretches 1 mm, what is Young's Modulus?

Solution:

$$\varepsilon = \Delta L/L$$

$$\varepsilon = 1\ \text{mm}/4{,}000\ \text{mm}$$

$$\varepsilon = 0.00025$$

$$\text{Area} = \left(0.5\ \text{cm}^2\right)\left(1\ \text{m}/100\ \text{cm}\right)\left(1\ \text{m}/100\ \text{cm}\right)$$

$$\text{Area} = 0.00005\ \text{m}^2$$

$$F = mg$$

$$F = \left(225\ \text{kg}\right)\left(9.81\ \text{m/s}^2\right)$$

$$F = 2{,}207\ \text{kg} - \text{m/s}^2 = 2{,}207\ \text{N}$$

$$F = 2.2\ \text{kN}$$

$$\sigma = F/A$$

$$\sigma = \left(2.2\ \text{kN}\right)/\left(0.00005\ \text{m}^2\right)$$

$$\sigma = 44{,}000\ \text{kPa}$$

$$\sigma = 44\ \text{MPa}$$

$$E = \sigma/\varepsilon$$

$$E = 44\ \text{MPa}/0.00025$$

$$E = 17{,}600\ \text{MPa}$$

$$E = 176\ \text{GPa}$$

Example 1.4.50

A laboratory press is used to test the compressional behavior of a timber crib (pack) with base dimensions of 800 mm by 900 mm. If the load of 425 kN was applied to the crib, what is the stress on the pack?

Solution:

$$\sigma = F/A$$

$$\sigma = \left(425\ \text{kN}\right)/\left[\left(0.8\ \text{m}\right)\left(0.9\ \text{m}\right)\right]$$

$$\sigma = \left(425\ \text{kN}\right)/\left(0.72\ \text{m}^2\right)$$

$$\sigma = 590{,}000\ \text{N/m}^2$$

$$\sigma = 590\ \text{kPa}$$

Example 1.4.51

A cylindrical rock specimen is 82.5 mm long and is placed in a press and loaded until it has deformed in the direction of loading by 0.145 mm. What is the longitudinal strain if tension is taken as positive or negative?

Solution:

Length after compressive load

$$82.5 - 0.145\ \text{mm} = 82.355\ \text{mm}$$

Assume that the compression is positive and the tension is negative:

$$\varepsilon = \Delta L / L_o$$

$$\varepsilon = (82.5 - 82.355\ \text{mm})/82.5\ \text{mm}$$

$$\varepsilon = (0.145\ \text{mm})/82.5\ \text{mm}$$

$$\varepsilon = 0.00176\varepsilon$$

$$\varepsilon = 1.76\text{m}\varepsilon$$

Now assume that the compression is negative and the tension is positive:

$$\varepsilon = \Delta L / L_o$$

$$\varepsilon = (82.355\ \text{mm} - 82.5\ \text{mm})/82.5\ \text{mm}$$

$$\varepsilon = (-0.145\ \text{mm})/82.5\ \text{mm}$$

$$\varepsilon = -0.00176\varepsilon$$

$$\varepsilon = -1.76\ \text{m}\varepsilon$$

The key is being consistent in the sign convention.

REFERENCES

Hall, M. (2015) What is anisotropy? Agile Scientific Blog-Geoscience & Technology. www.agilescientific.com/blog/2015/2/9/what-is-anisotropy

Journal of Petroleum Technology. (1977) The SI Metric System of Units and SPE's Tentative Metric Standard, Vol. 29, Issue 12, 1575–1616.

NASA. (2022). Motion of Free Falling Object. NASA. https://www1.grc.nasa.gov/beginners-guide-to-aeronautics/motion-of-free-falling-object/

2 Basic Rock Mechanics

The following is reproduced and modified from Spearing (1995) with kind permission of the South African Institute of Mining & Metallurgy with modifications.

2.1 INTRODUCTION

Rock mechanics is the theoretical and applied science of the mechanical behavior of rock. It is the branch of mechanics concerned with the response of rock to the force fields within its physical environment. Rock mechanics in its broadest sense includes the whole range of mine design, support, blasting, extraction, and safety.

Solid mechanics is based on continuum theory, which is a basic mathematical approximation. The molecular structure of any material is replaced by an equivalent continuum whose overall behavior is representative mathematically to the original solid. The equivalent continuum is usually assumed to be homogenous and isotropic. Discontinuum methods are used for representing complex physical systems, such as broken or caving rock, and are approached if continuum methods are unsuitable for the desired solution.

2.2 STRESS COMPONENTS AND NOTATION IN TWO DIMENSIONS

The sign convention is shown in Figure 2.1. The shear components (τ) have two subscripts: the first indicates the coordinate axis normal to the face of the element on which it acts, and the second indicates the direction in which it acts.

Prior to mining, all rock stresses are compressive, and so it is generally taken, or assumed, as positive which is unique to technical mining problems. Most important, however, is consistency regardless of the sign convention used.

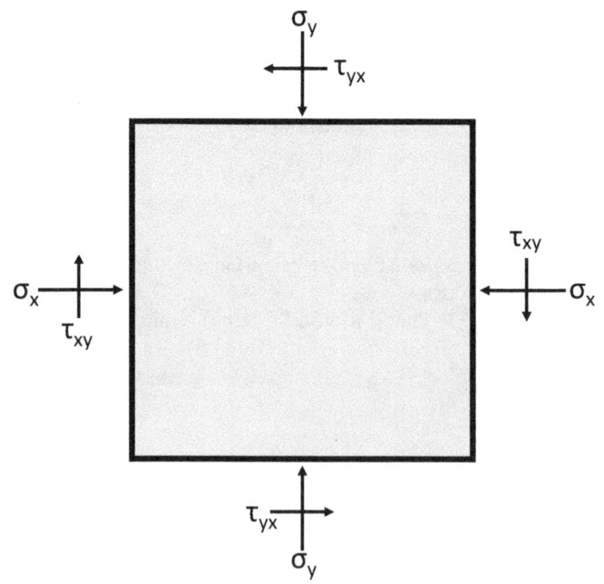

FIGURE 2.1 The typical sign convention used in mining on an arbitrary square element.

DOI: 10.1201/9781032704845-2

2.3 STRESSES ON AN INCLINED PLANE

Consider the following to determine stress components on an oblique plane. As shown in Figure 2.2, the orientation of a plane is specified by θ which is the angle at which the normal to the oblique face must be rotated to bring its side parallel to the positive x-axis.

If a triangle, ABO, is defined with lengths as follows:

$$AB = dl$$

$$OA = dy$$

$$OB = dx$$

then the resultant condition is as shown in Figure 2.3, where **n** is a unit vector normal to the tangent plane AB and ρ_n is the resultant of forces ρ_{nx} and ρ_{ny}.

Resolving forces along OX,

$$dl(\rho_{nx}) = (\sigma_x)dy + (\tau_{yx})dx \tag{2.1}$$

Resolving forces along OY,

$$dl(\rho_{ny}) = (\sigma_y)dx + (\tau_{xy})dy \tag{2.2}$$

Using the basic geometry, as shown in Figure 2.4, the following can be derived and supplanted into Equations 2.1 and 2.2:

$$\angle OZB = 90^\circ$$

$$\angle OBZ = 180^\circ - 90^\circ - \theta = 90^\circ - \theta$$

$$\angle OAB = 180^\circ - \angle AOB - \angle OBA = 180^\circ - 90^\circ - (90^\circ - \theta) = \theta$$

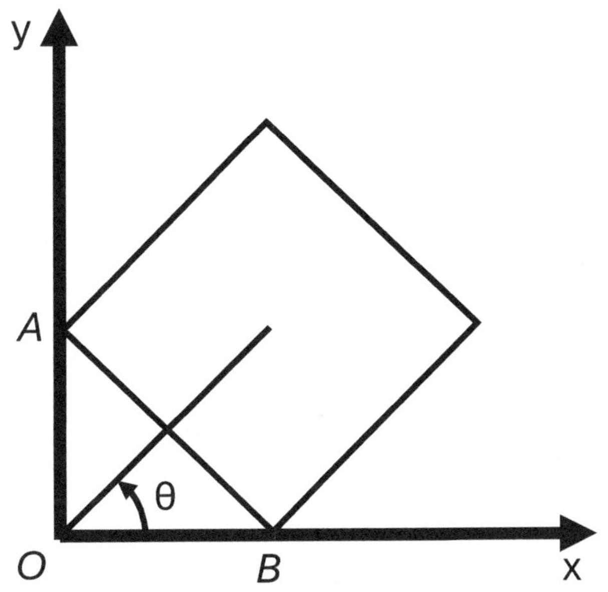

FIGURE 2.2 An oblique (angled) plane.

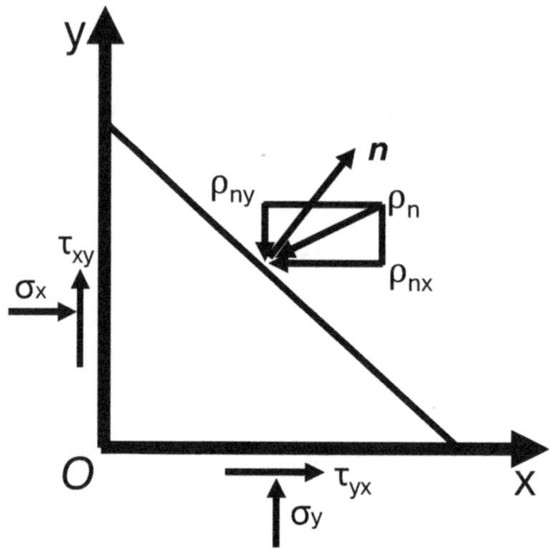

FIGURE 2.3 Resolving stresses on the oblique plane.

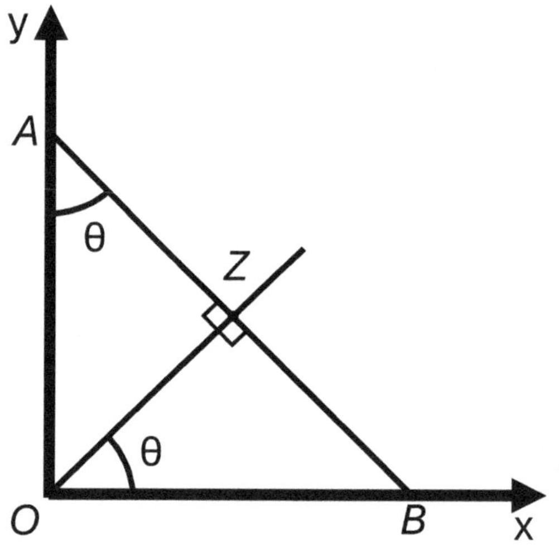

FIGURE 2.4 The basic geometry to solve for the stresses on the oblique plane (AB).

$$\sin\theta = OB/AB = dx/dl$$

$$\cos\theta = OA/AB = dy/dl$$

Therefore,

$$\rho_{nx} = \sigma_x\cos\theta + \tau_{yx}\sin\theta$$

$$\rho_{ny} = \sigma_y\sin\theta + \tau_{xy}\cos\theta$$

The resultant force ρ_n can be broken into a normal σ_n and shear stress τ_{nm} components, as shown in Figure 2.5, and resolving in the directions of σ_n and τ_{nm} provides the following:

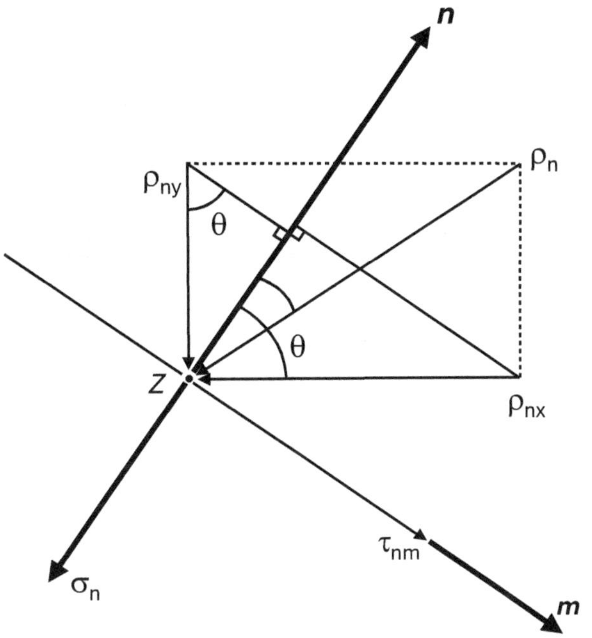

FIGURE 2.5 Resolving the stresses on the oblique plane.

$$\sigma_n = \rho_n \cos\gamma = \rho_{nx}\cos\theta + \rho_{ny}\sin\theta \quad (2.3)$$

$$\tau_{nm} = \rho_n \sin\gamma = \rho_{ny}\cos\theta - \rho_{nx}\sin\theta \quad (2.4)$$

Equations 2.3 and 2.4 are obtained by substitution of the expressions for ρ_{nx} and ρ_{ny}:

$$\sigma_n = \rho_{nx}\cos\theta + \rho_{ny}\sin\theta$$

$$\sigma_n = \left(\sigma_x\cos\theta + \tau_{yx}\sin\theta\right)\cos\theta + (\tau_{xy}\cos\theta + \sigma_y\sin\theta)\sin\theta$$

$$\sigma_n = \sigma_x\cos^2\theta + (\tau_{xy} + \tau_{yx})\sin\theta\cos\theta + \sigma_y\sin^2\theta \quad (2.5)$$

Similarly,

$$\tau_{nm} = \rho_{ny}\cos\theta - \rho_{nx}\sin\theta$$

$$\tau_{nm} = (\tau_{xy}\cos\theta + \sigma_y\sin\theta)\cos\theta - \left(\sigma_x\cos\theta + \tau_{yx}\sin\theta\right)\sin\theta$$

$$\tau_{nm} = (\sigma_y - \sigma_x)\sin\theta\cos\theta + \tau_{xy}\cos^2\theta - \tau_{yx}\sin^2\theta \quad (2.6)$$

2.4 STRESS ON ELEMENTS

Simplifications are possible for an element in two dimensions, as shown in Figure 2.6, where

$$AB = CD = dx \text{ and } CB = AD = dy$$

In many cases, rock mechanics problems consider square elements, and therefore, dx will equal dy.

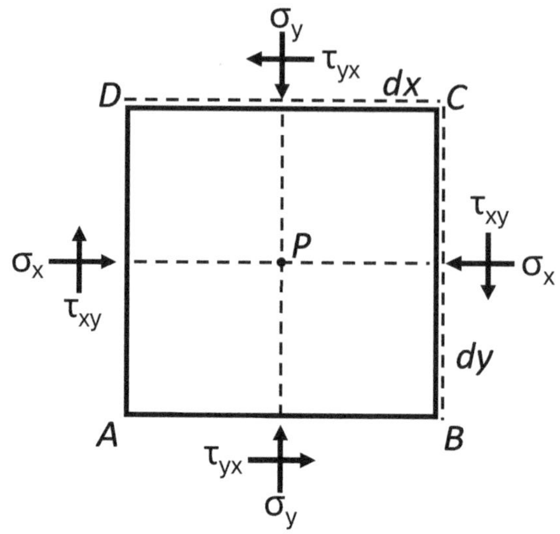

FIGURE 2.6 The stress tensor in 2D.

For the elements to remain in two-dimensional equilibrium, the moments around the centroid P must be zero. Note that $\tau_{yx}dy$ is a force:

$$dx\left(2\tau_{yx}\right)dy/2 = dy\left(2\tau_{xy}\right)dx/2$$

$$\tau_{yx} = \tau_{xy}$$

Equations 2.5 and 2.6 can be simplified with the following trigonometric identities:

$$\sin 2\theta = 2\sin\theta\cos\theta$$

$$\sin^2\theta = \tfrac{1}{2}(1-\cos 2\theta)$$

$$\cos^2\theta = \tfrac{1}{2}(1+\cos 2\theta)$$

$$\sigma_n = \sigma_x\cos^2\theta + (\tau_{xy}+\tau_{yx})\sin\theta\cos\theta + \sigma_y\sin^2\theta$$

$$\sigma_n = \sigma_x\cos^2\theta + \sigma_y\sin^2\theta + 2\tau_{xy}\sin\theta\cos\theta$$

$$\sigma_n = \sigma_x\left[\tfrac{1}{2}(1+\cos 2\theta)\right] + \sigma_y[\tfrac{1}{2}(1-\cos 2\theta)] + \tau_{xy}\sin 2\theta$$

$$\sigma_n = \sigma_x\left[\tfrac{1}{2}+\tfrac{1}{2}\cos 2\theta\right] + \sigma_y[\tfrac{1}{2}-\tfrac{1}{2}\cos 2\theta] + \tau_{xy}\sin 2\theta$$

$$\sigma_n = \tfrac{1}{2}(\sigma_x+\sigma_y) + \tfrac{1}{2}(\sigma_x-\sigma_y)\cos 2\theta + \tau_{xy}\sin 2\theta \qquad (2.7)$$

$$\tau_{nm} = (\sigma_y-\sigma_x)\sin\theta\cos\theta + \tau_{xy}\cos^2\theta - \tau_{yx}\sin^2\theta$$

$$\tau_{nm} = \tfrac{1}{2}(\sigma_y-\sigma_x)\sin 2\theta + \tfrac{1}{2}\tau_{xy}(1+\cos 2\theta) - \tfrac{1}{2}\tau_{yx}(1-\cos 2\theta)$$

$$\tau_{nm} = \tfrac{1}{2}(\sigma_y - \sigma_x)\sin 2\theta + \tfrac{1}{2}\tau_{xy}\cos 2\theta + \tfrac{1}{2}\tau_{yx}\cos 2\theta$$

$$\tau_{nm} = \tfrac{1}{2}(\sigma_y - \sigma_x)\sin 2\theta + \tau_{xy}\cos 2\theta \tag{2.8}$$

2.5 STRESSES AND STRAINS IN POLAR COORDINATES

In the previous sections, stresses were derived in rectangular coordinates. However, in the polar system of coordinates, the position of a point is fixed by two variables: r and θ. Here, θ is an angle from the x–y origin and r is the distance from the point to the origin. A representation is shown in Figure 2.7.

From the previous formula for stresses in Cartesian coordinates, it is possible to show that

$$\sigma_r = \sigma_x \cos^2\theta + 2\tau_{xy}\sin\theta\cos\theta + \sigma_y \sin^2\theta \tag{2.9}$$

$$\sigma_\theta = \sigma_x \sin^2\theta - 2\tau_{xy}\sin\theta\cos\theta + \sigma_y \cos^2\theta \tag{2.10}$$

$$\tau_{r\theta} = \tfrac{1}{2}(\sigma_y - \sigma_x)\sin 2\theta + \tau_{xy}\cos 2\theta \tag{2.11}$$

The addition of Equations 2.9 and 2.10 results in the following:

$$\sigma_r + \sigma_\theta = \sigma_x + \sigma_y \tag{2.12}$$

Strain transformations produce equations of a similar form:

$$\varepsilon_r = \varepsilon_x \cos^2\theta + \gamma_{xy}\sin\theta\cos\theta + \varepsilon_y \sin^2\theta \tag{2.13}$$

$$\varepsilon_\theta = \varepsilon_x \sin^2\theta - \gamma_{xy}\sin\theta\cos\theta + \varepsilon_y \cos^2\theta \tag{2.14}$$

$$\gamma_{r\theta} = (\varepsilon_y - \varepsilon_x)\sin 2\theta + \gamma_{xy}\cos 2\theta \tag{2.15}$$

It is also evident that $2\tau_{xy}$ will equate to γ_{xy}.

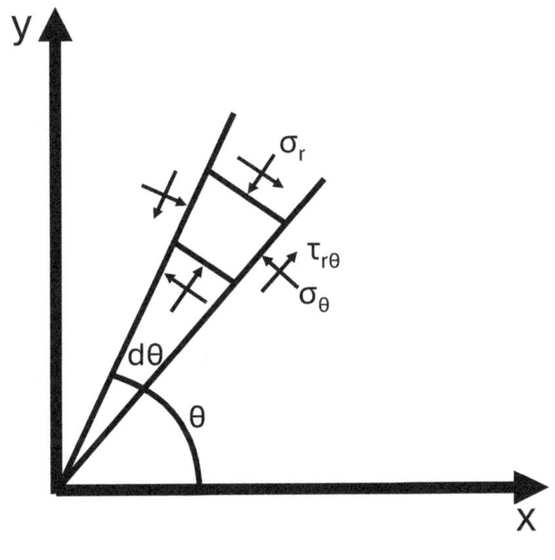

FIGURE 2.7 An oblique plane and element in polar coordinates.

Similarly,

$$\varepsilon_r + \varepsilon_\theta = \varepsilon_x + \varepsilon_y \tag{2.16}$$

2.6 PRINCIPAL STRESSES

The concept of principal stresses is useful for practitioners especially in two dimensions because it effectively reduces the problem to only normal stresses (no shear stresses), which is often easier to visualize. The same is true of principal strains.

In three dimensions, the same is also true except now, and the stress (or strain) components are reduced from nine to three components.

If a plane is rotated about a point, both the magnitude and the direction of the stresses change. It is possible to find a position where a vector coincides with the vector normal to the plane (**n**). At this position therefore, $\tau_{nm}=0$, and Equation 2.8 becomes

$$\tau_{nm} = \tfrac{1}{2}(\sigma_y - \sigma_x)\sin 2\theta + \tau_{xy}\cos 2\theta$$

$$0 = \tfrac{1}{2}(\sigma_y - \sigma_x)\sin 2\theta + \tau_{xy}\cos 2\theta$$

$$0 = (\sigma_y - \sigma_x)\sin 2\theta + 2\tau_{xy}\cos 2\theta$$

$$(\sigma_x - \sigma_y)\sin 2\theta = 2\tau_{xy}\cos 2\theta$$

$$(\sin 2\theta / \cos 2\theta) = 2\tau_{xy}/(\sigma_x - \sigma_y)$$

$$\tan 2\theta = 2\tau_{xy}/(\sigma_x - \sigma_y) \tag{2.17}$$

There are four possible solutions for the orientation of the normal (**n**) specifying two mutually perpendicular axes for which the stress has no shear component. These directions are known as the principal directions, or orientations, θ_1 and θ_2. The corresponding maximum and minimum principal stresses – in two dimensions – are defined as σ_1 and σ_2, respectively. In some cases, the notations σ_{max} and σ_{min} are used instead. A convention is adopted in which $\sigma_1 > \sigma_2$. In three dimensions, σ_1 and σ_3 represent the maximum and minimum stresses, respectively.

The following is the relationship between σ_1 and σ_2 and any other stress vector, and can be proven by the Mohr circle as shown by Spearing (1995):

$$\sigma_1 = \tfrac{1}{2}(\sigma_x + \sigma_y) + \tfrac{1}{2}[(\sigma_x - \sigma_y)^2 + 4(\tau_{xy})^2]^{\frac{1}{2}} \tag{2.18}$$

$$\sigma_2 = \tfrac{1}{2}(\sigma_x + \sigma_y) - \tfrac{1}{2}[(\sigma_x - \sigma_y)^2 + 4(\tau_{xy})^2]^{\frac{1}{2}} \tag{2.19}$$

If these equations are added, the result is the most important formula for checking calculations involving stresses:

$$\sigma_1 + \sigma_2 = \sigma_x + \sigma_y \tag{2.20}$$

2.7 ROCK STRENGTH AND STRAIN

2.7.1 Strength of Rock

Rock strength is dependent on a myriad of factors because it is naturally occurring and not engineered. The basic factors include the following:

- Genesis of the rock.
- Intact rock strength, such as tensile and compressive strengths.
- Degree of rock alteration due to weathering.
- The presence of major geological features such as bedding planes, joints, and their spacing and orientation.
- Hydrologic considerations such as the presence of water from groundwater or flooded workings.

Factors extend well beyond these basic considerations depending on the rock mechanics requirements and can include rock strength at extreme temperatures, strength with time, strength at varying scales, failure criterion, and even post-failure strength of rock. In terms of scale, rock strength is dependent on the size of the sample tested. The term effective strength is used to express the strength of rock considering any fluid, gas, or water present.

By far the most universal test, but not always the most appropriate, is the unconfined compressive strength (UCS) test. The International Society for Rock Mechanics and Rock Engineering (ISRM) has compiled a complete list of suggested methods for rock strength and characterization. A generalized depiction of a rock sample under uniaxial stress for the determination of peak strength is shown in Figure 2.8.

Laboratory-scale rock tests allow only for the derivation of certain intact properties, meaning that there are often limitations to lab results that should be well understood. In the case of a UCS test, one can determine the compressive strength of a rock sample at a certain scale.

For the information to be used reliably in practice, methods such as larger-scale tests or rock classification are used to characterize the rock behavior under compression before it is widely applied for design purposes. For more complex rock parameters such as friction angle or cohesion, a more appropriate test is a triaxial strength test.

2.7.2 Analysis of Strain in Rock

A solid body in equilibrium, under external forces, undergoes deformation. This causes displacement of points in the body from its original position. By convention, the components of the displacements along the x, y, and z axes are denoted by u, v, and w, respectively.

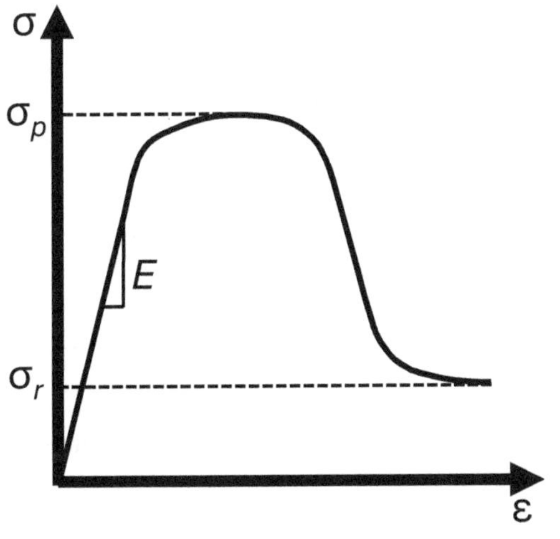

FIGURE 2.8 Generalized stress–strain curve for unconfined compressive strength (UCS) test on a rock sample illustrating peak stress (σ_p) and residual stress or post-failure stress (σ_r).

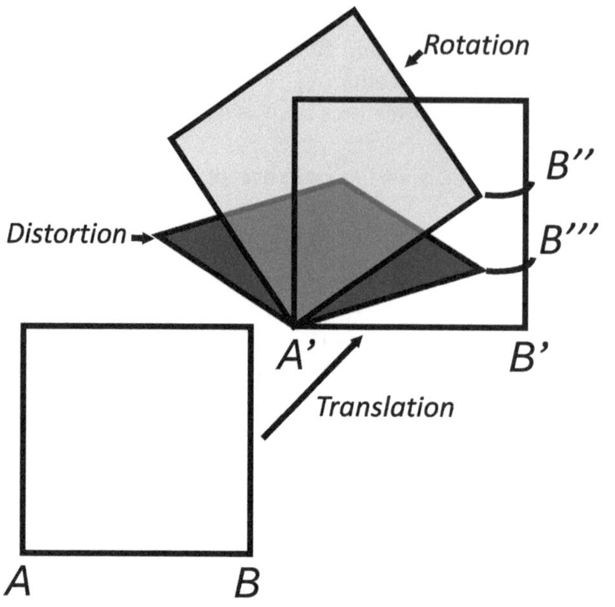

FIGURE 2.9 A force on A–B causing translation, rotation, and distortion.

Under the common convention, positive displacement represents movement toward the negative direction of the coordinate axes, reiterating that compressional forces are usually taken as positive in mining.

Consider the element in Figure 2.9, which, due to an external force system, moved from A to A′. The displacements of points in the element could be a result of any combination of (i) rigid-body translation of the element from A to A′, (ii) rigid-body rotation of the element from B′ to B″, and (iii) distortion of the element.

Owing to rigid-body motions, the relative positions of points in the element remain unchanged. Therefore, they do not induce any deformation in the element and can be disregarded in the subsequent analysis. Explained in a different way, the treatment of such elements in engineering practice is to assume that small displacements and strains to assure assumptions of rigid-body mechanics are not invalidated.

The compressive stress (σ) causes the material to decrease in its original length from L_o to L (Figure 2.10). The association was discussed in the previous chapter but reprinted here for convenience:

$$\varepsilon_n = (L_o - L)/L_o \tag{2.21}$$

where ε_n is the normal strain, taken as positive when it results in shortening, compressive strain (ε_c), and negative, when it results in lengthening, tensile strain (ε_t).

Just as a note, when analyzing rock bolt behavior, the bolts are generally in tension and/or shear, rather than compression.

Consider instead the simple shear mode of deformation as shown in Figure 2.11. The external shear forces introduce deformation on OB, producing a rotation of $\gamma/2$. By convention, γ represents the shearing strain measured in radians.

A simple way to remember this concept is to imagine holding opposing corners of the square element with two fingers. To get the sign convention correct, imagine one corner is at point 0 in Figure 2.11 and the other corner is diametrically opposite. The diametrically opposite corner is referred to as the right upper quadrant of the element, which theoretically is the region on which

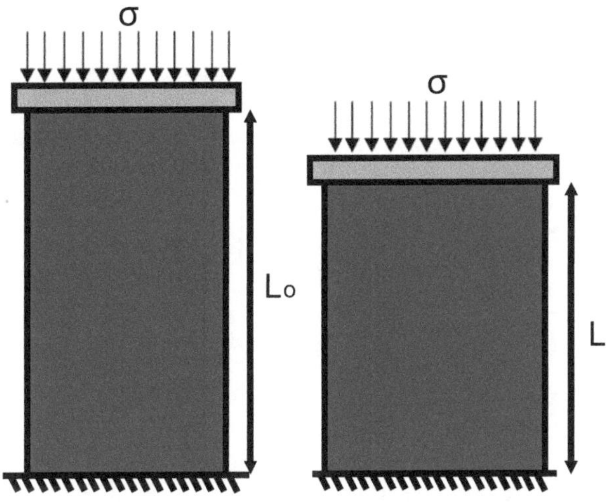

FIGURE 2.10 The effect of stress on a body resulting in small strain and displacement.

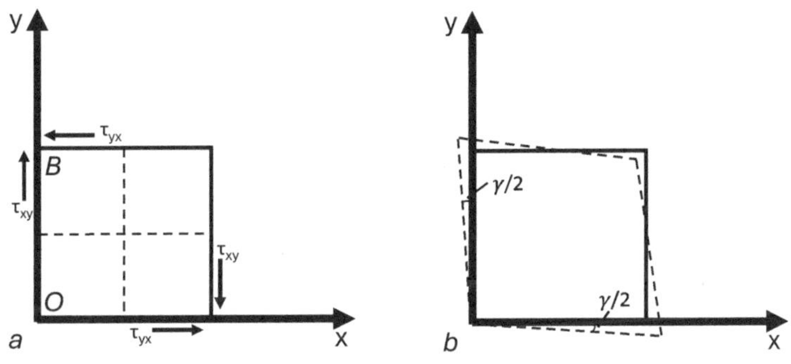

FIGURE 2.11 Deformation caused by strain: (a) an applied shear stress to a block in 2D and (b) the corresponding strain resulting.

sign conventions are based for simple strain problems. Now, if we pinch and pull our fingers away from one another, the element will decrease its angle in the upper right quadrant, thereby resulting in a positive shear strain. Likewise, pushing the fingers toward one another will increase the angle, resulting in a negative shear strain.

When a sample cube of material is stretched in one direction, it tends to contract in the other two orthogonal directions and vice versa. This is called Poisson's effect denoted by the Greek letter nu (ν), shown again in Equation 2.22 for convenience. Assuming that a material is compressed along the axial direction,

$$\nu = -\varepsilon_{lateral}/\varepsilon_{longitude} = -\varepsilon_{transverse}/\varepsilon_{axial} = -\varepsilon_x/\varepsilon_y \tag{2.22}$$

where $\varepsilon_{transverse}$ is the transverse strain and ε_{axial} is the axial strain expressed as negative for tension, or extensions, and positive for compression, or contraction. Figure 2.12 expands further on the interrelationships between stress and strain.

From geometrical considerations similar to those for the stress case, it can be shown that

$$\varepsilon_n = \varepsilon_x \cos^2\theta + \gamma_{xy} \sin\theta\cos\theta + \varepsilon_y \sin^2\theta \tag{2.23}$$

$$\gamma_{nm} = (\varepsilon_y - \varepsilon_x)\sin 2\theta + \gamma_{xy}\cos 2\theta \tag{2.24}$$

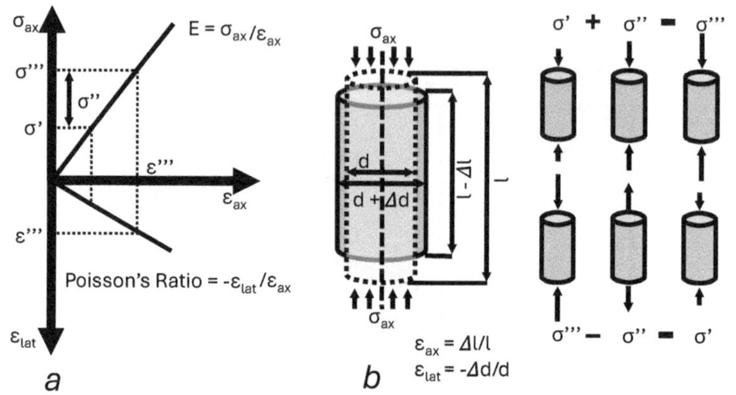

FIGURE 2.12 The inter-relationship between stress and strain (Patchet, 1970): (a) the axial stress and radial strain performance of a rock core resulting from an applied axial strain and (b) diagram of how the core behaves due to the applied axial strain (in the elastic zone).

In general, the stress and strain equations have similarities because ε_x, ε_y, and $\frac{1}{2}\gamma_{xy}$ correspond to σ_x, σ_y, and τ_{xy}, respectively.

2.8 EXAMPLES

Example 2.8.1

The axial strain of a rock specimen is 15 με, and Poisson's Ratio is 0.35. Determine the lateral strain.

Solution:

$$\nu = -\varepsilon_{\text{transverse}}/\varepsilon_{\text{axial}}$$

$$0.35 = -\varepsilon_{\text{transverse}}/15\ \mu\varepsilon$$

$$\varepsilon_{\text{transverse}} = -5.25\ \mu\varepsilon$$

Example 2.8.2

An elastic body is placed under a compressive stress of 150 psi and then under a tension of 40 psi (on the same axis). Determine the total stress and if it is compression or tension.

Solution:

$$\sigma_T = \sigma_c - \sigma_t$$

$$\sigma_T = 150\text{ psi} - 40\text{ psi}$$

$$\sigma_T = 110\text{ psi in compression}$$

Example 2.8.3

An elastic body is placed under a compressive stress of 5 MPa in the x- and z-directions. If the shear stress is initially zero (Case 1), but then increased to 3 MPa (Case 2), what is the maximum principal stress for both cases?

Solution:

$$\text{Case 1}: \tau_{xz} = 0$$

$$\sigma_1 = ½(\sigma_x + \sigma_z) + ½\left[(\sigma_x - \sigma_z)^2 + 4(\tau_{xz})^2\right]^{½}$$

$$\sigma_1 = ½(5+5) + ½\left[(5-5)^2 + 4(0)^2\right]^{½}$$

$$\sigma_1 = 5 \text{ MPa}$$

$$\text{Case 2}: \tau_{xz} = 3 \text{ MPa}$$

$$\sigma_1 = ½(\sigma_x + \sigma_z) + ½\left[(\sigma_x - \sigma_z)^2 + 4(\tau_{xz})^2\right]^{½}$$

$$\sigma_1 = ½(5+5) + ½\left[(5-5)^2 + 4(3)^2\right]^{½}$$

$$\sigma_1 = 5 + ½[36]^{½}$$

$$\sigma_1 = 5 + 3$$

$$\sigma_1 = 8 \text{ MPa}$$

Example 2.8.4

Convert the rectangular coordinates (x,y) in Table 2.1 to polar coordinates (r,θ) shown in Table 2.2.

Solution:

The following is needed to convert into polar coordinates:

- Determine which quadrant each point is located in.
- Calculate the angle θ.
- Calculate the radius r from the origin (i.e., 0,0).

TABLE 2.1
The x and y Coordinates

x	y
5	5
5	3
−5	−5
5	−1

TABLE 2.2
The x and y Coordinates

x	y	Quadrant	$\theta = \tan^{-1}(y/x)$	$r = (x^2+y^2)^{1/2}$
5	5	1	45.0	7.1
5	3	1	30.9	5.8
−5	−5	3	225.0	7.1
5	−1	4	348.7	5.1

Example 2.8.5

A square object is placed under equal elastic compressive stress in the y- and x directions as shown in Figure 2.13. The object is freed of stress, and then rotated 90° in the x–y plane, while the stresses maintain their original orientation. Describe its expected deformation if the material of the object is homogenous and isotropic. Figure 2.13 shows the object and stresses.

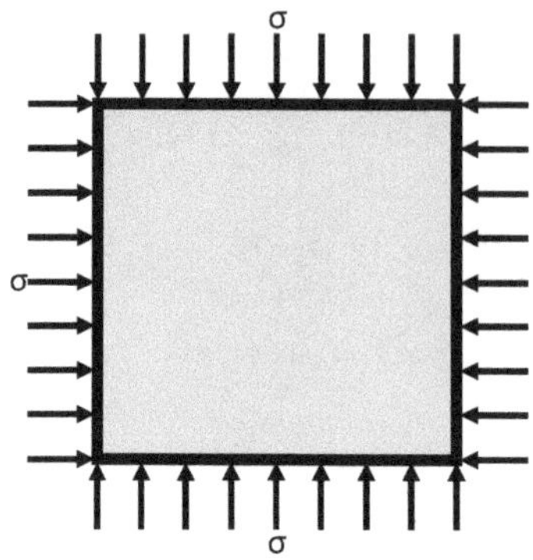

FIGURE 2.13 The object with the equal applied stresses applied on both orthogonal axes.

Solution:

The deformation of the object will be equal in all directions because material properties and stresses are the same in all directions even after the rotation of 90°.

Example 2.8.6

Assume a weak band of material is within the body as shown in Figure 2.14 which causes the material to be weaker in the direction parallel to the band. For the case in the figure, the weaker direction, parallel to the band, is in the y-direction. Describe the orientation of the block so that the greatest strength aligns with the greatest stress.

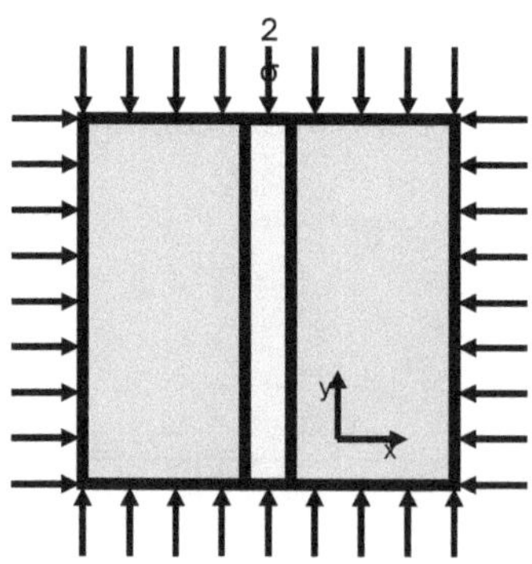

FIGURE 2.14 The block with the weak band shown parallel to the y-axis.

Solution:

If the object is rotated 90° from the figure, the greatest strength is in the direction of the greatest stress (2σ).

Example 2.8.7

If the elastic compressive stress in Example 2.8.5 is suddenly removed in the x-direction, describe the deformation. The applied stress is shown in Figure 2.15a, and the resulting deformation is shown in Figure 2.15b.

Problem and Solution:

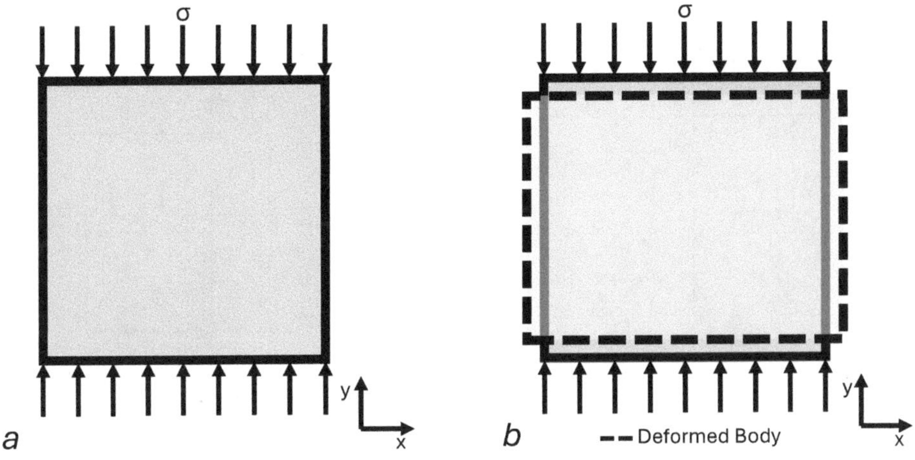

FIGURE 2.15 (a) The applied stress and (b) the resulting deformation.

In this case, the sudden removal of stress in the x-direction will cause the object to compress in the y-direction and expand in the x-direction. The magnitude of the deformation in the x-direction depends on Poisson's Ratio.

Example 2.8.8

If the base of the square object in Figure 2.16 is fixed from displacement, what is the anticipated deformation given the applied stresses? Assume that the object is homogenous and isotropic.

Problem and Solution:

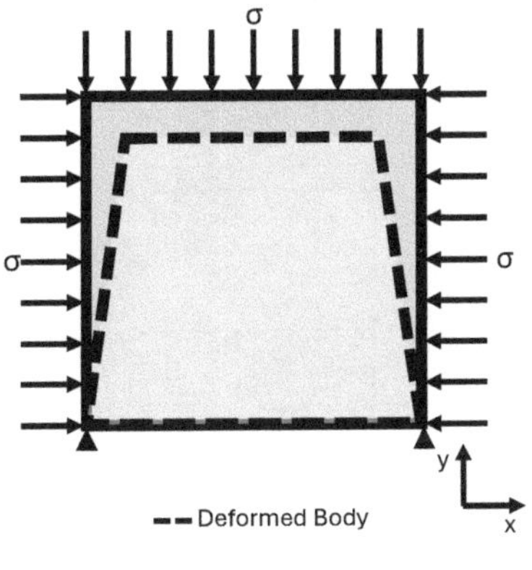

FIGURE 2.16 The applied stresses on the square body and the resulting deformed body (shown by the dashed shape).

Example 2.8.9

The following example is abstract in nature. Given the figures, stresses, and fixities shown below in Figure 2.17, determine the anticipated deformation (ignoring Poisson's Ratio for simplicity).

Problems and solutions:

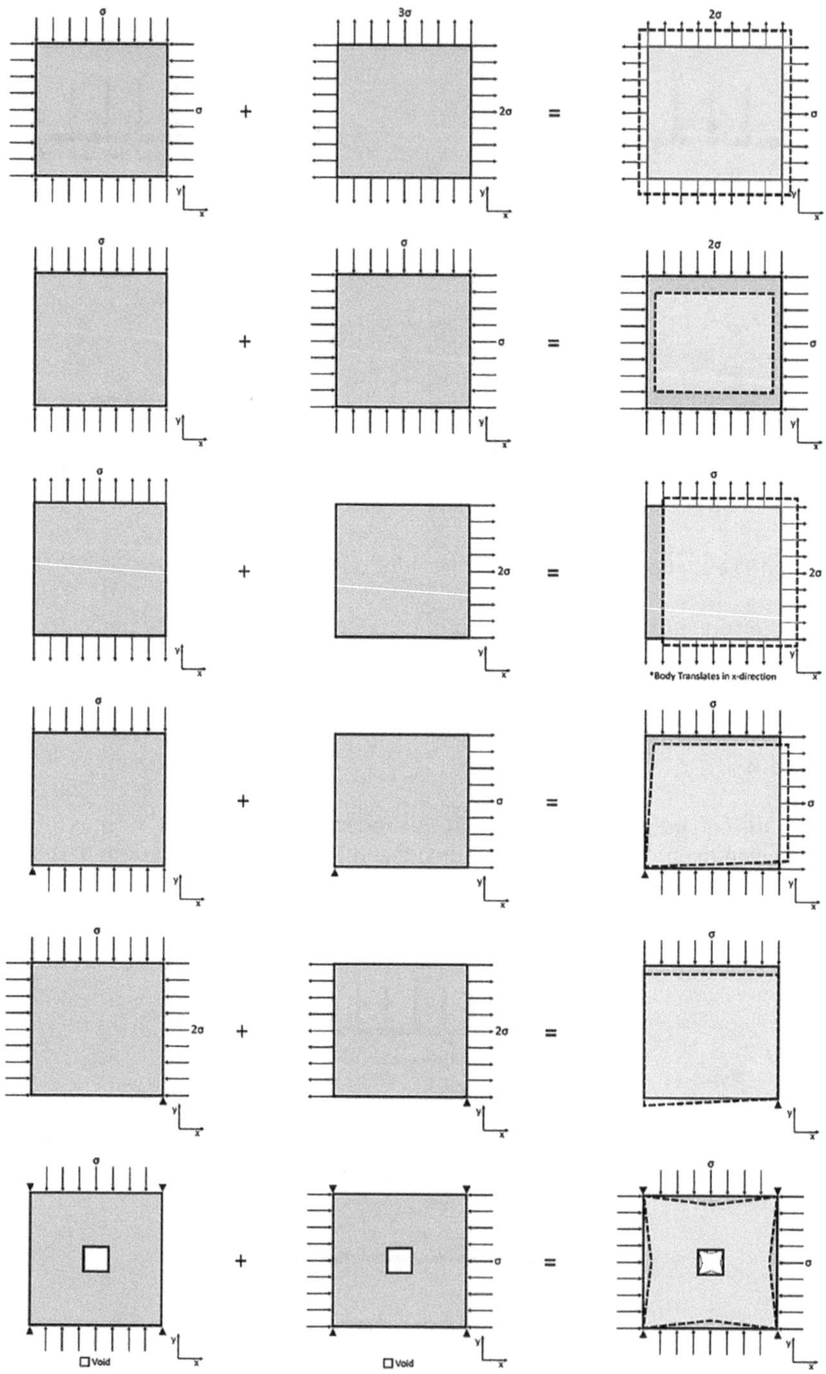

FIGURE 2.17 The applied stresses and the resulting deformation.

Example 2.8.10

Assume that two rock specimens have similar properties. Each has a compressive strength of 2,000 psi and a tensile strength of 300 psi. If one rock specimen is placed under tensile load and the other is placed under compressive load, determine if the specimens reach yield, and if so, under which loading condition. The modulus of elasticity is 5,000 psi; the strain in tension (ε_t) is 6%; and in compression, the strain (ε_c) is 5%. Assuming a 5% strain in tension and compression what must be the modulus in compression and tension to yield at the yield strength?

$$E = \sigma_c / \varepsilon_c$$

$$5{,}000\ \text{lbs/in}^2 = \sigma_c / (5/100)$$

$$\sigma_c = 250\ \text{psi}$$

$$E = \sigma_t / \varepsilon_t$$

$$5{,}000\ \text{lbs/in}^2 = \sigma_t / (6/100)$$

$$\sigma_t = 300\ \text{psi}$$

The specimen has reached yield in tension:

$$E = \sigma_c / \varepsilon_c$$

$$E = (2{,}000\ \text{psi}) / (5/100)$$

$$E = 40{,}000\ \text{psi}$$

$$E = \sigma_t / \varepsilon_t$$

$$E = (300\ \text{psi}) / (5/100)$$

$$E = 6{,}000\ \text{psi}$$

Example 2.8.11

Determine the minimum and maximum principal stresses if the stresses in the x- and y-directions are 10 and 4 MPa, respectively. Also find the maximum shear stress.

Given:

$$\sigma_x = 10\ \text{MPa}$$

$$\sigma_y = 4\ \text{MPa}$$

$$\tau_{xy} = 0\ \text{MPa}$$

Solution:

$$\sigma_1 = \tfrac{1}{2}(\sigma_x + \sigma_y) + \tfrac{1}{2}\left[(\sigma_x - \sigma_y)^2 + 4(\tau_{xy})^2\right]^{1/2}$$

$$\sigma_1 = \tfrac{1}{2}(10 + 4) + \tfrac{1}{2}\left[(10 - 4)^2 + 4(0)^2\right]^{1/2}$$

$$\sigma_1 = \tfrac{1}{2}(14) + \tfrac{1}{2}\left[(6)^2\right]^{1/2}$$

$$\sigma_1 = \sigma_{max} = 7 + 3 = 10 \text{ MPa}$$

$$\sigma_2 = \tfrac{1}{2}(\sigma_x + \sigma_y) - \tfrac{1}{2}\left[(\sigma_x - \sigma_y)^2 + 4(\tau_{xy})^2\right]^{1/2}$$

$$\sigma_2 = \sigma_{min} = 7 - 3 = 4 \text{ MPa}$$

$$\tau_{max} = (\sigma_{max} - \sigma_{min})/2$$

$$\tau_{max} = (10 - 4)/2$$

$$\tau_{max} = 3 \text{ MPa}$$

The following relationship can be used as a check:

$$\sigma_1 + \sigma_2 = \sigma_x + \sigma_y$$

Example 2.8.12

Determine the minimum and maximum principal stresses on a body where normal stresses are zero and shear stress is 12 psi.

Given:

$$\tau_{xy} = 12 \text{ psi}$$

Solution:

$$\sigma_1 = \tfrac{1}{2}(\sigma_x + \sigma_y) + \tfrac{1}{2}\left[(\sigma_x - \sigma_y)^2 + 4(\tau_{xy})^2\right]^{1/2}$$

$$\sigma_1 = \tfrac{1}{2}(0) + \tfrac{1}{2}\left[(0)^2 + 4(12)^2\right]^{1/2}$$

$$\sigma_1 = \tfrac{1}{2}[\left(4(12)^2\right]^{1/2}$$

$$\sigma_1 = 12 \text{ psi}$$

$$\sigma_2 = \tfrac{1}{2}(\sigma_x + \sigma_y) - \tfrac{1}{2}\left[(\sigma_x - \sigma_y)^2 + 4(\tau_{xy})^2\right]^{1/2}$$

$$\sigma_2 = \tfrac{1}{2}(0) - \tfrac{1}{2}\left[(0)^2 + 4(12)^2\right]^{1/2}$$

$$\sigma_2 = 0 - \tfrac{1}{2}\left[4(12)^2\right]^{1/2}$$

$$\sigma_2 = -12 \text{ psi}$$

Example 2.8.13

Find the maximum and minimum principal stresses given the following:

Given:

$$\sigma_x = 120 \text{ MPa}$$

$$\sigma_y = -250 \text{ MPa}$$

$$\tau_{xy} = 45 \text{ MPa}$$

Solution:

$$\sigma_1 = ½(\sigma_x + \sigma_y) + ½\left[\left(\sigma_x - \sigma_y\right)^2 + 4\left(\tau_{xy}\right)^2\right]^{½}$$

$$\sigma_1 = ½\left[(120 + (-250 \text{ MPa}))\right] + ½\left[(120 + 250)^2 + 4(45)^2\right]^{½}$$

$$\sigma_1 = ½(-130) + ½\left[(370)^2 + 4(45)^2\right]^{½}$$

$$\sigma_1 = -65 + ½\left[(136{,}900 + 8{,}100)\right]^{½}$$

$$\sigma_1 = -65 + ½[145{,}000]^{½}$$

$$\sigma_1 = -65 + 190$$

$$\sigma_1 = 125 \text{ MPa}$$

$$\sigma_2 = ½(\sigma_x + \sigma_y) - ½\left[\left(\sigma_x - \sigma_y\right)^2 + 4\left(\tau_{xy}\right)^2\right]^{½}$$

$$\sigma_2 = -65 - 190$$

$$\sigma_2 = -255 \text{ MPa}$$

Example 2.8.14

Find the total normal and shear strain for any orientation θ given the following:
Given:

$$\varepsilon_x = 120\ \mu\varepsilon$$

$$\varepsilon_y = -250\ \mu\varepsilon$$

$$\gamma_{xy} = 45\ \mu\varepsilon$$

Solution:

$$\varepsilon_n = \varepsilon_x\cos^2\theta + \gamma_{xy}\sin\theta\cos\theta + \varepsilon_y\sin^2\theta$$

$$\varepsilon_n = 120\cos^2\theta + 45\sin\theta\cos\theta - 250\sin^2\theta$$

$$\gamma_{nm} = \left(\varepsilon_y - \varepsilon_x\right)\sin2\theta + \gamma_{xy}\cos2\theta$$

$$\gamma_{nm} = (-250 - 120)\sin2\theta + 45\cos2\theta$$

$$\gamma_{nm} = -370\sin2\theta + 45\cos2\theta$$

Example 2.8.15

Find the radial and tangential stresses on a body if the stress field is rotated by 15° and the stresses in rectangular coordinates are $\sigma_x = 230$ MPa and $\sigma_y = 178$ MPa.

Solution:

$$\sigma_r = \sigma_x \cos^2\theta + 2\tau_{xy}\sin\theta\cos\theta + \sigma_y \sin^2\theta$$

$$\sigma_r = 230\cos^2(15°) + 178\sin^2(15°)$$

$$\sigma_r = 215 + 12$$

$$\sigma_r = 227 \text{ MPa}$$

$$\sigma_\theta = \sigma_x \sin^2\theta - 2\tau_{xy}\sin\theta\cos\theta + \sigma_y \cos^2\theta$$

$$\sigma_\theta = 230\sin^2(15°) + 178\cos^2(15°)$$

$$\sigma_\theta = 15 + 166$$

$$\sigma_\theta = 181 \text{ MPa}$$

As a check:

$$\sigma_r + \sigma_\theta = \sigma_x + \sigma_y$$

Example 2.8.16

Determine the average between the minimum and maximum principal stresses assuming zero shear stress when:

Given:

$$\sigma_x = 160 \text{ psi}$$

$$\sigma_y = -160 \text{ psi}$$

Solution:

$$\sigma_1 = ½(\sigma_x + \sigma_y) + ½\left[(\sigma_x - \sigma_y)^2 + 4(\tau_{xy})^2\right]^{½}$$

$$\sigma_1 = ½[(160 + (-160))] + ½[(160 + 160)^2 + 4(0)^2]^{½}$$

$$\sigma_1 = 0 + ½\left[(320)^2\right]^{½}$$

$$\sigma_1 = 160 \text{ psi}$$

$$\sigma_2 = ½(\sigma_x + \sigma_y) - ½\left[(\sigma_x - \sigma_y)^2 + 4(\tau_{xy})^2\right]^{½}$$

$$\sigma_2 = ½[(160 + (-160))] - ½\left[(160 + 160)^2 + 4(0)^2\right]^{½}$$

$$\sigma_2 = 0 - ½\left[(320)^2\right]^{½}$$

$$\sigma_2 = -160 \text{ psi}$$

$$\sigma_{avg} = (\sigma_1 + \sigma_2)/2$$

$$\sigma_{avg} = [160 + (-160)]/2$$

$$\sigma_{avg} = 0 \text{ psi}$$

Example 2.8.17

Consider the following stresses. Find the radial and tangential stress at an orientation of 0°. If the stress in the y-direction in removed causing a reorientation of 45° find the radial and tangential stress. The initial stress state is as follows:

Given:

$$\sigma_x = 120 \text{ MPa}$$

$$\sigma_y = -250 \text{ MPa}$$

$$\tau_{xy} = 45 \text{ MPa}$$

Solution:

Part 1:

$$\sigma_r = \sigma_x\cos^2\theta + 2\tau_{xy}\sin\theta\cos\theta + \sigma_y\sin^2\theta$$

$$\sigma_r = 120\cos^2 0° + 2(45)\sin 0°\cos 0° + (-250)\sin^2 0°$$

$$\sigma_r = 120 \text{ MPa}$$

$$\sigma_\theta = \sigma_x\sin^2\theta - 2\tau_{xy}\sin\theta\cos\theta + \sigma_y\cos^2\theta$$

$$\sigma_\theta = 120\sin^2 0° - 2(45)\sin 0°\cos 0° + (-250)\cos^2 0°$$

$$\sigma_\theta = -250 \text{ MPa}$$

Part 2:

$$\sigma_x = 120 \text{ MPa}$$

$$\sigma_y = 0 \text{ MPa}$$

$$\tau_{xy} = 45 \text{ MPa}$$

$$\sigma_r = \sigma_x\cos^2\theta + 2\tau_{xy}\sin\theta\cos\theta + \sigma_y\sin^2\theta$$

$$\sigma_r = \left[120\cos^2(45°)\right] + \left[2(45)\sin(45°)\cos(45°)\right] + \left[0\sin^2(45°)\right]$$

$$\sigma_r = 60 + 45$$

$$\sigma_r = 105 \text{ MPa}$$

$$\sigma_\theta = \sigma_x \sin^2\theta - 2\tau_{xy}\sin\theta\cos\theta + \sigma_y\cos^2\theta$$

$$\sigma_\theta = 120\sin^2(45°) - 2(45)\sin(45°)\cos(45°) + (0)\cos^2(45°)$$

$$\sigma_\theta = 60 - 45$$

$$\sigma_\theta = 15 \text{ MPa}$$

Example 2.8.18

If the modulus of elasticity of a material is 80 MPa, then determine the principal strains only if the yield stress has not been exceeded by one of the principal stresses. The yield stress of the material is 300 MPa in compression and 30 MPa in tension. The following normal stresses are given:

Given:

$$\sigma_x = 40 \text{ MPa}$$

$$\sigma_y = -15 \text{ MPa}$$

$$\tau_{xy} = 45 \text{ MPa}$$

Solution:

$$\sigma_1 = ½(\sigma_x + \sigma_y) + ½\left[(\sigma_x - \sigma_y)^2 + 4(\tau_{xy})^2\right]^{½}$$

$$\sigma_1 = ½[40 + (-15)] + ½[(40 + 15)^2 + 4(45)^2]^{½}$$

$$\sigma_1 = ½[25] + ½\left[(55)^2 + 4(45)^2\right]^{½}$$

$$\sigma_1 = 12.5 + ½[3{,}025 + 8{,}100]^{½}$$

$$\sigma_1 = 12.5 + 52.7$$

$$\sigma_1 = 65 \text{ MPa}$$

$$\sigma_2 = ½(\sigma_x + \sigma_y) - ½\left[(\sigma_x - \sigma_y)^2 + 4(\tau_{xy})^2\right]^{½}$$

$$\sigma_2 = ½\left[40 + (-15)\right] - ½\left[(40 + 15)^2 + 4(45)^2\right]^{½}$$

$$\sigma_2 = 12.5 - 52.7$$

$$\sigma_2 = -40 \text{ MPa}$$

The secondary principal stress exceeds the tensile strength of the material.

Example 2.8.19

If the maximum and minimum principal stresses are 10 MPa and 5 MPa, respectively, and the maximum shear stress can be determined using the formula below, determine the shear strain corresponding to the maximum shear stress if the modulus of rigidity is 50 MPa:

$$\tau_{max} = (\sigma_{max} - \sigma_{min})/2$$

$$\sigma_1 = 10 \text{ MPa}$$

$$\sigma_2 = 5 \text{ MPa}$$

$$\tau_{max} = (\sigma_{max} - \sigma_{min})/2$$

$$\tau_{max} = (10 - 5)/2$$

$$\tau_{max} = 2.5 \text{ MPa}$$

$$G = \tau_{max}/\gamma_{max}$$

$$50 \text{ MPa} = 2.5 \text{ MPa}/\gamma$$

$$\gamma = 0.05$$

Example 2.8.20

Show that $\sigma_r+\sigma_\theta=\sigma_x+\sigma_y$. If $\sigma_x=7$ psi, $\sigma_y=9$ psi, and $\tau_{xy}=3$ psi, check that the polar and rectangular coordinates match.

Solution:

$$\sigma_r = \sigma_x\cos^2\theta + 2\tau_{xy}\sin\theta\cos\theta + \sigma_y\sin^2\theta$$

$$\sigma_\theta = \sigma_x\sin^2\theta - 2\tau_{xy}\sin\theta\cos\theta + \sigma_y\cos^2\theta$$

$$\sigma_r + \sigma_\theta = \left[\sigma_x\cos^2\theta + 2\tau_{xy}\sin\theta\cos\theta + \sigma_y\sin^2\theta\right] + \left[\sigma_x\sin^2\theta - 2\tau_{xy}\sin\theta\cos\theta + \sigma_y\cos^2\theta\right]$$

$$\sigma_r + \sigma_\theta = \left[\sigma_x\cos^2\theta + \sigma_x\sin^2\theta\right] + \left[\sigma_y\sin^2\theta + \sigma_y\cos^2\theta\right]$$

$$\sigma_r + \sigma_\theta = \sigma_x\left(\cos^2\theta + \sin^2\theta\right) + \sigma_y(\cos^2\theta + \sin^2\theta)$$

$$\text{Note}: \cos^2\theta + \sin^2\theta = 1 \text{ therefore}:$$

$$\sigma_r + \sigma_\theta = \sigma_x + \sigma_y$$

As a check:

$$\sigma_r = \sigma_x\cos^2\theta + 2\tau_{xy}\sin\theta\text{Cos}\theta + \sigma_y\sin^2\theta$$

$$\sigma_r = (7)\cos^2 0° + 2(3)\sin 0°\cos 0° + (9)\sin^2 0°$$

$$\sigma_r = 7 \text{ psi}$$

$$\sigma_\theta = \sigma_x \sin^2\theta - 2\tau_{xy}\sin\theta\cos\theta + \sigma_y\cos^2\theta$$

$$\sigma_\theta = (7)\sin^2 0° - 2\tau_{xy}\sin 0° \cos 0° + (9)\cos^2 0°$$

$$\sigma_\theta = 9 \text{ psi}$$

$$\sigma_r + \sigma_\theta = 7 \text{ psi} + 9 \text{ psi} = 16 \text{ psi}$$

$$\sigma_x + \sigma_y = 7 \text{ psi} + 9 \text{ psi} = 16 \text{ psi}$$

Example 2.8.21

Determine the stresses in polar coordinates and the maximum and minimum principal stresses in rectangular coordinates given the following:
Given:

$$\sigma_x = 50 \text{ MPa}$$

$$\sigma_y = 55 \text{ MPa}$$

$$\tau_{xy} = 5 \text{ MPa}$$

$$\theta = 0°$$

Solution:

$$\sigma_r = \sigma_x \cos^2\theta + 2\tau_{xy}\sin\theta\cos\theta + \sigma_y \sin^2\theta$$

$$\sigma_r = \sigma_x \cos^2\theta$$

$$\sigma_r = 50 \text{ MPa}$$

$$\sigma_\theta = \sigma_x \sin^2\theta - 2\tau_{xy}\sin\theta\cos\theta + \sigma_y\cos^2\theta$$

$$\sigma_\theta = \sigma_y \cos^2\theta$$

$$\sigma_\theta = 55 \text{ MPa}$$

$$\sigma_1 = ½(\sigma_x + \sigma_y) + ½\left[(\sigma_x - \sigma_y)^2 + 4(\tau_{xy})^2\right]^{½}$$

$$\sigma_1 = ½(50 + 55) + ½\left[(50 - 55)^2 + 4(5)^2\right]^{½}$$

$$\sigma_1 = 52.5 + ½[25 + 100]^{½}$$

$$\sigma_1 = 58 \text{ MPa}$$

$$\sigma_2 = ½(\sigma_x + \sigma_y) - ½\left[(\sigma_x - \sigma_y)^2 + 4(\tau_{xy})^2\right]^{½}$$

$$\sigma_2 = ½(50 + 55) - ½\left[(50 - 55)^2 + 4(5)^2\right]^{½}$$

$$\sigma_2 = 52.5 - ½[25 + 100]^{½}$$

$$\sigma_2 = 47 \text{ MPa}$$

Example 2.8.22

Repeat the previous example for a shear stress of the same magnitude as the normal stress of the x-direction.

Given:

$$\sigma_x = 50 \text{ MPa}$$

$$\sigma_y = 55 \text{ MPa}$$

$$\tau_{xy} = 50 \text{ MPa}$$

$$\theta = 0°$$

Solution:

$$\sigma_r = \sigma_x \cos^2\theta + 2\tau_{xy}\sin\theta\cos\theta + \sigma_y \sin^2\theta$$

$$\sigma_r = \sigma_x \cos^2\theta$$

$$\sigma_r = 50 \text{ MPa}$$

$$\sigma_\theta = \sigma_x \sin^2\theta - 2\tau_{xy}\sin\theta\cos\theta + \sigma_y \cos^2\theta$$

$$\sigma_\theta = \sigma_y \cos^2\theta$$

$$\sigma_\theta = 55 \text{ MPa}$$

$$\sigma_1 = ½(\sigma_x + \sigma_y) + ½\left[(\sigma_x - \sigma_y)^2 + 4(\tau_{xy})^2\right]^{½}$$

$$\sigma_1 = ½(50 + 55) + ½\left[(50 - 50)^2 + 4(50)^2\right]^{½}$$

$$\sigma_1 = 52.5 + ½[0 + 10{,}000]^{½}$$

$$\sigma_1 = 102.5 \text{ MPa}$$

$$\sigma_2 = ½(\sigma_x + \sigma_y) - ½\left[(\sigma_x - \sigma_y)^2 + 4(\tau_{xy})^2\right]^{½}$$

$$\sigma_2 = ½(50 + 55) - ½\left[(50 - 50)^2 + 4(50)^2\right]^{½}$$

$$\sigma_2 = 52.5 - ½[0 + 10{,}000]^{½}$$

$$\sigma_2 = 2.5 \text{ MPa}$$

Example 2.8.23

Principal strains are expressed similarly as principal stresses as given below:

$$\varepsilon_1 = ½(\varepsilon_x + \varepsilon_y) + ½\left[(\varepsilon_x - \varepsilon_y)^2 + (\gamma_{xy})^2\right]^{½}$$

$$\varepsilon_2 = ½(\varepsilon_x + \varepsilon_y) - ½\left[(\varepsilon_x - \varepsilon_y)^2 + (\gamma_{xy})^2\right]^{½}$$

Given the following, calculate the minimum and maximum principal strains:

$$\varepsilon_x = 120\ \mu\varepsilon$$

$$\varepsilon_y = -250\ \mu\varepsilon$$

$$\gamma_{xy} = 45\ \mu\varepsilon$$

Solution:

$$\varepsilon_1 = ½(\varepsilon_x + \varepsilon_y) + ½\left[(\varepsilon_x - \varepsilon_y)^2 + (\gamma_{xy})^2\right]^{½}$$

$$\varepsilon_1 = ½[(120 + (-250)] + ½\left[(120 + 250)^2 + (45)^2\right]^{½}$$

$$\varepsilon_1 = -65 + ½[136900 + 2025]^{½}$$

$$\varepsilon_1 = -65 + 186.4$$

$$\varepsilon_1 = 121.4\ \mu\varepsilon$$

$$\varepsilon_2 = ½(\varepsilon_x + \varepsilon_y) - ½\left[(\varepsilon_x - \varepsilon_y)^2 + (\gamma_{xy})^2\right]^{½}$$

$$\varepsilon_2 = ½[120 + (-250)] - ½\left[(120 + 250)^2 + (45)^2\right]^{½}$$

$$\varepsilon_2 = -65 - 186$$

$$\varepsilon_2 = -251.4\ \mu\varepsilon$$

Example 2.8.24

Resolve the stresses below in polar coordinates. Provide the answer in Cartesian (rectangular) coordinates at zero degrees.
Given:

$$\sigma_x = 12\ \text{MPa}$$

$$\sigma_y = 24\ \text{MPa}$$

$$\tau_{xy} = 5\ \text{MPa}$$

$$\theta = 90°$$

Solution:

In polar coordinates:

$$\sigma_r = \sigma_x \cos^2\theta + 2\tau_{xy}\sin\theta\cos\theta + \sigma_y \sin^2\theta$$

$$\sigma_r = \left[12\cos^2(90°)\right] + \left[2(5)\sin(90°)\cos(90°)\right] + \left[24\sin^2(90°)\right]$$

$$\sigma_r = 24 \text{ MPa}$$

$$\sigma_\theta = \sigma_x \sin^2\theta - 2\tau_{xy}\sin\theta\cos\theta + \sigma_y \cos^2\theta$$

$$\sigma_\theta = \left[12\sin^2(90°)\right] - \left[2(5)\sin(90°)\cos(90°)\right] + \left[24\cos^2(90°)\right]$$

$$\sigma_\theta = 12 \text{ MPa}$$

$$\tau_{r\theta} = ½(\sigma_y - \sigma_x)\sin 2\theta + \tau_{xy}\cos 2\theta$$

$$\tau_{r\theta} = ½(24 - 12)\sin 2(90°) + 5\cos 2(90°)$$

$$\tau_{r\theta} = -5 \text{ MPa}$$

In rectangular coordinates:

$$\sigma_r = \sigma_x \cos^2\theta + 2\tau_{xy}\sin\theta\cos\theta + \sigma_y \sin^2\theta$$

$$24 = \sigma_x \cos^2(0°) + 2\tau_{xy}\sin(0°)\cos(0°) + \sigma_y \sin^2(0°)$$

$$\sigma_x = 24 \text{ MPa}$$

$$\sigma_\theta = \sigma_x \sin^2\theta - 2\tau_{xy}\sin\theta\cos\theta + \sigma_y \cos^2\theta$$

$$12 = \sigma_x \sin^2(0°) - 2\tau_{xy}\sin(0°)\cos(0°) + \sigma_y \cos^2(0°)$$

$$\sigma_y = 12 \text{ MPa}$$

$$\tau_{r\theta} = ½(\sigma_y - \sigma_x)\sin 2\theta + \tau_{xy}\cos 2\theta$$

$$-5 = ½(\sigma_y - \sigma_x)\sin 2(0°) + \tau_{xy}\cos 2(0°)$$

$$\tau_{xy} = -5 \text{ MPa}$$

Example 2.8.25

Determine the radial and tangential normal and shear strains at 90°, 180°, and 270° given the following strains in rectangular coordinates:

Given:

$$\varepsilon_x = 5\ \mu\varepsilon$$

$$\varepsilon_y = 10\ \mu\varepsilon$$

$$\gamma_{xy} = -3\ \mu\varepsilon$$

$$\theta = 0°$$

Solution:

Part 1 where $\theta = 90°$:

$$\varepsilon_r = \varepsilon_x \cos^2\theta + \gamma_{xy}\sin\theta\cos\theta + \varepsilon_y \sin^2\theta$$

$$\varepsilon_r = (5)\cos^2(90°) + (-3)\sin(90°)\cos(90°) + (10)\sin^2(90°)$$

$$\varepsilon_r = 10\ \mu\varepsilon$$

$$\varepsilon_\theta = \varepsilon_x \sin^2\theta - \gamma_{xy}\sin\theta\cos\theta + \varepsilon_y \cos^2\theta$$

$$\varepsilon_\theta = (5)\sin^2(90°) - (-3)\sin(90°)\cos(90°) + (10)\cos^2(90°)$$

$$\varepsilon_\theta = 5\ \mu\varepsilon$$

$$\gamma_{r\theta} = (\varepsilon_y - \varepsilon_x)\sin 2\theta + \gamma_{xy}\cos 2\theta$$

$$\gamma_{r\theta} = (10 - 5)\sin 2(90°) + (-3)\cos 2(90°)$$

$$\gamma_{r\theta} = 3\ \mu\varepsilon$$

Part 2 where $\theta = 180°$:

$$\varepsilon_r = \varepsilon_x \cos^2\theta + \gamma_{xy}\sin\theta\cos\theta + \varepsilon_y \sin^2\theta$$

$$\varepsilon_r = (5)\cos^2(180°) + (-3)\sin(180°)\cos(180°) + (10)\sin^2(180°)$$

$$\varepsilon_r = 5\ \mu\varepsilon$$

$$\varepsilon_\theta = \varepsilon_x \sin^2\theta - \gamma_{xy}\sin\theta\cos\theta + \varepsilon_y \cos^2\theta$$

$$\varepsilon_\theta = (5)\sin^2(180°) - (-3)\sin(180°)\cos(180°) + (10)\cos^2(180°)$$

$$\varepsilon_\theta = 10\ \mu\varepsilon$$

$$\gamma_{r\theta} = (\varepsilon_y - \varepsilon_x)\sin 2\theta + \gamma_{xy}\cos 2\theta$$

$$\gamma_{r\theta} = (10 - 5)\sin 2(180°) + (-3)\cos 2(180°)$$

$$\gamma_{r\theta} = -3\ \mu\varepsilon$$

Part 3 where $\theta = 270°$:

$$\varepsilon_r = \varepsilon_x \cos^2\theta + \gamma_{xy}\sin\theta\cos\theta + \varepsilon_y \sin^2\theta$$

$$\varepsilon_r = (5)\cos^2(270°) + (-3)\sin(270°)\cos(270°) + (10)\sin^2(270°)$$

$$\varepsilon_r = 10\ \mu\varepsilon$$

$$\varepsilon_\theta = \varepsilon_x \sin^2\theta - \gamma_{xy}\sin\theta\cos\theta + \varepsilon_y \cos^2\theta$$

$$\varepsilon_\theta = (5)\sin^2(270°) - (-3)\sin(270°)\cos(270°) + (10)\cos^2(270°)$$

$$\varepsilon_\theta = 5\ \mu\varepsilon$$

$$\gamma_{r\theta} = (\varepsilon_y - \varepsilon_x)\sin 2\theta + \gamma_{xy}\cos 2\theta$$

$$\gamma_{r\theta} = (10 - 5)\sin 2(270°) + (-3)\cos 2(270°)$$

$$\gamma_{r\theta} = 3\ \mu\varepsilon$$

Example 2.8.26

Calculate the principal stresses for the two cases described. For Case 1, use the sign conventions in mining where compression is taken as positive. For Case 2, use the traditional engineering convention where compression is negative. For both cases, assume that the normal stresses given below are compressive.

The stresses are as follows:

$$\sigma_x = 50 \text{ MPa}(\text{compression})$$

$$\sigma_y = 50 \text{ MPa}(\text{compression})$$

$$\tau_{xy} = -35 \text{ MPa}$$

Solution:

Case 1:

$$\sigma_1 = ½(\sigma_x + \sigma_y) + ½\left[(\sigma_x - \sigma_y)^2 + 4(\tau_{xy})^2\right]^{½}$$

$$\sigma_1 = ½(50 + 50) + ½\left[(50 - 50)^2 + 4(-35)^2\right]^{½}$$

$$\sigma_1 = 50 + 35$$

$$\sigma_1 = 85 \text{ MPa compression}$$

$$\sigma_2 = ½(\sigma_x + \sigma_y) - ½\left[(\sigma_x - \sigma_y)^2 + 4(\tau_{xy})^2\right]^{½}$$

$$\sigma_2 = ½(50 + 50) - ½\left[(50 - 50)^2 + 4(-35)^2\right]^{½}$$

$$\sigma_2 = 50 - 35$$

$$\sigma_2 = 15 \text{ MPa compression}$$

Case 2:

$$\sigma_1 = ½(\sigma_x + \sigma_y) + ½\left[(\sigma_x - \sigma_y)^2 + 4(\tau_{xy})^2\right]^{½}$$

$$\sigma_1 = ½(-50 + -50) + ½\left[(-50 - -50)^2 + 4(-35)^2\right]^{½}$$

$$\sigma_1 = -50 + 35$$

$$\sigma_1 = -15 \text{ MPa compression}$$

Since compression was taken as negative, this is –15 MPa in compression which is the same answer as Case 1:

$$\sigma_2 = ½(\sigma_x + \sigma_y) - ½\left[(\sigma_x - \sigma_y)^2 + 4(\tau_{xy})^2\right]^{½}$$

$$\sigma_2 = ½(-50 + -50) - ½\left[(-50 - -50)^2 + 4(-35)^2\right]^{½}$$

$$\sigma_2 = -50 - 35$$

$$\sigma_2 = -85 \text{ MPa compression}$$

Since compression was taken as negative, this is −85 MPa in compression which is the same answer as Case 1.

REFERENCES

Patchet, S.J. (1970) Rock mechanics studies associated with the development of a deep potash mine. Ph.D. Thesis, University of Newcastle-upon-Tyne, United Kingdom.

Spearing, A.J.S. (1995) *Handbook on Hard Rock Strata Control*. SAIMM Special Publication Series SP6. ISBN 1-874832-45-5.

3 The Earth and Structural Geology

3.1 GEOLOGIC TIME

The earth is at least 4.6 billion years old (Nelson, 2024). The geologic time scale of the earth (equivalent to a calendar for rock formation) is shown in Figure 3.1 and consists of four broad time periods, with the Eon being the largest:

- Eon.
- Era.
- Period.
- Epoch.

They each represent significant geological events relative to the landscape, diversity, and climate.

A section classifying the layers, or the geosphere, of the earth's crust is shown in Figure 3.2. According to Christiansen and Hamblin (2015), the layers are defined by their physical properties and thicknesses as follows:

- Solid inner core (6,370–5,150 km).
- Liquid outer core (5,150–2,900 km).
- Solid mesosphere (2,900 to 350–500 km).

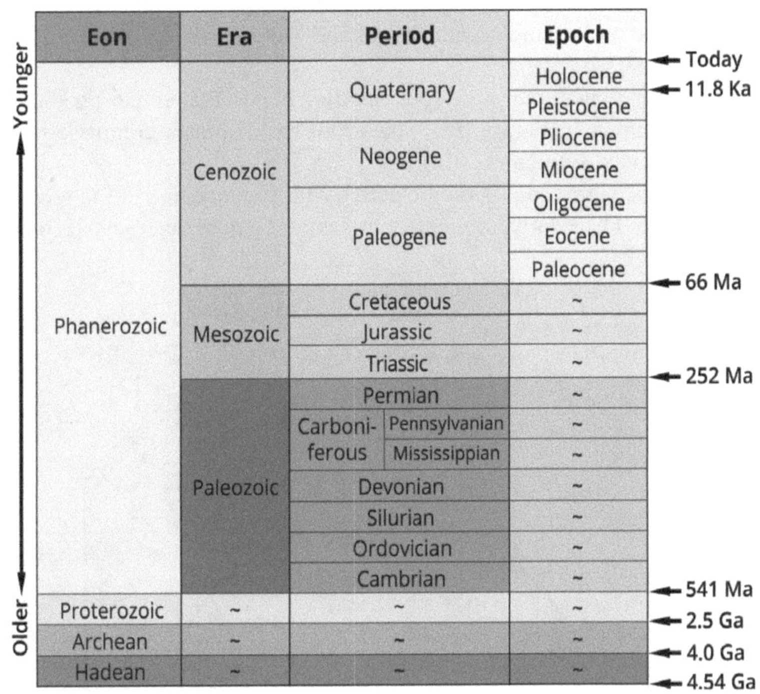

FIGURE 3.1 Geologic time scale (Hendricks, 2019).

DOI: 10.1201/9781032704845-3

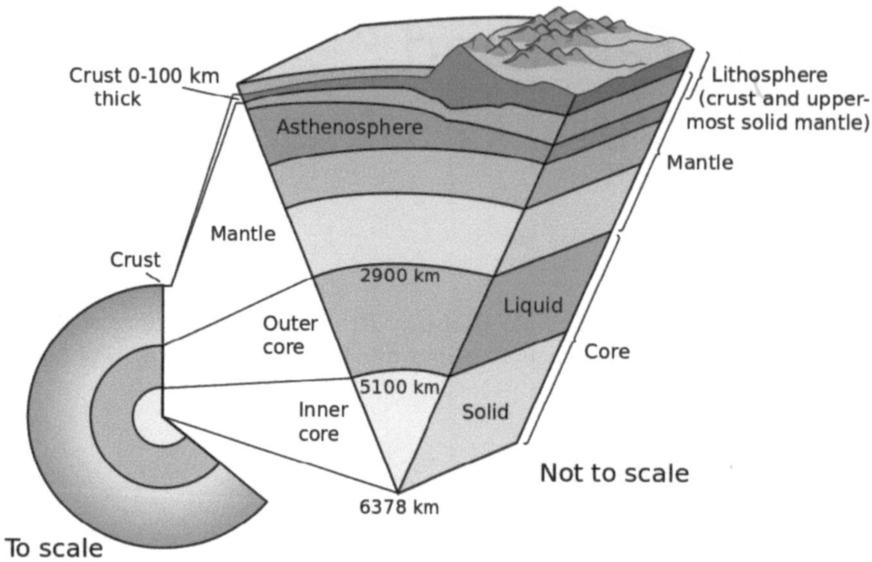

FIGURE 3.2 Section through the earth (USGS, 1999).

- Plastic asthenosphere (350–500 to 100–300 km).
- Rigid lithosphere (100–300 to 0 km).

3.2 PLATE TECTONICS

Plate tectonics is a theory explaining the changing structure of the earth's crust and many associated phenomena resulting from the interaction of rigid lithospheric plates which move very slowly in geological time over the underlying mantle (Read and Watson, 1975). An example of these movements is shown in Figure 3.3.

There are seven major plates and numerous smaller plates that make up the crust and upper mantle of the earth as shown in Figure 3.4. The seven major plates comprise nearly 94% of the earth's surface area (Tarbuck & Lugens 2014).

At the plate boundaries, significant geologic activity takes place due to the relative movement of the plates with each other. The relative boundary movement can be characterized as follows:

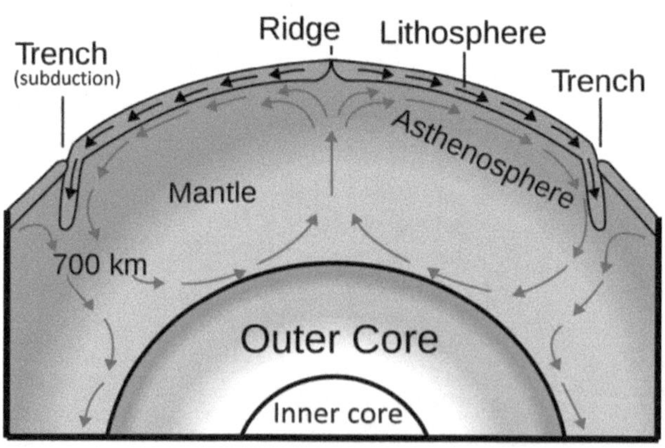

FIGURE 3.3 Section of the earth's crust showing the basic mechanics of tectonic plates (Kious and Tilling, 1996).

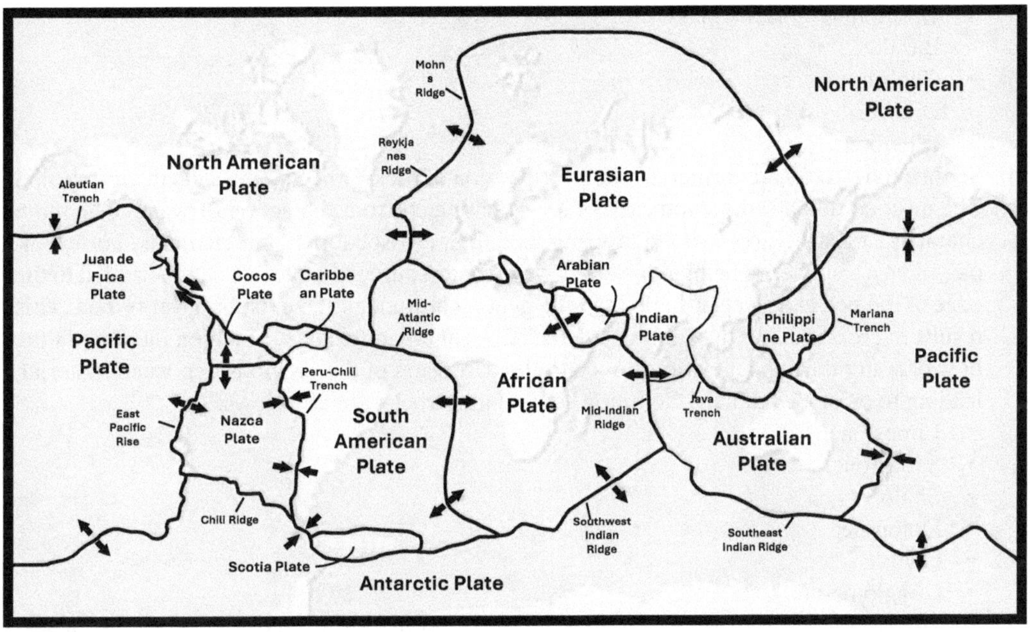

FIGURE 3.4 The major tectonic plates (McCarthy, 1996).

- Divergent plates moving away from each other.
- Convergent plates moving into each other.
- Plates grinding past each other, which are also known as transform plate boundaries.

Tectonics are responsible for the following particularly important conditions and geological features:

- Continents.
- Mountain ranges.
- Faults (mainly in brittle rock).
- Folds (mainly in ductile rock).
- Earthquake belts (zones).

3.3 ROCK TYPES AND THE ROCK CYCLE

A mineral is a naturally occurring inorganic substance formed through geological processes that has a characteristic chemical composition, a highly ordered atomic structure, and specific physical properties. In terms of matter, minerals are solids. By comparison, a rock is an aggregate, or assemblage, of minerals and generally does not have a specific chemical composition.

3.3.1 Fundamental Rock Classes

There are three fundamental classes of rock:

i. Igneous rocks are initially formed by the crystallization and solidification of molten magma that has been intruded into the earth's crust. The rock forms as magma cools, which also causes shrinkage and thus jointing in the rock. Common igneous rocks are as follows:
   - Granite.
   - Lava rock.

- Obsidian or "black" glass.
- Basalt.
- Gabbro.
- Diorite.
- Rhyolite.

ii. Sedimentary rocks are formed by the precipitation of rock-forming materials in air or water. The most common sedimentary rocks are precipitated from the action of water. The main characteristic structures include layering, bedding, crossbedding, and channels. For example, when a sand bed is being deposited away from a source of material, sand is added to the edge of the bed at a steeper angle than the general attitude of the existing layer or bed. This results in crossbedding in sedimentary rocks. As another example, a sudden increase in the flow of water during deposition can cause local erosion of the previous deposited material, leading to erosion channels. Common sedimentary rocks are as follows:

- Limestone.
- Sandstone.
- Shale.
- Dolomite.
- Coal.
- Conglomerate.

iii. Metamorphic rocks are transformed after initial formation by the influence of heat and pressure. For example, shales transform into slates, limestones into marble, and sandstone into quartzite. Generally, a metamorphic rock is stronger and more massive than the originally formed rock. The main metamorphic feature of interest is the strong anisotropy imparted to the rock. Some common metamorphic rocks are as follows:

- Gneiss.
- Slate.
- Marble.
- Schist.
- Quartzite.

3.3.2 The Rock Cycle

The rock cycle describes the relationship of the three rock types and how external changes can disturb the existing equilibrium conditions and hence transform rocks from one type to another over geological time. External changes include plate tectonics, pressure, heat, temperature, wind, rain, and ice.

Rocks exposed to high temperatures and pressures can be altered physically and/or chemically to form a metamorphic rock. Surface rocks exposed to the atmosphere are subject to weathering and hence erosion, which fragments the rock and can then be transported from the source by gravity, wind, and/or water. Upon compaction, this can form a sedimentary rock. When rocks are buried over geological time, they can melt and form magma. If the conditions subsequently change and the magma cools, it solidifies into an igneous rock.

Cold also influences the rock. For example, ice formation can freeze and crack rock, or ice can form glaciers which can erode rock. Snow and ice also affect erosion because as they melt, they turn to water which is an effective erosion agent and can transport materials a great distance.

A typical representation of the rock cycle is shown in Figure 3.5.

3.4 GEOLOGICAL STRUCTURES AND MECHANICAL ALTERATIONS IN ROCK

The following are the more important geological features for rock mechanics:

- Faults.
- Folds.

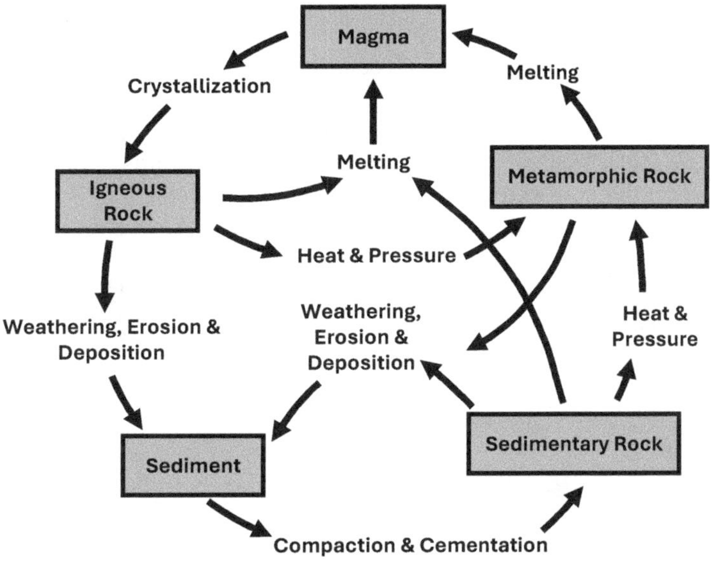

FIGURE 3.5 The rock cycle (redrawn from Kindred, 2022. Creative Commons by 4.0 Deed).

- Joints.
- Bedding planes.
- Dykes and sills.

The following is a brief description of some significant geological and structural features :

- Cleavage: The property of a rock where splitting occurs in a particular direction along parallel surfaces. This typically occurs during pressure metamorphism which imparts strong anisotropic mechanical properties to the rock.
- Dyke: An igneous wall-like intrusion that cuts across bedding or other layered structure in a rock.
- Erosion channel: Sediment-filled channel or stream that was cut into existing layers of rock during deposition.
- Fault: A break in a rock mass where observable displacement has occurred.
- Fault breccia: An assemblage of rock fragments associated with the formation of a fault (or subsequent movement) and located between the two fault surfaces.
- Fault gouge: A clay-like material occurring between the walls of a fault as a result of relative movement between the fault surfaces.
- Fissure: A gapped fracture.
- Foliation: A structure (or texture) due to the parallel disposition in layers (or lines) of one or more conspicuous minerals.
- Fold: A bend in a surface or layer of rock.
- Joint: A plane that separates, or tends to separate, two parts of a once-continuous rock mass (or block). Unlike a fault, no noticeable displacement has occurred.
- Sill: An igneous intrusive sheet of approximately uniform thickness and large lateral extent. It tends to force itself between sedimentary beds during its formation.
- Slickensides: The striations, grooves, and polish on faults and shear surfaces, caused by friction and movement of one rock mass against the other (Nelson and Nelson, 1967).
- Unconformity: A period in geological history during which new strata were not deposited. The strata formed before this period may be uplifted or tilted by earth's movements.

3.4.1 Typical Faulting of Rock

Faults are considered planes of weakness in rock, and an example is shown in Figure 3.6. They cause potentially large unstable blocks of rock. In high-stress conditions, they can be a source of seismicity, potentially causing rockbursts, earthquakes, or both. Depending on their relative orientation to rock surfaces, such as high walls, pillars, and excavation walls, they can introduce serious instability. They generally need extra or special support considerations.

Figure 3.7 shows the most common faults, created in typically brittle rocks, which are:

- Normal fault caused by tensile stresses creating a "loss of ground".
- Reverse fault caused by compressive stresses creating a "gain in ground".
- Strike-slip fault caused by shear stresses.

FIGURE 3.6 Example of a fault. Downloaded with kind permission from the International Society of Rock Mechanics (ISRM): website isrm.net, slides donated by Prof. John Franklin to the ISRM slide collection.

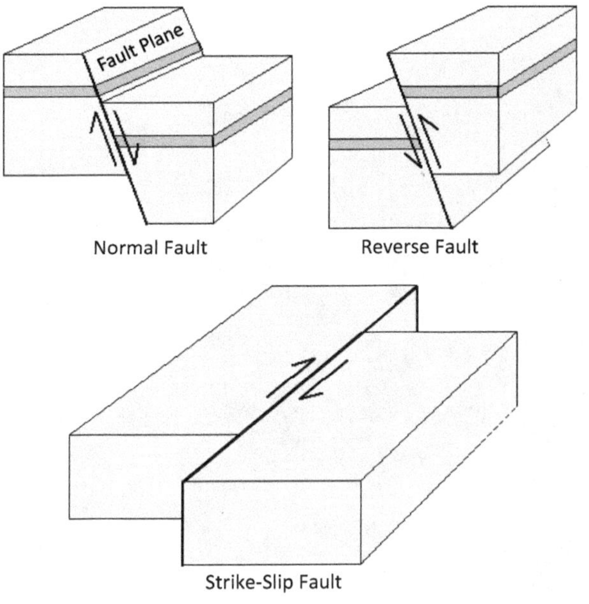

FIGURE 3.7 Diagrammatic representations of the common faults.

3.4.2 Typical Jointing in Rock

Joints are created by fracture in brittle rock due to tensile stresses, and rock is much weaker in tension than compression. The tensile stress may arise from the stretching of rock layers, increase in pore pressure, or originate from shrinkage caused by the cooling of the rock. Joints are considered localized weaknesses, particularly in igneous rock. They can form key blocks that could result in gravity-induced collapse. Properly aligned bolting relative to the joint planes can be effective.

Some massive joints are shown in Figure 3.8.

Where there is little topsoil cover and low rains, the rock joints are clearly visible in the ground as shown in Figure 3.9. Joints can be a natural path for water, which makes the rock and excavations less stable.

3.4.3 Folding in Layered Rocks

Folds can be small or large (creating mountains) and typically occur in ductile rocks. They are caused by high tectonic compressive stresses. They typically cause small friable and generally weak rock. Mesh, membranes, and shotcrete are effective supports, and the optimum support choice depends on the site-specific conditions (Figure 3.10).

3.4.4 Stratigraphy

Stratigraphy is mainly associated with sedimentary rocks. The fundamental premise is embodied in the Law of Superposition. That is, within a sequence of stratum, the oldest stratum is at the bottom and the youngest is at the top.

Rocks can be classified stratigraphically in three ways:

- Lithostratigraphic: according to the character of the rocks.
- Biostratigraphic: according to their fossil content (paleontological character).
- Chronostratigraphic: according to the age of the rock.

FIGURE 3.8 Typical joint sets. Downloaded with kind permission from the ISRM: website isrm.net, slides donated by Prof. John Franklin to the ISRM slide collection.

FIGURE 3.9 Rock joints visible below soil. Downloaded with kind permission from the ISRM: website isrm.net, slides donated by Prof. John Franklin to the ISRM slide collection.

The following are the principal lithostratigraphic terms used:

- Group describes a rock mass consisting of two or more related formations.
- Formation refers to the base rock unit.

FIGURE 3.10 An example of tightly folded rock strata. Downloaded with kind permission from the ISRM: website isrm.net, slides donated by Prof. John Franklin to the ISRM slide collection.

- Member refers to an entity within a formation.
- Bed describes a distinctive stratum (layer) within a member or formation.

3.5 CHEMICAL ALTERATIONS OF ROCK

Chemical defects are due to the alteration or removal of rock constituents resulting from a chemical reaction between these constituents and water (Terzaghi, 1946). Water transports via the various physical defects discussed in the previous sections and can be ascending (from depth) or descending (from surface). The source of the fluid can dictate its temperature. Of course, at low-enough temperature, water can freeze and fracture rock, and at high-enough temperature, it can weather rock and dissolve minerals. Depending on the rock and mineral type, fluids can generally be expected to lessen the homogeneity of the rock structure, thereby creating voids (e.g., natural caves) or cause swelling and disintegration. Physical properties are clearly altered. Friction and strength loss should be anticipated.

3.6 THE IMPORTANCE OF GEOLOGY IN MINING AND ROCK MECHANICS

Geology plays a vital role in the stability of underground excavations. Geological features significantly reduce the strength of the rock mass compared to a small intact rock lab test.

The importance of minor geological features, such as bedding planes and joints, somewhat diminishes at increasing depth because the effects of stresses become large enough to cause overriding stress-induced rock failure around the excavation periphery. An understanding of the interaction between the rock mass and the excavation is vital to create a safe and cost-effective structure in the rock.

Faults, joints, and bedding planes create natural planes of weakness and boundaries for potential rock failure. The relative orientation of an excavation with geological features is a critical consideration for the stability of the excavation and support.

During initial, large-scale exploration, geology associated with the orebody and surrounding rock mass is estimated to generate a geological model that requires information regarding the following:

- Stratigraphy.
- Rock types.
- Rock properties.
- Regional and local geology.
- Hydrology.
- Orebody extents and properties.
- Estimation of stresses.

A resource model determines the ore tonnage and grade of a geological deposit using the geological model.

3.7 GENERAL TERMINOLOGY FOR DESCRIBING A ROCK MASS

Deere (1964) suggested the terminology given in Table 3.1 for describing joint spacing and thickness of bedding units. There seems to be no agreement on an exact terminology, especially between hard and soft rocks. In general, terms with metrics are meant to be used as a guide and refinement may be necessary to suit specific conditions.

A description of the suggested terminology is given in Table 3.1.

Generalized terminology by Terzaghi (1946) was developed for describing in situ rock conditions in tunnels. Many quantitative classification methods have been introduced over the years which further characterize a rock mass to a greater detail.

Table 3.2 gives suggested rock mass conditions in tunnels.

3.8 STEREONETS

Modified from Heasley (2009, Personal Communication based on his course notes).

3.8.1 Hemispherical Projection of a Plane

Hemispherical projection is a method of representing and analyzing the three-dimensional relations between planes and lines on a two-dimensional diagram. It has long been a widely used tool in the field of structural geology and has recently found increasing use for solving engineering problems.

The basis of the method and its classic geological applications are described by Philipps (1971). Rock engineering applications are described in detail by Goodman (1976), Hoek and Brown (1980), Priest (1985), and Hoek and Bray (1981).

TABLE 3.1
Descriptive Terms for Spacing of Joints and Thickness of Beds (Deere, 1964)

Descriptive Terms		Spacing of Joints or Thickness of Beds	
Joints	Beds	English	Metric
Very close	Very thin	Less than 2 in	Less than 5 cm
Close	Thin	2 in–1 ft	5–30 cm
Moderately close	Medium	1–3 ft	30 cm–1 m
Wide	Thick	3–10 ft	1–3 m
Very wide	Very thick	Greater than 10 ft	Greater than 3 m

TABLE 3.2
Category and Descriptions for Rock Mass in Tunnels Originally by Terzaghi (1946) and Summarized by Deere, Merritt, and Coon (1969)

Category	Description
Intact	Rock containing no joints
Stratified	Rock consisting of individual strata with little or no resistance against separation along the planes of contact
Moderately jointed	Rock containing joints with blocks interlocked so that no lateral support is necessary for vertical walls
Blocky and seamy	A jointed rock mass composed of separated intact rock fragments which would require support for vertical walls
Crushed	Chemically unweathered rock reduced to small particle size
Squeezing	Rock that advances into a tunnel without perceptible volume increase
Swelling	Rock that exhibits a volume increase caused by minerals having a high swelling capacity

The method uses a sphere which is free to move in space so that it can be centered on an inclined plane as illustrated in Figure 3.11. The intersection of the plane and the surface of the sphere is a great circle, shown at the perimeter of the shaded area in Figure 3.11. A line perpendicular to the plane and passing through the center of the sphere intersects the sphere at two diametrically opposite points called the poles of the plane.

The great circle and the pole representing the plane appear on both the upper and lower parts of the sphere, and therefore, only one hemisphere needs to be used to plot and manipulate structural data. In rock mechanics, the lower hemisphere projection is almost always used. The upper-hemisphere projection is often used in textbooks on structural geology and can be used for rock mechanics studies if required; refer to Goodman (1976) for an example.

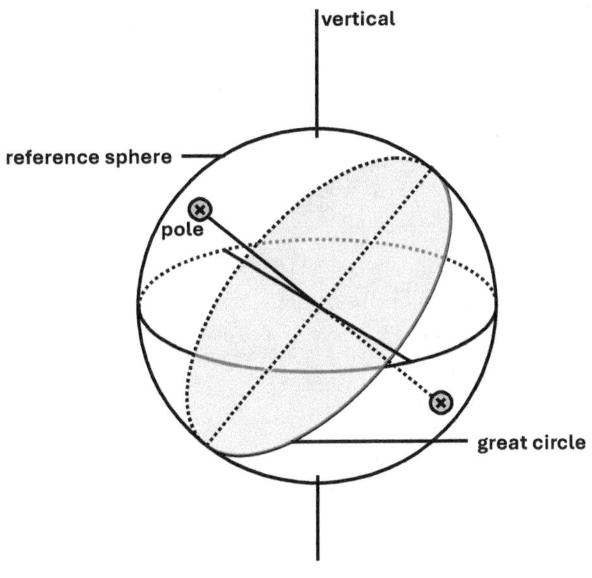

FIGURE 3.11 The great circle and its poles which define the orientation of the plane (Hoek and Brown, 1980).

The hemispherical projection provides a means of representing the great circle (a circular arc) and pole (a point perpendicular to the plane) which is easier to visualize in Figure 3.11 on a horizontal plane. As shown in Figure 3.12, this is achieved by connecting all points on the great circle and the pole with the zenith or point at which a vertical through the center of the sphere intersects the top of the sphere. The hemispherical projections of the great circle and the pole are then given by the intersections of these projection lines with the horizontal plane.

The plotting of planes and their poles is carried out with the aid of a stereonet as shown in Figure 3.13. The great circles representing planes of constant dip are constructed as circular arcs centered on extensions of the east–west axis of the net. The stereonet also contains a series of small

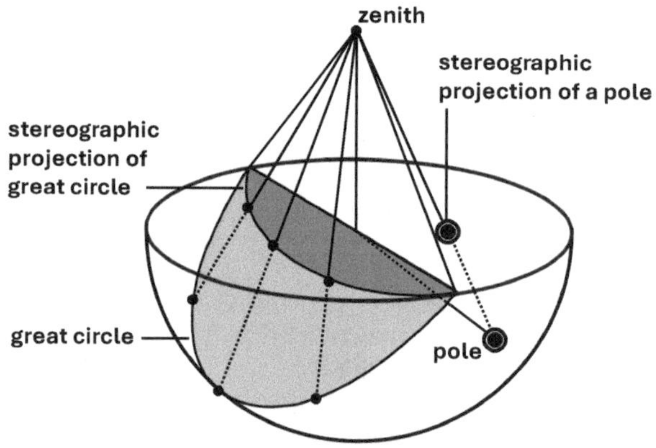

FIGURE 3.12 Stereographic projection of a great circle and its pole on to the horizontal plane from the lower reference hemisphere (Hoek and Brown, 1980).

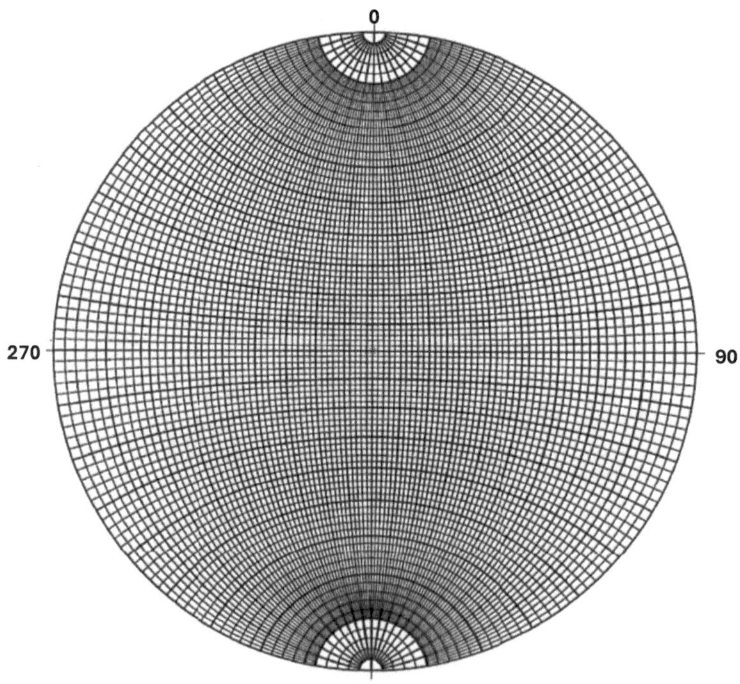

FIGURE 3.13 Meridional stereographic or equal area net.

circles centered on extensions of the north–south axis. The angle between any two points on a great circle is determined by counting the small circle divisions along the great circle between the two points concerned.

3.8.2 Typical Uses of Stereonets

Stereonets are used for:

- Plotting the trend and plunge of a line.
- Plotting the strike and dip of a plane.
- Plotting the orientation of a line in a plane.
- Measuring the angle between two lines.
- Measuring the angle between two planes.
- Determining the orientation of the intersection of two planes.
- Stereographic projection as a statistical tool.

The above applications can then be used to determine the average dip and strike of joint sets and the orientation of faults and bedding planes which affect the stability of underground excavations.

3.8.3 Steps to Constructing a Plane on a Stereonet

Generally, a stereonet requires an underlying net on a sheet of cardboard or paper for reference and an overlying tracing paper, preferably held in place with a pin in the center of the circle. First, draw the primitive on the tracing paper which is the perimeter of the stereonet. Label north, south, east, and west on the primitive as shown in Figure 3.14.

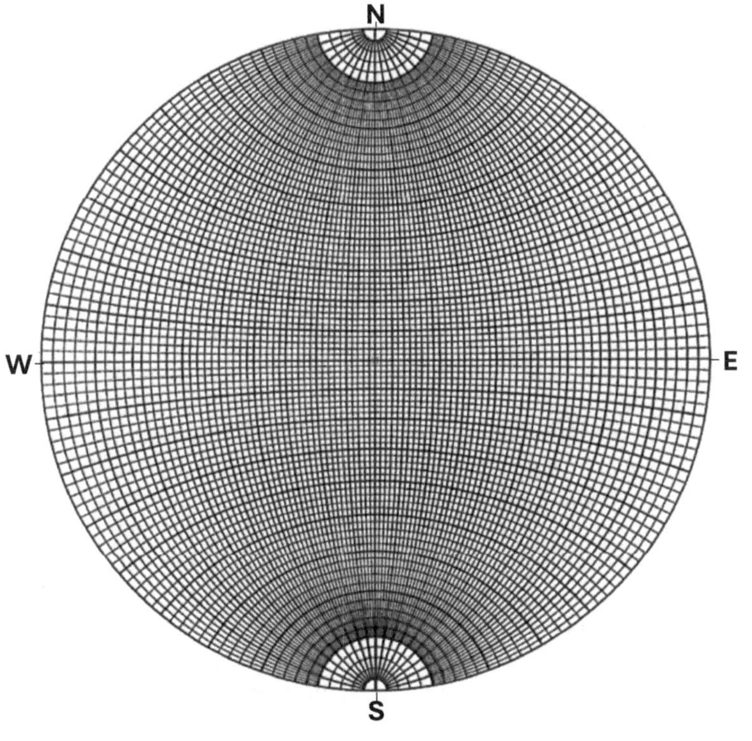

FIGURE 3.14 Initial labeling of a stereonet.

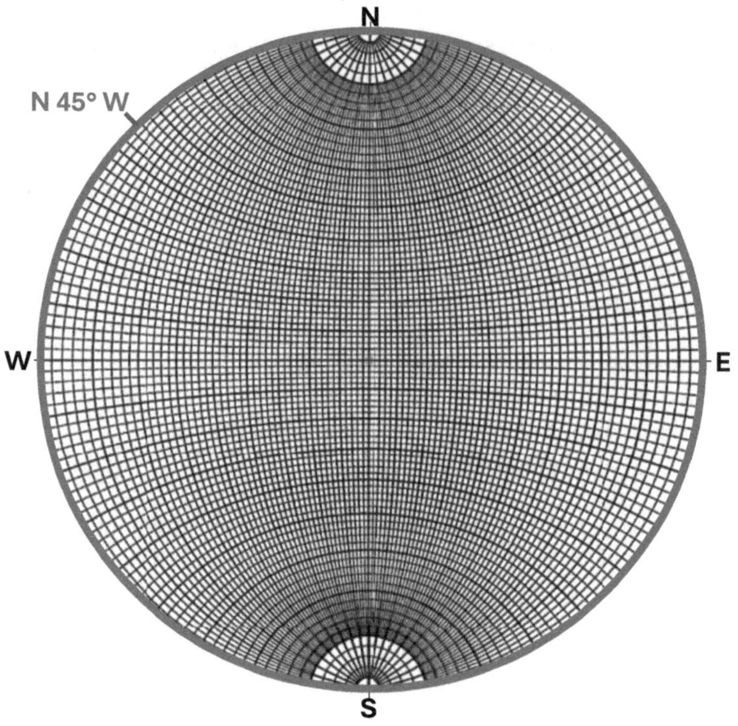

FIGURE 3.15 Marking the strike on a stereonet.

Next, find the strike on the primitive and make a tick mark. For this example, a strike azimuth of 315° or a bearing of N45°W (same direction but in quadrant form) is used (Figure 3.15).

Rotate the tick mark on the trace paper to the north without rotating the stereonet. Essentially, the N45°W tick should rotate 45° clockwise as shown in Figure 3.16.

Mark the dip by working inward from the primitive along the equator. Make sure that the dip is dipping in the correct direction when rotated to normal. For this example, let's assume a dip angle of 60° in the southwest direction. Southwest is on the left-hand side of the circle, so we start at the left-hand equator. Once the dip is marked, draw the great circle with the strike at north and through the dip (Figure 3.17).

Finally, return the primitive to the original position as shown in Figure 3.18. Adjacent to the following figure is the representation on the trace paper. Note the peak of the great circle which is dipping in the southwest direction.

There can be ambiguity in how features are plotted depending on the convention used. For example, the strike of a plane may be given as a bearing which ranges from 0° to 360° or be accompanied with the compass quadrant with angles only measuring between 0° and 90°, such as N45°W as in the example just given.

Additionally, the strike can be measured in two directions in the field. For the example above, those strike directions would be N45°W or S45°E. Obviously, both can be found on the great circle.

To avoid confusion, inclusion of the quadrants is useful for engineers. Reliance on the right-hand rule can also be helpful and consistent, whereby a plane should dip downward to your right when facing the strike direction of the plane. Alternatively, imagine placing your right hand on the surface of a dipping plane, and all fingers except your thumb will point in the downward direction of the planes dip; your thumb would be 90° counterclockwise from the dip and pointing in the direction of the strike.

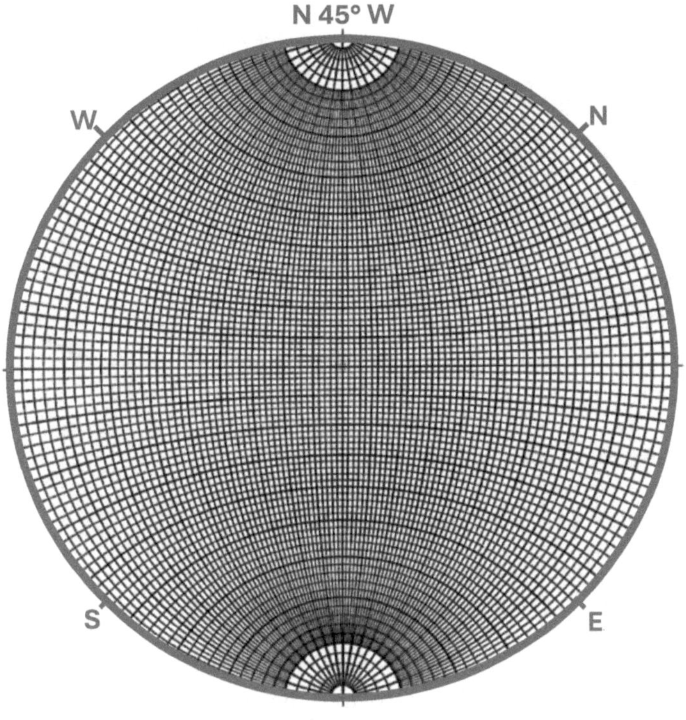

FIGURE 3.16 Rotation of the strike on a stereonet.

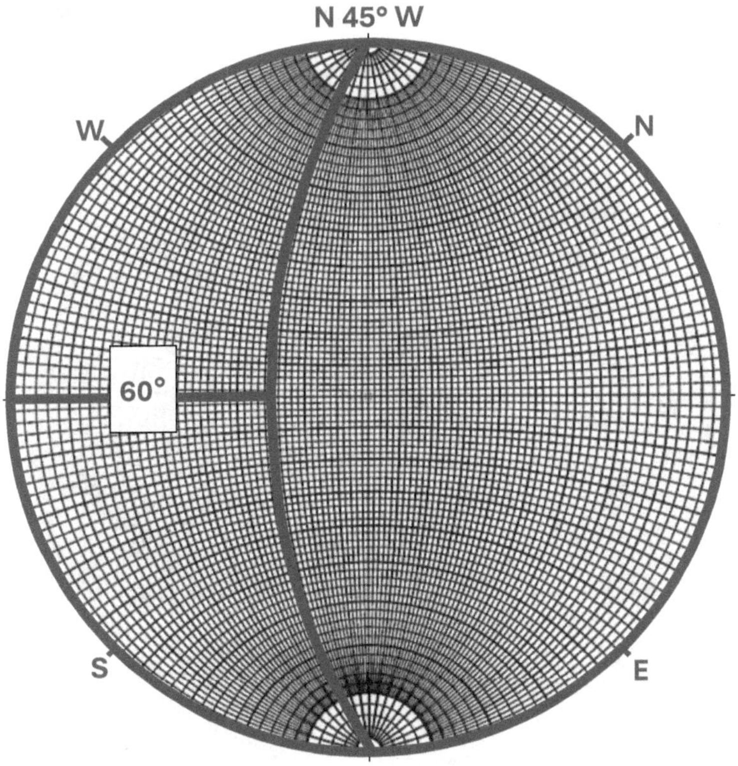

FIGURE 3.17 Marking the dip on a stereonet.

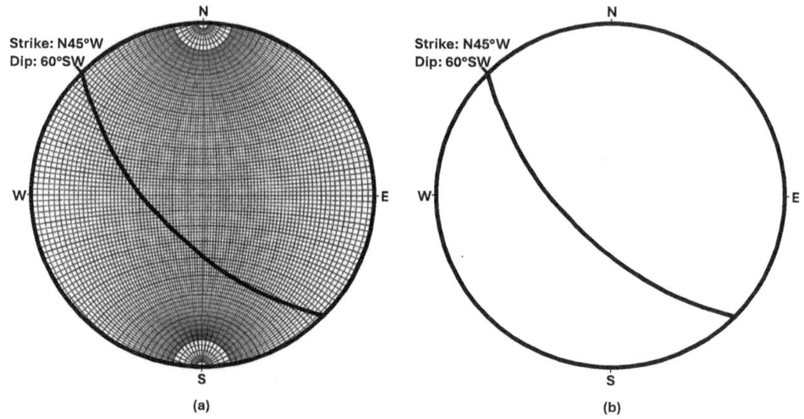

FIGURE 3.18 A plotted plane on a stereonet.

3.8.4 Steps to Constructing a Pole on a Stereonet

Poles are necessary when there are many planes, joint sets, or faults on a single stereonet, and hence, they are expressed as a single point (a pole) to describe a plane. Modern computing systems allow import of many data points from field measurements, so it is only practical to have a means to portray such data without overwhelming the stereonet (and the user) and therefore the interpretation. The first step in plotting a pole to the plane is to draw the plane as shown in the previous section and rotate the plane to the north–south direction as shown in Figure 3.19.

The pole is then marked by counting 90° along the equator (as shown in Figure 3.20). Note this can be done in either direction as the whole circle is 180°.

The stereonet is then rotated back to the original position as shown in Figure 3.21.

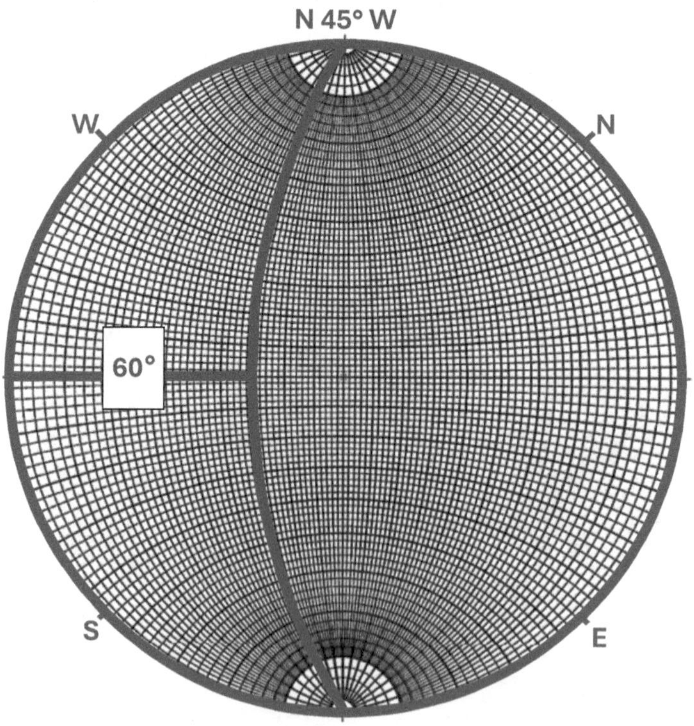

FIGURE 3.19 Rotation of stereonet for plotting of poles.

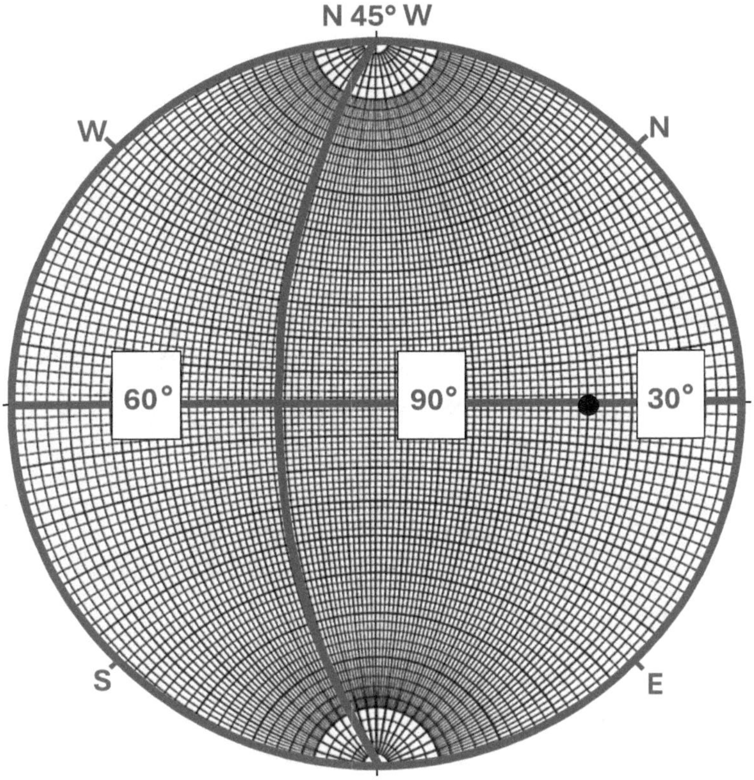

FIGURE 3.20 Plotting a pole on a stereonet.

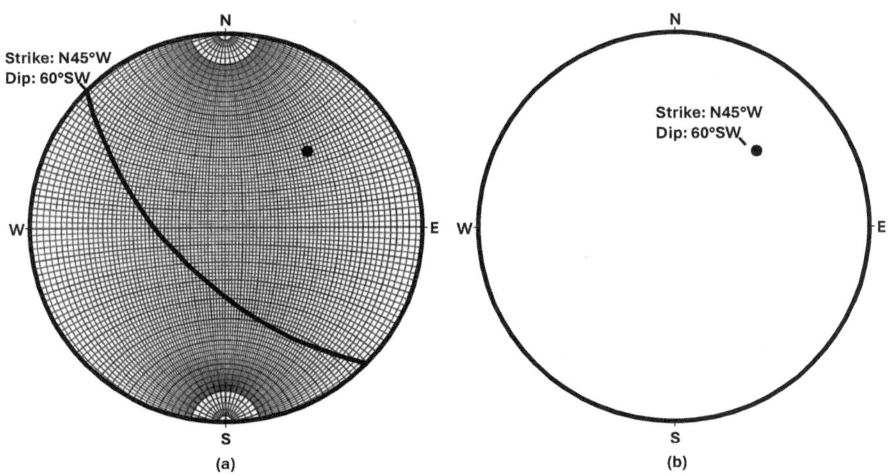

FIGURE 3.21 Final plotted pole on a stereonet.

3.9 WORKED STEREONET EXAMPLES

Example 3.9.1

Plot the following great circles of the following planes on a stereonet:

- Strike: N40°E, Dip: 60°SE.
- Strike: N55°E, Dip: 55°SE.
- Strike: N75°E, Dip: 38°SE.

The solution is given in Figure 3.22.

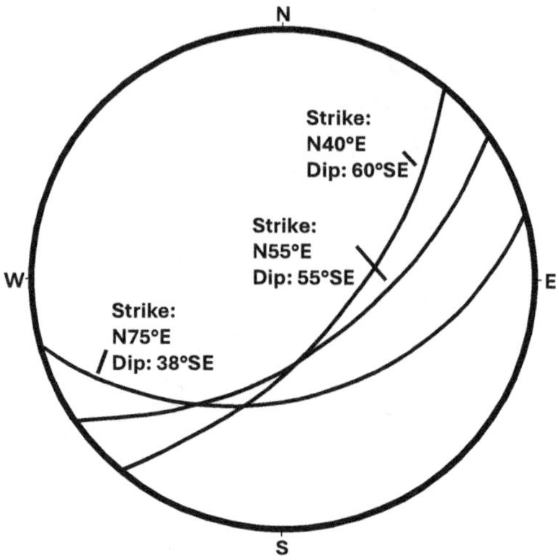

FIGURE 3.22 Solution to Example 3.9.1.

Example 3.9.2

Plot the great circles of the following planes on a stereonet:

- Strike: 152°, Dip: 20°, Direction: SW.
- Strike: 154°, Dip: 60°, Direction: SW.
- Strike: 330°, Dip: 30°, Direction: NE.
- Strike: 320°, Dip: 80°, Direction: NE.

The solution is given in Figure 3.23.

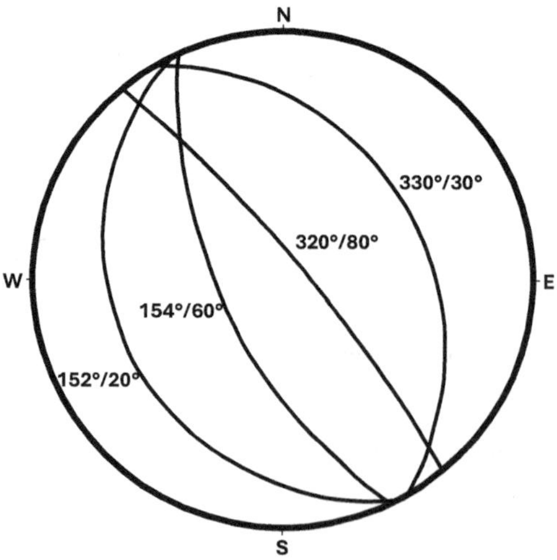

FIGURE 3.23 Solution to Example 3.9.2.

Example 3.9.3

Draw the following vertical planes on a stereonet:

- 030/0, NW or SE.
- 115/0, NE or SW.
- 180/0, E or W.
- 270/0, N or S.

The solution is given in Figure 3.24.

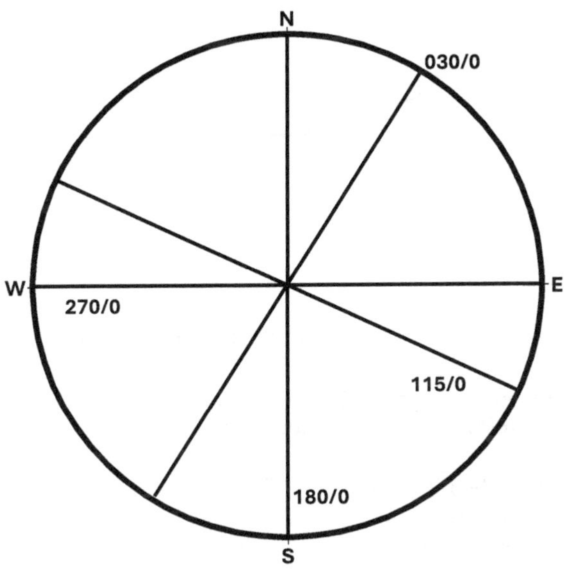

FIGURE 3.24 The solution to Example 3.9.3.

Example 3.9.4

It has been observed that at a coal mine a bed of coal is slightly uplifted. The average attitude of the coal bed has been measured by field geologists striking at 315°, dipping 60° to the southwest. Plot the plane on a stereonet.

The solution is given in Figure 3.25.

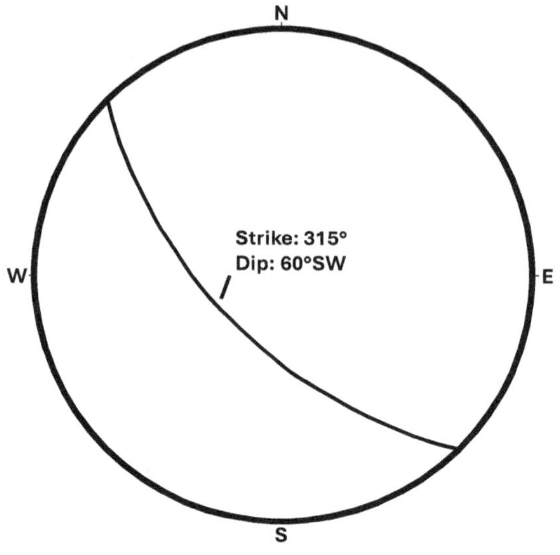

FIGURE 3.25 The solution to Example 3.9.4.

Example 3.9.5

Two geologists provide you a field measurement on the same feature:

- Geologist 1: Strike: 000°, Dip: 50°, Direction: W.
- Geologist 2: Strike: 180°, Dip: 50°, Direction: W.

Plot the planes, and explain the difference in the field measurements provided.

The solution is given in Figure 3.26.

Geologist 2 is using the right-hand rule, whereas Geologist 1 is not; however, the plane described is still the same.

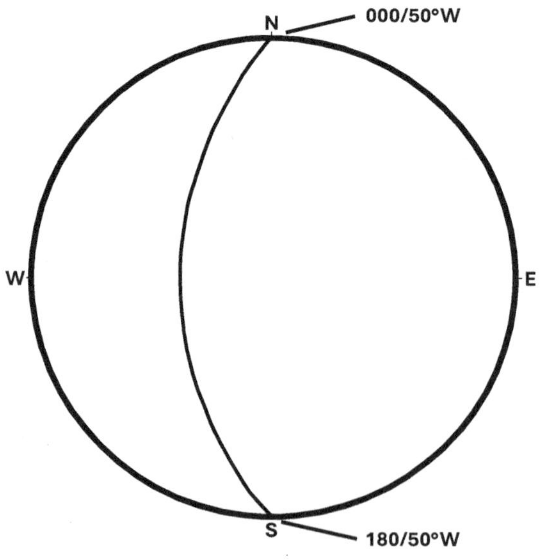

FIGURE 3.26 The solution to Example 3.9.5.

Example 3.9.6

The manager at a mine asks for a bedding plane orientation as described. This is best understood by showing the pole to the bedding plane because it can determine the orientation with a single point on the stereonet. The bed has an attitude of the following:

- N74°E/80°N.

The solution is given in Figure 3.27.

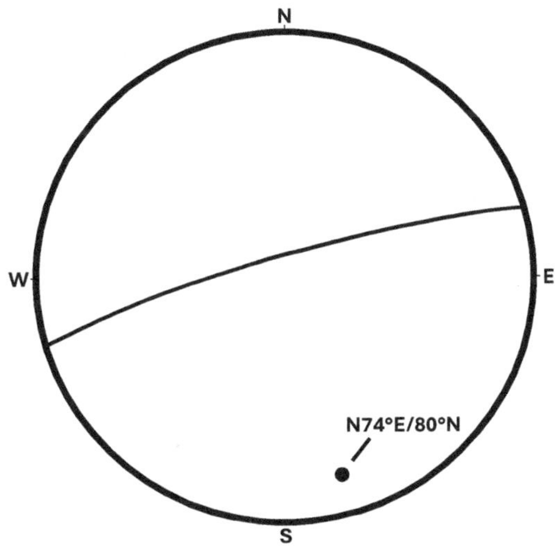

FIGURE 3.27 Solution to Example 3.9.6.

Example 3.9.7

Suppose that a plane strikes north–south and dips 40° west. Represent this situation using an equal area stereonet and plot the pole to plane.

The solution is given in Figure 3.28.

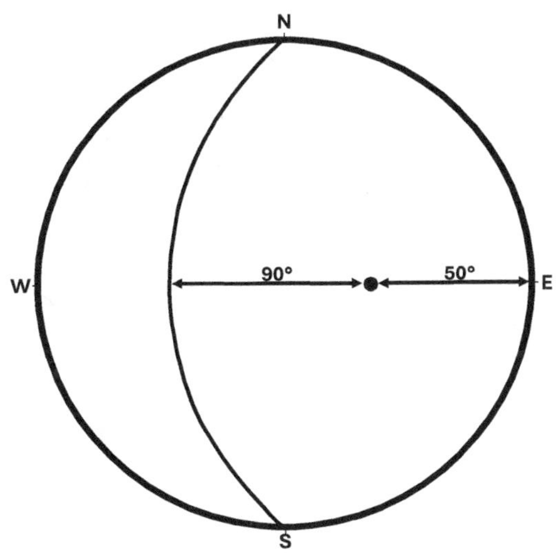

FIGURE 3.28 Solution to Example 3.9.7.

Example 3.9.8

Plot the following line on a stereonet. Unlike planes that rotate to the north and south once the strike is plotted, lines rotate to the equator once the trend is plotted on the primitive. Lines will be represented as a point only.

Trend: 230°, Plunge: 20°

The solution is given in Figure 3.29.

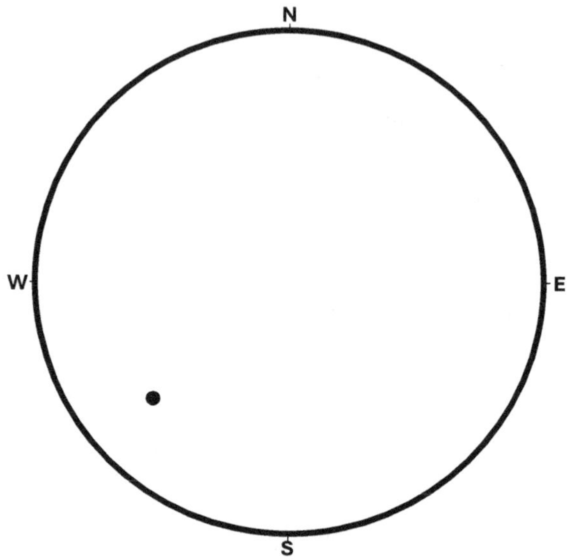

FIGURE 3.29 The solution to Example 3.9.8.

Example 3.9.9

Plot the following planes shown in Figure 3.30, on a stereonet. Assume that the planes strike in the north–south direction. Plane C is horizontal.

The solution to Example 3.9.9 is given in Figure 3.31.

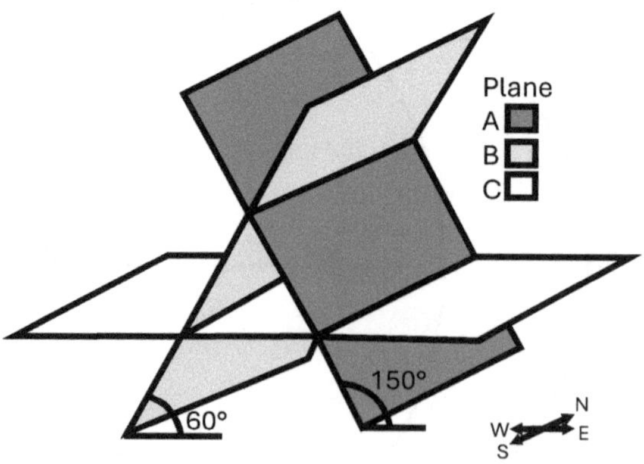

FIGURE 3.30 Diagram of the three planes.

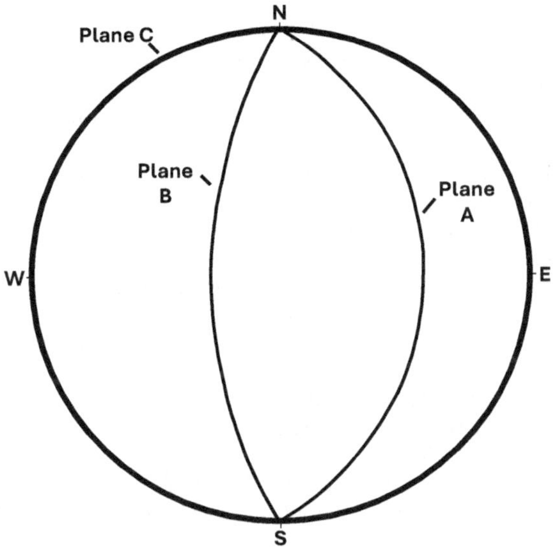

FIGURE 3.31 The solution to Example 3.9.9.

Example 3.9.10

Using the stereonet from the previous example, what is the attitude of the lines at the intersection of the planes?

Using the right-hand rule and plane A as our dipping plane, line intersecting A–C has a trend at 0° (or north–south) and a plunge of 0°. Similarly using the right-hand rule again and plane B as the dipping plane, line intersecting B–C has a trend of 180° (or NS) and a plunge of 0°. If the same approach is used for the intersection of A–B, then the line could be described with a trend of 0° or 180° with a plunge of 0°. For practical purposes, a north–south trend is likely appropriate to describe all the intersecting lines (Figure 3.32).

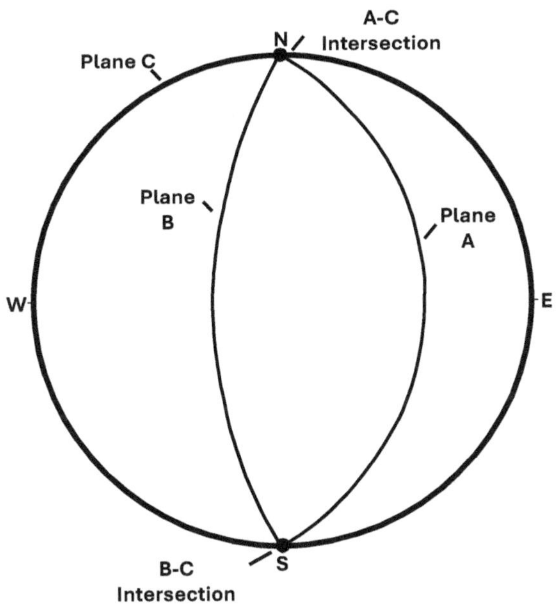

FIGURE 3.32 The solution to Example 3.9.10.

REFERENCES

Christiansen, E.H. and Hamblin, W.K. (2015) *Dynamic Earth: An Introduction to Physical Geology*. Jones and Bartlett Learning, LLC, Burlington.

Deere, D.U. (1964) Technical description of rock cores for engineering purposes. *Rock Mech. Eng. Geol.* 1(1), 17–22.

Deere, D.U., Merritt, A.H. and Coon, R.F. (1969) *Engineering Classification of In-Situ Rock* (Vol. 67). Air Force Weapons Laboratory, Air Force Systems Command, Kirtland Air Force Base, Montgomery County, OH.

Goodman, R.E. (1976) *Methods of Geological Engineering*. West Publications, New York.

Hendricks, J.R. (2019) *Geological Time. Digital Atlas of Ancient Life*. Paleontological Research Institution, New York. www.digitalatlasofancientlife.org/learn/geological-time/geological-time-scale/.

Hoek, E. and Brown, E.T. (1980) *Underground Excavations in Rock*. E. & F.N. Spon, London.

Hoek, E. and Bray, J.W. (1981) *Rock Slope Engineering (*3rd Edition). Institution of Mining and Metallurgy, London, UK.

Kindred, G. (2022) Creative Commons by 4.0 Deed.

Kious, W.J. and Tilling, R.I. (1996) *This Dynamic Earth: The Story of Plate Tectonics*. U.S. Geological Survey, Washington, DC. DOI: 10.3133/7000097.

McCarthy, D.D. (1996) *Technical Note 21*. International Earth Rotation Services (IERS), Paris France.

Nelson, A. and Nelson, K.D. (1967) *Dictionary of Applied Geology, Mining & Civil Engineering*. Philosophical Library Inc., New York.

Nelson, S.A. (2024) The Age of the Earth. Geological Time, Earth Science Australia. Retrieved from: www.earthsci.org/education/teacher/basicgeol/geotim/geotim.html.

Philipps, F. (1971) *The Use of Stereographic Projection in Structural Geology*. Edward Arnold, London.

Priest, S.D. (1985) *Hemispherical Projection Methods in Rock Mechanics*. Allen and Unwin, New York.

Read, H.H. and Watson, J. (1975) *Introduction to Geology*. Halsted: New York.

Tarbuck, E.J. and Lutgens, F.K. (2014) *Illustrated by Dennis Tasa. Earth An Introduction to Physical Geology, Custom Edition for Southern Illinois University – Carbondale*. Pearson, Upper Saddle River, NJ.

Terzaghi, K. (1946) Chemical defects of rocks. In: Proctor, R.V. and White, T.L. (eds) *Rock Tunneling with Steel Supports with an Introduction to Tunnel Geology*. The Commercial Shearing and Stamping Co., New York, pp 39–46.

USGS (1999) Cutaway view showing the internal structure of the earth. Volcano Hazards Program. Retrieved from: www.usgs.gov/media/images/cutaway-views-showing-internal-structure-earth-left

4 Stress and Strain

4.1 STRESS TRANSFORMATIONS USING MOHR CIRCLE

Consider a square element, as shown in Figure 4.1, having dimensions Δx and Δy in their respective coordinate directions. Assuming that the normal and shear stresses are known, the graphical technique of Mohr circle can be used to transform the stresses to a desired orientation in two- and three-dimensional spaces.

The principal stress planes, for a given stress state, orientate where the shear stresses are zero, leaving only normal stresses acting on the elemental square, referred to as the principal stresses. These principal stresses are critical for design purposes as is the maximum shear stress.

4.1.1 Principal Stress Orientation

Stress transformation Equations 4.1 and 4.2 have been resolved as follows:

$$\sigma_n = 1/2(\sigma_x + \sigma_y) + 1/2(\sigma_x - \sigma_y)\cos 2\theta + \tau_{xy} \sin 2\theta \tag{4.1}$$

$$\tau_{nt} = -1/2(\sigma_x - \sigma_y)\sin 2\theta + \tau_{xy} \cos 2\theta \tag{4.2}$$

The principal stress direction is found by taking the derivative of the normal stress, σ_n, in Equation 4.1, with respect to θ. With the inclusion of terms sin 2θ and cos 2θ in Equations 4.1 and 4.2, the stresses are periodic in π and must have maximum and minimum values or be constant (Obert and Duvall, 1967).

The following derivatives are needed:

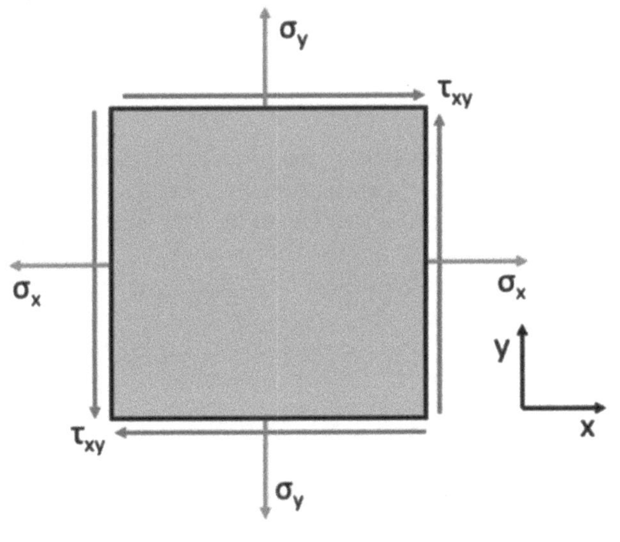

FIGURE 4.1 Stresses acting on an element in equilibrium.

DOI: 10.1201/9781032704845-4

$$d/d\theta(\sin 2\theta) = 2\cos 2\theta$$

$$d/d\theta(\cos 2\theta) = -2\sin 2\theta$$

The simple meaning of a derivative with respect to a certain factor, such as in the case of $d\sigma_n/d\theta$ shown here, is that any factors not including θ are treated as constants, and therefore,

$$\sigma_n = \tfrac{1}{2}(\sigma_x + \sigma_y) + \tfrac{1}{2}(\sigma_x - \sigma_y)\cos 2\theta + \tau_{xy}\sin 2\theta$$

$$d\sigma_n/d\theta = -(\sigma_x - \sigma_y)\sin 2\theta + 2\tau_{xy}\cos 2\theta$$

Equating to zero, this can now be readily solved for θ:

$$\theta = 1/2\tan^{-1}\left[\left(2\tau_{xy}/(\sigma_x - \sigma_y)\right)\right] \tag{4.3}$$

Technically, there are two directions: one for the minimum (θ+90°) and the other for the maximum (θ) principal stress. Depending on the preference, these stresses take the notations of P, σ_1, σ_p, or σ_{max} for the maximum and Q, σ_2, σ_q, or σ_{min} for the minimum principal stress. The terms major and minor principal stresses may also be used.

For simplicity, the directions follow the subscript notation such that the principal stress direction is θ_1, θ_p, or θ_{max}. For three-dimensional space, an intermediate principal stress, σ_2, is introduced, and the minimum principal stress becomes σ_3. It is important to always keep the notations as simple and consistent as possible.

4.1.2 Mohr Circle Derivation

The minimum and maximum principal stresses are resolved by substituting Equation 4.3 into Equation 4.1 and simplifying (Obert and Duvall, 1967). The following equations are obtained:

$$\sigma_1 = \tfrac{1}{2}(\sigma_x + \sigma_y) + \tfrac{1}{2}\left[(\sigma_x - \sigma_y)^2 + 4(\tau_{xy})^2\right]^{1/2} \tag{4.4}$$

$$\sigma_2 = \tfrac{1}{2}(\sigma_x + \sigma_y) - \tfrac{1}{2}[(\sigma_x - \sigma_y)^2 + 4(\tau_{xy})^2]^{1/2} \tag{4.5}$$

The diameter of the Mohr circle is bounded by the principal stresses as shown in Figure 4.2. The circle represents all possible normal and shear stresses on an element at any orientation in two dimensions. The stresses on any plane as a function of the principal stresses and θ can be resolved:

$$\sigma_x = \tfrac{1}{2}(\sigma_1 + \sigma_2) + \tfrac{1}{2}(\sigma_1 - \sigma_2)\cos 2\theta \tag{4.6}$$

$$\sigma_y = \tfrac{1}{2}(\sigma_1 + \sigma_2) - \tfrac{1}{2}(\sigma_1 - \sigma_2)\cos 2\theta \tag{4.7}$$

$$\tau_{xy} = -\tfrac{1}{2}(\sigma_1 - \sigma_2)\sin 2\theta \tag{4.8}$$

The above equations are further simplified by replacing 2θ by φ:

$$\sigma_x = \tfrac{1}{2}(\sigma_1 + \sigma_2) + \tfrac{1}{2}(\sigma_1 - \sigma_2)\cos\varphi \tag{4.9}$$

$$\sigma_y = \tfrac{1}{2}(\sigma_1 + \sigma_2) - \tfrac{1}{2}(\sigma_1 - \sigma_2)\cos 2\varphi \tag{4.10}$$

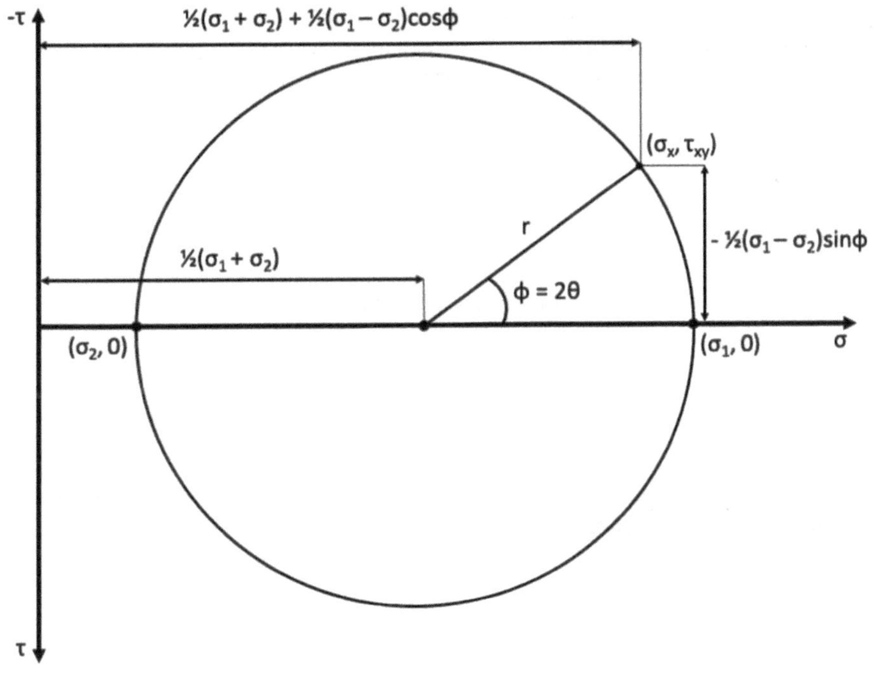

FIGURE 4.2 Depiction of a Mohr circle.

$$\tau_{xy} = -\tfrac{1}{2}(\sigma_1 - \sigma_2)\sin\varphi \tag{4.11}$$

The circle is centered, c, at an average stress, σ_{avg}, on the σ-axis of the σ–τ space:

$$c = \sigma_{avg} = \tfrac{1}{2}(\sigma_1 + \sigma_2) \tag{4.12}$$

The radius, r, of the Mohr circle represents the maximum shear stress, τ_{max}:

$$r = \tau_{max} = \tfrac{1}{2}(\sigma_1 - \sigma_2) \tag{4.13}$$

The planes of maximum shear stress are orientated at θ=45° from the principal stress axes when using the coordinate axes system. In the Mohr circle, orientations are represented by 2θ, and thus, the same stress planes will be represented by 90° during graphical construction. Similarly, the principal stress planes are mutually perpendicular but represented by 2θ in the Mohr circle, and therefore shown as 180° from one another in Figure 4.2.

4.2 STRESSES IN MINING

4.2.1 State of Stress in Rock

It can generally be assumed that prior to mining underground, rock is in equilibrium and acted upon by virgin (primitive) stress. When an excavation is developed, the stresses are redistributed and become concentrated in the vicinity of the opening since stress cannot be transferred across a free face, and the surfaces of the excavation tend to converge. This displacement is responsible for the work done on the rock to produce the excavation. The energy dissipates in heating, fracturing, and bursts of the rock surrounding the excavation.

To design the support for an excavation, it is essential to understand the rocks' initial state of stress (pre-mining stresses) and the stress on the excavation due to mining (mining-induced stresses). Generally, unless failure occurs, no support is required in an excavation.

4.2.2 Pre-Mining Stresses

According to Jaeger and Cook (1969), the state of stress, in the absence of an excavation, can be grouped accordingly:

- Stresses (both vertical and horizontal) due to gravity.
- Tectonic stresses.
- Structural stresses.
- Residual stresses.

The stress due to gravity, which occurs in the vertical direction, is the effect of the mass of the superincumbent rock at any specified depth. If one considers a homogeneous and elastic rock mass with density (ρ) the vertical component of the stress, σ_z in Pascals (N/m^2) will be

$$\sigma_z = \rho g z \tag{4.14}$$

where (in SI units)
ρ is the rock density typically in kg/m^3.
g is the gravitational constant 9.81 m/s^2.
z is the depth in meters.

This equation can be modified for the more general case of a stratified rock mass by a summation process in which ρ_i and z_i are taken to denote the density and thickness of the individual strata, respectively. Equation 4.14 becomes

$$\sigma_z = \sum_{i=1}^{n} \rho_i g z_i \tag{4.15}$$

For example, if there are three main strata (1, 2, and 3), then Equation 4.15 takes the following form:

$$\sigma_z = \rho_1 g z_1 + \rho_2 g z_2 + \rho_3 g z_3$$

The horizontal component of stress due to gravity (the effect of Poisson's ratio) is also found to be dependent on the depth below surface (z). In a relatively geologically undisturbed rock mass, the two horizontal stress components are found to be approximately equal in magnitude (i.e., $\sigma_x \approx \sigma_y$). The constant of proportionality relating the value of σ_x and σ_y, to σ_z, is specified by k, where

$$\sigma_x = \sigma_y = k\sigma_z \tag{4.16}$$

In practice, it is difficult to fix a value to k, but it is generally approximated using Poisson's ratio (ν)

$$k = \nu/1-\nu \tag{4.17}$$

Typically, ν varies between 0.15 and 0.35, giving the corresponding value of k ranging from 0.17 to 0.54.

Studies carried out in South Africa by Pallister et al. (1970) and later by Gay (1975) suggest that in deep-level gold mines, the relationship between horizontal and vertical stresses is as follows (Gay, 1975):

$$\sigma_h \approx 0.6\sigma_v \tag{4.18}$$

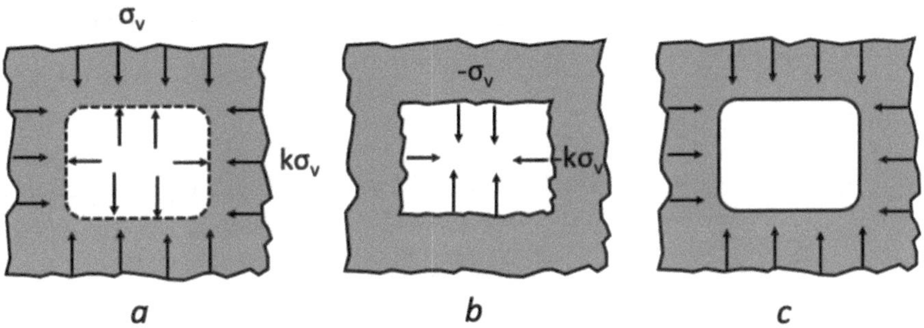

FIGURE 4.3 The effect of an excavation on the stress field: (a) the equilibrium virgin state of stress (prior to mining), (b) the mining-induced stress caused by the excavation, and (c) the resultant stress (Budavari, 1983).

Tectonic stresses are due to the straining of the earth's crust by regional uplift, down warping, faulting, folding, or surface irregularities. The superimposition of these stresses on the gravity-induced stresses can yield substantial changes from the expected magnitudes and directions of the principal stresses.

Residual stresses remain in the rock mass even after the cause has been removed, by, for example, massive erosion. Jaeger and Cook (1969) consider the residual stresses as "locked-in" stresses associated with the history of rock, and structural stresses are associated with the inhomogeneity of the rock, such as differences in properties. Generally, the aid of in situ stress measurements are needed to predict the magnitudes of stresses beyond the stresses due to gravity.

4.2.3 Mining-Induced and Resultant Stresses

Once an excavation has been created, the original equilibrium does not exist and the normal stresses on the periphery of the excavation become reduced to zero. For the "new" equilibrium, a stress field of the same magnitude but opposite in direction to the initial stress must be superimposed around the excavation (refer to Figure 4.3). These are referred to as the mining-induced stresses (or the induced stresses).

These induced stresses are of great importance to the stability of the excavation. Figure 4.3 illustrates the effect of an excavation on the stress field. After mining, the sum of the induced stress and the virgin stress is in practice known as the resultant stress, because the tectonic and residual stresses if present are unknown. It is important, however, if stress measurements are not done at or near a mine site and the in situ excavation stability is significantly worse than expected, it is worth assuming that either the initial analysis input was incorrect or there are significant tectonic and residual stresses. If this is the case, judiciously located stress measurements should be seriously considered, or the analysis should be calibrated using another acceptable method based on the in situ mining actual behavior. Any deformation caused by the excavation creation is due to the induced stress only, and thus the deformation is called the induced deformation. To reiterate, if the deformation is significant, it can be responsible for the partial closure of the excavation and for the straining of the surrounding rock mass.

4.2.4 Stress–Strain Relationships

As mentioned in the above sub-section, the mining-induced stresses created by an excavation result in mining-induced strains. The relationship between stress and strain is known as Hooke's law. It has been shown that the relationship between the Force (F) and displacement (x) is related by

$$F = kx \tag{4.19}$$

For a homogenous and isotropic body, subject to elastic stress in the z-component direction, the stress (σ_z) and strain (ε_z) are related by the elastic constant (E) such that

$$\varepsilon_z = \sigma_z / E \tag{4.20}$$

If the body is in tension, it will result in contraction in the x- and y-component directions due to Poisson's ratio (ν) such that

$$\varepsilon_y = -\nu\sigma_z / E \tag{4.21}$$

$$\varepsilon_x = -\nu\sigma_z / E \tag{4.22}$$

If the body is instead subject to stress in three orthogonal directions (x, y, and z), then the estimate of the strains must become a function of all the stresses (σ_x, σ_y, and σ_z) due to the Poisson effect of each component such that:

$$\varepsilon_x = (1/E)[\sigma_x - \nu(\sigma_y + \sigma_z)] \tag{4.23}$$

$$\varepsilon_y = (1/E)[\sigma_y - \nu(\sigma_z + \sigma_x)] \tag{4.24}$$

$$\varepsilon_z = (1/E)[\sigma_z - \nu(\sigma_x + \sigma_y)] \tag{4.25}$$

The shear stress (γ) is related to the shear strain by the shear modulus (G) and can be resolved by the following:

$$\gamma_{xy} = \tau_{xy} / G \tag{4.26}$$

$$\gamma_{yz} = \tau_{yz} / G \tag{4.27}$$

$$\gamma_{xz} = \tau_{xz} / G \tag{4.28}$$

4.2.5 Plane Stress

If a thin plate is externally loaded by forces, as shown in Figure 4.4a, then the stress components of σ_z, τ_{xz}, and τ_{yz} are zero on both faces and within the plate. The state of stress is defined by σ_x, σ_y, and τ_{xy} only (Timoshenko and Goodier, 1951). A two-dimensional stressed body deforms in the plane of loading and in the direction normal to the plane by Poisson's ratio such that $\varepsilon_z \neq 0$, even though σ_z equals zero. Both plane stress and plane strain are assumptions made to simplify more complex problems. For the case of plane stress,

$$\sigma_z = \tau_{xz} = \tau_{yz} = 0 \tag{4.29}$$

$$\varepsilon_z \neq 0 \tag{4.30}$$

$$\varepsilon_x = (1/E)(\sigma_x - \nu\sigma_y) \tag{4.31}$$

$$\varepsilon_y = (1/E)(\sigma_y - \nu\sigma_x) \tag{4.32}$$

$$\varepsilon_z = -(\nu/E)(\sigma_x + \sigma_y) \tag{4.33}$$

$$\gamma_{xy} = \tau_{xy} / G \tag{4.34}$$

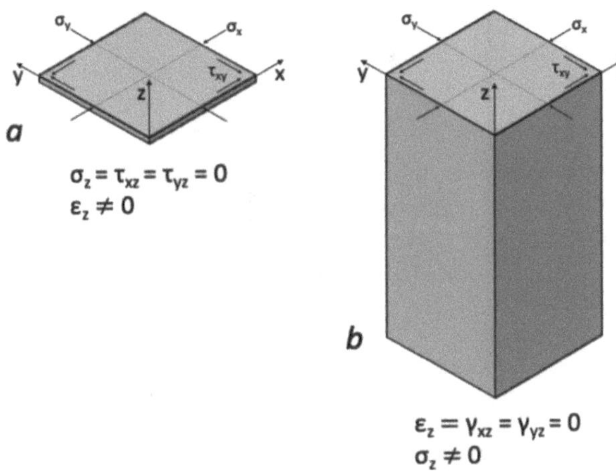

FIGURE 4.4 Plane stress (a) and plane strain (b).

4.2.6 Plane Strain

If a body has a significant dimension in a direction, here assume the z-direction, which is larger than the other directions (x and y axes), then a condition, plane strain, can be assumed as shown in Figure 4.4b. Unlike plane stress, ε_z will equal zero and $\sigma_z \neq 0$. Therefore, in the case of plane strain, the general case is

$$\varepsilon_z = \gamma_{xz} = \gamma_{yz} = 0 \tag{4.35}$$

$$\sigma_z \neq 0 \tag{4.36}$$

If ε_z is set to zero in Equation 4.25, then the following relationships can be resolved:

$$\sigma_z = \nu(\sigma_x + \sigma_y) \tag{4.37}$$

$$\varepsilon_x = (1/E)\left[(1-\nu^2)\sigma_x - \nu(1+\nu)\sigma_y\right] \tag{4.38}$$

$$\varepsilon_y = (1/E)[(1-\nu^2)\sigma_y - \nu(1+\nu)\sigma_x] \tag{4.39}$$

$$\gamma_{xy} = \tau_{xy}/G \tag{4.40}$$

A plain strain example relevant to mining would be a long and relatively straight horizontal tunnel in a rock mass, whose behavior could therefore be approximated in 2D.

It should be noted that plane stress and plane strain cannot coexist.

4.2.7 Strain Energy

When an elastic body is strained by the application of a stress, energy is stored in the form of strain energy density as shown in Figure 4.5. The strain energy density refers to the strain energy per unit volume, expressed as follows:

$$W = \int_0^{\varepsilon} \sigma \, d\varepsilon \tag{4.41}$$

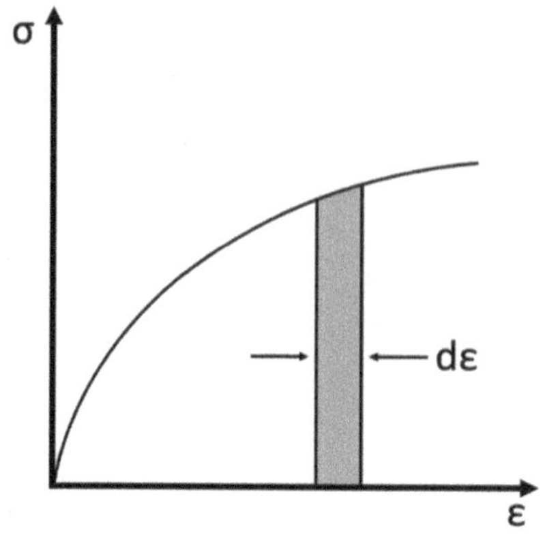

FIGURE 4.5 Strain energy density resulting from a stress on an object.

Units of work, W, or strain energy, take the form of N-m for example. Strain energy density has units of the form N-m/m³.

For a linear elastic substance under uniaxial loading, the strain energy density is

$$W = \tfrac{1}{2}\sigma\varepsilon \tag{4.42}$$

Under triaxial conditions, the equation takes the form of

$$W = \tfrac{1}{2}(\sigma_1\varepsilon_1 + \sigma_2\varepsilon_2 + \sigma_3\varepsilon_3) \tag{4.43}$$

The equation can also be expressed in stress and strain components:

$$W = \tfrac{1}{2}(\sigma_x\varepsilon_x + \sigma_y\varepsilon_y + \sigma_z\varepsilon_z + \tau_{xy}\gamma_{xy} + \tau_{yz}\gamma_{yz} + \tau_{zx}\gamma_{zx}) \tag{4.44}$$

4.3 THEORIES OF FAILURE

Rock is not infinitely strong, and the failure of most brittle materials depends on the mechanical conditions existing in the rock, when a certain failure strength is exceeded for example. The following are failure theories, summarized originally by Obert and Duvall (1967), and as such assume tension is positive.

4.3.1 Coulomb–Navier Theory of Failure

This theorizes that failure will occur in a material when the maximum shear stress at a point reaches a specific value (S_0), which is referred to as the shear strength (Jaeger, 1960). If σ_1, σ_2, and σ_3 are the principal stresses, whereby $\sigma_1 > \sigma_2 > \sigma_3$, then the maximum shear stress is given by

$$\tau_{max} = \tfrac{1}{2}(\sigma_1 - \sigma_3) \tag{4.45}$$

According to Coulomb (1776), failure will occur when shear stress has the following magnitude:

$$S_0 = \tfrac{1}{2}(\sigma_1 - \sigma_3) \tag{4.46}$$

Moreover, the failure plane will bisect the angle between the maximum and minimum principal stresses. For a triaxial test, the plane of failure should be at 45° degrees to the axial load, which is in fact not the case in practice nor is it borne out of experiment.

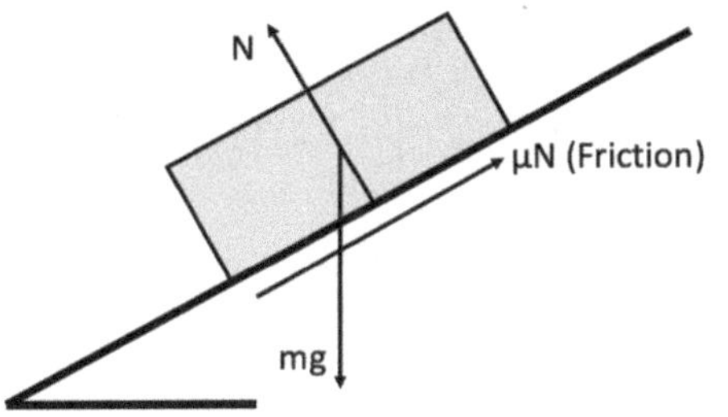

FIGURE 4.6 Friction from an inclined body.

Navier (1823) modified the theory, because he postulated that the normal stress to the failure plane increased the shear resistance of the material proportionally to the magnitude of the applied normal stress.

For a two-dimensional case, if σ_F and τ_F are the normal and shear stresses acting on the plane of failure, respectively, then the modified theory states that failure will occur when (depending on the sign convention)

$$|\tau_F| = S_0 - \mu\sigma_F \tag{4.47}$$

As defined earlier, S_0 is the maximum shear stress causing failure.

The concept can be taken as analogous to the friction against a body on an inclined plane caused by the normal reaction, as shown in Figure 4.6 for the limiting case where a body of mass, m, is about to slip. In the Coulomb–Navier criterion, μ is the coefficient of internal friction.

The normal and shear stresses at any orientation are as follows:

$$\sigma_F = \tfrac{1}{2}(\sigma_1 + \sigma_3) + \tfrac{1}{2}(\sigma_1 - \sigma_3)\cos 2\theta \tag{4.48}$$

$$\tau_F = -\tfrac{1}{2}(\sigma_1 - \sigma_3)\sin 2\theta \tag{4.49}$$

Equation 4.47 becomes

$$S_0 = \mu\sigma_F + |\tau_F| = \mu\left[\tfrac{1}{2}(\sigma_1 + \sigma_3) + \tfrac{1}{2}(\sigma_1 - \sigma_3)\right](\sin 2\theta + \mu\cos 2\theta) \tag{4.50}$$

Equation 4.50 is maximum when $dS_0/d\theta = 0$, that is when

$$\tan 2\theta = 1/\mu \tag{4.51}$$

The magnitude of S_0 at this angle is

$$S_0 = \tfrac{1}{2}\sigma_1\left[\mu + \left(\mu^2 + 1\right)^{1/2}\right] + \tfrac{1}{2}\sigma_3\left[\mu - \left(\mu^2 + 1\right)^{1/2}\right] \tag{4.52}$$

As μ increases, the plane of failure moves toward the direction of the algebraically least stress (compression in this case being negative).

When the criterion of failure is expressed in terms of T_0 (tensile strength) and C_0 (compressive strength), the equations are as follows by treating $\sigma_1 = T_0$ and $\sigma_3 = 0$ in Equation 4.52:

Failure in tension:

$$S_0 = \tfrac{1}{2}T_0\left[\left(\mu^2 + 1\right)^{1/2} + \mu\right] \tag{4.53}$$

Failure in compression is obtained by treating $\sigma_3 = -C_0$ and $\sigma_1 = 0$:

$$S_0 = \tfrac{1}{2} C_0 \left[\left(\mu^2 + 1\right)^{1/2} - \mu \right] \tag{4.54}$$

When Equation 4.54 is divided by Equation 4.53, the following is derived:

$$C_0/T_0 = \left[\left(\mu^2 + 1\right)^{1/2} + \mu \right] / \left[\left(\mu^2 + 1\right)^{1/2} - \mu \right] \tag{4.55}$$

4.3.2 Mohr Theory of Failure (1870)

According to Obert and Duvall (1967), Mohr postulated that a material will fail (fracture or permanently deform) when the shear stress (τ_θ) on the fracture, or slip plane, has increased to a value that depends on the normal stress (σ_θ) acting on the same plane, or when the largest tensile principal stress has reached a limiting value T_0 (the tensile strength of the material). Therefore, failure occurs when either of the following two conditions are met:

$$\tau_\theta = f(\sigma_\theta) \tag{4.56}$$

$$\sigma_1 = T_0 \tag{4.57}$$

The relationship in Equation 4.56 must be determined experimentally, and is represented in Figure 4.7, by the envelope defined by curves AB and CD. For any state of stress represented by a Mohr circle that lies entirely inside the envelope, failure will not occur. This criterion also predicts the angle of the fracture plane.

For trona rock specimens tested in triaxial compression, τ_θ increases in a positively monotonic manner with σ_θ. Therefore, as one parameter increases, so does the other but not in proportion.

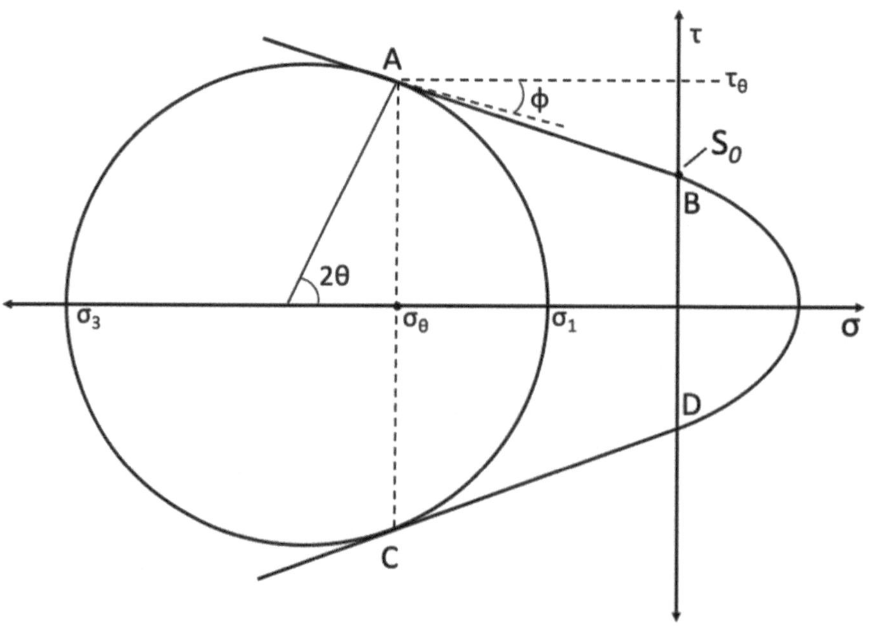

FIGURE 4.7 Failure envelope developed using the Mohr circle (Obert and Duvall, 1967).

Hence, under this theory, a material will not fail in hydrostatic compression, which is also borne out of experiments (Obert and Duvall, 1967).

4.3.3 Griffith's Theory of Brittle Fracture

Crystalline materials contain micro-fractures, and Griffith's hypothesis (Griffith, 1921) states that stress concentrations develop at the end of these cracks, causing the cracks to propagate and ultimately create a macroscopic fracture (after Obert and Duvall, 1967). The McClintock and Walsh modification (1962) of the Griffith criterion was developed to include the closing of the cracks during compression and the development of friction along the cracks. It predicts a linear relationship between σ_1 and σ_3 as does the Coulomb–Navier theory:

$$\left[\sigma_1\left(1+\mu^2\right)^{1/2}+\mu\right]/\left[C_0\left(1+\mu^2\right)^{1/2}-\mu\right]=1+(\sigma_3/C_0) \quad (4.58)$$

4.4 EFFECTS OF WEAKNESS PLANES

On a large mine wide (megascopic) scale, the rock masses contain numerous planes of weakness created by jointing, other fractures, faults, bedding planes, or partings. Consider a material containing a plane of weakness whose failure envelope has been determined by direct shear or triaxial tests, line AB in Figure 4.8a (Obert and Duvall, 1967). Assume that this material is subjected to principal compressive stresses σ_1 and σ_3 with the plane of weakness orientated at an angle of γ with respect to the σ_3 axis as shown in Figure 4.8b. If on the Mohr circle, the angle 2γ, is such that the radius OC touches or intersects the failure envelop (between D and E), then failure will result. For all other values of 2γ, there will be no failure. It should be noted that for these examples, tension is assumed as positive by Obert and Duvall (1967).

In Figure 4.8, φ is the angle of internal friction and S_0 is the inherent (system) shear strength.

4.5 EFFECTS OF MOISTURE AND PORE PRESSURE

The mechanical behavior of rock depends not only on the state of stress and strain and on the rate of stressing and straining, but also on the moisture and pore pressure.

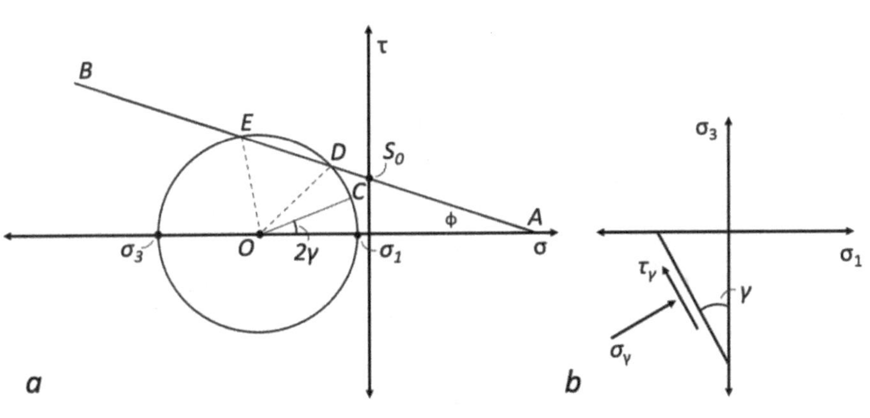

FIGURE 4.8 Using the Mohr circle to determine if failure potentially occurs. (Modified after Obert and Duvall (1967).)

Many underground rocks contain moisture ranging from less than 1% to over 35% (Obert and Duvall, 1967), depending on the rock type. The presence of moisture in porous rocks can have varying negative effects on the behavior, such as:

- If the water is migrating, it creates a mechanical (abrasive) action.
- Chemical action may take place and dissolve the material or a component of that material.
- Chemical hydration or dehydration may cause a volumetric change.
- Exposure to varying humidity (or wetting and drying) can cause certain rocks to separate on bedding planes for example.

Moisture is not always fully understood in a specific mine site, but it negatively affects the compressive strength and elastic properties of rock especially.

Pore water pressure is the pressure produced by water contained in the gaps (pores) in soils or rocks. Pore pressure reduces principal stress to an effective principal stress. For example, the Coulomb–Navier condition for failure becomes

$$\tau_\theta = S_0 - \mu(\sigma_\theta - p_0) \quad (4.59)$$

$$T_0 = \sigma_1 + p_0 \quad (4.60)$$

Therefore, the pore pressure reduces the normal stress on the failure plane by an amount p_0.

The presence of pore water also alters the location of the Mohr circle.

4.6 WORKED EXAMPLES

Example 4.6.1

Label the following figures as positive or negative for the shear stress and strain as shown in Figure 4.9. Given the direction provided, indicate if they plot below or above the normal stress and strain axis on the Mohr circle.

The solutions are given in Figure 4.10.

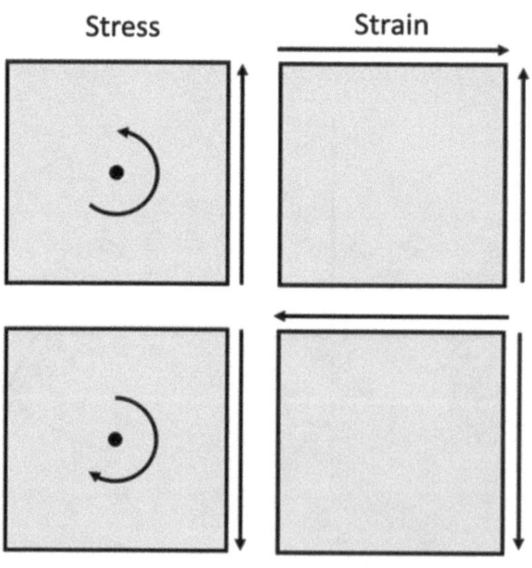

FIGURE 4.9 The stress and strain conditions.

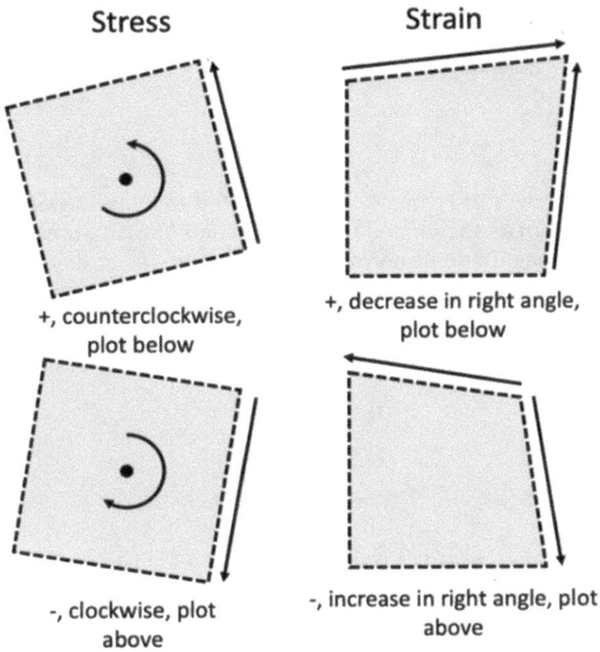

FIGURE 4.10 The solutions to Example 4.6.1.

Example 4.6.2

Given the following relationship discussed in this chapter, what value of ν is necessary for a lithostatic virgin stress. Use the relationship, $\sigma_x = \sigma_y = k\sigma_z$.

For lithostatic condition, stress is the same in all directions, and therefore, $\sigma_x = \sigma_y = \sigma_z$.

Solution:

This means k = 1, and therefore,

$$k = \nu/1 - \nu$$

$$1 = \nu/1 - \nu$$

$$1 - \nu = \nu$$

$$1 = 2\nu$$

$$\nu = 0.5$$

Example 4.6.3

Using the relationship by Pallister et al. (1970) and Gay (1975) ($\sigma_h \approx 0.6\sigma_v$), if it is assumed that $\sigma_h = \sigma_x$ and $\sigma_v = \sigma_z$, what would be the value of Poisson's ratio, ν?

Solution:

$$k = \nu/1 - \nu$$

$$0.6 = \nu/1 - \nu$$

$$0.6 - 0.6\nu = \nu$$

$$0.6 = 1.6\nu$$

$$\nu = 0.375$$

Example 4.6.4

A square element (as shown in Figure 4.11) is subjected to a stress state in an arbitrary rectangular coordinate system as shown and then subdivided into an array of smaller elements of equal size. Assign the magnitude of normal and shear stresses to its appropriate symbol and subscript for element t as shown. Determine the minimum and maximum principal stresses at element t.

Solution:

$$\sigma_x = \sigma_p = 50 \text{ MPa}$$

$$\sigma_y = \sigma_q = 50 \text{ MPa}$$

$$\tau_{xy} = \tau_{pq} = -35 \text{ MPa}$$

$$\sigma_1 = ½(\sigma_x + \sigma_y) + ½[(\sigma_x - \sigma_y)^2 + 4(\tau_{xy})^2]^{½}$$

$$\sigma_1 = ½(\sigma_p + \sigma_q) + ½[(\sigma_p - \sigma_q)^2 + 4(\tau_{pq})^2]^{½}$$

$$\sigma_1 = ½(50 + 50) + ½[(50 - 50)^2 + 4(-35)^2]^{½}$$

$$\sigma_1 = 50 + 35$$

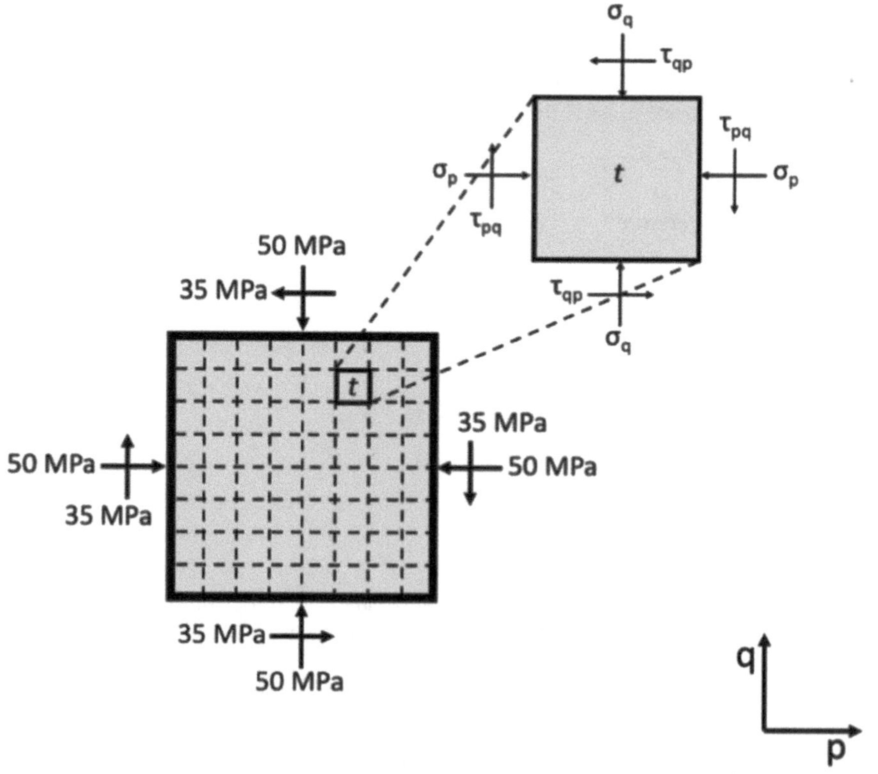

FIGURE 4.11 The stress state on an element and the stresses on element t.

$$\sigma_1 = 85 \text{ MPa}$$

$$\sigma_2 = ½(\sigma_x + \sigma_y) - ½[(\sigma_x - \sigma_y)^2 + 4(\tau_{xy})^2]^{½}$$

$$\sigma_2 = ½(\sigma_p + \sigma_q) - ½[(\sigma_p - \sigma_q)^2 + 4(\tau_{pq})^2]^{½}$$

$$\sigma_2 = ½(50 + 50) - ½[(50 - 50)^2 + 4(-35)^2]^{½}$$

$$\sigma_2 = 50 - 35$$

$$\sigma_2 = 15 \text{ MPa}$$

Example 4.6.5

Repeat previous Example 4.6.4, but now assume it is of interest to determine the stress on an imaginary plane which acts through the same element at an angle of 45° (in Figure 4.12). Determine the normal and shear stresses on the imaginary plane.

Solution:

$$\sigma_x = \sigma_p = 50 \text{ MPa}$$

$$\sigma_y = \sigma_q = 50 \text{ MPa}$$

$$\tau_{xy} = \tau_{pq} = -35 \text{ MPa}$$

Check stresses before orientation:

The first normal stress (σ_t) is orientated at $\theta=0°$:

$$\sigma_t = ½(\sigma_x + \sigma_y) + ½(\sigma_x - \sigma_y)\cos 2\theta + \tau_{xy}\sin 2\theta$$

$$\sigma_t = ½(50 + 50) + ½(50 - 50)\cos(0°) + (-35)\sin(0°)$$

$$\sigma_t = 50 \text{ MPa}$$

The second normal stress (σ_n) is orientated at $\theta=90°$:

$$\sigma_n = ½(\sigma_x + \sigma_y) + ½(\sigma_x - \sigma_y)\cos 2\theta + \tau_{xy}\sin 2\theta$$

$$\sigma_n = ½(50 + 50) + ½(50 - 50)\cos(180°) + (-35)\sin(180°)$$

$$\sigma_n = 50 \text{ MPa}$$

$$\tau_{nt} = -½(\sigma_x - \sigma_y)\sin 2\theta + \tau_{xy}\cos 2\theta$$

$$\tau_{nt} = -½(50 - 50)\sin(0°) - 35\cos(0°)$$

$$\tau_{nt} = -35 \text{ MPa}$$

Orientate at 45°, $\theta=0°+45°$:

$$\sigma_t = ½(\sigma_x + \sigma_y) + ½(\sigma_x - \sigma_y)\cos 2\theta + \tau_{xy}\sin 2\theta$$

$$\sigma_t = ½(50 + 50) + ½(50 - 50)\cos(90°) + (-35)\sin(90°)$$

$$\sigma_t = 50 - 35$$

$$\sigma_t = 15 \text{ MPa}$$

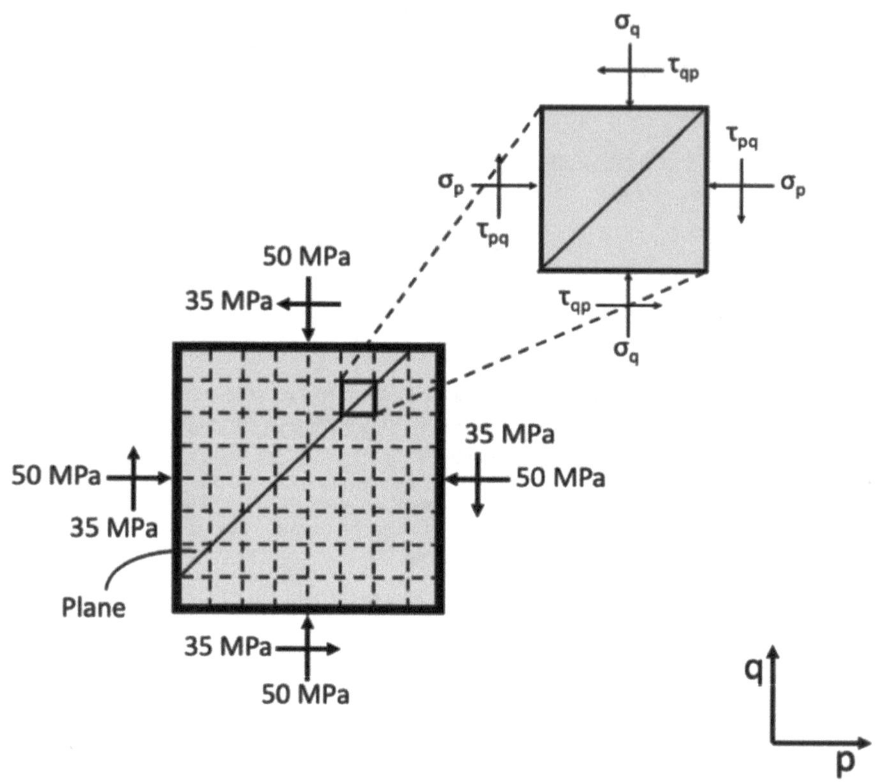

FIGURE 4.12 The stress state on the re-orientated element.

The second normal stress (σ_n) is orientated at $\theta = 90° + 45°$:

$$\sigma_n = ½(\sigma_x + \sigma_y) + ½(\sigma_x - \sigma_y)\cos 2\theta + \tau_{xy}\sin 2\theta$$

$$\sigma_n = ½(50 + 50) + ½(50 - 50)\cos(270°) + (-35)\sin(270°)$$

$$\sigma_n = 50 + 35$$

$$\sigma_n = 85 \text{ MPa}$$

$$\tau_{nt} = -½(\sigma_x - \sigma_y)\sin 2\theta + \tau_{xy}\cos 2\theta$$

$$\tau_{nt} = -½(50 - 50)\sin(90°) - 35\cos(90°)$$

$$\tau_{nt} = 0 \text{ MPa}$$

Example 4.6.6

Show and describe the stepwise construction of the Mohr circle for the following state of stress on the square element as shown in Figure 4.13. Determine the maximum and minimum principal stress magnitudes and directions. A compass, protractor, ruler, and engineering paper are recommended for this graphical exercise.

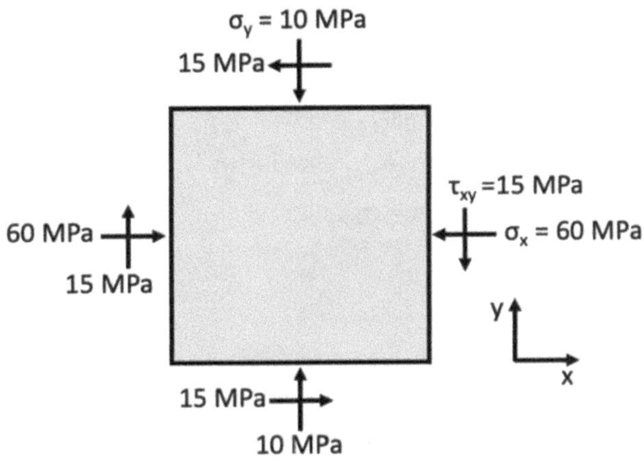

FIGURE 4.13 The problem stress state.

Solution:

First Step: Determine the state of stress on a square element in x–y space, and construct orthogonal axes of normal stress (σ) and shear stress (τ) as shown in Figure 4.14. The axes should be of equal scale, and a special note is made of the location of negative and positive shear stresses on the axes. After construction, the magnitudes of σ_x and σ_y can be marked on the σ axis.

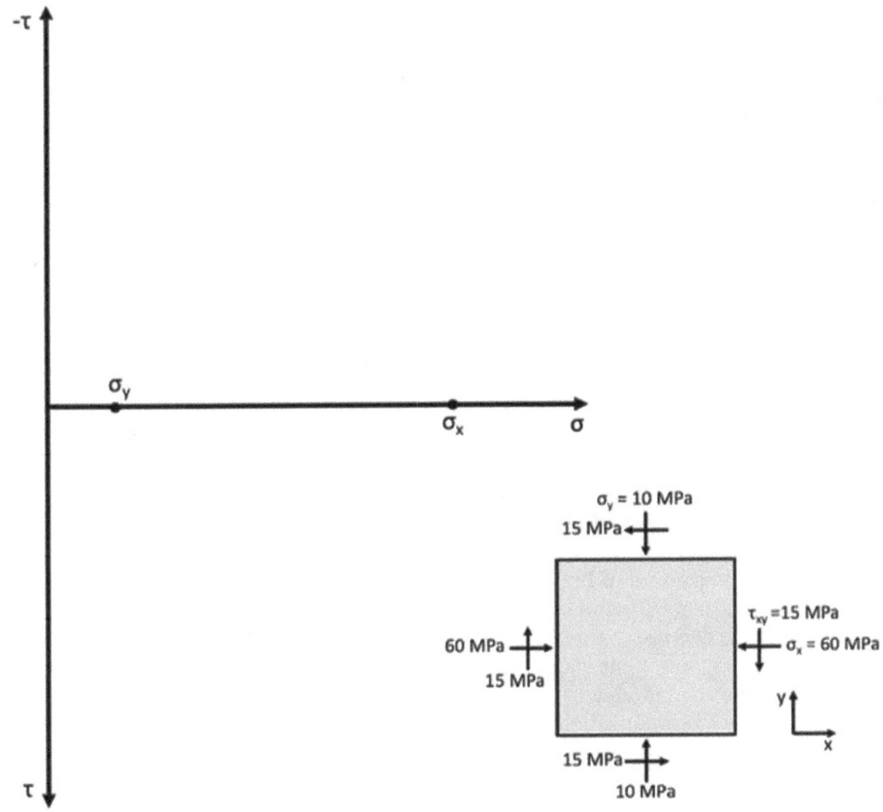

FIGURE 4.14 Construction of the σ–τ axes and the location of σ_x and σ_y.

Second Step: Calculate the average normal stress which will represent the center of the Mohr circle and plot on the normal stress axis (Figure 4.15):

$$c = \sigma_{avg} = ½(\sigma_x + \sigma_y)$$

$$c = \sigma_{avg} = ½(60 + 10)$$

$$c = \sigma_{avg} = 35 \text{ MPa}$$

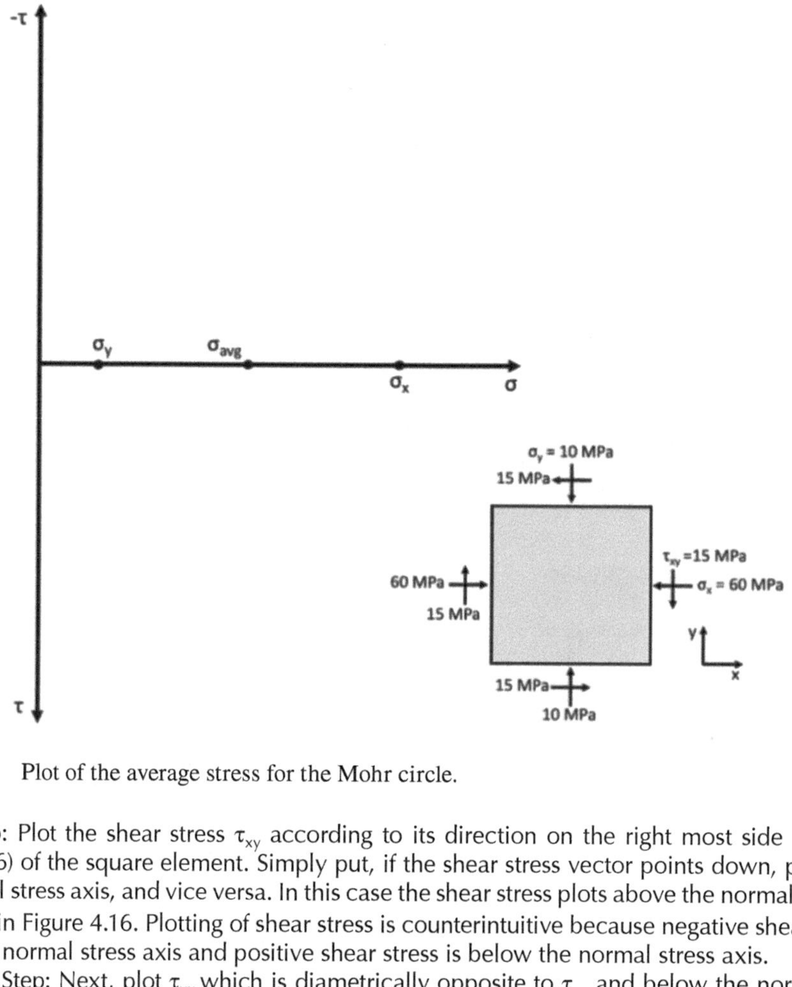

FIGURE 4.15 Plot of the average stress for the Mohr circle.

Third Step: Plot the shear stress τ_{xy} according to its direction on the right most side (note a in Figure 4.16) of the square element. Simply put, if the shear stress vector points down, plot above the normal stress axis, and vice versa. In this case the shear stress plots above the normal stress (σ_x) as shown in Figure 4.16. Plotting of shear stress is counterintuitive because negative shear stress is above the normal stress axis and positive shear stress is below the normal stress axis.

Fourth Step: Next, plot τ_{yx} which is diametrically opposite to τ_{xy} and below the normal stress axis (in Figure 4.17). This corresponds to the stress on the top of the element as shown by note b in Figure 4.17. A straight line should be formed from τ_{yx} through σ_{avg} to τ_{xy}. The distance from σ_{avg} to either shear stress forms the radius, r, of the Mohr circle.

At this point, it is important to highlight that angle c is expressed by two angles. One angle is in the rectangular space at 90°, and the other is 180° as shown in the Mohr circle in Figure 4.18.

Fifth Step: Using a compass, centered on σ_{avg} with a radius, r, draw a complete circle. The circle should pass through τ_{yx} and τ_{xy} as shown in Figure 4.19.

With this step, the magnitudes of the maximum (σ_1 or σ_p) and minimum (σ_2 or σ_q) principal stresses as well as the maximum shear stress (τ_{max}) will be known graphically (see Figure 4.20).

The values can also be calculated as $\sigma_p = 64$ MPa, $\sigma_q = 6$ MPa, and $\tau_{max} = 29$ MPa below:

$$\sigma_1 = ½(\sigma_x + \sigma_y) + ½[(\sigma_x - \sigma_y)^2 + 4(\tau_{xy})^2]^{½}$$

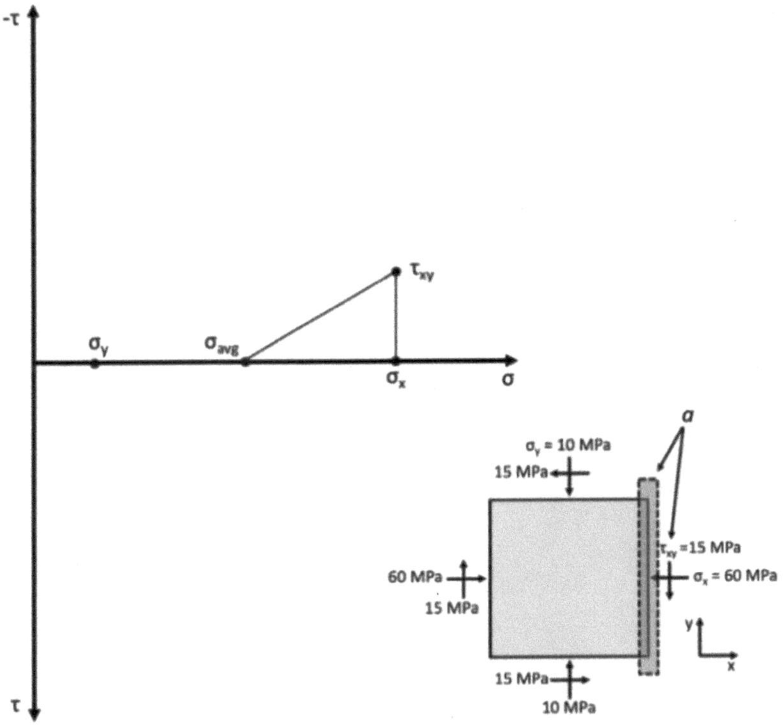

FIGURE 4.16 Plotting of shear stress for the Mohr circle.

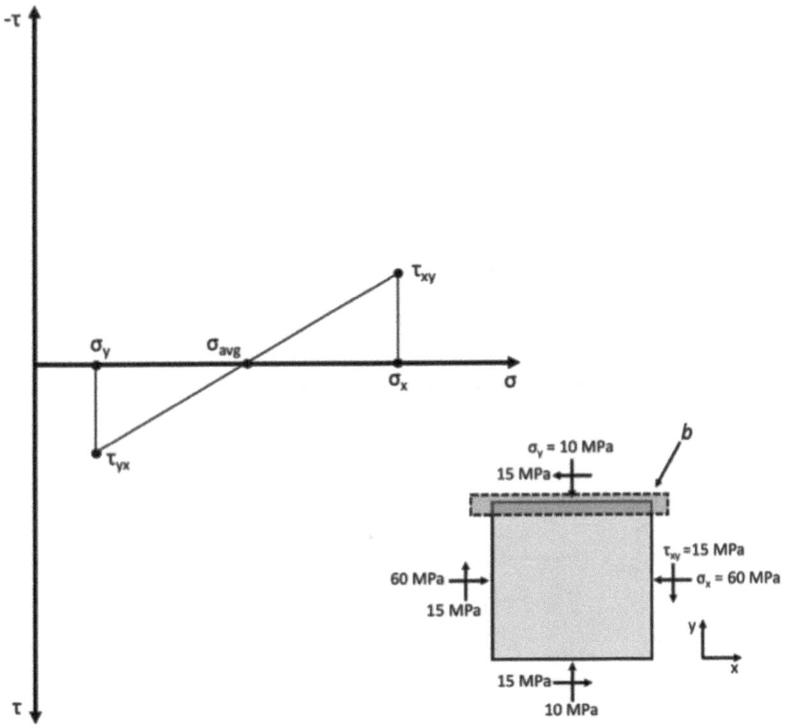

FIGURE 4.17 Plotting of the second shear stress for the Mohr circle.

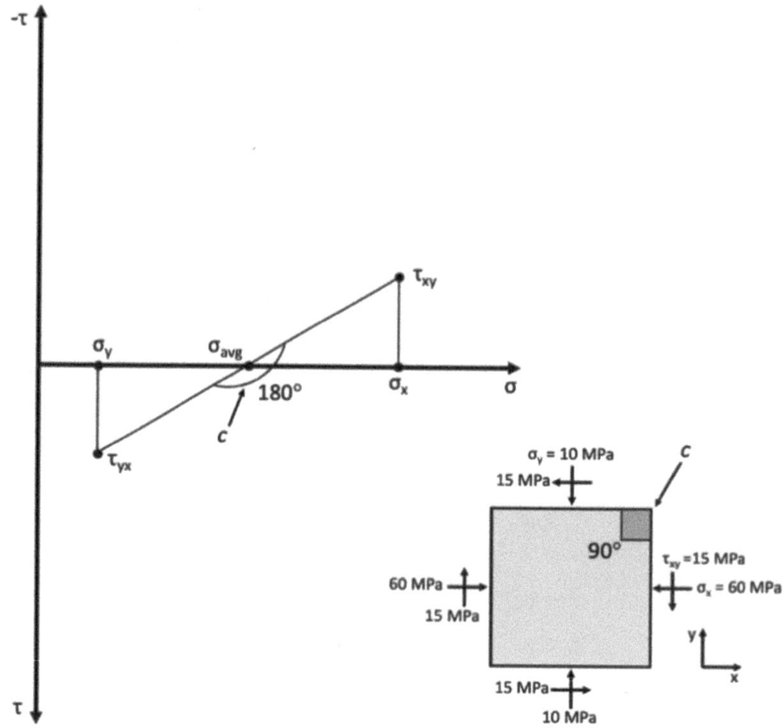

FIGURE 4.18 Angles in the Mohr circle and rectangular space.

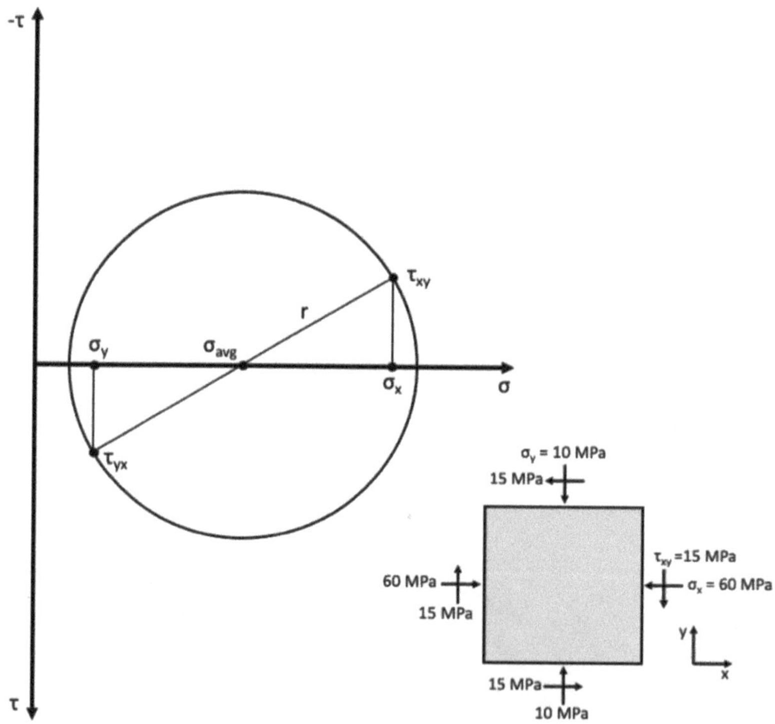

FIGURE 4.19 Drawing the Mohr circle.

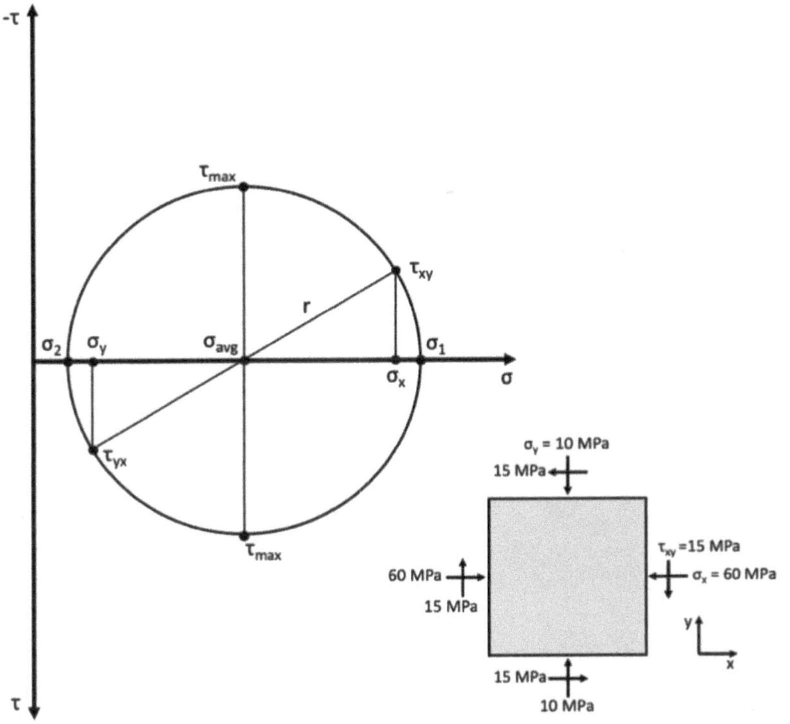

FIGURE 4.20 Location of maximum and minimum principal stresses and maximum shear stress on the Mohr circle.

$$\sigma_1 = \tfrac{1}{2}(60+10)+\tfrac{1}{2}[(60-10)^2 + 4(-15)^2]^{\frac{1}{2}}$$

$$\sigma_1 = 35+\tfrac{1}{2}[2{,}500+900]^{\frac{1}{2}}$$

$$\sigma_1 = 35+29$$

$$\sigma_1 = 64 \text{ MPa}$$

$$\sigma_2 = \tfrac{1}{2}(\sigma_x+\sigma_y)-\tfrac{1}{2}[(\sigma_x-\sigma_y)^2 + 4(\tau_{xy})^2]^{\frac{1}{2}}$$

$$\sigma_2 = 35-29$$

$$\sigma_2 = 6 \text{ MPa}$$

$$r = \tau_{max} = \tfrac{1}{2}(\sigma_1-\sigma_2)$$

$$r = \tau_{max} = \tfrac{1}{2}(64-6)$$

$$r = \tau_{max} = 29 \text{ MPa}$$

A simplified option for calculating the maximum and minimum principal stresses uses the center and radius of the circle:

$$\sigma_1 = c+r$$

$$\sigma_1 = 35 \text{ MPa} + 29 \text{ MPa}$$

$$\sigma_1 = 64 \text{ MPa}$$

$$\sigma_2 = c-r$$

$$\sigma_2 = 35 \text{ MPa} - 29 \text{ MPa}$$

$$\sigma_2 = 6 \text{ MPa}$$

Sixth Step: The final step is determining the orientation of the principal stress axes relative to the element orientation and plotting the square element into our original set of x–y axes. The angle, ϕ, in the Mohr circle is equal to 2θ, or –31° (refer to Figure 4.21):

$$\theta = ½\tan^{-1}[(2\tau_{xy}/(\sigma_x - \sigma_y)]$$
$$\theta = ½\tan^{-1}[2(-15)/(60-10)]$$
$$\theta = ½\tan^{-1}(-30/50)$$
$$\theta = -15.5°$$

The direction of rotation of the element follows the same angular sense as the rotation in the Mohr circle for the normal stress to become the principal stress as shown in Figure 4.22. This is important as there are instances in mine design where the plane of zero shear stress and the plane of maximum shear stress are useful to determine.

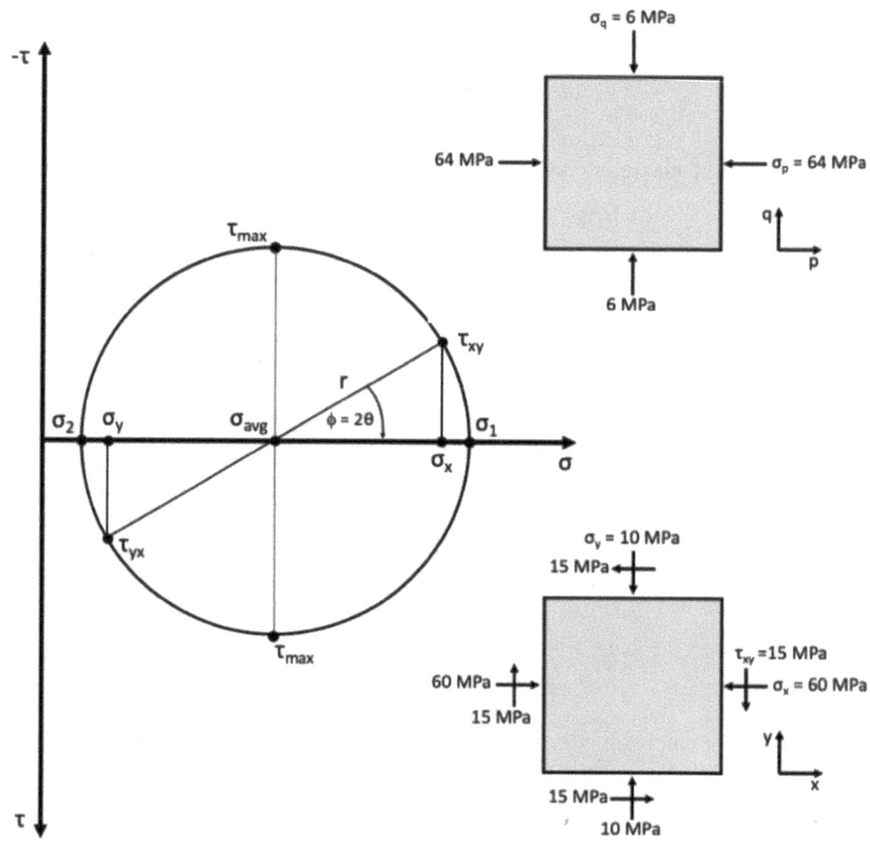

FIGURE 4.21 Resolving orientations in the Mohr circle.

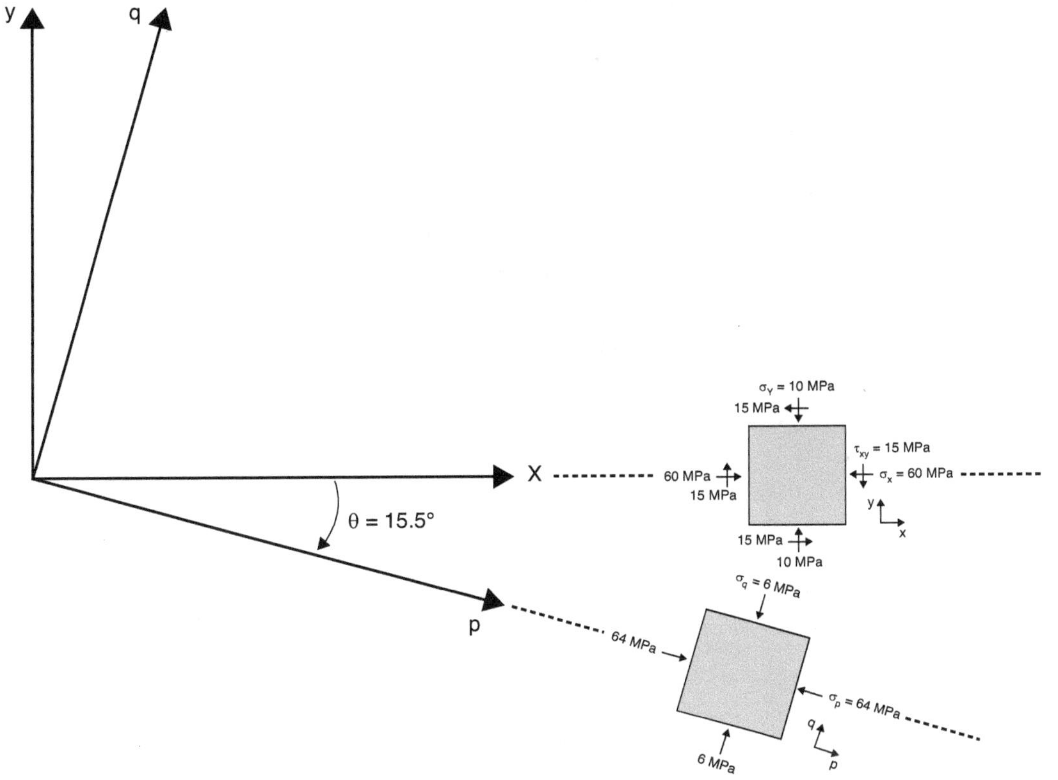

FIGURE 4.22 Final rotation of the square element from the Mohr circle.

Example 4.6.7

Calculate the vertical stress at 850 ft deep given the following rock weights and thicknesses as shown in Figure 4.23.

Solution:

$$\sigma_z = \gamma_1 z_1 + \gamma_2 z_2 + \gamma_3 z_3 + \gamma_4 z_4 + \gamma_5 z_5$$

$$\sigma_z = (147 \text{ lbs/ft}^3)(200 \text{ ft}) + (170 \text{ lbs/ft}^3)(200 \text{ ft}) + (152 \text{ lbs/ft}^3)(50 \text{ ft}) + (140 \text{ lbs/ft}^3)(100 \text{ ft}) + (164 \text{ lbs/ft}^3)(300 \text{ ft})$$

$$\sigma_z = 29{,}400 \text{ lbs/ft}^2 + 34{,}000 \text{ lbs/ft}^2 + 7{,}600 \text{ lbs/ft}^2 + 14{,}000 \text{ lbs/ft}^2 + 49{,}200 \text{ lbs/ft}^2$$

$$\sigma_z = 134{,}200 \text{ lbs/ft}^2 \text{ or } 932 \text{ psi}$$

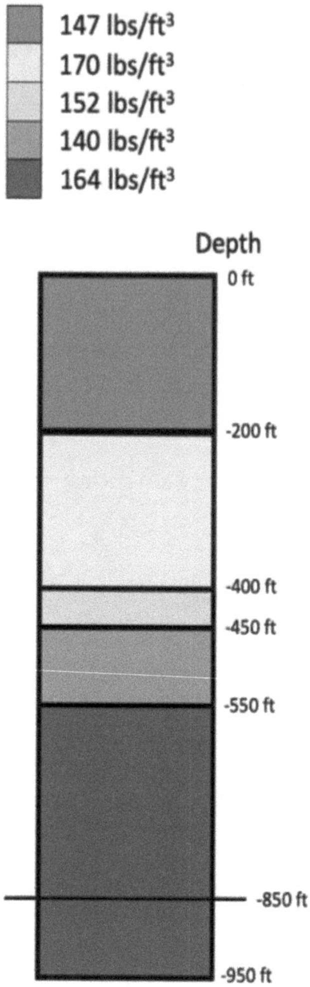

FIGURE 4.23 The unit weights and rock mass thickness for Example 4.6.7.

Example 4.6.8

Calculate the vertical stress at the 850 ft depth mark used in Example 4.6.7 as indicated in Figure 4.24 for the additional Conditions 2 and 3. Values are in elevation.

Solution:

Condition 2:

$$\sigma_z = \gamma_1 z_1 + \gamma_2 z_2 + \gamma_3 z_3 + \gamma_4 z_4 + \gamma_5 z_5$$

$$\sigma_z = \left(147 \text{ lbs/ft}^3\right)(400 \text{ ft}) + \left(170 \text{ lbs/ft}^3\right)(200 \text{ ft}) + \left(152 \text{ lbs/ft}^3\right)(50 \text{ ft}) + \left(140 \text{ lbs/ft}^3\right)(100 \text{ ft}) + \left(164 \text{ lbs/ft}^3\right)(300 \text{ ft})$$

$$\sigma_z = 58{,}800 \text{ lbs/ft}^2 + 34{,}000 \text{ lbs/ft}^2 + 7{,}600 \text{ lbs/ft}^2 + 14{,}000 \text{ lbs/ft}^2 + 49{,}200 \text{ lbs/ft}^2$$

$$\sigma_z = 163{,}600 \text{ lbs/ft}^2 \text{ or } 1{,}136 \text{ psi}$$

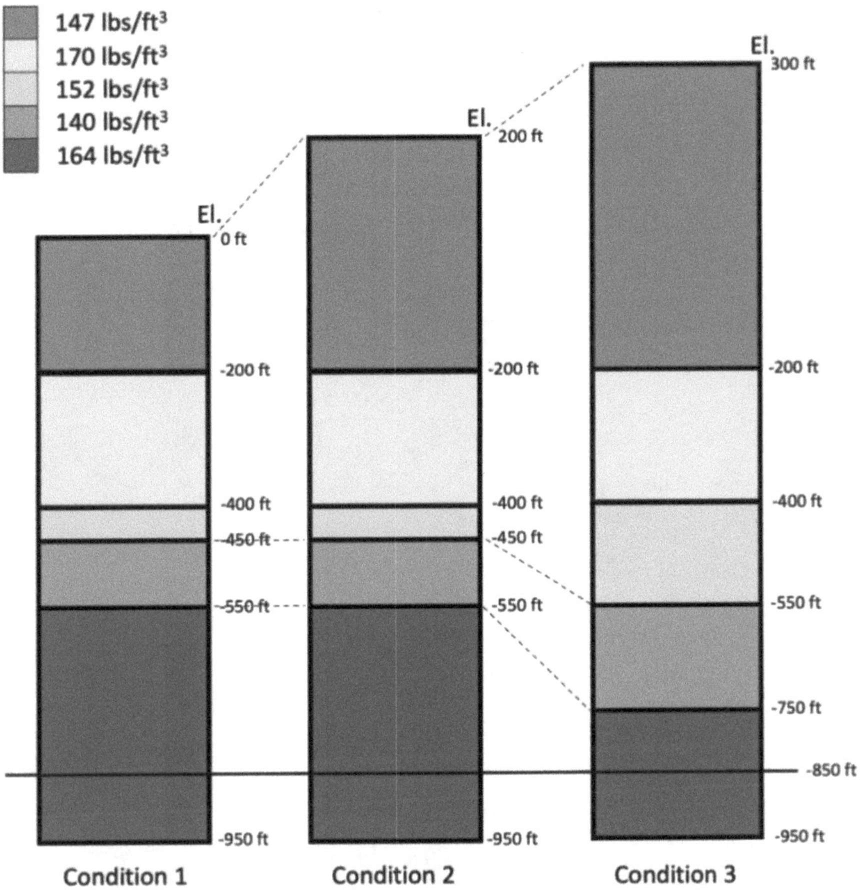

FIGURE 4.24 The unit weights and rock mass thickness for Example 4.6.8.

Condition 3:

$$\sigma_z = \gamma_1 z_1 + \gamma_2 z_2 + \gamma_3 z_3 + \gamma_4 z_4 + \gamma_5 z_5$$

$$\sigma_z = (147 \text{ lbs/ft}^3)(500 \text{ ft}) + (170 \text{ lbs/ft}^3)(200 \text{ ft}) + (152 \text{ lbs/ft}^3)$$
$$(150 \text{ ft}) + (140 \text{ lbs/ft}^3)(200 \text{ ft}) + (164 \text{ lbs/ft}^3)(100 \text{ ft})$$

$$\sigma_z = 73{,}500 \text{ lbs/ft}^2 + 34{,}000 \text{ lbs/ft}^2 + 22{,}800 \text{ lbs/ft}^2 + 28{,}000 \text{ lbs/ft}^2 + 16{,}400 \text{ lbs/ft}^2$$

$$\sigma_z = 174{,}700 \text{ lbs/ft}^2 \text{ or } 1{,}213 \text{ psi}$$

Example 4.6.9

Given the following element in Figure 4.25, determine the principal stresses and maximum shear stress by calculations and by using the Mohr circle:

$$\sigma_x = -100 \text{ MPa}$$

$$\sigma_y = -200 \text{ MPa}$$

$$\tau_{xy} = -60 \text{ MPa}$$

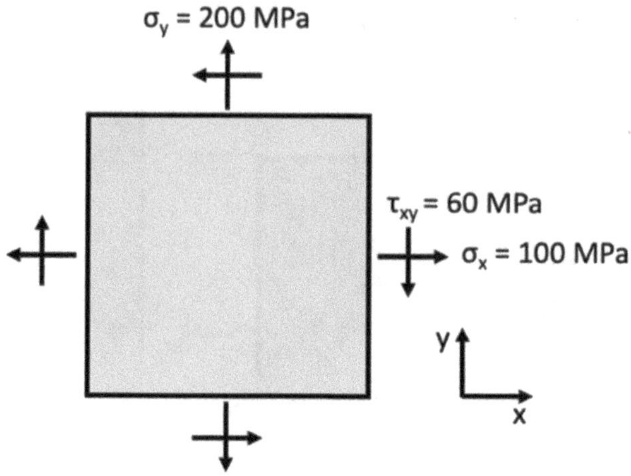

FIGURE 4.25 The stress state.

Solution:

$$\sigma_1 = ½(\sigma_x + \sigma_y) + ½[(\sigma_x - \sigma_y)^2 + 4(\tau_{xy})^2]^{½}$$

$$\sigma_1 = ½(-100 - 200) + ½[(-100 + 200)^2 + 4(-60)^2]^{½}$$

$$\sigma_1 = -150 + ½[10{,}000 + 14{,}400]^{½}$$

$$\sigma_1 = -150 + 78$$

$$\sigma_1 = -72 \text{ MPa}$$

$$\sigma_2 = ½(\sigma_x + \sigma_y) - ½[(\sigma_x - \sigma_y)^2 + 4(\tau_{xy})^2]^{½}$$

$$\sigma_2 = -150 - 78$$

$$\sigma_2 = -228 \text{ MPa}$$

$$\tau_{max} = ½(\sigma_1 - \sigma_2)$$

$$\tau_{max} = ½(-72 + 228)$$

$$\tau_{max} = 78 \text{ MPa}$$

$$\theta = ½\tan^{-1}[(2\tau_{xy}/(\sigma_x - \sigma_y))]$$

$$\theta = ½\tan^{-1}[-120/(-100 + 200)]$$

$$\theta = ½\tan^{-1}(-120/100)$$

$$\theta = -25°$$

The Mohr circle solution is shown in Figure 4.26 with the approximated graphical values of stress.

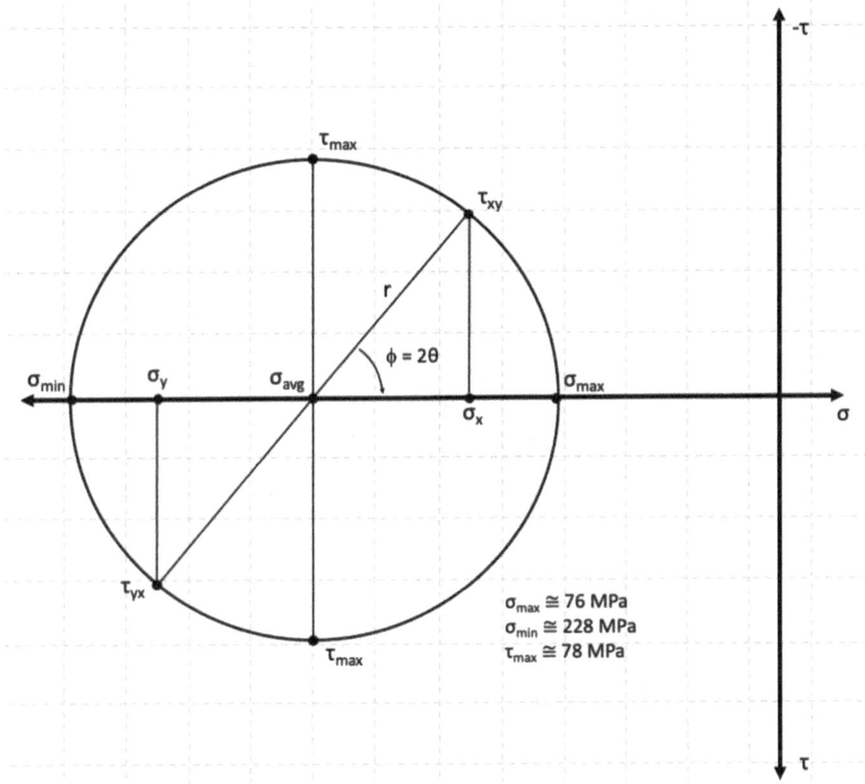

FIGURE 4.26 The Mohr circle solution for Example 4.6.9.

Example 4.6.10

Construct the Mohr circle given the following:

$$\sigma_x = 100 \text{ MPa}$$

$$\sigma_y = -100 \text{ MPa}$$

$$\tau_{xy} = 50 \text{ MPa}$$

Solution:

$$r = ½[(\sigma_x - \sigma_y)^2 + 4(\tau_{xy})^2]^{1/2}$$

$$r = ½\{(100 + 100)^2 + [2(50)]^2\}^{1/2}$$

$$r = ½[40{,}000 + 10{,}000]^{1/2}$$

$$r = 112 \text{ MPa}$$

$$c = ½(\sigma_x + \sigma_y)$$

$$c = ½(100 - 100)$$

$$c = 0 \text{ MPa}$$

Calculation included for completeness:

$$\sigma_1 = c + r$$

$$\sigma_1 = 0 \text{ MPa} + 112 \text{ MPa}$$

$$\sigma_1 = 112 \text{ MPa}$$

$$\sigma_2 = c - r$$

$$\sigma_2 = 0 \text{ MPa} - 112 \text{ MPa}$$

$$\sigma_2 = -112 \text{ MPa}$$

$$\tau_{max} = r$$

$$\tau_{max} = 112 \text{ MPa}$$

$$\theta = ½\tan^{-1}[(2\tau_{xy}/(\sigma_x - \sigma_y)]$$

$$\theta = ½\tan^{-1}[(100/(100+100)]$$

$$\theta = ½\tan^{-1}(100/200)$$

$$\theta = 13°$$

The Mohr circle is given in Figure 4.27.

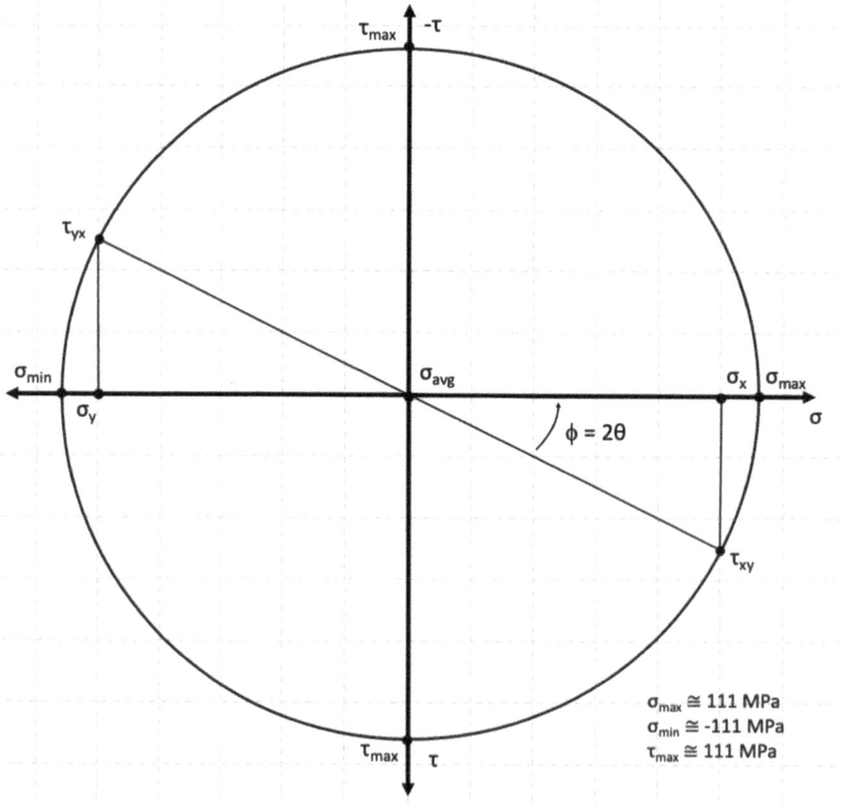

FIGURE 4.27 The Mohr circle solution for Example 4.6.10.

Example 4.6.11

Illustrate an orientation of 45° using the elemental square from the previous Example 4.6.10 and the Mohr circle. Calculate the stresses on the new element.

Solution:

The Mohr circle is shown in Figure 4.28.

$$\sigma_1 = 112 \text{ MPa}$$

$$\sigma_2 = -112 \text{ MPa}$$

Note that θ is the angle from the principal stress axis because we see the equations below are a function of σ_1 and σ_2 so account for the change from the principal stress axis:

$$\theta = 45° - 13° = 33°$$

$$\sigma_x = ½(\sigma_1 + \sigma_2) + ½(\sigma_1 - \sigma_2)\cos 2\theta$$

$$\sigma_x = 0 + ½(224)\cos 66°$$

$$\sigma_x = 46 \text{ MPa}$$

$$\sigma_y = ½(\sigma_1 + \sigma_2) - ½(\sigma_1 - \sigma_2)\cos 2\theta$$

$$\sigma_y = 0 - ½(224)\cos 66°$$

$$\sigma_y = -46 \text{ MPa}$$

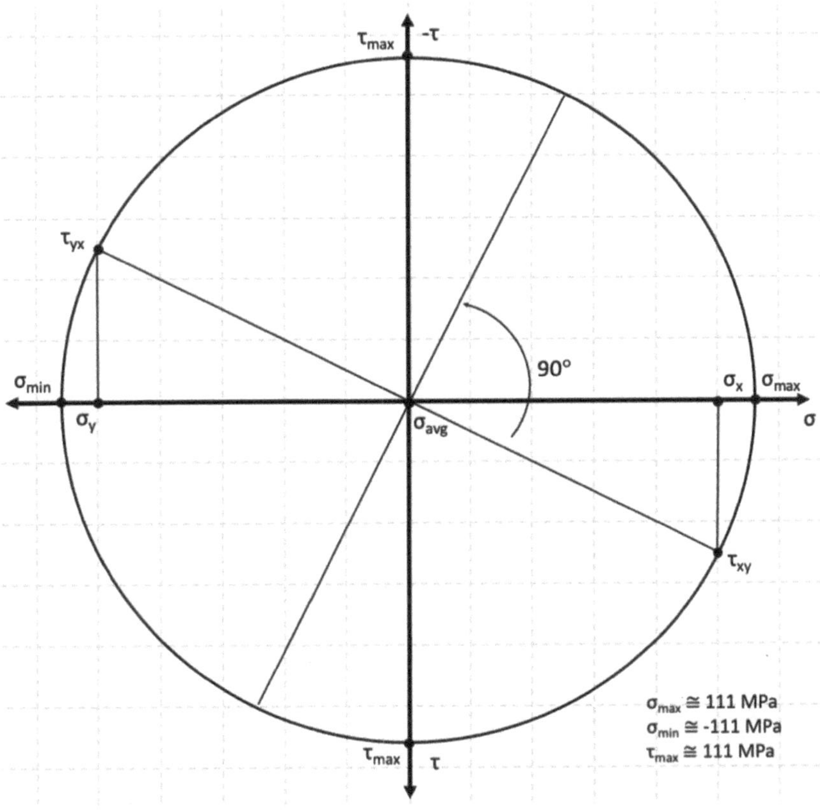

FIGURE 4.28 The Mohr circle solution for Example 4.6.11.

$$\tau_{xy} = -\tfrac{1}{2}(\sigma_1 - \sigma_2)\sin 2\theta$$

$$\tau_{xy} = -\tfrac{1}{2}(112 + 112)\sin 66^\circ$$

$$\tau_{xy} = -102 \text{ MPa}$$

Example 4.6.12

The following data approximated from Farmer (1968) provide the results of triaxial tests on dry and fully saturated quartzite by Colback and Wiid (1965). Determine the reduction of the shear intercept constant S_0 due to saturation using the Mohr circle and the approximated principal stresses in Table 4.1.

TABLE 4.1
The Stress Data Approximated from Farmer (1968)

	Unsaturated Quartzite (kg/cm²)	Saturated Quartzite (kg/cm²)
σ_1	2,750	1,780
σ_3	500	500
σ_1	1,825	950
σ_3	50	50

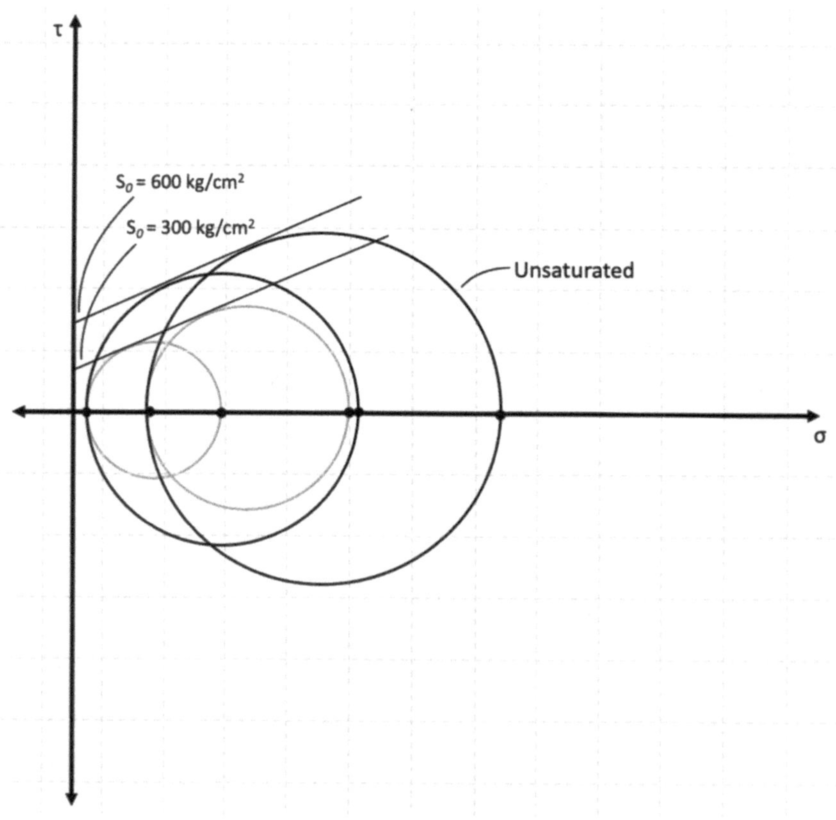

FIGURE 4.29 The Mohr circle.

Solution:

After plotting the Mohr circles, the reduction of the intercept can be determined graphically or using the Coulomb–Navier criterion. After Farmer (1968),

$$\tau = S_0 + (\sigma - \sigma_w)\tan\varphi$$

$$\text{where, } \sigma_w \tan\varphi = 300 \text{ kg/cm}^2$$

The answer should show an approximately 40%–60% reduction in unsaturated S_0 based on the differences in the values for the unsaturated and saturated values in Table 4.1.

Example 4.6.13

Pore water pressure (p_0) in rock reduces the overall principal stress to an effective principal stress (σ_1'). If the pore water pressure is measured in a pressure relief hole as 150 psi and the mine stresses are measured at 1,500 psi for the maximum principal stress and 500 psi for the minimum principal stress, determine the effective principal stresses.

Solution:

$$\sigma_1' = \sigma_1 - p_0$$

$$\sigma_1' = 1{,}500 - 150$$

$$\sigma_1' = 1{,}350 \text{ psi}$$

$$\sigma_3' = \sigma_3 - p_0$$

$$\sigma_3' = 500 - 150$$

$$\sigma_3' = 350 \text{ psi}$$

Example 4.6.14

Given the following principal stresses (refer to Figure 4.30), determine where the stress exceeds the strength if the rock has a cohesion of 100 psi and a friction angle of 12°. Use the Mohr circle.

Solution:

The solution is given in Figure 4.31. Failure is defined on the σ_1–σ_3 axis (the Mohr circle associated with σ_1–σ_3) within the grayed area in the figure. This is predicted because the Mohr circle of stress exceeds the failure envelope defined by the cohesion and friction.

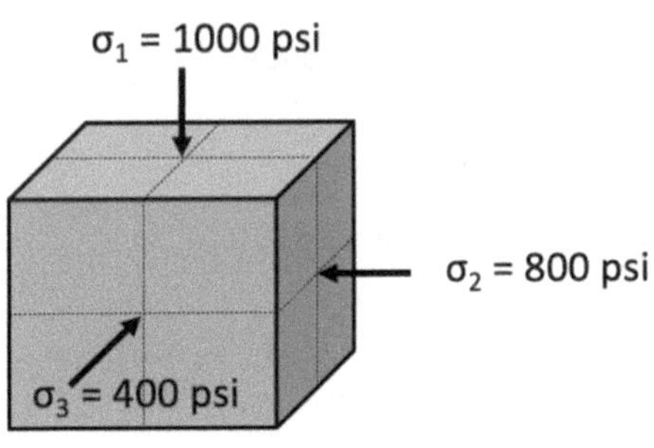

FIGURE 4.30 The principal stresses on the rock element.

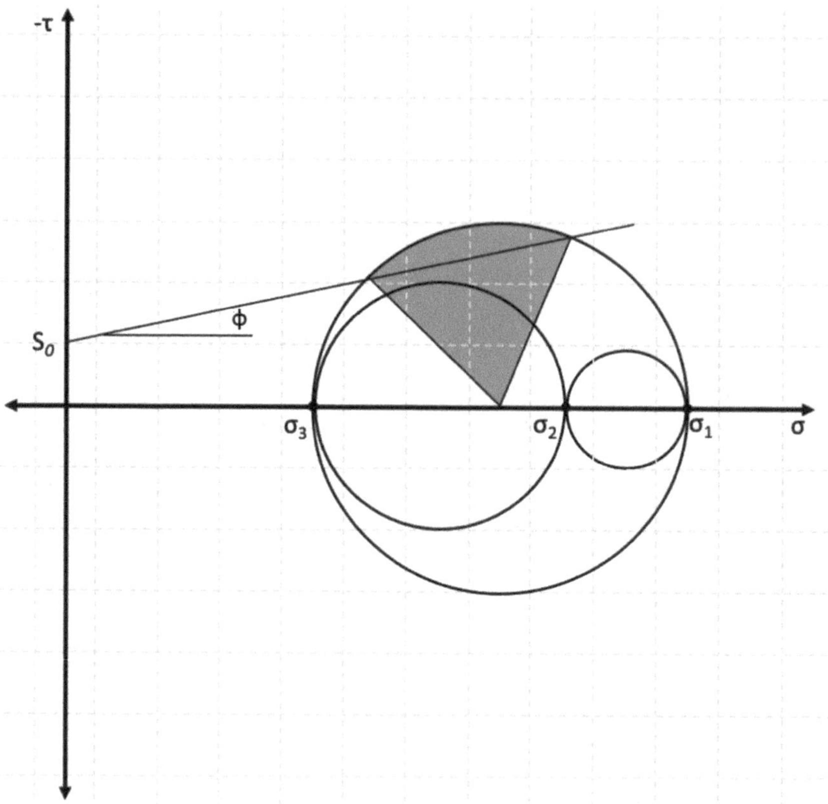

FIGURE 4.31 The Mohr circle solution for Example 4.6.14.

Example 4.6.15

Given the following strain components, plot the Mohr circle of strain to determine the principal strains. It should be noted that the process is the same for the Mohr circle of stress; however, the y-axis is half the shear strain (γ/2). Determine the maximum shear strain:

$$\varepsilon_x = 500\varepsilon$$

$$\varepsilon_y = 500\varepsilon$$

$$\gamma_{xy} = 500\varepsilon$$

Solution:

The solution is given in Figure 4.32.

From the Mohr circle of strain, we see that $\gamma/2 = 250\ \mu\varepsilon$, so $\gamma_{max} = 500\ \mu\varepsilon$.

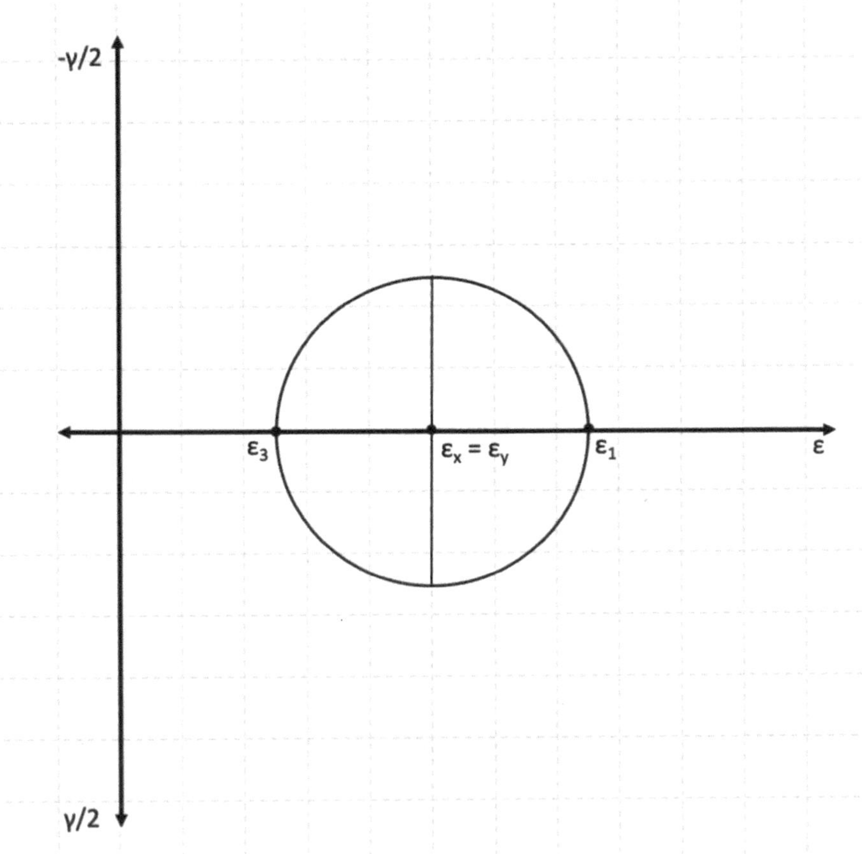

FIGURE 4.32 The Mohr circle solution for Example 4.6.15.

Example 4.6.16

The densities and thicknesses of three rock units are provided below. Calculate the total vertical stress at the bottom of the rock column. If Poisson's ratio is determined to be 0.25, find the horizontal stress due to gravity.

Rock unit 1: $\rho_1 = 3$ t/m³, thickness = 200 m.
Rock unit 2: $\rho_1 = 2.8$ t/m³, thickness = 300 m.
Rock unit 3: $\rho_1 = 2.5$ t/m³, thickness = 700 m.

Solution:

$$\sigma_v = \rho_1 g z_1 + \rho_2 g z_2 + \rho_3 g z_3$$

$$\sigma_v = (3\ \text{t/m}^3)(1{,}000\ \text{kg/m}^3)(9.81\ \text{m/s}^2)(200\ \text{m}) + (2.8\ \text{t/m}^3)(1{,}000\ \text{kg/m}^3)(9.81\ \text{m/s}^2)(300\ \text{m}) + (2.5\ \text{t/m}^3)(1{,}000\ \text{kg/m}^3)(9.81\ \text{m/s}^2)(700\ \text{m})$$

$$\sigma_v = 5.89\ \text{MPa} + 8.24\ \text{MPa} + 17.17\ \text{MPa}$$

$$\sigma_v = 31.3\ \text{MPa}$$

$$k = \nu/1 - \nu$$

$$k = 0.25/1 - 0.25$$

$$k = 0.33$$

$$\sigma_H = k\sigma_v$$
$$\sigma_H = (0.33)31.3 \text{ MPa}$$
$$\sigma_H = 10.3 \text{ MPa}$$

Example 4.6.17

If the stresses calculated in Example 4.6.16, represent the stress at point *"a"* in Figure 4.33, and calculate the stresses at point b if the rock is represented by rock unit 3. What are the principal stresses at points *"a"* and *"b"*? Are these virgin stresses or mining-induced stresses?

Solution:

$$\sigma_v = \rho_1 g z_1 + \rho_2 g z_2 + \rho_3 g z_3$$

$$\sigma_v = (3 \text{ t/m}^3)(1{,}000 \text{ kg/m}^3)(9.81 \text{ m/s}^2)(200 \text{ m}) + (2.8 \text{ t/m}^3)(1{,}000 \text{ kg/m}^3)(9.81 \text{ m/s}^2)(300 \text{ m}) + (2.5 \text{ t/m}^3)(1{,}000 \text{ kg/m}^3)(9.81 \text{ m/s}^2)(700+8 \text{ m})$$

$$\sigma_v = 5.89 \text{ MPa} + 8.24 \text{ MPa} + 17.36 \text{ MPa}$$
$$\sigma_v = 31.5 \text{ MPa}$$

$$k = \nu/1 - \nu$$

$$k = 0.25/1 - 0.25$$

$$k = 0.33$$

$$\sigma_H = k\sigma_v$$
$$\sigma_H = (0.33)31.5 \text{ MPa}$$
$$\sigma_H = 10.4 \text{ MPa}$$

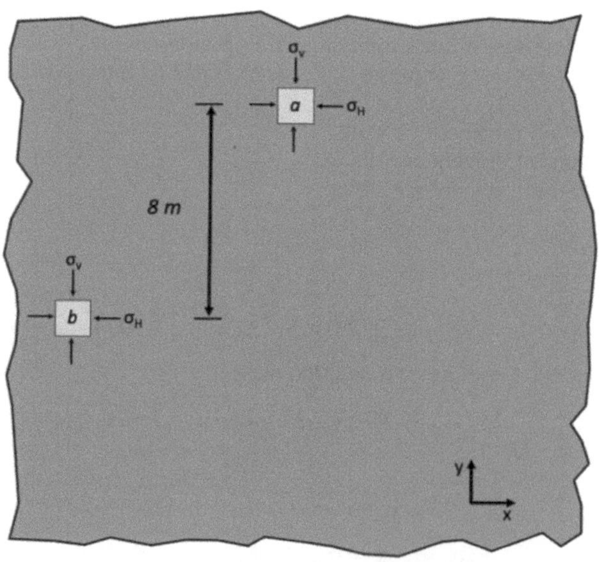

FIGURE 4.33 The excavations in the rock.

Because there is no shear indicated, the principal stresses at point "a" are as follows:

$$\sigma_1 = 31.3 \text{ MPa}$$

$$\sigma_3 = 10.3 \text{ MPa}$$

At point "b", the principal stresses are as follows:

$$\sigma_1 = 31.5 \text{ MPa}$$

$$\sigma_3 = 10.4 \text{ MPa}$$

These are the virgin stresses.

Example 4.6.18

Estimate the in situ vertical stress using the following rules of thumb and unit weight of shale found by Hudson and Harrison (1997). There should be three completed parts to this example. Assume a depth of 1,000 ft. Convert the final answers to units of psi and bar.

Determining vertical stress using rules of thumb:

(Part a) 1 psi induced by 1 ft of overlying rock.
(Part b) 1 MPa induced by 40 m of overlying rock.

Solution:

Determining vertical stress using unit weights:

$\sigma_v = \gamma z$
(Part c) $\gamma = 0.023$ MN/m^3 for some shales

Part a

$$\sigma_v = (1 \text{ psi/1 ft})(1{,}000 \text{ ft}) = 1{,}000 \text{ psi}$$

$$\sigma_v = (1{,}000 \text{ psi})(0.0689 \text{ bar/1 psi}) = 69 \text{ bar}$$

Part b

$$z = (1{,}000 \text{ ft})(0.3048 \text{ m/1 ft}) = 304.8 \text{ m}$$

$$\sigma_v = (1 \text{ MPa/40 m})(304.8 \text{ m}) = 7.62 \text{ MPa}$$

$$\sigma_v = (7.62 \text{ MPa})(145 \text{ psi/1 MPa}) = 1{,}105 \text{ psi}$$

$$\sigma_v = (1{,}105 \text{ psi})(0.0689 \text{ bar/1 psi}) = 76 \text{ bar}$$

Part c

$$\sigma_v = \gamma z$$

$$\sigma_v = (0.023 \text{ MN/m}^3)(304.8 \text{ m}) = 7.01 \text{ MN/m}^2$$

$$\sigma_v = (7.01 \text{ MN/m}^2)(145 \text{ psi/1 MPa}) = 1{,}017 \text{ psi}$$

$$\sigma_v = (1{,}017 \text{ psi})(0.0689 \text{ bar/1 psi}) = 70 \text{ bar}$$

Example 4.6.19

Assume a Poisson's ratio of 0, 0.25, and 0.50. Find the relationship between the horizontal and vertical gravitational stresses.

Part a:

$$\nu = 0$$

$$\sigma_h = (\nu/1 - \nu)\sigma_v$$

$$\sigma_h = 0$$

$$\nu = 0.25$$

$$\sigma_h = (\nu/1 - \nu)\sigma_v$$

$$\sigma_h = (0.25/1 - 0.25)\sigma_v$$

$$\sigma_h = 0.33\sigma_v$$

$$\nu = 0.50$$

$$\sigma_h = (\nu/1 - \nu)\sigma_v$$

$$\sigma_h = (0.50/1 - 0.50)\sigma_v$$

$$\sigma_h = \sigma_v$$

Example 4.6.20

The following relationship by Hoek and Brown (1980) estimates the vertical stress using a generalized unit weight. Another relationship estimates the proportionality constant (as a range) for approximating the horizontal stress. Estimate the vertical and horizontal stresses at a depth of 100 m and again at 2,500 m. The example and information within it were created from the textbook by Hudson and Harrison (1997).

Solution:

Vertical stress by Hoek and Brown (1980),

$$\sigma_v = 0.027z$$

0.027 is a generic unit weight.

It should be noted that this is a generalized expression to which Hudson and Harrison (1997) caution, there are cases at shallow depths, where the prediction can be dramatically different from measured, that is, up to five times greater than predicted, whereas at depths of 500–1,500 m, there are cases where the measured vertical stress is five times less than predicted.

Vertical stress at 100 m.

$$\sigma_v = 0.027z$$

$$\sigma_v = 0.027(100\ \text{m})$$

$$\sigma_v = 2.7\ \text{MPa}$$

Proportionality constant by Hoek and Brown (1980)

$$100/z + 0.3 < k < 1{,}500/z + 0.5$$

where

k is the ratio of the average horizontal stress to the average vertical stress.

z is the depth in meters.

$$100/(100\ \text{m}) + 0.3 < k < 1{,}500/(100\ \text{m}) + 0.5$$

$$1.3 < k < 15.5$$

If the vertical stress is taken as 2.7 MPa, then the horizontal stress estimate is

$$3.5 \text{ MPa} < \sigma_h < 42 \text{ MPa}$$

At shallow depths, the bounded range is very wide.
Vertical stress at 2,500 m.

$$\sigma_v = 0.027z$$

$$\sigma_v = 0.027(2{,}500 \text{ m})$$

$$\sigma_v = 68 \text{ MPa}$$

$$100/z + 0.3 < k < 1{,}500/z + 0.5$$

$$100/2{,}500 \text{ m} + 0.3 < k < 1{,}500/2{,}500 \text{ m} + 0.5$$

$$0.34 < k < 1.1$$

$$(0.34)68 \text{ MPa} < \sigma_h < (1.1)68 \text{ MPa}$$

$$23 \text{ MPa} < \sigma_h < 75 \text{ MPa}$$

REFERENCES

Budavari, S. (1983) Rock Mechanics in Mining Practice. SAIMM monograph Series 5, Johannesberg.

Colback, P. and Wiid, B. (1965) Influence of moisture content on the compressive strength of rock. *Proceedings Rock Mechanics Symposium Canadian Department of Mines and Technology Surveys*, Ottawa.

Coulomb, C.A. (1776) Essai sur une application des regies de maximis & minimis a quelques problemes de statique, relatifs a l'architecture, Memoires de Mathematique et de Physique. Presentes a V Academie Roy ale des Sciences par divers Savans, 1, 343–382 (written, 1773).

Farmer, I.W. (1968) *Engineering Properties of Rocks*. E. & F.N. Spon Ltd, London.

Gay, N.C. (1975) In situ stress measurements in southern Africa. *Tectonophys*. 29, 1–4.

Griffith, A.A. (1921) The phenomena of rupture and flow in solids. *Philos. Trans. R. Soc. London. A*. 221, 163–198.

Hoek, E. and Brown, E.T. (1980) Underground excavations in rock. Institution of Mining and Metallurgy, London, p. 527.

Hudson, J. and Harrison, J. (1997) *Engineering Rock Mechanics: An Introduction to the Principles, Part One*. Elsevier Science Ltd. Kidlington, Oxford.

Jaeger, J.C. (1960) Shear failure of anisotropic rock. *Geol. Magazine*. 97(1). 65–72. https://doi.org/10.1017/S0016756800061100.

Jaeger, J.C. and Cook, N.G.W. (1969) *Fundamentals of Rock Mechanics*. Methuen & Co Ltd, London.

McClintock, F.A. and Walsh, J. (1962) Friction on Griffith cracks in rock under pressure. Int. Cong. Appl. Mech., Berkeley, Cal.

Mohr, O.C. (1870) Beitrag zur Theorie der elastischen Bogentrager. *Zeitschrift des Architekten und Ingenieur-Vereins zu Hannover*. 16, 389.

Navier, C.L. (1823) Rapport à Monsieur Becquey, Conseiller d'État, Directeur Général des Ponts et Chaussées et des Mines; et mémoire sur les ponts suspendus. Paris, France: L'Imprimerie Royale

Obert, L. and Duvall, W.I. (1967) Rock Mechanics and the Design of Structures in Rock. John Wiley and Sons, New York.

Pallister, G.F., Gay, N.C. and Cook, N.G.W. (1970) Measurement of the virgin state of stress in rock at depth. *Proceedings of 2nd Congress of the International Society for Rock Mechanics (ISRM)*, Belgrade, Serbia.

Timsoshenko, S, & Goodier, J.,N. (1951). *Theory of Elasticity*. McGraw-Hill Book Company, Inc. 2nd ed.

5 Rock Properties and Testing

Rocks generally exhibit some form of brittle behavior when unconfined and become more plastic as the confinement increases. Rock is unconfined at surfaces, such as the periphery of an excavation, but becomes confined further away from the surface of the excavation. The concept known as lateral confinement generally restricts the rock mass from deforming and thus can effectively increase its stress-bearing capacity. Some rocks, such as salt, potash, and clays, are, however, ductile (exhibiting some plastic behavior) and tend to rather squeeze or creep under stress.

Rock behavior is dependent on many factors and can be variable when occurring naturally. The factors include the following:

- Formation of the rock.
- Intact tensile strength.
- Intact compressive strength.
- Stress state.
- Degree of weathering.
- Geological features.
- The presence of water and whether it is mobile.

5.1 PHYSICAL ROCK PROPERTIES

In rock mechanics, the physical properties of rock are relatively less important than its mechanical properties. Physical rock properties of smaller samples include the following:

- Luster.
- Texture.
- Shape.
- Size.
- Weight.
- Color.

5.1.1 Luster

Luster is defined as the state or quality of shining by reflecting light.

5.1.2 Texture

Texture is the visual or tactile quality of a surface. For example, a surface can be described as rough, smooth, or bumpy. The texture of surfaces and fractures such as bedding planes, faults, and joints are particularly important in rock mechanics.

5.1.3 Shape

Shape is the specific form, boundary, or outline of a body. For example, shape is important for aggregates used in concrete and shotcrete.

DOI: 10.1201/9781032704845-5

5.1.4 Size

Size is the specific dimensions or proportions of a body. Size is important in rock mechanics, particularly excavations, because the larger the excavation the less inherently stable.

5.1.5 Weight

The weight of an object is the force acting on the object because of gravity. Weight is obviously critical in support design, especially strata control.

5.1.6 Color

Color describes how light reflects off an object with respect to the visible-light portion of the electromagnetic spectrum. Dark-colored objects are less easy to see than lighter-colored ones. For example, coal being dark makes illumination and visibility underground poor compared to limestone which reflects more light.

The physical properties of large samples of rocks include the following:

- Mineral composition (the micro-structure within a rock mass).
- Specific gravity.
- Unit weight.
- Density.
- Void ratio.
- Porosity.
- Moisture content.
- Swelling.
- Degree of saturation.
- Coefficient of permeability.
- Electrical and thermal properties.
- Anisotropy.
- Durability.

5.2 MECHANICAL ROCK PROPERTIES

Mechanical rock properties are critical in mining and include the following:

- Compressive strength.
- Tensile strength.
- Young's modulus.
- Poisson's ratio.

Large-scale measurements underground are generally difficult to arrange and costly. They can be applicable for a significant project where the absolute data are important. Laboratory testing using small-scale specimens, extracted from diamond drill cores, is the usual approach for obtaining rock properties. Results from small samples do not directly represent that of the larger rock mass because geological features at scale, e.g., beds, faults, and joints, can alter the lab-tested properties in situ.

There are three types of rock deformation that occur naturally:

- Elastic deformation
  - Reversible if the stress is removed.
- Ductile deformation

    - Irreversible resulting in a permanent change to the shape or size of the rock.
- Brittle deformation
    - Irreversible generally resulting in the breakage of the rock.

5.3 ROCK TESTING

Complete test methods can be obtained from the International Society for Rock Mechanics (2007 & 2015). Only brief descriptions and important aspects are considered here.

5.3.1 Unconfined Compression Strength (UCS) Testing

UCS tests are undertaken using a compression-testing machine that is sometimes equipped with strain gauges that provide additional information about how a rock specimen deforms. A typical machine for determining the UCS is shown in Figure 5.1, and results of a test are shown in Figure 5.2.

The UCS strength of small rock specimens will always be higher than that of the rock mass, because of geological features and water for example. Cognizance must always be taken of this when designing mines, for excavations and pillars for example.

Important aspects of the UCS test as suggested in the ISRM guidelines (1981) are as follows:

- As a rule, at least five samples should be tested to obtain representative results and the standard deviation.
- Rock specimens should be straight, circular cylinders with a length-to-diameter ratio between 2.5 and 3.
- The diameter of the specimen should be related to the largest grain size in the rock and no less than ten times the diameter of the largest grain.
- The ends of the specimen should be cut parallel to each other and at right angles to the longitudinal axis.
- Stress loading should be constant at a rate between 0.5 and 1.0 MPa/s, or the rock should fail within 5–10 minutes after starting the test.

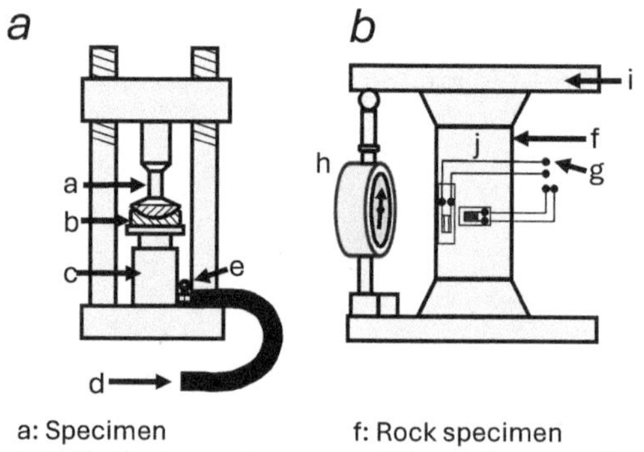

FIGURE 5.1 Equipment for UCS testing: (a) diagrammatic sketch of compression-testing machine and (b) strain-measuring devices on rock specimen. Note that the strain gauges are situated in two directions: one in the direction of load, and the other at 90° to the load (Patchet, 1970).

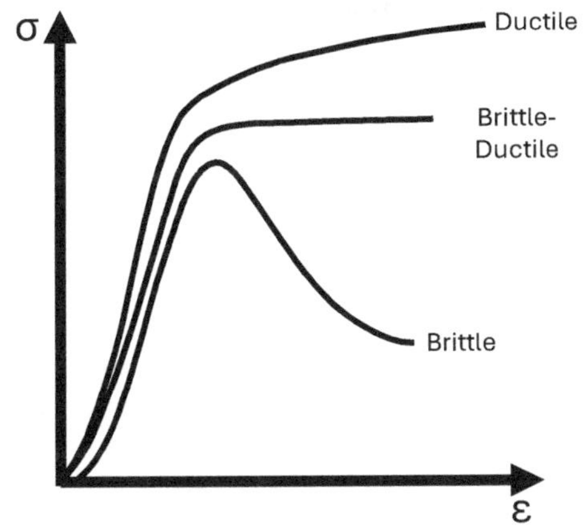

FIGURE 5.2 Examples of lab UCS test results for rock (Mogi, 1966).

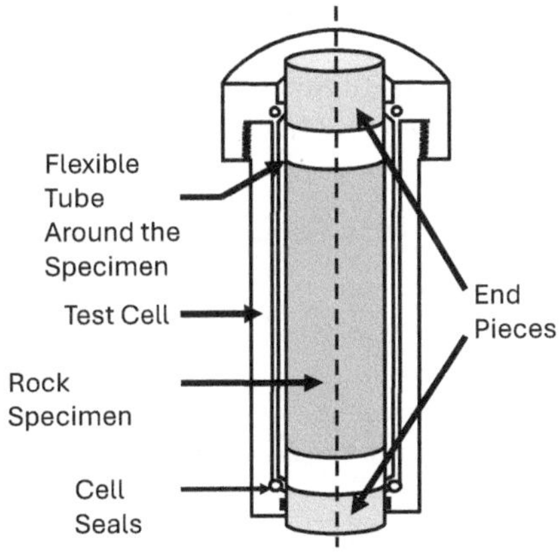

FIGURE 5.3 A circular triaxial cell which applies pressure equally around the rock specimen hydraulically.

5.3.2 Triaxial Compression Testing

Rock is generally under a triaxial stress state, so triaxial testing can be useful, for example, to determine a Mohr–Coulomb failure envelope for the specific rock. The rock specimen is generally tested in the same type of machine as for a UCS test, but the specimen is enclosed in a triaxial cell as shown in Figure 5.3.

When testing under triaxial conditions, the major principal stress (σ_1) is applied along the axis of the specimen by the compression-testing machine. The confining stresses are applied hydraulically using oil pressure, and in most triaxial cells, the minor principal stresses (σ_2 and σ_3) are equal.

A typical set of results is shown in Figure 5.4 after Patchet (1970).

5.3.3 Uniaxial Tensile Testing (UTS)

Direct tensile testing of rock is simple in concept but difficult and time consuming. This is mainly because the carefully prepared sample must be adhered to the platens and the epoxy (or similar) often fails first during the test. If the resin holds, for the result to be acceptable, the failure of the

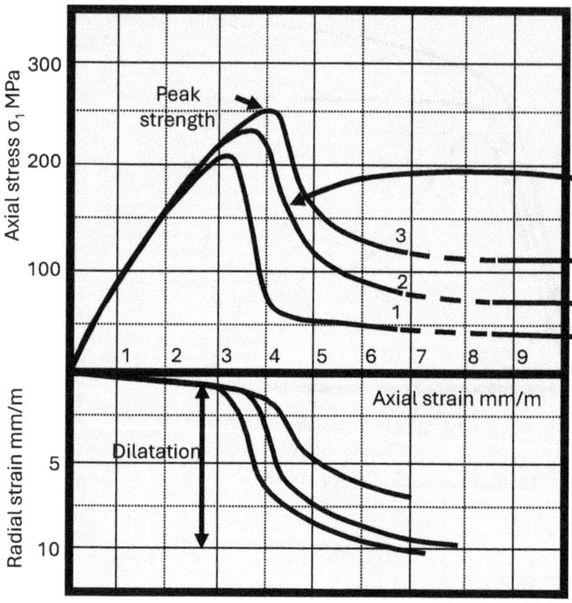

FIGURE 5.4 Example of triaxial test results. Illustrated are the axial stress as a function of the axial strain and radial strain as a function of axial strain for rock specimens from a single piece of Witwatersrand quartzite (Patchet, 1970).

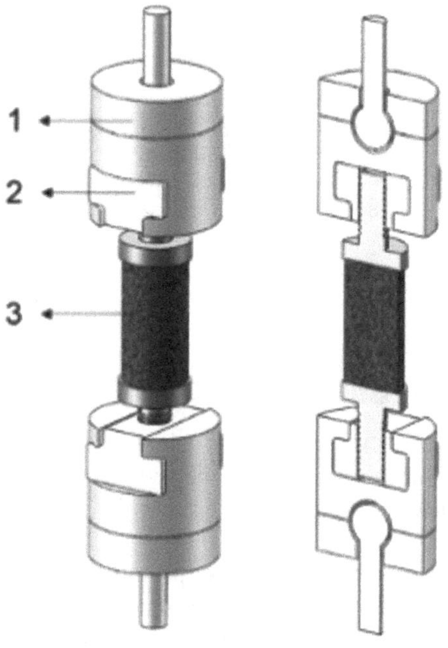

FIGURE 5.5 Direct tensile testing schematic and section (Zhang et al, 2021).

rock sample should be in the middle third of the sample which again does not always happen. The indirect tensile test (Brazilian test) is therefore much more commonly used to obtain the UCS tensile strength of rocks. The UTS test involves placing a cylindrical rock specimen and loading it in tension as shown in Figure 5.5.

In Figure 5.5, (i) shows the ball hinge to try and ensure axial loading, (ii) is the T-shape slide block to attach the sample platens with the core securely bonded with typically an epoxy, and (iii) is the rock core for tensile testing.

The tensile strength is calculated as follows:

$$T_o = F_t / A \tag{5.1}$$

where

T_o is the tensile strength.
F_t is the failure load.
A is the specimen cross section.

An alternative for obtaining the tensile strength of rock is the Brazilian (disc) method by introducing an indirect tensile failure to a disc of rock by compression. The diameter of the disc is twice the length of the sample.

The Brazilian test is shown in Figure 5.6, and the method of calculation, for plane stress, to arrive at the tensile strength is given below:

$$\sigma_1 = P/\pi R \tag{5.2}$$

$$\sigma_2 = 0 \tag{5.3}$$

$$\sigma_3 = -P/\pi R\left(\text{the tensile strength}\right) \tag{5.4}$$

So, the indirect tensile strength is calculated as follows:

$$T_i = P/\pi R, \left(\text{unit thickness}\right) \tag{5.5}$$

See Figure 5.6a for the variables.

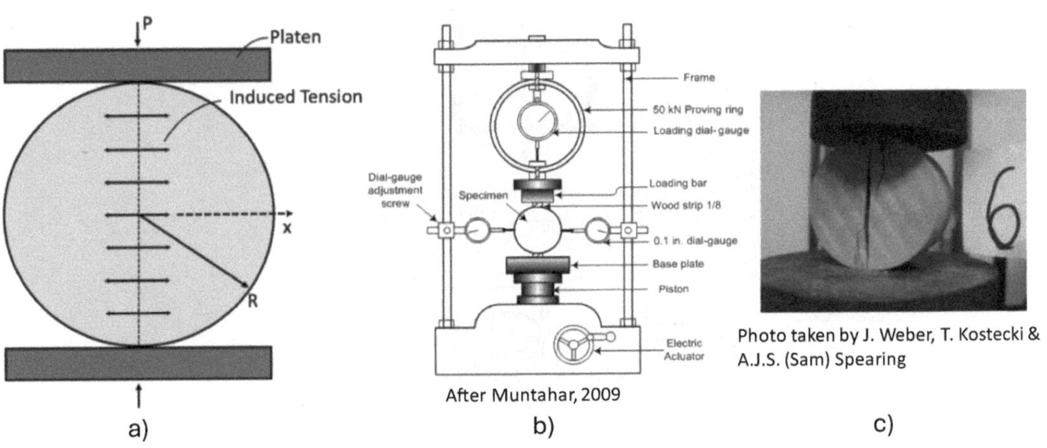

FIGURE 5.6 (a) Indirect tensile testing failure mode and important parameters, (b) equipment for a Brazilian indirect tensile test (after Muntahar 2009), and (c) a rock specimen after failure after J. Weber and A.J.S Spearing.

5.3.4 Point Load Testing

This test was designed as a useful field or lab tool to estimate the rock strength of specimens in the form of rock core or random irregular lumps. It is a quicker alternative to the UCS test but less repeatable (Akazawa, 1943; Carneiro, 1943).

The machine consists of a simple portable hydraulic compression-testing rig with specially designed platens as shown in Figure 5.7. The various ways of testing core or lumps are shown in Figure 5.8.

The point load strength index (uncorrected) is defined as follows:

$$I_s = P/D_e^2 \tag{5.6}$$

where

P is the load applied.

D_e is the equivalent core diameter.

As it is known, the size of a rock influences the strength, so to normalize results for the point load index test, the size is often corrected to a common size using a correction chart.

Therefore, to allow for size influences of the rock samples tested, the point load strength index should always be reported as the $I_s(50)$, which is the index if a 50 mm diameter sample was used. An example of a correction chart is given in Figure 5.9.

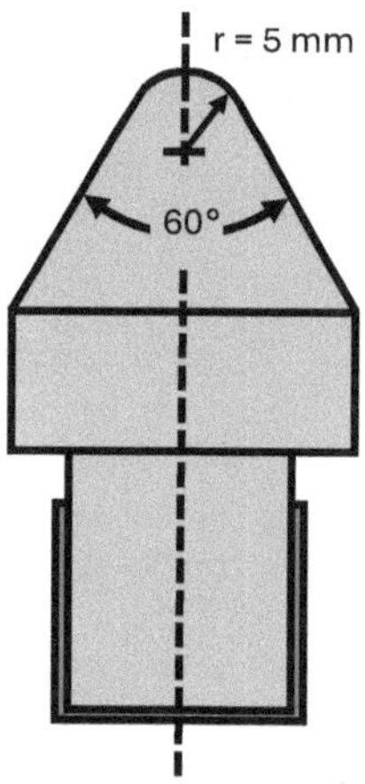

FIGURE 5.7 The point load application platen (ASTM D5731-95, 1995).

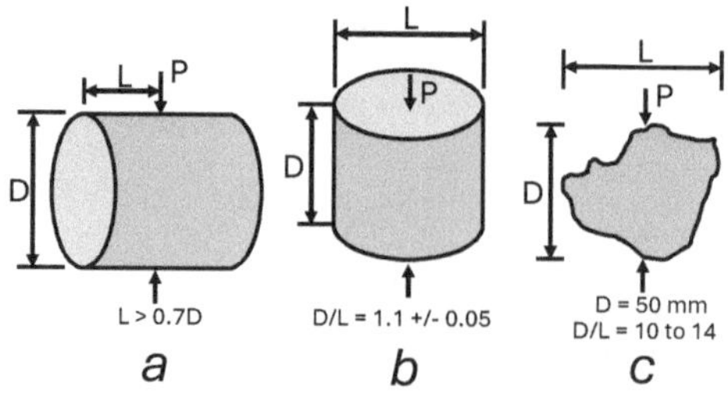

FIGURE 5.8 Typical-point load-testing samples and recommended dimensions for (a) diametral test, (b) axial test, and (c) irregular lump test (Broch and Franklin, 1972).

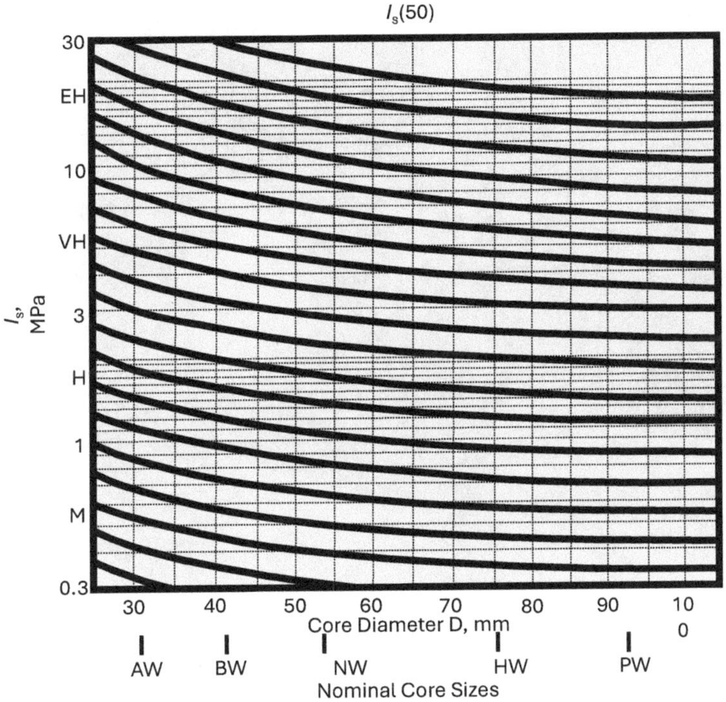

FIGURE 5.9 Point load correction chart (Broch and Franklin, 1972). Note the strength designations on the y-axis as EH (extremely high), VH (very high), H (high), and M (medium). Nominal core sizes on the x-axis refer to a drill classification.

Broch and Franklin (1972) developed a relationship between the point load strength index and the uniaxial compressive strength of a rock specimen as follows:

$$UCS = 24 \times I_s(50) \tag{5.7}$$

According to the ISRM (2007), the range of the constant in Equation 5.7 for isotropic rocks is between 20 and 25, but for anisotropic rocks, the range is more, typically 15–50.

Figure 5.10 shows a point load test being conducted.

5.4 THE IMPORTANCE OF ROCK PROPERTIES

The stability of all excavations is influenced by the properties of the surrounding rock mass, and hence, the reliable determination of these properties from lab tests, or in situ, is important. Tests at a small lab scale reflect different strength and performance than that of the larger-scale rock mass. In the rock mass, the strength is affected by bedding planes, faults, joint sets, and water for example. Rock classification systems relate lab-scale properties to those of the rock mass as discussed in Chapter 6.

The measure of a mine design's success is its safety, functionality, and cost-effectiveness. Engineering judgment based on reliable information, such as rock tests, is essential, and the design must be continually revisited, and at times modified, as more information becomes available. Any assumptions and data used for a particular design must be clearly recorded, such that when more information is available, the design can be effectively updated and the rationale for the update understood.

FIGURE 5.10 Conducting a point load test. Downloaded with kind permission from the ISRM website isrm.net, slides donated by Prof. John Franklin to the ISRM slide collection.

5.5 WORKED EXAMPLES

Example 5.5.1

A rock specimen of dolerite has the following properties. Based on the magnitudes alone, determine which property is for tension and which is for compression and why?

$$S = 3{,}000\ \text{kg/cm}^2$$
$$S = 300\ \text{kg/cm}^2$$

Solution:

Rocks are always weaker in tension compared to its compressive strength. The compression property is 3,000 kg/cm^2, and the tension property is 300 kg/cm^2. A rule of thumb for the tensile strength is 1/10 (10%) of the compressive strength – clearly this doesn't always apply.

Example 5.5.2

A target of ten tests at minimum is good practice, unless guidelines such as those by ISRM or other societies specify otherwise, and the more tests which can be conducted the better. Less than ten tests will reduce the level of confidence in the result, but this does not mean that the results are nullified. Some tests are expensive and time consuming but have supreme levels of accuracy, whereas others are cruder, inexpensive, and rapid.

Part a

Ten UCS tests are conducted on a specimen of marble. The results are given in Table 5.1. Calculate the average and standard deviation of the results:

$$\text{Standard deviation} = \sigma = \sqrt{\frac{\sum_{i=1}^{n}(x_i - \bar{x})^2}{n-1}}$$

TABLE 5.1
The Test Data with the Average Strength and Standard Deviation

Parameter	Test No. (i)	S_c (kg/cm²)	$(x_i - \bar{x})^2$
	1	1,260	$(1{,}260 - 1{,}422)^2 = 26{,}147$
	2	1,315	11,385
	3	1,617	38,142
	4	1,435	177
	5	1,553	17,240
	6	1,302	14,328
	7	1,620	39,323
	8	1,521	9,860
	9	1,323	9,742
	10	1,271	22,710
N		10	
Sum		14,217	189,054
$\bar{x}$		1,422	
n − 1			9
Standard deviation (σ)			145

Solution:

Part b

What is the standard deviation if only the first three tests are conducted?

$$N = 3$$

$$\text{Sum} = 1{,}260 + 1{,}315 + 1{,}617 = 4{,}192 \text{ Average} = 4{,}192/3 = 1{,}397$$

$$n - 1 = 2$$

$$(x_1 - \bar{x})^2 = (1{,}260 - 1{,}397)^2 = (-137)^2 = 18{,}769$$

$$(x_2 - \bar{x})^2 = (1{,}315 - 1{,}397)^2 = (-82)^2 = 6{,}724$$

$$(x_3 - \bar{x})^2 = (1{,}617 - 1{,}397)^2 = (-220)^2 = 48{,}400$$

$$(x_1 - \bar{x})^2 + (x_2 - \bar{x})^2 + (x_3 - \bar{x})^2 = 18{,}769 + 6{,}724 + 48{,}400 = 73{,}893$$

$$\sigma = \sqrt{\frac{\sum_{i=1}^{n}(x_i - \bar{x})^2}{n-1}} = \sqrt{\frac{73{,}893}{2}} = 192$$

Example 5.5.3

Based on the following information, determine the $I_s(50)$ and strength designation. Inputs and calculations for this example are found in Broch and Franklin (1972); however, the usage of the correction chart is not detailed in the source, so how it may have been used to arrive at the $I_s(50)$ answer is shown. Note there are other methods of creating and using correction factors as detailed in ISRM (2007):

$$D = 41.5 \text{ mm}$$

$$P = 4.00 \text{ kN}$$

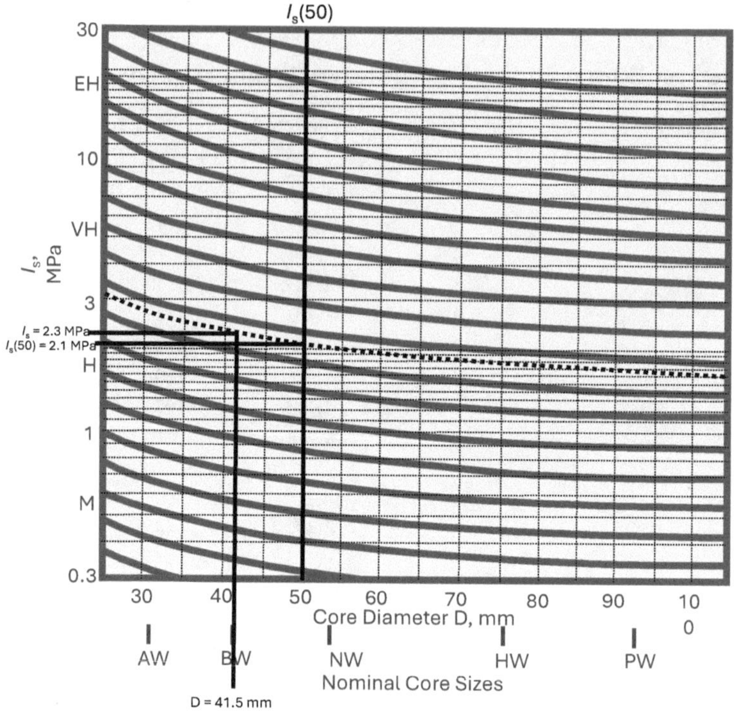

FIGURE 5.11 The solution using the correction chart.

Solution:

$$I_s = P/D^2$$

$$I_s = (0.004\ MN)/(0.0415\ m)^2$$

$$I_s = 2.3\ MN/m^2$$

From the chart (Figure 5.11), the $I_s(50)$ vertical line can be determined and chart an approximate curve to determine an $I_s(50)$ of 2.1 MPa. The strength designation is High (H).

Example 5.5.4

Find the UCS from the previous example using the $I_s(50)$ value.

Solution:

$$UCS = 24 \times I_s(50)$$

$$UCS = 24 \times 2.1\ MPa$$

$$UCS = 50.4\ MPa$$

Examples 5.5.5–5.5.16 are produced using equations, processes, and recommendations found in the ISRM guidelines of the 2007 version (blue book).

Example 5.5.5

An axial point load test is conducted on a rock sample of diameter 50 mm. If the length-to-diameter ratio is 1.0 and the failure load is 10 kN, what is the uncorrected point load strength? Does the recorded strength need to be corrected?

Solution:

$$I_s = P/D_e^2$$

$$\text{For diametral tests } D_e^2 = D^2$$

$$I_s = P/D^2$$

$$I_s = (10{,}000\ \text{N})/(50\ \text{mm})^2$$

$$I_s = 4\ \text{N/mm}^2 = 4\ \text{MPa}$$

Since the test is a diametral test and the diameter is 50 mm, it does not need to be corrected.

Example 5.5.6

An irregular lump test is completed on a rock sample. Prior to the test, it was determined that the width of the sample was 18 mm and the depth was 8 mm. The failure load was 1.0 kN. Find the corrected point load strength index if the correction factor (F) is estimated as 0.56.

Solution:

For an irregular lump test:

$$A = WD$$

$$A = (18\ \text{mm})(8\ \text{mm})$$

$$A = 144\ \text{mm}^2$$

where A is the minimum cross-sectional area of plane through platen contact points:

$$D_e^2 = 4A/\pi$$

$$D_e^2 = 4(144\ \text{mm})/\pi$$

$$D_e^2 = 183\ \text{mm}$$

$$D_e = 13.5\ \text{mm}$$

$$I_s = P/D^2$$

$$I_s = 1{,}000\ \text{N}/(13.5\ \text{mm})^2$$

$$I_s = 5.49\ \text{MPa}$$

$$I_{s(50)} = (I_s)(F)$$

$$I_{s(50)} = (5.49)(0.56)$$

$$I_{s(50)} = 3.07$$

Example 5.5.7

A point load index test is completed on a block sample of width 42 mm and a depth of 32 mm. If the failure load is 6 kN, what is the corrected point value index strength if the correction factor is estimated at 0.90?

Solution:

For block samples:

$$A = WD$$
$$A = (42\ \text{mm})(32\ \text{mm})$$
$$A = 1{,}344\ \text{mm}^2$$
$$D_e^2 = 4A/\pi$$
$$D_e^2 = 4(1{,}344\ \text{mm})/\pi$$
$$D_e^2 = 1{,}711\ \text{mm}$$
$$D_e = 41.4\ \text{mm}$$
$$I_s = P/D^2$$
$$I_s = 6{,}000\ \text{N}/(41.4\ \text{mm})^2$$
$$I_s = 3.50\ \text{MPa}$$
$$I_{S(50)} = (I_s)(F)$$
$$I_{S(50)} = (3.50)(0.90)$$
$$I_{S(50)} = 3.15\ \text{MPa}$$

Example 5.5.8

Table 5.2 gives the results of ten valid point load index tests. Using ISRM's approach, determine the mean value for $I_{S(50)}$ for the test results parallel to the rocks planes of weakness, and perpendicular to the rocks planes of weakness. Values are in MPa.

TABLE 5.2
The Test Series Results

Test No.	$I_{s(50)}$, Parallel	$I_{s(50)}$, Perpendicular
1	5.04	10.32
2	4.85	8.40
3	6.23	9.28
4	5.28	10.10
5	3.23	8.73
6	6.98	9.20
7	4.32	9.54
8	7.05	9.89
9	5.52	10.56
10	4.21	8.25

Solution:

The highest and lowest values of each test can be negated for this number of tests:

Mean $I_{S(50)}$, Parallel = (5.04 + 4.85 + 6.23 + 5.28 + 6.98 + 4.32 + 5.52 + 4.21)/8 = 5.30 MPa

Mean $I_{S(50)}$, Perpendicular = (10.32 + 8.40 + 9.28 + 10.10 + 8.73 + 9.20 + 9.54 + 9.89)/8 = 9.43 MPa.

Example 5.5.9

The strength anisotropy index, $I_{a(50)}$, is a measure of the rock's anisotropic nature. It is found using the ratio of $I_{S(50)}$, Perpendicular to $I_{S(50)}$, Parallel. Values approaching 1.0 are closer to quasi-isotropic and increasing in anisotropy as the value increases. Find the strength anisotropy index for the rock in the previous example.

Solution:

$$I_{a(50)} = I_{S(50)}\text{, Perpendicular/ } I_{S(50)}\text{, Parallel}$$

$$I_{a(50)} = 9.43\ \text{MPa}/5.30\ \text{MPa}$$

$$I_{a(50)} = 1.78$$

Example 5.5.10

A cylindrical rock sample of gypsum has a diameter of 54 mm and a length-to-diameter ratio of 3.0. If the ultimate compressive load on test was 60 kN, what is the UCS? Note that during testing, the cross-sectional area was found to increase by 2%.

Solution:

Use the original cross-sectional area:

$$\text{Area} = \pi(D)^2/4$$

$$\text{Area} = \pi(54\ \text{mm})^2/4$$

$$\text{Area} = 2{,}290\ \text{mm}^2$$

$$\text{UCS} = P/A$$

$$\text{UCS} = (60{,}000\ \text{N})/(2{,}290\ \text{mm}^2)$$

$$\text{UCS} = 26.2\ \text{MPa}$$

Example 5.5.11

To measure the deformability of rock, a cylindrical sample is prepared from drill core, with a diameter of 54 mm and a length-to-diameter ratio of 3.0. The sample is tested in uniaxial compression and affixed with strain gauges to measure the average axial and diametric strain. Using ISRM suggested methods, determine the volumetric strain (ε_v) diametric strain (ε_d) and axial strain (ε_a), if the length decreases by 8 mm and the diameter increases 1 mm during testing. Assume that the point of interest is at 50% the ultimate uniaxial compressive strength.

Solution:

$$\varepsilon_d = \Delta d/d_o$$

$$\varepsilon_d = (-1\ \text{mm})/(54\ \text{mm})$$

$$\varepsilon_d = -0.019\ \text{mm/mm or } -1.9\%$$

$$\varepsilon_a = \Delta l / l_o$$

$$\varepsilon_a = (8 \text{ mm})/(162 \text{ mm})$$

$$\varepsilon_a = 0.049 \text{ mm/mm or } 4.9\%$$

At a stress level of 50% UCS, the volumetric strain is as follows:

$$\varepsilon_v = \varepsilon_a + 2\varepsilon_d$$

$$\varepsilon_v = 0.049 + 2(-0.019)$$

$$\varepsilon_v = 0.011 \text{ mm/mm or } 1.1\%$$

Example 5.5.12

If a sample of marble has a slope of the diametric stress strain curve at 0.15 to 150 (strain as a percent to stress as MPa) and the axial stress strain curve is 0.5 to 150, using the sign conventions of the previous example, what is Poisson's ratio of the sample?

Solution:

Axial Young's modulus (E_a):

$$E_a = \Delta\sigma_a / \Delta\varepsilon_a$$

$$E_a = (150 \text{ MPa})/(0.005 \text{ mm/mm})$$

$$E_a = 30{,}000 \text{ MPa}$$

$$\nu = -\text{slope of the axial stress - strain curve/slope of the diametric stress strain curve}$$

$$\nu = -30{,}000 \text{ MPa}/(-150 \text{ MPa}/0.0015)$$

$$\nu = -30{,}000 \text{ MPa} / -100{,}000 \text{ MPa}$$

$$\nu = 0.30$$

Example 5.5.13

A sample of sandstone is tested to find that the UCS is 180 MPa reaching this magnitude at 0.006 mm/mm strain. What is the secant modulus which is taken from 0% to 50% the UCS and the tangent modulus? The tangent modulus is also taken at 50% but between a strain of 0.003 and 0.005 across a $\Delta\sigma$ of 80 MPa.

Solution:

$$\text{Secant Modulus, } E_s = \Delta\sigma / \Delta\varepsilon_o$$

$$\text{Secant Modulus, } E_s = 90 \text{ MPa}/(0.006 \text{ mm/mm})$$

$$\text{Secant Modulus, } E_s = 15 \text{ GPa}$$

$$\text{Tangent Modulus, } E_t = \Delta\sigma / \Delta\varepsilon$$

$$\text{Tangent Modulus, } E_t = 80 \text{ MPa}/(0.005 - 0.003)$$

$$\text{Tangent Modulus, } E_t = 40 \text{ GPa}$$

Example 5.5.14

A direct tensile test is complete on a rock sample. The relevant sample dimensions are a diameter of 55 mm. If the ultimate load was 9.5 kN, what is the tensile strength?

Solution:

$$\text{Area} = \pi D^2/4$$

$$\text{Area} = \pi(55\text{ mm})^2/4$$

$$\text{Area} = 2{,}376\text{ mm}^2$$

$$\sigma_t = P/A$$

$$\sigma_t = (9{,}500\text{ N})/(2{,}376\text{ mm}^2)$$

$$\sigma_t = 4.0\text{ MPa}$$

Example 5.5.15

An indirect tensile test is completed using a jaw-type apparatus on the same rock type as the previous example. The sample has a diameter of 55 mm and a thickness equal to the radius, and reaches a load of 9.5 kN. What is the tensile strength?

Solution:

$$\sigma_t = (2/\pi)(P/Dt)$$

$$\sigma_t = (2/\pi)\left[(9{,}500\text{ N})/(55\text{ mm})(22.5\text{ mm})\right]$$

$$\sigma_t = (0.636)\left[(9{,}500\text{ N})/(55\text{ mm})(22.5\text{ mm})\right]$$

$$\sigma_t = 4.9\text{ MPa}$$

Example 5.5.16

A triaxial test is completed in the lab, and it's determined that the friction angle and cohesion of the rock sample are 30° and 900 psi, respectively. Show the reduced form of the σ–p relationship using the friction angle and cohesion.

Solution:

The failure envelope can be represented by

$$\sigma = m_i p + b_i$$

where

b_i is an ordinate on the axial stress (σ) vs confining stress (p) plot represented on the y-axis

m_i is the slope angle of the strength envelope on the axial stress (σ) vs confining stress (p) plot.

The internal friction angle (ϕ) is expressed as

$$\phi_i = \arcsin\left[(m_i - 1)/(m_i + 1)\right]$$

And the apparent cohesion (c) is given by

$$c_i = b_i(1 - \sin\phi_i)/(2\cos\phi_i)$$

Step 1 solve for m_i

$$\phi_i = \arcsin\left[(m_i - 1)/(m_i + 1)\right]$$

$$\sin(\phi_i) = \left[(m_i - 1)/(m_i + 1)\right]$$

$$(m_i + 1)\sin(\phi_i) = (m_i - 1)$$

$$(m_i)\sin(\phi_i) + \sin(\phi_i) = m_i - 1$$

$$(m_i)\sin(\phi_i) - m_i = -1 - \sin(\phi_i)$$

$$(m_i)\left(\sin(\phi_i) - 1\right) = -1 - \sin(\phi_i)$$

$$m_i = \left[-1 - \sin(\phi_i)\right] / \left(\sin(\phi_i) - 1\right)$$

$$m_i = \left[-1 - \sin(30)\right]/(\sin(30) - 1)$$

$$m_i = [-1 - 0.5]/(0.5 - 1)$$

$$m_i = (-1.5)/(-0.5)$$

$$m_i = 3$$

Step 2 solve for b_i

$$c_i = b_i\left(1 - \sin\phi_i\right) / \left(2\cos\phi_i\right)$$

$$2c_i(\cos\phi_i)/(1 - \sin\phi_i) = b_i$$

$$(2)(900)[1.73] = b_i$$

$$3{,}114 = b_i$$

Resolve:

$$\sigma = m_i p + b_i$$

$$\sigma = 3p + 3{,}114$$

REFERENCES

Akazawa, T. (1943) New test method for evaluating internal stress due to compression of concrete: the splitting tension test. *J. Japan Soc. Civil Eng.* 29, 777–787.

ASTM D 5731-95 (1995) *Standard Test Method for Determination of the Point Load Strength Index of Rock.* American Society for Testing and Materials, West Conshohocken, PE.

Broch, E. and Franklin, J.A. (1972) The point load test. *Inst. J. Rock Mech. Mining Sci.* 1972, 669–697.

Carneiro F.L.L.B. (1943) A new method to determine the tensile strength of concrete. *Paper presented at the Proceedings of the 5th meeting of the Brazilian Association for Technical Rules* ("Associação Brasileire de Normas Técnicas—ABNT"), 3d. section

International Society of Rock Mechanics (1981) *Rock Characterization, Testing & Monitoring: ISRM Suggested Methods.* Pergamon Press, Oxford.

International Society of Rock Mechanics (2007) *The Complete ISRM Suggested Methods for Rock Characterization, Testing & Monitoring: 1974–2006.* Elsevier, Amsterdam, Netherlands.

International Society of Rock Mechanics (2015) *The ISRM Suggested Methods for Rock Characterization, Testing and Monitoring: 2007–2014.* Ulusay, R. (ed). Springer, Switzerland.

Mogi, K. (1966) Pressure dependence of rock strength and transition from brittle fracture to ductile flow. *Bull. Earthquake Res. Inst.* 44, 215–232.

Muntahar, A.S. (2009) Influence of plastic waste fibers on the strength of lime-rice husk ash stabilized clay soil. *Civil Eng. Dimen.* 11(1), 32–40.

Patchet, S.J. (1970) Rock mechanics studies associated with the development of a deep potash mine. Ph.D Thesis, University of Newcastle-upon-Tyne, United Kingdom.

Zhang, Y., Zhang, Q-Y., Zhou, X-Y. and Xiang, W. (2021) Direct tensile tests of red sandstone under different loading rates with self-developed centering device. *Geotech. Geol. Eng.* 39, 709–718.

6 Strength Criteria in Rock and Constitutive Relationships

6.1 INTRODUCTION TO CONSTITUTIVE RELATIONSHIPS

A constitutive relationship is a relationship between two physical quantities that are specific to a material or substance. It approximates the response of that material to external stimuli, usually applied forces. A constitutive relationship approximates the observed physical behavior of a material under specific conditions of interest (Wikiwand).

The way in which stress and strain are related in a material under load (underground) is described qualitatively as constitutive behavior (Brady and Brown, 2004). Constitutive relationships are useful in rock mechanics as they aid in the analysis of complex underground conditions and situations.

To be useful, constitutive models should ideally be:

- Widely applicable with the conditions and constraints clearly specified.
- Easy and simple to use.

Typical examples include (only the first two are, however, discussed):

- Kirsch Equations (1898).
- Mohr Coulomb Model (1870).
- Goodman's Model (1976).
  - This is used to model joints or fractures in rock masses. It can be used to model both elastic and plastic deformation.
- Barton–Bandis model (Bandis et al., 1981).
  - This model allows the shear strength and normal stiffness of rock joints.

6.2 STRESSES AROUND EXCAVATIONS

6.2.1 Introduction

Consider a cube of rock subjected to a vertical compressive stress σ_z as shown in Figure 6.1a. The strain in the z-direction due to σ_z is as follows (modified from Pillalamarry and Spearing, 2009):

$$\varepsilon_z = \sigma_z/E \tag{6.1}$$

The compressive stress results in the extension of the body in the x- and y-directions producing the following strains due to σ_z:

$$\varepsilon_x = -(\nu\sigma_z/E) \tag{6.2}$$

$$\varepsilon_y = -(\nu\sigma_z/E) \tag{6.3}$$

If the simultaneous action of σ_x, σ_y, and σ_z occurs as shown in Figure 6.1b, the resultant normal strains are related to the stress in all directions such that the z-direction normal strain is

$$\varepsilon_z = (1/E)[\sigma_z - \nu(\sigma_x + \sigma_y)] \tag{6.4}$$

DOI: 10.1201/9781032704845-6

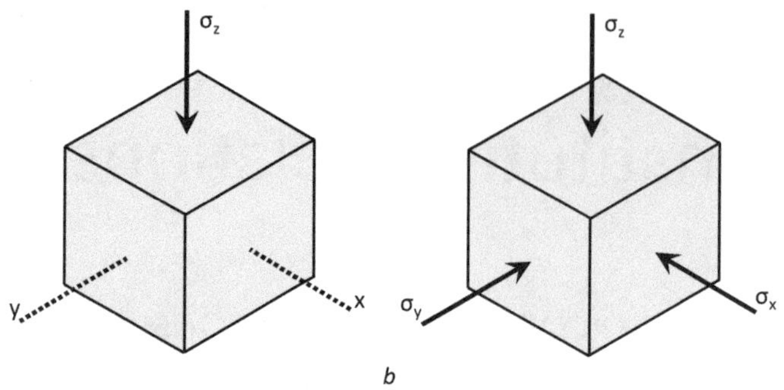

FIGURE 6.1 Stresses acting on a cube of rock underground.

Similarly, the resultant strain in the x-direction is

$$\varepsilon_x = (1/E)[\sigma_x - \nu(\sigma_y + \sigma_z)] \tag{6.5}$$

The resultant strain in the y-direction is

$$\varepsilon_y = (1/E)[\sigma_y - \nu(\sigma_x + \sigma_z)] \tag{6.6}$$

In the case of sedimentary rocks in geologically undisturbed areas, where the strata were built up in horizontal layers in such a way that the horizontal dimensions remained basically unchanged, the lateral stresses σ_x and σ_y are basically equal and $\varepsilon_x = \varepsilon_y = 0$ in Equations 6.5 and 6.6, and therefore,

$$\sigma_x = \sigma_y = [\sigma_z/(1-\nu)] \tag{6.7}$$

where
ν is Poisson's ratio.

6.2.2 Stress and Displacement Distributions around Single Circular Excavations

When a circular opening is made in a rock mass, as shown in Figure 6.2, equilibrium is disturbed causing the stresses in the rock to distribute and concentrate in the rock surrounding the opening. Stresses eventually diminish with distance from the opening to the in situ stress based on the Saint Venants Principle (1855). The principle states that the difference between the effects caused by two different, but statically equivalent, loads become small at sufficiently large distances from the load.

To calculate the stresses, strains, and displacements induced around excavations in elastic materials, knowledge of theory of elasticity is required. According to Brady and Brown (2004), conditions to satisfy any solution for the stress and displacement distributions for a problem geometry and loading conditions are boundary conditions, differential equations of equilibrium, constitutive equations for the material, and strain compatibility equations. For the case of a circular opening in a rock mass, the solutions are known as the Kirsch equations (Kirsch, 1898).

Stress components at a point based on radial distance from the center of the excavation and orientation (r,θ) are as follows (Brady and Brown, 2004):

Radial stress

$$\sigma_r = (\sigma_v/2)\left[(1+k)\left(1-a^2/r^2\right)-(1-k)\left(1-4a^2/r^2+3a^4/r^4\right)\cos 2\theta\right] \tag{6.8}$$

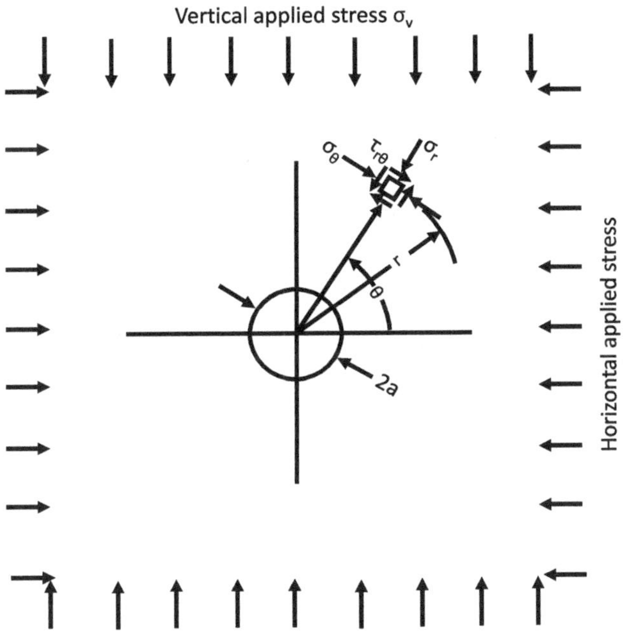

FIGURE 6.2 Stress changes caused by an opening in rock. (Modified after Hoek and Brown, 1980a.)

Tangential stress

$$\sigma_\theta = (\sigma_v/2)\left[(1+k)\left(1+a^2/r^2\right)+(1-k)\left(1+3a^4/r^4\right)\cos 2\theta\right] \tag{6.9}$$

Shear stress

$$\tau_{r\theta} = (\sigma_v/2)\left[(1-k)\left(1+2a^2/r^2-3a^4/r^4\right)\sin 2\theta\right] \tag{6.10}$$

Note that σ_v is represented instead by p, in Brady and Brown (2004), and is one component of a biaxial stress field where $p = p_{yy}$, and $p_{xx} = kp$ such that k is the ratio of horizontal stress to vertical stress.

The displacements can be calculated as follows:

Radial displacement (Brady and Brown, 2004)

$$u_r = \left[-pa^2/4Gr\right]\left\{(1+k)-(1-k)[4(1-\nu)-a^2/r^2]\cos 2\theta\right\} \tag{6.11}$$

Tangential displacement

$$u_\theta = \left[-pa^2/4Gr\right]\left\{(1-k)\left[2(1-2\nu)+a^2/r^2\right]\sin 2\theta\right\} \tag{6.12}$$

where G represents the shear modulus and p is as described above.

The calculations for the principal stresses and directions are given by Hoek and Brown (1980b):

Maximum stress

$$\sigma_1 = (1/2)(\sigma_r+\sigma_\theta)+\left[(1/4)(\sigma_r-\sigma_\theta)^2+\tau_{r\theta}^2\right]^{1/2} \tag{6.13}$$

Minimum stress

$$\sigma_2 = (1/2)(\sigma_r+\sigma_\theta)-\left[(1/4)(\sigma_r-\sigma_\theta)^2+\tau_{r\theta}^2\right]^{1/2} \tag{6.14}$$

Inclination relative to the radial direction

$$\tan 2\alpha = 2\tau_{r\theta}/(\sigma_\theta - \sigma_r) \tag{6.15}$$

where

σ_v is the vertical stress equivalent to γh.
a is the inside radius of opening.
k is the ratio of horizontal stress to vertical stress.

6.2.3 Stresses at the Excavation Boundary

At the boundary of the excavation or when r=a, the radial stress σ_r and the shear stress $\tau_{r\theta}$ are both zero and the tangential stress of Equation 6.9 reduces as follows:

$$\sigma_\theta = \sigma_v[(1+k)+2(1-k)\cos 2\theta] \tag{6.16}$$

At the sidewalls of the opening, θ=0° and 180°, Equation 6.16 reduces to

$$\sigma_\theta = \sigma_v(3-k) \tag{6.17}$$

Similarly, at the crown and floor of the opening, θ=90° and 270°, Equation 6.16 becomes

$$\sigma_\theta = \sigma_v(3k-1) \tag{6.18}$$

From Figure 6.3, it can be observed that for k=0, the stresses in the roof and floor of the opening are tensile. For k=0.33, the stresses in the roof and floor are zero, and for increasing k values, i.e., 0.33<k<1, all stresses on the boundary of the opening are compressive. The sidewall stresses decrease from a maximum of $3\sigma_v$ for k=0 to a value of $2\sigma_v$ for k=1.

Only tangential stresses (no radial stress) exist at the boundary of the excavation when free of internal loading. If the inside surface of the opening is loaded by means of water pressure, concrete lining, or the loads applied through Rockbolts, then these must be factored in the stress redistribution in the rock surrounding the opening.

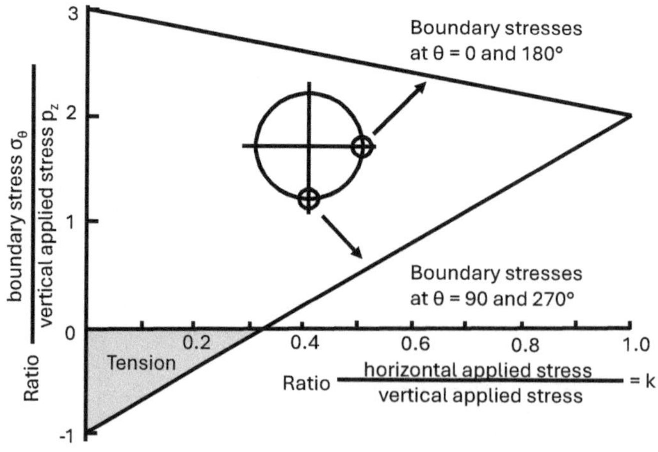

FIGURE 6.3 Boundary stresses around a circular excavation. (Modified from Hoek and Brown, 1980a.)

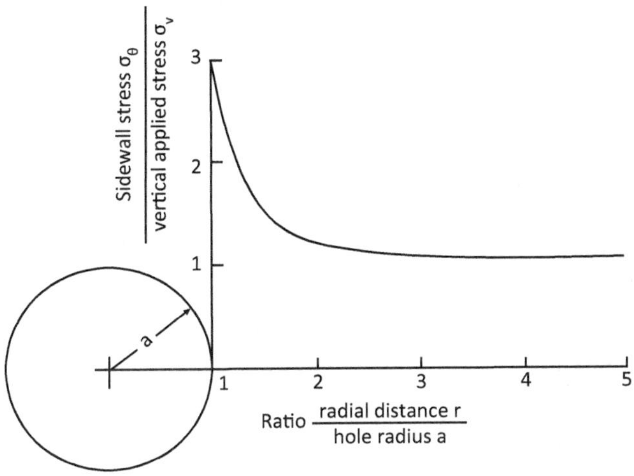

FIGURE 6.4 Stress concentrations along the horizontal from the edge of a circular excavation (based on Terzaghi and Richart, 1952).

6.2.4 Stress Remote from the Excavation Boundary

As the distance (r) from the excavation increases, the influence of the opening upon the stresses in the rock decreases (assuming elastic behavior). For example, at the crown, $\theta=90°$, Equation 6.9 becomes

$$\sigma_0 = (\sigma_v/2)[(1+k)(1+a^2/r^2)-(1-k)(1+3a^4/r^4) \tag{6.19}$$

For a special case where $k=0$, then Equation 6.19 becomes

$$\sigma_\theta = (\sigma_v/2)(a^2/r^2 - 3a^4/r^4) \tag{6.20}$$

Figure 6.4 plots the ratios σ_θ/σ_v against r/a, illustrating that the stress concentrating effect of the hole reduces and at $r/a=3$, the ratio of induced tangential to applied vertical stress is close to unity. Therefore, the mining-induced stresses in the rock diminish with distance. An example of the radial and tangential stresses at the crown (top), invert (bottom), and sidewall (springline) of a circular excavation when k is 0.25 is shown in Figure 6.5.

For multiple excavations, each will have mining-induced stresses, and in the elastic theory, the principle of superposition holds, indicating that excavations placed too close in proximity will have elevated stresses even beyond those shown in Figure 6.4.

6.2.5 Excavation Shape

For an elliptical excavation with axes orientated in the directions of the in situ principal stresses as shown in Figure 6.6, the circumferential stresses at the boundary can be determined according to closed-form solutions found by Hudson and Harrison (1997).

After Hudson and Harrison (1997), the tangential stress at point A is

$$\sigma_A = p(1-k+2q) = p\left[1-k+(2W/\rho_A)^{0.5}\right] \tag{6.21}$$

At point B,

$$\sigma_B = p(k-1+2k/q) = p\left[k-1+k(2H/\rho_B)^{0.5}\right] \tag{6.22}$$

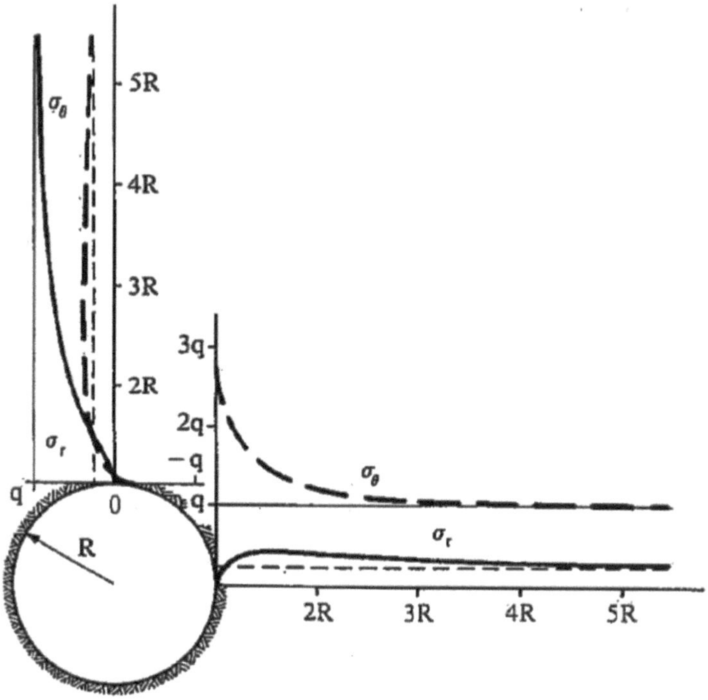

FIGURE 6.5 Stress distribution around a circular excavation where the stress k ratio is 0.25 (based on Terzaghi and Richart, 1952).

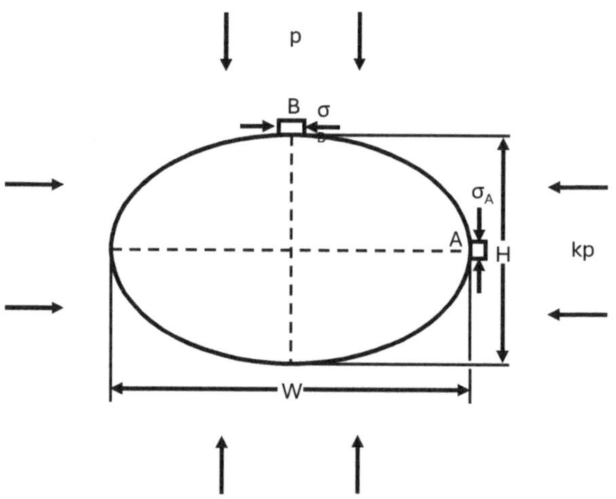

FIGURE 6.6 An elliptical excavation under stress. (Modified after Brady and Brown, 2004.)

The radii of curvature are

$$\rho_A = \left(H^2/2W\right) \tag{6.23}$$

$$\rho_B = \left(W^2/2H\right) \tag{6.24}$$

The aspect ratio of the ellipse is

$$q = W/H \tag{6.25}$$

Compare the circular excavation in Figure 6.5 to an elliptical excavation in Figure 6.7 also giving the radial and tangential stresses at the crown (top) and invert (bottom) as well as the sidewall (springline) when k is again 0.25.

Elliptical (or modified) cross sections can be useful in highly stressed shafts and tunnels.

Square excavations presented by Hudson and Harrison (1997) are summarized here. For a square opening with round corners in a hydrostatic stress field, the highest stress concentration should occur where the radius of curvature is smallest which is at the corners.

For square corners, the stress concentrations approach infinity in theory, and thus, rounded corners are used for closed-form approximations (Obert and Duvall, 1967). Taking the case in Figure 6.8, reorienting the square allows for an approximation to be made using Equation 6.21.

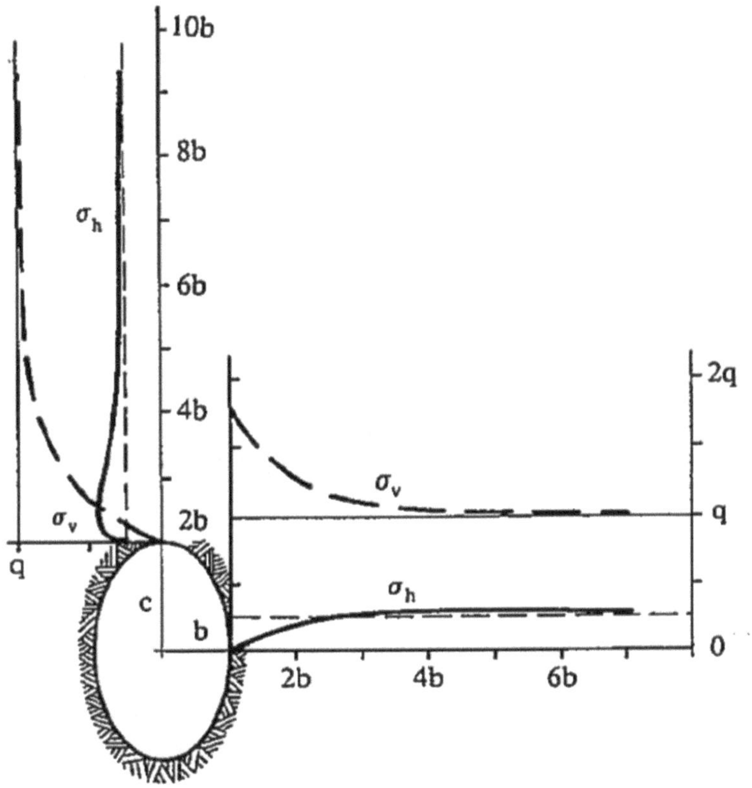

FIGURE 6.7 Stress distribution around an elliptical excavation where the stress k ratio is 0.25 (based on Terzaghi and Richart, 1952).

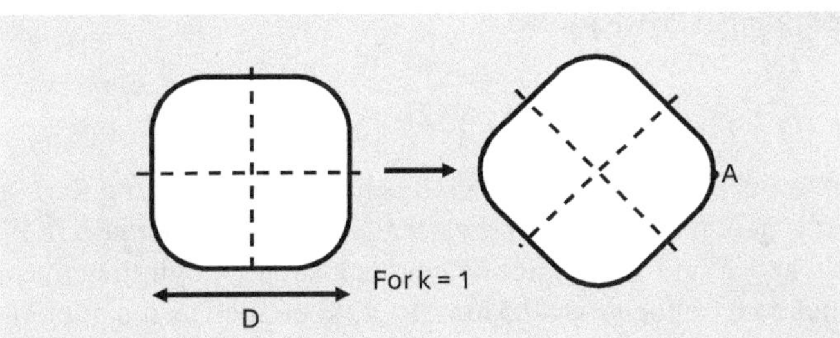

FIGURE 6.8 Square excavation with rounded corners. (Modified after Hudson and Harrison, 1997.)

As an example, Hudson and Harrison (1997) treat the radius of curvature as 0.2D and the width equal to 1.25D. Therefore, the reduced form for stress at A is

k = 1 for hydrostatic stress field and assuming D = 1:

$$\sigma_A = p\left[1 - k + (2W/\rho_A)^{0.5}\right]$$

$$\sigma_A = p\left\{1 - 1 + [2(1.25D)/(0.2D)]^{0.5}\right\}$$

$$\sigma_A = 3.53p$$

Numerical modeling is highly beneficial alongside the approaches in the preceding sections, because it can easily expand on point estimates, such as shown in Figure 6.8, and allow for greater complexity and multiple scenarios to be analyzed quickly and efficiently. Calibration is always an important consideration, as is underground monitoring, and designs should be revisited frequently as more data becomes available.

6.2.6 Stresses versus Excavation Size

The equations for stresses around a circular excavation use the dimensionless ratio, r/a, which implies that the boundary stresses are independent of the excavation radius. For example, the same stress concentrations occur in the walls of 1 and 10 m diameter tunnels in elastic rock. However, stress magnitudes and concentrations are critical especially if there are other excavations in the vicinity.

For all practical purposes, rock masses that are fractured, and therefore not elastic, the stability will generally be controlled by the ratio of excavation size to the size of the discrete blocks in the rock mass. For example, increasing the size of an excavation in a weakly jointed rock mass may not cause a significant increase in stress, but it will almost certainly give rise to a decrease in stability.

6.3 ROCK MASS CLASSIFICATION SYSTEMS

A rock mass does not behave as a small core sample tested in the laboratory as the in situ strength is affected by geological features such as bedding planes, faults, and joint sets and water can be another major strength reducer. Classification systems are designed as an adjustment for such circumstances. For greenfield mine development (i.e., an area or region where no similar type of mines exist), classification systems represent the only initial information available using diamond core.

The following classification systems will be covered:

- Rock quality designation (RQD).
- Q system.
- Mine rock mass rating system (MRMR) and design rock mass strength (DRMS).
- Coal mine roof rating (CMRR).

6.3.1 Rock Quality Designation (RQD)

The RQD was first introduced by Deere in 1964. It is based on measuring the core recovery percentage which incorporates only pieces that are greater than 100 mm in length. In this respect, pieces of core that are not hard and sound should not be counted, even though they may be 100 mm or more in length. The optimal core diameter is 47.5 mm. RQD is defined as the quotient as follows:

$$RQD = 100(\text{Sum of } 10)/L_{tot} \tag{6.26}$$

TABLE 6.1
The Relationship of RQD to Rock Quality (Deere et al., 1967)

Rock Quality Designation	Description of Rock Quality
Very poor	0%–25%
Poor	25%–50%
Fair	50%–75%
Good	75%–90%
Excellent	90%–100%

where

(Sum of 10) is the sum of competent core pieces defined as having lengths ≥10 cm

L_{tot} is the total length of core run.

RQD has value as an initial estimate for stability as it can identify low-quality rock zones and can guide support requirements for excavations. It is considered a standard parameter in drill core logging and forms a basic parameter for most of the major rock mass classification systems. Table 6.1 summarizes the relationship between RQD and basic rock quality.

The only caution is that fractures caused by the diamond drilling itself, for example, from poor operation or handling, should not be included if possible. However, if these fractured cores are included, they introduce more conservatism into the design, making it safer not more hazardous. An example of RQD is provided in Figure 6.9.

6.3.2 The Q System

The Q system, after Barton et al. (1974), is a detailed approach to rock mass classification and builds on the RQD system by integrating a comprehensive classification of joints and support pressure, based on over 200 case studies in Norway. Figure 6.10 illustrates a jointed rock mass. The overall rock quality, Q, is expressed by six parameters:

$$Q = (RQD/J_n)(J_r/J_a)(J_w/SRF) \qquad (6.27)$$

where

RQD is the rock quality designation (Table 6.1).

J_n is a joint set number (Table 6.2).

J_r is a roughness number (Table 6.3).

J_a is a joint alteration number (Table 6.4).

J_w is a joint water reduction factor (Table 6.5).

SRF is the stress reduction factor (Table 6.6).

The assembled terms of the Q value can be explained as follows:

- (RQD/J_n) is an indication of block size or degree of jointing in the rock mass.
- (J_r/J_a) is the joint friction or an indication of the shear strength between the blocks in the rock mass.
- (J_w/SRF) is a function of the active stress acting on the rock mass.

Tables 6.2 and 6.3 reflect the rankings in the Norwegian Geotechnical Institute's handbook for usage of the Q system which has changed from the original document on the Q system in Barton et al. (1974).

Joint roughness is quite subjective and difficult to determine in situ. Figure 6.11 shows a "rough/uneven" joint and a Barton comb used to measure the joint roughness.

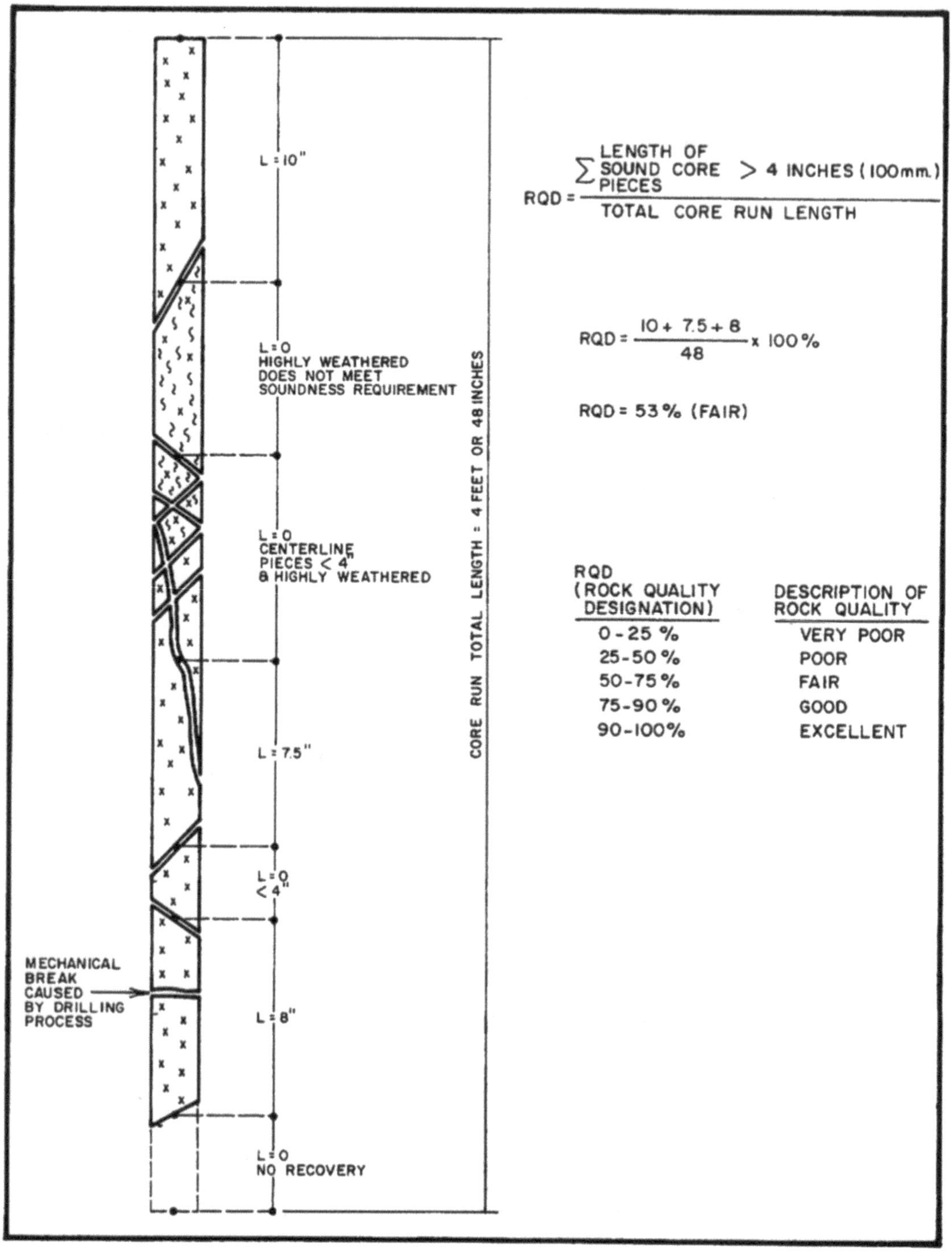

FIGURE 6.9 RQD example after Deere (1989).

Figure 6.12 shows the weakening effects of water flowing through joints. The white mineralization has been removed from the host rock by the action of flowing water and forming stalactites.

The RQD can also be estimated from the number of joints per unit volume. The updated relationship is given as follows (Palmström, 2005):

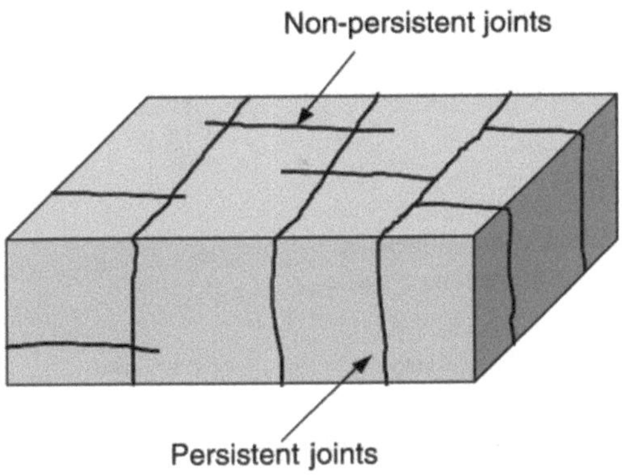

FIGURE 6.10 A rock mass schematic with joints sets (Kim et al., 2007).

TABLE 6.2
Joint Set Number (J_n) (NGI, 2022)

A	Massive, no or few joints	0.5–1.0
B	One joint set	2
C	One joint set plus random joints	3
D	Two joint sets	4
E	Two joint sets plus random joints	6
F	Three joint sets	9
G	Three joint sets plus random joints	12
H	Four or more joint sets, random, heavily jointed, "sugar cube", etc.	15
J	Crushed rock, earth like	20

Note: For tunnel intersections, use ($3.0\times J_n$).
For portals, use ($2.0\times J_n$).

TABLE 6.3
Joint Roughness Number (J_r) (NGI, 2022)

a. Rock-wall contact, and
b. Rock-wall contact before 10cm shear movement

A	Discontinuous joints	4
B	Rough or irregular, undulating	3
C	Smooth, undulating	2
D	Slickensided, undulating	1.5
E	Rough or irregular, planar	1.5
F	Smooth, planar	1.0
G	Slickensided, planar	0.5

Note: Descriptions refer to small-scale and intermediate-scale features, in that order.

c.No rock-wall contact when sheared

H	Zone containing clay minerals thick enough to prevent rock-wall contact when sheared	1.0

Note: Add 1.0 if the mean spacing of the relevant joint set is greater than 3 m (dependent on the size of the underground opening).
$J_r=0.5$ can be used for planar slickensided joints having lineations, provided that the lineations are orientated in the estimated sliding direction.

TABLE 6.4
Joint Alteration Number (J_a) (NGI, 2022)

		J_a	ϕ_r Approx. Residual Friction
a. Rock-wall contact (no mineral fillings, only coatings)			
A	Tightly healed, hard, non-softening, impermeable filling, i.e., quartz or epidote.	0.75	
B	Unaltered joint walls, surface staining only.	1.0	25°–35°
C	Slightly altered joint walls. Non-softening mineral coatings, sandy particles, clay-free disintegrated rock, etc.	2.0	25°–30°
D	Silty or sandy-clay coatings, small clay fraction (non-softening).	3.0	20°–25°
E	Softening or low-friction clay mineral coatings, i.e., kaolinite or mica. Also, chlorite, talc, gypsum, graphite, etc. and small quantities of swelling clays.	4.0	8°–16°
b. Rock-wall contact before 10 cm shear (thin mineral fillings)			
F	Sandy particles, clay-free disintegrated rock, etc.	4.0	25°–30°
G	Strongly over-consolidated non-softening clay material fillings (continuous but <5 mm thickness).	6.0	16°–24°
H	Medium or low over-consolidation, softening, clay mineral fillings (continuous but <5 mm thickness).	8.0	12°–16°
J	Swelling clay fillings, i.e., montmorillonite (continuous, but <5 mm thickness). Value J_a depends on the percentage of swelling clay-sized particles.	8–12	6°–12°
c. No rock-wall contact when sheared (thick mineral fillings)			
K	Zones or bands of disintegrated or crushed rock. Strongly over-consolidated.	6	16°–24°
L	Zones or bands of clay, disintegrated or crushed rock. Medium or low over-consolidation of softening fillings.	8	12°–16°
M	Zones or bands of clay, disintegrated or crushed rock. Swelling clay. J_a depends on percent of swelling clay-size particles.	8–12	6°–12°
N	Thick continuous zones or bands of clay. Strongly over-consolidated.	10	12°–16°
O	Thick continuous zones or bands of clay. Medium to low over-consolidation.	13	12°–16°
P	Thick continuous zones or bands with clay. Swelling clay. J_a depends on percent of swelling clay-size particles.	13–20	6°–12°

TABLE 6.5
Joint Water Reduction Factor (J_w) (NGI, 2022)

		J_w
A	Dry excavations or minor inflow (humid or a few drips).	1.0
B	Medium inflow, occasional outwash of joint fillings (many drips/"rain").	0.66
C	Jet inflow of high pressure in competent rock with unfilled joints.	0.5
D	Large inflow or high pressure, considerable outwash of joint fillings.	0.33
E	Exceptionally high inflow or water pressure decaying with time. Causes outwash of material and perhaps cave in.	0.2–0.1
F	Exceptionally high inflow or water pressure continuing without noticeable decay. Causes outwash of material and perhaps cave in.	0.1–0.05

Note: Factors C to F are crude estimates. Increase J_w if the rock is drained or grouting is carried out. Special problems caused by ice formation are not considered.

TABLE 6.6
Stress Reduction Factor (SRF) (NGI, 2022)

a. Weak zones intersecting the underground opening, which may cause loosening of rock mass			**SRF**
A Multiple occurrences of weak zones within a short section containing clay or chemically disintegrated, very loose surrounding rock (any depth), or long sections with incompetent (weak) rock (any depth). For squeezing, see L and M.			10
B Multiple shear zones within a short section in competent rock (clay-free) with loose surrounding rock (any depth).			7.5
C Single weak zones with or without clay or chemically disintegrated rock (depth of excavation ≤ 50 m).			5
D Loose, open joints, heavily jointed or "sugar cube", etc. (any depth).			5
E Single weak zones with or without clay or chemically disintegrated rock (depth of excavation > 50 m).			2.5
Note: Reduce these values of SRF by 25%–50% if the weak zones only influence but do not intersect the underground opening.			
b. Competent, mainly massive rock, stress problems	σ_c/σ_1	σ_θ/σ_c	**SRF**
F Low stress, near surface, open joints	>200	<0.01	2.5
G Medium stress, favorable stress condition	200–10	0.01–0.3	1
H High stress, very tight structure. Usually favorable to stability. May be unfavorable to stability depending on stress orientations compared to jointing/weak planes.*	10–5	0.3–0.4	0.5–2 2–5*
J Moderate spalling and/or slabbing after > 1 hour in massive rock	5–3	0.5–0.65	5–50
K Spalling or rock burst after a few minutes in massive rock	3–2	0.65–1	50–200
L Heavy rock burst and immediate dynamic deformation in massive rock	<2	>1	200–400
Note: For strongly anisotropic virgin stress field (if measured): when $5 \le \sigma_1/\sigma_3 \le 10$, reduce σ_c to $0.75\sigma_c$. When $\sigma_1/\sigma_3 > 10$, reduce σ_c to $0.5\sigma_c$, where σ_c = unconfined compression strength, σ_1 and σ_3 are the major and minor principal stresses and σ_θ = maximum tangential stress (estimated from elastic theory). When depth of crown below surface is less than span, SRF increase from 2.5 to 5 for such cases (see F).			
c. Squeezing rock: plastic flow of incompetent rock under the influence of high pressure		σ_θ/σ_1	**SRF**
M Mild squeezing rock pressure		1–5	5–10
N Heavy squeezing rock pressure		>5	10–20
Note: Determination of squeezing rock conditions can be made according to literature such as Singh et al. (1992) and Bhasin and Grimstad (1996)			
d. Swelling rock: chemical swelling activity depending on presence of water			**SRF**
O Mild swelling rock pressure			5–10
P Heavy swelling rock pressure			10–15

$$RQD = 110 - 2.5J_v \left(\text{for } J_v \text{ between 4 and 44}\right) \tag{6.28}$$

where

J_v is the total number of joints per m^3.

In order to relate the calculated Q value to the support requirement underground, Barton et al. (1974) developed a further concept of the equivalent dimension:

$$D_e = \left(\text{Excavation span, diameter, or height}\right)/\text{Excavation Support Ratio}\left(\text{ESR}\right) \tag{6.29}$$

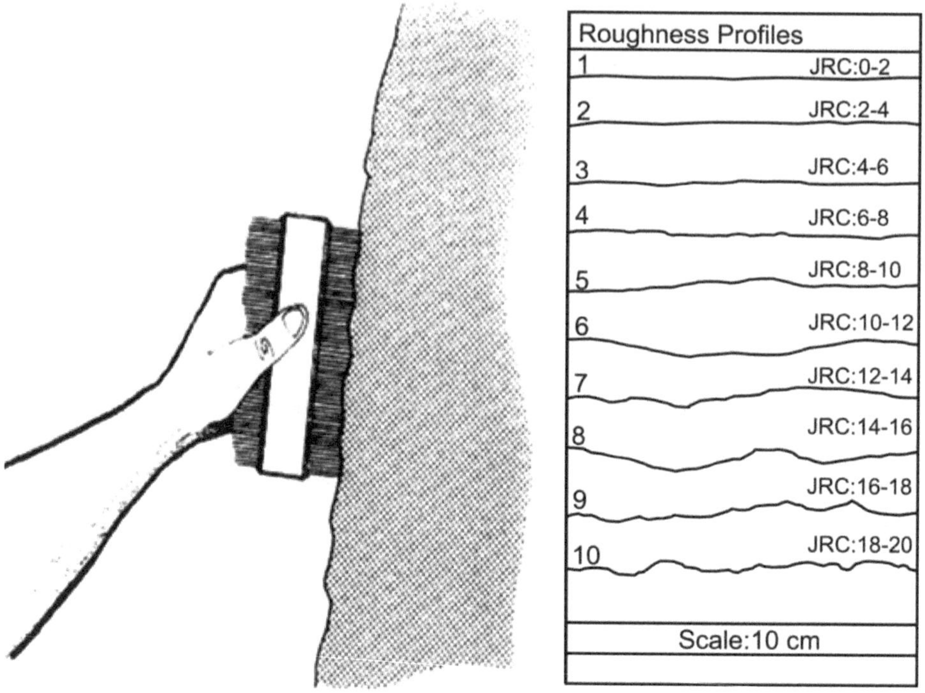

FIGURE 6.11 Joint roughness being measured using a Barton comb (Morelli, 2013).

FIGURE 6.12 The effects of water flowing through joints on a rock Island in the South China Sea. The white are minerals dissolved out of the rock by water. Photo taken by Sam Spearing.

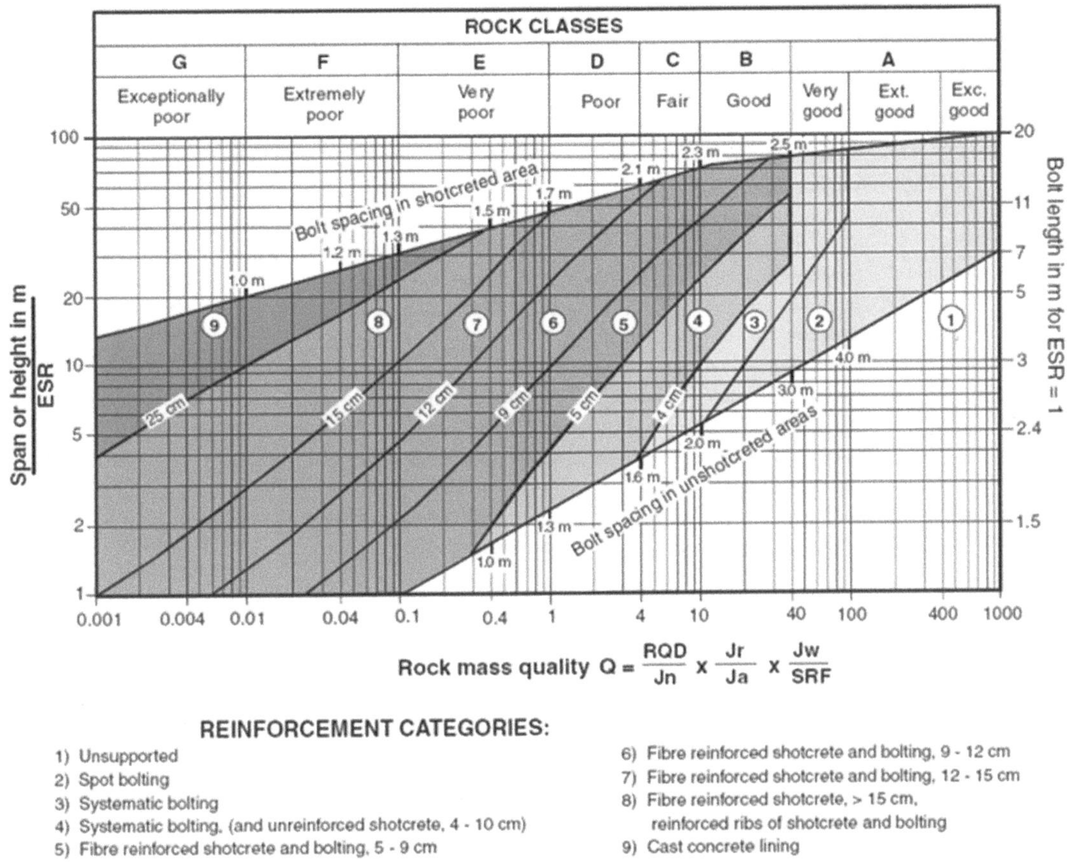

FIGURE 6.13 The relationship between the Q value and the equivalent dimension (D_e). (After Barton et al., 1974.)

The ESR recognizes the variable support practices in construction, for example, the use of the excavation, or the time the excavation is required to be stable for. The Norwegian Geotechnical Institute (NGI) provides the updated metrics for ESR which were originally found in Barton et al. (1974).

Figure 6.13 (Barton et al., 1974) gives a derived empirically based support requirement guide.

6.3.3 Mine Rock Mass Rating (MRMR)

The MRMR system was first introduced by Laubscher in 1975 as a further refinement of the Rock Mass Rating by Bieniawski (1973). Jakubec and Laubscher (2000) described how it can be especially useful for the design of cave mining and sub-level cave mining designs.

There are many similarities between the Q system and the MRMR system as they use the same base parameters but give the individual parameters different weighting systems.

Figure 6.14 gives an overview of MRMR (modified after Jakubec and Laubscher, 2000).

Table 6.8 compares the main differences between the three important rock mass classification systems used in mining (Jakubec and Esterhuizen, 2007).

TABLE 6.7
ESR Values (NGI, 2022)

Ranking	Excavation Category	ESR
A	Temporary mine openings, etc.	3–5
B	Vertical shafts[a]:	
	i. circular sections	2.5
	ii. rectangular/square section	2.0
C	Permanent mine openings, water tunnels for hydro power (exclude high-pressure penstocks), water supply tunnels, pilot tunnels, drifts, and headings for large excavations.	1.6
D	Minor road and rail tunnels, surge chambers, access tunnels, sewage tunnels, etc.	1.3
E	Power houses, storage rooms, water treatment plants, major road and rail tunnels, civil defense chambers, portals, intersections, etc.	1.0
F	Underground nuclear power stations, railway stations, sports and public facilities and factories, etc.	0.8
G	Very important caverns and underground openings with a long lifetime, such as 100 years or without access for maintenance.	0.5

[a] Dependent on the purpose may be lower than given values.

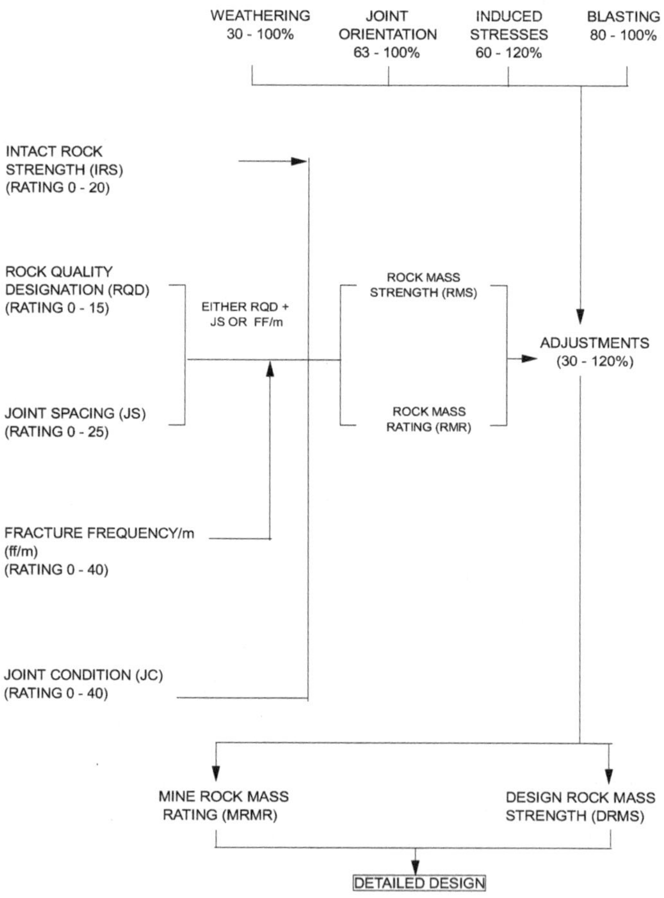

FIGURE 6.14 Overview of the MRMR components (Jakubec and Laubscher, 2000).

TABLE 6.8
Comparison of the RMR, Q, and MRMR Rock mass Classification Systems

Category	Parameter	RMR	Q	MRMR[a]	MRMR[b]
Intact rock strength	UCS	X	X	X	X
Open joint frequency	RQD	X	X	X	–
	FF/m	X	-	X	X
	Joint set (J_n)	X	X	X	X
Open joint strength	Roughness (J_r)	X	X	X	X
	Alteration (J_a)	X	X	X	X
	Infill (J_a)	X	X	X	X
Cemented joints (CJ)	CJ/m	–	–	–	X
quantity and strength	CJ strength	–	–	–	X

[a] No emLaubscher (1990).
[b] No emLaubscher and Jakubec (2001).

6.3.4 Design Rock Mass Strength (DRMS)

DRMS relates to the strength of an unconfined rock mass in a unique mining environment. It is typically expressed in megapascals and can be related to mining-induced stresses. DRMS can be used to estimate the support needed before mining commences.

As in rock mechanics in general, the support design must be regularly revisited as more actual data become available. The designs must include all the assumptions made and the input parameters used.

Figure 6.15 gives an overview of DRMS from Laubscher (1990) and how it relates to the mining environment stress.

Table 6.9 gives a relationship between MRMR and different mining methods and can be used as a rough guideline.

6.3.5 Coal Mine Roof Rating

CMRR employs a similar format to that of Bieniawki's RMR, by summing the individual rating categories to obtain a final CMRR on a 0–100 scale (Mark and Molinda, 2005). It is also designed so that the CMRR, the unsupported span, and the standup time relationships are roughly the same as that determined for the RMR. The following Sections is reproduced directly from Mark and Molinda (2005).

6.3.5.1 Data Collection for the Calculation of CMRR

The data required for CMRR can be determined either from underground exposures such as roof falls and overcasts, or from exploratory drill core. In either case, the main parameters measured are as follows:

- The uniaxial compressive strength (UCS) of the intact rock.
- The intensity (spacing and persistence) of bedding and other discontinuities.
- The shear strength (cohesion and roughness) of bedding and other discontinuities.
- The moisture sensitivity of the rock.
- The presence of a strong bed in the bolted interval.

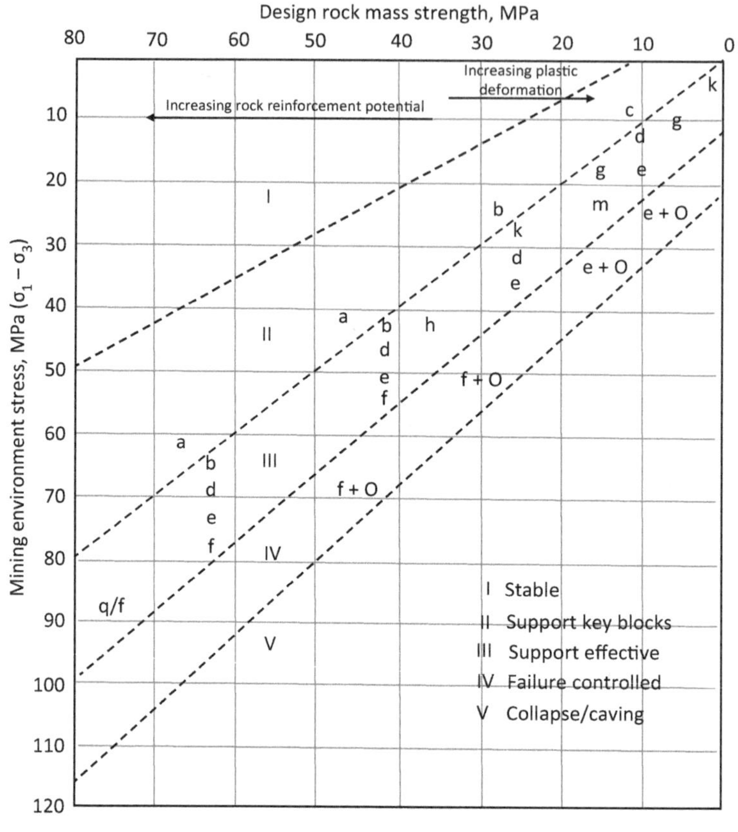

FIGURE 6.15 Support requirements as a function of DRMS for maximum stress conditions (Laubscher, 1990).

Other factors include the following:

- Number of rock layers.
- Presence of groundwater.
- Surcharge from overlying typically weak beds.

CMRR is calculated in a two-step process. First, the mine roof is divided into structural units, and unit ratings are determined for each. A structural unit generally contains one lithologic layer, but several rock layers may be lumped together if their engineering properties are similar.

In the second step, CMRR is determined by averaging all the unit ratings within the bolted interval (with the contribution of each unit weighted by its thickness) and applying appropriate adjustment factors. This second step is the same regardless of whether the unit ratings were from the data collected underground or from core. Figure 6.16 outlines the process.

6.3.5.2 Parameters Needed to Calculate CMRR

6.3.5.2.1 Uniaxial Compressive Strength (UCS)

The UCS of the rock material affects roof strength in several ways. First, it determines the ease with which new fracturing (as opposed to movement along pre-existing discontinuities) will take place. Second, the compressive strength of the rock is a factor in the shear strength of discontinuities.

TABLE 6.9
Mining Methods Related by MRMR Classification (Laubscher, 1990)

Class	**5**	**4**	**3**	**2**	**1**
Rating	**5–20**	**21–40**	**41–60**	**61–80**	**80–100**
		Block Caving			
Undercut (m)	1–8	8–18	18–32	32–50	+50
Cavability	Very good	Good	Fair	Poor	Very poor
Fragmentation (m)	0.01-0.3	0.1-0.2	0.4-5	1.5-9	3-20
2nd lay-on blast/drill (g/t)	0–50	50–150	150–400	400–700	+700
	0–20	20–60	60–150	150–250	+250
Hangups as % of tonnage	0	15	30	45	>60
Dia. of draw zone (m)	6–7	8–9	10–11.5	12–13.5	15
Drawpoint span (m)					
Grizzly	5–7	7–10	9–12		
Slusher	5–7	7–10	9–12		
LHD (m)	9	9–13	11–15	13–18	
Brow support	Steel and concrete		Concrete	Blast protection	
	Reinf. concrete				
Drift support	Lining, rock reinf., repair techniques		Lining reinf.	Rock reinf.	
Width of point (m)	1.5–2.4	2.4–3.5	2.4–4	4	
Direction of advance	Toward low stress		Toward high stress		
Comments	Fine fragmentation, poor ground, heavy support, repairs	Medium fragmentation, good ground, fair support	Medium coarse fragmentation, good drill hangups	Coarse fragmentation, large LHDs, drill hangups	
		Sub-level Caving			
Loss of holes	Excessive	Fair	Negligible	Nil	Nil
Brow wear	Excessive	Fair	Low	Nil	Nil
Support	Heavy	Medium	Low	Localized	Nil
Dilution	Very high	High	Medium	Low	Very low
Cave SI (m)	1–8	8–18	18–32	32–50	+50
Comments	Not practicable	Applicable	Suitable	Suitable	Suitable, large hangingwall cave area
Sub-level Open Stoping	1–5	5–20	20–30	30–80	100
Minimum span (m)	N/A	1–8	8–16	16–35	+35
Stable area, i.e., SI (m)					
N/A = Not applicable					

Approximately one-third of the CMRR is determined by the compressive strength rating, which is approximately twice the weight given to the UCS to the original RMR as a comparison.

Laboratory testing is generally considered the standard method of determining the UCS. In coal mining, the variability in the results can be high, with the standard deviation typically about one-third of the average for a coal seam.

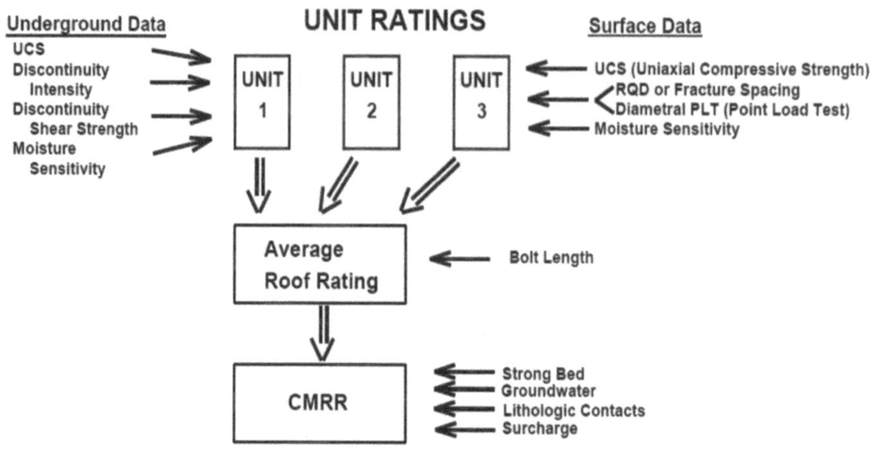

FIGURE 6.16 Steps involved in calculation of CMRR (Mark and Molinda, 2005).

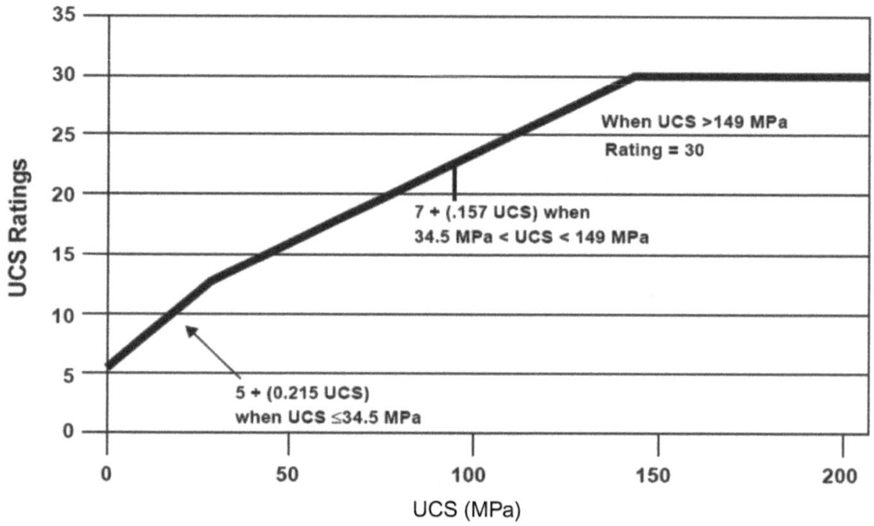

FIGURE 6.17 CMRR rating scale for axial point load or UCS tests (Mark and Molinda, 2005).

As an alternative, the CMRR recommends the Point Load Test (PLT) for drill core. When the axial PLT is used to estimate the UCS, the Point Load Index (Is_{50}) is converted using the following equation:

$$UCS = K(Is_{50}) \tag{6.30}$$

where K is the conversion factor.

A comprehensive study involving more than 10,000 tests of coal measure rocks reported that K=21 fits the data well for the range of coal measures in all geographic regions (Rusnak and Mark, 2000). The UCS rating scale used (in place of the actual UCS) in the CMRR calculation is shown in Figure 6.17.

TABLE 6.10
CMRR Rating Scale for Ball Peen Hammer Test (Mark and Molinda, 2005)

Ball Peen Hammer Class	Williamson (1984) UCS Range (MPa)	CMRR UCS Range (MPa)	CMRR Rating
Molds	<7	<14	5
Craters	7–21	14–35	10
Dents	21–56	35–70	15
Pits	56–105	70–120	22
Rebounds	>105	>120	30

TABLE 6.11
Bedding and Discontinuity Intensity Table for Underground Data

	Spacing				
Persistence (m)	>1.8 (m)	0.6–1.8 (m)	0.2–0.6 (m)	60–200 (mm)	<60 (mm)
0–1	35	30	24	17	9
1–3	32	27	21	15	9
>3	30	25	20	13	9

Underground, the CMRR employs an indentation test to estimate UCS, proposed by Williamson (1984). The exposed rock face is struck with the round end of a ball peen hammer, and the impact reaction is used to calculate the CMRR rating as given in Table 6.10. The nature of the reaction (indentation), not its magnitude, is what is important.

6.3.5.2.2 Discontinuity Intensity

Intensity is determined by the spacing between bedding planes or other discontinuities, and the persistence, or extent, of each individual discontinuity. The more closely spaced a set of discontinuities, the greater weakening effect it has on the rock mass. Persistence is more important for discontinuities that are widely spaced. Like UCS, intensity accounts for about one-third of the total CMRR.

Underground, both spacing and persistence can be measured directly, using the standard methods for rock mass characterization (ISRM, 1981). Table 6.11 shows the Bedding/Discontinuity Rating Scale for underground data. The matrix shows what point value is added for each combination of spacing and persistence of discontinuities.

Most standard geotechnical core logging procedures include some measure of the natural breaks in the core. The two most commonly employed procedures are fracture spacing and RQD. Fracture spacing is easily determined by counting the core breaks in a particular unit, and then dividing by the thickness of the unit.

6.3.5.2.3 Shear Strength of Discontinuities

Bedding plane shear strength is a critical parameter for coal mine ground control, because the most severe loading applied to coal mine roof is normally lateral, caused by horizontal stress (Mark and Barczak, 2000). The shear strength of discontinuities rating is evaluated using two parameters: cohesion of bedding surfaces and roughness of surfaces.

Underground, the cohesion of bedding surfaces is evaluated by using a 9 cm mason chisel and a hammer to split hand samples of rock. Cohesion can also be estimated by observing the nature of the fractured wall of a roof fall.

TABLE 6.12
Bedding/Discontinuity Shear Strength Rating Table for Underground Data

Roughness	Cohesion			
	Strong	Moderate	Weak	Slickensided
Jagged	35	29	24	10
Wavy	35	27	20	10
Planar	35	25	16	10

The roughness of a surface is estimated visually and classified into "jagged", "wavy", or "planar" (Barton et al., 1974). This measure is to be applied on a scale which ranges from hand sample size to several feet across a fall exposure. The CMRR assumes that roughness significantly affects shear strength only when cohesion is in the middle range. Table 6.12 shows the shear strength rating for underground data.

6.3.5.2.4 Moisture Sensitivity Deduction

The maximum deduction for moisture sensitivity is 15 points in CMRR. The data sheet used in the immersion test data sheet is shown in Figure 6.18. If immersion test results are not available, moisture sensitivity can sometimes be estimated visually in underground exposures.

Usually, some time is required for contact with humid mine air to affect rock strength. In short-term applications, therefore, it may not be appropriate to apply the moisture sensitivity deduction.

Research by Mark and Molinda (2005) was conducted to explore the relationship between the Slake Durability Test (SDT) and the immersion test. The correlation is less reliable for distinguishing "moderately sensitive" rocks from "severely sensitive" rocks. Table 6.13 indicates how the results from either test can be used for input to the CMRR.

6.3.5.2.5 Multiple Discontinuity Adjustment

CMRR contains a multiple discontinuity adjustment so that the weakening effects of slickensides and other discontinuities can be considered.

6.3.5.3 Calculation of CMRR

Figure 6.19 shows the components used in CMRR (Molinda and Mark, 1994).

When using underground data, the equation for calculating the Unit Rating is

$$\text{Unit Rating} = \begin{pmatrix} \text{UCS Rating} + \text{Discontinuity Intensity Rating} + \text{Discontinuity Shear Strength Rating} \\ +\text{Multiple Discontinuity Adjustment} + \text{Moisture Sensitivity Deduction} \end{pmatrix} \tag{6.31}$$

For drill core data, the equation is even simpler:

$$\text{Unit Rating} = \left(\text{UCS Rating} + \text{Discontinuity Rating} + \text{Moisture Sensitivity Deduction}\right) \tag{6.32}$$

The next step in the calculation of the CMRR is to determine the thickness-weighted average of the unit ratings of all the units within the bolted interval. For example, assume that the roof consists of three units (from top down):

IMMERSION TEST

Mine: ____________________ Date: ____________________

Unit No.: ____________________ Tester: ____________________

Sample Description (Lithology, bedding, etc.):

____

Immersion

Observation	Rating
Appearance of Water	
Clear = 0	_____
Misty = -1	
Cloudy = -3	
Talus Formation	
None = 0	
Minor = -1	
Major = -3	_____
Cracking of Sample	
None = 0	
Minor/Random = -1	
Major/Preferred Orientation =-3	_____
Specimen Breakdown = -9	
Total Immersion Index	_____

Breakability

Observation	Rating
No Change 0	
Small Change-2	
Large Change -6	
Breakability Index	_____

Procedure for Immersion Test

1. Select sample(s) - ~ hand sized.
2. Test for hand breakability.
3. Rinse specimen (to remove surface dirt, dust, etc.).
4. Immerse in water for 1 hour.
5. Observe and rate water appearance, talus formation, and cracking of sample.
6. *Sum Rating for Immersion Index.*
7. Retest for hand breakability.
8. *Determine Breakability Index.*
9. The final Immersion Test Index is the greater of the Breakability Index or the Immersion Index.

FIGURE 6.18 Data sheet for immersion test (Mark and Molinda, 2005).

TABLE 6.13
Moisture Sensitivity Classes and Rating from Both Immersion and Slake Durability Tests (Mark and Molinda, 2005)

Moisture Sensitivity Class	Rating Adjustment	Immersion Index	Slake Durability Index
Not sensitive	0	0–1	100–98
Slightly sensitive	−3	2–4	98–92
Moderately sensitive	−7	5–9	92–80
Severely sensitive	−15	>9	<80

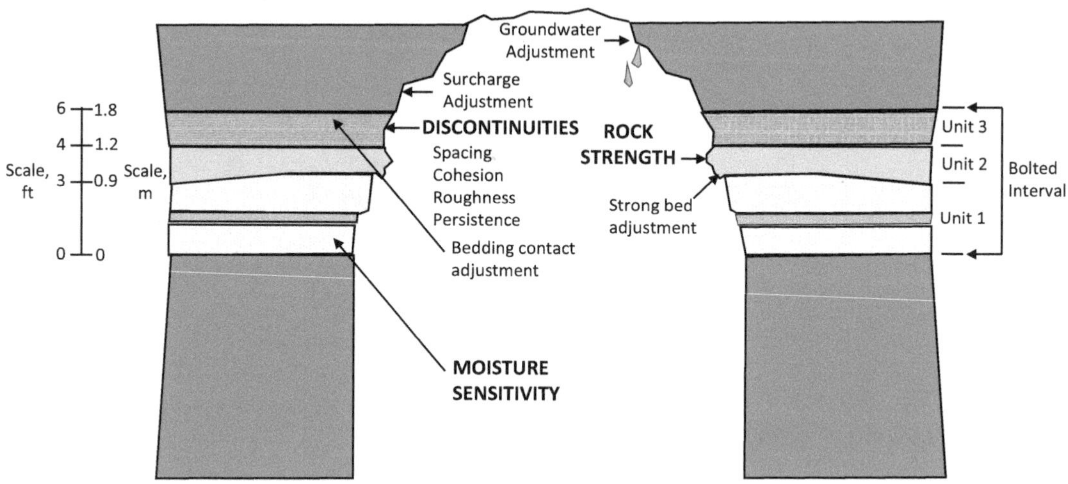

FIGURE 6.19 The components used to calculate the CMRR (Molinda and Mark, 1994).

- 2 m Sandstone, Unit Rating = 60.
- 0.8 m Siltstone, Unit Rating = 50.
- 0.4 m Shale, Unit Rating = 40.

According to Mark and Molinda (2005), if the length of the roof bolts is 1.8 m, then the Roof Rating thickness-Weighted average (RRW) is

$$\text{RRW} = \left[(0.4)(40) + (0.8)(50) + (0.6)(60)\right]/1.8 = 51.1$$

Note that although the uppermost layer was 2 m thick, only the lowest 0.6 m (the distance to the top of the bolts) was used in the calculation.

The CMRR is now determined by applying several adjustment factors to the RRW.

6.3.5.3.1 Strong Bed Adjustment (SBADJ)

The strongest rock layer within the length of the rock bolts installed often determines the performance and stability of the immediate rock. This is considered in the CMRR classification system. The following therefore can be important:

- The effective strength of the strongest layer in relation to the strength of the other layers in the immediate roof that is rock bolted.
- The strongest layer needs to be at least 0.3 m thick to have a significant effect.

- The installed rock bolts need to be embedded in at least 0.3 m of the strongest layer.
- Generally, the higher into the immediate roof the strongest layer is located, the less its positive effect.

In the original CMRR, the SBADJ was determined using a table (Mark and Molinda, 2005). For improved accuracy, the following equation was derived using multiple regression:

$$SBADJ = \left[(0.72\ SBD)(THSB) - 2.5\right]\left[1 - 0.33(THWR) - 0.5\right] \tag{6.33}$$

where

- SBD is the Strong Bed Difference, the difference between the strong bed's Unit Rating and the thickness-weighted average of all the unit ratings within the bolted interval.
- THSB is the thickness of the strong bed (m).
- THWR is the thickness of the weak rock suspended from the strong bed (m).

Note that if the strong bed is at the top of the bolted interval, its full thickness is used in the calculation of the SBADJ (up to a maximum of 1.2 m).

- Other Adjustments
  - Number of Units

    When depositional processes change and deposit distinctly different material, there is generally, but not always, a sharp contact between units. Since gradational contacts do not weaken the roof, the characteristics of major bedding contact surfaces (cohesion and roughness) should be noted. The maximum deduction from the CMRR is 5 points when more than four weak contacts are present.
  - GroundWater Adjustment

    Groundwater is most prevalent in shallow mines, particularly beneath stream valleys, but it can also be introduced by leakage from pooled water in abandoned mines or fracturing of overlying aquifers during high extraction mining. The CMRR maintains the RMR system's rating scale, with a maximum deduction for flowing groundwater of 10 points.
  - Surcharge

    The strength of rocks overlying the bolted interval is only considered when they are significantly weaker than the rocks within it. An example was a Western mine where 1.2 m of relatively strong top coal was overlain by 6 m of weak, rooted claystone. Because the roof beam needed to carry some of the surcharge (extra weight) of the incompetent claystone, stability was reduced. The CMRR accounts for the surcharge with a 3-point deduction.

6.4 HOEK AND BROWN ROCK FAILURE CRITERIA

6.4.1 The Basic Criteria

Introduced in 1980, the Hoek–Brown failure criteria (Hoek and Brown, 1980a) has been widely adopted for modeling a wide range of rock types. The criterion was derived from the results of research into the brittle failure of intact rock by Hoek (1968) and studies on joint rock mass behavior by Brown (1970) that related geological conditions to observations. A version of the failure criterion for jointed rock masses, described by Hoek and Brown (1997), is as follows:

$$\sigma_1' = \sigma_3' + \sigma_{ci}\left[m_b(\sigma_3'/\sigma_{ci}) + s\right]^a \tag{6.36}$$

where

σ_1' is the major effective stress at failure.
σ_3' is the minor effective stress at failure.
σ_{ci} is the UCS of the intact rock material.
s and a are constants which depend upon the properties of the rock mass.

The input m_b is a reduced material constant of the Hoek–Brown constant m_i, both for the rock mass and expressed as

$$m_b = m_i \exp\left[(GSI - 100)/28\right] \tag{6.37}$$

where

GSI is the Geological Strength Index (GSI).

6.4.2 Determining the Hoek–Brown Criteria Parameters

The intact UCS σ_{ci}, and Hoek–Brown constant, m_i, are determined from triaxial tests. The tests should be carried out over a confining stress range from zero to one-half of the UCS. Five or more tests should be conducted, and the values are determined using the following equations:

$$\sigma^2_{ci} = (\Sigma y/n) - \{[\Sigma xy - (\Sigma x\Sigma y/n)]/[\Sigma x^2 - (\Sigma x)^2/n]\}(\Sigma x/n) \tag{6.38}$$

$$m_i = (1/\sigma_{ci})\{[\Sigma xy - (\Sigma x\Sigma y/n)]/[\Sigma x^2 - (\Sigma x)^2/n]\} \tag{6.39}$$

The accuracy of σ_{ci} and m_i is determined using coefficient determination, r^2, that is given as

$$r^2 = \left[\Sigma xy - (\Sigma x\Sigma y/n)\right]^2 / \left\{\left[\Sigma x^2 - (\Sigma x)^2/n\right]\left[\Sigma y^2 - (\Sigma y)^2/n\right]\right\} \tag{6.40}$$

where

n is the number of tests:

$$x = \sigma_3'$$

$$y = (\sigma_1' - \sigma_3')^2$$

6.4.3 Geological Strength Index (GSI)

The strength of jointed rock mass depends on the properties of the intact rock and on the ability of the intact rock to rotate and slide under different applied stress conditions. This freedom of movement is controlled by the geometric shape of the rock and the condition of surfaces between these pieces (Hoek and Brown, 1997). Hoek (1994) introduced the GSI to provide a system for estimating the reduction in strength for different geological conditions. The system is outlined in Figure 6.20, from which the GSI value can be determined.

Once the GSI has been estimated, the parameters s and a are determined using the following relationships:

For GSI > 25, i.e., rock masses of good reasonable quality (Hoek and Brown, 1997):

$$s = \exp\left[(GSI - 100)/9\right] \tag{6.41}$$

$$a = 0.5 \tag{6.42}$$

For GSI < 25, i.e., rockmasses of very poor quality:

$$s = 0 \tag{6.43}$$

$$a = 0.65 - GSI/200 \tag{6.44}$$

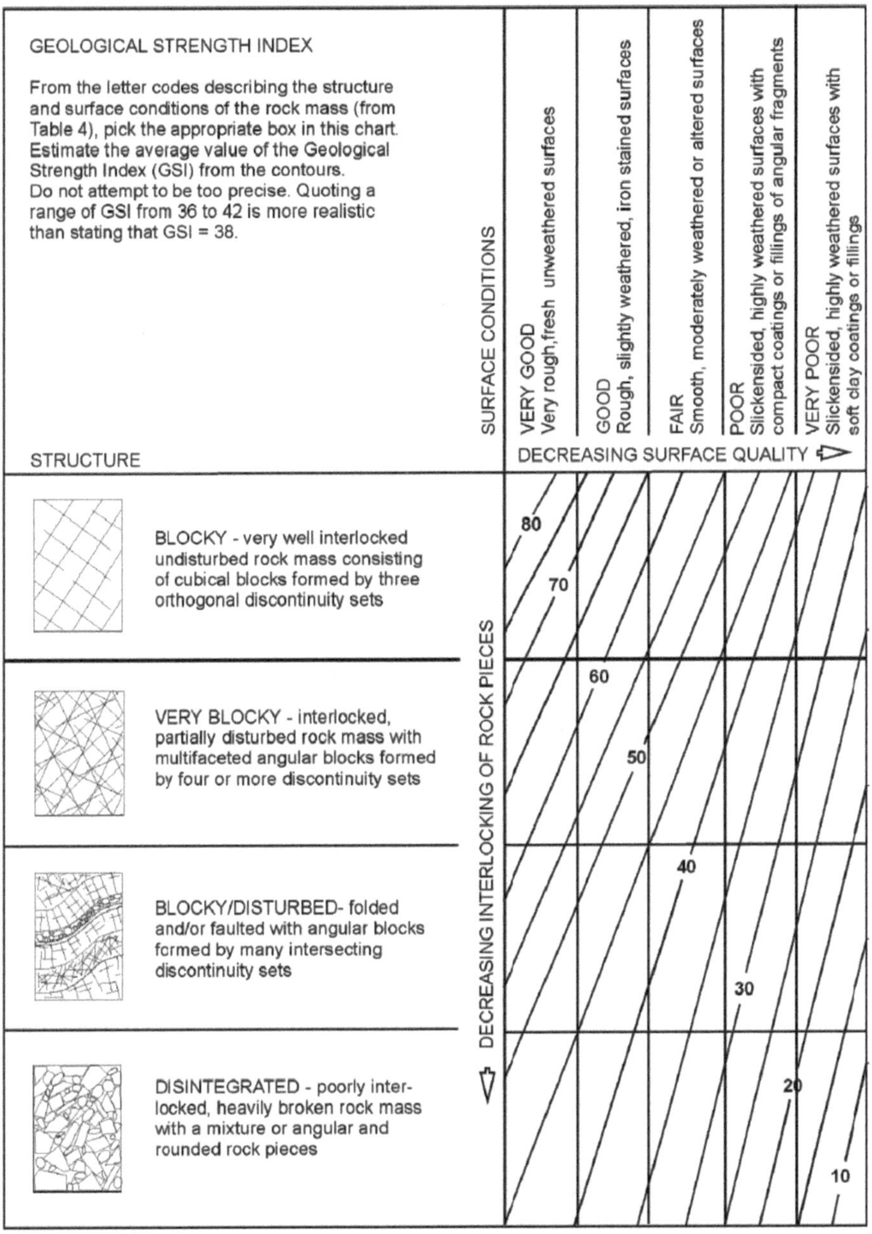

FIGURE 6.20 The geologic strength index (Marinos and Hoek, 2000).

6.4.3.1 Influence of Various Parameters on the Results

6.4.3.1.1 Effect of Water

The strength of rock decreases with increasing pore water pressure. Before using the Hoek–Brown criteria, it is necessary to determine the effective stress. The stress parameters in Equation (6.45) should be taken as the effective stress rather than absolute stress. The effective stress for a porous rock mass is determined using following relation:

$$\sigma' = \sigma - u \tag{6.45}$$

where

σ' is the effective stress.
σ is the stresses in rock mass.
u is the pore water pressure.

The pore water pressure is determined using piezometers or from manually constructed or numerically generated flow nets.

6.4.3.1.2 Effect of Sample Size

In general, there is significant reduction in the strength of a sample with increasing sample size. Hoek and Brown (1997) suggested that the UCS σ_{cd} of a rock specimen with a certain diameter in mm is related to the UCS of σ_{c50} of a 50 mm diameter sample by following equation:

$$\sigma_{cd} = \sigma_{c50}(50/d)^{0.18} \tag{6.46}$$

Figure 6.21 illustrates the situations when the Hoek–Brown failure criteria can be applied (Hoek and Brown, 1988). They suggested that these situations can be divided into three situations:

1. When the volume under consideration is small enough that it does not contain any structural discontinuities, the Hoek–Brown failure equation can be applied, using the m and s values for the intact rock. This condition would apply to small-scale specimens which has been extracted for laboratory testing or to the analysis of concentrated forces such as those which may be exerted by an individual pick on a tunnel-boring machine cutter head.
2. When the volume of rock being considered is such that only a few structural discontinues are contained in this volume, the Hoek–Brown failure criteria should not be used.

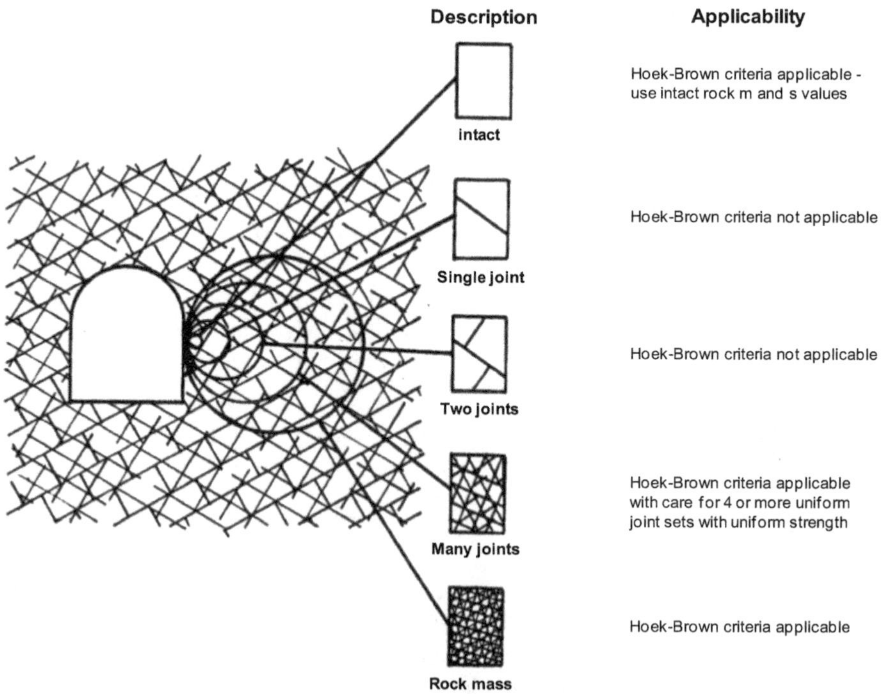

FIGURE 6.21 Applicability of the Hoek–Brown criteria to different scales of rock mass (Hoek and Brown, 1988).

The behavior of this rock is likely to be highly anisotropic, and the Hoek–Brown failure equation, which is only applicable to isotropic rock, will give erroneous results.

3. When the volume under consideration contains four or more closely spaced discontinuity sets and none of these discontinuity sets is weaker than any of the others, the Hoek–Brown criteria can be used. If one of the discontinuities is very weak as compared to the other, the rock mass should be treated as anisotropic and the Hoek–Brown criteria should not be applied unless allowance is made for this anisotropy.

Table 6.14 gives suggested values for the empirical parameters (Hoek and Brown, 1988).

6.5 WORKED EXAMPLES

Example 6.5.1

Assume a circular excavation is made in rock. If the radial stress is determined to be 0 MPa and the tangential stress is 20 MPa, find the stress concentration factors if the vertical stress is 10 MPa.

Traditionally, Stress Concentration Factors (SCF) are expressed in the form of σ_r/σ_v and σ_θ/σ_v.

Solution:

Radial Stress Concentration Factor

$$SCF = \sigma_r/\sigma_v$$

$$SCF = 0\ MPa/10\ MPa$$

$$SCF = 0$$

Tangential Stress Concentration Factor

$$SCF = \sigma_\theta/\sigma_v$$

$$SCF = 20\ MPa/10\ MPa$$

$$SCF = 2$$

Example 6.5.2

Assume that a circular excavation is made in rock. If the radial stress is determined to be 0 MPa and the tangential stress is determined to be –10 MPa, determine the SCF if the vertical stress is 10 MPa.

Solution:

Radial Stress Concentration Factor

$$SCF = \sigma_r/\sigma_v$$

$$SCF = 0\ MPa/10\ MPa$$

$$SCF = 0$$

Tangential Stress Concentration Factor

$$SCF = \sigma_\theta/\sigma_v$$

$$SCF = -10\ MPa/10\ MPa$$

$$SCF = -1$$

TABLE 6.14

Suggested Values for Hoek and Brown Constants (Hoek and Brown, 1988)

Disturbed Rock Mass m and s Values		**Undisturbed Rock Mass m and s Values**				
Empirical Failure Criterion $\sigma_1' = \sigma_3' + (m\sigma_c \sigma_3' + s\sigma_c^2)^{0.5}$ σ_1' = major principal effective stress σ_3' = minor principal effective stress σ_c = uniaxial compressive strength of the intact rock m and s are empirical constants		Carbonate Rocks with Well Developed Crystal Cleavage *dolomite, limestone, and marble*	Lithified Argilaceous Rock *mudstone, siltstone, shale, and slate (normal to cleavage)*	Arenaceous Rocks with Strong Crystals and Poorly Developed Crystal Cleavage *sandstone and quartzite*	Fine Grained Polymineralic Igneous Crystalline Rocks *andesite, dolerite, diabase, and rhyolite*	Coarse Grained Polymineralic Igneous & Metamorphic Crystalline Rocks amphibolite, gabbro gneiss, granite, norite, quartz-diorite
Intact Rock Samples Laboratory-size specimens free from discontinuities	m	7.00	10.00	15.00	17.00	25.00
CSIR rating: RMR = 100	s	1.00	1.00	1.00	1.00	1.00
NGI rating: Q = 500	m	7.00	10.00	15.00	17.00	25.00
	s	1.00	1.00	1.00	1.00	1.00
Very Good Quality Rock Mass						
Tightly interlocking undisturbed rock with	m	2.40	3.43	5.14	5.82	8.56
unweathered joints at 1–3 m	s	0.082	0.082	0.082	0.082	0.082
CSIR rating: RMR = 85	m	4.10	5.85	8.78	9.95	14.63
NGI rating: Q = 100	s	0.189	0.189	0.189	0.189	0.189
Good-Quality Rock Mass						
Fresh to slightly weathered rock, slightly	m	0.575	0.821	1.231	1.395	2.052
disturbed with joints at 1–3 m	s	0.00293	0.00293	0.00293	0.00293	0.00293
CSIR rating: RMR = 65	m	2.006	2.865	4.298	4.871	7.163
NGI rating: Q = 10	s	0.0205	0.0205	0.0205	0.0205	0.0205

(Continued)

TABLE 6.14 (*Continued*)
Suggested Values for Hoek and Brown Constants (Hoek and Brown, 1988)

Disturbed Rock Mass m and s Values		Undisturbed Rock Mass m and s Values				
Fair-Quality Rock Mass						
Several sets of moderately weathered joints	m	0.128	0.183	0.275	0.311	0.458
spaced at 0.3–1 m	s	0.00009	0.00009	0.00009	0.00009	0.00009
CSIR rating: RMR=44	*m*	0.947	1.353	2.030	2.301	3.383
NGI rating: Q=1	*s*	0.00198	0.00198	0.00198	0.00198	0.00198
Poor-Quality Rock Mass						
Numerous weathered joints at 30–500 mm,	m	0.029	0.041	0.061	0.069	0.102
some gouge. Clean compacted waste rock	s	0.000003	0.000003	0.000003	0.000003	0.000003
CSIR rating: RMR=23	m	0.447	0.639	0.959	1.087	1.598
NGI rating: Q=0.1	s	0.00019	0.00019	0.00019	0.00019	0.00019
Very Poor Quality Rock Mass						
Numerous heavily weathered joints spaced	m	0.007	0.010	0.015	0.017	0.025
<50 mm with gouge. Waste rock with	s	0.0000001	0.0000001	0.0000001	0.0000001	0.0000001
fines.	m	0.219	0.313	0.469	0.532	0.782
CSIR rating: RMR=3	s	0.00002	0.00002	0.00002	0.00002	0.00002
NGI rating: Q=0.01						

Example 6.5.3

If, at a point, an SCF for the tangential stress is determined to be –1 in one case, but +2 in another case, what does this suggest of the stress at this point if the vertical stress has a magnitude of 20 MPa?

Solution:

In the case where SCF = –1, σ_θ = 20 MPa. The magnitudes are equal; however, the sign is opposite. So, for a compressive vertical stress, the tangential stress at the point is tensile.

In the case where SCF = +2, σ_θ = 40 MPa. The magnitude of stress is double the vertical stress, and the sign is the same. So, for a compressive vertical stress, the tangential stress at the point is also compressive.

Example 6.5.4

Given the following conditions, for a circular excavation, determine the radial, tangential, and shear stress at a point.

$$\sigma_v = 10 \text{ MPa}$$

$$\sigma_h = 5 \text{ MPa}$$

$$a = 5 \text{ m}$$

$$r = 5 \text{ m}$$

$$\theta = 0°$$

Solution:

$$k = \sigma_h/\sigma_v = 5 \text{ MPa}/10 \text{ MPa} = 0.5$$

$$\sigma_r = (\sigma_v/2)[(1+k)\left(1-a^2/r^2\right)-(1-k)\left(1-4a^2/r^2+3a^4/r^4\right)\cos 2\theta]$$

$$\sigma_r = (10/2)\left[(1+0.5)\left(1-5^2/5^2\right)-(1-0.5)\left(1-4\left(5^2/5^2\right)+3\left(5^4/5^4\right)\right)\cos(0°)\right]$$

$$\sigma_r = (5)\left[(1.5)(1-1)-(0.5)(1-4+3)(1)\right]$$

$$\sigma_r = (5)\left[(1.5)(0)-(0.5)(0)(1)\right]$$

$$\sigma_r = 0 \text{ MPa}$$

$$\sigma_\theta = (\sigma_v/2)[(1+k)\left(1+a^2/r^2\right)+(1-k)\left(1+3a^4/r^4\right)\cos 2\theta]$$

$$\sigma_0 = (10/2)\left[(1+0.5)\left(1+5^2/5^2\right)+(1-0.5)\left(1+3\left(5^4/5^4\right)\right)\cos(0°)\right]$$

$$\sigma_0 = (5)\left[(1.5)(1+1)+(0.5)(1+3)(1)\right]$$

$$\sigma_0 = (5)[3+2]$$

$$\sigma_0 = 25 \text{ MPa}$$

$$\tau_{r\theta} = (\sigma_v/2)[(1-k)\left(1+2a^2/r^2-3a^4/r^4\right)\sin 2\theta]$$

$$\tau_{r\theta} = (10/2)\left[(1-0.5)\left(1+2\left(5^2/5^2\right)-3\left(5^4/5^4\right)\right)\sin(0°)\right]$$

$$\tau_{r\theta} = (5)\left[(0.5)(1+2-3)(0)\right]$$

$$\tau_{r\theta} = 0 \text{ MPa}$$

Example 6.5.5

Repeat the previous example. Note the change of a and r. Comment on the results and their significance.

$$\sigma_v = 10 \text{ MPa}$$

$$\sigma_h = 5 \text{ MPa}$$

$$a = 10 \text{ m}$$

$$r = 10 \text{ m}$$

$$\theta = 0°$$

Solution:

$$k = \sigma_h/\sigma_v = 5 \text{ MPa}/10 \text{ MPa} = 0.5$$

$$\sigma_r = (\sigma_v/2)[(1+k)\left(1-a^2/r^2\right)-(1-k)\left(1-4a^2/r^2+3a^4/r^4\right)\cos 2\theta]$$

$$\sigma_r = (10/2)\left[(1+0.5)\left(1-10^2/10^2\right)-(1-0.5)\left(1-4\left(10^2/10^2\right)+3\left(10^4/10^4\right)\right)\cos(0°)\right]$$

$$\sigma_r = (5)\left[(1.5)(1-1)-(0.5)(1-4+3)(1)\right]$$

$$\sigma_r = (5)\left[(1.5)(0)-(0.5)(0)(1)\right]$$

$$\sigma_r = 0 \text{ MPa}$$

$$\sigma_\theta = (\sigma_v/2)[(1+k)\left(1+a^2/r^2\right)+(1-k)\left(1+3a^4/r^4\right)\cos 2\theta]$$

$$\sigma_\theta = (10/2)\left[(1+0.5)\left(1+10^2/10^2\right)+(1-0.5)\left(1+3\left(10^4/10^4\right)\right)\cos(0°)\right]$$

$$\sigma_\theta = (5)\left[(1.5)(1+1)+(0.5)(1+3)(1)\right]$$

$$\sigma_0 = (5)[3+2]$$

$$\sigma_0 = 25 \text{ MPa}$$

$$\tau_{r\theta} = (\sigma_v/2)[(1-k)\left(1+2a^2/r^2-3a^4/r^4\right)\sin 2\theta]$$

$$\tau_{r\theta} = (10/2)\left[(1-0.5)\left(1+2\left(10^2/10^2\right)-3\left(10^4/10^4\right)\right)\sin(0°)\right]$$

$$\tau_{r\theta} = (5)\left[(0.5)(1+2-3)(0)\right]$$

$$\tau_{r\theta} = 0 \text{ MPa}$$

The answers are the same as in Example 6.5.4 because the ratio a/r, being a dimensionless parameter, assures that the magnitude of stress in the rock surrounding the excavation is not only affected by the size of the excavation (a) but also the distance from the center (r). This results in identical SCF for a 5 m excavation as a 10 m excavation, at the sidewall of the opening.

Example 6.5.6

If the vertical stress is determined to be 50 MPa, find the tangential stress at the sidewall for the following horizontal to vertical stress ratios. Provide the SCF.

$$k = 0$$

$$k = 0.25$$

$$k = 0.33$$

$$k = 0.50$$

$$k = 1.0$$

Solution:

$$k = 0$$

$$\sigma_\theta = \sigma_v(3 - k)$$

$$\sigma_\theta = 50(3)$$

$$\sigma_\theta = 150 \text{ MPa}$$

$$SCF = \sigma_\theta/\sigma_v$$

$$SCF = 150/50$$

$$SCF = 3$$

$$k = 0.25$$

$$\sigma_\theta = \sigma_v(3 - 0.25)$$

$$\sigma_0 = 50(2.75)$$

$$\sigma_\theta = 137.5 \text{ MPa}$$

$$SCF = \sigma_\theta/\sigma_v$$

$$SCF = 137.5/50$$

$$SCF = 2.75$$

$$k = 0.33$$

$$\sigma_\theta = \sigma_v(3 - 0.33)$$

$$\sigma_\theta = 50(2.67)$$

$$\sigma_0 = 133.5 \text{ MPa}$$

$$SCF = \sigma_\theta/\sigma_v$$

$$SCF = 133.5/50$$

$$SCF = 2.67$$

$$k = 0.50$$

$$\sigma_\theta = \sigma_v(3 - 0.5)$$

$$\sigma_\theta = 50(2.5)$$

$$\sigma_\theta = 125 \text{ MPa}$$

$$SCF = \sigma_\theta/\sigma_v$$

$$SCF = 125/50$$

$$SCF = 2.5$$
$$k = 1.0$$
$$\sigma_\theta = \sigma_v(3-1)$$
$$\sigma_\theta = 50(2)$$
$$\sigma_\theta = 100 \text{ MPa}$$
$$SCF = \sigma_\theta/\sigma_v$$
$$SCF = 100/50$$
$$SCF = 2$$

Example 6.5.7

Is the change in k and SCF linear at the sidewall of a circular opening?

Solution:

It can be seen from the equation, $\sigma_\theta=\sigma_v(3-k)$, that the relationship is linear. A plot also confirms the relationship (Figure 6.22).

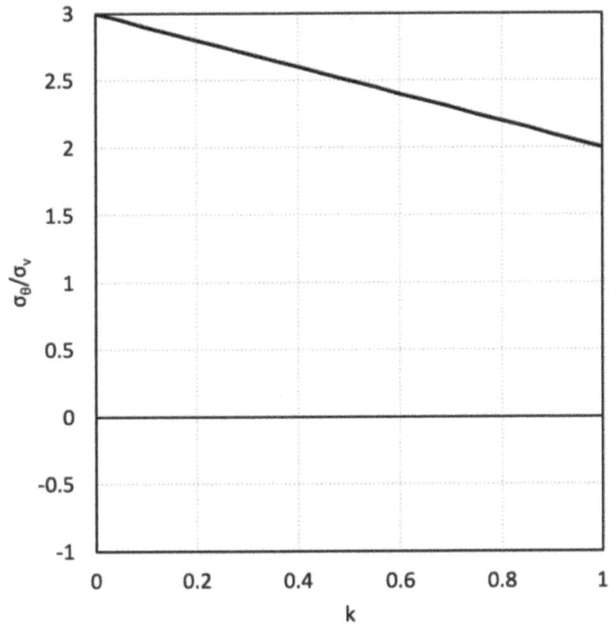

FIGURE 6.22 Plot of SCF versus k in the sidewall.

Example 6.5.8

If the vertical stress is determined to be 50 MPa, determine the tangential stress at the crown for the following horizontal to vertical stress ratios. Provide the SCF with the following k values:

$$k = 0$$
$$k = 0.25$$

$$k = 0.33$$
$$k = 0.50$$
$$k = 1.0$$

Solutions:

$$k = 0$$
$$\sigma_\theta = \sigma_v(3k - 1)$$
$$\sigma_\theta = 50[3(0) - 1]$$
$$\sigma_\theta = -50 \text{ MPa}$$
$$SCF = \sigma_\theta/\sigma_v$$
$$SCF = -50/50$$
$$SCF = -1$$
$$k = 0.25$$
$$\sigma_\theta = \sigma_v(3k - 1)$$
$$\sigma_0 = 50[3(0.25) - 1]$$
$$\sigma_\theta = -12.5 \text{ MPa}$$
$$SCF = \sigma_\theta/\sigma_v$$
$$SCF = -12.5/50$$
$$SCF = -0.25$$
$$k = 0.33$$
$$\sigma_\theta = \sigma_v(3k - 1)$$
$$\sigma_\theta = 50[3(0.33) - 1]$$
$$\sigma_\theta = -0.5 \text{ MPa}$$
$$SCF = \sigma_\theta/\sigma_v$$
$$SCF = -0.5/50$$
$$SCF = -0.01$$
$$k = 0.5$$
$$\sigma_\theta = \sigma_v(3k - 1)$$
$$\sigma_0 = 50[3(0.5) - 1]$$
$$\sigma_0 = 25 \text{ MPa}$$
$$SCF = \sigma_\theta/\sigma_v$$
$$SCF = 25/50$$
$$SCF = 0.5$$
$$k = 1.0$$
$$\sigma_\theta = \sigma_v(3k - 1)$$
$$\sigma_\theta = 50[3(1.0) - 1]$$
$$\sigma_0 = 100 \text{ MPa}$$
$$SCF = \sigma_0/\sigma_v$$
$$SCF = 100/50$$
$$SCF = 2.0$$

Example 6.5.9

Is the change in k and SCF linear at the crown?

Solution:

It can be seen from the equation, $\sigma_\theta = \sigma_v(3k - 1)$, that the relationship is linear.
A plot again confirms the relationship (Figure 6.23).

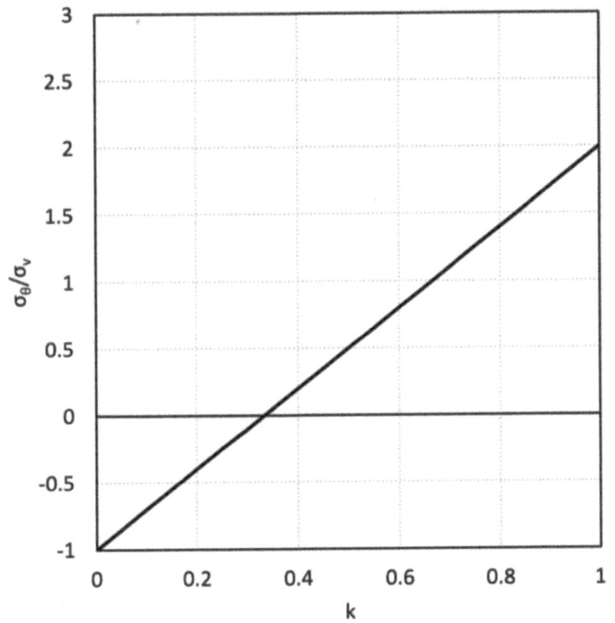

FIGURE 6.23 Plot of SCF versus k in the crown.

Example 6.5.10

Provide a plot of the SCF at the sidewall of a circular excavation for a distance of four times the radius of the opening. The radius (a) is 5 m. The plot should have the SCF on the y-axis and the ratio (r/a) on the x-axis. Assume a vertical stress of 3,000 psi and a value of k=1.0. As a reference, two calculations at r/a=4 have been completed (Figure 6.24).

Solution:

$$\sigma_r = (\sigma_v/2)[(1+k)(1-a^2/r^2) - (1-k)(1-4a^2/r^2+3a^4/r^4)\cos 2\theta]$$

$$\sigma_r = (3{,}000/2)\left[(1+1)(1-5^2/20^2) - (1-1)\left(1-4*(5)^2/(20)^2+3*(5)^4/(20)^4\right)\cos 0\right]$$

$$\sigma_r = 1{,}500\left[(2)(1-25/400)\right]$$

$$\sigma_r = 2{,}813 \text{ psi}$$

$$\text{SCF} = 2{,}813/3{,}000 = 0.94$$

$$\sigma_\theta = (\sigma_v/2)[(1+k)(1+a^2/r^2) + (1-k)(1+3a^4/r^4)\cos 2\theta]$$

$$\sigma_\theta = (3{,}000/2)\left[(1+1)\left(1+(5)^2/(20)^2\right) + (1-1)\left(1+3*(5)^4/(20)^4\right)\cos 0\right]$$

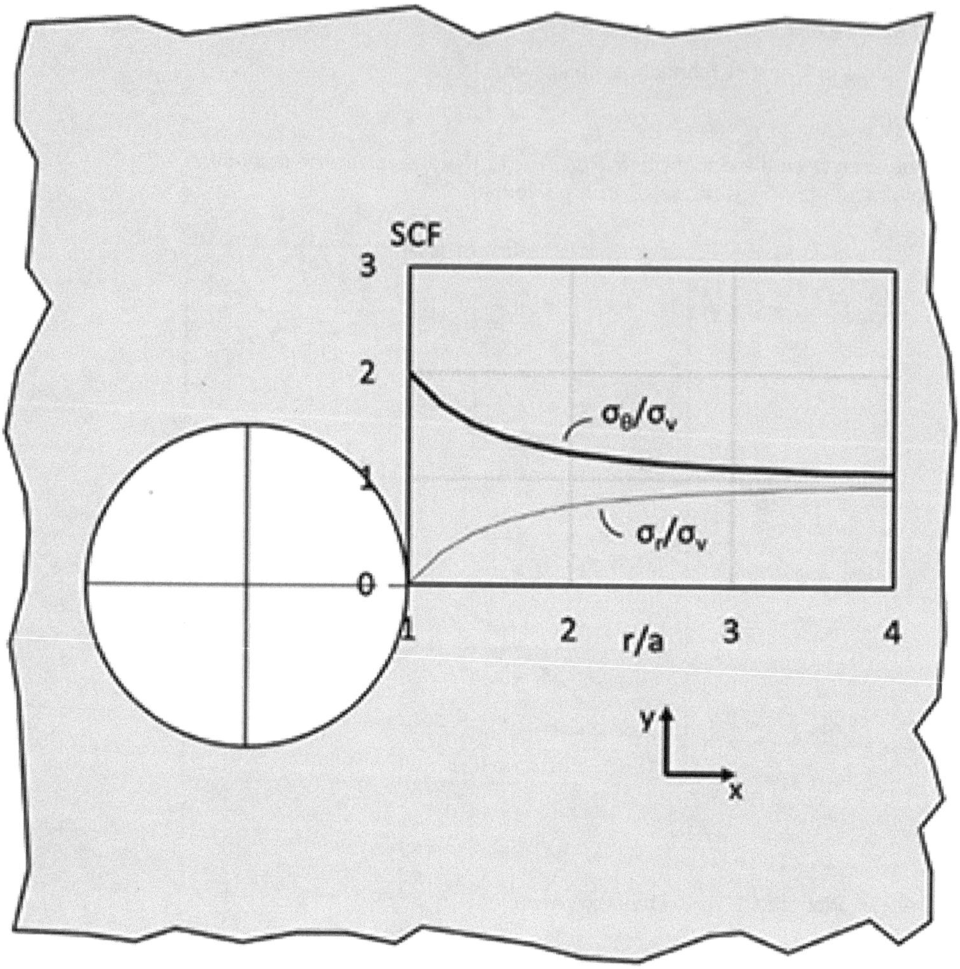

FIGURE 6.24 Plot of sidewall stress concentrations.

$$\sigma_\theta = (1{,}500)\left[(2)(1+25/400)\right]$$

$$\sigma_\theta = 3{,}188 \text{ psi}$$

$$\text{SCF} = 3{,}188/3{,}000 = 1.06$$

Example 6.5.11

Repeat the previous example, but produce plots for the following values of k (Figure 6.25).

$$k = 0$$

$$k = 0.33$$

$$k = 0.50$$

$$k = 1.0$$

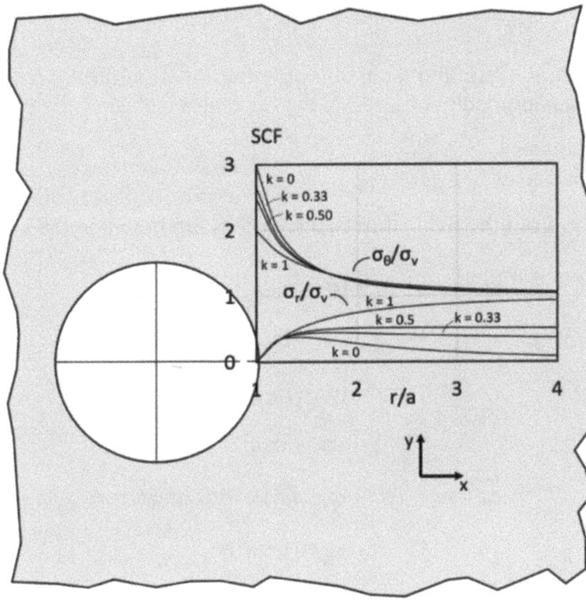

FIGURE 6.25 Plot of sidewall concentrations as a function of k.

Example 6.5.12

Determine the stresses around a circular opening for a hydrostatic stress field. Let $\theta=45°$, $a=3$ m, and $r=18$ m.

Solution:

For a case of hydrostatic stress, $k=1$.

$$\sigma_r = (\sigma_v/2)[(1+k)(1-a^2/r^2)-(1-k)(1-4a^2/r^2+3a^4/r^4)\cos 2\theta]$$

$$\sigma_r = (\sigma_v/2)\left[(1+1)(1-3^2/18^2)-(1-1)(1-4*3^2/18^2+3*3^4/18^4)\cos 90°\right]$$

$$\sigma_r = (\sigma_v/2)\left[(2)(1-3^2/18^2)\right]$$

$$\sigma_r = 0.97(\sigma_v)$$

$$\sigma_\theta = (\sigma_v/2)[(1+k)(1+a^2/r^2)+(1-k)(1+3a^4/r^4)\cos 2\theta]$$

$$\sigma_\theta = (\sigma_v/2)\left[(1+1)(1+3^2/18^2)+(1-1)(1+3*3^4/r^4)\cos 90°\right]$$

$$\sigma_\theta = (\sigma_v/2)\left[(2)(1+3^2/18^2)\right]$$

$$\sigma_\theta = 1.0(\sigma_v)$$

$$\tau_{r\theta} = (\sigma_v/2)[(1-k)(1+2a^2/r^2-3a^4/r^4)\sin 2\theta]$$

$$\tau_{r\theta} = (\sigma_v/2)\left[(1-1)(1+2*3^2/18^2-3*3^4/18^4)\sin 90°\right]$$

$$\tau_{r\theta} = 0$$

Example 6.5.13

Assume a hydrostatic stress field and a circular opening. At what distance from the center of the opening, do the stresses approach within ±5% the field stresses? After the content found in Brady and Brown (2004).

Solution:

Stresses approach ±5% from the field stresses if σ_r and σ_θ are between 0.95 σ_v and 1.05 σ_v:

$$\sigma_r = (\sigma_v)\left(1 - a^2/r^2\right)$$

$$\sigma_r = \sigma_v - \sigma_v a^2/r^2$$

$$\sigma_\theta = (\sigma_v)\left(1 + a^2/r^2\right)$$

$$\sigma_\theta = \sigma_v + \sigma_v a^2/r^2$$

$$\sigma_\theta - \sigma_r = (\sigma_v + \sigma_v a^2/r^2) - (\sigma_v - \sigma_v a^2/r^2)$$

$$\sigma_\theta - \sigma_r = 2\sigma_v a^2/r^2$$

$\sigma_\theta - \sigma_r$ must approach a value of $0.1\sigma_v$, and therefore,

$$0.1\sigma_v = 2\sigma_v a^2/r^2$$

$$0.1 = 2a^2/r^2$$

$$0.05 = a^2/r^2$$

Preferably, a distance r from the center of the opening is needed, and therefore,

$$0.05^{1/2} = \left(a^2/r^2\right)^{1/2}$$

$$0.224 = a/r$$

$$0.224r = a \text{ or}$$

$$r = 4.5\ a$$

The distance between opening centers is therefore r=4.5a+1a=5.5a.

From Brady and Brown (2004), the general rule of thumb for these conditions is a distance between centers of D≥6a.

Example 6.5.14

Assume that two circular openings are to be excavated. The stress is to be assumed hydrostatic. The first opening has a radius of 4 m, and the second has a radius of 8 m. The edge-to-edge distance between the openings is 20 m. Determine if one opening will influence the stresses of the other.

Solution:

From Example 6.5.13, the rule of thumb was determined, according to Brady and Brown (2004), to be D≥6a between centers to minimize stress influence on an adjacent excavation. Thus, for the smaller opening where a=4 m, D≥24 m. Therefore, the smaller opening should not affect the larger opening because the distance between centers is 32 m.

However, since the diameter of the second opening is a=8 m, then, D≥48 m. Because the distance between centers is 32 m, which is less than the minimum required distance of 48 m, the second opening introduces a stress interaction between the two openings.

Example 6.4.15

A circular opening is to be excavated as in Figure 6.26. If the tensile strength is zero and the crack initiation stress, σ_{ci}, is 10 MPa, determine the regions around the excavation periphery where compressive and tensile fracture could initiate. This example and the process for solving it are found from a similar example in Brady and Brown (2004). Note that the stress at which stable crack propagation initiates is considered the crack initiation stress (σ_{ci}).

Solution:

$$\sigma_\theta = (\sigma_v/2)[(1+k)\left(1+a^2/r^2\right)+(1-k)\left(1+3a^4/r^4\right)\cos 2\theta]$$

$$\sigma_\theta = (9/2)\left[(1+0.3)(1+1)+(1-0.3)(1+3(1))\cos 2\theta\right]$$

$$\sigma_\theta = 4.5(2.6+2.8\cos 2\theta)$$

For compressive failure, then

$$4.5(2.6+2.8\cos 2\theta) \geq 10 \text{ MPa}$$

Ranges of θ which satisfy this condition are
Compressive failure at $-49° \leq \theta \leq 49°$ and $131° \leq \theta \leq 229°$ (Figure 6.27).
For tensile failure, then

$$4.5(2.6+2.8\cos 2\theta) \leq 0 \text{ MPa}$$

Ranges of θ which satisfy this condition are
Tensile failure at $79° \leq \theta \leq 101°$ and $259° \leq \theta \leq 281°$.

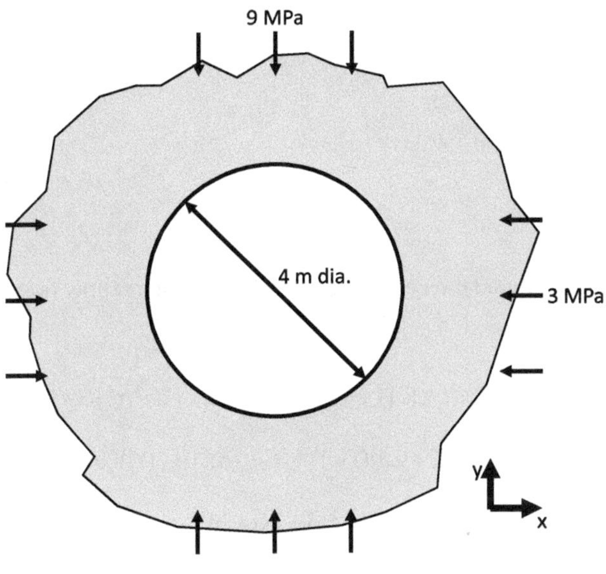

FIGURE 6.26 Sketch of the problem.

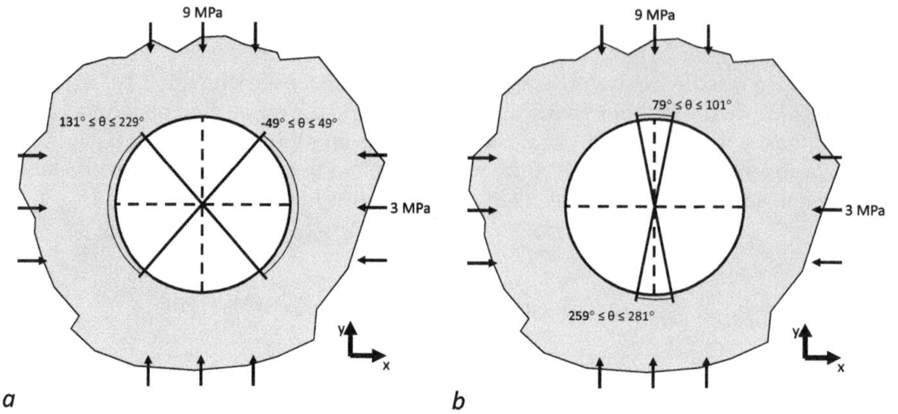

FIGURE 6.27 Zones compressive failure (a) and zones of tensile failure (b).

Example 6.5.16

If k=1 for Example 6.5.15 and the vertical stress remains constant, how does the solution change?

Solution:

$$\sigma_\theta = (\sigma_v/2)[(1+k)(1+a^2/r^2)+(1-k)(1+3a^4/r^4)\cos 2\theta]$$

$$\sigma_\theta = (9/2)\left[(1+1)(1+1)+(1-1)(1+3(1))\cos 2\theta\right]$$

$$\sigma_\theta = 4.5(4)$$

$$\sigma_\theta = 18 \text{ MPa}$$

Therefore,
$\sigma_\theta \geq 10$ MPa for all values of θ

Example 6.5.17

If k=0.5 for Example 6.5.15, and the vertical stress is held constant, how does the answer change?

Solution:

$$\sigma_\theta = (\sigma_v/2)[(1+k)(1+a^2/r^2)+(1-k)(1+3a^4/r^4)\cos 2\theta]$$

$$\sigma_\theta = (9/2)\left[(1+0.5)(1+1)+(1-0.5)(1+3(1))\cos 2\theta\right]$$

$$\sigma_\theta = 4.5(3+2\cos 2\theta)$$

For compressive failure, then

$$4.5(3+2\cos 2\theta) \geq 10 \text{ MPa}$$

Ranges of θ which satisfy this condition are
Compressive failure at $-56° \le \theta \le 56°$ and $124° \le \theta \le 236°$.
For tensile failure, then

$$4.5(3 + 2\cos 2\theta) \le 0 \text{ MPa}*$$

*Notice that no values of θ will satisfy this condition as cosine oscillates between 1 and –1; thus, no range exists less than 0 MPa.

Example 6.5.18

If the field stress of Example 6.5.15 is σ_{xx}=–4.5 MPa and the vertical stress is held constant, how does the answer change?

Solution:

$$\sigma_\theta = (\sigma_v/2)[(1+k)(1+a^2/r^2)+(1-k)(1+3a^4/r^4)\cos 2\theta]$$

$$\sigma_\theta = (9/2)\left[(1-0.5)(1+1)+(1+0.5)(1+3(1))\cos 2\theta\right]$$

$$\sigma_0 = 4.5(1+6\cos 2\theta)$$

For compressive failure, then

$$4.5(1+6\cos 2\theta) \ge 10 \text{ MPa}$$

Ranges of θ which satisfy this condition are
Compressive failure at $-39° \le \theta \le 39°$ and $141° \le \theta \le 219°$
For tensile failure, then

$$4.5(1+6\cos 2\theta) \le 0 \text{ MPa}$$

Ranges of θ which satisfy this condition are
Tensile failure at $50° \le \theta \le 130°$ and $230° \le \theta \le 310°$

Example 6.5.19

An elliptical excavation will have a height of 2 m and a width of 4 m. If the vertical stress is 8 MPa and k=0.5, what are the stresses at the crown and sidewall?

$$\rho_A = \left(H^2/2W\right)$$

$$\rho_A = \left(2^2/2(4)\right)$$

$$\rho_A = 0.5$$

$$\rho_B = \left(W^2/2H\right)$$

$$\rho_B = \left(4^2/2(2)\right)$$

$$\rho_B = 4$$

$$\sigma_A = p[1-k + (2W/\rho_A)^{0.5}]$$

$$\sigma_A = 8\left\{1 - 0.5 + \left[2(4)/0.5\right]^{0.5}\right\}$$

$$\sigma_A = 36 \text{ MPa}$$

$$\sigma_B = p[k - 1 + k(2H/\rho_B)^{0.5}]$$

$$\sigma_B = 8\left\{0.5 - 1 + 0.5\left[2(2)/(4)\right]^{0.5}\right\}$$

$$\sigma_B = 0 \text{ MPa}$$

Example 6.5.20

Three circular excavations are proposed as shown in Figure 6.28. The diameters are a=2.3 units, b=3 units, and c=2 units. The x, y centers for each excavation are in Figure 6.28. Determine any potential zones of stress interaction due to these excavations.

Solution:

Drawing to scale, the zones of potential stress interaction can be readily identified with the rule of thumb of 6a from each excavations center (marked by the dotted line for each excavation). Excavation a will influence part of excavation b and all of excavation c. Excavation b will influence all of excavation a and c. Excavation c will influence part of excavation a and b. Hence, stress interactions should be expected (Figure 6.29).

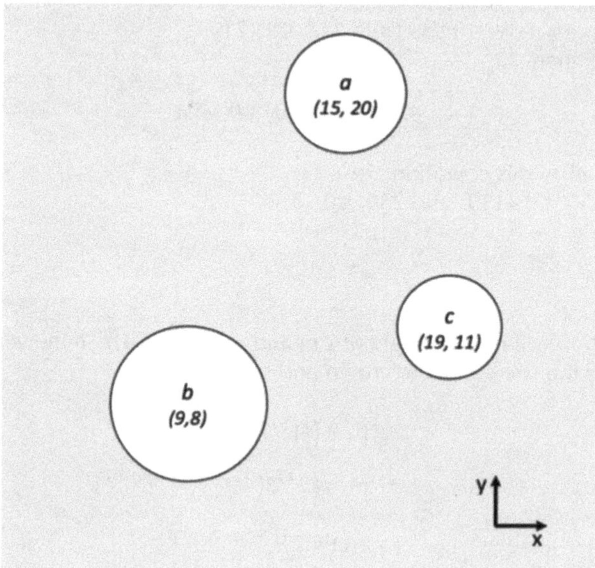

FIGURE 6.28 Section of the problem.

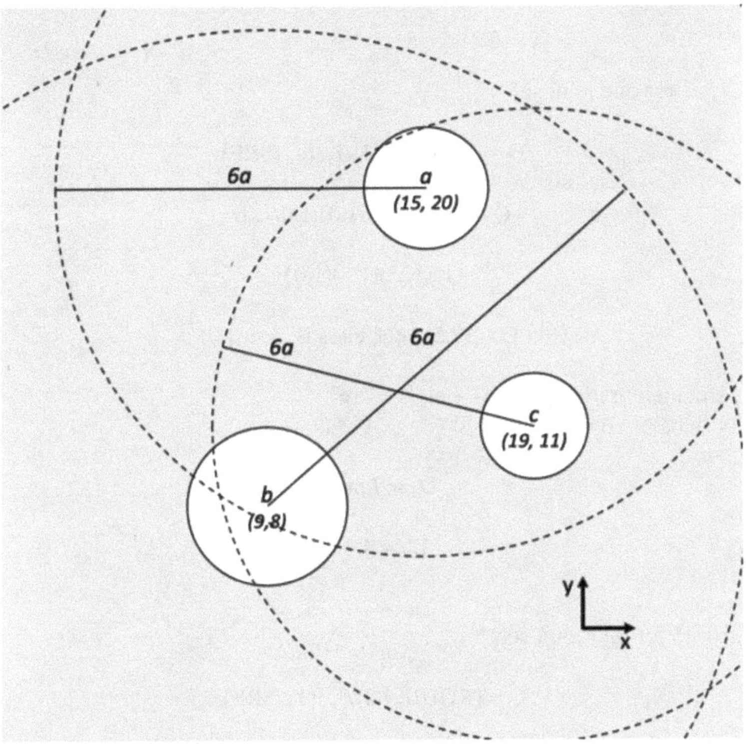

FIGURE 6.29 The areas of influence from each of the three excavations.

Example 6.5.21

Determine the Q value for good rock with an RQD of 75 and given the following information. Categorize the rock class based on Figure 6.13.

Rock Mass Quality, $Q=(RQD/J_n)(J_r/J_a)(J_w/SRF)$

RQD=75 (good rock)

$J_n=2$ (Category B/one joint set)

$J_r=1.5$ (Category E/rough irregular, planar)

$J_a=1.0$ (Category B/unaltered joint walls, surface staining only)

$J_w=1.0$ (Category A/dry or minor inflow)

SRF=1.0 (Category G/medium stress, favorable stress condition)

Solution:

$$Q = (75/2)(1.5/1.0)(1.0/1.0)$$

$$Q = (37.5)(1.5)(1.0)$$

$$Q = 56(\text{Rock class A, very good})$$

Example 6.5.22

Based on the previous example, if the approximations J_r/J_a and J_w/SRF remain unchanged, determine if shotcrete needs to be considered if the RQD is rated at 50 with two possible degrees of jointing, one joint set, and four joint sets. Assume that the span will be a permanent mine opening with a height of 7 m.

Solution:

Case 1

RQD=50, J_n=2 for one joint set

$$Q = (RQD/J_n)(J_r/J_a)(J_w/SRF)$$

$$Q = (50/2)(1.5/1.0)(1.0/1.0)$$

$$Q = (25)(1.5)(1.0)$$

$$Q = 37.5(\text{Rock class B, good})$$

ESR=1.6 for permanent mine openings, height=7 m

D_e=Span or height in m/ESR

$$D_e = 7\text{ m}/1.6$$

$$D_e = 4.4\text{ m}$$

Case 2

RQD=50, J_n=15 for four joint sets

$$Q = (RQD/J_n)(J_r/J_a)(J_w/SRF)$$

$$Q = (50/15)(1.5/1.0)(1.0/1.0)$$

$$Q = (3.3)(1.5)(1.0)$$

$$Q = 5(\text{Rock class C, fair})$$

ESR=1.6 for permanent mine openings, height=7 m

D_e=Span or height in m/ESR

$$D_e = 7\text{ m}/1.6$$

$$D_e = 4.4\text{ m}$$

As shown in Figure 6.30, only Case 2 approaches category 4 where shotcrete with systematic bolting is recommended. Depending on site conditions, shotcrete may need to be implemented.

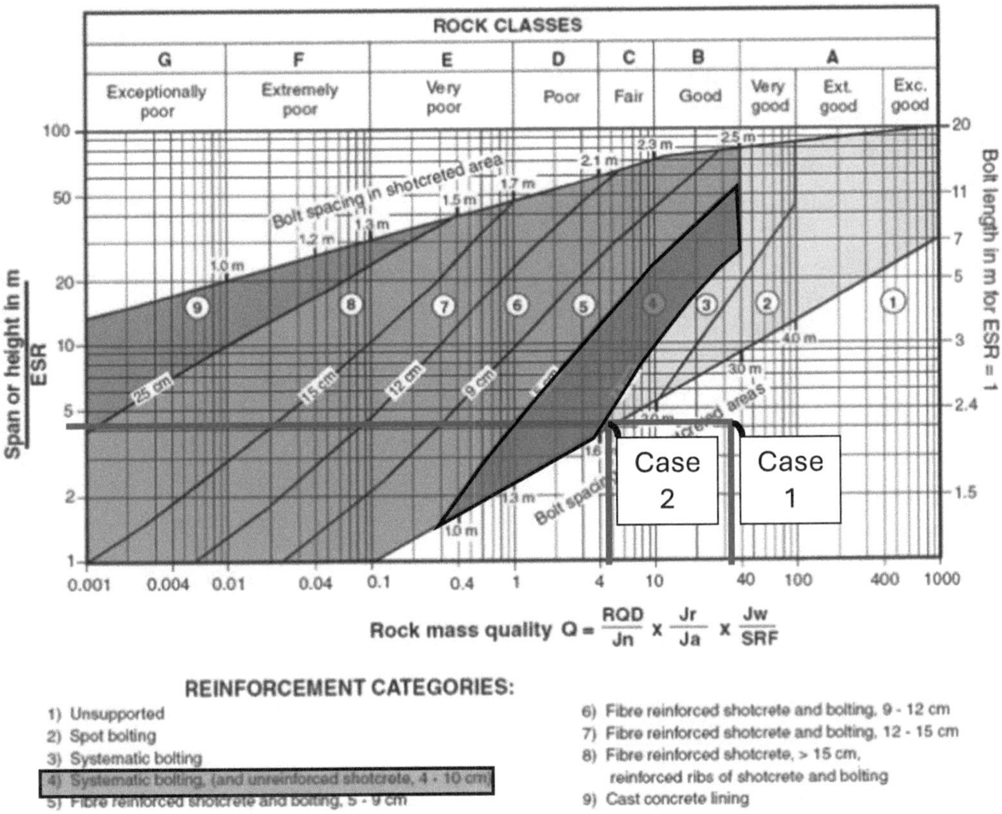

FIGURE 6.30 Selecting reinforcement with the Q system (Barton et al., 1974).

Example 6.5.23

Find the Hoek–Brown constant m_i, the UCS σ_{ci}, and the coefficient of determination r^2 based on the data from five triaxial tests with results below. Calculations for row one are shown. Complete the table of calculations. An example similar to this is found in the Appendix of Hoek and Brown (1997) (Tables 6.15 and 6.16).

TABLE 6.15
Triaxial Test Data (Incomplete Table)

	x		$y = (\sigma_1' - \sigma_3')^2$	xy	x^2	y^2
	σ_3'	σ_1'				
	0	40	$(40 - 0)^2 = 1{,}600$	$(1{,}600)(0) = 0$	0	2,560,000
	6	60	–	–	–	–
	12	70	–	–	–	–
	18	100	–	–	–	–
	24	120	–	–	–	–
Sum	–	–	–	–	–	–

TABLE 6.16
Triaxial Test Data (Complete Table)

	x		$y = (\sigma_1' - \sigma_3')^2$	xy	x^2	y^2
	σ_3'	σ_1'				
	0	40	1,600	0	0	2,560,000
	6	60	2,916	17,496	36	8,503,056
	12	70	3,364	40,368	144	11,316,496
	18	100	6,724	121,032	324	45,212,176
	24	120	9,216	221,184	576	84,934,656
Sum	60	390	23,820	400,080	1,080	152,526,384

n is the number of tests

$$x = \sigma_3'$$

$$y = (\sigma_1' - \sigma_3')^2$$

Solution:

$$\sigma^2_{ci} = (\Sigma y/n) - \{[\Sigma xy - (\Sigma x \Sigma y/n)]/[\Sigma x^2 - (\Sigma x)^2/n]\}(\Sigma x/n)$$

$$\sigma^2_{ci} = (23{,}820/5) - \{[400{,}080 - (60)(23{,}820)/5]/[1{,}080 - (60)^2/5]\}(60/5)$$

$$\sigma^2_{ci} = (4{,}764) - \{[400{,}080 - 285{,}840]/[1{,}080 - 720]\}(12)$$

$$\sigma^2_{ci} = (4{,}764) - \{[114{,}240]/[360]\}(12)$$

$$\sigma^2_{ci} = 4{,}764 - 3{,}808$$

$$\sigma_{ci} = 31\text{ MPa}$$

$$m_i = (1/\sigma_{ci})\{[\Sigma xy - (\Sigma x \Sigma y/n)]/[\Sigma x^2 - (\Sigma x)^2/n]\}$$

$$m_i = (1/31)\{[400{,}080 - (60)(23{,}820)/5]/[1{,}080 - (60)^2/5]\}$$

$$m_i = (1/31)\{[114240]/[360]\}$$

$$m_i = 10.2$$

$$r^2 = [\Sigma xy - (\Sigma x \Sigma y/n)]^2/\{[\Sigma x^2 - (\Sigma x)^2/n][\Sigma y^2 - (\Sigma y)^2/n]\}$$

$$r^2 = [(400{,}080 - (60)(23{,}820)/5)]^2/\{[1{,}080 - (60)^2/5][152{,}526{,}384 - (23{,}820)^2/5]\}$$

$$r^2 = [1.31 \times 10^{10}]/\{[360][39047904]\}$$

$$r^2 = [1.31 \times 10^{10}]/[1.41 \times 10^{10}]$$

$$r^2 = 0.932$$

Example 6.5.24

Estimate the Hoek–Brown constants m_b, s, and a if $m_i = 11.3$ and the rock is described by a geologist as very blocky with a smooth surface condition. Use Figure 6.20 to estimate the GSI.

If the same rock is considered weathered and disturbed by blasting, estimate the Hoek–Brown constant m_b. See Hoek and Brown (1997) for additional information. Note that GSI can be reflected as a range rather than a precise value. For the sake of the example, an average will be used.

Using the contours of Figure 6.20 for very blocky structure with smooth surface condition, the GSI can be roughly estimated from 37 to 54.

Solution:

For GSI = 37,

$$m_b = m_i \exp\left[(GSI - 100)/28\right]$$

$$m_b = (11.3)\exp\left[(37 - 100)/28\right]$$

$$m_b = 1.2$$

For GSI > 25,

$$s = \exp\left[(GSI - 100)/9\right]$$

$$s = \exp\left[(37 - 100)/9\right]$$

$$s = 9.1 \times 10^{-4}$$

$$a = 0.5$$

According to Hoek and Brown (1997), weathering and blasting could reduce the GSI by as much as one column and one row. Thus, the GSI could be instead represented by Very Blocky/Disturbed and Poor GSI = 22–37.

Example 6.5.25

The following example, in part, can be found in Barton et al. (1974) for a 20 m span machine hall in phyllite. The following rock mass classification is given, and determine the Q value. The unconfined compressive strength of phyllite is 400 kg/cm^2, the major principal stress is 30 kg/cm^2, and the minor principal stress is 10 kg/cm^2.
Given:
Minor water inflow: $J_w = 1.0$

Solution:

Joint Set 1
Strongly developed foliation
Smooth-planar ($J_r = 1$)
Chlorite coatings ($J_a = 4$)
15 joints/meter
Joint Set 2
Smooth undulating ($J_r = 2$)
Slightly altered joint walls ($J_a = 2$)
5 joints/meter
J_v = total number of joints per m^3 (note that RQD = 100 for $J_v < 4.5$)

$$J_v = 5 + 15 = 20$$

$$RQD = 115 - 3.3J_v$$

Note this does not reflect the updated approach by Palmström (2005).

$$RQD = 115 - 3.3(20) = 49$$

$J_n = 4$ (for two joint sets)

$$\sigma_1/\sigma_3 = 30\ kg/cm^2/10\ kg/cm^2 = 3.0$$

$$\sigma_c/\sigma_3 = 13.3$$

SRF for medium stress = 1.0

For J_r/J_a, the minimum is used, $J_r/J_a = ¼$.

$Q = (RQD/J_n)(J_r/J_a)(J_w/SRF)$

$Q = (49/4)(1/4)(1/1)$

$Q = 3.1$ (poor rock)

REFERENCES

Bandis, S., Lumsden, A. and Barton, N. (1981) Experimental studies of scale effects on the shear behaviour of rock joints. *Int. J. Rock Mech. Min. Sci. Geomech. Abstr.* 18(1), 1–21. https://doi.org/10.1016/0148-9062(81)90262-X.

Barton, N., Lien, R. and Lunde J. (1974) Engineering classification of rock masses for the design of tunnel support. *Rock Mech.* 6, 189–236.

Bhasin, R. and Grimstad, E. (1996): The use of stress – strength relationships in the assessment of tunnel stability. Tunnelling and Underground Space Technology, 11(1): 93–98.

Bieniawski, Z.T. (1973) Engineering classification of jointed rock masses. *Trans. S. Afr. Civ. Engrs*, 15.

Brady, B.H. and Brown, E.T. (2004) *Rock Mechanics: For Underground Mining* (3rd Edition). Springer Science and Business Media, German.

Brown, E.T. (1970) Strength of models of rock with intermittent joints. *J. Soil Mech. Found. Div., ASCE* 96, 1935–1949.

Deere, D.U. (1964) Technical description of rock cores for engineering purposes. *Rock Mech. Eng. Geol.* 1(1), 16–22.

Deere, D.U. (1989) Rock Quality Designation (RQD) after Twenty Years. U.S. Army Corps of Engineers Contract Report GL-89-1. Waterways Experiment Station, Vicksburg, MS, p. 67.

Deere, D.U., Hendron, A.J., Jr, Patton, F.D. and Cording, E.J. (1967) Design of surface and near surface construction in rock. Failure and breakage of rock. In: Fairhurst, C. (ed) *Society of Mining Engineers of AIME*. AIME, New York, pp. 237–302.

Goodman, R.E. (1976) *Methods of Geological Engineering in Discontinuous Rocks*. West Publishing Co., St Paul.

Hoek, E. (1968) Brittle failure of rock. In: Stagg, K.G. and Zienkiewicz, O.C. (eds) *Rock Mechanics in Engineering Practice*. Wiley, New York, pp. 99–124

Hoek, E. (1994) Strength of rock and rock masses. *ISRM News Journal*, 2 (2), 4–16.

Hoek, E. and Brown, E.T. (1980a). *Underground Excavations in Rock*. Instn Min. Metall, London, p. 527

Hoek, E. and Brown, E.T. (1980b) Empirical strength criterion for rock masses. *J. Geotech. Eng. Div.* 1980, 1013–1035.

Hoek, E. and Brown, E.T. (1988) The Hoek-Brown failure criteria – A 1988 update. In: Curran, J.H. (ed) *15th Canadian Rock Mechanics Symposium*. University of Toronto, Toronto, Canada, pp. 31–38

Hoek, E. and Brown, E.T. (1997) Practical estimates of rock mass strength. *J. Rock Mech. Min. Sci. Geomech.* 34, 1165–1186.

Hudson, J.A. and Harrison, J.P. (1997) *Engineering Rock Mechanics: An Introduction to the Principles*. Elsevier Science, Amsterdam, The Netherlands.

International Society of Rock Mechanics (1981) *Suggested Methods for Rock Characterization, Testing, and Monitoring*. Pergamon Press, Oxford, UK.

Jakubec, J. and Esterhuizen, G.S. (2007) *Use of the Mining Rock Mass Rating (MRMR) Classification: Industry Experience*. CDC, Canada.

Jakubec, J. and Laubscher, D.H. (2000) *The MRMR Rock Mass Classification System in Mining Practice.* MassMin, Brisbane, Australia.

Kim, B.H., Cai, M., Kaiser, P.K. and Yang, H.S. (2007) Estimation of block sizes for rock masses with nonpersistent joints. *Rock Mech. Rock Eng.* 40(2), 169–192.

Kirsch, E.G. (1898) Die Theorie der Elastizität und die Bedürfnisse der Festigkeitslehre. *Zeitschrift des Vereines deutscher Ingenieure.* 42, 797–807.

Laubscher, D.H. (1975) Class distinction in rock masses. *Coal Gold Base Min. S. Afr.* 23, 119–128.

Laubscher, D.H. (1990) A geomechanics classification system for the rating of rock mass in mine design. *Trans. S Afr. Inst. Min. Metal.* 9(10), 257–272.

Laubscher, D.H. and Jakubec, J. (2001) The MRMR rock mass classification for jointed rock masses. In: Hustrulid, W.A. and Bullock, R.L. (eds) *Underground Mining Methods: Engineering Fundamentals and International Cases Studies.* Society for Mining, Metallurgy, and Exploration, Inc., Littleton, CO, pp. 475–481.

Marinos, P. and Hoek, E. (2000) GSI: A geologically friendly tool for rock mass strength estimation. *Proceedings of the Eighth International Conference on Deep and High Stress Mining*, Australian Centre for Geomechanics, Perth, pp. 1422–1442

Mark, C. and Barczak, T.M. (2000) Fundamentals of coal mine roof support. In: Mark, C., Dolinar, D.R., Tuchman, R.J., Barczak, T.M., Signer, S.P., Wopat, P.F. (eds) *Proceedings: New Technology for Coal Mine Roof Support.* U.S. Department of Health and Human Services, Public Health Service, Centers for Disease Control and Prevention, National Institute for Occupational Safety and Health, DHHS (NIOSH) Publication No. 2000–151, IC 9453, Pittsburgh, PA, pp. 23–42.

Mark, C. and Molinda, G.M. (2005) The coal mine roof rating (CMRR)—A decade of experience. *Int. J. Coal Geol.* 64, 85–103.

Mohr, O.C. (1870) Beitrag zur Theorie der elastischen Bogentrager. *Zeitschrift des Architekten und Ingenieur-Vereins zu Hannover*, 16, 389.

Molinda, G.M. and Mark, C. (1994) Coal Mine Rock Rating (CMRR): A practical rock mass classification for coal mines. Bureau of Mines Information Circular (IC) 9387, Pittsburgh, PA.

Morelli, G.H. (2013) On joint roughness: measurements and use in rock mass characterization. *Geotech. Geol. Eng.* 32, 345–362.

Norwegian Geotechnical Institute (2022) Using the Q-System. www.ngi.no.

Obert, L. and Duvall, W.I. (1967) *Rock Mechanics and the Design of Structures in Rock.* John Wiley and Sons, New York.

Palmström, A. (2005) Measurements of and correlations between block size and rock quality designation (RQD). *Tunn. Underg. Space Technol.* 20(4), 362–377.

Pillalamarry, M. and Spearing, A.J.S. (2009) *Internal Notes for Rock Mechanics Class.* Southern Illinois University Carbondale, Carbondale, IL.

Rusnak, J. and Mark, C. (2000) Using the point load test to determine the uniaxial compressive strength of coal measure rock. In: Peng, S.S. and Mark, C. (eds) *Proceedings of the 19th International Conference on Ground Control in Mining.* West Virginia University, Morgantown, WV, pp. 362–371.

Saint-Venant, A.J. (1855) Memoire sur la Torsion des Prismes. *Memoires Divers Savants.* 14, 233–560.

Singh, B., Jethwa, J.L., Dube, A.K. and Singh, B. (1992): Correlation between observed support pressure and rock mass quality. Tunnelling and Underground Space Technology 7(1): 59–74.

Terzaghi, K. and Richart, F.E. (1952) Stresses in rock around cavities. *Geotechnique.* 3(2), 57–90.

Williamson, D.A. (1984) Unified rock classification system. *Bull. Assoc. Eng. Geol.* 21, 345–354.

Wikiwand – Constitutive relationships. www.wikiwand.com/en/articles/Constitutive_relationship

7 Basic Hydrology and Soil Mechanics

7.1 BASIC HYDROLOGY

7.1.1 Introduction to Hydrology

Hydrology is the study of the movement, distribution, quantity, and quality of water throughout the Earth, and addresses the hydrologic cycle and water resources (modified from Chem Europe, 2025). Hydrology has been of study and interest for thousands of years and is still very relevant and important, especially in rock mechanics. For example, in about 4000 BC, the Nile was dammed to improve water availability throughout the year, decrease the effects of drought, and increase the overall agricultural output. Ancient Mesopotamian (west Asian) towns were protected from flooding by building high earthen walls. Aqueducts were built by the ancient Greeks and Romans to supply water, while the Chinese used similar systems to irrigate fields and control flood.

7.1.2 The Evolution of Modern Hydrology

The pioneers of the modern science of hydrology include the following:

- Pierre Perrault (1674), who developed the hydrologic cycle.
- Edme Mariotte (1679), who independently formulated Boyle's law and designed Newton's cradle (a pendulum of ball bearings hung in contact with each other).
- Edmund Halley (1686), who, apart from his huge contribution to astronomy (Halley's comet), also developed the relationship between barometric pressure and height above sea level.

In addition, Perrault measured rainfall, runoff, and drainage areas around the Seine River (France) and showed that the rainfall was sufficient to explain the flow in this river. Mariotte combined velocity and river cross-section measurements to obtain discharge, again in the Seine River. Halley showed that the evaporation from the Mediterranean Sea was sufficient to account for the outflow of rivers flowing into the sea.

Advances in the 18th century included Bernoulli's piezometer, equation, and the Pitot tube. The 19th century saw development in groundwater hydrology, including:

- Darcy's law (1856) which describes the flow of liquid through a porous material.
- The Dupuit–Thiem well formula (1863) which establishes the flow of liquid toward a pumping well in unconfined aquifers.
- The Hagen–Poiseuille capillary flow equation (1839) which gives the pressure drop of a gas or liquid in laminar flow conditions traveling in a long pipe with a constant circular cross-sectional area.

7.2 WATER DISTRIBUTION ON THE EARTH

The distribution of water on the Earth is given in Figure 7.1 (US Geological Survey, n.d.).

The total amount of water on Earth is constant. Water is continually renewed while circulating between oceans, land, and atmosphere by the processes of evaporation, precipitation, and runoff. A concept known as the hydrological cycle is shown in Figure 7.2.

DOI: 10.1201/9781032704845-7

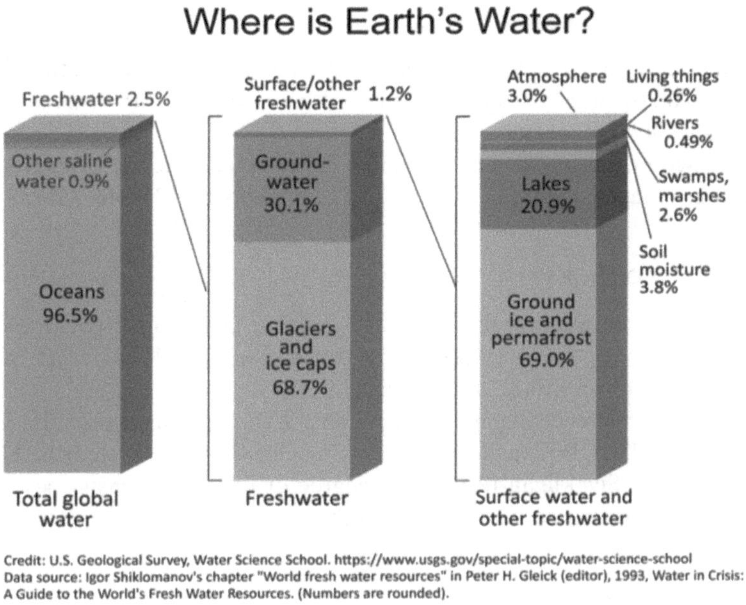

FIGURE 7.1 The Earth's water.

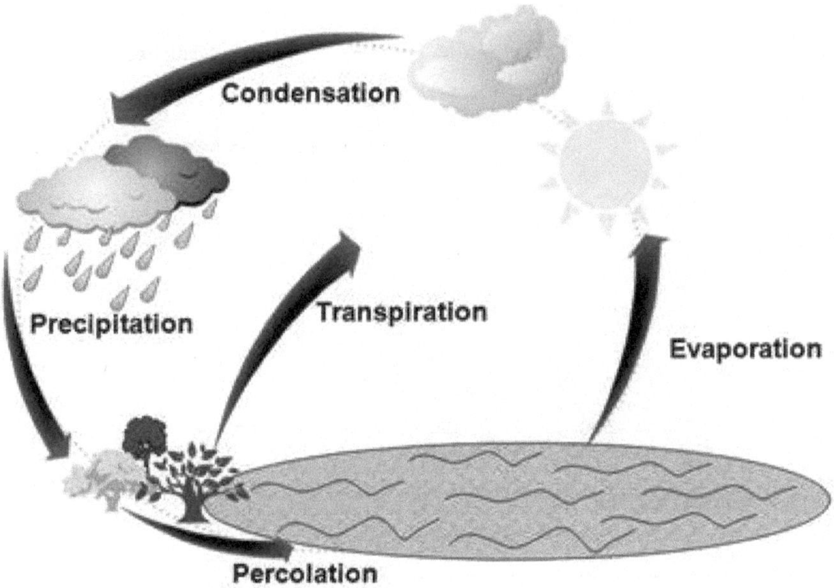

FIGURE 7.2 The hydrological cycle (Volkholz, 2021).

7.3 POROUS MEDIA

Porous media is comprised of:

- Solids such as mineral material.
- Liquids such as water in pores.
- Gases such as air.

7.3.1 Typical Media Profile from Surface

After chemical weathering, the resulting products remain in their original location. The soil created is termed a residual soil (referred to as just soil in Figure 7.3) and appears in areas where there have never been conditions of soil transport.

Unlike soils created by erosion that are not found in the same location, residual soils do not consolidate or over-consolidate and are not subjected to a stress history (Wesley, 2013).

7.3.2 Particle Sizes for Soils

Soils are a conglomerate of mineral particles, and the size and relative packing of the particles have a significant influence on the properties. Particles in a soil sample can range in size from very coarse, such as boulders, to very fine, such as clays.

Soils are classified by reference to particle size. Table 7.1 compares the size distributions according to British Standard 5930 (2007) and the Unified Soil Classification System ASTM 2487 (2006).

According to Das and Sivakugan (2019), the degree of firmness (consistency), i.e., soft, firm, stiff of a fine-grained soil, varies significantly with the water content. As the water content of

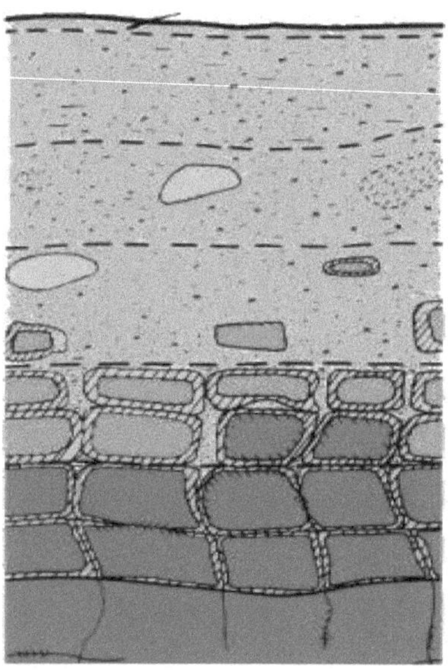

FIGURE 7.3 Soil profile (Fredlund, 1993).

TABLE 7.1
Soil Particle Size Definitions

Material	BS 5930 Size Definitions	USCS Size Definitions (U.S. Standard Sieve)
Clay	Less than 0.002 mm	Passing 0.075, use Atterberg Limits
Silt	0.002–0.063 mm	Passing 0.075, use Atterberg Limits
Sand	0.063–2 mm	0.075–4.75 mm
Gravel	2–63 mm	4.75–75 mm
Cobbles	63–200 mm	75–300 mm
Boulders	Greater than 200 mm	Greater than 300 mm

a fine-grained soil is increased slowly from 0%, it goes through different consistencies: solid to semi-solid to plastic and finally the liquid state (Figure 7.4). Atterberg limits are the borderline water contents between four such states. They were developed in the early 1900s by a Swedish soil scientist Atterberg (1911), working in the ceramics industry. Later, Terzaghi (1925) modified them to suit geotechnical problems.

Table 7.2 converts U.S. Standard Sieve sizes, which are commonly used in the United States, to a size in mm.

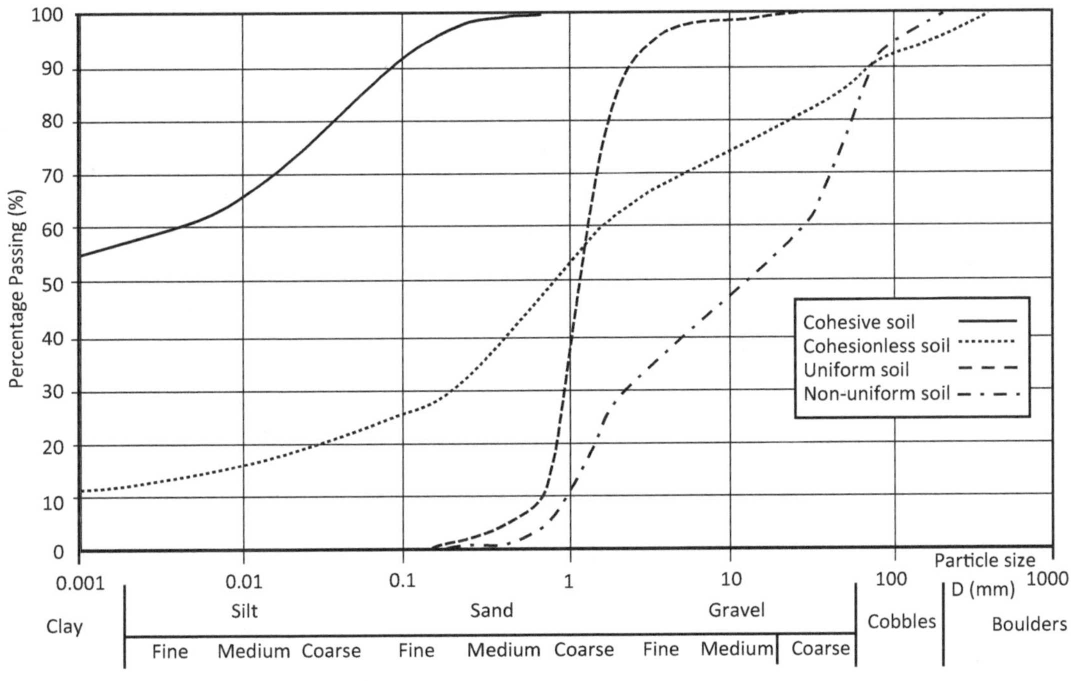

FIGURE 7.4 Particle size distribution for soils (Elementary Engineering, 2019).

TABLE 7.2
U.S. Standard Sieve Sizes (Das and Sivakugan, 2019)

Sieve No.	Opening, mm	Sieve No.	Opening, mm
4	4.75	35	0.500
5	4.00	40	0.425
6	3.35	50	0.300
7	2.80	60	0.250
8	2.36	70	0.212
10	2.00	80	0.180
12	1.70	100	0.150
14	1.40	120	0.125
16	1.18	140	0.106
18	1.00	170	0.090
20	0.850	200	0.075
25	0.710	270	0.053
30	0.600		

The particle size distribution is defined as a list of values or a mathematical function that defines the relative amount, typically by mass, of the particles according to a size, in a powder, granular material, or particles dispersed in a liquid (Jillavenkatesa et al., 2001).

Figure 7.4 gives examples of several soil distributions.

From Figure 7.4, the soils can be described as follows (ASTM D653-21, 2001; D2488-23, 2023):

- Cohesive soil is typically considered a clay (a finely grained soil), or soil with a high clay content, which has cohesive strength. Cohesive soil does not crumble and behaves plastically when moist. Cohesive soil is hard to break up when dry and exhibits significant cohesion when wet.
- Cohesionless soil is free flowing like relatively coarse sand or gravel. The strength depends solely on the friction between the particles.
- Uniform soil is poorly graded soil and typically consists of particles that are nearly the same size. This type of soil has generally poor compaction, high permeability, and low shear strength.
- Non-uniform soil (heterogeneous soil) is well-graded soil that generally has relatively high shear strength, good compaction, and low permeability.

7.3.3 Soil Properties and Relationships

The following are important relationships that govern soil behavior.

7.3.3.1 Coefficient of Uniformity

The coefficient of uniformity, C_u, classifies soils. It is the ratio of the particle diameter at 60% finer (D_{60}) to the particle diameter at 10% finer (D_{10}) found from the particle size distribution curve. It is used in determining whether a soil is well graded, poorly graded, or gap graded. A higher coefficient indicates a greater range of particle sizes and less uniformity:

$$C_u = D_{60}/D_{10} \tag{7.1}$$

where (as above)

D_{10} is the size such that 10% of the particles are smaller than that size.

D_{60} is similarly defined for 60%.

The D_{10} is known as the effective size and represents the water flow in soils and sands. The coefficient of uniformity C_u may be assumed as the inverse slope of the particle size distribution curve, taken in the middle section considered as the most representative for the soil. If a curve has a slope of "m", the inverse slope is then 1/"m". A value of C_u greater than 4–6 is considered well graded and less than 4 is considered poorly graded (or uniform).

7.3.3.2 Coefficient of Curvature

The general shape (slope) of a particle distribution curve can be described by using the coefficient of curvature (C_c) defined as follows:

$$C_c = D_{30}^{\,2}/D_{60}D_{10} \tag{7.2}$$

where

The significance of the diameters is the same presented for the coefficient of uniformity above (in Equation 7.1).

For a well-graded soil, sand and gravel (not clay), the coefficient of curvature is typically between 1 and 3, but coefficient of uniformity also has an influence and must also be considered. For values outside of this range, the soil is considered poorly graded.

7.3.3.3 Porosity

The particle size, shape, and distribution control all the mechanical and hydraulic properties of a porous media such as the conductivity (the ability of a material to transfer heat), deformation behavior, and anisotropy. All of these characteristics of the pore systems are mainly derived from the texture of the soil or of the rock masses.

The porosity (n) is defined as the ratio of the volume of voids (V_v) to the total volume (V_t) of the soil sample:

$$n = V_v / V_t \tag{7.3}$$

The porosity can also be expressed by percentage:

$$n = (V_v / V_t)(100)\% \tag{7.4}$$

7.3.3.4 Effective Porosity

Porosity indicates the maximum amount of water that rock contains when saturated. If water is stored in the ground, a portion of the water is available and will drain under gravity, termed specific yield (S_y), while the remaining portion is retained in the rock as a film on surfaces or very small openings, termed specific retention (S_r) (Heath, 1983). The expression according to Heath (1983) is

$$n = S_y + S_r = V_d / V_t + V_r / V_t \tag{7.5}$$

where

V_d is the volume of water that drains from the total volume V_t.

V_r is the volume of water that retains from the total volume V_t.

In some cases, S_y is referred to as the effective porosity (Bear, 1979). In underground flow situations, only the effective porosity permits the flow of water.

Effective porosity can be divided into macroporosity and microporosity. Macroporosity is responsible for water flow, and microporosity is associated with the movement of capillary water. Fracture systems, such as those from mining, should increase the initial macroporosity, thus increasing the potential for water inflow.

7.3.3.5 Void Ratio

The void ratio (e) represents the ratio of the volume of voids (V_v) to the volume of solids (V_s):

$$e = V_v / V_s \tag{7.6}$$

As a function of porosity and vice versa:

$$n = e/(1 + e) \tag{7.7}$$

$$e = n/(1 - n) \tag{7.8}$$

7.3.3.6 Soil Particle Density

According to McCarty et al. (2016), the soil particle density and bulk (dry) density can be described as follows.

The soil particle density (ρ_s) does not include space occupied by air or water and is defined as

$$\rho_s = M_s / V_s \tag{7.9}$$

where

M_s is the mass of the soil particles.

V_s is the unit volume of the soil particles.

TABLE 7.3
Typical Densities and Porosities for Some Materials

Soil or Rock Type	Density Range (g/cm^3)	Porosity (%)
Soil example	1.20–2.40	30–70
Clay	1.63–2.60	40–70
Gravel	1.70–2.40	25–40
Sand	1.70–2.30	25–50
Sandstone	1.61–2.76	5–30
Shale	1.77–3.20	–
Limestone	1.93–2.90	0–20
Dolomite	2.28–2.90	1–5

7.3.3.7 Bulk (Dry) Density

The dry bulk density (ρ_b) is defined as the dry soil per unit volume; thus, it includes pores occupied by air and water:

$$\rho_b = M_s / V_t \tag{7.10}$$

Soils with higher bulk densities will have slower water infiltration and percolation than noncompacted soils (McCarty et al., 2016).

Table 7.3 provides typical maximum porosities and density of various soils (Yu et al., 1993).

7.3.3.8 Unit Weight

According to Das (2010), the dry unit weight (γ_d) of the soil is related to the dry density of the soil:

$$\gamma_d \left(kN/m^3\right) = \rho_d g \left(kg/m^3\right)/1,000 \tag{7.11}$$

Similarly, the unit weight (γ) is related to the density of the soil:

$$\gamma \left(kN/m^3\right) = \rho g \left(kg/m^3\right)/1,000 \tag{7.12}$$

where the unit weight of water (γ_w) is 9.81 kN/m^3, 62.4 lb/ft^3, or 1,000 kgf/m^3.

7.3.3.9 The Compaction of Soil

Void ratio can be used to characterize different degrees of compaction in situ. Using the actual void ratio (e) between the limit values e_{max} and e_{min}, the relative density (D_r) is

$$D_r = \left(e_{max} - e\right)/\left(e_{max} - e_{min}\right) \tag{7.13}$$

If the soil is in its densest possible state (void ratio of e_{min}), then the relative density of a soil is 1, or 100%, while in its loosest possible state (void ratio of e_{max}), the relative density is 0, or 0%.

7.3.3.10 The Liquid Phase

Water density (ρ_w) is defined as the ratio between the mass of water (M) and the volume occupied by water (V):

$$\rho_w = M/V \tag{7.14}$$

The unit weight of a liquid (γ) is defined as the ratio between the weight of the liquid and its volume. For water, the unit weight is about 9,810 N/m^3. It may also be obtained by multiplying the density by acceleration due to gravity (g):

$$\gamma = W/V = \rho g \tag{7.15}$$

where

W is the weight of the water.
g is the acceleration of gravity.

7.3.3.11 Viscosity

The viscosity of a fluid is the property where it develops resistance forces between surfaces with relative motion. It is also considered a fluid's resistance to flow. The viscosity decreases as the temperature increases.

The following relationship gives the dynamic coefficient of viscosity in terms of the shear stress:

$$\mu = \tau/\left(dv/dy\right) \tag{7.16}$$

where

μ is the dynamic coefficient of viscosity.
ν is the velocity.
τ is the shear stress.
$d\nu/dy$ is a velocity gradient.

The dynamic coefficient of viscosity decreases with increase in the temperature. For water, the dynamic coefficient of viscosity is of about 10^{-3} Pa-s at 20°C.

7.3.3.12 Kinematic Viscosity

The dynamic coefficient of viscosity is not the only method to quantify the viscosity of a fluid. The kinematic coefficient of viscosity (ν) is defined by the following relationship:

$$\nu = \mu/\rho \tag{7.17}$$

where

ν is the kinematics' coefficient of viscosity.
ρ is the density of the fluid.
For water, the kinematic coefficient of viscosity is 10^{-6} m^2/s at 20°C.
Viscosity is important in rock as it influences their flow and deformation.

7.4 SUBSURFACE WATER

The origin of subsurface water can be explained by taking into consideration the two main ways of subsurface water accumulation (Galperin et al., 1993):

- Exogenic via infiltration and sedimentation.
- Endogenic by metamorphic and magmatic processes.

In the upper levels of the Earth's crust, the subsurface water is collected mainly from infiltration (seeping into the soil and rock) and precipitation (rain). The natural recharge by infiltration is the main source of water in underground formations. This type of water is most common, and significant in the hydrologic cycle. In some cases, infiltration from surface waters into the ground may also occur.

Subsurface formations containing water may be divided vertically into a zone of aeration in which the pores contain both air and water and an underlying zone of saturation in which the pores are filled up with water. The boundary between the two zones is referred to as the water table.

The underground water may exist in the following forms (refer to Figure 7.5):

- Bound water, as chemically bounded water and physically bounded water.
- Free water as capillary water and gravitational water.
- Gaseous water.

7.4.1 Subsurface Water Distribution

The subsurface water distribution is diagrammatically shown in Figure 7.6. The top layer is mainly wet soils followed by the Vadose zone, which is the undersaturated zone before the phreatic (capillary) zone.

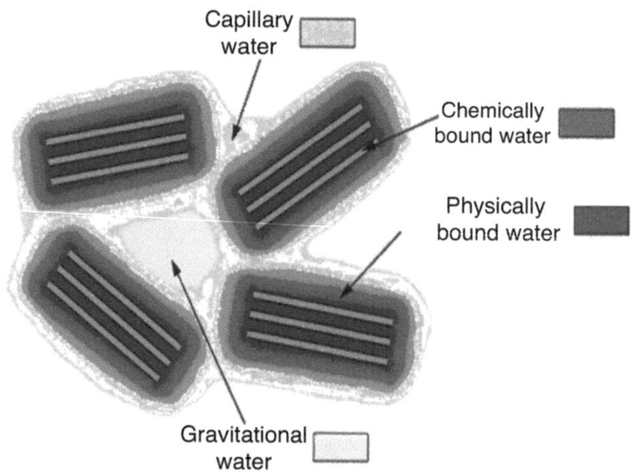

FIGURE 7.5 Underground water (Galperin et al., 1993).

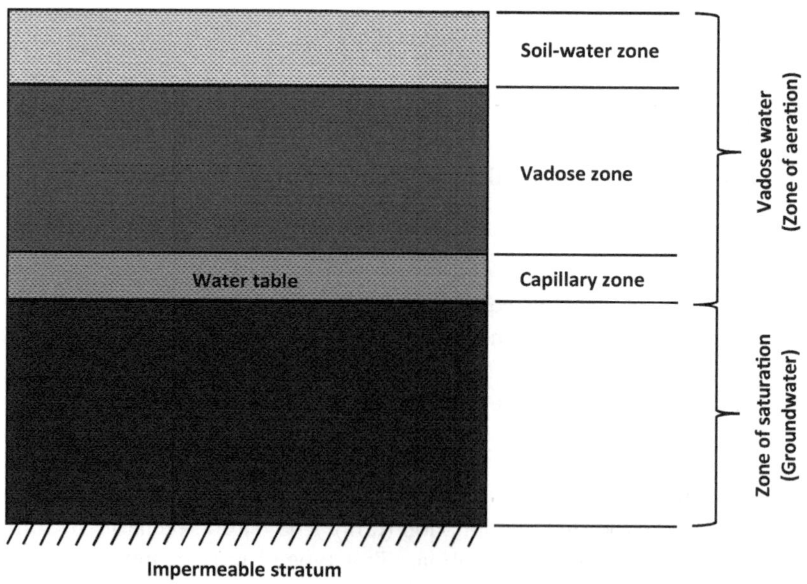

FIGURE 7.6 Subsurface water distribution (Galperin et al., 1993).

7.4.2 Groundwater

Groundwater denotes water found beneath the ground surface. In hydrogeology, groundwater is beneath the water table in the saturation zone of Figure 7.6. Figure 7.7 shows a phase diagram representing the typical mass–volume relationship for soil.

According to Fredlund and Rahardjo (1993), the total density (ρ) can also be considered the bulk density (ρ_b) contrary to the bulk (dry) density expression in Equation 7.10. Using the expressions in Figure 7.7, in natural conditions, the ratio of the total mass to the total volume can be expressed as follows (ignoring the mass of the air – M_a):

$$\rho_b = M/V_t$$

Using $M = M_a + M_w + M_s$, where $M_a = 0$, and $V_t = V_v + V_s$ and plugging in

$$\rho_b = (M_s + M_w)/(V_s + V_v)$$

Dividing through by V_s

$$\rho_b = \left[(M_s + M_w)/V_s\right]/\left[(V_s + V_v)/V_s\right]$$

Factoring and simplifying

$$\rho_b = \left[M_s\left(1 + M_w/M_s\right)/V_s\right]/\left[(V_s + V_v)/V_s\right]$$

$$\rho_b = \left[\left(M_s/V_s\right)\left(1 + M_w/M_s\right)\right]/\left[1 + V_v/V_s\right]$$

If $\rho_s = M_s/V_s$, then the following expression is resolved:

$$\rho_b = \rho_s(1 + w)/(1 + e) \tag{7.18}$$

where
w is the moisture content.

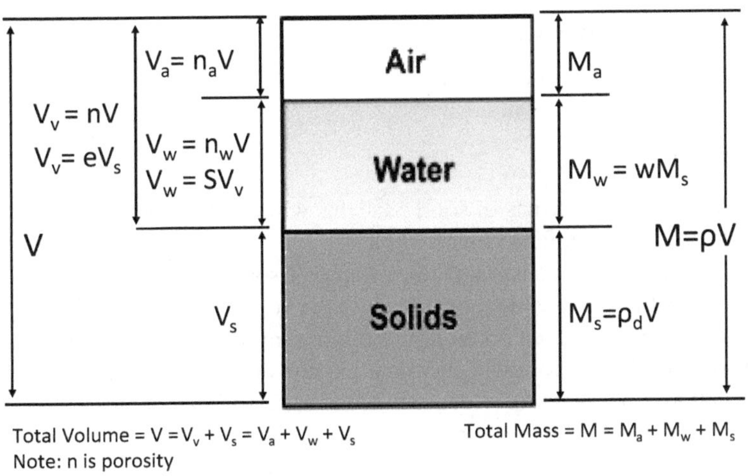

FIGURE 7.7 Phase diagram (Fredlund and Rahardjo, 1993).

The bulk density can also be written in terms of porosity:

$$\rho_b = \rho_s(1+w)/(1+e)$$

$$\rho_b = \rho_s(1+w)/[1+n/(1-n)]$$

$$\rho_b = \rho_s(1+w)(1-n)/[1+n/(1-n)](1-n)$$

$$\rho_b = \rho_s(1+w)(1-n)/[(1-n)+n]$$

$$\rho_b = \rho_s(1-n)(1+w) \tag{7.19}$$

The dry density of the soil (ρ_d) can be obtained similarly:

$$\rho_d = M_s/V_t$$

$$\rho_d = M_s/(V_s+V_v)$$

$$\rho_d = (M_s/V_s)/[(V_s+V_v)/V_s]$$

$$\rho_d = (M_s/V_s)/[(1+V_v/V_s]$$

$$\rho_d = \rho_s/(1+e)$$

$$\rho_d = \rho_s/[1+n/(1-n)]$$

$$\rho_d = \rho_s(1-n)/[1+n/(1-n)](1-n)$$

$$\rho_d = \rho_s(1-n)/[(1-n)+n]$$

$$\rho_d = \rho_s(1-n) \tag{7.20}$$

7.5 BASIC SOIL MECHANICS

Modified after Marais and Spearing (1990).

7.5.1 Introduction

As previously discussed, soil consists of solid particles with voids in between which are filled with water, air, or partly with both (refer to Figure 7.8).

The total volume (V_t) of soil consists of the volume of solids (V_s) and the volume of voids (V_v). The volume of voids is the volume of water (V_w) and the volume of air (V_a).

Schematically it can be represented as a phase diagram as shown in Figure 7.7, which relates masses, M, and volumes, V. The following sections expand on the various weight, mass, and volume relationships as well as other relevant relationships for basic soil mechanics.

7.5.2 The Effect of an External Load on the Void Ratio

This is best demonstrated by considering backfill. After backfill has been placed in a mine void, the vertical stress on the fill will typically increase due to stope closure. This will cause a decrease

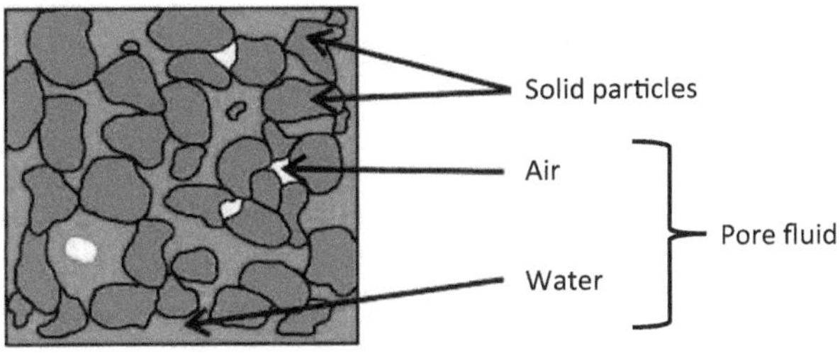

FIGURE 7.8 Typical soil made up of solid particles, water, and air gaps (Dutto, 2014).

in the void ratio of the backfill. The over-consolidation ratio (OCR) is an important parameter. The OCR can be defined as follows:

$$\text{OCR} = \sigma'_p/\sigma'_{vo} \tag{7.21}$$

where

σ'_p is the preconsolidation pressure of the material (maximum past effective pressure)

σ'_{vo} is the present effective overburden pressure.

The following classifications generally apply:

OCR = 1 the soil is normally consolidated.
OCR < 1 the soil is under-consolidated.
OCR > 1 the soil is over-consolidated.

Often backfill is placed as a slurry, so it is normally consolidated or slightly over-consolidated (saturated or overly saturated). If the material is saturated and pore water pressure does exist, the material is under-consolidated. This phenomenon can be explained by considering the following.

Consider a rock that exerts a pressure of 500 kPa on the soil underneath it. The load from the rock is considered the preconsolidation pressure, and while the rock is in place, the soil is considered normally consolidated. If the rock is removed and replaced by a building structure that exerts a pressure of 250 kPa, the OCR changes. The pressure due to the structure is called the overburden pressure, and the soil is now over-consolidated. The calculation is as follows:

$$\text{OCR} = \sigma'_p/\sigma'_{vo} = 500\ \text{kPa}/250\ \text{kPa} = 2.0$$

Consolidation curves are plotted as void ratio versus log pressure as shown in Figure 7.9 and can be obtained from standard consolidation tests.

According to Suryakanta (2015), the curves represent the following:

Curve A: This curve is obtained from a loose to medium cohesionless soil. This type of curve shows no clear point of failure.

Curve B: This curve is obtained from a cohesive soil. As the load increases, the curve trends toward a settlement curve.

Curve C: This curve is obtained from a partially cohesive soil. This type of curve also does not show a clear point of failure.

Curve D: This curve is obtained from a dense cohesionless soil. This type of curve is common for dense sand or gravel deposits.

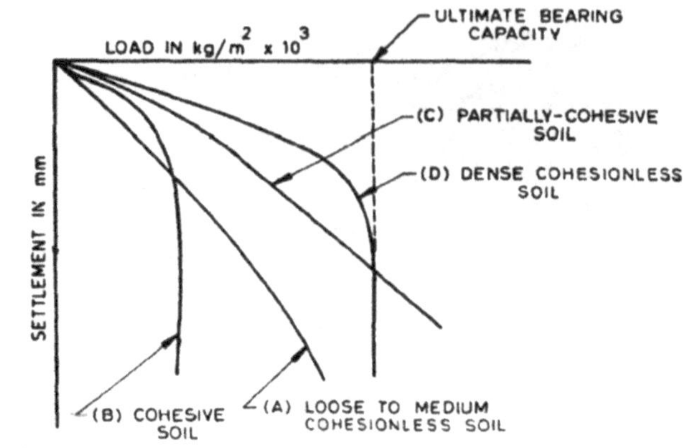

FIGURE 7.9 Typical load settlement curve (Suryakanta, 2015). (Note: The y-axis could be settlement or void ratio.)

The initial flatter portion of the void–log pressure curve is termed the over-consolidated portion, and the part after the "break" in the curve is called the virgin compression curve (reconsolidation stress). In the case of backfill, the reconsolidation curve does not exist until it has been drained to the extent that there is no more pore water pressure.

In the case of backfill, for example, the preconsolidation pressure is small compared to the overburden pressure, and hence, significant settlement can occur. Once backfills are loaded and deform considerably, the backfill can become more resistant to deformations due to lateral confinement and the consolidation mechanisms discussed above.

7.5.3 Degree of Saturation

The degree of saturation (S) is defined as the ratio of the volume of water to the volume of voids:

$$S = V_w / V_v \tag{7.22}$$

Degree of saturation can be expressed as a percentage or as a decimal. If the soil is completely dry, then S = 0%, and if the pores are completely full of water, then the soil is fully saturated and S = 100%.

7.5.4 Water Content

The water content can be defined as the ratio of the mass of water to the mass of solids:

$$w = M_w / M_s \tag{7.23}$$

7.5.5 The Effect of Load on Water Content and the Degree of Saturation

When any soil or backfill is subject to load, there is a decrease in void ratio.

In the case of the degree of saturation, the situation is different. As the void ratio (e) decreases, there is an increase in the degree of saturation. The effect of a decrease in void ratio on the water content and the degree of saturation can be illustrated as in Figure 7.10.

The point e_x represents the void ratio where water is forced out of the voids (drainage) to allow a further decrease in void ratio. This can lead to build up in pore water pressure which can affect the stability of the fill material.

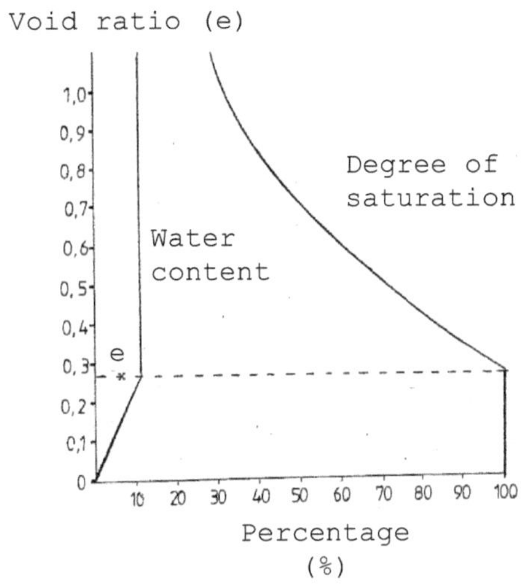

FIGURE 7.10 Theoretical curve for a sample with an initial void ratio of 1 and a water content of 10% (reference unknown).

Placing or obtaining dry backfill is not possible due to practical problems such as the method of placement and dust, as well as the relatively high humidity and wet conditions in most mines.

7.5.6 Specific Gravity

The specific gravity (G_s) can be defined as the ratio of the density of the substance to the density of water:

$$G_s = (M_s/V_s)/\rho_w = (W_s/V_s)/\gamma_w \tag{7.24}$$

where

M_s is the mass of solids.
V_s is the volume of solids.
ρ_w is the density of water.
W_s is the weight of solids.
γ_w is the unit weight of water.

7.5.7 Bulk Unit Weight

Bulk unit weight, γ_b, is a measure of the solid particles per unit volume of a soil or backfill similarly expressed as Equation 7.15.

7.5.8 Saturated Unit Weight

The saturated unit weight, γ_{sat}, is equal to the bulk unit weight of the soil when the total voids are filled with water, that is, when S = 100%:

$$\gamma_{sat} = W/V_t \tag{7.25}$$

In terms of specific gravity and void ratio the saturated unit weight is

$$\gamma_{sat} = \gamma_w (G_s + e)/(1 + e) \tag{7.26}$$

7.5.9 Dry Unit Weight

The dry unit weight, γ_d, is the weight of the soil solids per unit volume, that is, when S = 0%:

$$\gamma_d = W_s / V_t \quad (7.27)$$

$$\gamma_d = \gamma_w G_s / (1+e) \quad (7.28)$$

7.5.10 Submerged Unit Weight

When a soil is below the water table, part of its weight is balanced by a buoyant effect equal to the weight of water displaced. Submerged, or buoyant, unit weight, γ_{sub}, is defined as the difference between the saturated unit weight and the unit weight of water:

$$\gamma_{sub} = \gamma_{sat} - \gamma_w$$

$$\gamma_{sub} = \left[\gamma_w (G_s + e)/(1+e)\right] - \gamma_w$$

$$\gamma_{sub} = \gamma_w (G_s - 1)/(1+e) \quad (7.29)$$

7.5.11 Density

Bulk density (ρ_b) was previously provided as follows:

$$\rho_b = \rho_s (1+w)/(1+e)$$

where $\rho_s = G_s \rho_w$:

$$\rho_b = G_s \rho_w (1+w)/(1+e)$$

From weight–volume relationships, the degree of saturation can be expressed as follows:

$$Se = wG_s$$

which resolves the bulk density in terms of degree of saturation and void ratio:

$$\rho_b = \rho_w (G_s + eS)/(1+e) \quad (7.30)$$

Saturated Density (ρ_{sat})

$$\rho_{sat} = \rho_w (G_s + e)/(1+e) \quad (7.31)$$

Dry Density (ρ_d)

$$\rho_d = \rho_w G_s / (1+e) \quad (7.32)$$

Submerged Density (ρ_{sub})

$$\rho_{sub} = \rho_w (G_s - 1)/(1+e) \quad (7.33)$$

The expressions G_s, e, and S are all dimensionless.

7.5.12 Capillarity

Capillarity arises from surface tension, which is a basic fluid property. The concept can be illustrated with the use of a small-diameter clean glass tube resting at the phreatic surface of a tank of pure water and represents the pores in a soil, as shown in Figure 7.11. Clean water tends to rise in the capillary tube, above the phreatic surface, due to forces between the water and glass, forming a meniscus between the walls as illustrated in Figures 7.11 and 7.12. The height of the rise is inversely proportional to the diameter of the tubing. Considering the meniscus geometry for water in a capillary tube, equations can be derived for the forces acting in the water column.

The force acting downward is the weight of the column of water:

$$F = h_c(\pi d^2/4)\rho_w g \tag{7.34}$$

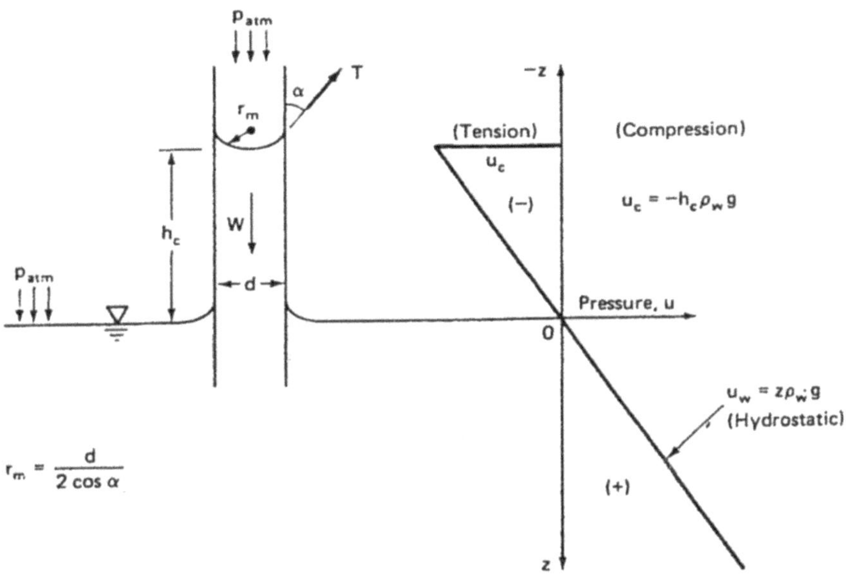

FIGURE 7.11 Meniscus geometry of water in a capillary tube (reference unknown).

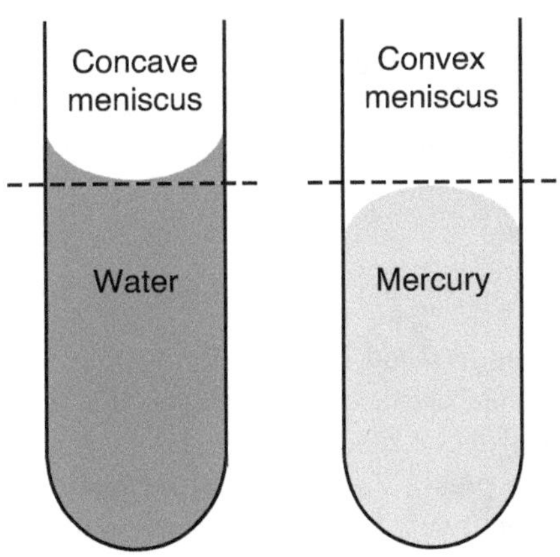

FIGURE 7.12 Meniscus examples in a glass tube (Ahlawat, 2018).

where

F is the force.

h_c is the height of capillary rise.

d is the diameter of the tube.

ρ_w is the density of water.

g is the acceleration due to gravity.

The upward force is the vertical component of the reaction of the meniscus against the tube circumference:

$$F = \pi d T \cos\alpha \tag{7.35}$$

where

T is the surface tension of the water–air interface, which acts on the circumference of the tube. The dimensions of the surface tension are force/length.

For equilibrium,

$$h_c(\pi d^2/4)\rho_w g = \pi d T \cos\alpha \tag{7.36}$$

where

α is the contact angle shown in Figure 7.11.

Solving for the height of capillary rise in clean glass tubes and pure water, where $\alpha = 0$,

$$h_c = 4T/(\rho_w g d) \tag{7.37}$$

The pressure increase below the phreatic surface is positive and hydrostatic as shown in Figure 7.11. Above the reservoir surface, the water pressure is negative, or less than atmospheric. From Equation 7.37, the pressure for pure water is

$$u_c = -h_c\rho_w g = -4T/d = -2T/r_m \tag{7.38}$$

where

r_m is the radius shown in Figure 7.11 expressed as d/2cosα.

The shape of the meniscus is approximately spherical with radius r_m. It is considered that the minimum energy condition as a sphere has the minimum surface area and hence the lowest energy state. The radius is greater than or equal to the radius of the tube depending on the contact angle. Therefore, when α is approximately zero, r_m = d/2. In larger diameter tubes, the surface will be nearly flat with only a concave or convex surface near the tube edges due to surface tension.

In soils, it is common to assume that the effective pore diameter is about 20% of the effective grain size (D_{10}). This assumption points out the importance of pore size, not grain size, as the controlling factor in capillary. Two particles connected by menisci of radius r_1 are shown in Figure 7.13 to illustrate this.

7.5.13 Shrinkage

Shrinkage of clay soils can be understood by considering the analogy of capillary stresses and a horizontal tube with compressible elastic walls. Assume that a tube, as shown in Figure 7.14a, is filled with water and the radii of the menisci have not reached their final shape and are very large. As evaporation occurs, the pressure in the water decreases and the menisci begin to form (Figure 7.14b). As evaporation continues, the radii become smaller. The compression in the compressible walls of the tube increases and the tube shrinks in length and diameter. The limiting case, shown in Figure 7.14c, is when the radii of the menisci are fully developed.

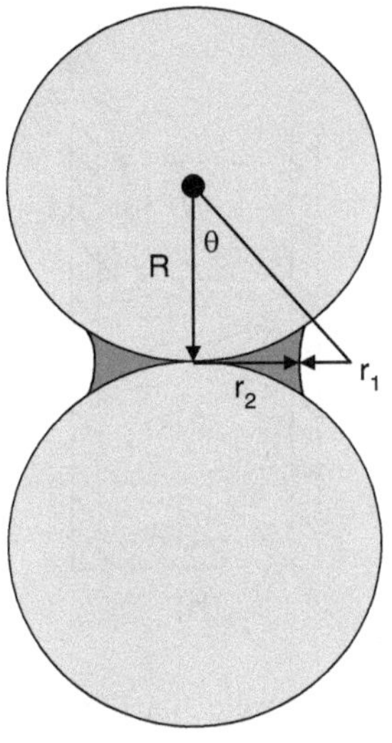

FIGURE 7.13 Two particles held together by a capillary film (Likos and Lu, 2004).

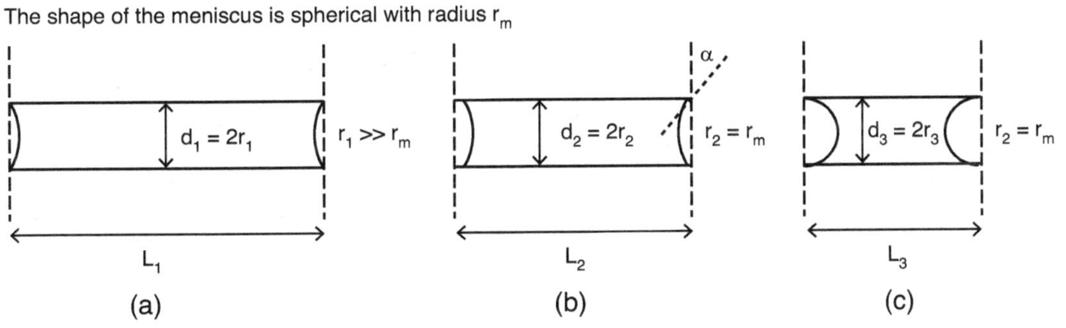

FIGURE 7.14 Capillary stresses causing shrinkage (reference unknown).

The negative pressure in the capillary tube causes the walls of the tube to shrink to an equilibrium condition between the rigidity of the walls and the capillary forces. If the tube is immersed in water, the menisci are destroyed, and the tube can expand because the capillary forces acting on the tube walls are eliminated. Unless the walls are perfectly elastic, the tube will not return completely to its original length and diameter.

In the case of backfill, the voids are too big to cause significant shrinkage. If any shrinkage tests are to be done, it is important to measure the horizontal, and not the vertical, shrinkage.

7.5.14 Consolidation Settlement

When load is placed on soil or backfill, it will compress because of the deformation of particle grains and compression and evacuation of air and water in the voids.

Backfill is a special case where initial settlement is caused by the self-weight of the backfill material. Load due to the overlying rock will generate after the roof in a mine has come down to the extent that it touches the fill. In addition, there can be settlement if the water is not allowed to drain from the backfill. Settlement and shrinkage are major issues when transporting and placing backfill as a slurry.

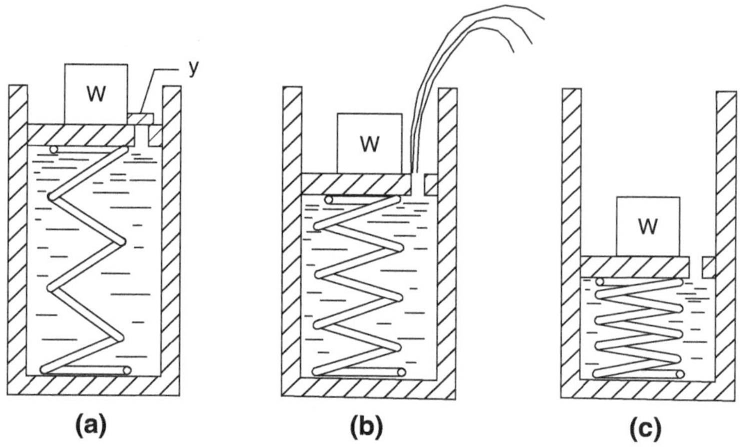

FIGURE 7.15 Spring analogy as applied to consolidation (Terzaghi, 1943).

Consider the consolidation of fine clay which is explained by the spring-cylinder analogy shown in Figure 7.15 (Terzaghi, 1943). A piston is loaded vertically and compresses a spring inside the chamber, which is filled with water. The spring is analogous to the soil mineral skeleton, while the water in the cylinder represents the pore water in the soil voids.

To simulate a fine-grained soil with low hydraulic conductivity, assume that the valve at "y" in 7.15a is opened. The water will slowly leave the cylinder under the initial excess pressure caused by the force (W) which acts on the cylinder's area (A). With time, as the water flows out, the water pressure (pore water pressure in soil) decreases, and the load is fully transferred to the spring. At equilibrium (Figure 7.15c), no further water is squeezed out of the cylinder and the water pressure equals zero, resulting in the spring being in equilibrium with the load.

7.6 THE EFFECT OF HYDROLOGY ON MINING

Subsidence and hydrology impacts occur at every underground mining operation bringing about changes to surface landforms, groundwater, and surface water.

Subsidence can be a natural or a man-made phenomenon associated with a variety of processes including compaction of natural sediments, groundwater dewatering, wetting, melting of permafrost, liquefaction and crustal deformation, withdrawal of petroleum and geothermal fluids, and general mining (Soliman, 1998).

Singh (1992) reports, "Subsidence is an inevitable consequence of underground mining – it may be small and localized or extend over large areas, it may be immediate or delayed for many years". Subsidence is discussed in the next chapter.

Drainage gradients are usually significantly altered by the disturbance of the strata around mine areas and the creation of additional fracturing. Rocks may also become weakened by saturation and erosion patterns could change.

Where surface water is present, it can readily migrate into fractures and fissures in the strata and the mine and may induce subsidence. Generally, the creation of major mining excavations often results in subsidence.

Another mining-related phenomenon that also creates subsidence is the withdrawal of water to facilitate underground mining. Water withdrawal also causes the formation of cavities (which were once filled with water) and, like cavities directly created by mining, may result in subsidence as the hydro-geological properties of the associated strata are changed.

7.7 WORKED EXAMPLES

Example 7.7.1

A leach solution is measured to have a surface tension of roughly 73 dyne/cm and a density of 1.08 g/cm³. Determine the capillary rise for this solution as a function of capillary diameter if the contact angle is zero (after Schlitt, 1984).

Solution:

Note that Schlitt (1984) uses a double-tilde operator suggesting that the relationship is approximately equal theoretically:

$$h_c(\pi d^2/4)\rho_w g = \pi d T \cos\alpha \tag{7.36}$$

$$h_c = \pi d T \cos\alpha / (\pi d^2/4)\rho_w g$$

$$h_c = 4T\cos\alpha / d\rho_w g$$

$$h_c = [4(73\ \text{dyne/cm})\cos(0)] / \left[(1.08\ \text{gm/cm}^3)(980.7\ \text{cm/s}^2)d\right]$$

$$h_c = 292\ \text{dyne/cm} / \left[(1{,}059\ \text{gm/cm}^2\text{s}^2)d\right]$$

$$h_c = 0.28/d$$

Example 7.7.2

If a relationship between the capillary diameter (d) and effective rock particle diameter (d_p) is determined to be $d_c = 0.8d_p$, determine the approximate theoretical capillary rise for two particle sizes. The first particle size is 0.11 cm and the other is 0.024 cm (after Schlitt, 1984).

Solution:

$$h_c = 0.28/d_c$$

$$d_c = 0.8d_p$$

$$h_c = 0.28/0.8d_p$$

$$h_c = 0.35/d_p$$

$$d_p = 0.11\ \text{cm}$$

$$h_c = 0.35/d_p$$

$$h_c = 0.35/0.11$$

$$h_c = 3.2\ \text{cm}$$

$$d_p = 0.024\ \text{cm}$$

$$h_c = 0.35/d_p$$

$$h_c = 0.35/0.024$$

$$h_c = 14.6\ \text{cm}$$

Example 7.7.3

A particle size distribution curve (PSD) is provided. From the information on the curve, determine the D_{60}, D_{30}, and D_{10}.

Solution:

The PSD at D_{60}, D_{30}, and D_{10} results are estimated from Figure 7.16 as shown in Figure 7.17.

The particle diameters at D_{60}, D_{30}, and D_{10} will be linearly interpolated between the known points.

Solution:

Slope (m) of a straight line with two points

$$m = (y_2 - y_1)/(x_2 - x_1)$$

Point-slope formula

$$y = y_1 + m(x - x_1)$$

Combining creates the linear interpolation formula for finding y:

$$y = y_1 + (y_2 - y_1)(x - x_1)/(x_2 - x_1)$$

D_{60}

$$(x_1, y_1) = (0.425, 69)$$

$$(x_2, y_2) = (0.25, 43.1)$$

$$(x, y) = (x, 60)$$

$$60 = 69 + (43.1 - 69)(x - 0.425)/(0.25 - 0.425)$$

$$-9 = (-25.9)(x - 0.425)/(-0.175)$$

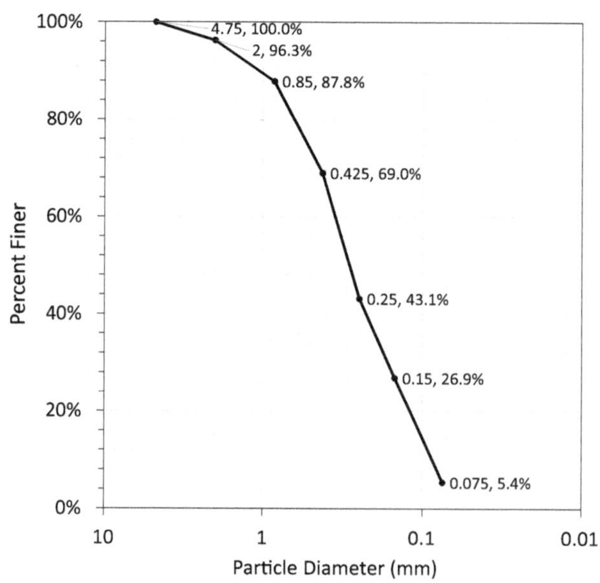

FIGURE 7.16 The PSD for the example.

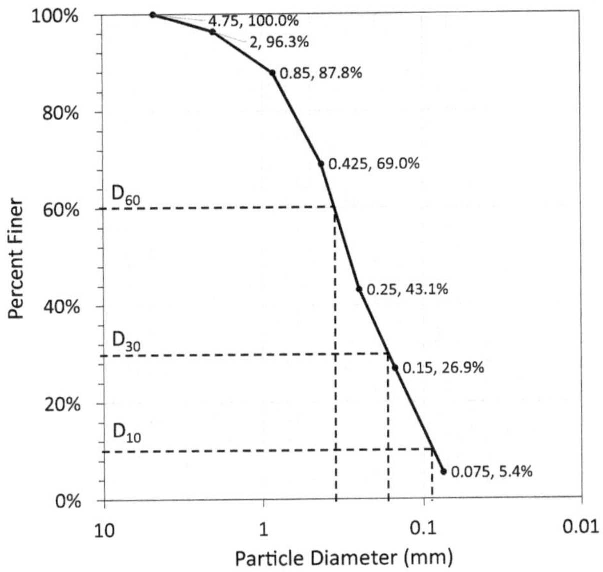

FIGURE 7.17 The PSD with the important particle diameters estimated.

$$1.575 = -25.9x + 11$$

$$-9.425 = -25.9x$$

$$0.363 = x = D_{60}$$

D_{30}

$$(x_1, y_1) = (0.25, 43.1)$$

$$(x_2, y_2) = (0.15, 26.9)$$

$$(x, y) = (x, 30)$$

Linear interpolation formula for finding x

$$x = x_1 + (y - y_1)(x_2 - x_1)/(y_2 - y_1)$$

$$x = 0.25 + (30 - 43.1)(0.15 - 0.25)/(26.9 - 43.1)$$

$$x = 0.25 + (-13.1)(-0.10)/(-16.2)$$

$$x = 0.25 + (-0.081)$$

$$D_{30} = x = 0.169$$

D_{10}

$$(x_1, y_1) = (0.15, 26.9)$$

$$(x_2, y_2) = (0.075, 5.4)$$

$$(x, y) = (x, 10)$$

$$x = x_1 + (y - y_1)(x_2 - x_1)/(y_2 - y_1)$$

$$x = 0.15 + (10 - 26.9)(0.075 - 0.15)/(5.4 - 26.9)$$

$$x = 0.15 + (-16.9)(-0.075)/(-21.5)$$

$$x = 0.15 + (-0.059)$$

$$D_{10} = x = 0.091$$

Example 7.7.4

Calculate the coefficient of uniformity (C_u) and the coefficient concavity (C_c) from the previous example.

Solution:

Coefficient of uniformity

$$C_u = D_{60}/D_{10}$$

$$C_u = 0.363/0.091$$

$$C_u = 3.99$$

Coefficient of concavity

$$C_c = D_{30}{}^2/D_{60}D_{10}$$

$$C_c = (0.169)^2/(0.363)(\ 0.091)$$

$$C_c = (0.0286)/(0.033)$$

$$C_c = 0.87$$

Example 7.7.5

The results of a sieve analysis are provided in Table 7.4. Construct the PSD curve.

TABLE 7.4
The Sieve Analysis

U.S. Sieve No. Column 1	Opening Size, mm Column 2	Mass Retained on Sieve, g Column 3
4	4.750	0
12	1.700	18.50
25	0.710	54.20
35	0.500	67.30
60	0.250	110.2
100	0.150	67.30
200	0.075	42.30
270	0.053	18.50
Pan	–	15.20

Solution:

Percent finer is solved as follows (Table 7.5):
% finer = (cumulative mass – column 4)/(cumulative mass)
% finer = (393.5 – 18.5)/(393.5)
% finer = 0.952
% finer = 95.2%
The complete PSD is given in Figure 7.18.

TABLE 7.5
The Information for the PSD

U.S. Sieve No.	Opening Size (mm)	Mass Retained on Sieve (g)	Cumulative Mass Retained on Sieve (g)	Percent Finer (%)
Column 1	Column 2	Column 3	Column 4 (new)	Column 5 (new)
4	4.750	0	0	100.0
12	1.700	18.50	18.50	95.2
25	0.710	54.20	72.70	81.5
35	0.500	67.30	140.0	64.4
60	0.250	110.2	250.2	36.4
100	0.150	67.30	317.5	19.3
200	0.075	42.30	359.8	8.6
270	0.053	18.50	378.3	3.9
Pan	–	15.20	393.5	0.0

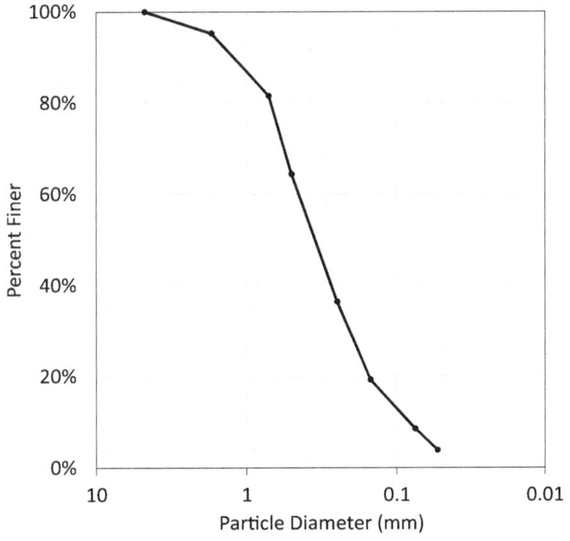

FIGURE 7.18 The PSD for Example 7.7.5.

Example 7.7.6

Given a soil sample of volume 1.2 m^3, total mass of 2,350 kg, moisture content of 8.6%, and specific gravity of 2.71, determine the moist density, dry density, void ratio, porosity, degree of saturation, and volume of water of the soil sample, using mass–volume relationships (after Das, 2006).

Refer to Figure 7.19.

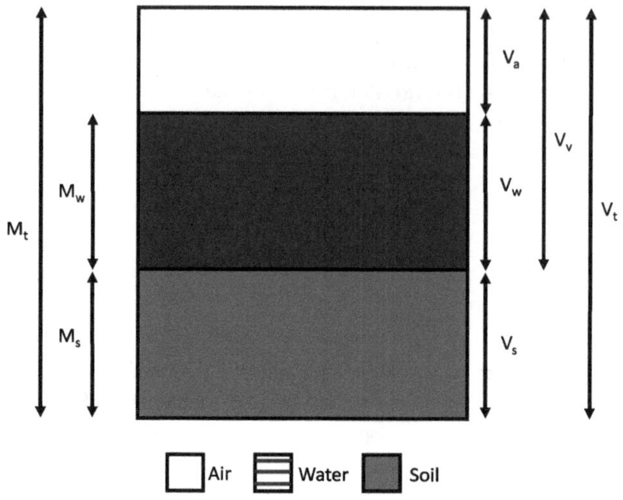

FIGURE 7.19 The makeup of a soil sample.

Solution:

Moist density can be expressed as, $\rho = M/V$,

$$\rho = M/V$$

$$\rho = 2{,}350 \text{ kg}/1.2 \text{ m}^3$$

$$\rho = 1{,}958 \text{ kg/m}^3$$

Dry density, $\rho_d = M_s/V$

$$M = M_s + M_w$$

$$M = M_s + \left(M_w/M_s\right)M_s$$

$$M = M_s + wM_s$$

$$M = M_s(1 + w)$$

$$M_s = M/(1 + w)$$

$$M_s = \left(2{,}350 \text{ kg}\right)/(1 + 8.6/100)$$

$$M_s = 2{,}163.9 \text{ kg}$$

$$\rho_d = 2{,}163.9 \text{ kg}/1.2 \text{ m}^3$$

$$\rho_d = 1{,}803 \text{ kg/m}^3$$

Void ratio, $e = V_v/V_s$

$$\rho_s = M_s/V_s$$

$$(\rho_w)(G_s) = M_s/V_s$$

$$(1{,}000\ \text{kg/m}^3)(2.71) = 2{,}163.9\ \text{kg/V}_s$$

$$V_s = 0.798\ \text{m}^3$$

$$V = V_v + V_s$$

$$1.2\ \text{m}^3 = V_v + 0.798\ \text{m}^3$$

$$0.402\ \text{m}^3 = V_v$$

$$e = V_v/V_s$$

$$e = 0.402\ \text{m}^3/0.798\ \text{m}^3$$

$$e = 0.504$$

Porosity, $n = V_v/V_t$

$$n = 0.402\ \text{m}^3/1.2\ \text{m}^3$$

$$n = 0.335$$

Degree of saturation, $S = V_w/V_v$

$$\rho_w = M_w/V_w$$

$$M = M_w + M_s$$

$$2{,}350\ \text{kg} = M_w + 2{,}163.9\ \text{kg}$$

$$M_w = 186.1\ \text{kg}$$

$$V_w = M_w/\rho_w$$

$$V_w = 186.1\ \text{kg}/1{,}000\ \text{kg/m}^3$$

$$V_w = 0.186\ \text{m}^3$$

$$S = V_w/V_v$$

$$S = 0.186\ \text{m}^3/0.402\ \text{m}^3$$

$$S = 0.463 \text{ or } 46.3\%$$

Example 7.7.7

The compressive principal stresses acting on a rock specimen of dolerite are determined to be 150 and 500 psi under dry conditions. Determine the normal and shear stresses on a plane of maximum shear for all values of theta for dry and saturated conditions. Comment on how the normal and shear stresses are affected by the pore water pressure.

The equations below are found in Farmer (1968), however the normal stress equation with pore water pressure uses a plus sign instead of a minus sign and the sign convention for shear stress is negative instead of positive.

Solution:

Dry conditions

$$\sigma = \tfrac{1}{2}(\sigma_1 + \sigma_2) + \tfrac{1}{2}(\sigma_1 - \sigma_2)\cos 2\theta$$

$$\sigma = \tfrac{1}{2}(500 + 150) + \tfrac{1}{2}(500 - 150)\cos 2\theta$$

$$\sigma = 325 + 175\cos 2\theta$$

$$\tau = -\tfrac{1}{2}(\sigma_1 - \sigma_2)\sin 2\theta$$

$$\tau = -\tfrac{1}{2}(500 - 150)\sin 2\theta$$

$$\tau = -175\sin 2\theta$$

Saturated conditions result in a pore-pressure, which counteracts the compressive principal stresses from dry conditions, and therefore,

$$\sigma = \tfrac{1}{2}[(\sigma_1 - \sigma_w) + (\sigma_2 - \sigma_w)] + \tfrac{1}{2}[(\sigma_1 - \sigma_w) - (\sigma_2 - \sigma_w)]\cos 2\theta$$

$$\sigma = \tfrac{1}{2}[(500 - \sigma_w) + (150 - \sigma_w)] + \tfrac{1}{2}[(500 - \sigma_w) - (150 - \sigma_w)]\cos 2\theta$$

$$\sigma = \tfrac{1}{2}(500 - 2\sigma_w + 150) + \tfrac{1}{2}(500 - 150)\cos 2\theta$$

$$\sigma = 325 + 175\cos 2\theta - \sigma_w$$

$$\tau = -\tfrac{1}{2}(\sigma_1 - \sigma_2)\sin 2\theta$$

$$\tau = -\tfrac{1}{2}[(\sigma_1 - \sigma_w) - (\sigma_2 - \sigma_w)]\sin 2\theta$$

$$\tau = -\tfrac{1}{2}[(500 - \sigma_w) - (150 - \sigma_w)]\sin 2\theta$$

$$\tau = -175\sin 2\theta$$

According to Farmer (1968) pore water will affect the normal stress but not the shear stress on the plane of maximum shear.

Example 7.7.8

The following mass–volume relationship is provided by Das (2006) for lab estimation of shrinkage limits of soils. Given the following information in the example, calculate the shrinkage limit. The lab test for shrinkage limit involves submersion in mercury (see ASTM D-427, 2004) or by a wax method (ASTM D-4943, 2018). An example is also provided by Das (2006):

$$\text{Shrinkage Limit (SL)} = [(M_1 - M_2)/M_2](100) - [(V_i - V_f)/M_2](\rho_w)(100)$$

where
M_1 is the mass of wet soil pat in dish at beginning of lab test.
M_2 is the mass of dry soil pat.
V_i is the initial volume of wet soil pat.
V_f is the volume of oven dried soil pat.
Note: A "soil pat" refers to a small, molded sample of soil that can be dried or wet enough to be shaped but not too wet that it seeps, flows, or deforms.

Given:
Initial volume of soil in saturated state: 16.2 cm^3
Final volume of soil in dry state: 10.8 cm^3
Initial mass in saturated state: 44.6 g
Final mass in dry state: 32.8 g

Solution:

$$\text{Shrinkage Limit (SL)} = \left[(M_1 - M_2)/M_2\right](100) - \left[(V_i - V_f)/M_2\right](\rho_w)(100)$$

$$SL = \left[(44.6 - 32.8)/32.8\right](100) - \left[(16.2 - 10.8)/32.8\right](1\ g/cm^3)(100)$$

$$SL = 35.97 - 16.46 = 19.5\%$$

Example 7.7.9

A spring is situated into a spring-cylinder model and filled with water. If a load (P) of 25 units is applied to the piston, in a cylinder which has an area of 25 cm^2, find the load carried by the spring and the water (a) if the valve to the model is closed and (b) if the valve to the model is open and the water is allowed to fully evacuate the cylinder. Also, find the hydrostatic pressure in the piston if the valve is closed. Concept and equations for the example are from Das (2006).

Solution:

For the spring-cylinder model, the load on the piston can be represented as follows:

$$P = P_s + P_w$$

P is the load on the piston.
P_s is the load carried by the spring.
P_w is the load carried by the water.
Part a: If the valve is closed to the system, the water cannot evacuate the soil, and therefore, the additional load will be carried by the incompressible fluid:
$P_s = 0$ units
$P_w = 25$ units

$$P = P_s + P_w$$

P = 0 units + 25 units
P = 25 units
The hydrostatic pressure is the load divided by the area of the cylinder:
$\Delta u = P/A = 25\ \text{units}/25\ cm^2 = 1\ \text{unit}/cm^2$
Part b: If the valve to the spring-cylinder model is open and the water is allowed to fully evacuate the cylinder,
$P_s = 25$ units
$P_w = 0$ units
P = 25 units + 0 units
P = 25 units

Example 7.7.10

The iron mineral, hematite, has a unit weight of 8,560 lbs/yd^3 and the gangue varies with respect to the hematite. If the mineral hematite is sampled at 40% by weight of the ore, the gangue samples at 60% by weight of ore, determine the volume occupied by 1,000 lbs of ore in a natural bed if the unit weight of gangue is 4,000 lbs/yd^3 (after Church 1981).

Solution:

Mineral portion

$$40\%\left(1{,}000 \text{ lbs}\right)/\left(8{,}560 \text{ lbs/yd}^3\right) = 0.0467 \text{ yd}^3$$

Gangue portion

$$60\%\left(1{,}000 \text{ lbs}\right)/\left(4{,}000 \text{ lbs/yd}^3\right) = 0.15 \text{ yd}^3$$

$$\text{Total Volume of Ore} = 0.0467 \text{ yd}^3 + 0.15 \text{ yd}^3 = 0.197 \text{ yd}^3$$

Example 7.7.11

Soil A has a liquid limit (LL) of 80% and a plastic limit (PL) of 42%. Soil B has an LL of 75% and a PL of 35%. Based on the plasticity index, which soil would be characterized with a higher measure of plasticity?

Solution:

Plasticity Index = Liquid Limit – Plastic Limit
Soil A:

$$\text{PI} = \text{Liquid Limit} - \text{Plastic Limit} = 80\% - 42\% = 38\%$$

Soil B

$$\text{PI} = \text{Liquid Limit} - \text{Plastic Limit} = 75\% - 35\% = 40\%$$

Soil B has a higher measure of plasticity.

Example 7.7.12

If a total volume of dry sand representing one cubic meter is combined with 0.3 m^3 of water, what is the porosity of the now-saturated sand after Heath (1983)?

Solution:

$$\text{Porosity, n} = \left(\text{Volume of voids, } V_v\right)/\left(\text{Total Volume, } V_t\right)$$

$$\text{Porosity, n} = 0.3 \text{ m}^3/1.0 \text{ m}^3 = 0.3$$

Example 7.7.13

If the same column of sand in the previous example is allowed to drain freely under the influence of gravity, which yields 0.2 m^3 of water, find the porosity, specific yield, and specific retention. Developed from the content in Heath (1983).

Solution:

Specific yield

$$S_y = V_d/V_t$$

$$S_y = 0.2 \text{ m}^3/1.0 \text{ m}^3$$

$$S_y = 0.2$$

Specific retention

$$S_r = V_r/V_t$$

$$S_r = \left(0.3\ m^3 - 0.2\ m^3\right)/1.0\ m^3$$

$$S_r = 0.1$$

Porosity

$$n = S_y + S_r$$

$$n = 0.2 + 0.1$$

$$n = 0.3 \text{ or } 30\% \text{ by volume}$$

Example 7.7.14

Given the following specific yield values for limestone, estimate the volume of water that yields from an area of one acre from a saturated limestone unit that is one meter thick. Provide the answer in gallons.

Solution:

$$S_y = V_d/V_t$$

$$V_t = (1\ \text{acre})(1\ m)$$

$$V_t = \left(43{,}560\ ft^2\right)(1\ m)(1\ ft/0.3048\ m)$$

$$V_t = 142{,}913\ ft^3$$

S_y = 18% by volume for limestone (Heath, 1983)

$$S_y = V_d/V_t$$

$$(18/100) = V_d/\left(142{,}913\ ft^3\right)$$

$$V_d = \left(25{,}724\ ft^3\right)\left(7.48\ g/1\ ft^3\right)$$

$$V_d = 192{,}418\ g$$

Example 7.7.15

Find the soil particle density, dry bulk density, and porosity for a soil where the mass of 1.0 cm^3 of soil particles is 2 g. The soil occupies a space of 1.5 cm^3 in a container.

Solution:

Soil particle density

$$\rho_s = \left(\text{Mass of soil particles, } M_s\right)/\left(\text{Volume of soil particles, } V_s\right)$$

$$\rho_s = 2\ \text{g}/1.0\ \text{cm}^3$$

$$\rho_s = 2\ \text{g/cm}^3$$

Bulk density

$$\rho_b = \left(\text{Mass of soil particles, M}_s\right)/\left(\text{Total volume, V}_t\right)$$

$$\rho_b = 2\ \text{g}/1.5\ \text{cm}^3$$

$$\rho_b = 1.3\ \text{g/cm}^3$$

Porosity

$$n = (1 - \rho_b/\rho_s)(100)$$

$$n = \left[1 - \left(1.3\ \text{g/cm}^3\right)/\left(2\ \text{g/cm}^3\right)\right](100)$$

$$n = 35\%$$

Example 7.7.16

A 1 m high column of soil is fully saturated with water. Above the column rests a pool of water that is replenished and holds a constant height of 0.1 m above the top of the soil. Use Darcy's law (below) to find the flow rate of water at the bottom of the soil if the area of the column is 2 m^2. For this simple case, treat the hydraulic conductivity of the soil as 10 cm/hr. This problem is similar to an example in McCarty et al. (2016).

Darcy's law

$$Q = kA\Delta h/L$$

where

Q is volume of water collected.
A is the area of the soil column.
L is the length of the soil column.
Δh is the difference in hydraulic head between the water surfaces.
k is the hydraulic conductivity of the soil.

Solution:

$$Q = kA\Delta h/L$$

$$Q = \left[(10\ \text{cm/h})(1\ \text{m}/100\ \text{cm})\left(2\ \text{m}^2\right)(1\ \text{m} + 0.1\ \text{m})\right]/(1\ \text{m})$$

$$Q = 0.22\ \text{m}^3/\text{h}$$

Example 7.7.17

Consider the column of soil in the previous example, but negate the pool of water above the sample. Instead, assume that the hydraulic head is measured at the top and bottom of the column with piezometers. If the elevation from a datum up to the second piezometer (located at the bottom of the soil) is h_2 = 0.28 m and the elevation to the first piezometer (located at the top of the soil) is h_1 = 1.43 m, find the flow rate using Darcy's law. How many gallons of water flow through the soil in 24 hours?

Darcy's law takes the same form as in the example above; however, a negative sign is used since the hydraulic head is taken differently:

$$Q = -kA\Delta h/L$$

$$\Delta h = h_2 - h_1$$

$$\Delta h = 0.28\ \text{m} - 1.43\ \text{m}$$

$$\Delta h = -1.15\ \text{m}$$

Solution:

$$Q = -kA\Delta h / L$$

$$Q = -\left[(10\ \text{cm/h})(1\ \text{m/100 cm})(2\ \text{m}^2)(-1.15\ \text{m})\right]/(1\ \text{m})$$

$$Q = 0.23\ \text{m}^3/\text{h}$$

Volume of water in 24 hours

$$V_w = Qt$$

$$V_w = (0.23\ \text{m}^3/\text{h})(24\ \text{h})(7.48\ \text{g/1 ft}^3)\left[(1\ \text{ft})^3/(0.3048\ \text{m})^3\right]$$

$$V_w = (5.52\ \text{m}^3)(7.48\ \text{g/0.0283 m}^3)$$

$$V_w = 1{,}458\ \text{gallons}$$

Example 7.7.18

The results of a sieve analysis are provided in Table 7.6. Construct the PSD curve, and determine the uniformity and curvature coefficients (Figure 7.20).

D_{60}

$$(x_1, y_1) = (4.75, 75.8)$$

TABLE 7.6
The Sieve Analysis

U.S. Sieve No.	Opening Size (mm)	Mass Retained on Sieve (g)	Cumulative Mass (g)	Percent Finer (%)
4	4.75	95	95	75.8
12	1.7	90	185	52.9
25	0.71	60	245	37.7
35	0.5	43	288	26.7
60	0.25	35	323	17.8
100	0.15	27	350	10.9
200	0.075	20	370	5.9
270	0.053	18	388	1.3
Pan	–	5	393	0.0

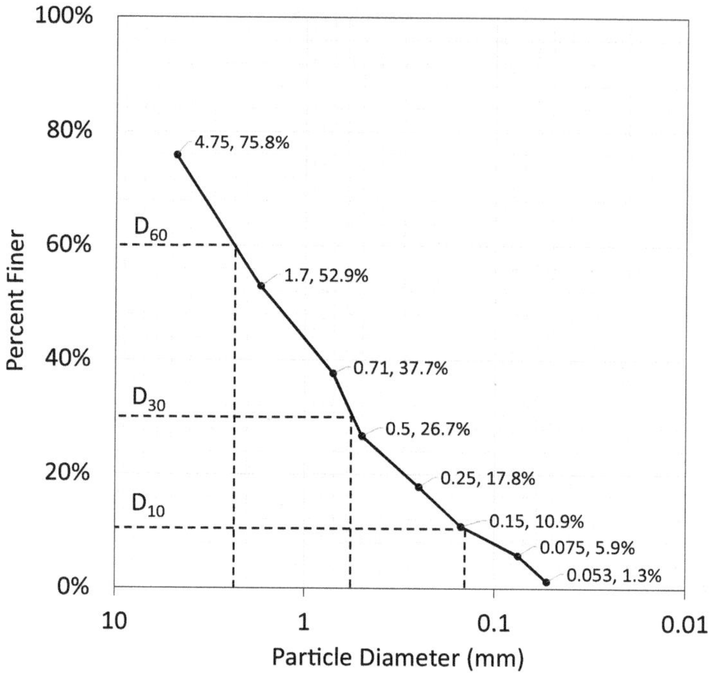

FIGURE 7.20 The PSD for Example 7.7.18.

$$(x_2,y_2) = (1.70, 52.9)$$

$$(x,y) = (x, 60.0)$$

$$x = x_1 + (y - y_1)(x_2 - x_1)/(y_2 - y_1)$$

$$x = 4.75 + (60 - 75.8)(1.70 - 4.75)/(52.9 - 75.8)$$

$$x = 4.75 + (-15.8)(-3.05)/(-22.9)$$

$$x = 4.75 + (-2.10)$$

$$D_{60} = x = 2.65$$

D_{30}

$$(x_1,y_1) = (0.71, 37.7)$$

$$(x_2,y_2) = (0.50, 26.7)$$

$$(x,y) = (x, 30.0)$$

$$x = x_1 + (y - y_1)(x_2 - x_1)/(y_2 - y_1)$$

$$x = 0.71 + (30 - 37.3)(0.50 - 0.71)/(26.7 - 37.3)$$

$$x = 0.71 + (-7.3)(-0.21)/(-10.6)$$

$$x = 0.71 + (-0.15)$$

$$D_{30} = x = 0.56$$

D_{10}

$$(x_1, y_1) = (0.15, 10.9)$$

$$(x_2, y_2) = (0.075, 5.9)$$

$$(x, y) = (x, 10.0)$$

$$x = x_1 + (y - y_1)(x_2 - x_1)/(y_2 - y_1)$$

$$x = 0.15 + (10 - 10.9)(0.075 - 0.15)/(5.9 - 10.9)$$

$$x = 0.15 + (-0.9)(-0.075)/(-5)$$

$$x = 0.15 + (-0.0135)$$

$$D_{10} = x = 0.14$$

Coefficient of uniformity

$$C_u = D_{60}/D_{10}$$

$$C_u = 2.65/0.14$$

$$C_u = 18.9$$

Coefficient of concavity

$$C_c = D_{30}^2/D_{60}D_{10}$$

$$C_c = (0.56)^2/(2.65)(0.14)$$

$$C_c = 0.85$$

Example 7.7.19

The results of a sieve analysis are provided in Table 7.7. Construct the PSD curve, and determine the percent gravel, sand, and fines using the United Soil Classification System (USCS).

According to the USCS Classification (Figure 7.21),
75 mm > Gravels > 4.75 mm
From the chart, gravels represent 100% – 93.2% = 6.8%:
4.75 mm > Sands > 0.075 mm
From the chart, sands represent 93.2% – 3.4% = 89.8%:
Fines: Passing 0.075 mm
From the chart, fines represent 3.4%.

TABLE 7.7
The Sieve Analysis

U.S. Sieve No.	Opening Size (mm)	Mass Retained on Sieve (g)	Cumulative Mass (g)	Percent Finer (%)
4	4.75	10	10	93.2
10	2	15	25	83.1
25	0.71	34	59	60.1
35	0.5	32	91	38.5
60	0.25	25	116	21.6
100	0.15	15	131	11.5
200	0.075	12	143	3.4
270	0.053	5	148	0.0
Pan	–	0	148	0.0

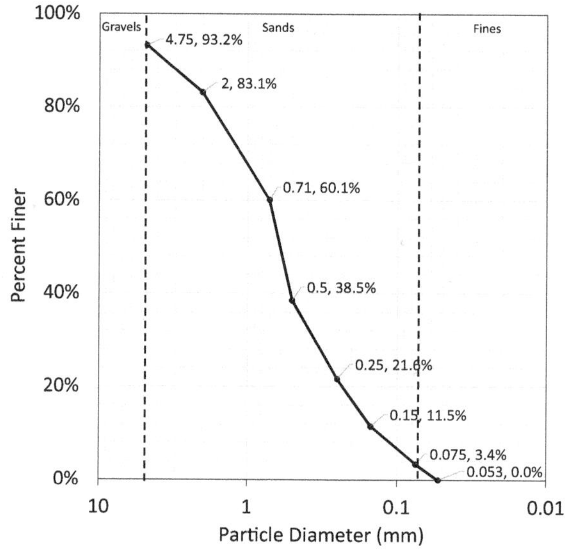

FIGURE 7.21 The PSD for Example 7.7.19.

Example 7.7.20

From the previous example, determine the percent gravel, sand, and fines using the British Standard Classification.

According to the British Standard Classification,

63 mm > Gravels > 2.00 mm

From the chart, gravels represent 100% – 83.1% = 16.9%:

2.00 mm > Sands > 0.063 mm

From the chart, sands represent 83.1% – 1.55% = 81.6%:

Percent passing 0.063 mm

$$(x_1, y_1) = (0.075, 3.4)$$

$$(x_2, y_2) = (0.053, 0.0)$$

$$(x, y) = (0.063, y)$$

$$y = y_1 + (y_2 - y_1)(x - x_1)/(x_2 - x_1)$$

$$y = 3.4 + (0 - 3.4)(0.063 - 0.075)/(0.053 - 0.075)$$

$$y = 3.4 + (-3.4)(-0.012)/(-0.022)$$

$$y = 3.4 + (-1.85)$$

$$y = 1.55$$

Fines: Passing 0.063 mm
From the chart, fines represent 1.55%.

Example 7.7.21

A thin layer between two surfaces contains a liquid. If one surface shifts due to a shear stress of 1 kPa, determine the velocity if the thickness between surfaces is 3.0 mm. If the area over which the movement occurs is reduced by half, what is the velocity? Treat the dynamic coefficient of viscosity of the liquid as 50 Pa-s.

Solution:

Part a

$$\mu = \tau/(dv/dy)$$

$$50\ Pa - s = [(1\ kPa)(1{,}000\ Pa / 1\ kPa)]/[(dv/(3\ mm)(1\ m/1{,}000\ mm)]$$

$$50\ N - s/m^2 = (1{,}000\ Pa)/(dv/3 \times 10^{-3} m)$$

$$(50\ N - s/m^2)(dv/3 \times 10^{-3} m) = (1{,}000\ N/m^2)$$

$$(16{,}667\ N - s/m^3)(dv) = (1{,}000\ N/m^2)$$

$$dv = 0.059\ m/s$$

Part b
If the area is reduced by half, the shear stress will double as the force is constant:

$$\mu = \tau / (dv/dy)$$

$$50\ Pa - s = [(2\ kPa)(1{,}000\ Pa/1\ kPa)]/[(dv/(3\ mm)(1\ m/1{,}000\ mm)]$$

$$50\ N - s/m^2 = (2{,}000\ Pa)/(dv/3 \times 10^{-3}\ m)$$

$$(50\ N - s/m^2)(dv/3 \times 10^{-3} m) = (2{,}000\ N/m^2)$$

$$(16{,}667\ N - s/m^3)(dv) = (2{,}000\ N/m^2)$$

$$dv = 0.119\ m/s$$

Example 7.7.22

A sample of soil has been pre-consolidated to a stress of 20 kN/m^2. Find the over-consolidation ratio if the present effective vertical stress is 20 kN/m^2. If the void ratio of the soil decreases, how will the OCR change?

Solution:

$$OCR = \sigma'_p / \sigma'_{vo}$$

$$OCR = (20\ kN/m^2)/(20\ kN/m^2)$$

$$OCR = 1, \text{normally consolidated.}$$

If the void ratio decreases, it is a result of an increase in the effective pressure, σ'_{vo}. Therefore, the soil would now be classified as under-consolidated.

REFERENCES

Ahlawat, A. (2018) Lecture 9-Calculation of capillarity rise and fall. Mechanical Education Hub. Retrieved from: www.modemech.blogspot.com

ASTM D2487 (2006) *Standard Practice for Classification of Soils for Engineering Purposes (Unified Soil Classification System)*. American Society for Testing and Materials (ASTM), West Conshohocken, PA.

ASTM D2488 (2023) *Description and Identification of Soils*. American Society for Testing and Materials (ASTM), West Conshohocken, PA.

ASTM D427 (2004) *Standard Test Method for Shrinkage Factors of Soils by the Mercury Method (Withdrawn 2008)*. American Society for Testing and Materials (ASTM), West Conshohocken, PA.

ASTM D4943 (2018) *Standard Test Method for Shrinkage Factors of Cohesive Soils by the Water Submersion Method. Annual Book of ASTM Standards, Vol. 04.08*. American Society for Testing and Materials (ASTM), West Conshohocken, PA.

ASTM D653 (2021) *Standard Terminology Relating to Soil, Rock and Contained Fluids*. American Society for Testing Materials (ASTM) Standards, West Conshohocken, PA.

Atterberg, A. (1911) Die Plastizität der Tone. *Int. Mitt. Bodenkd.* 1, 4–37.

Bear, J. (1979) *Hydraulics of Groundwater*. McGraw-Hill, New York, p. 569.

British Standards Institution (2007). *BS 5930:1999 Code of Practice for Site Investigations*. BSI, London.

Chem Europe (2025) Hydrology. Retrieved from: https://www.chemeurope.com/en/encyclopedia/Hydrology.html

Church, H.K. (1981) *Excavation Handbook*. McGraw-Hill Book Company, New York.

Darcy, H. (1856) *Les fontaines publiques de la ville de Dijon*. Dalmont, Paris.

Das, B.M. (2006) *Principles of Geotechnical Engineering* (6th Edition). Cengage Learning Inc., Stamford, CT.

Das, B.M. (2010) *Principles of Geotechnical Engineering* (7th Edition). Cengage Learning Inc., Stamford, CT.

Das, B.M. and Sivakugan, N. (2016) *Introduction to Geotechnical Engineering* (2nd Edition). Cengage Learning Inc., Stamford, CT.

Das, B.M. and Sivakugan, N. (2019) *Principles of Foundation Engineering* (9th Edition). Cengage Learning Inc., Stamford, CT.

Dupuit, J. (1863) *Etudes Théoriques et Pratiques sur le Mouvement des Eaux dans les Canaux Découverts et à Travers les Terrains Perméables [Theoretical and Practical Studies on the Movement of Water in Open Channels and Permeable Ground]*. Dunod, Paris, France.

Dutto, P. (2014) Modelling of landslides propagation with SPH: Effects of rheology and pore water pressure. Thesis. Universidad Politecnica De Madrid, Spain.

Elementary Engineering (2019) Particle size distribution curve-Civil Engineering. Retrieved from: www.elementaryengineeringlibrary.com/civil-engineering/soil-mechanics/particle-size-distribution-curve

Farmer, I.W. (1968) *Engineering Properties of Rocks*. E. & F.N. Spon Ltd, London.

Fredlund, D.G. (1993) *Soil Mechanics for Unsaturated Soils*. Wiley, New York.

Fredlund, D.G. and Rahardjo, H. (1993) *Soil Mechanics for Unsaturated Soils*. John Wiley and Sons, New York.

Galperin, M., Zayetsev, V.S. and Normatov, Y.A. (1993) *Hydrogeology and Engineering Geology: Geotechnika*. CRC Press, Taylor & Francis Group, Boca Raton, FL.

Hagen, G.H.L. (1839) Uber die Bewegung des Wassers in engen cylindrischen R6hren. *Poggendorf's Annalen der Physik und Chemie*. 46, 423–442.

Halley, E. (1686) An historical account of the trade winds, and monsoons, observable in the seas between and near the Tropics, with an attempt to assign the physical cause of the said winds. *Philos. Trans.* 16, 153–168.

Heath, R.C. (1983). Basic ground-water hydrology, U.S. Geological Survey Water-Supply Paper 2220, p. 86.

Jillavenkatesa, A., Dapkunas, S.J. and Lum, L-S.H. (2001) *Particle Size Characterization*. NIST Special Publication, Washington, DC, pp. 960–961.

Likos, W.J. and Lu, N. (2004) *Unsaturated Soil Mechanics*. John Wiley & Sons Inc., New Jersey

Marais, C.S and Spearing, A.J.S. (1990) *Internal Training Module*. Anglogold Division, Anglo American Corp, South Africa.

Mariotte, E. (1679) Essais de physique, ou Mémoires pour servir à la science des choses naturelles attribué à Roberval (1679–1681).

McCarty, L.B., Hubbard, L.R. and Quisenberry, V.L. (2016) *Applied Soil Physical Properties, Drainage, and Irrigation Strategies*. Springer International Publishing, Switzerland.

Perrault, P. (1966). *On the Origin of Springs*. Translation of "De l'origine des fontaines" (1674). Hafner. ASIN B0026MBIHE

Schlitt, W. J. (1984) *The Role of Solution Management in Heap and Dump Leaching. Au and Ag Heap and Dump Leaching Practice*. SME-AIME, Denver, CO, pp. 69–83.

Singh, M.M. (1992) *Mine Subsidence (Chapter 10.6) in SME Mining Engineering Handbook*. SME-AIME, Denver, CO, pp. 938–971

Soliman, M. (1998) *Environmental Hydrogeology*. CRC Press LLC, Boca Raton, FL.

Suryakanta, P. (2015) *How to calculate bearing capacity of soil from plate load test?* Retrieved from: CivilBlog.org.

Terzaghi, K. (1925) *Erdbaumechanik auf Bodenphysikalischer Grundlage*. Franz Deuticke, Liepzig-Vienna, Austria.

Terzaghi, K. (1943) *Theoretical Soil Mechanics*. John Wiley & Sons Inc., New York, p. 510.

US Geological Survey (n.d.) Water Science School. Retrieved from: www.usgs.gov/special-topic/water-science-school

Volkholz, J. (2021) *River Flood Modelling*. ISIpedia, Germany.

Wesley, L. D. (2013). Residual soils and the teaching of soil mechanics. In *Proceedings of the 18th International Conference on Soil Mechanics and Geotechnical Engineering* (pp. 3479–3482).

Yu, C., Loureiro, C., Cheng, J., Jones, L.G., Wang, Y.Y., Chia, Y.P. and Faillace, E. (1993) *Data Collection Handbook to Support Modeling Impacts of Radioactive Material in Soil*. Environmental Assessment and Information Sciences Division-Argonne National Laboratory, Argonne, IL.

8 Subsidence

8.1 INTRODUCTION TO SUBSIDENCE

Subsidence refers to a surface elevation sinking to a lower elevation. With the vertical ground movement is horizontal displacement particularly at the edges of the subsidence zone. Subsidence can be considered mining-induced or natural and can be planned, because of the mining method, or unplanned. Impacts of subsidence are potentially severe and can affect surface water, groundwater, roads, rail, buildings, vegetation, and future mining.

Subsidence occurs due to the creation of a void, or space, from the intentional or unintentional removal of gases, solids, or liquids. If the space is too great relative to the competency of the rock, subsidence can initiate. Subsidence without warning is a major safety concern.

Factors related to mining-induced subsidence include the following:

- Mine design and method.
- Orebody depth, dip, shape, and size.
- Extent of mining.
- Topography.
- Thickness of the deposit.
- Properties of the rock.
- Regional and local geology.
- Hydrology (surface and groundwater).

Natural subsidence occurs because of the following (Whittaker and Reddish, 1989):

- Soil compaction.
- Soil shrinkage.
- Lowering of the water table.
- Subterranean voids.
- Tectonic and volcanic activity.

Figure 8.1 shows a typical mine subsidence cycle (Pennsylvania Department of Environmental Protection, n.d.).

Underground mining can cause surface subsidence, and the main concerns are as follows:

- Magnitude of the vertical subsidence.
- Lateral extent of surface subsidence.
- Time over which the subsidence will occur.
- Nature of the subsidence (i.e., continuous or sporadic).

Countries tend to have their own regulations regarding mining-induced subsidence and interference with the surface and groundwater. In the United States, for example, subsidence is regulated by the Federal Surface Mining Control and Reclamation Act (SMCRA) of 1977 for coal. As part of the regulation, prospective mines develop subsidence and hydrologic impact studies and plans before mining begins. Solutions are provided to deal with any potentially negative environmental impacts identified. The potential impacts are typically covered by Environmental Impact Studies (or other equivalent studies) in many countries.

DOI: 10.1201/9781032704845-8

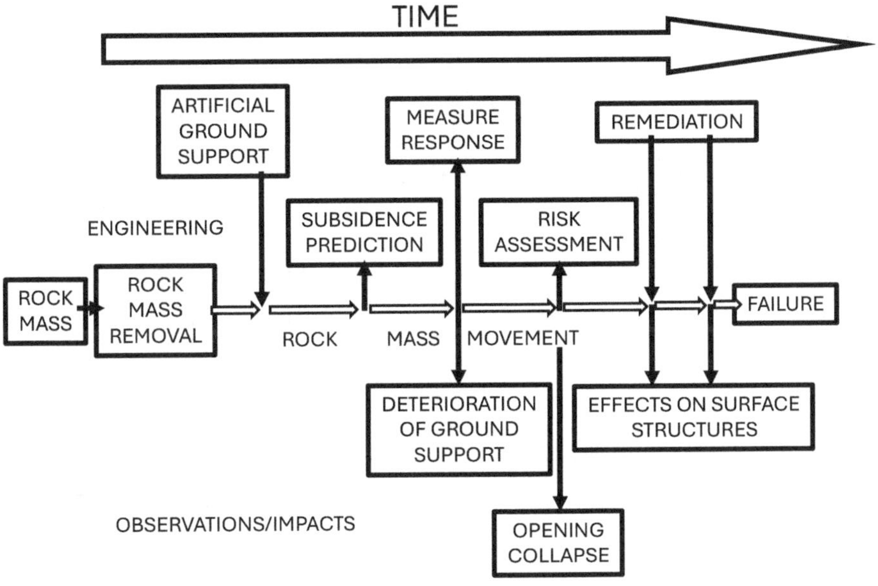

FIGURE 8.1 The mine subsidence cycle (PA Department of Environmental Protection, n.d.).

The source for much of the information below is taken from the Pennsylvania Department of Environmental Protection website.

8.2 NATURAL SUBSIDENCE

Many natural factors cause damage to structures other than mine subsidence. Most structures have some degree of damage even prior to mining. Property owners are not always aware of the structural condition of their homes and properties since they do not generally conduct routine and thorough inspections. It is therefore important for a mine operator to assess the stability of structures prior to any mining as it can help justify claims from property owners that may arise later. Some of the common reasons for structural damages other than mine subsidence, as shown in Figures 8.2 and 8.3, are as follows:

- Settlement due to drying of soils or the weight of surface structures.
- Landslides and soil creep.
- Shrinking and/or swelling of soils.
- Freezing and thawing of soils.
- Surface and subsurface erosion.
- Poor construction design, methods, and products.
- Structural deterioration with age, lack of maintenance, or misuse.
- Structural movements.

8.3 TYPICAL MINING-INDUCED SUBSIDENCE

8.3.1 Room-and-Pillar Coal Mine–Induced Subsidence

Room-and-pillar coal mining is typically suitable and economic in shallow coal deposits with strong immediate roof rock. Room-and-pillar mining transfers the overburden load to pillars, while the roof is supported using rock anchors for the duration of extraction. Some pillars can be totally or partially recovered on a retreat basis before secondary mining is completed, thereby leaving a condition for the remaining remnant pillars to possibly fail under the increased pillar load. Pillar

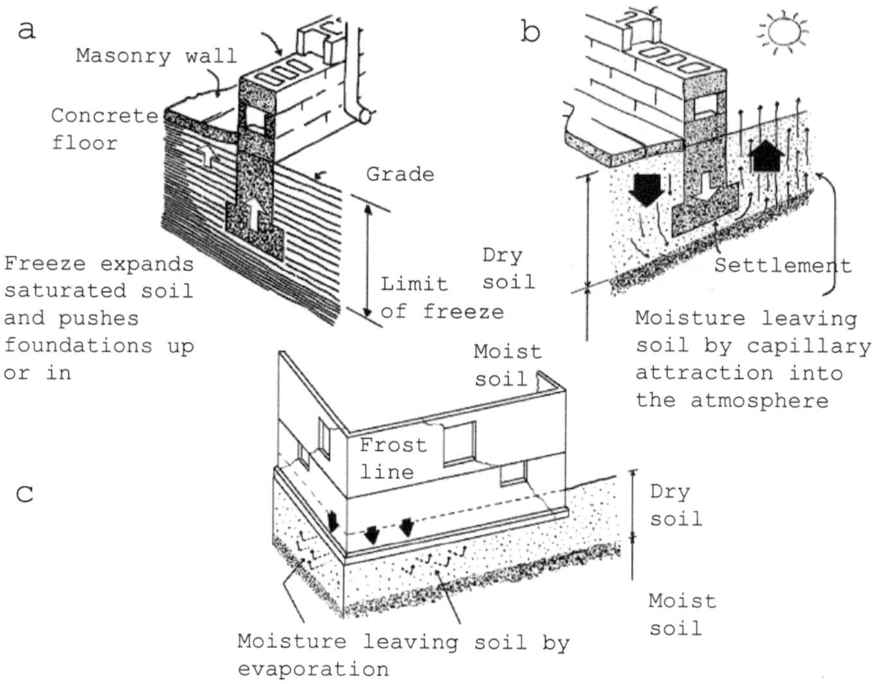

FIGURE 8.2 Building damage due to frost action (a), settlement due to drying of soil (b), and foundation failures due to soil shrinkage (c). (Retrieved from PADEP (n.d.), modified after American Insurance Services Group, Inc. (1972) and published in American Insurance Services Group, Inc. (1990).)

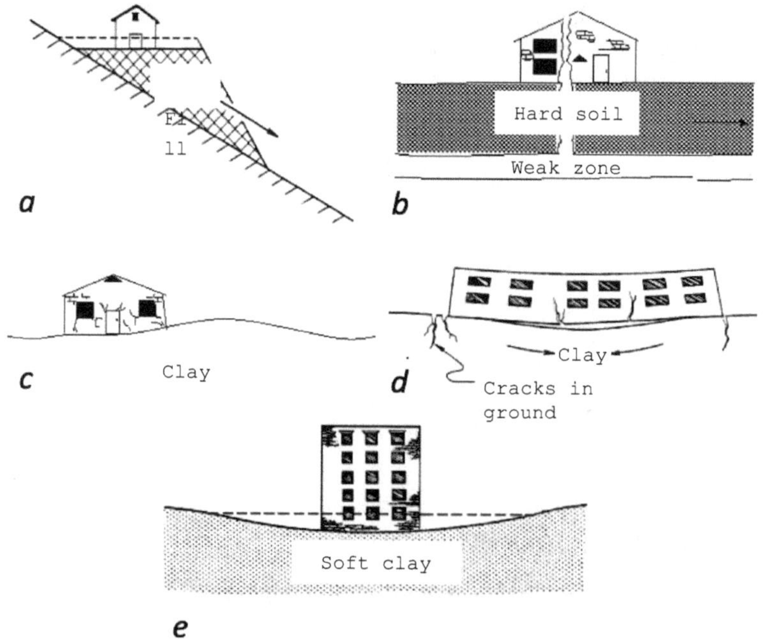

FIGURE 8.3 Translational ground movement by landslide from fill placement (a) and earthquake (b). Distortional movement caused by swelling by wetting of soils (c), shrinkage of clay under the center of a building (d), and settlement under weight of a structure (e). (Retrieved from PADEP (n.d.), modified after Tschebotarioff (1951).)

extraction has a much higher hazard associated with the extraction, so it must be carefully designed and implemented, and it has a larger chance for surface subsidence.

Depending on the pillar design if mining is shallow, and especially if the roof and/or floor rock is weak, it is easy for subsidence to progress to surface during the years after extraction. The nature and magnitude of these effects depend on the overburden geology and properties, related rock mechanics features, and the mine design. The two major forms of subsidence mechanisms from room-and-pillar coal mining are sinkholes and subsidence troughs.

According to the Pennsylvania Department of Environmental Protection (PADEP), sinkholes can be common in areas overlying shallow room-and-pillar mines. Sinkholes occur from the generally time-dependent collapse of the mine roof into a mine opening, resulting in caving of the overlying strata and a depression at the ground surface. This type of subsidence is generally localized in extent, affecting a relatively small area on the overlying surface. However, structures and surface features affected by sinkhole subsidence tend to experience extensive and costly damage. Sinkholes can be associated with active or abandoned mine workings. Modern underground mines operate at depths sufficient to generally preclude the development of sinkhole subsidence. Figure 8.4 shows the typical modes of surface subsidence above shallow room (bord) and pillar workings.

Subsidence troughs from room-and-pillar mining can occur above active or abandoned mines. The resultant surface impacts and damages can be similar; however, the mechanisms that trigger the subsidence are dramatically different. In abandoned mines, troughs usually occur when the overburden sags downward due to the failure of remnant mine pillars, or by punching of the pillars into a soft mine floor or roof. It is difficult to predict if or when failure in an abandoned mine might occur, since abandoned mines may collapse many decades after the mining is completed, especially if the mine workings were not designed to provide long-term support (Whittaker and Reddish, 1989). Water ingress into abandoned workings over time is clearly a major contributing factor.

The following controlling factors make it difficult to predict with accuracy the time frame and extent of subsidence. The influencing factors include the following:

- Seam depth below surface.
- Mining height.
- Presence of multiple seams that have been or will be mined.
- Strength, dimensions, and layout of support pillars.

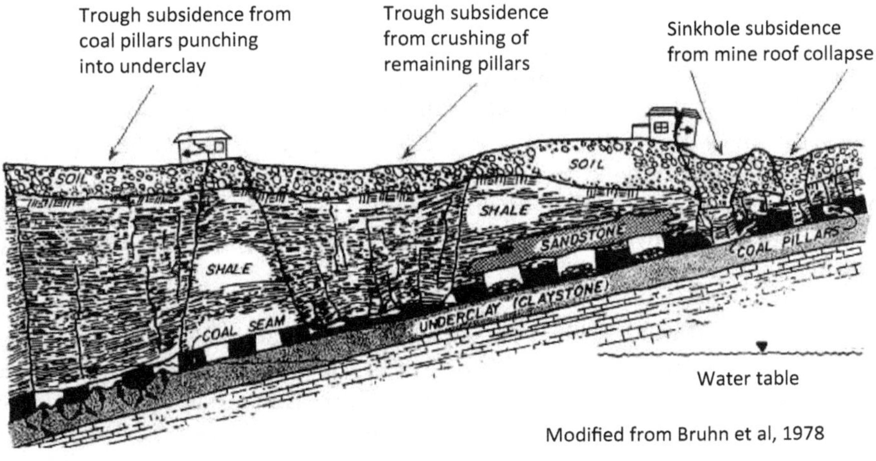

FIGURE 8.4 Modes of surface subsidence caused by shallow room-and-pillar coal mining. (Retrieved from PADEP (n.d.), modified from Bruhn et al. (1978).)

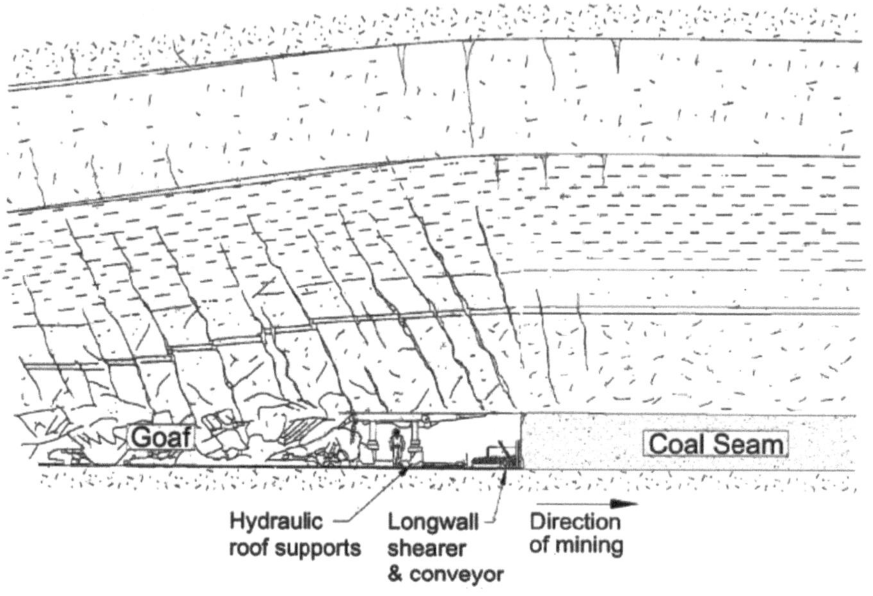

FIGURE 8.5 Section through an advancing longwall panel showing the strata collapse behind the shields (Mine Subsidence Engineering Consultants, 2007).

- Variation in the strength properties within a rock unit.
- Water inflow.
- Mining method, especially deliberate caving.
- Rate of mining.
- Fill support, if used.

8.3.2 Typical Subsidence Caused by Longwall Coal Mining

Longwall mining is the safest and most productive and generally has the highest extraction (approximately 80%) of any other underground coal mining method. It does, however, induce significant ground failure and subsidence as the overlying strata collapses (gobs or goafs) behind the longwall shields. A typical layout is shown in Figure 8.5.

A principle behind the longwall mining system is that the roof close to the advancing face is supported by huge hydraulic supports (immediately adjacent to the worked seam). Behind these supports (at the rear of them), the roof is allowed to fall and break up, and with the swell of the material as it fractures, the void where the seam was is filled. Eventually this establishes equilibrium in the strata behind the mined area (in the collapsed, goafed, or gobbed area).

Depending mainly on the mining height, the depth, and the overburden rock geology and properties (that control the goafing behind the longwall shields), longwall surface subsidence may be one of the following:

- Subcritical longwall width: This occurs at a relatively small longwall span as the goaf is able to arch and limit the surface subsidence to a level less than S_{max} as shown in Figure 8.6c.
- Critical longwall width: This is a very specific situation as shown in Figure 8.6a, where the conditions are such that the maximum possible surface subsidence (S_{max}) occurs but just in the center of the longwall.
- Supercritical longwall width: Any longwall width wider than the critical width will not create a larger maximum subsidence (S_{max}), but the extent of the area that has the maximum subsidence increases as shown in Figure 8.6b.

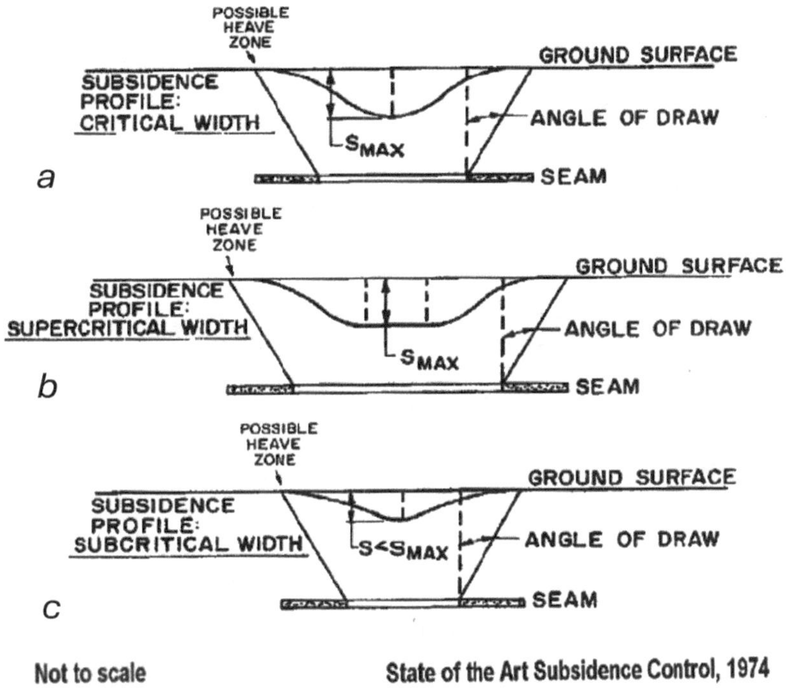

FIGURE 8.6 Surface subsidence profiles above longwall mines showing critical (a), supercritical (b), and subcritical (c) widths of panels (Gray et al., 1974).

Jeran and Trevits (1995) determined that the total subsidence for longwall panels in the Pittsburgh Coalbed was complete about a year after mining. Above the center of the panel, where subsidence is greatest, the subsidence reaches 90% complete by the time the face has advanced a distance equal to one overburden thickness beyond a particular surface point.

Figure 8.7 shows the surface tension and compression conditions and the angle of draw above a longwall panel. The Inflection Point defines where the strain on surface transitions from tension to compression (see the inset of Figure 8.7). The Inflection Point is generally considered to occur slightly in from the panel edge toward the gob. The Angle of Draw or Limit Angle is the outward angle between the vertical line at the panel edge and a line connecting the panel edge and the point on the surface where the observed subsidence is zero (Michaud, 2018). All of the surface subsidence occurs within the angle of draw. In the case of Figure 8.7, the angle of draw is shown as 25°.

Figure 8.8 shows the subsidence trough on the surface above a longwall panel and the respective movements and strains. Buildings can tolerate vertical movement, but tilt (or horizontal movement) can cause significant cracks and structural damage. When a structure has to be undermined by longwall mining, damage is reduced if it is located near the center of the longwall panel. This is because it will only tilt briefly as the longwall advances past it, but then, the movement will be mainly vertical. If the structure is close to the longwall edge, it will be subjected to near-continual tilt.

8.3.3 Subsidence Typically Caused by Salt, Potash, and Trona Mining

Subsidence caused by underground salt mining depends on the mining method used which is typically done by:

- Solution mining

  This method can cause severe subsidence typically as water is injected into the salt deposit to dissolve it and the saline-rich solution is pumped to the surface. This causes underground voids that can ultimately cause the overlying strata to collapse.

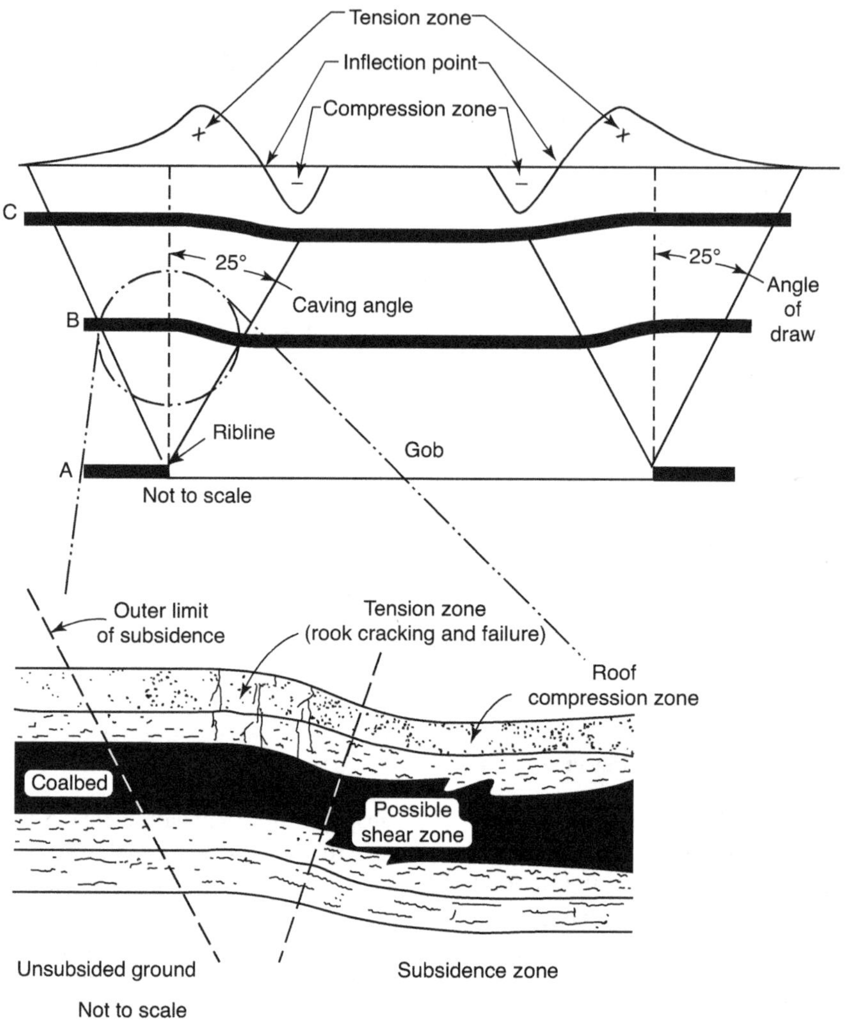

FIGURE 8.7 The subsidence profile above a longwall coal mine showing the inflection point and angle of draw. After Michaud (2018).

Solution cavities' size and extent are generally unknown, which can create significant issues once abandoned and especially if water is allowed to enter the cavities and interconnect with other voids (Whittaker and Reddish, 1989).

- Room-and-pillar mining

  Mining salt using room and pillar methods tends to create much more common surface subsidence than coal because the salt tends to creep, and after mining, any water ingress creates solution voids that exacerbate subsidence.
- Longwall mining

  Trona and potash can be mined using longwall mining in the same way as coal. The surface subsidence tends to be similar to that when coal is longwall mined.

8.3.4 Subsidence Caused by Hard-Rock Mining

Subsidence is dependent on the mining method and whether backfill is used. The overall subsidence principles are the same for hard- and soft-rock mining, although the magnitude and extent of the subsidence may be different.

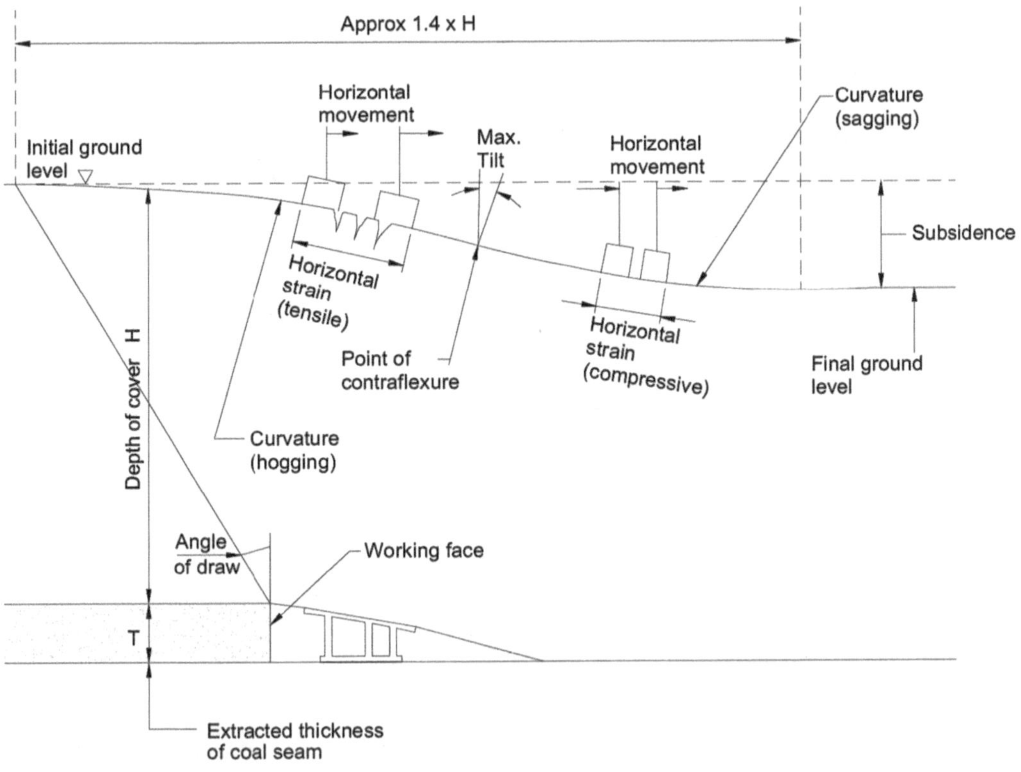

FIGURE 8.8 The development of the subsidence trough above an advancing longwall panel (Mine Subsidence Engineering Consultants, 2007).

The main hard-rock mining methods that can create significant subsidence issues are as follows:

- Block caving.
- Sub-level caving.
- Room-and-pillar mining.
- Solution mining.
- Sub-level stoping (without backfill). Backfill is frequently used with sub-level stoping, but this is not common practice in the United States currently.

Block caving and sub-level caving generally create the most damaging and extensive subsidence that cannot be cost effectively repaired. The huge volume of rock extracted and the near-total extraction achieved result in massive ground subsidence. The rock generally fails in tension and shear. Typically, subsidence above a block cave mine in shown in Figure 8.9 (Van As, 2003).

8.3.5 Potential Hydrologic Impacts of Underground Mining

Underground mine excavations often intercept aquifers or can create new ones due to mining, which can transport surface water and groundwater into mined openings. When mining below the water table, mine excavations become effectively low-pressure sinks so that groundwater flows freely into the excavations from the surrounding water-bearing rocks or geological feature. This can cause the partial or total dewatering of the overlying rock strata. Depending on the geology, hydrology, stresses, and fracturing, there is also potential to negatively impact remote water-bearing strata and surface water too.

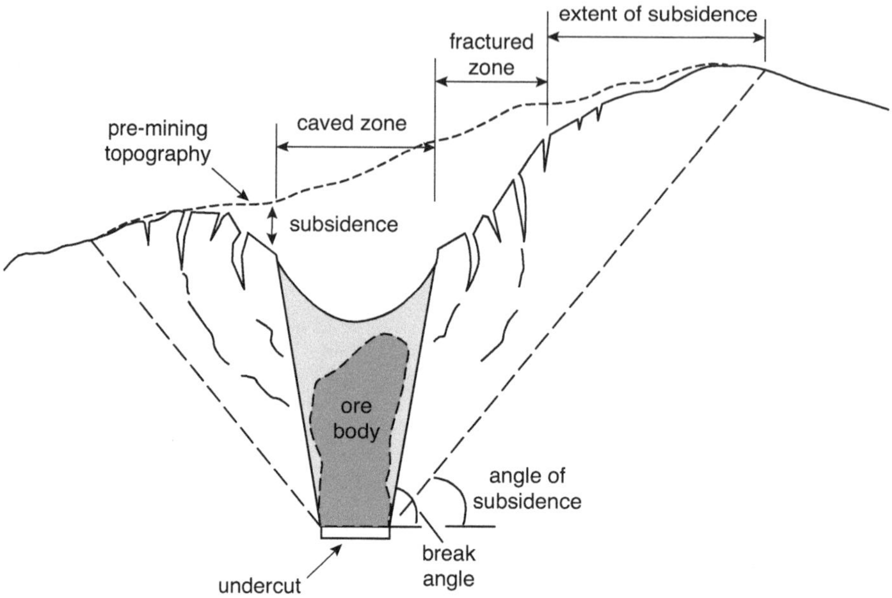

FIGURE 8.9 Subsidence above a typical block cave mine (Van As, 2003).

The impact on localized water resources varies considerably mainly depending on the mining method used. For example, the subsidence impacts of room-and-pillar mining tend to be localized, irregular, and often long after mining, whereas longwall subsidence is much more immediate, pervasive, systematic, and ultimately predictable (Booth, 1998).

8.4 PREDICTING SUBSIDENCE

There are four main techniques for predicting mine subsidence (Sepehri et al., 2017):

- Empirical methods (graphical methods, profile functions, and influence functions).
- Analytical methods.
- Numerical methods.
- Hybrid methods.

8.4.1 Graphical Methods

- These are empirically derived graphs and charts which are used to determine the subsidence at specific points on the surface.
- They are simple but have limited use as they are only applicable to the area where the graphs and charts were derived, using whatever measurements were available.

8.4.2 Profile Functions

- These are functions or mathematical equations developed to match the actual measured surface profile.
- They are also simple but are again fairly limited in application.
- Profile functions are generally used only for single coal seam applications.

8.4.3 Influence Functions

- The subsidence of specific areas of a seam is numerically integrated in order to calculate the total surface subsidence, using a derived formula. They use the Superposition Principle (Bals, 1931).
- The functions are moderately difficult to calculate but can be applied in many applications.
- Numerous parameters must be known such as the influence angle of the longwall to apply the formula.
- The functions can handle many situations such as irregular panels, variable seam dips, and irregular surface topography.

8.4.4 Numerical Methods

- The Fundamental Laws of Mechanics are used in complex mathematical approximations typically using Finite-Element, Finite-Difference, Boundary-Element, and Discrete-Element for the orebody and the overburden for calculating the subsidence.
- They can be difficult to set up especially in 3-D but are extremely flexible.
- They need reliable material properties, representative failure criteria, and calibration with real measurement data. The method is useful as it can easily produce qualitative (comparative) results and sensitivity analyses if the input parameters are not well known.

There are many packages available that can be used for predicting subsidence including:

- Subsidence Engineers Handbook – (Graphical Method) National Coal Board in the United Kingdom.
- Subside – (Influence Function) Beulah Engineering.
- SPASID – (Influence Function) Penn State University.
- SUBPRO – (Profile Function) U.S. Bureau of Mines (now defunct).
- Surface Deformation Prediction System – (Influence Function) Office of Surface Reclamation and Enforcement.
- LWSUB – (Influence Function) West Virginia University, US.
- LAMODEL – (Boundary Element) National Institute of Occupational Safety and Health, US.
- FLAC – Itasca Consulting Group.
- UDEC and 3DEC – Itasca Consulting Group.

8.5 WORKED EXAMPLES

Example 8.5.1

A point load of 24 kN acts on the surface of an elastic mass of very great extent. This is an unworked example found in Terzaghi et al. (1996).

What is the intensity of vertical pressure, due to load:

Part a: At a depth of 1 m directly below the load?
Part b: At a depth of 2 m?
Part c: At a depth of 5 m?
Part d to f: What is the intensity of vertical pressure at the same depths at a horizontal distance of 1 m from the line of action of the load?

After Terzaghi et al. (1996), one of Boussinesq's equations for determining the state of stress at point, N, of Figure 8.10:

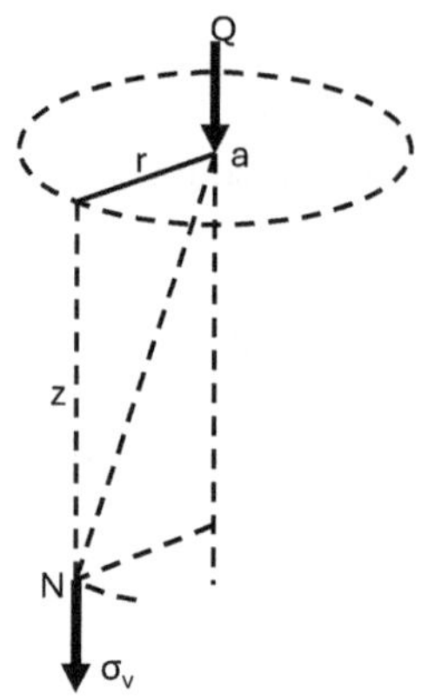

FIGURE 8.10 Vertical pressure at point N in interior of a semi-infinite solid. (Redrawn after Terzaghi et al. (1996).)

$$\Delta\sigma_v = (3Q/2\pi z^2)\left\{1/\left(1+(r/z)^2\right)\right\}^{5/2}$$

where

$\Delta\sigma_v$ is the change in vertical pressure.
Q is a concentrated vertical load.
r is the horizontal distance from N to the line of action of the point load Q.
z is the vertical distance between N and the surface of the mass.

Solution:

Part a

$$\Delta\sigma = (3Q/2\pi z^2)\left\{1/\left[1+(r/z)^2\right]\right\}^{5/2}$$

$$\Delta\sigma = \left[(3)(24)/2\pi(1\ \text{m})^2\right]\left\{1/\left[1+(0/1)^2\right]\right\}^{5/2}$$

$$\Delta\sigma = 72\ \text{kN}/6.28\ \text{m}^2$$

$$\Delta\sigma = 11.5\ \text{kN/m}^2$$

Part b

$$\Delta\sigma = (3Q/2\pi z^2)\left\{1/\left[1+(r/z)^2\right]\right\}^{5/2}$$

$$\Delta\sigma = \left[(3)(24)/2\pi(2\ \text{m})^2\right]\left\{1/\left[1+(0/2)^2\right]\right\}^{5/2}$$

$$\Delta\sigma = 72\ \text{kN}/25.1\ \text{m}^2$$

$$\Delta\sigma = 2.86\ \text{kN/m}^2$$

Part c

$$\Delta\sigma = (3Q/2\pi z^2)\left\{1/\left[1+(r/z)^2\right]\right\}^{5/2}$$

$$\Delta\sigma = \left[(3)(24)/2\pi(5\text{ m})^2\right]\left\{1/\left[1+(0/5)^2\right]\right\}^{5/2}$$

$$\Delta\sigma = 72\text{ kN}/157.1\text{ m}^2$$

$$\Delta\sigma = 0.46\text{ kN/m}^2$$

Part d

$$\Delta\sigma = (3Q/2\pi z^2)\left\{1/\left[1+(r/z)^2\right]\right\}^{5/2}$$

$$\Delta\sigma = \left[(3)(24)/2\pi(1\text{ m})^2\right]\left\{1/\left[1+(1/1)^2\right]\right\}^{5/2}$$

$$\Delta\sigma = \left[72\text{ kN}/6.28\text{ m}^2\right](0.177)$$

$$\Delta\sigma = 2.0\text{ kN/m}^2$$

Part e

$$\Delta\sigma = (3Q/2\pi z^2)\left\{1/\left[1+(r/z)^2\right]\right\}^{5/2}$$

$$\Delta\sigma = \left[(3)(24)/2\pi(2\text{ m})^2\right]\left\{1/\left[1+(1/2)^2\right]\right\}^{5/2}$$

$$\Delta\sigma = \left[72\text{ kN}/25.1\text{ m}^2\right](0.572)$$

$$\Delta\sigma = 1.64\text{ kN/m}^2$$

Part f

$$\Delta\sigma = (3Q/2\pi z^2)\left\{1/\left[1+(r/z)^2\right]\right\}^{5/2}$$

$$\Delta\sigma = \left[(3)(24)/2\pi(5\text{ m})^2\right]\left\{1/\left[1+(1/5)^2\right]\right\}^{5/2}$$

$$\Delta\sigma = \left[72\text{ kN}/157.1\text{ m}^2\right](0.907)$$

$$\Delta\sigma = 0.42\text{ kN/m}^2$$

Example 8.5.2

A circular area on the surface of an elastic mass of great extent carries a uniformly distributed load of 120 kPa. The radius of the circle is 3 m. What is the intensity of vertical pressure at points 3, 6, and 12 m beneath the center of the circle and at points at the same depth beneath the edge of the circle? This is an unworked example found in Terzaghi et al. (1996).

Figure 8.11 is necessary to complete the above problem in closed form. The figure is originally from Lambe and Whitman (1969) but reproduced in Terzaghi et al. (1996).

After Terzaghi et al. (1996) for a uniform load applied over a circular area at the ground surface:

$$\Delta\sigma = I(r/a, z/a)q$$

where

$\Delta\sigma_v$ is the change in vertical pressure.
I(r/a, z/a) is a dimensionless influence factor found in Figure 8.11.
a is the radius of the circular area.
z is the depth.
q is a uniformly distributed load.

Solution:

Part a

a=3 m and z=3 m therefore I(r/a, z/a)=0.65 from figure 8.11.

$$\Delta\sigma = I(r/a, z/a)q$$

$$\Delta\sigma = 0.65(120 \text{ kPa})$$

$$\Delta\sigma = 78 \text{ kPa}$$

Part b

$$\Delta\sigma = I(r/a, z/a)q$$

$$\Delta\sigma = 0.29(120 \text{ kPa})$$

$$\Delta\sigma = 35 \text{ kPa}$$

Part c

$$\Delta\sigma = I(r/a, z/a)q$$

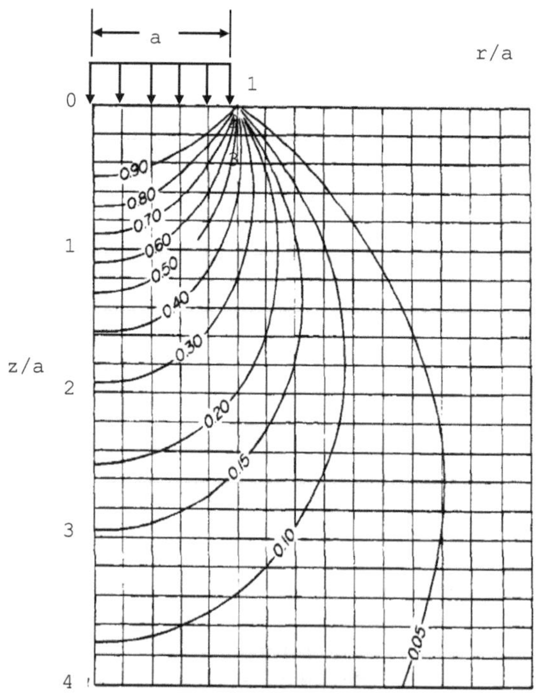

FIGURE 8.11 Influence factors for the uniform load over a circular area. (Originally from Lambe, T.W. and R.V. Whitman (1969) found in Terzaghi et al. (1996).)

$$\Delta\sigma = 0.09(120\text{ kPa})$$

$$\Delta\sigma = 11\text{ kPa}$$

Part d

$$\Delta\sigma = I(r/a, z/a)q$$

$$\Delta\sigma = 0.33(120\text{ kPa})$$

$$\Delta\sigma = 40\text{ kPa}$$

Part e

$$\Delta\sigma = I(r/a, z/a)q$$

$$\Delta\sigma = 0.2(120\text{ kPa})$$

$$\Delta\sigma = 24\text{ kPa}$$

Part f

$$\Delta\sigma = I(r/a, z/a)q$$

$$\Delta\sigma = 0.08(120\text{ kPa})$$

$$\Delta\sigma = 10\text{ kPa}$$

Example 8.5.3

A gravity retaining wall has a base width of 2.5 m. The line of action of the resultant vertical and horizontal forces intersects the base at a point of 1 m from the toe of the wall. The vertical component of the resultant force is 175 kN/m. What is the subgrade reaction at the toe? At the heel? This is an unworked example found in Terzaghi et al. (1996).

Through the subject of statics, the following equations are produced and are related to Figure 8.12 after Terzaghi et al. (1996).

$$Q = (\tfrac{1}{2})B(p_a + p_b)$$

$$Qa = (1/6)B^2p_a + (1/3)B^2p_b$$

where
B is the width of the base.
Q is the resultant of the vertical loads on the base.
a is the distance from the toe.
p_a is the subgrade reaction at the toe.
p_b is the subgrade reaction at the heel.

Solution:

Given two equations and two unknowns, the procedure is as follows:

$$Q = (\tfrac{1}{2})B(p_a + p_b)$$

$$175\text{ kN/m} = (\tfrac{1}{2})(2.5\text{ m})(p_a + p_b)$$

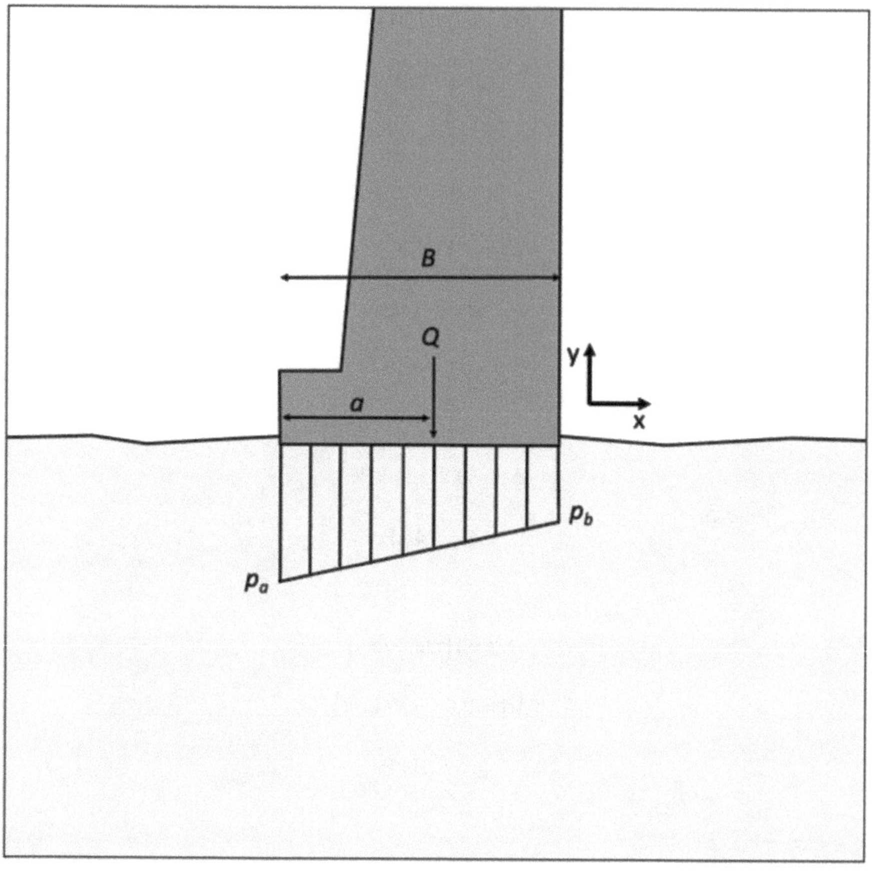

FIGURE 8.12 Section showing the problem. (Modified from Terzaghi et al. (1996).)

$$175 \text{ kN/m} = 1.25 \text{ m}\left(p_a + p_b\right)$$

$$140 \text{ kN/m}^2 = p_a + p_b$$

$$140 \text{ kN/m}^2 - p_a = p_b$$

$$Qa = \left(1/6\right)B^2 p_a + \left(1/3\right)B^2 p_b$$

$$\left(175 \text{ kN/m}\right)\left(1 \text{ m}\right) = \left(1/6\right)\left(2.5 \text{ m}\right)^2 p_a + \left(1/3\right)\left(2.5 \text{ m}\right)^2 p_b$$

$$175 \text{ kN} = \left(1.04 \text{ m}^2\right)p_a + \left(2.08 \text{ m}^2\right)p_b$$

$$175 \text{ kN} = \left(1.04 \text{ m}^2\right)p_a + 2.08 \text{ m}^2\left(140 \text{ kN/m}^2 - p_a\right)$$

$$175 \text{ kN} = \left(1.04 \text{ m}^2\right)p_a + 291.2 \text{ kN} - \left(2.08 \text{ m}^2\right)p_a$$

$$-116.2 \text{ kN} = -1.04 \text{ m}^2 p_a$$

$$p_a = 112 \text{ kN/m}^2$$

$$p_b = 28 \text{ kN/m}^2$$

Example 8.5.4

The following example pertains to chimney caving using limit equilibrium analysis. This example is found in Brady and Brown (2004).

Determine the factor of safety (F) against vertical sliding for a shallow abandoned room-and-pillar coal mine. The overburden is considered very poor with the following properties:

$$\gamma = 0.023\ \text{MN/m}^3$$

$$\sigma_c = 15\ \text{MPa}\left(\text{uniaxial compressive strength of intact rock}\right)$$

$$a = b = 10\ \text{m}\left(\text{room span}\right)$$

$$h = 30\ \text{m}\left(\text{depth to roof from surface}\right)$$

$$k = 1\left(\text{constant relating the horizontal to vertical stress}\right)$$

Constants for the given rock mass:

$$A = 0.05$$

$$B = 0.539$$

Solution:

$$F = \left[2(a+b)k^B h^B A\sigma_c^{\,1-B}\right]/\left[\gamma^{1-B}ab(1+B)\right]$$

$$F = \left[2(10+10)\left(1^{0.539}\right)\left(30^{0.539}\right)(0.05)\left(15^{1-0.539}\right)\right]/\left[0.023^{1-0.539}(10)(10)(1+0.539)\right]$$

$$F = \left[2\left(20\ \text{m}\right)(1)(6.25)(0.05)(3.48)\right]/\left[0.176(10)(10)(1.539)\right]$$

$$F = (43.5)/(27.1)$$

$$F = 1.61$$

Example 8.5.5

The following example also pertains to chimney caving using limit equilibrium analysis. This example is found in Brady and Brown (2004).

In order to back-calculate the critical span of an underground opening (a) the factor, F, is treated as 1. This is not meant to be a recommended factor of safety. The following properties are provided for a mine on the Zambian copper belt. The image for problems 8.5.5 to 8.5.7 can be found in Brady and Brown (2004). The shear strength properties for the weathered rock are as follows:

$$c' = 50\ \text{kPa}$$

$$\phi' = 31°$$

$$\gamma = 23.5\ \text{kN/m}^3$$

$$k = 0.33\left(\text{material is residual soil}\right)$$

Criteria related to the stopes and pillars:
Stopes were separated by 2-5 m wide rib pillars.

$$h = 140 \text{ m}$$

$$a = 12 - 20 \text{ m}$$

$$b = 80 - 95 \text{ m} \left(\text{taken as } 90 \text{ m}\right)$$

$$\alpha = 45°$$

For the case of a water table below the ground surface (d ≥ h), the following equation is provided:

$$F = \left\{\left[2c'(a + b\cos\alpha)\right]/(\gamma ab\cos\alpha)\right\} + \{k\tan\phi'/(2h - b\sin\alpha)\}\{[h^2 + (h - b\sin\alpha)^2/(b\cos\alpha)] + (2/a)[h(h - b\sin\alpha) + b^2\sin^2\alpha/3]\}$$

Solution:

All values are substituted into the above equation, except for "a":

$$1.0 = \left\{\left[2(50)(a + 90\cos 45°)\right]/(23.5(a)(90)\cos 45°)\right\} + \left\{0.33\tan 31°/(2(140) - 90\sin 45°)\right\} \left\{\left[140^2 + (140 - 90\sin 45°)^2/(90\cos 45°)\right] + (2/a)\left[140(140 - 90\sin 45°) + 90^2\sin^2 45°/3\right]\right\}$$

Solving for a:

$$1.0 = \left\{\left[100(a + 63.6)\right]/1,495.6a\right\} + \{0.198/216\}\left\{\left[19,600 + (76.4)^2/(63.6)\right] + (2/a)\left[140(76.3) + 1,350\right]\right\}$$

$$1.0 = \left\{[100a + 6,364]/1,495.6a\right\} + \left\{9.15 \times 10^{-4}\right\}\left\{[401.8] + (24,081/a)\right\}$$

$$1.0 = 0.067 + (4.26/a) + 0.3676 + (22.03/a)$$

$$0.5654 = 4.26/a + 22.03/a$$

$$0.5654a = 26.29$$

$$a = 46.5 \text{ m}$$

(in the reference, it is given 46.7 due to rounding).

For the next example, a solution for F is needed in terms of a. This is found in the above calculations as follows:

$$1.0 = 0.067 + (4.26/a) + 0.3676 + (22.03/a)$$

Using F in place of 1.0,

$$F = 0.067 + (4.26/a) + 0.3676 + (22.03/a)$$

$$F = 0.4346 + 26.29/a$$

Example 8.5.6

Repeat the previous example but for a case when the water table is 50 m below the ground surface, after Brady and Brown (2004).

For this case, $0 \le d \le h - b\text{Sin}\alpha$, the following equation is provided:

$$F = F_1 - \{[(2\gamma_w \tan\phi')]/[3\gamma(2h - b\sin\alpha)]\}\{[h^2 + (h - b\sin\alpha)^2 - 2d(2h - b\sin\alpha - d)]/(b\text{Cos}\alpha)$$

$$+(2/3a)[3h(h - b\sin\alpha) + b^2\sin^2\alpha - 3d(2h - b\sin\alpha - d)]\}$$

Solution:

$$F = F_1 - \{[(2(9.81)\text{Tan}31°)/[3(23.5)(2(140) - (90)\text{Sin}45°)]\}\{[140^2 + (140 - (90)\text{Sin}45°)^2 -$$

$$2(50)(2(140) - (90)\text{Sin}45° - 50)]/((90)\text{Cos}45°) + (2/3a)[3(140)(140 - (90)\text{Sin}45°) + 90^2\text{Sin}^2 45°$$

$$-3(50)(2(140) - (90)\text{Sin}45° - 50)]\}$$

$$F = F_1 - [11.8/15{,}253.4][138.2 + 7{,}444.9/a]$$

$$F = F_1 - 0.1068 - 5.754/a$$

From Example 8.5.5, it is known that $F_1 = 0.4346 + 26.29/a$ for a critical span $F = 1$,

$$F = 0.4346 + 26.29/a - 0.1068 - 5.754/a$$

$$1 = 0.3278 + 20.536/a$$

$a = 30.5$ m (in the reference, it is given as 30.8 m).

Example 8.5.7

Repeat the previous example but for a case when the water table is at a depth of 120 m below surface.

For this case, $h \ge d \ge h - b\text{Sin}\alpha$, the following equation is provided by Brady and Brown (2004):

$$F = F_1 - \{[2\gamma_w(h - d)^2\tan\phi']/[3\gamma b(2h - b\sin\alpha)]\}\{\sec\alpha + [2(h - d)]/3a\sin\alpha]\}$$

Solution:

$$F = F_1 - \{[2(9.81)(140 - 120)^2\tan 31°]/[3(23.5)(90)(2(140) - (90)\sin 45°)]\}$$

$$\{\sec 45° + [2(140 - 120)] + /[3a\sin 45°]\}$$

$$F = F_1 - \{[4{,}715.6/1{,}372{,}806.7]\}\{1.41 + 40/2.12a\}$$

$$F = F_1 - 4.843 \times 10^{-3} - 0.0648/a$$

From Example 8.6.5, we know that $F_1 = 0.4346 + 26.29/a$ and for a critical span $F = 1$,

$$1 = 0.4346 + 26.29/a - 4.843 \times 10^{-3} - 0.0648/a$$

$$a = 45.9 \text{ m}$$

Example 8.5.8

The principle of volume constancy states that subsidence and the associated trough which forms are related linearly to the thickness of the seam mined (M) and a factor (a) which is related to the manner of stowing (Kratzsch, 1983):

- Part a: Estimate the dimensions of the subsidence trough for a coal seam which has a mined thickness of 10 ft, panel width of 1,000 ft, and panel length of 10,000 ft, for a subsidence factor of a = 0.95 (total caving). Assume a depth to surface of 1,000 ft and a conservative angle of draw of approximately 27 gradians. Estimate a conservative acreage influence on surface.
- Part b: If the mining operation uses partial backfilling and has a subsidence factor instead of a = 0.45, what is the extent of the trough?
- Part c: Determine the vertical subsidence for each case.

For this basic principle, the rib-line of the entry projects to surface (see Figure 8.13), and therefore, the total extent of the trough is the width or length of the panel plus horizontal distances related to the angle of draw from each rib-line (for this simple example). Note that the figure has been modified from Kratzsch (1983) to include the angle of draw as the internal angle delta. In his work, the external angle is given in the figure and treated as gamma.

Solution:

Part a

Convert gradians to degrees: (27 gr)(1°/1.11 gr) = 24°

Tan 24° = horizontal distance/1,000 ft

Horizontal distance = 445 ft

Total trough width = 1,000 ft + 2(445 ft) = 1,890 ft

Total trough length = 10,000 ft + 2(445 ft) = 10,890 ft

The subsidence trough will possess rounded corners, but for a conservative estimate, assume that a rectangular area on surface is influenced. Therefore,

(1,890 ft)(10,890 ft)(1 acre/43,560 ft²) = 473 acres.

Part b

Even though the subsidence factor reduces by more than half, the extent of the trough is the same under this principle.

Part c:

For a = 0.95

Vertical convergence = aM = 0.95(10 ft) = 9.5 ft

For a = 0.45

Vertical convergence = aM = 0.45(10 ft) = 4.5 ft

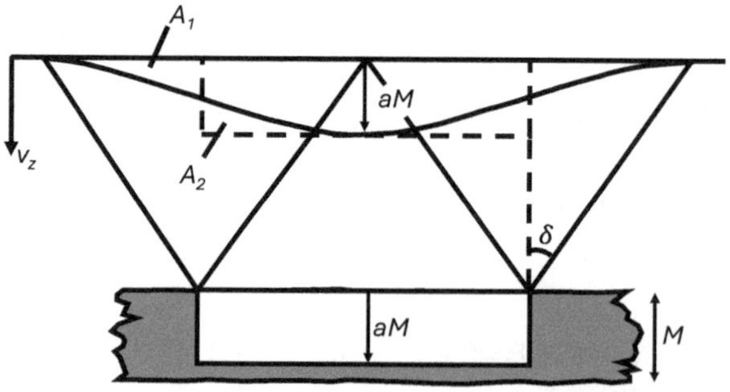

FIGURE 8.13 Section through the mined coal seam. (Modified after Kratzsch (1983).)

Example 8.5.9

As an exercise, approximate the maximum mine subsidence using the principle of superposition for the multiple-seam scenario described. Assume that the principle of volume constancy also applies (see the previous example). For the scenario described, assume that the mid-line of the panels is within a reasonable tolerance.

Mine A was a retreat mine in a coal seam at approximately 1,200 ft depth. The seam had a mined height of 6 ft. Pillars at this mine were 60 ft square, and entries were 20 ft wide. The panels were six pillars wide. The angle of draw is approximated at 24 gradians.

Mine B was a room-and-pillar mine in a coal seam at approximately 800 ft depth. Although the mine was room and pillar, remnants were left significantly smaller in panel due to the era in which they were extracted. The pillars averaged 35 ft square, and entries were wide, at 25 ft. Mapping shows that the panels were seven pillars wide in the area of interest. The seam had a mined height of 5.5 ft. Angle of draw is approximated at 27 gradians.

For this example, estimate the subsidence factor, a, using the following relationship by Peng et al. (1995). Refer to Figure 8.14:

$$a = 0.6815519\left(0.9997398^{h}\right)$$

where

h is the overburden depth in feet.

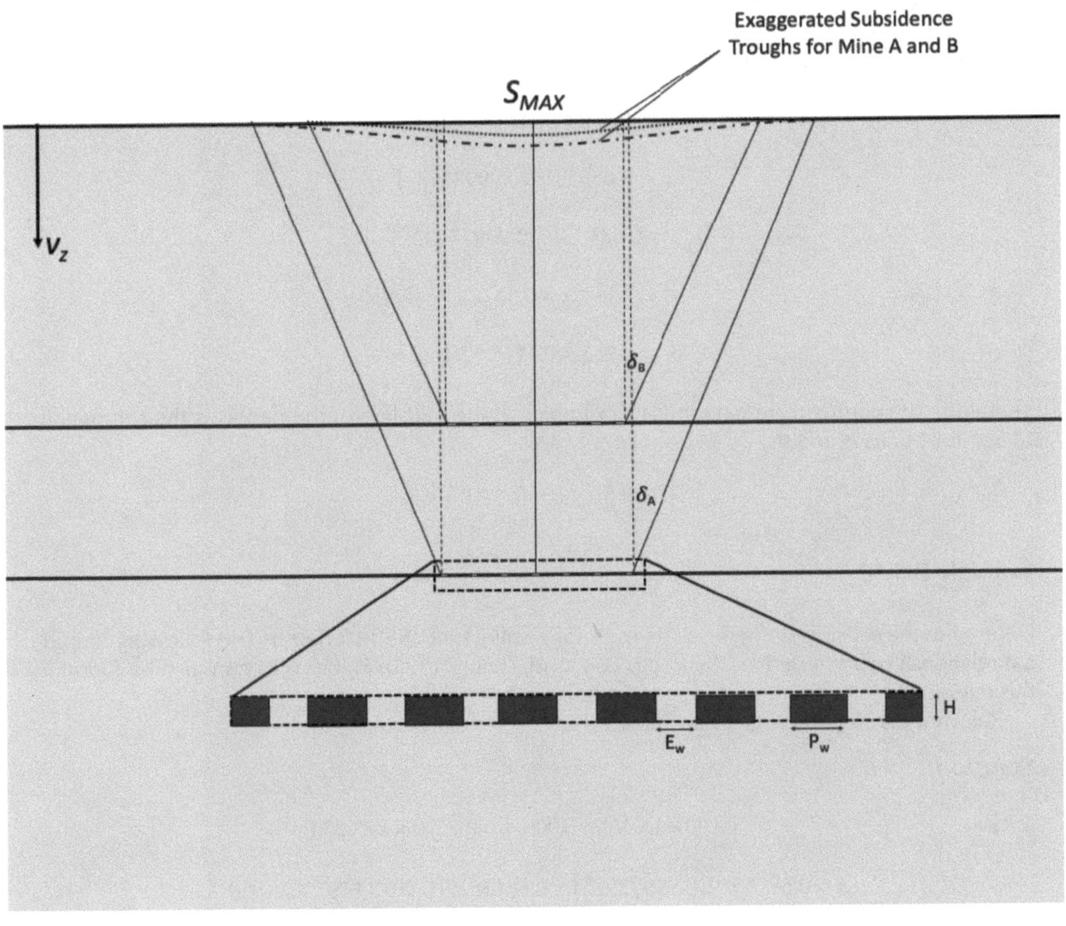

FIGURE 8.14 Section of the mining subsidence problem.

Solution:

Kratzsch (1983) reviewed the principle of superposition as it relates to subsidence in detailed terms but suggests that a trough is calculated for each working panel and the subsidence of each is added together to give a combined trough. It is discussed in the text for multiple extraction areas of a single seam or of multiple seams. The following example combines the principle of superposition and the principle of volume constancy (shown in the previous example) for approximation of the maximum subsidence at a single point. Clearly there are many assumptions made in the following approach.

Mine A

Rib-to-rib panel width is approximated as follows:

$$60\text{ ft}(6) + 20\text{ ft}(7) = 500\text{ ft}$$

$$a = 0.6815519\left(0.9997398^{h}\right)$$

$$a = 0.6815519\left(0.9997398^{1,200}\right)$$

$$a = 0.50$$

$$S_{max} = aM = 0.50\left(6\text{ ft}\right) = 3\text{ ft}$$

Mine B

Mine B was a room-and-pillar mine, but because of the era of mining and size of the pillars, a conservative estimate would be to assume that the pillars have crushed, and therefore, we treat the panel as gob, knowing that if the pillars haven't failed, our estimate is still conservative:

$$35\text{ ft}(7) + 25\text{ ft}(8) = 445\text{ ft}$$

$$a = 0.6815519\left(0.9997398^{h}\right)$$

$$a = 0.6815519\left(0.9997398^{800}\right)$$

$$a = 0.55$$

$$S_{max} = aM = 0.55\left(5.5\text{ ft}\right) = 3\text{ ft}$$

Therefore, the approximate maximum subsidence at the mid-line of the panels is the summation of S_{max} for Mines A and B:

$$\text{Total } S_{max} = 3\text{ ft} + 3\text{ ft} = 6\text{ ft}$$

Example 8.5.10

Determine the subsidence parameters for a longwall panel at 500 ft depth. The following empirical relationships were derived by Peng, Luo, and Zhang (1995). Figure 8.15 can also be found in this reference.

Relevant information is given in Table 8.1.

Solution:

$$W_c = 100 + 1.046h = 100 + 1.046\left(500\text{ ft}\right) = 623\text{ ft}$$

$$a = 0.6815519\left(0.9997398^{h}\right) = 0.6815519\left(0.9997398^{500}\right) = 0.60$$

$$d = 0.38123he^{-0.000699h} = 0.38123\left(500\text{ ft}\right)e^{-0.000699(500)} = 134\text{ ft}$$

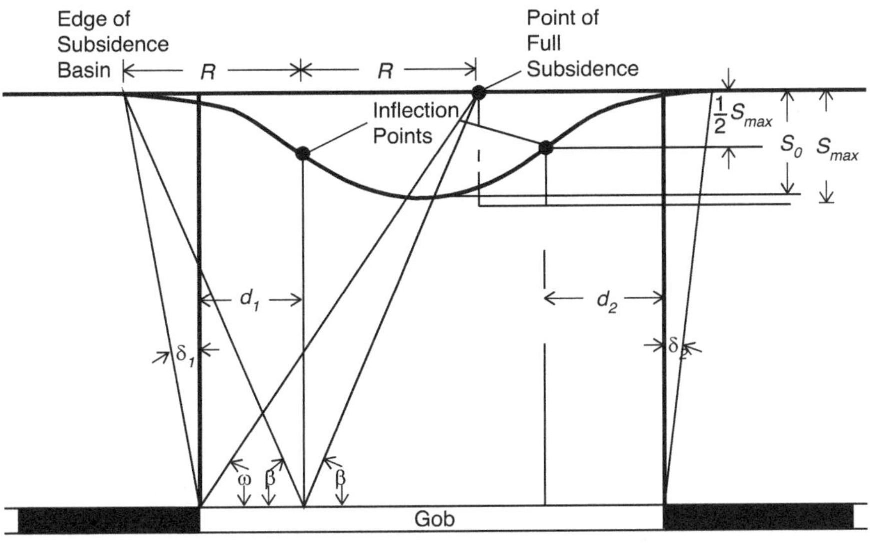

FIGURE 8.15 Subsidence above a longwall coal mine (Peng et al., 1995).

TABLE 8.1
Relevant Subsidence Related Information

Parameter	Definition	Approximation
Critical panel width	Panel width when the maximum subsidence is reached	$W_c = 100 + 1.046h$
Subsidence factor	Ratio of maximum subsidence to mining height	$a = 0.6815519\ (0.9997398^h)$
Offset distance	Horizontal distance between inflection point and panel edge	$d = 0.38123he^{-0.000699h}$
Angle of draw	Angle between the vertical line at the panel edge and a line connecting the edge of the subsidence and the panel edge	$\delta_o = 6.87 - 0.0072h + 8.872 \times 10^{-6}(h^2)$
Angle of major influence	Angle between the horizontal line at the mining depth to the point of full subsidence and the vertically projected point of inflection on the coal seam	$\beta = 58.89 + 0.03089h - 1.84 \times 10^{-5}(h^2)$
Angle of full subsidence	Angle between a horizontal line at the mining level and the line connecting the panel edge and the point of full subsidence	$\omega = 42.55 + 0.0417h - 2.16 \times 10^{-5}(h^2)$
Angle of critical deformation	Angle between the vertical line at the panel edge and the line connecting the panel edge and the point of critical deformation on surface	$\phi = 27.96 - 0.02426h + 6.9 \times 10^{-6}(h^2)$

Source: Summarized from Peng et al. (1995).

$$\delta_o = 6.87 - 0.0072h + 8.872 \times 10^{-6}(h^2) = 6.87 - 0.0072(500) + 8.872 \times 10^{-6}(500^2) = 5.5°$$

$$\beta = 58.89 + 0.03089h - 1.84 \times 10^{-5}(h^2) = 58.89 + 0.03089(500) - 1.84 \times 10^{-5}(500^2) = 69.7°$$

$$\omega = 42.55 + 0.0417h - 2.16 \times 10^{-5}(h^2) = 42.55 + 0.0417(500) - 2.16 \times 10^{-5}(500^2) = 58°$$

$$\phi = 27.96 - 0.02426h + 6.9 \times 10^{-6}(h^2) = 27.96 - 0.02426(500) + 6.9 \times 10^{-6}(500^2) = 17.6°$$

Example 8.5.11

What must be the panel width and length to be deemed a supercritical panel at 500 ft depth? A subcritical panel at 500 ft depth?

Solution:

$$W_c = 100 + 1.046h = 100 + 1.046(500 \text{ ft}) = 623 \text{ ft}$$

A supercritical panel is one where the width and length exceed the critical width (W_c) of 623 ft. A subcritical panel is one where the width and length do not exceed the critical width of 623 ft.

Example 8.5.12

Determine a rough, initial estimate of the cave angle for a block cave mine using the Laubscher method (2000). Limitations of Laubscher's method are summarized in Contreras et al (1996), and the following inputs are found in Contreras (2022).

Density of caved material: 1.87 t/m^3
Height of caved material: 355 m
Depth: 460 m
MRMR: 25
Minimum span: 105 m
Factor = (Density of caved material/1.5)(height of caved material/100)(depth/min. span).

Solution:

$$\text{Factor} = (1.87/1.5)(355/100)(460/105)$$

$$\text{Factor} = (1.25)(3.55)(4.38)$$

$$\text{Factor} = 19.4$$

Using Figure 8.16 from Laubscher (2000), the cave angle is determined to be 74°.

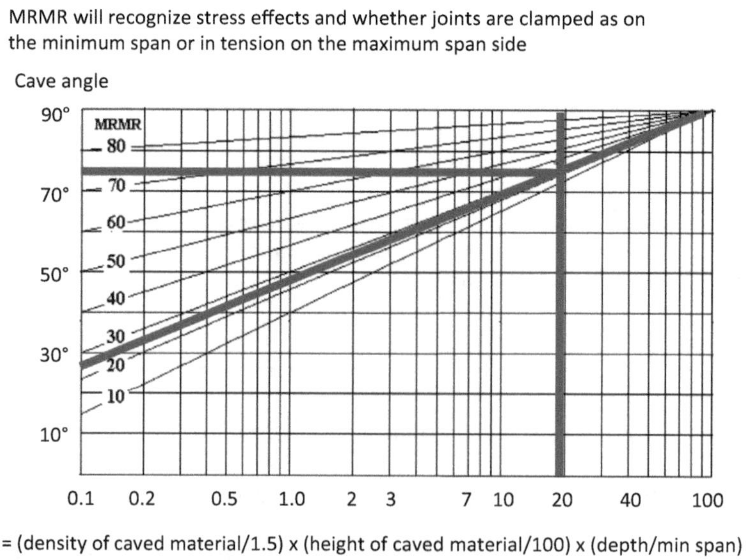

FIGURE 8.16 MRMR figure (Laubscher, 2000).

Example 8.5.13

Contreras (2022) expands on the previous example and shows the method for determining the overall cave angle for various rock units of different MRMR values. Given the following table and equations, determine the overall cave angle. Calculations are for values in bold and shown in Table 8.2. This is for example purposes only and does not include any buffer zone or similar on surface.

TABLE 8.2
The Given Parameters and the Calculated Parameters (in Bold)

i	Depth (m)	HCM (m)	MRMR	Min Span (m)	Factor	Cave Angle	Increase in Footprint (m)
1	460	**355**	25	**105**	**19.4**	**74°**	**60.2**
2	250	**195**	43	**225**	**2.7**	**65°**	**51.3**
3	140	**117.5**	25	**328**	**0.63**	**44°**	**46.6**
4	95	**85**	16	**425**	**0.24**	**30°**	**34.6**
5	75	**37.5**	6	**494**	**0.1**	**9°**	**473.5**

Note: Some results differ from Contreras (2022) due to usage of Laubscher's chart versus the use of the derived functions from the chart used by Contreras.

Solution:

Height of caved material $(HCM)_i = Depth_i - [(Depth_i - Depth_{i+1})/2]$

Factor = (Density of caved material/1.5) (height of caved material/100) (depth/min. span)

Increase in $footprint_{i=}(Depth_i - Depth_{i+1})/\tan(cave\ angle_i)$

Min $span_{i+1}$ = Min $span_{i+2}$ (Increase in $footprint_i$)

Overall cave angle $= \tan^{-1}$ (Maximum depth/Σ Increase in footprint)

Iteration i = 1:

Height of caved material $(HCM)_1 = Depth_1 - [(Depth_1 - Depth_2)/2]$

Height of caved material $(HCM)_1 = 460 - [(460 - 250)/2] = 355$ m

Factor = (Density of caved material/1.5) (height of caved material/100) (depth/min. span)

Factor = (1.87/1.5) (355/100) (460/105)

Factor = (1.25) (3.55) (4.38)

Factor = 19.4

Increase in $footprint_1 = (Depth_1 - Depth_2)/\tan(cave\ angle_1)$

Increase in $footprint_1 = (460 - 250)/\tan(74°) = 60.2$ m

Iteration i = 2:

Min $span_{i+1}$ = Min $span_i + 2$(Increase in $footprint_i$)

Min $span_2$ = Min $span_1 + 2$(Increase in $footprint_1$)

Min $span_2 = 105 + 2(60.2) = 225$ m

Height of caved material $(HCM)_2 = 250 - [(250 - 140)/2] = 195$ m

Factor = (Density of caved material/1.5)(height of caved material/100)(depth/min. span)

Factor = (1.87/1.5)(195/100)(250/225)

Factor = (1.25)(1.95)(1.11)

Factor = 2.7, from the chart in Figure 8.16, cave angle is approximately 65°

Increase in $footprint_2 = (Depth_2 - Depth_3)/\tan(cave\ angle_2)$

Increase in $footprint_2 = (250 - 140)/\tan(65°)$

Increase in $footprint_2 = 51.3$ m

Iteration i = 3:

Min $span_3$ = Min $span_2 + 2$(Increase in $footprint_2$)

Min $span_3 = 225 + 2(51.3) = 328$ m

Height of caved material $(HCM)_3 = 140 - [(140 - 95)/2] = 117.5$ m

Factor = (Density of caved material/1.5) (height of caved material/100) (depth/min. span)

Factor = (1.87/1.5) (117.5/100) (140/328)

Factor = (1.25) (1.175) (0.43)

Factor=0.63, from the chart in Figure 8.16, cave angle is approximately 44°
Increase in $footprint_3 = (Depth_3 - Depth_4)/\tan\ (cave\ angle_3)$
Increase in $footprint_3 = (140 - 95)/\tan\ (44°)$
Increase in $footprint_3 = 46.6\,m$
Iteration i=4:
Min $span_4$=Min $span_3$+2(Increase in $footprint_3$)
Min $span_4 = 332 + 2(46.6) = 425\,m$
Height of caved material $(HCM)_4 = 95 - [(95 - 75)/2] = 85\,m$
Factor=(Density of caved material/1.5)(height of caved material/100)(depth/min. span)
Factor=(1.87/1.5)(85/100)(95/425)
Factor=(1.25)(0.85)(0.22)
Factor=0.24, from the chart in Figure 8.16, cave angle is approximately 30°
Increase in $footprint_4 = (Depth_4 - Depth_5)/\tan\ (cave\ angle_4)$
Increase in $footprint_4 = (95 - 75)/\tan(30°)$
Increase in $footprint_4 = 34.6\,m$
Iteration i=5:
Min $span_5$=Min $span_4$+2(Increase in $footprint_4$)
Min $span_5 = 425 + 2(34.6) = 494\,m$
Height of caved material $(HCM)_5 = 75 - [(75 - 0)/2] = 37.5\,m$
Factor=(Density of caved material/1.5) (height of caved material/100) (depth/min. span)
Factor=(1.87/1.5) (37.5/100) (95/494)
Factor=(1.25) (0.375) (0.19)
Factor=0.1, from the chart in Figure 8.16, cave angle is approximately 9°
Increase in $footprint_5 = (Depth_5 - Depth_6)/\tan\ (cave\ angle_5)$
Increase in $footprint_5 = (75 - 0)/\tan(9°)$
Increase in $footprint_5 = 473.5\,m$
Overall cave angle:
Overall cave angle=$\tan^{-1}$(Maximum depth/ΣIncrease in footprint)
Overall cave angle=$\mathrm{Tan}^{-1}[460/(60.2+51.3+46.6+34.6+473.5)]$
Overall cave angle=35°
(the angle between, the horizontal line defining the maximum depth of the workings, and the surface).

Example 8.5.14

The following example and image are found in Kratzsch (1983) and pertains to an introductory example on the use of the influence function method and the determination of subsidence at a point (P) above extracted workings. The extraction height (M) is to be taken as 1.8 m and the subsidence factor (a) of 0.9.

Solution:

The first step involves the division of circular grid shown in Figure 8.17 into five concentric rings of equal influence. Each ring is further subdivided into eight sectors so that a total of 40 zones are created. The zones are clearly of different sizes, but each represents the same percentage of the maximum subsidence, which is still to be calculated. Of the maximum subsidence ($v_{z,\,max}$) each of the 40 zones represents 2.5%.

The integration grid is constructed on a transparent paper. The grid, as shown in the figure, is a diameter $2R = 2h\cot\gamma$, which represents the "critical area". The parameter (γ) represents the limit angle. The transparent paper is overlain on the seam of interest so that the central point of the grid (P) overlays the point of interest (P′).

Solution:

The influence factor (e) is obtained by counting the number (n) grids covering the workings. This is also shown in the figure:

$$e = 0.025n$$

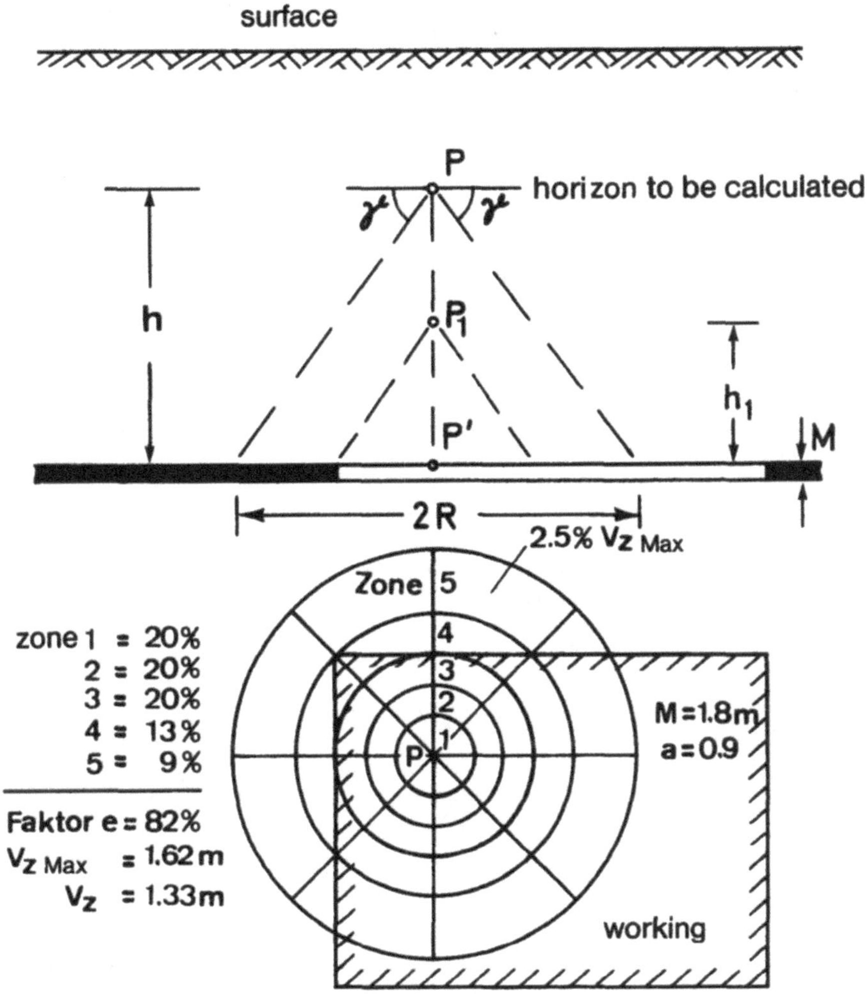

FIGURE 8.17 Subsidence zones using the influence function method (Kratzsch, 1983).

Zone 1: 8 zones @ 2.5% each = 20%

Zone 2: 8 zones @ 2.5% each = 20%

Zone 3: 8 zones @ 2.5% each = 20%

Zone 4: (4 zones @ 2.5% each) + (remaining fraction of zones @ 2.5% each) = 13%

Zone 5: (2 zones @ 2.5% each) + (remaining fraction of zones @ 2.5% each) = 9%

Factor, e, = (20 + 20 + 20 + 13 + 9) = 82%

The amplitude of subsidence is resolved from

$$v_z = e v_{z,\,max}$$

where $v_{z,\,max}$ is resolved using the relationship Ma as shown in Example 8.5.8:

$$v_{z,\,max} = Ma$$

$$v_{z,\,max} = (1.8\text{ m})(0.9)$$

$$v_{z,\,max} = 1.62\text{ m}$$

The associated magnitude of subsidence (v_z) at point (P) for the horizon shown is

$$v_z = ev_{z,max}$$

$$v_z = 0.82(1.62 \text{ m})$$

$$v_z = 1.33 \text{ m}$$

Though not stated explicitly in the example, associated tensile and compressive strains with expected subsidence magnitudes at specific points are generally necessary to compare with limiting strains for specific structures on surface which could be affected.

REFERENCES

American Insurance Services Group, Inc (1972) *Blasting Damage and Other Structural Cracking: A Guide for Adjusters and Engineers*. American Insurance Services Group. New York.

American Insurance Services Group, Inc (1990). *Blasting Damage and Other Structural Cracking: A Guide for Adjusters and Engineers* (3rd Edition). American Insurance Services Group Inc., New York.

Bals, R. (1931) *Beitrag zur Frage der Vorausberechnung Bergbaulicher Senkugen (A contribution to the Problem of Precalculating Mine Subsidence)*. Mitteilungen aus dem Markscheidewesen.

Booth, C.J. (1998) Impacts of mine subsidence on groundwater. In: Hooks, C.L., Vories, K.C. and Throgmorton, D. (eds). *Proceedings of Prime, Farmland Interactive Forum*. Department of Agronomy, Illinois Agricultural Experiment Station, University of Illinois at Urbana-Champaign, University of Southern Indiana, Evansville, pp. 143–148.

Brady, B.H. and Brown, E.T. (2004) *Rock Mechanics: For Underground Mining* (3rd Edition). Springer Science & Business Media, New York.

Bruhn, R.W., Magnuson, M.O. and Gray, R.E. (1978) Subsidence over the mined-out Pittsburgh coal. ASCE Preprint 3293, American Society of Civil Engineers Spring Convention, Pittsburgh, pp. 26–55.

Contreras, C, Elmo, D, Jakubec, J and Thomas, A (2022) Reviewing Laubscher's empirical method to estimate subsidence limits. In Potvin, Y. (ed) *Caving 2022: Proceedings of the Fifth International Conference on Block and Sublevel Caving, Australian Centre for Geomechanics*, Perth, pp. 805–818, https://doi.org/10.36487/ACG_repo/2205_55

Contreras, C.P. (2022) Reviewing Laubscher's empirical method to estimate subsidence limits (T). University of British Columbia. Retrieved from: https://open.library.ubc.ca/collections/ubctheses/24/items/1.0421707

Gray, R.E., Gamble, J.C., McLaren, R.J. and Rogers, D.J. (1974) State of the Art of Subsidence Control. Appalachian Reg. Comm., ARC-73-111-2550.

Jeran, P.W. and Trevits, M.A. (1995) Mining Publication: Timing and Duration of Subsidence Due to Longwall Mining. US Bureau of Mines (USBM RI 9552). January.

Kratzsch, H. (1983) *Mining Subsidence Engineering*, Springer, Berlin, Germany.

Lambe, T.W. and Whitman, R.V. (1969) *Soil Mechanics*. John Wiley & Sons, Inc., New York, p. 553.

Laubscher, D.H. (2000) *Block Caving Manual. International Caving Study*. JKMRC and Itasca Consulting Group, Inc, Brisbane.

Michaud, D. (2018) Longwall mining method and design. 911 Metallurgist. Retrieved from: www.911metallurgist.com/longwall-mining-methods/

Mine Subsidence Engineering Consultants. (2007) Introduction to longwall mining and subsidence. Retrieved from: https://www.minesubsidence.com/index_files/files/Intro_Longwall_Mining_and_Subs.pdf

Peng, S.S., Luo, Y. and Zhang, Z.M. (1995) Subsidence parameters – their definitions and determination. In *Society for Mining, Metallurgy & Exploration Annual Meeting Transactions*, Vol. 300, pp. 60–65, Littleton, CO.

Pennsylvania Department of Environmental Protection. (n.d.). Retrieved from: https://files.dep.state.pa.us/Mining/District%20Mining/DistrictMinePortalFiles/California/Underground%20Mining/MineSubsidence/

Sepehri, M., Apel, D.B. and Hall, R.A. (2017) Prediction of mining-induced surface subsidence and ground movements at a Canadian diamond mine using an elastoplastic finite element model. *Inter. J. Rock Mech. Mining Sci.* 100, 73–82.

Terzaghi, K., Peck, R.B. and Mesri, G. (1996) *Soil Mechanics in Engineering Practice*. John Wiley & Sons, New York.

Tschebotarioff, G.P. (1951) *Soil Mechanics, Foundations and Earth Structures*. McGraw-Hill Book Company, New York.

Van As, A. (2003) Subsidence definitions for block caving mines. Technical Report. Rio Tinto Technical Services, Sydney, p. 59.

Whittaker, B.N. and Reddish, D.J. (1989) *Subsidence: Occurrence, Prediction and Control*. Elsevier Science Publishers, Amsterdam, The Netherlands.

9 Pillars and Classical Pillar Design *(Mainly Modified After Spearing et al., 2023)*

9.1 THE USE AND NEED FOR PILLARS

The entire Chapter is mainly modified after Spearing et al, 2023.

Pillars are used in many underground mining methods, and their design is critical to ensure safe and productive mining. Over-sized pillars reduce the percentage extraction, whereas under-sized pillars can squeeze, or collapse suddenly, which is very hazardous.

Pillars are used for many underground applications. The following are examples of pillars:

- Internal pillars.
- Barrier pillars.
- Rib pillars.
- Crown or sill pillars.
- Shaft protection pillars.
- Water barrier pillars.
- Yield pillars.

Pillars can also serve a vital function to protect the mine from failure along major geological features. In most underground mining operations, pillars can be broadly divided into support pillars and protective pillars.

9.1.1 Support and Local-Support Pillars

Local-support pillars often have temporary use and can be extracted once they have fulfilled their initial support purpose, which is mainly after the time when active mining has taken place in their vicinity. Subsequent failure of these pillars can take place, provided that the mode of failure is stable and they are not being used for subsidence control. The use of yielding support pillars falls into this category.

Barrier and wide inter-panel pillars are typical examples of pillars that provide regional support and will generally have a higher factor of safety than local-support pillars. Pillars in seismically active areas may need even higher factors of safety and/or yield pillars.

9.1.2 Protective Pillars

It is essential to protect underground and surface structures from the effects of mining. One of the practical means of achieving this is to leave portions of a seam or orebody unmined to form protective pillars. The design criteria for protective pillars depend largely on the nature of the structure that needs to be protected. In the case of surface structures, the design comes down mainly to limit strain and tilt that can be tolerated by the structure above without significant damage. In the case of underground structures, such as bunkers, pump stations, and service excavations, it is mainly the magnitude of the stresses that determines the size of protective pillars.

DOI: 10.1201/9781032704845-9

9.2 LOCAL PILLAR DESIGN METHODOLOGIES

9.2.1 Basic Pillar Design Principles

Classical pillar design (typically using tributary area theory) compares the strength of the pillar to its stress to meet a desired safety factor suitable for the specific mining application. A higher safety factor can allow for uncertainty regarding the following:

- Strength of the rock in the roof, pillar, and floor.
- Pillar stress.
- Size and shape of the pillar.
- Pillar height.
- Geological and hydrological conditions.

The dimensions and rock properties of a pillar affect the strength and the post-failure behavior of a pillar as shown in Figure 9.1. Most classical pillar designs only consider the elastic portion of the curves.

Based on Figure 9.1, the width to height (w/h) ratio of a pillar influences the peak strength and the yield behavior of the pillar. The depth at which pillars are developed is important as the design approach can be different. For example, rigid or yielding pillars may be required depending on the stress conditions and mining method, and the post-failure (yielding) behavior of pillars can perform in a controlled manner if well designed or uncontrolled if poorly designed. This is illustrated in Figure 9.2.

9.2.2 Strength of Pillars

For determining the in situ strength of a pillar, empirical formula are used, which are derived from specific criteria. For example, in coal, they may be deduced from pillar case histories from specific coal basins. Extrapolation of the empirical approach is generally considered a weakness of the method as the approach is theoretically bounded by the database from which it is derived.

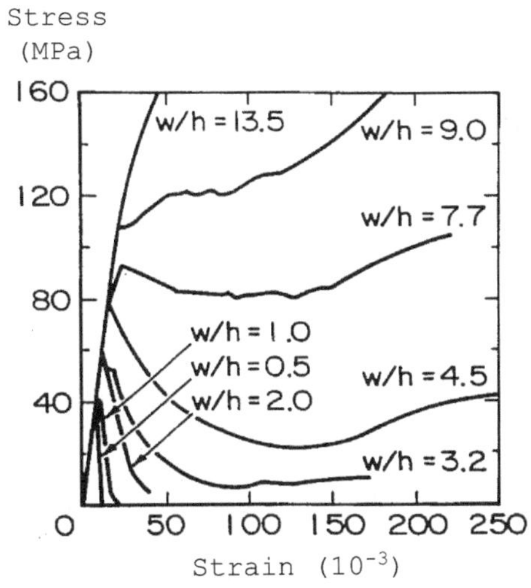

FIGURE 9.1 The stress versus strain behavior of lab samples representing pillars at different width to height ratios (Das, 1986).

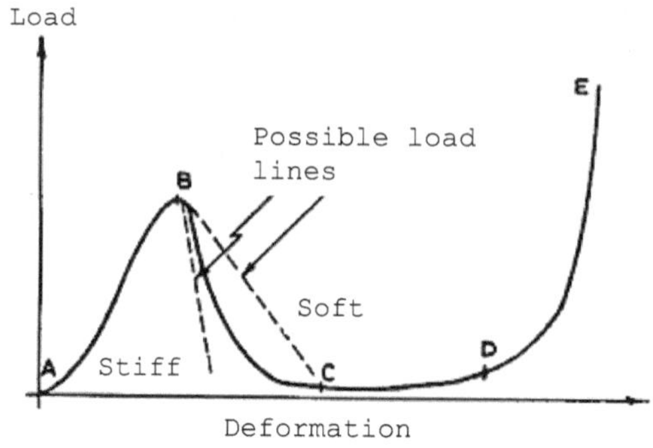

FIGURE 9.2 The post-peak performance of pillars (Ozbay and Roberts, 1988).

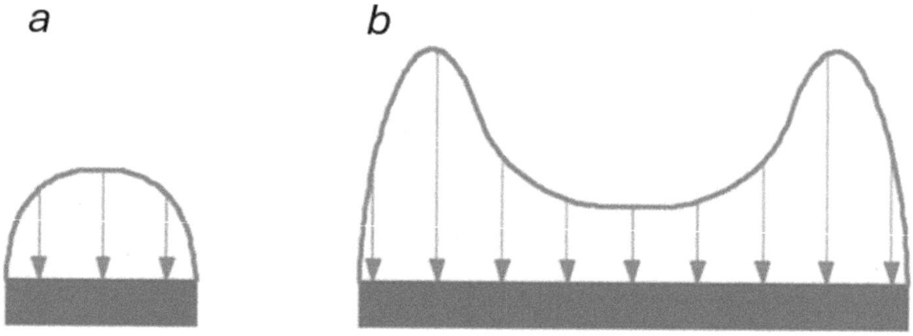

FIGURE 9.3 Examples of stresses across (a) a yielding pillar and (b) a wide pillar (Pan et al., 2020).

9.2.2.1 Factors Affecting the Pillar Strength

At the fundamental level, the behavior of a pillar depends on the distribution of stresses in the pillar. Stress distributions for two different types of pillars are shown in Figure 9.3.

As it relates to Figure 9.3, for a yielding pillar, the width of a pillar is smaller than a wide stable pillar, and the edges gradually crush, transferring the highest load to the center (core) as it crushes in a stable manner. For the wide pillar, the higher stresses are carried nearer the pillar edges.

As shown in Figure 9.1, pillar strength tends to increase with width to height ratio due to the lateral confining effect (Salamon and Wagner, 1985). For example, the effect of the width to height ratio on the strength of pillars can be expressed by the following relationship, which, according to Salamon and Wagner (1985), was proposed by Greenwald et al. (1939), Steart (1954), Holland and Gaddy (1957), and Skinner (1959):

$$\text{Pillar Strength } P_s = Kh^{\alpha}w^{\beta} \tag{9.1}$$

where

K is the uniaxial compressive strength of the pillar material based on lab cube tests (MPa)
h is the effective pillar height (m).
w is the pillar width (m).
α and β are appropriately chosen constants based on back analyses and statistical work. Note that the early use of Equation 9.1 associated the exponent alpha, with h, and the exponent beta, with w. This is sometimes reversed as shown in Equation 9.14. Hence the constants must be carefully chosen and confirmed before use.

An examination of the formula for pillar strength published by the authors above shows that the value of constants α varies from −1.0 for laboratory pillar tests (Steart, 1954) to −0.83 for moderately large model pillars (Greenwald et al, 1939). The value of constant β was found to be 0.5 in all the model pillar tests, while a value of 0.46 was obtained from the statistical analysis of many case histories (Salamon & Munro, 1967). The effect of the pillar width (always considered the width as being less than or the same as the length) on its strength tends to be independent of the length of the pillar, while the effect of pillar height is greatest in the case of small pillars but tends to be less significant as the pillar size increases.

Table 9.1 tracks the development of some of the numerous pillar strength empirical formula in chronological order.

TABLE 9.1
Some Commonly Used Empirical Coal Pillar Strength (Prassetyo et al., 2019)

Year	Developer	Pillar Formula
1876	Bauschinger[a]	$S_p = S_{cube}$ [0.778 + 0.222(w/h)]
1911	Bunting[a]	S_p = 6.9 [0.7 + 0.3(w/h)]
1939	Greenwald et al.[a]	$S_p = 19.2\ (w^{1/2}/h^{5/6})$
1957	Holland and Gaddy[a]	$S_p = K\ [(\sqrt{w})/h]$
1967	Salamon and Munro[a]	$S_p = 7.2\ (w^{0.46}/h^{0.66})$
1968	Bieniawski	$S_p = S_{cube}$ [0.64 + 0.36(w/h)]
1973	Holland	$S_p = S_{cube}\ \sqrt{(w/h)}$
1984	Wagner and Madden[b]	$S_p = 7.2\ (R_0^{\,b}/V^a)\{(b/\varepsilon)[(R/R_0)^\varepsilon - 1] + 1\}$
1985	Salamon and Wagner	$S_p = 0.00786[(R^{2.5} + 181.6)/V^{0.0667}]$
1987	Sheorey et al.	$S_p = 0.27\sigma_c h^{-0.36} + (z/160)[(w/h) - 1]$
1991	Madden	$S_p = 5.24(w^{0.63}/h^{0.78})$
1997	Mark and Chase[a]	$S_p = S_{cube}\{0.64 + [0.54(w/h) - (0.18w^2/Lh)]\}$
1999	Galvin et al.[a]	$S_p = 6.88(w^{0.5}/h^{0.7})$
2003	van der Merwe[c]	S_p = 1.5(w/h)
2003	van der Merwe[d]	S_p = 3.5(w/h)

[a] Shown in Figure 9.4.
[b] Squat pillar formula.
[c] Weak coal.
[d] Normal coal.

where typically

S_p = pillar strength (MPa).
S_{cube} = strength of a 1:1 width to height ratio coal cubic sample in MPa.
σ_c = uniaxial compressive strength of the coal sample in MPa.
w = pillar width (m).
h = pillar height (m).
K = Gaddy factor that depends on the characteristic of the coal.
R = width to height ratio.
R = critical width to height ratio.
V = pillar volume (m^3).
L = pillar length (m).
z = depth to the coal seam (m).
ε = rate of strength increase.

a = constant assumed to be 0.0667.
b = constant assumed to be 0.5933.

Figure 9.4 compares several pillar strength formula among those listed in Table 9.1. This comparison uses a pillar height of 2.1 m. The values of S_{cube} (the strength of a 1:1 ratio coal sample) are varied by the researchers and are shown in Table 9.2.

There are many pillar strength formula derived empirically, analytically, or numerically. These formula and the methods from which they were derived are compared in Figure 9.5, which can be divided into the three categories as follows (Michaud, 2017):

- Empirical formulas, comprising those of Bieniawski (1968), Sheorey et al (1987), and Obert and Duvall (1967), who take a middle road (around the average), implying a linear increase in pillar strength.
- Analytical models, comprising the Salamon and Munro (1967) and Holland and Gaddy (1956 and 1964) empirical formulas and Barron's (1992) analytic formula, indicate that the core stress and the pillar strength tend toward some maximum, limiting value.

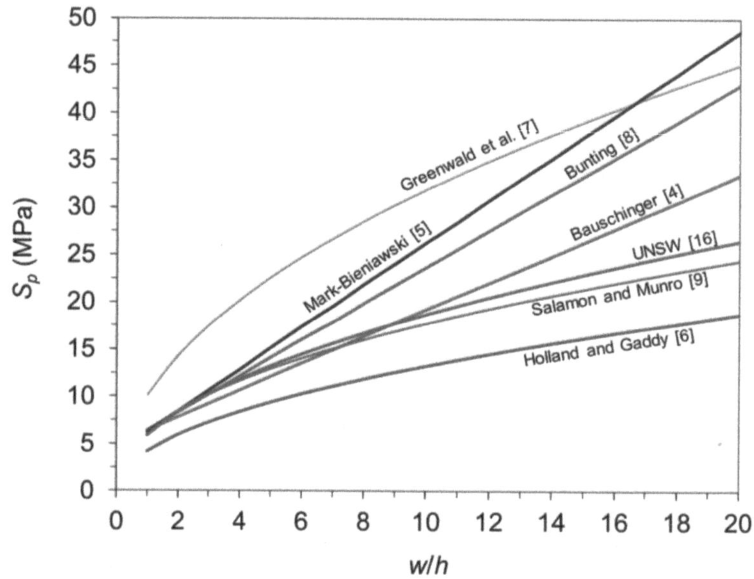

FIGURE 9.4 A comparison of a few of the empirical formula for a specific pillar geometry (Prassetyo et al., 2019).

TABLE 9.2
Values of the Cube Strength Used by Various Researchers (Prassetyo et al., 2019)

Math Formula	Researcher(s)	K Value	S_{cube} (MPa)
Linear	Bauschinger	–	6.4
	Bunting	–	6.9
	Mark and Bieniawski	–	6.4
Power	Greenwald et al.	–	19.2
	Holland and Gaddy	5,095	4.2
	Salamon and Munro	–	7.2
	Galvin et al.	–	6.9

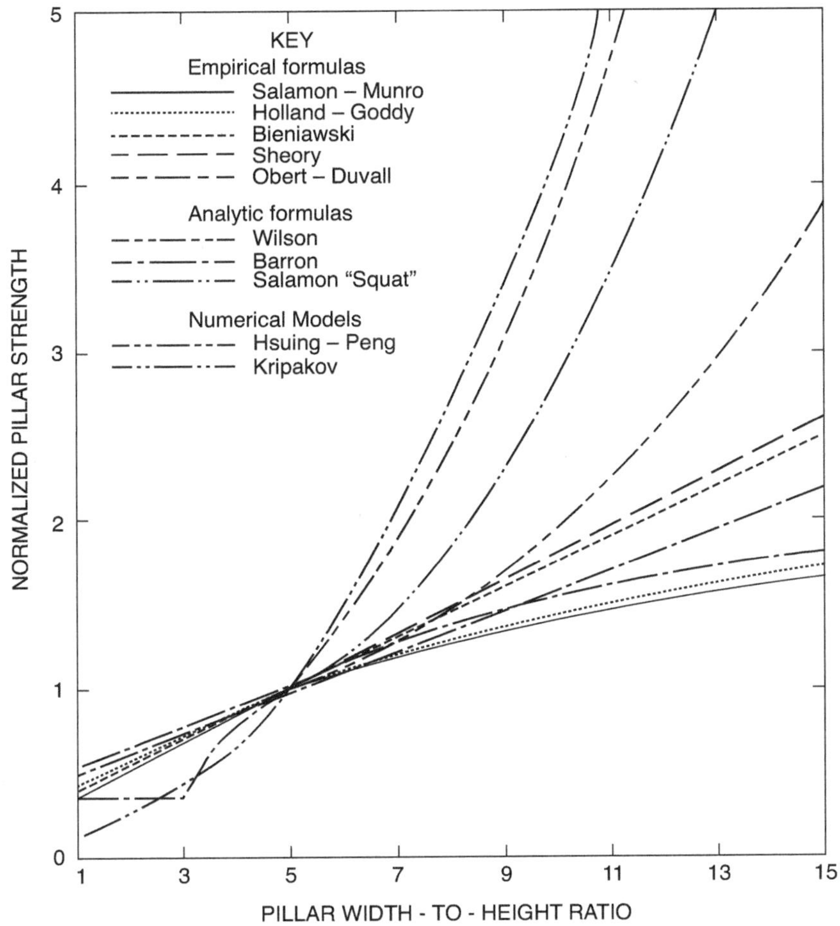

FIGURE 9.5 Comparison of different pillar strength formula and models (Michaud, 2017).

- Numerical models, comprising Hsuing and Peng's (1985), Kripakov's (1981), Wilson's (1981) formulas and Salamon's "squat" formula, indicate that the pillar strength increases exponentially as the pillar width is increased. It assumes that confinement grows rapidly within squat pillars, allowing the core stress to build strength infinitely.

It has been suggested that the strength of rectangular and irregular pillars is about the same as that of a square pillar having the same cross-sectional area (Wagner, 1980). On the basis of this assumption, an effective pillar width (w_{eff}) can be defined as follows:

$$w_{eff} = \left(A_p\right)^{0.5} \tag{9.2}$$

where

A_p is the cross-sectional area of the pillar.

The effective width formula is useful because not all local pillars have a square cross section, but some are rectangular. Rectangular pillars are generally used for examples, when the orebody is dipping, orienting the long pillar axis along the dip direction or along main entries/drives/gates to reduce the number of cross-cuts that will need a ventilation stopping to control the air flow.

TABLE 9.3
Exponents for Pillar Strengths of Some Coal Regions (Hustrulid, 1976)

Coal Region	α	β
South African coal in situ failures	−0.66 ± 0.16	0.46
Pittsburgh (US) coal from model tests	−0.83	0.50
West Virginia (US) coal from lab tests	−1.00	0.50

In a detailed study of the failure process of coal pillars, it was found that the failure commences at the circumference of the pillar and migrates inward (Wagner, 1974). Based on these observations, Wagner (1980) suggests that the ratio of the area (A_p) to the circumference (C) of a coal pillar has a strong influence on the pillar strength. Accordingly, the effective width, in meters, of a pillar of irregular shape is defined as follows:

$$w_{eff} = \left(4A_p\right)/C \tag{9.3}$$

Table 9.3 gives some values that have been found for the two parameters α and β for different coal seams.

9.2.2.2 Strength of Roof and Floor

An important aspect in pillar mining is the strength of the roof and floor which can affect the following:

- Immediate roof support.
- Standing support.
- Working conditions.
- Pillar stability.
- Ventilation.
- Subsidence.

Considering immediate roof control in room-and-pillar panels, the stability decreases as the rock strength decreases, the density of the bedding planes in the immediate roof strata increases, and the strength of the rock composing the individual beds decreases. In addition, the presence of water is always deleterious. In principle, there are three ways of improving the behavior of the immediate roof:

- Reducing the room width and/or increasing the pillar width.
- Increasing the apparent strength of the roof by typically installing more support.
- Reducing the induced stress in the roof.

Wagner (1980) states that the most important parameters that control the magnitude of the induced stresses are the size of the pillars and the room width. In general, the magnitude of the induced stresses decreases with increasing pillar size and decreasing room width. In laminated or bedded roof strata, marked improvements in the quality of the roof can often be achieved with relatively small reductions in room width. Multiple seam mining adds significant additional complexity to any mine and support design.

Wagner (1980) further states that the presence of a relatively thin, soft shale band or coal stringer in the roof or floor strata of the seams that are being worked can have a significant adverse influence on the quality of the immediate roof or floor, even if these are composed of relatively competent sandstone. An example of excessive floor heave caused by a very soft shale band, beneath

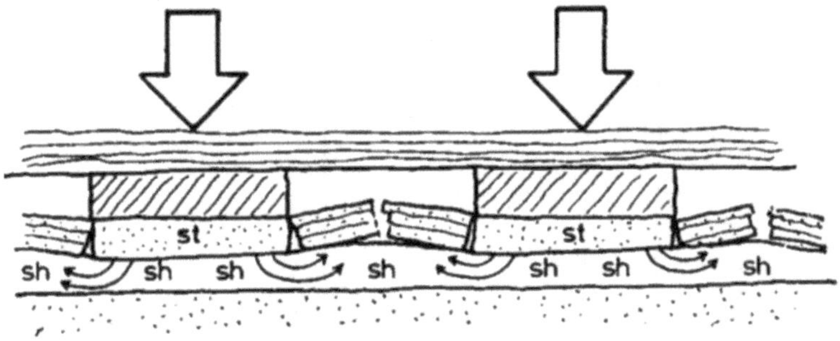

FIGURE 9.6 Example of excessive floor heave caused by weak immediate floor strata (Wagner, 1980).

an otherwise competent sandstone bed, is shown in Figure 9.6. Phenomena of this kind are usually confined to relatively deep-lying coal seams. This type of floor heave is often observed in pillar-extraction sections where the situation is aggravated by the high abutment stresses that act on the pillars in the retreat line. In the case of very narrow seams, floor heave can be so severe that it becomes a serious production problem.

9.2.3 Pillar Stress

The prediction of pillar stress is one of the principal problems in the design of pillar layouts. The only theoretical method of determining pillar stress was based on the tributary area theory. This basic theory applies to horizontal workings of which the lateral extent is large in relation to the depth below surface. Under these conditions, the average vertical pillar stress (p_m) is approximated by the following equation:

$$p_m = \gamma H\left(A/A_p\right) \tag{9.4}$$

where

γ is the unit weight (mass of rock per unit volume).
H is the depth to seam below surface.

The area of the pillar is denoted by A_p, and the corresponding tributary area (Figure 9.7) is denoted by A. The average pillar stress can also be expressed in terms of the extraction ratio (e):

$$e = \left(A - A_p\right)/A \tag{9.5}$$

Therefore,

$$p_m = \gamma H/(1-e) \tag{9.6}$$

The pillar stress from Equations 9.4 and 9.6 represents the maximum average pillar stress in any horizontal working of constant height supported by uniform pillars (as it assumes an infinite span in both directions and the room-and-pillar panel spans are always limited by unmined abutments or barrier pillars between panels).

9.2.4 Tributary Area Theory for Panel Pillar Design

The theoretical pillar stress is usually obtained by either computer analysis or tributary area theory particularly for coal seams. This theory ignores the effects of abutments and hence is only really applicable where the panel span of the room-and-pillar mining is at least equal to the depth of the

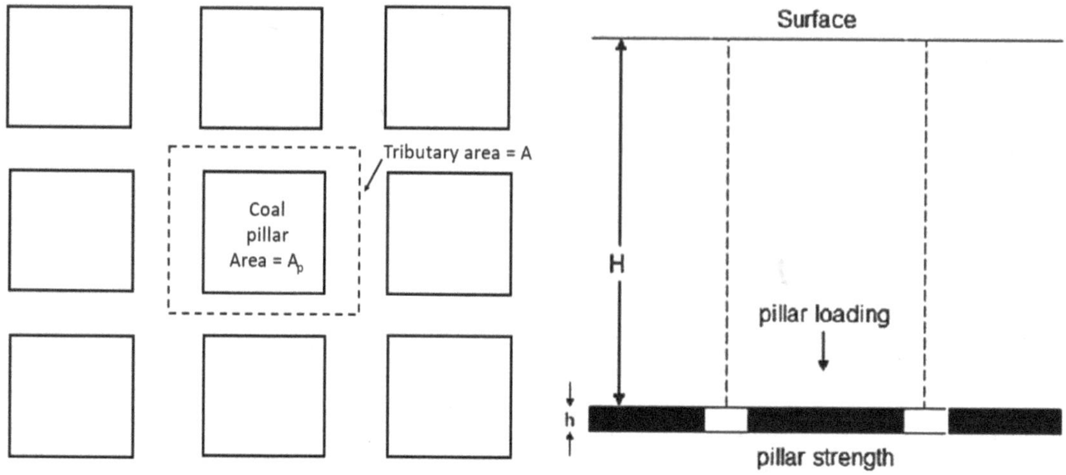

FIGURE 9.7 The tributary area and pillar area.

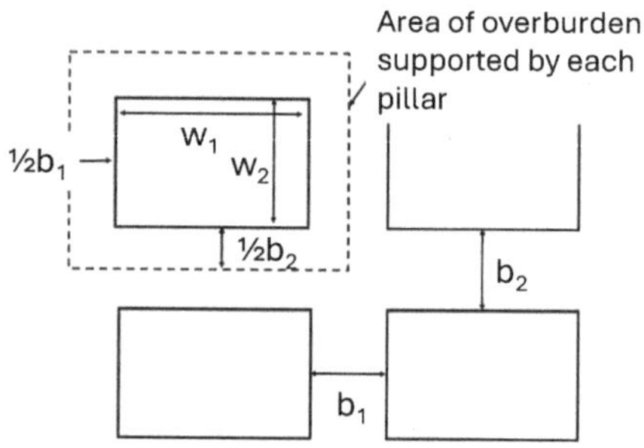

FIGURE 9.8 The basis of tributary area theory.

mining below surface. Should this not be the case, the pillar loads calculated will be substantially higher than the actual pillar loads resulting in a very conservative approach. Stress transfer and arching are also not considered, which brings in an even more conservative approach. Tributary Area Theory is expanded in Figure 9.8, assuming a simple pillar layout.

The total tributary area supported by a pillar is as follows:

$$A = (w_1 + b_1)(w_2 + b_2) \tag{9.7}$$

The area of the pillar is as follows:

$$A_p = w_1 w_2 \tag{9.8}$$

Assuming the depth of the pillars to be h and the average rock density of the overburden to be ρ, the overburden stress is considered to be

$$\sigma_v = \rho g h \tag{9.9}$$

The pillar stress is a function of the overburden stress, and therefore,

$$\sigma_p = \sigma_v[(w_1 + b_1)(w_2 + b_2)]/(w_1 + w_2) \tag{9.10}$$

The percentage extraction (e) in percent and in terms of Figure 9.8 is equal to

$$e = (100)\{1-(w_1 w_2)/[(w_1+b_1)(w_2+b_2)\} \quad (9.11)$$

The strength of pillars (P_s) according to Salamon and Munro (1967), is approximated as follows:

$$P_s = k\left(w^{0.46}/h^{0.66}\right) \quad (9.12)$$

Equation 9.12 was based on research in coal mines, and k was found to be 7.2 MPa for South African coal.

For hard-rock room-and-pillar mining sections, the following has been derived by Hedley and Grant (1972):

$$P_s = k\left(w^{0.50}/h^{0.75}\right) \quad (9.13)$$

where

k is a constant depending on the rock type.
w is the pillar width (assumed square).
h is the effective pillar height.

The effective pillar height includes the additional height of the immediate roof and floor which is highly fractured and this can be significantly more than the original mined height especially as a result of local collapses and extensive fracturing in the immediate roof.

Equation 9.13 was based on research into uranium mines in Canada, and k was found to be 133 MPa. Back analyses on typical quartzites on South African tabular gold mines by Spearing (internal AAC report, 1987) found that k was between 110 and 140 MPa.

9.2.5 Limestone Room-and-Pillar Design

Underground stone, mainly limestone, mines are relatively flat and utilize the room-and-pillar mining method (Esterhuizen et al., 2011). The roof is generally strong with few significant geological features, and little or no support is used except spot bolting where necessary. Table 9.4 shows the average and range of pillar conditions found in a survey undertaken by Esterhuizen et al. (2011). The unconfined compression strength (UCS) of limestone is typically in the range of 13,000–30,000 psi (90–210 MPa).

TABLE 9.4
Summary of Mining Dimensions and Depth of Mines Included in the Study by Esterhuizen et al. (2011)

Dimension	Average	Minimum	Maximum
Pillar width (ft)	42.9	15.1	70.5
Pillar height (ft)	36.4	15.7	131.2
Width to height ratio	1.41	0.29	3.52
Room width (ft)	44.3	29.9	52.8
Cover depth (ft)	383.9	75.1	2,198.2

According to Iannacchione et al. (2002), room widths average about 43 ft (13 m) with a range from 20 to 60 ft (6–18 m). Overall extraction averages 75% with a range from 55% to 90%.

Most empirically based hard-rock formula are derived from the basic formula below, as detailed earlier:

$$\text{Pillar strength} = \sigma_p = k\left(w^a/h^b\right) \tag{9.14}$$

where again

k is a constant related to the material strength.

h is the pillar height.

w is the pillar width (square pillar assumed).

Clearly if the orebody is dipping, this adds a constraint into the pillar design as a shear component is added. Generally, pillars under these conditions can be made rectangular with large dimensions in the dip direction. The strength of the roof and floor and the presence of water must also be considered. Mining multiple seams increases the design complexity significantly.

According to Esterhuizen et al. (2011), the strength equation for stone mine pillars with the large discontinuity and rectangular pillar adjustments can be rewritten as follows:

$$P_s = 0.65(\text{UCS})(\text{LDF})(w^{0.3}/h^{0.59}) \tag{9.15}$$

where

UCS is the uniaxial compressive strength of the intact rock in MPa.

LDF is the large discontinuity factor as given in Equation 9.16.

h is the height in meters.

w is the pillar width in meters.

For rectangular pillars, w is replaced by the equivalent width w_e, which can be calculated using

$$w_e = w + \left(4A/C - w\right)(\text{LBR}) \tag{9.16}$$

where

w is the minimum width of the pillar.

A is the pillar plan area.

C is the circumference of the pillar.

LBR is the length benefit ratio.

Values for LBR for rectangular pillars at various w/h ratios are given in Table 9.5.

The value of LDF can be determined from Equation 9.17. If no large discontinuities are present, LDF will equal 1.0:

$$\text{LDF} = 1 - (\text{DDF})(\text{FF}) \tag{9.17}$$

where

DDF is the discontinuity dip factor.

FF is the discontinuity frequency factor.

Full details are available in Esterhuizen et al. (2011 and 2019).

TABLE 9.5
Values of LBR for Rectangular Pillars (Esterhuizen et al., 2011)

Width to Height Ratio	Length Benefit Ratio (LBR)
0.5	0.00
0.6	0.06
0.7	0.22
0.8	0.50
0.9	0.76
1.0	0.89
1.1	0.96
1.2	0.98
1.3	0.99
1.4	1.00

TABLE 9.6
Values for Finding the DDF (Esterhuizen et al., 2011)

	Pillar Width to Height Ratio								
Dip (°)	≤0.5	0.6	0.7	0.8	0.9	1.0	1.1	1.2	>1.2
30	0.15	0.15	0.15	0.15	0.16	0.16	0.16	0.16	0.16
40	0.23	0.26	0.27	0.27	0.25	0.24	0.23	0.23	0.22
50	0.61	0.65	0.61	0.53	0.44	0.37	0.33	0.30	0.28
60	0.94	0.86	0.72	0.56	0.43	0.34	0.29	0.26	0.24
70	0.83	0.68	0.52	0.39	0.30	0.24	0.21	0.20	0.18
80	0.53	0.41	0.31	0.25	0.20	0.18	0.17	0.16	0.16
90	0.31	0.25	0.21	0.18	0.17	0.16	0.16	0.15	0.15

TABLE 9.7
Values for the FF (Esterhuizen et al., 2011)

Average Frequency of Large Discontinuities per Pillar	Frequency Factor (FF)
0.0	0.00
0.1	0.10
0.2	0.18
0.3	0.26
0.5	0.39
1.0	0.63
2.0	0.86
3.0	0.95
>3.0	1.00

Table 9.6 shows how the value of DDF is obtained, and Table 9.7 shows how FF is obtained.

Based on an extensive investigation into stone mine pillar stability, Esterhuizen et al. (2011) have suggested the design chart in Figure 9.9.

This has been incorporated into S-Pillar Software for ease of use (NIOSH, 2024).

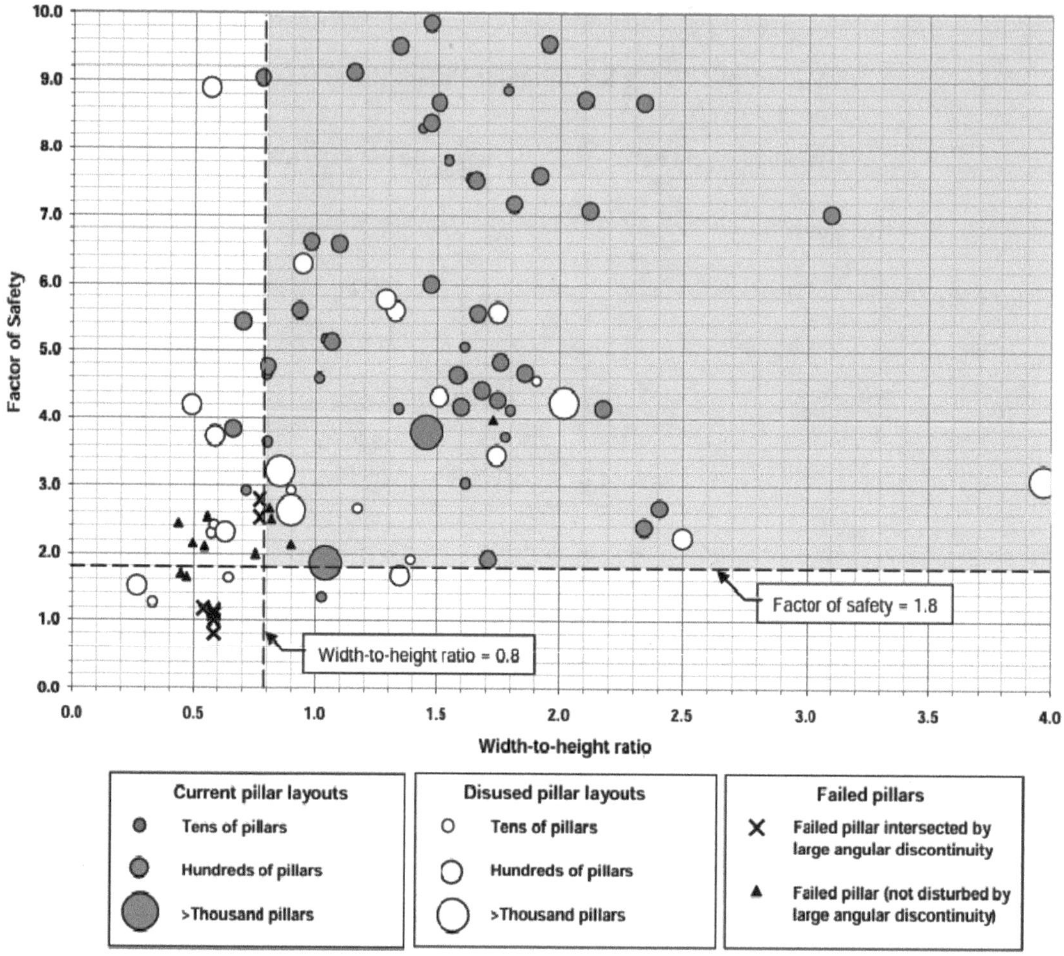

FIGURE 9.9 Limestone pillar design chart (Esterhuizen et al., 2011).

9.2.6 Pillar Factor of Safety

The greatest hazard associated with pillar mining is the potential of violent pillar failure and collapse. Sudden failure is not always preceded by pillar spalling, and once the failure starts, it can uncontrollably propagate. Massive collapses are also associated with hazardous air blasts.

This really depends on the design approach. In some cases a higher value may be needed. However the conditions and design equation may dictate a greater factor of safety. A compliment to the aforementioned pillar design methods, albeit used less frequently, include variability in the input and output using an example such as Monte Carlo simulations where variability of parameters is fundamental.

9.2.6.1 The Use of Reliability Concepts in Pillar Design

The approximate strength of pillars depends on rock strength and pillar dimensions and discontinuities. For a standard pillar design approach, these parameters are taken as point estimates (without variation), whereas in practice, the parameters can vary considerably. For example, the pillars are often cut to variable sizes and the height of each pillar is seldom the same due to stoping width variations. Similarly, the stresses on pillars are assumed to be the same, but this is also not the case in practice, due to the different mining spans and the varying overburden. Care must also be taken that the design is not too conservative as, for example, using the worst-case scenario may not be realistic.

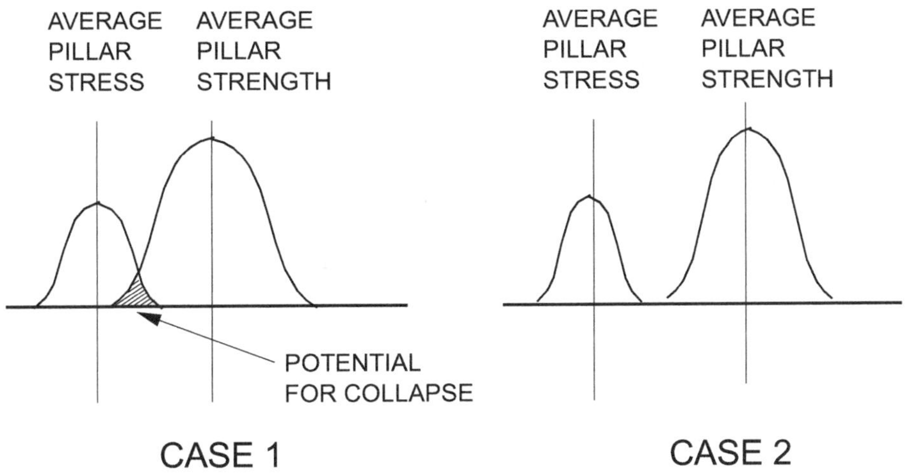

FIGURE 9.10 Pillar probabilistic examples.

Probabilistic concepts consider these variations in the FOS of pillar design as illustrated in Figure 9.10. Since variations of inputs are included, a variation in the FOS is also resolved in the process.

The FOS is determined in the same way for both cases:

$$\text{FOS} = \text{Pillar Strength}/\text{Pillar Stress} \tag{9.18}$$

The bell-shaped curves in Figure 9.10 account for variability in pillar strength and stress. Using a classical deterministic approach, which only uses the averages (vertical lines in Figure 9.10), both cases show a FOS that is acceptable because the average pillar strength exceeds the stress in both cases. However, using a probabilistic approach whereby the whole bell-shaped curve is considered (i.e., variability is considered), there exists an overlap whereby the pillar stress exceeds the strength.

To undertake such an investigation on a mine, databases need to be prepared, and the following provides examples of data that can be gathered:

- The average UCS of the orebody, by taking multiple samples at numerous locations, and a standard deviation would be useful. Young's modulus and Poisson's ratio are also useful parameters for elastic models if they can be collected.
- The stratigraphic column at various locations above and below the orebody to determine the average overburden density.
- Actual pillar dimensions.
- Mining dimensions and spans (room width, etc.)
- In multiple seam environments, maps of all seams, especially abandoned ones, must be determined. An assessment of interburden competency and any potential groundwater (flooded areas particularly) is very important.

The use of datasets can be beneficial when designing pillars by identifying potential instabilities that average stresses and strengths would not reveal. This technique can, therefore, be used to formalize more representative FOS required or encountered in past mining.

9.3 BARRIER PILLAR DESIGN

Barrier pillars are typically used:

- To protect the main entries in longwall and room-and-pillar mines.
- To isolate mining districts (panels) especially room-and-pillar panels from each other so to limit the possibility of massive pillar runs from one panel to another.
- When mining close to other mines, tapped gas or oil fields, especially considering the potential for water and/or gas inrushes.
- To limit surface subsidence.
- To create safe ventilation and escapeways.
- To control the effect of mining on major geological structures especially those that could be burst-prone.
- To control potential inrushes from surface and groundwater.

Clearly under-designed pillars can lead to massive potential hazards, while over-designed pillars lead to a loss in mineable reserves and hence revenue. Additional consideration and care must be taken if the conditions could be seismically active.

Again depends on several factors. Not at all a hard rule. This is probably more a minimum.

Appropriate computer-aided numerical design packages, if correctly modeled and validated, are useful, especially because sensitivity analyses can be undertaken. Field data can also be effectively used to calibrate the computer model. An empirically designed method for coal that can be used within the data parameters used to establish the method is the Analysis of Coal Pillar Stability (ACPS). According to Mark and Agioutantis (2018), this method combined the Analysis of Longwall Pillar Stability (ALPS), the Analysis of Retreat Mining Stability (ARMPS), and the Analysis of Multiple Seam Stability (AMSS) design methods, but this is limited to two seams. LaModel (Heasley, 2014) can handle four seams successfully.

In the United States, longwall barrier pillars ranged between 152 and 183 m wide, but with the application of effective computer software packages and analyses, they have been safely reduced from 61 to 92 m (Peng, 2020).

Continuously monitoring and observing the behavior of pillars is essential, and as more data is made available, the design should be revisited and amended if necessary.

Should pillars start to spall, squeeze, and fail unexpectedly (unless designed as yield pillars), the effect of local pillar failure must be assessed, and if the pillar damage can be contained, then the area can be cordoned off and abandoned. If any collapse is a possibility, miners safety is paramount and they should be removed from the area immediately, but only implemented if safe to do so:

- Cease mining in the area and remove personnel until the reason for the pillar problem has been established.
- Backfill the affected area as even if the fill is only partial, it reduces the effective pillar height and hence increases the pillar strength. Waste rock or hydraulic fill can be used for this.
- Mesh and bolt the pillars.
- Install a breaker line of free-standing support around the affected area to limit falls of ground associated with the unplanned pillar failure.

Thorough and regular mine examinations should be considered essential.

9.4 SPECIAL CONDITIONS

Pillar design (in hard-rock mines) usually assumes a relatively flat seam, with stable roof and floor, no blast damage and no significant water or adverse geological features. Typically, the in situ conditions are different to the "ideal" ones assumed in pillar design and must be considered when deciding on a final pillar design. It is also essential to monitor the actual pillars as they are mined and revisit the design assumptions depending on the actual pillar behavior. Aspects to consider include the following:

- Actual pillar dimensions, e.g., as-mined are they larger or smaller than the design?
- Room widths, e.g., are they larger or smaller than the design?
- Rock condition, e.g., is there unexpected pillar spalling?
- Pillar movement, e.g., are pillars punching into the roof or floor?

9.4.1 The Effect of an Inclined Seam

An increased pillar inclination increases the potential for sliding on weak planes. This can cause a reduction in the load that pillars can sustain (Jessu and Spearing, 2018). As a first estimate, it was found that the maximum pillar load reduction results in an inclined situation is

$$\text{Pillar strength reduction factor} = 1 - 0.0077\left(\text{orebody dip in degrees}\right) \quad (9.19)$$

The formula was developed using both laboratory testing and numerical modeling, which showed good correlation, as well as the Lunder and Palkalnis (1997) theoretical results. The current limitations of this relationship are as follows:

- A maximum orebody inclination of 20° was tested.
- Width to height ratios were between 0.5 and 2.0.

These results tend to show that if pillars are designed using traditional approaches (like tributary area theory), a higher FOS should be used. It is also well known that at dip, rectangular pillars (with the longer axis on dip) are more effective as they can resist the shearing tendency.

9.4.2 The Effect of a Weak Floor or Roof

As it relates to pillar design, the rock behavior of floor and roof is of equal importance toward mine stability. Historically, rock mechanics literature demonstrates methods to control discontinuities and/or rocks of unfavorable chemical make-up, or competency, and how they could compromise an opening for miner, material, or ventilation. In coal for example, even in low-stress environments, a seam underlain by a weak clay can fail when saturated. In most cases, but not always, this failure will manifest slowly, as the floor gradually creeps into the open entry. Examples of the indirect consequences of this type of failure are as follows:

- Pillar displacement into the floor.
  - At shallow depth, the pillar FOS in classical design would be high, but the foundation could still fail. This could also represent the cause of surface subsidence.
- Floor displacement into the room.
  - At times, this can be remediated, but if not, it represents a ventilation hazard in gassy mines.
- Deterioration of roof due to floor displacement.
  - In fact, floor failure, especially in the presence of significant water inflow, can make relatively strong roof (e.g., sandstones or limestones) perform like weaker roof rock (e.g., weak shale).

At depth, strong roof and floor can represent other hazards, such as rockburst conditions which in coal can manifest at pillar ribs or even from the floor. Clearly the unpredictable nature of rock bursts, versus slow gradual floor failure, is the reason why rock burst prevention is of critical priority.

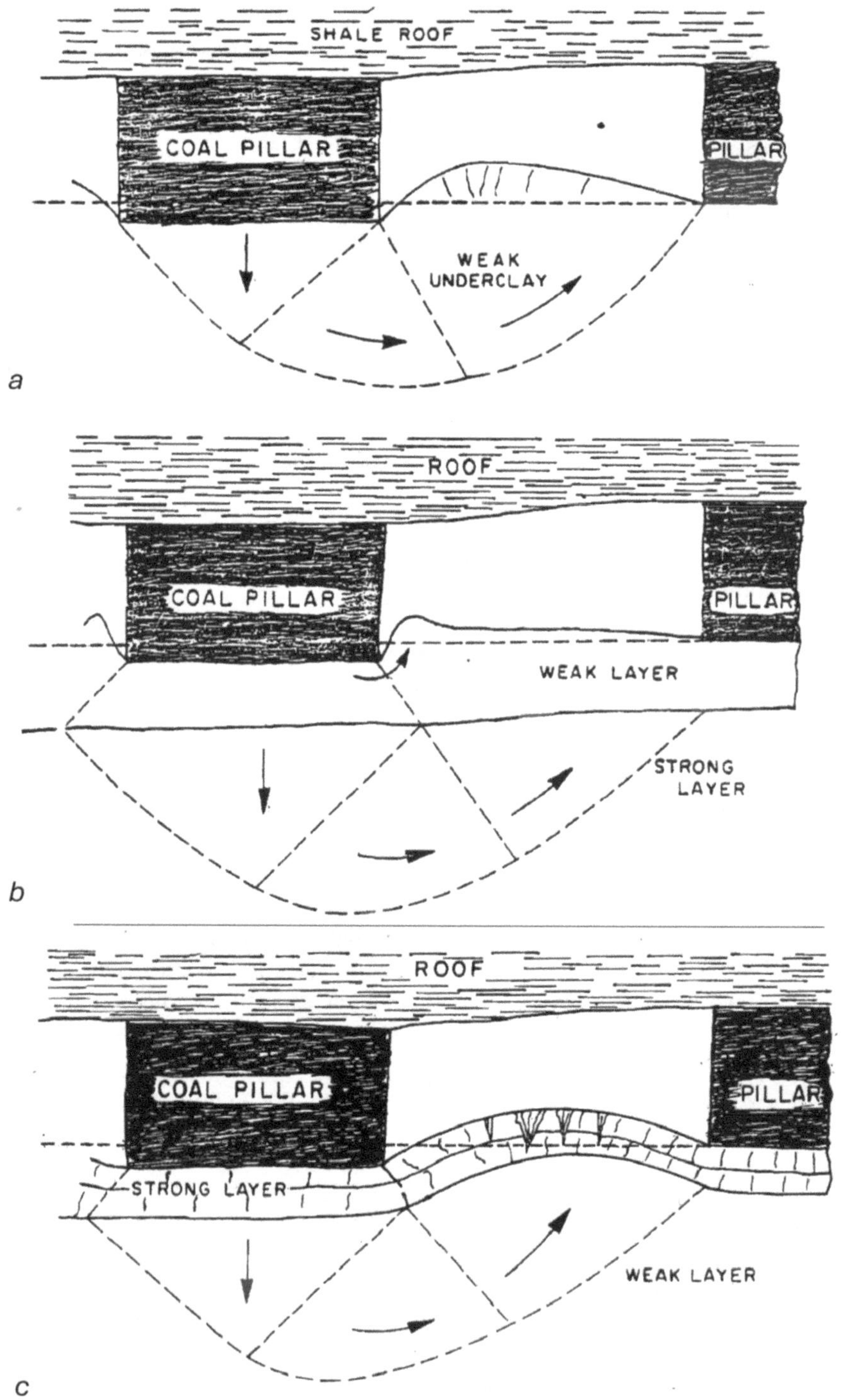

FIGURE 9.11 Three forms of floor failure for coal pillars: general shear failure (a), punching pillar failure (b), and buckling failure (c) (Rockaway and Stephenson, 1979).

In multiple seam mining in coal, floor rock is hardly considered except when the interburden is thin, and even in those conditions, there are few tools and guidance that provide a high level of confidence toward how to design these types of mines. Similarly in stone and hard rock, floor stability is not necessarily considered directly into the classical pillar design. Currently available are adaptations from soil mechanics and foundation design, but they are not widely used proactively or are empirically bounded for rather specific mining regions.

9.4.3 The Effect of Blast Damage

Where pillars are blasted, the blast-induced fractures reduce the strength of the pillar rock mass around its immediate perimeter, by effectively softening it (Hoek, 1994; Jessu et al., 2018). This can be done by either reducing the pillar rock strength (Hoek, 1994) or by reducing the dimensions of an equivalent "un-blasted" pillar, when calculating its strength (Jessu et al., 2018). Jessu et al.'s (2018) research showed that the effect of blasting needs to be considered in pillar design as it can decrease the strength of the pillars. Therefore, the damage factor and damage thickness are important parameters.

It was generally found that the larger the width to height ratio of the pillar, the larger the percentage strength reduction generally caused by blasting.

9.4.4 The Effect of Adverse Geology

Discontinuities in pillars can have a significant adverse effect on the strength of the pillar. Bedding planes are not often a problem in horizontal pillars in orebodies (or seams) but can become serious in inclined deposits. Geological features (mainly faults but also joints) can also be a serious issue depending on their orientation and properties.

9.4.5 The Effect of Water

Water is always unfavorable when mining, and from a rock mechanics perspective, it can:

- Cause weathering of the rock.
- Lubricate faults, bedding planes, and joints, thus reducing their effective shear strength.
- Create adverse pore pressure.
- Corrode the installed support.

9.5 WORKED EXAMPLES

Example 9.5.1

Given the following square pillar layout (in Figure 9.12) for a room-and-pillar mine, calculate the percent extraction for development mining. The pillars are at 80 ft centers, and the rooms are 20 ft spans.

Solution:

Percent extraction (%) = $(100)[1 - (w_1)(w_2)/(w_1 + b_1)(w_2 + b_2)]$

$$w_1 = 80\text{ ft} - 20\text{ ft} = 60\text{ ft}$$

$$w_2 = 80\text{ ft} - 20\text{ ft} = 60\text{ ft}$$

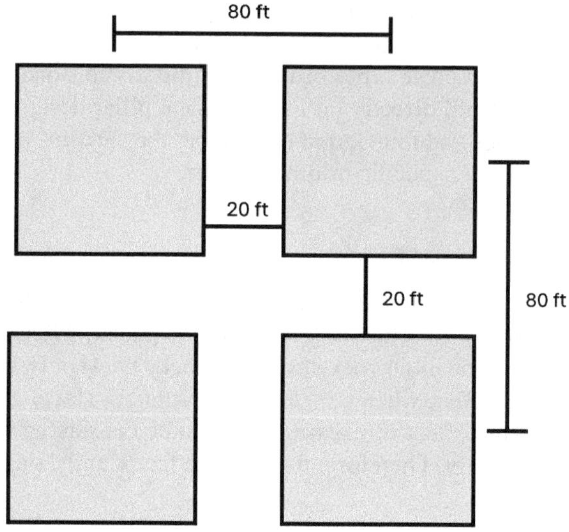

FIGURE 9.12 The room-and-pillar layout for Example 9.5.1.

$$b_1 = 20 \text{ ft}$$

$$b_2 = 20 \text{ ft}$$

Percent extraction, % = (100)[1 – (60 ft)(60 ft)/(60 ft + 20 ft)(60 ft + 20 ft)]
Percent extraction, % = (100)[1 – (3,600 ft^2)/(6,400 ft^2)]
Percent extraction, % = 44%

Example 9.5.2

Back calculate the approximate room spans for the square pillars on 60 ft centers as shown in Figure 9.13. The percentage extraction is 40%.

Solution:

Percent extraction (%) = (100)[1 – $(w_1)(w_2)/(w_1 + b_1)(w_2 + b_2)$]

$$w_1 = w_2 = 60 \text{ ft} - x$$

$$b_1 = b_2 = x$$

$$40\% = (100)(1 - (60 \text{ ft} - x)(60 \text{ ft} - x) / \{[(60 \text{ ft} - x) + x][(60 \text{ ft} - x) + x]\}$$

$$0.40 = 1 - (60 \text{ ft} - x)^2 / (60 \text{ ft})(60 \text{ ft})$$

$$0.40 - 1 = -(60 \text{ ft} - x)^2 / (3{,}600 \text{ ft}^2)$$

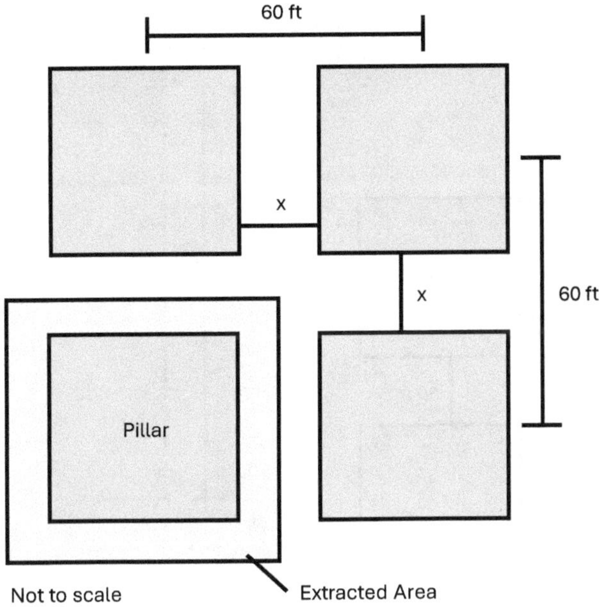

FIGURE 9.13 The room-and-pillar layout for Example 9.5.2.

$$-0.60 = -(60\text{ ft} - x)^2/(3{,}600\text{ ft}^2)$$

$$2{,}160\text{ ft}^2 = (60\text{ ft} - x)^2$$

$$(2{,}160\text{ ft}^2)^{0.5} = 60\text{ ft} - x$$

$$46.5\text{ ft} = 60\text{ ft} - x$$

$$x = 13.5\text{ ft}$$

These examples are not necessarily real and practical, and are just shown as examples.

Example 9.5.3

The proposed layout of pillars for a room-and-pillar mine is given alongside an overburden detail in Figure 9.14. Determine the pillar stress, the percent extraction, pillar strength, and the FOS. Use Hedley and Grant's equation (1972) assuming that the constant k is 130 MPa and the pillar height is 5 m.

Solution:

$$w_1 = 12\text{ m}$$

$$w_2 = 12\text{ m}$$

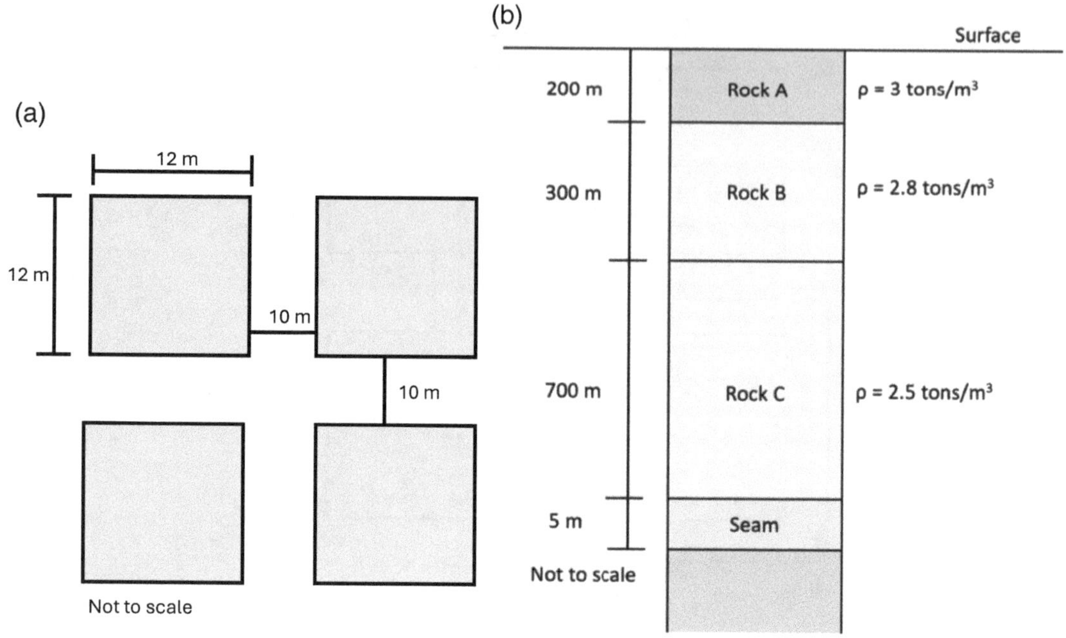

FIGURE 9.14 The hard-rock room-and-pillar layout and overburden.

$$b_1 = 10 \text{ m}$$

$$b_2 = 10 \text{ m}$$

Vertical Stress:

$$\sigma_v = \rho g h = \left(3 \text{ t/m}^3\right)\left(1{,}000 \text{ kg/t}\right)\left(9.81 \text{ m/s}^2\right)\left(200 \text{ m}\right) + \left(2.8 \text{ t/m}^3\right)\left(1{,}000 \text{ kg/t}\right)$$
$$\left(9.81 \text{ m/s}^2\right)\left(300 \text{ m}\right) + \left(2.5 \text{ t/m}^3\right)\left(1{,}000 \text{ kg/t}\right)\left(9.81 \text{ m/s}^2\right)\left(700 \text{ m}\right)$$

$$\sigma_v = 5{,}886{,}000 \text{ N/m}^2 + 8{,}240{,}400 \text{ N/m}^2 + 17{,}167{,}500 \text{ N/m}^2$$

$$\sigma_v = 31{,}293{,}900 \text{ N/m}^2$$

$$\sigma_v = 31.3 \text{ MPa}$$

Pillar Stress:

$$\sigma = \sigma_v[(w_1 + b_1)(w_2 + b_2)]/(w_1 + w_2)$$

$$\sigma = 31.3 \text{ MPa}[(12 \text{ m} + 10 \text{ m})(12 \text{ m} + 10 \text{ m})]/[(12 \text{ m})(12 \text{ m})]$$

$$\sigma = 31.3 \text{ MPa}[(22 \text{ m})(22 \text{ m})]/\left(144 \text{ m}^2\right)$$

$$\sigma = 105 \text{ MPa}$$

Pillar Strength:

Convert m to ft for width and height:

$$w = 12\ \text{m}\left(1\ \text{ft}/0.3048\ \text{m}\right) = 39.4\ \text{ft}$$

$$h = 5\ \text{m}\left(1\ \text{ft}/0.3048\ \text{m}\right) = 16.4\text{ft}$$

$$\sigma_p = k(w^{0.5}/h^{0.75})$$

$$\sigma_p = 130\ \text{MPa}(39.4^{0.5}/16.4^{0.75})$$

$$\sigma_p = 130\ \text{MPa}(6.28/8.15)$$

$$\sigma_p = 130\ \text{MPa}(0.77)$$

$$\sigma_p = 100\ \text{MPa}$$

FOS for a Single Pillar:

$$F = \sigma_p/\sigma_s$$

$$F = 100\ \text{MPa}/105\ \text{MPa}$$

$$F = 0.95$$

Percent Extraction:

Percentage extraction, % = (100)[1 – $(w_1)(w_2)/(w_1 + b_1)(w_2 + b_2)$]
Percentage extraction, % = (100)[1 – (12)(12)/(12 + 10)(12 + 10)]
Percentage extraction, % = (100)[1 – (144)/(22)(22)]
Percentage extraction, % = 70%

Example 9.5.4

Determine the pillar FOS for a square pillar at a depth of 200 ft. Pillars are on 80 ft centers with 20 ft spans. Use Holland's formula (1973) where seam height is 6 ft and the compressive strength of a 4 in rock specimen is 2,000 psi. Assume that the weight of the overburden can be approximated at 1.1 psi per foot of cover.

Solution:

$$w_1 = 60\ \text{ft}$$

$$w_2 = 60\ \text{ft}$$

$$b_1 = 20\ \text{ft}$$

$$b_2 = 20\ \text{ft}$$

Approximate the vertical stress since the stratigraphic column is not given:

$$\sigma_v = [(1.1\text{psi})/\text{ft}](200\ \text{ft})$$

$$\sigma_v = 220\ \text{lb/in}^2(\text{psi})$$

Pillar Stress:

$$\sigma = \sigma_v[(w_1 + b_1)(w_2 + b_2)]/(w_1 + w_2)$$

$$\sigma = 220\ \text{lb / in}^2[(60\ \text{ft} + 20\ \text{ft})(60\ \text{ft} + 20\ \text{ft})]/[(60\ \text{ft})(60\ \text{ft})]$$

$$\sigma = 220\ \text{lb/in}^2(6{,}400\ \text{ft}^2)/(3{,}600\ \text{ft}^2)$$

$$\sigma = 392\ \text{lb/in}^2(\text{psi})$$

Holland's Pillar Strength Formula (1973) takes the following form:

$$\sigma_p = \sigma_{cc}(w/h)^{0.5}$$

where σ_{cc} is the strength at critical size of coal specimens such that (Hustrulid, 1976)

$$\sigma_{cc} = k/(36)^{0.5}\ \text{cubical pillars with height, h} > 36\ \text{in}$$

$$\sigma_{cc} = k/(h)^{0.5}\ \text{cubical pillar with height, h} < 36\ \text{in}$$

where
the constant k, after Gaddy (1956), is

$$k = \sigma_c(d)^{0.5}$$

where

σ_c = uniaxial compressive strength of a specimen having a diameter or cube dimension "d"

$$k = \sigma_c(d)^{0.5}$$

$$k = (2{,}000\ \text{lb / in}^2)(4\ \text{in})^{0.5}$$

$$k = 4{,}000\ \text{lb/in}^2$$

$$\sigma_{cc} = k/(36)^{0.5}\ \text{cubical pillars with height, h} > 36\ \text{in}$$

$$\sigma_{cc} = 4{,}000\ \text{lb/in}^2/(36)^{0.5}$$

$$\sigma_{cc} = 667\ \text{lb/in}^2$$

$$\sigma_p = \sigma_{cc}\left(w/h\right)^{0.5}$$

$$\sigma_p = 667\ \text{lb/in}^2\left(60\ \text{ft}/6\ \text{ft}\right)^{0.5}$$

$$\sigma_p = 667\ \text{lb/in}^2(10)^{0.5}$$

$$\sigma_p = 2{,}109\ \text{lb/in}^2$$

FOS for a single pillar:

$$F = \sigma_p/\sigma$$

$$F = 2{,}109\ \text{lb/in}^2/392\ \text{lb/in}^2$$

$$F = 5.4$$

Note that this is very conservative for development-only panel pillars in coal due to the pillar strength and depth. The FOS is more appropriate for barrier pillars but may even be still too high.

Example 9.5.5

Determine the FOS for the previous example using Salamon and Munro's formula for pillar design (1967).

Solution:

$$w_1 = 60\ \text{ft}\ \left(0.3048\ \text{m}/1\ \text{ft}\right) = 18.3\ \text{m}$$

$$w_2 = 60\ \text{ft} = 18.3\ \text{m}$$

$$b_1 = 20\ \text{ft} = 6.1\ \text{m}$$

$$b_2 = 20\ \text{ft} = 6.1\ \text{m}$$

$$h = 6\ \text{ft} = 1.8\text{m}$$

Pillar stress: σ = 392 psi (1 MPa/145 psi) = 2.7 MPa
Pillar Strength for Salamon and Munro in SI units

$$\sigma_p = 7.2(w^{0.46}/h^{0.66})$$

where
h and w are pillar dimensions in meters.
Pillar Strength:

$$\sigma_p = k(w^{0.46}/\ h^{0.66})$$

$$\sigma_p = 7.2(18.3^{0.46}/1.8^{0.66})$$

$$\sigma_p = 7.2(2.58)$$

$$\sigma_p = 18.6 \text{ MPa}$$

Factor of Safety:

$$F = \sigma_p/\sigma$$

$$F = 18.6 \text{ MPa}/2.7 \text{ MPa}$$

$$F = 6.9$$

Example 9.5.6

Determine the FOS for a pillar at 1,000 ft depth using the Obert, Duvall (1967) and Wang et al (1977) equation. Compare the answer using Bieniawski's (1968) pillar formula. Assume a pillar height of 10 ft. The pillar is square at 40 ft with rooms of 18 ft. Average unit weight of the overburden is roughly 145 lbs/ft^3. Assume that $\sigma_{cc.}$ is 315 lbs/in^2 (psi).

Solution:

$$w_1 = 40 \text{ ft}$$

$$w_2 = 40 \text{ ft}$$

$$b_1 = 18 \text{ ft}$$

$$b_2 = 18 \text{ ft}$$

Vertical stress where stratigraphic column is not given, so reasonable estimate is

$$\sigma_v = \gamma_{rock}(H)$$

$$\sigma_v = (145 \text{ lbs/ft}^3)(1{,}000 \text{ ft})$$

$$\sigma_v = 145{,}000 \text{ lbs/ft}^2 (\text{psf})$$

Pillar Stress:

$$\sigma = \sigma_v[(w_1 + b_1)(w_2 + b_2)]/(w_1 + w_2)$$

$$\sigma = 145{,}000 \text{ lbs/ft}^2[(40 \text{ ft} + 18 \text{ ft})(40 \text{ ft} + 18 \text{ ft})]/[(40 \text{ ft})(40 \text{ ft})]$$

$$\sigma = 145{,}000 \text{ lbs/ft}^2(3{,}364 \text{ ft}^2)/\left(1{,}600 \text{ ft}^2\right)$$

$$\sigma = 304{,}863 \text{ lb/ft}^2\left(\text{psf}\right)$$

Pillar Strength from Obert and Duvall (1967) and Wang et al. (1977) for w/h of 0.25–4.0:

$$\sigma_p = \sigma_{cc}\left[0.788 + 0.222\left(w/h\right)\right]$$

$$\sigma_p = \left(315 \text{ lbs/in}^2\right)\left(144 \text{ in}^2/1 \text{ ft}^2\right)\left[0.788 + 0.222\left(40/10\right)\right]$$

$$\sigma_p = 45{,}360 \text{ lbs/ft}^2\left[0.788 + 0.222(4)\right]$$

$$\sigma_p = 76{,}023 \text{ lbs/ft}^2\left(\text{psf}\right)$$

where

σ_{cc} is the uniaxial compressive strength of cubical specimen.

Pillar strength from Bieniawski (1968):

$$\sigma_p = \sigma_{cc}\left[0.64 + 0.36\left(w/h\right)\right]$$

$$\sigma_p = \left(315 \text{ lbs/in}^2\right)\left(144 \text{ in}^2/1 \text{ ft}^2\right)\left[0.64 + 0.36\left(40/10\right)\right]$$

$$\sigma_p = 94{,}348 \text{ lbs/ft}^2\left(\text{psf}\right)$$

where

σ_{cc} is the strength of a cubical specimen of critical size or greater.

FOS for Obert, Duvall, and Wang:

$$F = \sigma_p/\sigma$$

$$F = 76{,}023 \text{ lbs/ft}^2/304{,}863 \text{ lb/ft}^2$$

$$F = 0.25$$

FOS for Bieniawski:

$$F = \sigma_p/\sigma$$

$$F = 94{,}348 \text{ lbs/ft}^2/304{,}863 \text{ lb/ft}^2$$

$$F = 0.31$$

Example 9.5.7

Determine the pillar strength of a limestone pillar using the Hedley and Grant (1972), Krauland and Soder (1987), and Lunder and Pakalnis (1997) equations for a limestone pillar with the following criteria. The inputs were found in Esterhuizen et al. (2019). Assume a pillar height of 17.7 m with a w/h ratio of 0.5. The UCS strength of the intact rock is 182 MPa. Also, the ratio k/σ_c is 0.58 for Hedley and Grant and 0.354 for Krauland and Soder.

Parameters:
k: pillar strength constant
w: pillar width
h: pillar height
σ_c: uniaxial compressive strength intact rock
κ: mine pillar friction term
C_{pav}: average pillar confinement

Solution:

Hedley and Grant Equation (1972):

$$\sigma_p = k(w^{0.5}/h^{0.75})$$

$$0.58 = k/\sigma_c$$

$$0.58 = k/182\ \text{MPa}$$

$$k = 106\ \text{MPa}$$

$$w/h = 0.5$$

$$w = 0.5(17.7\ \text{m})$$

$$w = 8.85\ \text{m}\left(1\ \text{ft}/0.3048\ \text{m}\right) = 29.0\ \text{ft}$$

$$h = 17.7\ \text{m}\left(1\ \text{ft}/0.3048\ \text{m}\right) = 58.1\ \text{ft}$$

$$w = 8.85\ \text{m}\left(1\ \text{ft}/0.3048\ \text{m}\right) = 29.0\ \text{ft}$$

$$h = 17.7\ \text{m}\left(1\ \text{ft}/0.3048\ \text{m}\right) = 58.1\ \text{ft}$$

$$\sigma_p = k(w^{0.5}/h^{0.75})$$

$$\sigma_p = 106\ \text{MPa}\ (29^{0.5}/58.1^{0.75})$$

$$\sigma_p = 106\ \text{MPa}\ (5.39/21.0)$$

$$\sigma_p = 27.2\ \text{MPa}$$

Krauland and Soder Equation (1987):

$$\sigma_p = k\left[0.778 \; + \; 0.222\left(w/h\right)\right]$$

$$0.354 = k/\sigma_c$$

$$0.354 = k/182 \text{ MPa}$$

$$k = 64.4 \text{ MPa}$$

$$\sigma_p = k\left[0.778 + 0.222\left(w/h\right)\right]$$

$$\sigma_p = 64.4\left[0.778 + 0.222(0.50)\right]$$

$$\sigma_p = 57 \text{ MPa}$$

Lunder and Pakalnis' Equations (1997):

$$\sigma_p = 0.44\sigma_c(0.68 + 0.52\kappa)$$

Average Pillar Confinement:

$$C_{pav} = 0.46\{\log[\left(w/h\right) + 0.75]\}^{1.4/(w/h)}$$

$$C_{pav} = 0.46\{\log[(0.5) + 0.75]\}^{1.4/(0.5)}$$

$$C_{pav} = 0.46\left\{[0.0969]\right\}^{2.8}$$

$$C_{pav} = 0.000667$$

Mine Pillar Friction Term:

$$\kappa = \tan\left\{\cos^{-1}\left[\left(1 - C_{pav}\right)/\left(1 + C_{pav}\right)\right]\right\}$$

$$\kappa = \tan\left\{\cos^{-1}\left[(1 - 0.000667)/(1 + 0.000667)\right]\right\}$$

$$\kappa = \tan\left\{\cos^{-1}\left[(0.999333)/(1.000667)\right]\right\}$$

$$\kappa = \tan\{2.96\}$$

$$\kappa = 0.052$$

Pillar Strength:

$$\sigma_p = 0.44\sigma_c(0.68 + 0.52\kappa)$$

$$\sigma_p = 0.44(182\text{ MPa})[0.68 + 0.52(0.052)]$$

$$\sigma_p = 56.6\text{ MPa}$$

Table 9.8 compares the results.

TABLE 9.8
Pillar Strength Comparisons

Formula	Pillar Strength, σ_p
Hedley and Grant	27.2 MPa
Krauland and Soder	57.0 MPa
Lunder and Pakalnis	56.6 MPa

Example 9.5.8

If the average pillar stress for the previous example was found to be between 11 and 13 MPa, estimate the FOS for each approach.

Solution:

Hedley and Grant (1972):

$$\sigma_p = 27.2\text{ MPa}$$

$$F = \sigma_p/\sigma_s = 27.2\text{ MPa}/11\text{ MPa} = 2.5$$

$$F = \sigma_p/\sigma_s = 27.2\text{ MPa}/13\text{ MPa} = 2.1$$

Krauland and Soder (1987):

$$\sigma_p = 57\text{ MPa}$$

$$F = \sigma_p/\sigma_s = 57\text{ MPa}/11\text{ MPa} = 5.2$$

$$F = \sigma_p/\sigma_s = 57\text{ MPa}/13\text{ MPa} = 4.4$$

Lunder and Pakalnis (1997):

$$\sigma_p = 56.6\text{ MPa}$$

$$F = \sigma_p/\sigma_s = 56.6\text{ MPa}/11\text{ MPa} = 5.1$$

$$F = \sigma_p/\sigma_s = 56.6\text{ MPa}/13\text{ MPa} = 4.4$$

Example 9.5.9

Account for large discontinuities at a dip of 60° in a limestone pillar with a spacing at 2 m using the approach and inputs by Esterhuizen et al. (2011). Also, find the pillar strength with no discontinuities.

Given:

$$\sigma_c = 182\text{MPa}$$

$$w = 9.5 \text{ m}$$

$$h = 17.7 \text{ m}$$

Solution:

$$P_s = 0.65(\text{UCS})(\text{LDF})(w^{0.3} / h^{0.59})$$

LDF is a large discontinuity factor

w is the width in meters.

h is the height in meters.

$$\text{LDF} = 1 - \text{DDF}(\text{FF})$$

where

DDF is a special discontinuity dip factor.

FF is a frequency factor of large discontinuities per pillar.

Width to height ratio (Table 9.9):

$$w/h = 9.5 \text{ m}/17.7 \text{ m} = 0.54$$

For a precise value of Table 9.5, we can interpolate:

$$y = y_0 + (x - x_0)\left[(y_1 - y_0)/(x_1 - x_0)\right]$$

$$y = 0.86 + (0.54 - 0.6)\left[(0.94 - 0.86)/(0.5 - 0.6)\right]$$

$$y = 0.86 + (-0.06)\left[(0.08)/(-0.1)\right]$$

$$y = 0.86 + (-0.06)(-0.8)$$

$$y = 0.86 + 0.048$$

$$y = 0.91$$

Average Frequency of Large Discontinuities per Pillar = width/discontinuity spacing

TABLE 9.9
DDF Factor by Esterhuizen et al. (2011)

Dip (°)	Pillar w/h Ratio								
	≤0.5	0.6	0.7	0.8	0.9	1.0	1.1	1.2	>1.2
30	0.15	0.15	0.15	0.15	0.16	0.16	0.16	0.16	0.16
40	0.23	0.26	0.27	0.27	0.25	0.24	0.23	0.23	0.22
50	0.61	0.65	0.61	0.53	0.44	0.37	0.33	0.30	0.28
60	0.94	0.86	0.72	0.56	0.43	0.34	0.29	0.26	0.24
70	0.83	0.68	0.52	0.39	0.30	0.24	0.21	0.20	0.18
80	0.53	0.41	0.31	0.25	0.20	0.18	0.17	0.16	0.16
90	0.31	0.25	0.21	0.18	0.17	0.16	0.16	0.15	0.15

TABLE 9.10
FF by Esterhuizen et al. (2011)

Average Frequency of Large Discontinuities per Pillar	Frequency Factor (FF)
0.0	0.00
0.1	0.10
0.2	0.18
0.3	0.26
0.5	0.39
1.0	0.63
2.0	0.86
3.0	0.95
> 3.0	1.00

Average Frequency of Large Discontinuities per Pillar = 9.5/2 = 4.8 (Table 9.10)

$$FF = 1.00$$

$$LDF = 1 - DDF(FF)$$

$$LDF = 1 - (0.91)(1.00)$$

$$LDF = 0.09$$

$$P_s = 0.65(\sigma_c)(LDF)(w^{0.3}/h^{0.59})$$

$$P_s = 0.65(182\text{ MPa})(0.09)(9.5^{0.3}/17.7^{0.59})$$

$$P_s = 0.65(182\text{ MPa})(0.09)(1.96/5.45)$$

$$P_s = 3.8\text{ MPa}$$

$$P_s = 0.65(\sigma_c)(LDF)(w^{0.3}/h^{0.59})$$

$$P_s = 0.65(182\ \text{MPa})(1.0)(9.5^{0.3}/17.7^{0.59})$$

$$P_s = 0.65(182\ \text{MPa})(1.0)(1.96/5.45)$$

$$P_s = 42.5\ \text{MPa}$$

Example 9.5.10

An underground coal mine, in the Western Shelf of the Illinois Coal Basin, mines at an average depth of 135 ft. The mine currently extracts coal from the Herrin #6 seam. For this example, the location and the coal seam are important for selecting the correct design equations. Determine the floor safety factor (FSF) using the Vesic–Gadde equation found in Gadde (2009) and the given parameters below. Data for this example are given in Gadde (2009). This is only one example and Gadde (2009) has additional procedures and criteria for pillar design in weak floor which is typically found in the Illinois coal basin.

Immediate floor thickness, H = 3.2 ft
Width of pillar, B = 20 ft
Length of pillar, L = 67 ft
Width of entry, W_e = 26 ft
Width of cross-cut, W_{cc} = 33 ft
Moisture content of immediate floor, MC_{if} = 13.3%
Moisture content of the main floor, MC_{mf} = 6.7%
Step 1: Determine the bearing capacity factors (N_c) and (N_q) and the shape factor (s_c):
Note that ϕ' is taken as zero due to the difficulty in sampling and testing for the friction angle of weak friable floor, because the data for friction angle are insufficient, and because Vesic's bearing capacity model from which Gadde's approach expands is based on undrained conditions, so the friction angle can be treated as such:

$$N_q = e^{\pi \tan'} \tan^2(45° + \varphi'/2)$$

$$N_q = e^{\pi \tan 0} \tan^2(45° + 0/2)$$

$$N_q = 1$$

$$\text{For } \varphi' = 0, N_c = 5.14$$

Shape factor:

$$s_c = 1 + \left[(B/L)(N_q/N_c)\right]$$

$$s_c = 1 + \left[(20/67)(1/5.14)\right]$$

$$s_c = 1.06$$

Step 2: Determine the undrained shear strength of the immediate floor (c_1). For Western Shelf Mines of the Illinois Coal Basin:
For Western Shelf Mines:

$$c_1 = 1905e^{(-0.1MCif)}/(N_c s_c)$$

$$c_1 = 1,905e^{[-0.1(13.3)]}/(5.14)(1.06)$$

$$c_1 = 93 \text{ psi}$$

Step 3: Determine the undrained shear strength of the main floor (c_2), which follows similar rules as c_1. For this example,
For the Herrin #6 Seam of the Illinois Basin:

$$c_2 = 6,443e^{(-0.34MCmf)}/2$$

$$c_2 = 6,443e^{[-0.34(6.7)]}/2$$

$$c_2 = 330 \text{ psi}$$

Step 4: Determine the ratio of the undrained shear strength of the main floor to the immediate floor (K), (Nc*), and the punching index (β):

$$K = c_2/c_1$$

$$K = 330/93$$

$$K = 3.56$$

$$N_c{*} = s_c N_c$$

$$N_c{*} = (1.06)(5.14)$$

$$\beta = BL/[2(B + L)H]$$

$$\beta = (20)(67)/[2(20 + 67)(3.2)]$$

$$\beta = 2.4$$

Step 5: Determine the modified bearing capacity factor (N_m):

$$N_m = \{KN_c{*}(N_c{*} + \beta - 1)[(K + 1)(N_c{*})^2 + (1 + K\beta)N_c{*} + \beta - 1]\}/$$

$$\{[K(K+1)N_c{*} + K + \beta - 1][(N_c{*} + \beta)N_c{*} + \beta - 1] - [(KN_c{*} + \beta - 1)(N_c{*} + 1)]\}$$

For the sake of readability, the above equation is broken into the numerator and denominator:
Numerator:

$$\left\{KN_c{}^*\left(N_c{}^*+\beta-1\right)\left[\left(K+1\right)\left(N_c{}^*\right)^2+\left(1+K\beta\right)N_c{}^*+\beta-1\right]\right\}$$

$$\left\{(3.56)(5.44)(5.44+2.4-1)\left[(3.56+1)(5.44)^2+\left[1+(3.56)2.4\right]5.44+2.4-1\right]\right\}$$

$$(132.5)\left[(134.9)+(53.3)\right]$$

$$=24{,}939$$

Denominator:

$$\left\{\left[K(K+1)N_c{}^*+K+\beta-1\right]\left[\left(N_c{}^*+\beta\right)N_c{}^*+\beta-1\right]-\left[\left(KN_c{}^*+\beta-1\right)\left(N_c{}^*+1\right)\right]\right\}$$

$$\{\left[3.56(3.56+1)5.44+3.56+2.4-1\right]\left[(5.44+2.4)5.44+2.4-1\right]-[\left[3.56(5.44)+2.4-1\right]$$

$$(5.44+1)]\}$$

$$\left\{[93.3][44]-\left[[20.8](6.44)\right]\right\}$$

$$=3{,}971$$

$$N_m=24{,}939/3{,}971$$

$$N_m=6.28$$

Step 6: Determine the bearing capacity of the immediate floor (q_{floor}):

$$q_{floor}=c_1N_m$$

$$q_{floor}=(93\text{ psi})(6.28)$$

$$q_{floor}=584\text{ psi}$$

Step 7: Determine the average pillar stress using tributary area theory:

$$\sigma=\left[(B+W_e)(L+W_{cc})/(BL)\right]\sigma_v$$

$$\sigma=\left[(20+26)(67+33)/\left[(20)(67)\right]\right](1.125\text{ psi/ft})(135\text{ ft})$$

$$\sigma=\left[(46)(100)/\left[(1{,}340)\right]\right](148.5\text{ psi})$$

$$\sigma=521\text{ psi}$$

Step 8: Determine the FSF:

$$FSF = q_{floor}/\sigma > 1.5$$

$$FSF = 584 \text{ psi}/521 \text{ psi}$$

$$FSF = 1.12$$

Example 9.5.11

Using the Salamon and Munro (1967) formula for square pillars and the Mark and Bieniawski (1986) formula for rectangular pillars (Mark and Chase, 1997), determine the pillar safety factor using inputs from the previous example to show how weak floor can reduce the overall pillar FOS. Assume a pillar height of 5 ft and an in situ coal strength of 900 psi for the Mark and Bieniawski formula.

Salamon and Munro's (1967) formula found in Hartman (1992):

$$\sigma_p = 1{,}320(w^{0.46}/h^{0.66})$$

The formula is for square pillars, and therefore, the equivalent width (w_e) must be used and substituted for w.

The equivalent width is

$$w_{eff} = (A_p)^{0.5}$$

$$w_{eff} = [(67)(20)]^{0.5}$$

$$w_{eff} = 37 \text{ ft}$$

Pillar Strength:

$$\sigma_p = 1{,}320(37)^{0.46}/(5 \text{ ft})^{0.66}$$

$$\sigma_p = 2{,}402 \text{ psi}$$

Mark and Bieniawski's formula from Mark and Chase (1997) using $\sigma_1 = 900$ psi

Note that L is the length:

$$\sigma_p = \sigma_1\left[0.64 + 0.54(w/h) - 0.18(w^2/Lh)\right]$$

$$\sigma_p = 900 \text{ psi}\left\{0.64 + 0.54(20/5) - 0.18\left[20^2/(67)(5)\right]\right\}$$

$$\sigma_p = 900 \text{ psi}\{0.64 + 0.54(4) - 0.18[1.19]\}$$

$$\sigma_p = 900 \text{ psi}(2.58)$$

$$\sigma_p = 2{,}327 \text{ psi}$$

Pillar Stress Using Tributary Area

$$\sigma = \left[(B + W_e)(L + W_{cc})/(BL)\right]\sigma_v$$

$$\sigma = \left[(20 + 26)(67 + 33)/\left[(20)(67)\right]\left(1.125 \text{ psi/ft}\right)(135 \text{ ft})\right]$$

$$\sigma = (46)(100)/\left[(1{,}340)(148.5 \text{ psi})\right]$$

$$\sigma = 521 \text{ psi}$$

Classical FOS for Pillar

$$\text{Salamon and Munro Formula:} \quad F = \sigma_p/\sigma = 2{,}402 \text{ psi}/521 \text{ psi} = 4.6$$

$$\text{Mark – Bieniawski Formula:} \quad F = \sigma_p/\sigma = 2{,}327 \text{ psi}/521 \text{ psi} = 4.5$$

Example 9.5.12

The following example applies an orebody dip correction to pillar strength formulations. Given a pillar strength of 50 MPa, an orebody dip of 10°, and a w/h ratio of 0.8, find the corrected pillar strength, using the approach by Jessu et al. (2022). Sensibly there are limitations on this approach which include a limited number of lab tests, limited to w/h ratios of 0.5–2.0, and for orebody orientation, it is limited to 0°–40°.

Orebody Dip Direction
For an angle > 10°,
RF orebody dip = 1 – 0.0077(orebody dip in degrees)
RF orebody dip = 1 – 0.0077(10)
RF orebody dip = 1 – 0.077 = 0.923

$$\sigma_p = (0.923)50 \text{ MPa} = 46.2 \text{ MPa}$$

Example 9.5.13

Determine the average pillar confinement (C_{pav}) for a hard-rock pillar in a flat-lying orebody for a w/h of 0.5, 1.0, and 2.0. The equation is from Lunder's (1994) thesis.

For extraction ratios between 65% and 95%, C_{pav} is

$$C_{pav} = 0.46[\log(W_p/h + 0.75)]^{1.4/(Wp/h)}$$

where
W is the pillar width.
H is the pillar height.

Solution:

$$\text{Average pillar confinement for w/h} = 0.5:$$

$$C_{pav} = 0.46[\log(W_p/h + 0.75)]^{1.4/(Wp/h)}$$

$$C_{pav} = 0.46[\log(0.5 + 0.75)]^{1.4/(0.5)}$$

$$C_{pav} = 0.46[\log(1.75)]^{2.8}$$

$$C_{pav} = 0.46[0.00145]$$

$$C_{pav} = 0.000667$$

For $w/h = 1.0$:

$$C_{pav} = 0.46\left[\log(1.75)\right]^{1.4}$$

$$C_{pav} = 0.46[0.138]$$

$$C_{pav} = 0.0635$$

For $w/h = 2.0$:

$$C_{pav} = 0.46\left[\log(2.75)\right]^{0.7}$$

$$C_{pav} = 0.46[0.56]$$

$$C_{pav} = 0.258$$

Example 9.5.14

Show how the pillar strength changes for a w/h ratio of 0.5 and 2.0 using the results of the previous example. Use the approach by Lunder and Pakalnis (1997).

Mine Pillar Friction Term for w/h = 0.5:

$$\kappa = \tan\left\{\cos^{-1}\left[(1 - C_{pav})/(1\ C_{pav})\right]\right\}$$

$$\kappa = \tan\left\{\cos^{-1}\left[(1 - 0.000667)/(1 + 0.000667)\right]\right\}$$

$$\kappa = \tan\left\{\cos^{-1}\left[(0.999333)/(1.000667)\right]\right\}$$

$$\kappa = \tan\{2.96\}$$

$$\kappa = 0.052$$

Pillar Strength for w/h = 0.5:

$$\sigma_p = 0.44\sigma_c(0.68 + 0.52\kappa)$$

$$\sigma_p = 0.44\sigma_c\left[0.68 + 0.52(0.052)\right]$$

$$\sigma_p = 0.31\sigma_c$$

Mine Pillar Friction Term for w/h = 2.0:

$$\kappa = \tan\left\{\cos^{-1}\left[\left(1 - C_{pav}\right)/\left(1 + C_{pav}\right)\right]\right\}$$

$$\kappa = \tan\left\{\cos^{-1}\left[(1 - .258)/(1 + .258)\right]\right\}$$

$$\kappa = \tan\left\{\cos^{-1}\left[(0.742)/(1.258)\right]\right\}$$

$$\kappa = \tan\{53.8\}$$

$$\kappa = 1.37$$

Pillar Strength for w/h = 2.0:

$$\sigma_p = 0.44\sigma_c(0.68 + 0.52\kappa)$$

$$\sigma_p = 0.44\sigma_c\left[0.68 + 0.52(1.37)\right]$$

$$\sigma_p = 0.61\sigma_c$$

Example 9.5.15

A 2.5 m thick horizontal coal seam is located at a depth of 80 m, with the rock cover having a unit weight of 25 kNm^{-3}. An initial mining layout is based on 6.0 m room spans and 5.0 m square pillars, with the full orebody thickness of 2.5 m being mined. Use the pillar strength formula by Salamon and Munro (1967) to determine if a minimum pillar FOS of 1.6 is met.

Pillar Strength Formula:

$$P_s = k(w^{0.46} / h^{0.66})$$

where

P_s is the pillar strength in MPa.
k is 7.2 MPa.
w is the pillar width (m).
h is the pillar height (m).
Pillar Stress Formula:

$$\sigma_{ps} = \gamma H\left(A/A_p\right)$$

where

σ_{ps} is the pillar stress in MPa.
γ is the density (unit weight) in kN/m^3.
H is the seam depth (km).

A is the tributary area (m²).
A_p is the pillar area (m²).

Solution:

Pre-mining Stress:

$$P_{zz} = \gamma H = \left(25\,\text{kN/m}^3\right)\left(0.08\,\text{km}\right) = 2.0\,\text{MPa}$$

Pillar Stress:

$$\sigma_{ps} = \gamma H\left(A/A_p\right)$$

$$\sigma_{ps} = 2\left[(11)(11)/(5)(5)\right] \quad \text{Note}: A = (5+6)(5+6)$$

$$\sigma_{ps} = 9.7\ \text{MPa}$$

Pillar Strength:

$$P_s = 7.2\left(5^{0.46}\right)/\left(2.5^{0.66}\right)$$

$$P_s = 7.2(2.1)/(1.8)$$

$$P_s = 8.4\ \text{MPa}$$

Factor of Safety:
F= Pillar strength/Pillar stress

$$F = 8.4/9.7 = 0.87$$

The low FOS provided by this prospective layout indicates that a redesign is necessary to achieve the required FOS.

REFERENCES

Barron, K. and Pen, K. (1992) A revised model for coal pillars. *Proceedings of the Workshop on Coal Pillar Mechanics and Design USBM IC 9315*, Pittsburgh, PA, pp 144–157.

Bauschinger, J. (1876) *Mitteilungen aus dem Mechanisch Technishen Laboratorium der Königlichen Technischen Hochschule in München, vols 1–23*. Theodor Ackermann, München.

Bieniawski, Z.T. (1968) In situ strength and deformation characteristics of coal. *Eng. Geol.* 2, 325–340.

Bieniawski, Z.T. (1968) The effect of specimen size on compressive strength of coal. *Int. J. Rock Mech. Min. Sci.* 5, 327–335.

Bunting, D. (1911) Chamber pillars in deep anthracite mines. *Trans AIME*. 42, 236–245.

Das, M.N. (1986) Influence of width to height ratio on post-failure behaviour of coal. *Int. J. Min. Geol. Eng.* 4, 79–87.

Esterhuizen, G.S., Dolinar, D.R. and Ellenberger, J.L. (2011) Pillar strength in underground stone mines in the United States. *Int. J. Rock Mech. Min. Sci.* 48(1), 42–50.

Esterhuizen, G.S., Tyrna, P.L. and Murphy, M.M. (2019) A case study of the collapse of slender pillars affected by through-going discontinuities at a limestone mine in Pennsylvania. *Rock Mech. Rock Eng.* 52(12), 4941–4952.

Gadde, M.M. (2009) *Weak Floor Stability in the Illinois Basin Underground Coal Mines*. West Virginia University, Virginia.

Gaddy, F.L. (1956) Study of the ultimate strength of coal as related to absolute size of the cubical specimens tested. Virginia Polytechnic Institute Bulletin, Series No. 112, pp. 1–27.

Galvin, J.M., Hebblewhite, B.K. and Salamon, M.D.G. (1999) *University of New South Wales Pillar Strength Determinations for Australian and South African Conditions*. A.A. Balkema, Rotterdamn, pp. 63–71.

Greenwald, H.P., Howarth, H.C. and Hartmann, I. (1939) *Experiments on Strength of Small Pillars of Coal in the Pittsburgh Bed*. U.S. Department of Interior, Bureau of Mines, Washington, DC.

Hartman, H. L. (1992) *SME Mining Engineering Handbook* (Vol. 2). Society for Mining, Metallurgy, and Exploration, Denver, CO.

Heasley, K.A. (2014) *LaModel – A Program for Coal Mine Modelling*. AusIMM, Melbourne.

Hedley, D.G.F. and Grant, F. (1972) Stope and pillar design for the Elliot Lake uranium mines. *Can. Inst. Min. Metall. Bull*. 65, 37–44.

Hoek, E. (1994) Strength of rock and rock masses. *ISRM News J*. 2, 4–16.

Holland, C.T. (1964) The strength of coal in mine pillars. *Proceedings of the Sixth Symposium on Rock Mechanics*. University of Missouri, Rolla, MO, pp. 450–456.

Holland, C.T. (1973) *Mine Pillar Design. SME Mining Engineering Handbook* (Vol. 1, Sec. 13–18). AIME, New York, pp. 97–118.

Holland, C.T. (1973). Pillar design for support of the over burden in coal mines. *Proceedings of the 9th Canadian Symposium on Rock Mechanics*. Mines Branch of the Department of Energy, Mines and Resources, Ottawa, pp. 114–139.

Holland, C.T. and Gaddy, F.L. (1957). Some aspects of permanent support of overburden on coal beds. *Proc. West Virginia Coal Min. Inst*. 5, 43–65.

Hsuing, S.M. and Peng, S.S. (1985) Chain pillar design for U. S. longwall panels. *Min. Sci. Technol*. 2, 279–305.

Hustrulid, W.A. (1976) A review of coal pillar strength formulas. *Rock Mech*. 8, 115–145.

Iannacchione, A.T., Dolinar, D.R. and Mucho, T.P. (2002) High stress mining under shallow overburden in underground U.S. stone mines. *1st International Seminar on Deep and High Stress Mining, Section 32*. Australian Center for Geomechanics, Nedlands, Australia.

Jessu, K.V. and Spearing, A.J.S. (2018) Effect of dip on pillar strength. *J. S. Afr. Inst. Min. Metall*. 118(7), 765–776.

Jessu, K.V., Spearing, A.J.S. and Sharifzadeh, M. (2018) A parametric study of blast damage on hard rock pillar strength. *Energies*. 11(7), 190–191.

Jessu, K.V., Spearing, A.J.S. and Sharifzadeh, M. (2022) An improved pillar design methodology. *J. S. Afr. Inst. Min. Metall*. 122(10), 587–596.

Krauland, N. and Soder, P.E. (1987) Determining pillar strength from pillar failure observations. *Eng. Min. J*. 8, 34–40.

Kripakov, N.P. (1981) Analysis of pillar stability on steeply pitching seam using the finite element method. USBM Report RI 8579.

Lunder, P.J. (1994) Hard rock pillar strength estimation an applied empirical approach, Master's Thesis, University of British Columbia, Department of Mining and Mineral Process Engineering.

Lunder, P.J. and Pakalnis, R. (1997) Determining the strength of hard rock mine pillars. *CIM Bulletin*. 90, 51–55.

Madden, B.J. (1991) A re-assessment of coal pillar design. *J. S. Afr. Inst. Min. Metall*. 91, 27–37.

Mark, C. and Agioutantis, Z. (2018) Analysis of coal pillar stability (ACPS): A new generation of pillar design software. *Int. J. Min. Sci. Technol*. 29, 87–91.

Mark, C. and Bieniawski, Z.T. (1986) Field measurements of chain pillar response to longwall abutment loads. *Proceedings of the Fifth Conference on Ground Control in Mining*, Morgantown, WV, pp. 114–122.

Mark, C. and Chase, F. (1997). Analysis of retreat mining pillar stability. *Proceedings of new technology for ground control in retreat mining, NIOSH IC 9446*. National Institute of Occupational Safety and Health, Pittsburgh, PA, pp. 17–34.

Michaud, D. (2017) *Longwall Pillar Design Method*. 911 Metallurgist, Canada.

National Institute of Occupational Safety and Health (NIOSH) (2024) S-Pillar Windows Software for Stone Mine Pillar Design. Retrieved from: www.cdc.gov/niosh/mining/tools/s-pillar.html

Obert, L. and Duvall, W.I. (1967) *Rock Mechanics and the Design of Structures in Rock*. Wiley, New York, pp 542–545.

Ozbay, M.U. and Roberts, M.K.C (1988) Yield pillars in stope support. *Proceedings of the SANGORM Symposium in Africa*. South African National Group on Rock Mechanics, Johannesburg, Swaziland, pp. 317–326.

Pan, R., Li, Y., Wang, H., Xu, Y. and Yang, H. (2020) Assessment of ground instability risk of lower seam longwall panels during crossing overlying remnant pillars. *R. Soc. Open Sci*. 7, 201249.

Peng, S. (2020) *Longwall Mining* (3rd Edition). CRC Press. https://doi.org/10.1201/9780429260049

Prassetyo, S.H., Irnawan, M.A., Simangunsong, G.M., Wattimena, R.K., Arif, I. and Rai, M.A. (2019) New coal pillar strength formulae considering the effect of interface friction. *Int. J. Rock Mech. Min. Sci.* 123, 104102.

Rockaway, J.D, Stephenson, R.W. (1979) *Investigation of the effects of weak floor conditions on the stability of coal pillars.* Report No. BUMINES-ofr-12-81.

Salamon, M.D.G. and Munro, A.H. (1967) A study of the strength of coal pillars. *J. S. Afr. Inst. Min. Metall.* 68, 55–67.

Salamon, M.D.G. and Wagner, H. (1985) Practical experiences in the design of coal pillars. *Proceedings of the 21st International Conference on Mining Software Repositories.* A.A. Balkema, Rotterdam.

Sheorey, P.R., Das, M.N., Barat, D., Prasad, R.K. and Singh, B. (1987) Coal pillar strength estimation from failed and stable cases. *Int. J. Rock. Mech. Min. Sci. Geomech. Abstr.* 24, 347–355.

Skinner, W.J. (1959). *Experiments on the Compressive Strength of Anhydrite.* The Engineer, London.

Spearing, A.J.S. (1987) A practical rock mechanics manual. Anglo American Corp. of SA. Internal Report.

Spearing, A.J.S., Ma, L. and Ma, C. (2023) *Mine Design, Planning and Sustainable Exploitation in the Digital Age.* CRC Press, London, UK.

Steart, F.A. (1954) Strength and stability of pillars in coal mines. *J. Chem. Metall. Min. Soc. S. Afr.* 54, 309–325.

van der Merwe, J.N. (2003) New pillar strength formula for South African coal. *J. S. Afr. Inst. Min. Metall.* 103, 281–292.

Wagner, D. (1980) Pillar design in coal mines. *J. SA. Inst. Min. Metall.* 1980, 37–45.

Wagner, H. (1974). Determination of complete load deformation characteristics of coal pillars. *Proceedings of the 3rd ISRM Congress*, Denver, CO, pp. 1076–1081.

Wagner, H. and Madden, B.J. (1984) Fifteen years' experience with the design of coal pillars in shallow South African collieries: an evaluation of the performance of the design procedure and recent improvements. Design and Performance of Underground Excavations. *ISRM International Symposium,* British Geotechnical Society, Cambridge, UK, pp. 391–399.

Wang, F.D., Skelly, W.A., and Wolgamott, J. (1977). In-situ coal pillar strength study. Proceedings 18th U.S. Symposium on Rock Mechanics, Keystone, CO, pp. 2B5-1:9.

Wilson, A.H. (1981) Stress and stability in coal ribsides and pillars. *Proceedings, 1st Conference on Ground Control in Mining*, Morgantown, WV, pp. 1–12.

10 Mine Seismicity and Support

10.1 INTRODUCTION TO SEISMICITY

A seismic event is a sudden energy release in a rock mass which manifests as deformation, crack propagation, rock failure, and/or heat generation. Mining-induced seismic events are usually in response to a stress change in a rock mass close to an excavation caused by mining activities. The most notable form of a mining-induced seismic event is a rockburst which can cause significant damage to excavations, equipment, and/or injury to personnel. Damage can vary from minor rock slabbing to complete collapse of support structures such as pillars.

Common and important waveforms are as follows:

- Compressional and shear waves

  These are the two main waveforms associated with seismic events. P waves or compressional waves can travel through gases, liquids, and solids. S waves or shear waves can only travel through solids. Figure 10.1 represents the typical waveforms of P and S waves.
- Rayleigh and Love waves

  These are both surface waves as a result of a seismic event. Rayleigh waves move in an elliptical motion on the surface. Love waves move the surface in a side-to-side motion perpendicular to the direction of propagation, as shown in Figure 10.2.

Figure 10.3 shows a sample seismogram with the common waveforms as they are detected by a seismometer, i.e., the waveform with the greater velocity arrives first.

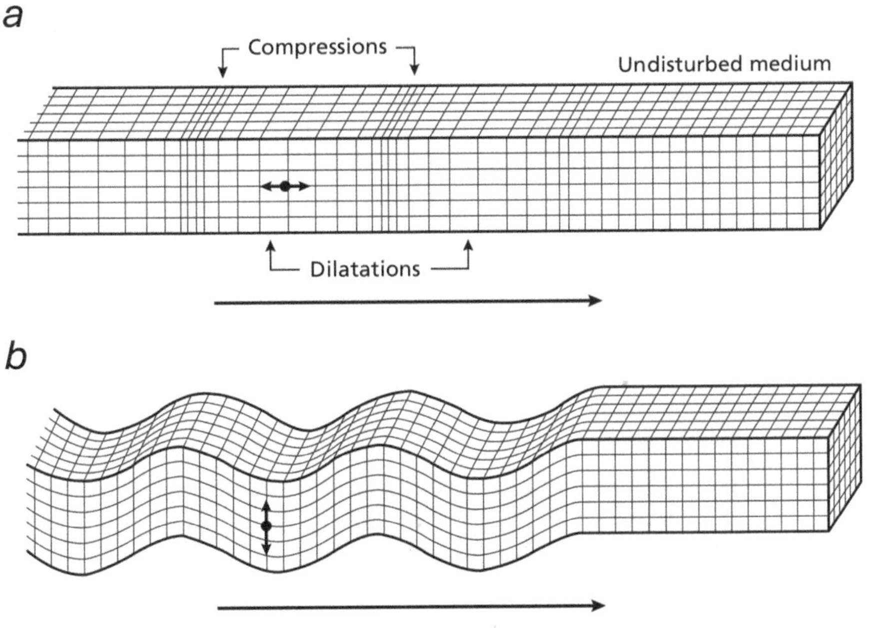

FIGURE 10.1 The two main body waves: (a) P wave and (b) S wave. Image found in Kearey et al. (2002), cited to Bolt (1982), which credits Bolt (1976).

DOI: 10.1201/9781032704845-10

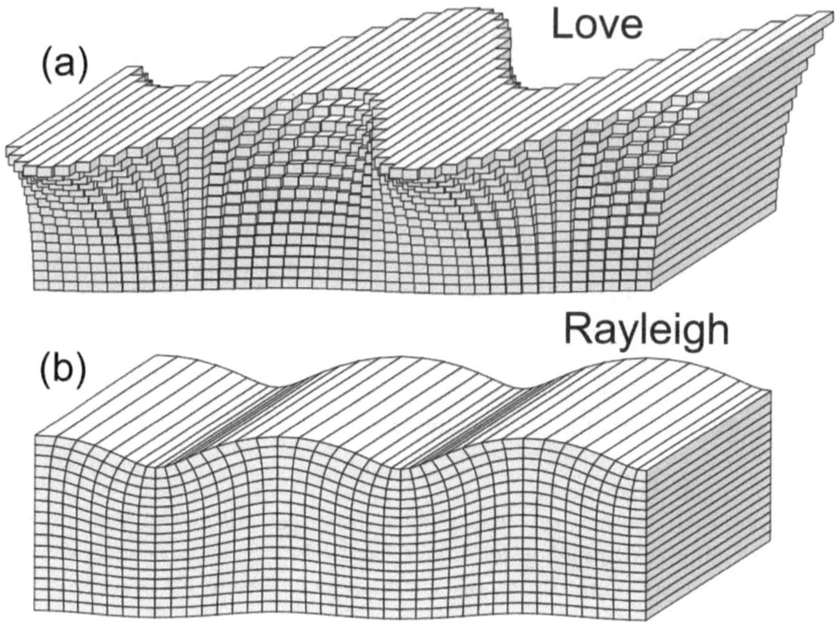

FIGURE 10.2 Surface ground motion from Rayleigh waves (a) and Love waves (b). Image found in Kearey et al. (2002), cited to Bolt (1982), which credits Bolt (1976).

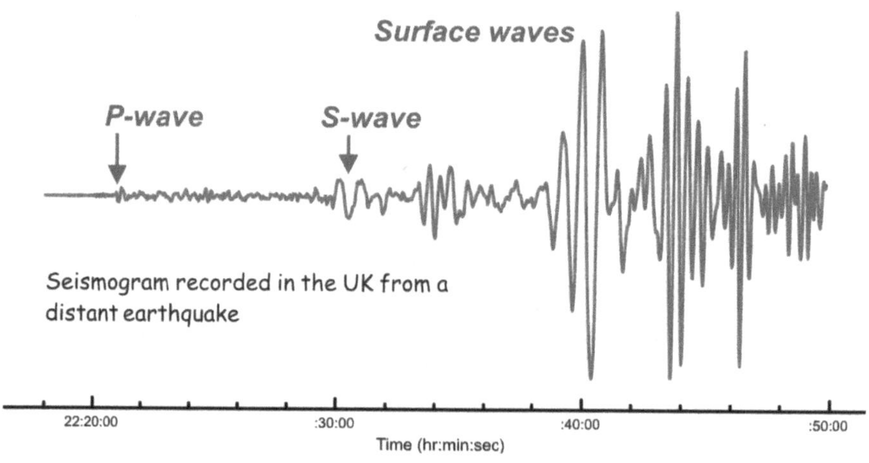

FIGURE 10.3 A sample seismogram (British Geological Survey, 2008).

Denkhaus (1964) postulated an energy balance associated with energy release rate (ERR) and seismicity as shown in Figure 10.4.

According to Figure 10.4, energy is either stored elastically or released. Of the released energy, it is dissipated plastically in solid rock, in fractured rock, absorbed by support, or it creates new fractures without violence. The remaining portion is violently released.

Much of this seismicity in hard-rock underground mines is also released immediately after blasting operations as the rock is removed almost instantaneously causing a shock loading to the surrounding rock which tends to cause seismicity.

Figure 10.5 shows a seismic event, with some terminology and how the energy is radiated.

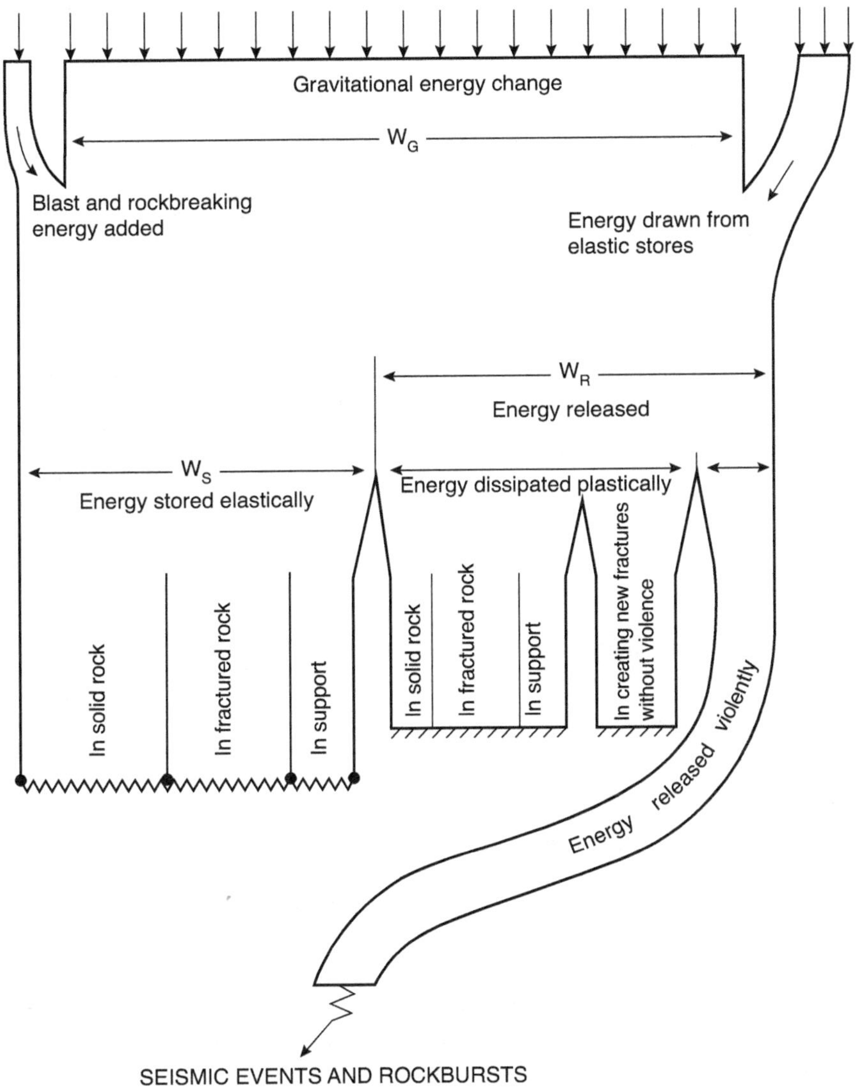

FIGURE 10.4 The energy balance in a mining cycle (Denkhaus, 1964).

10.2 SEISMIC MAGNITUDE

The Richter scale (and the Gutenberg–Richter scale) is a measure of the strength (energy or magnitude) of earthquakes, developed by Charles Richter (1935) in collaboration with Beno Gutenberg. The local magnitude, denoted by M_L, is determined by taking the log of the amplitude of the seismic waves determined from seismograms with appropriate corrections. Each whole number increase in the Richter scale represents a ten-fold increase in the seismic amplitude and about 31.6 times more energy.

The relationship between the amplitude of an earthquake and the Richter scale is defined as follows:

$$M_L = \log(A/A_0) \tag{10.1}$$

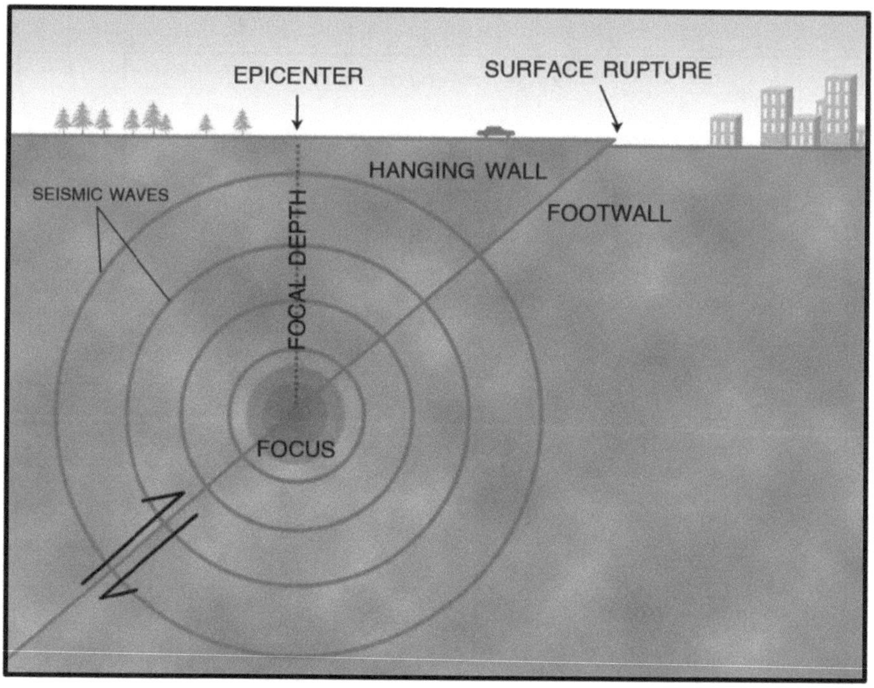

FIGURE 10.5 A section showing a seismic event (Yeats, 2020).

where

M_L is the magnitude on the Richter scale.

A is the maximum amplitude of the seismic event in mm as measured by a Wood-Anderson seismograph.

A_0 is the amplitude of a standard earthquake, defined as the amplitude from a earthquake 100 km from the earthquake epicenter.

The moment magnitude scale (M_w) is another approach to describe a seismic event and is considered more reliable because the moment magnitude scale avoids the issue around magnitude saturation as it is based on the total energy. Magnitude saturation is the tendency for earthquake magnitude calculations to cluster around certain numerical thresholds on the logarithmic scale which can create errors.

The basic formula for calculating the moment magnitude is as follows (Hanks & Kanamori., 1979):

$$M_w = 2/3\ \log(M_o) - 10.7 \tag{10.2}$$

where

M_o is the seismic moment, which is measured in dyne-cm, where 1 dyne-cm = 1×10^{-7} J.

According to the United States Geological Survey (n.d.), the seismic moment, M_o, is a measure of the total energy released by an earthquake. It is calculated by multiplying the shear modulus of the rocks along a fault, G, by the area of the fault, A, and the slip over the fault, D (Aki, 1966):

$$M_o = GAD \tag{10.3}$$

The graphic representation of the Richter scale is given in Figure 10.6.

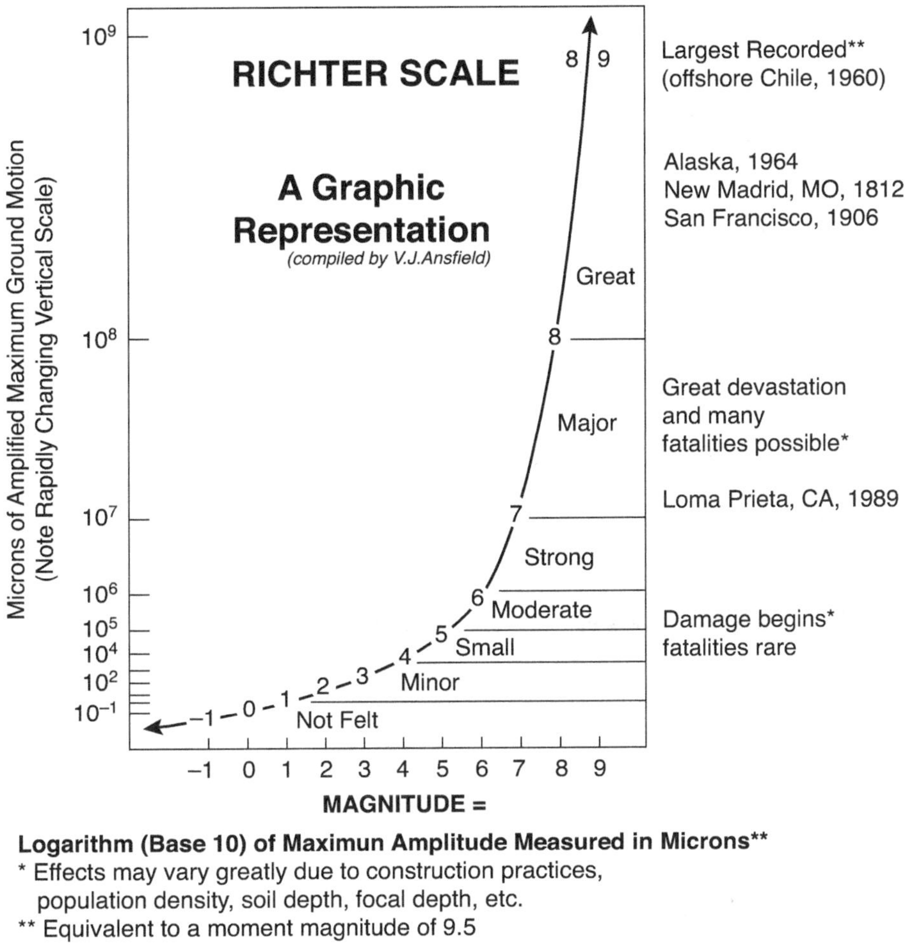

FIGURE 10.6 Graphic representation of the Richter scale (Ansfield, 1992).

10.3 ROCKBURST SUPPORT DESIGN PRINCIPLES MAINLY FOR HARD-ROCK MINES

10.3.1 Support Requirements for Rockbursts

Support for rockburst (dynamic/rapid deformation) conditions is different from quasi-static support because energy absorption is now the critical performance parameter. The behavior of underground excavations under seismic loading is different from that of surface structures because underground excavations are encased by rock. Hence, the design and support approaches for underground excavations with seismic potential are unique.

Seismicity affects underground structures in two main ways:

- Ground shaking which can cause excavation perimeter rocks to be ejected violently.
- Ground and support failure caused by the seismic event as the energy waves meet an excavation.

Figure 10.7 shows an example of rockburst damage in a deep and highly stressed tunnel developed in brittle rock.

FIGURE 10.7 Failure due to a rockburst in a deep gold mine tunnel in South Africa. (Photograph taken by W.D. Ortlepp and reproduced with the kind permission of the South African Institute of Mining & Metallurgy.)

Factors that influence the magnitude of rockburst effects are as follows:

- Overall mine design.
- Excavation shape in relation to the seismic source.
- Excavation dimensions.
- Energy associated with the seismic event.
- Rock mass properties.
- Stored energy accumulated in the system (rock mass and reinforcement).
- Regional and local geology.
- Installed support.

Wagner (1984) stated that "Rockbursts are the most serious and least understood problem facing deep mining operations all around the world". This comment is still relevant in spite of considerable advancements in seismicity, monitoring, dynamic computer modeling, preventative mine designs, and support that can better withstand rapid movements.

Wagner (1984) recommends the following requirements for rockburst support in excavations:

- A support resistance of at least 60 kN/m^2.
- The ability to deform rapidly and maintain a reactive load after the rapid deformation. The support should be able to sustain a closure rate of at least 2 m/s but preferably 3 m/s.
- The support system should be able to stably withstand the closure of at least 60 mm in tunnels.
- A means to keep the broken rock between the rockbolts in place (i.e., area or liner effective coverage is important, such as mesh or shotcrete).

10.3.2 Energy Release Rate (ERR) and Excess Shear Stress (ESS)

10.3.2.1 Energy Release Rate (ERR)

In mining, ERR is the rate at which energy is changed as a rock undergoes fracture. ERR is a calculated rather than a physical parameter. In narrow deep tabular mining, it takes into account the volumetric closure taking place during stoping, and the stope abutment stress. According to Brummer et al. (1988), >97% of the ERR is released in a non-violent way in the form of fracturing, crushing, and sliding of blocks in the fracture zone in front of the immediate face. Figure 10.8 illustrates the effects of ERR on mining conditions for low- and high-stress situations.

The ERR is defined after Cook et al. (1966) as follows:

$$\text{ERR} = \Delta U_{SA}/\Delta a \text{ in MJ/m}^2 \tag{10.4}$$

where

ΔU_{SA} is the available strain energy which is therefore related to the (abutment) rock stresses before the rock is removed (excavated)

Δa is the area being removed (excavated).

The relationship between increased ERR and increased seismicity is only partially reliable, but ERR is still a useful qualitative tool in the comparison and optimization of mining layouts as it relates directly to the magnitude of the stresses (Cook and Salamon, 1966).

In narrow tabular, deep, and hard-rock mines, stabilizing pillars are often left to control the ERR by reducing the stope face stresses and the stope closure. Where stabilizing pillars are used, the Average Pillar Stress (APS) is often considered as a guideline, and it should be limited as follows (COMRO, 1988):

$$\text{APS} \leq 2.5 * \text{UCS} \tag{10.5}$$

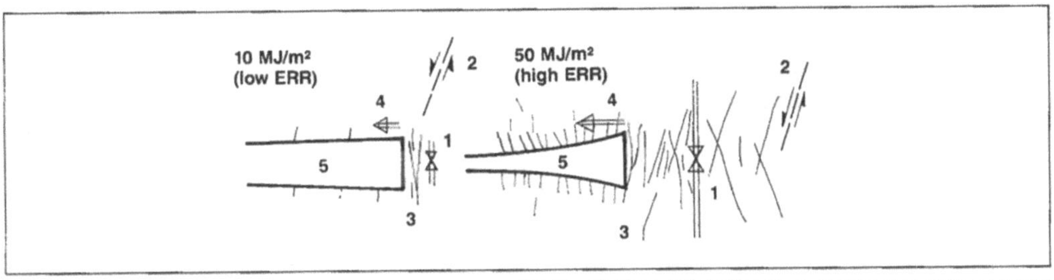

	ERR correlates with	Practical significance
1	Stresses in front of face	Drilling difficulties, crush bursting
2	Shear stresses on planes of weakness	Face-parallel shear seismic events
3	Depth & height of face fracturing	"Hard patches" on face. Seismic "cushion" of fractured rock
4	Dilation of fracture zone & thrust on h/w beam	Hangingwall stability
5	Stope convergence behind face	Mining conditions, generated support resistance

FIGURE 10.8 Energy release rate effects on mining conditions (Brummer et al., 1988).

where
UCS is the unconfined compressive strength of a core sample tested in the laboratory.

Obviously other considerations are important such as:

- The strength of the immediate hangingwall (roof) and footwall (floor).
- The effective height of any pillar includes the pillar height plus the immediate fractured roof (hangingwall) and floor height (footwall) and should be used in place of the actual mining height. This can be much larger than the actual mined height.

Figure 10.8 (Brummer et al., 1988) explains how ERRs affect hard-rock, deep, and tabular mining.

10.3.2.2 Excess Shear Stress (ESS) and Seismic Potential

The ESS is the theoretical shear stress in excess of that required to initiate slip on a fault or fracture (COMRO, 1988). According to Brummer et al. (1988), studies have confirmed that seismic events and accompanying rockbursts are greatly increased when mining in the vicinity of faults, dykes, and joint sets. ESS is based on the Coulomb–Navier failure criterion. It is mainly applied to faulting in the roc kmass which are arbitrarily oriented with strength anisotropies and subject to non-Andersonian stress systems (i.e., with none of the principal stresses acting in a vertical direction). The concept of ESS is shown diagrammatically in Figure 10.9 (Brummer et al., 1988).

ESS is qualitative aid when designing bracket pillars against potentially seismically active geological features such as faults. Various layouts using typically rectangular pillars (with the long axis perpendicular to the fault) can be modeled and the resulting ESS on the fault compared, albeit qualitatively.

Where a stiffer rock intrusion exists like a dyke, another basic strength-to-stress criterion can be useful (COMRO, 1988):

$$\left(\sigma_1/\sigma_c\right) \leq 1 \tag{10.6}$$

where
σ_1 is the maximum principal stress acting on the feature.
σ_c is the UCS of the stiff intrusion.

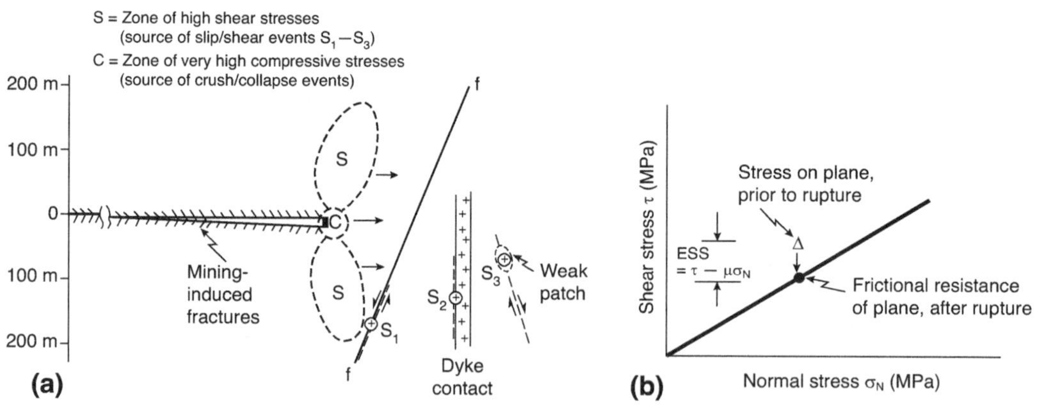

FIGURE 10.9 The concept of ESS specifically mining-induced shear stress creating seismic events (Brummer et al., 1988).

There are three main modes of rock mass instability leading to rockbursts:

- Fault-slip events that resemble natural earthquakes and usually occur on a mine panel or mine scale.
- Stress-induced fracturing or crushing of pillars, or fracturing at or near the mining face, can lead to local instabilities, sometimes called strain bursts, which occur on a stope or excavation scale.
- Stiffer rock such as dykes can store more energy than the surrounding rock mass, and this can lead to rockbursts due to the higher stresses in the stiffer rock mass.

10.4 TUNNEL SUPPORT DESIGN FOR ROCKBURSTS (SEISMICITY)

Unsupported rock and tunnel walls can collapse due to keyblock, shear, buckling, compression, or tensile failure. This can result in fatalities or injuries, damaging equipment, and delaying production because the excavation needs to be re-supported and re-habilitated. This is even more important in seismically active areas or areas of severe rock movement (dilation/convergence) due to the geotechnical and mining conditions.

Conventional rockbolt (used in quasi-static conditions) would yield a small amount (typically <10%) from the elastic and plastic deformation associated with the ductility of the steel which is generally used to make rockbolts and then suddenly fail without warning. This is because the support is deformed more than it can sustain and thus cannot absorb enough energy or quasi-static rock movement (dilation). Basically, the support design for large deformation or violent deformation is not the same as for quasi-static relatively small deformation conditions.

As an example, typical cable tendons used for rock reinforcements usually have a 10%–15% elongation capacity of the working element before failure, which depending on anchorage capacity, results in a limited energy absorption capability. If the cable bolt is fully grouted, then its total and stable elongation capacity is significantly reduced.

The compression wave (P wave) associated with a seismic event travels at about 5,200 m/s, and the shear wave (S wave) travels at about 4,800 m/s in stressed Witwatersrand quartzites in South Africa (Nkosi, 2016). However, due mainly to the fractured nature of the rock around highly stressed excavations, peak rock ejection velocities seldom exceed a few m/s, but this still creates very hazardous ground control situations. The key to rockburst control is therefore controlling the kinetic energy, and this is achieved by using support that can absorb energy by maintaining a support resistance while yielding.

The cost of any installed support (reinforcement) system must always be considered as it should be cost effective, even though safety is still the overriding concern. The total installed cost is made up of direct and indirect costs. The indirect costs are difficult to quantify but can be significant. An optimum support system is one that minimizes the total costs (direct and indirect).

The direct costs include the following:

- Labor costs.
- Material costs.
- Material transportation.
- Drilling costs.

The indirect costs include the following:

- Material wastage.
- Lost production due to an inappropriate support installation cycle.
- Loss of future production due to key access interference.
- Health, safety, and rehabilitation costs (if the system fails), and this must be a significant consideration in rockburst support design.

10.5 STOPE SUPPORT DESIGN FOR ROCKBURSTS

Many seismic events are induced immediately after a production blast when all personnel are safely out of the area. This is because a volume of rock is effectively removed almost instantaneously and the surrounding rock mass is effectively overstressed or shock loaded.

Reducing stresses in the rock mass around active mining can effectively reduce the energy available for seismicity. This can be achieved by:

- Optimizing the extraction sequence.
- Leaving strategically placed and sized pillars against potentially hazardous geological features.
- Designing yielding pillars that are sized such that they are in a post peak failure zone as they are developed (mined).
- Preconditioning (destressing) which involves typically drilling and blasting the structure which could be a tight fault or stiff dyke.
- By controlling the progression of excavations (rate of mining), the evolution of mining induced stresses can be partially controlled, in order to reduce seismicity and mitigate damages.

10.6 SUPPORT SYSTEMS FOR SEISMIC CONDITIONS

There are many commercially available rock anchors and standing supports (such as hydraulic props) that can yield rapidly and absorb the required energy during a rockburst. The liner support holds the broken rock between the installed support after a rockburst and is very important. The following can be considered depending on the severity of the anticipated seismicity:

- Mesh or screen (steel or synthetic).
- Fiber-reinforced or mesh-reinforced shotcrete. Steel arches (possibly yielding) or girders can also be used with shotcrete.
- Fiber-reinforced shotcrete with mesh on the shotcrete surface (not in the shotcrete) as the mesh extends the stable deformation after the shotcrete has failed.
- Light arches backfilled with lightweight or foamed concrete. This can be useful when excavating through a stiff dyke, for example, that may become seismically active.

Details of rock anchors suitable for seismic conditions are given in Chapter 17.

10.7 SEISMIC MONITORING ON MINES

10.7.1 The Benefits of Seismic Monitoring

Seismic monitoring can be useful on mines for the following reasons:

- Help focus rescue efforts quickly and in the right area on a mine after a rockburst has occurred.
- Identify areas of higher seismic risk and take extra precautions and install additional appropriate support.
- Identify potentially hazardous geological features so that appropriate pillars can be left for example.
- Identify any unexpected changes in the seismicity in an area.
- Create a database, so different mining layouts, different pillar designs, and different mining areas can be compared, and more effectively manage risk.

10.7.2 Seismic Event Location and Monitoring

10.7.2.1 Seismic Event Location

The distance from the epicenter can be readily determined using the time-travel seismograph, but the direction cannot be determined from a single seismic record. A minimum of three different seismographs are needed ideally around the epicenter.

Simplistically, a circle from each seismograph is drawn based on the arrival times, and hence, the distance the waves traveled from the source; the epicenter is located at the single intersection of all three circles.

According to Mendecki et al. (1999), "A seismic event is considered to be described quantitatively when apart from its timing, t, and location, X = (x, y, z), at least two independent parameters pertaining to the seismic source namely, seismic moment, M, which measures coseismic inelastic deformation at the source and radiated seismic energy, E, or seismic moment and stress drop, $\Delta\sigma$, are determined reliably."

For a mine wide or area seismic system, an array of seismometers are installed both inside and outside the mining area. Determining the x and y location of an event is relatively easy and accurate, but getting a reliable z (vertical) location can be more challenging due to underground access constraints especially if the orebody is narrow and tabular.

10.7.2.2 Seismic Monitoring

A basic seismic monitoring system consists of:

- An array of transducers.
- Data acquisition hardware.
- Data processing and analysis software.
- Data storage.

On the most seismically active mines, measuring all seismic events from −2 or smaller to a maximum of say 6 is desirable. Negative events to say −4 result in small fractures, but if they cluster, it can help indicate areas of high stress and potential location for large rockbursts. It should be noted that negative magnitudes are possible because the magnitude scale is a logarithmic measurement.

According to Mendecki et al. (1999), the selection of instruments to obtain a useful mine seismograph depends on the desired result:

- Geophones are ideal for large and sparse networks, where sensors are separated by distances of the order of 1 km on average.
- Piezoelectric accelerometers are ideal for dense networks where sensors are separated by the order of 100 m on average.
- It should be noted, however, that neither miniature geophones nor piezoelectric accelerometers cover the low frequencies generated by the larger magnitude and lower stress drop seismic events. For these cases, they report that it is useful to add at least one short period earthquake monitoring instrument to a mine network in the form of a 1 Hz geophone or strong motion force balance accelerometer which provides for the recovery of moment for the largest events.

For the same reasons as when measuring the absolute stress or stress change underground, seismic sensors should be installed in boreholes, outside the zone of mining-induced and other fractures. The grouting needs to be of a high quality and totally fill the hole to ensure good contact with the rock. The grout should be as dense and strong as possible to try to have the same wave flow properties as the surrounding rock mass. The orientation of the sensors in the borehole should also be known and recorded to obtain reliably oriented seismic waves in the x and y axes, which are at right angles to the z (vertical or depth axis).

Once the reasons for needing seismic monitoring have been determined and the goals clearly documented, the system needs to be designed and installed by experts as it is often not a trivial task.

10.8 WORKED EXAMPLES

Example 10.8.1

A mass is to be supported by a set of springs on the left and right sides. On the left side, there are two springs with a stiffness of 5 N/m each, connected in parallel. How many springs are needed on the right side to meet, or exceed, the effective stiffness on the left side? The springs are to be connected in series, and their stiffness is equal. The equations for effective stiffness of series and parallel spring systems are given below.

For parallel systems,

$$k = k_1 + k_2 + \ldots + k_n$$

For series systems,

$$1/k = 1/k_1 + 1/k_2 + \ldots + 1/k_n$$

Solution:

Left side

$$k = k_1 + k_2$$

$$k = 5\ \mathrm{N/m} + 5\ \mathrm{N/m}$$

$$k = 10\ \mathrm{N/m}$$

Right side

$$1/k = 1/k_1 + 1/k_2 + \ldots + 1/k_n$$

where

1/k must meet or exceed 10 N/m.

For series systems,

$$1/k = 1/k_1 + 1/k_2 + \ldots + 1/k_n$$

Therefore,

$$10\ \mathrm{N/m} = n\left(1/k_1\right)$$

$$10\ \mathrm{N/m} = n\left[1/\left(5\ \mathrm{N/m}\right)\right]$$

$$10/0.2 = n$$

$$n = 50\ \text{springs}$$

Example 10.8.2

The example illustrates the explosive force of a pressurized vessel. It is found as a worked example in the handbook on explosions in air (after Department of the Army, 1974).

A pressure vessel is used as part of the Apollo Service Module. The pressure vessel stores 7×10^4 in^3 of helium at a pressure of 3,500 psia (pounds per square in absolute). NASA was concerned about the possible blast effects of the rupture of this vessel during vibration testing in lab tests. Find the trinitrotoluene (TNT) equivalent to the vessel bursting.

Equation for computing energy, E, for this specific example is as follows:

$E = [p_oV/\gamma - 1][(p/p_o) - (p/p_o)^{1/\gamma}]$

p_o = ambient pressure = 14.7 psia which is the ambient temperature at Houston, Texas

p = initial absolute pressure = 3,500 psi

V = total volume = 7×10^4 in^3

γ = ratio of specific heats for the expanding gas = 5/3 (value assigned in text, stated because Helium is monatomic gas).

Solution:

$$E = [p_oV/\gamma - 1][(p/p_o) - (p/p_o)^{1/\gamma}]$$

$$E = \left[(14.7)(70{,}000)/(5/3) - 1\right]\left[(3{,}500/14.7) - (3{,}500/14.7)^{3/5}\right]$$

$$E = (1{,}543{,}500)(211)$$

$$E = 325.7 \times 10^6 \text{in} - \text{lbf}$$

Note a slight difference from solution given which was E = 3.27×10^8 in-lbf.

Given the detonation energy of TNT which is 1000 cal/g or 18.3×10^6 in-lbf/lbm, the following relationship is used to estimate a "TNT Equivalent":

$$W_{TNT} = 325.7 \times 10^6 / 18.13 \times 10^6$$

$$W_{TNT} = 17.9 \text{ lbm of TNT}$$

*Note: Slight difference from solution given in the text of 18.02 lbm because of the above quantity of 325.7×10^6 in-lbf.

Example 10.8.3

The following is a portion of a longer worked example in the *Handbook on Explosions in Air* by the Department of the Army (1974). Given the weight specific energy (E/W) of a 1 lbm sphere of pentolite, determine the total explosive energy, E.

Solution:

Weight Specific Energy, E/W, of Pentolite 50/50 is 20.50×10^6 in-lbf/lbm:

$$E/W = 20.50 \times 10^6 \text{in} - \text{lbf/lbm}$$

$$E = (1 \text{ lbm}) 20.50 \times 10^6 \text{in} - \text{lbf/lbm}$$

$$E = 20.50 \times 10^6 \text{in} - \text{lbf}$$

Example 10.8.4

For a linear elastic block of rock given in Figure 10.10 with a volume of 60 m^3 and a modulus of elasticity of 35 GPa, determine the strain energy, U, if the block is subject to a stress of 0.034 GPa. What is the strain energy if the modulus is reduced by half? The approach below is after Isaacson (1962).

Strain energy, U, for the block given as follows:

$$U = P^2V/2E$$

where

P is the compressive acting normal to the cross section.
V is the volume of the strained block = la^2.
E is the modulus of elasticity.

$$V = 60\ m^3$$

$$E = 35\ GPa$$

$$P = 0.034\ GPa.$$

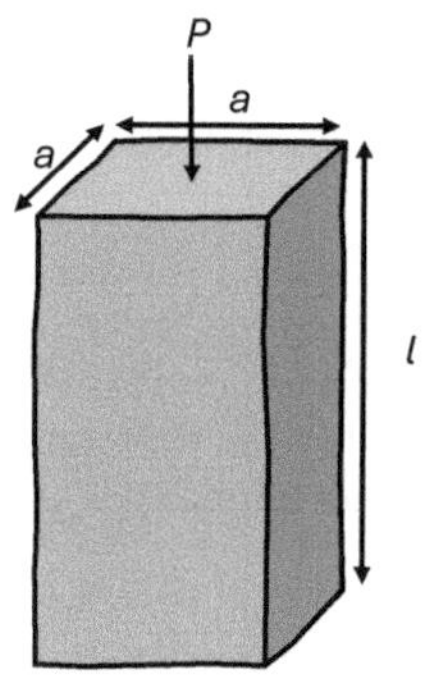

FIGURE 10.10 The block (Issacsson, 1962).

Solution:

$$U = P^2V/2E$$

$$U = \left[\left(3.4\times10^7 Pa\right)^2\left(60\ m^3\right)\right]/\left[2\left(3.5\times10^{10} Pa\right)\right]$$

$$U = 990{,}857\ N - m$$

For E reduced by half,

$$U = \left[\left(3.4\times10^7 Pa\right)^2\left(60\ m^3\right)\right]/\left[2\left(1.75\times10^{10} Pa\right)\right]$$

$$U = \left(6.94\times10^{16} N^2/m\right)/\left(3.5\times10^{10} N/m^2\right)$$

$$U = 1{,}981{,}714\ N - m$$

Example 10.8.5

How does the stress change if the modulus is halved in the previous example? Assume that strain energy is constant at 57,000 N-m.

Solution:

$$U = P^2V/2E$$

$$P = \left(U2E/V\right)^{0.5}$$

$$P = \left[(57{,}000\ \text{N} - \text{m})(2)\left(3.5\times10^{10}\,\text{Pa}\right)/\left(60\ \text{m}^3\right)\right]^{0.5}$$

$$P = 8.15\times10^{6}\,\text{Pa}$$

For E reduced by half,

$$U = P^2V/2E$$

$$P = \left(U2E/V\right)^{0.5}$$

$$P = \left[(57{,}000\ \text{N} - \text{m})(2)\left(1.75\times10^{10}\,\text{Pa}\right)/\left(60\ \text{m}^3\right)\right]^{0.5}$$

$$P = 5.77\times10^{6}\,\text{Pa}$$

Example 10.8.6

The following example is from Isaacson (1962). For a case of a cubic foot of rock, at a depth of 2,500 ft, a specific gravity of 3.0, Young's modulus of 6×10^6 psi, and Poisson's ratio of 0.25, find the total strain energy per unit volume, W. Assume that a horizontal stress of one-third the vertical stress and the resolved stresses are the principal stresses. The relationship for strain energy per unit volume uses the expression by Jaeger and Cook (1969) which is simplified to common symbols.

Solution:

$$W = 1/2E\left[\sigma_1^{\,2} + \sigma_2^{\,2} + \sigma_3^{\,2} - 2\nu(\sigma_2\sigma_3 + \sigma_3\sigma_1 + \sigma_1\sigma_2)\right]$$

Find the unit weight of rock:

$$\gamma_{\text{rock}} = (\text{S.G.})\gamma_{\text{water}}$$

$$\gamma_{\text{rock}} = (3.0)\left(62.4\ \text{lbs/ft}^3\right)$$

$$\gamma_{\text{rock}} = 187.2\ \text{lbs/ft}^3$$

Find the vertical stress:

$$\sigma_v = \sigma_1 = (\text{depth})\gamma_{\text{rock}}$$

$$\sigma_v = \sigma_1 = (2{,}500 \text{ ft})\left(187.2 \text{ lbs/ft}^3\right)$$

$$\sigma_v = \sigma_1 = \left(468 \times 10^3 \text{lbs/ft}^2\right)$$

$$\sigma_v = \sigma_1 = \left(468 \times 10^3 \text{lbs/ft}^2\right)\left(1 \text{ t/2,200 lbs}\right)\left(1 \text{ ft/12 in}\right)\left(1 \text{ ft/12 in}\right) = 1.47 \text{ t/in}^2$$

Find the horizontal stress:

$$\sigma_h = \sigma_2 = \sigma_3 = 1/3\sigma_v = (1/3)\left(1.47 \text{ t/in}^2\right) = 0.49 \text{ t/in}^2$$

$$W = (1/2)\left(6 \times 10^6 \text{psi}\right)\left\{1.47^2 + 0.49^2 + 0.49^2 - 2(0.25)\left[0.49^2 + 0.49(1.47) + (1.47)0.49\right)\right\}$$

$$W = 8.3 \times 10^{-8} \text{in}^2/\text{lbs}\left[2.16 + 0.24 + 0.24 - (0.5)(0.24 + 0.72 + 0.72)\right]$$

$$W = 8.3 \times 10^{-8} \text{in}^2/\text{lbs}[2.16 + 0.24 + 0.24 - 0.84]$$

$$W = \left[8.3 \times 10^{-8} \text{in}^2/\text{lbs}\right]\left[2200 \text{ lbs/1 t}\right]\left[1.8 \text{ t}^2/\text{in}^4\right]$$

$$W = \left[1.826 \times 10^{-4} \text{in}^2/\text{t}\right]\left[1.8 \text{ t}^2/\text{in}^4\right]$$

$$W = 3.29 \times 10^{-4} \text{t/in}^2$$

$$W = 0.047 \text{ t/ft}^2 \text{or } 0.047 \text{ ft} - \text{t/ft}^3$$

Example 10.8.7

Determine the compression and shear velocity through an elastic body given the following properties of dynamic modulus of elasticity, E, of 2×10^5 kg/cm² (part a) and 6×10^5 kg/cm² (part b).

Equations are from Farmer (1968) and have been derived specifically for Poison's ratio = 0.25 and a density of 2.6 gm/cc.

Compression Wave:

$$C_p = 6.8(E)^{0.5} \text{ m/s}$$

Shear Wave:

$$C_s = 3.9(E)^{0.5}$$

where

E is the dynamic modulus of elasticity.

Solution:

Part a:

Compression Wave:

$$C_p = 6.8(E)^{0.5} \text{ m/s}$$

$$C_p = 6.8\left(2 \times 10^5\right)^{0.5} \text{ m/s}$$

$$C_p = 3{,}041 \text{ m/s}$$

Shear Wave:

$$C_s = 3.9(E)^{0.5}$$

$$C_s = 3.9\left(2 \times 10^5\right)^{0.5}$$

$$C_s = 1{,}744 \text{ m/s}$$

Part b:

Compression Wave:

$$C_p = 6.8(E)^{0.5} \text{ m/s}$$

$$C_p = 6.8\left(6 \times 10^5\right)^{0.5} \text{ m/s}$$

$$C_p = 5{,}268 \text{ m/s}$$

Shear Wave:

$$C_s = 3.9(E)^{0.5}$$

$$C_s = 3.9\left(6 \times 10^5\right)^{0.5}$$

$$C_s = 3{,}021 \text{ m/s}$$

Example 10.8.8

Given a pillar with an area of 64 m^2 and a length of 2 m, determine the increase in strain energy in the pillar if its modulus of elasticity is 0.65 GPa. Assume a load is gradually applied to the pillar of 70 kN.

According to Brady and Brown (2004), the increase in strain energy can be expressed in terms of stress (in the direction of loading) and the elastic material properties. The increase in strain energy, ΔU_{st}, is expressed as follows:

$$\Delta U_{st} = 1/2 P_z u_z$$

where

ΔU_{st} is the increase in strain energy in a pillar.
P_z is the induced load to the pillar = $\sigma_{zz}A$.
u_z is the induced deflection to the pillar = $\varepsilon_{zz}L = (\sigma_{zz}/E)L$.
σ_{zz} is the induced static stress.
A is the pillar plan area.
L is the pillar length.
E is the modulus of elasticity.

Solution:

From the equation above, it follows that

$$\Delta U_{st} = 1/2 P_z u_z$$

$$\Delta U_{st} = ½(\sigma_{zz} A)(\sigma_{zz}/E)L$$

$$\Delta U_{st} = ½(\sigma_{zz}^2/E)AL$$

$$\sigma_{zz} = 70{,}000\ N/64\ m^2$$

$$\sigma_{zz} = 1{,}094\ N/m^2$$

$$\Delta U_{st} = ½\left[\left(1{,}094\ N/m^2\right)^2/\left(0.65\times10^9 N/m^2\right)\right]\left(64\ m^2\right)(2\ m)$$

$$\Delta U_{st} = 0.12\ N-m$$

Example 10.8.9

Find the body waves, i.e., P and S waves, velocities through anhydrite given the elastic modulus, Poisson's ratio, and density. Provide the answer in m/s. Properties are found in Birch et al. (1942). The original source for the equations is unknown.

Given:

$$E = 7.2\times10^{11} dynes/cm^2$$

$$\nu = 0.295$$

$$\rho = 2.957\ gm/cm^3$$

Solution:

P-Wave Velocity:

$$V_p = \left[(K + 4G/3)/\rho\right]^{0.5}$$

S-Wave Velocity:

$$V_s = (G/\rho)^{0.5}$$

Bulk Modulus:

$$K = \left[E/3(1 - 2\nu)\right]$$

$$K = \left[\left(7.2\times10^{11} dynes/cm^2\right)\right]/3\left[1-2(0.295)\right]$$

$$K = 5.85\times10^{11} dynes/cm^2$$

Shear Modulus:

$$G = E/\left[2(1 + \nu)\right]$$

$$G = \left(7.2\times10^{11} dynes/cm^2\right)/\left[2(1+0.295)\right]$$

$$G = 2.8 \times 10^{11} \text{dynes/cm}^2$$

$$K = \left(5.85 \times 10^{11} \text{dynes/cm}^2\right)\left(1 \text{ Pa/10 dynes/cm}^2\right) = 5.85 \times 10^{10} \text{N/m}^2$$

$$G = \left(2.8 \times 10^{11} \text{dynes/cm}^2\right)\left(1 \text{ Pa/10 dynes/cm}^2\right) = 2.8 \times 10^{10} \text{N/m}^2$$

$$\rho = 2.957 \text{ gm/cm}^3 \left(1 \text{ kg/1,000 gm}\right)\left(100 \text{ cm/1 m}\right)^3 = 2{,}957 \text{ kg/m}^3$$

P-Wave Velocity:

$$V_P = \left[\left(K + 4G/3\right)/\rho\right]^{0.5}$$

$$V_P = \left[\left(5.85 \times 10^{10} \text{dynes/cm}^2 + 4 * 2.8 \times 10^{10} \text{dynes/cm}^2/3\right)/2{,}957 \text{ kg/m}^3\right]^{0.5}$$

$$V_P = 5{,}686 \text{ m/s}$$

S-Wave Velocity:

$$V_s = (G/\rho)^{0.5}$$

$$V_s = \left(2.8 \times 10^{10} \text{N/m}^2/2{,}957 \text{ kg/m}^3\right)^{0.5}$$

$$V_s = 3{,}066 \text{ m/s}$$

Example 10.8.10

An earthquake is measured at a station where the P and S waves arrive 1 minute apart. If the P-wave velocity is 4,750 m/s and the S-wave velocity is 2,650 m/s, determine the distance from the station to the earthquake source.

Solution:

For the distance from source to a seismograph station.

$$d = \left(t_s - t_P\right)/\left(1/V_s - 1/V_P\right)$$

where

- t_s is the S-wave travel time.
- t_p is the P-wave travel time.
- $t_s - t_p$ is the time lag between the S and P waves.
- V_s is the S-wave velocity.
- V_p is the P-wave velocity.

$$d = \left[(1 \text{ min})\left(60 \text{ s/1 min}\right)\right]/\left[\left(1/2{,}650 \text{ m/s}\right) - \left(1/4{,}750 \text{ m/s}\right)\right]$$

$$d = (60 \text{ s})/\left(1.67 \times 10^{-4} \text{s/m}\right)$$

$$d = 359{,}281 \text{ m}$$

$$d = 359 \text{ km}$$

Example 10.8.11

A fault is estimated to have a surface area of 10 km by 20 km. If the fault dislocates by an average of 5 m and the modulus of rigidity is 2.8×10^{11} dyne/cm^2, find the seismic moment, M_o, in ergs.

Seismic moment after Aki (1966)

$$M_o = GAD$$

Solution:

Area of rupture surface, A = (10 km)(20 km) = (200 km^2)(1,000 m/1 km)^2(100 cm/1 m)^2 = 2×10^{12} cm^2

Fault displacement, D = (5 m)(100 cm/1 m) = 500 cm

Seismic Moment:

$$M_o = GAD$$

$$M_o = \left(2.8 \times 10^{11} \text{dyne/cm}^2\right)\left(2 \times 10^{12} \text{cm}^2\right)(500 \text{ cm})$$

$$M_o = 2.8 \times 10^{26} \text{ergs}$$

Example 10.8.12

Estimate the moment magnitude, M_w, from the previous example.

Moment magnitude, M_w after Hanks and Kanamori (1979):

$M_w = (2/3)\log_{10}M_o - 10.73$.

*Note that this is specific for M_o in dyne-cm.

Solution:

$$M_w = (2/3)\log_{10}M_o - 10.73$$

$$M_w = (2/3)\log_{10}\left(2.8 \times 10^{26}\right) - 10.73$$

$$M_w = (2/3)(26.45) - 10.73$$

$$M_w = 16.6 - 10.73$$

$$M_w = 6.90$$

Example 10.8.13

An arbitrary seismogram from an earthquake is presented in Figure 10.11. If the maximum amplitude of the S wave is 40 mm from the seismogram, determine the Richter local magnitude of an earthquake using Richter's (1958) nomogram for Southern California as shown in Figure 10.12. An example is already completed on the nomogram for reference.

Solution:

From the seismogram in Figure 10.11, the travel time difference = 18 – 5 = 13 seconds. Plotting on the nomogram with the max amplitude of 40 mm and the travel time difference of 13 seconds shows an intersection at a Richter local magnitude, M_L, of approximately 4.7 as shown in Figure 10.12.

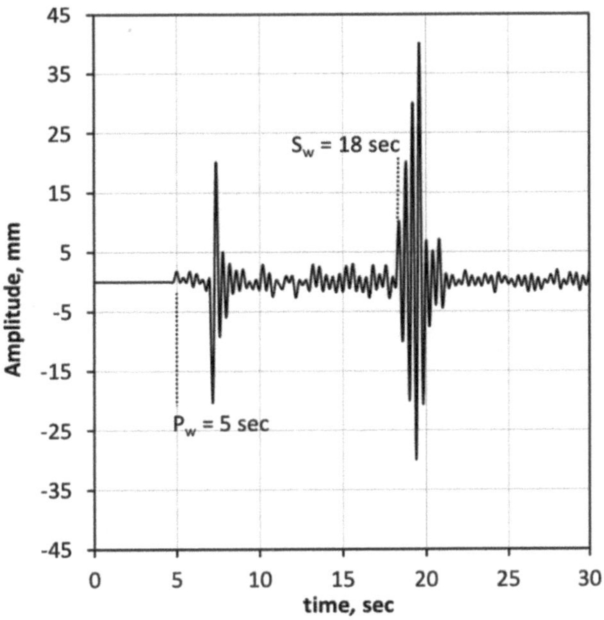

FIGURE 10.11 The seismogram.

Example 10.8.14

The differences between P-wave and S-wave travel times for three stations are provided in Table 10.1. Using the three-circle method, detailed in Villaverde (2009) and shown below, estimate the epicenter of the earthquake. Assume that a P wave travels at 5 km/s and the S wave travels at 3 km/s.

Distance from earthquake focus to observation point Villaverde (2009):

$$\Delta = (t_{sp})/(1/V_s - 1/V_p)$$

where

t_{sp} is the time lag between P and S waves.

Solution:

$$\Delta_1 = (t_{sp})/(1/V_s - 1/V_p)$$

$$\Delta_1 = (5.6 \text{ s})/(1/3 \text{ km/s} - 1/5 \text{ km/s})$$

$$\Delta_1 = (5.6 \text{ s})/(0.133 \text{ s/km})$$

$$\Delta_1 = 42.1 \text{ km}$$

$$\Delta_2 = (t_{sp})/(1/V_s - 1/V_p)$$

$$\Delta_2 = (7.2 \text{ s})/(1/3 \text{ km/s} - 1/5 \text{ km/s})$$

$$\Delta_2 = (7.2 \text{ s})/(0.133 \text{ s/km})$$

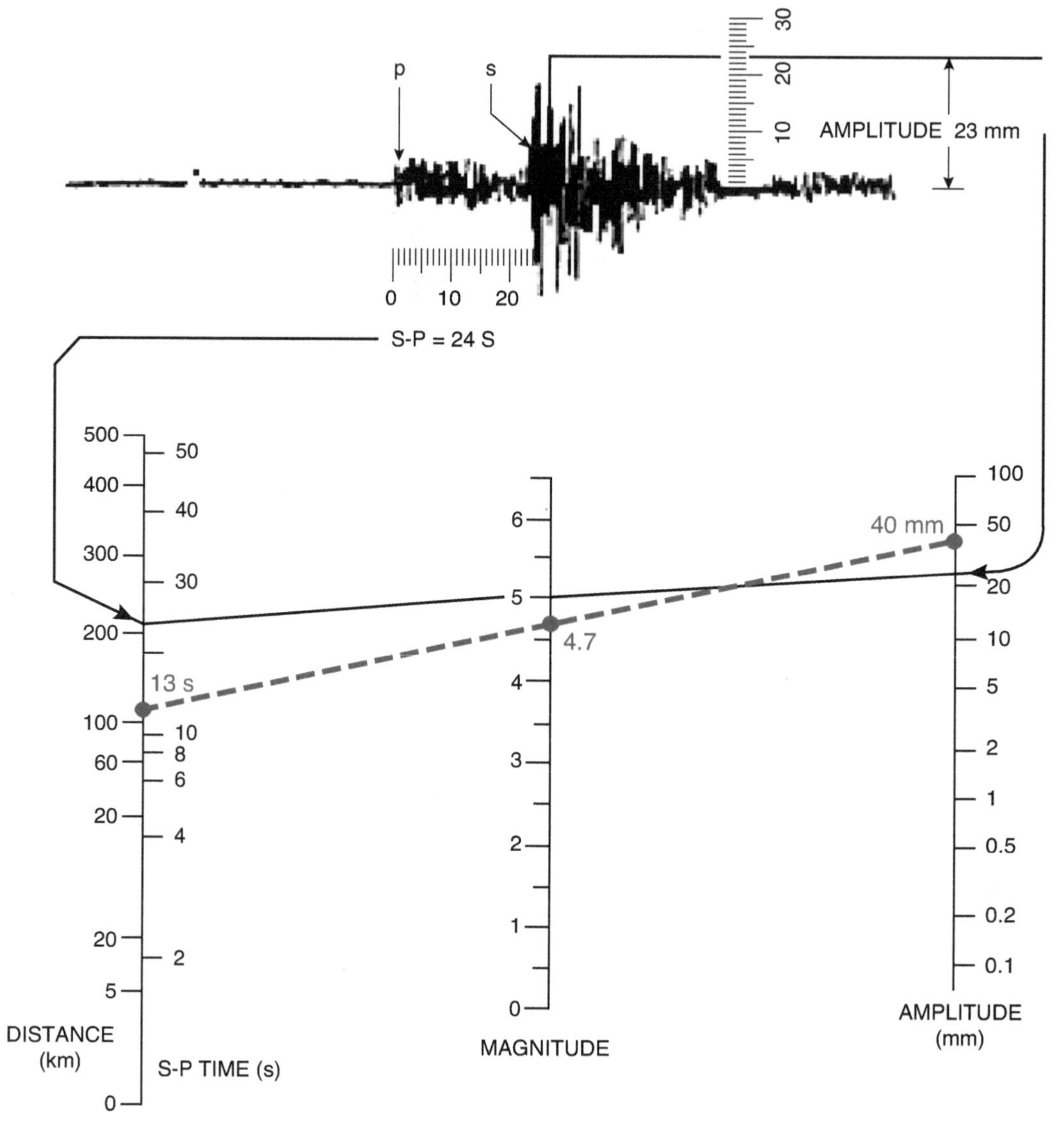

FIGURE 10.12 The nomogram for the example. Image was retrieved from U.S. Geological Survey USGS (n.d.), where it is attributed originally to Richter (1958).

TABLE 10.1
Time Lag between P and S Waves and Locations of Seismographic Stations

Seismographic Stations	Location of Station (x, y)	Time Lag	Δ_i
Station 1	(60 km, 45 km)	5.6 seconds	42.1 km
Station 2	(72 km, 72 km)	7.2 seconds	54.1 km
Station 3	(32 km, 87 km)	4.2 seconds	31.6 km

$$\Delta_2 = 54.1 \text{ km}$$

$$\Delta_3 = (t_{sp})/(1/V_s - 1/V_p)$$

$$\Delta_3 = (4.2 \text{ s})/(1/3 \text{ km/s} - 1/5 \text{ km/s})$$

$$\Delta_3 = (4.2 \text{ s})/(0.133 \text{ s/km})$$

$$\Delta_3 = 31.6 \text{ km}$$

The epicenter can now be constructed as shown in Figure 10.13 because the station locations are known as given in Table 10.2.

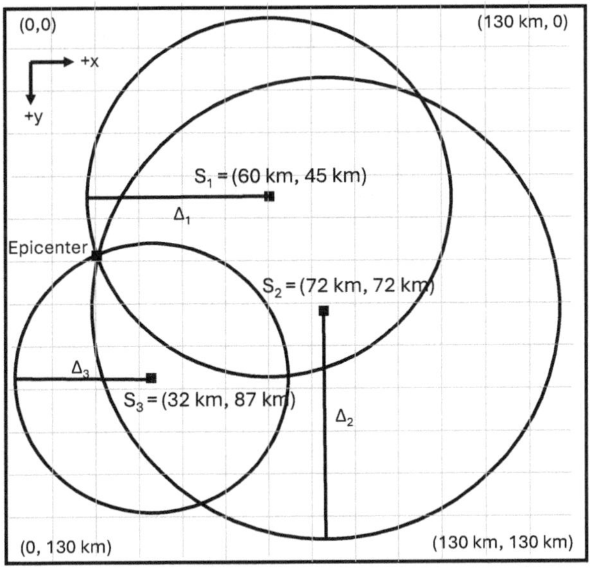

FIGURE 10.13 The geometric solution.

TABLE 10.2
Microseismic Record Ordered for P-Wave Arrival (from 0 to 4) Alongside Station Coordinates (x, y, z) from Blake (1971)

Location	x, a_i	y, b_i	z, c_i
0	800	250	305
1	723	190	160
2	815	468	305
3	424	278	305
4	277	576	305

Example 10.8.15

Determine the volume of rock effected by a seismic energy release from a source, for a micro-crack (part a) and for a rockburst (part b). The example and approach are found in Blake et al. (1974). Assume an applied stress field of 10,000 psi and a modulus of rigidity of 4 x 10^6 psi.

The following equation and definitions are found in Blake et al. (1974) and Duvall and Stephenson (1965):

$$W_s = 3/8\left(p^2/G\right)V_{ac}$$

where

W_s is the seismic energy released at the source.
p is the applied stress field at the source.
G is the modulus of rigidity of the rock.
V_{ac} is the volume of rock effected.

Solution:

Part a:

For a micro-crack, $W_s = 10^{-5}$ ft-lb. Blake, Leighton, and Duvall (1974) show this as ft/lb instead of ft-lb:

$$p = \left(10{,}000 \text{ lb/in}^2\right)\left(144 \text{ in}^2/\text{ft}^2\right) = 1.44 \times 10^6 \text{psf}$$

$$G = \left(4 \times 10^6 \text{lb/in}^2\right)\left(144 \text{ in}^2/\text{ft}^2\right) = 576 \times 10^6 \text{psf}$$

$$W_s = 3/8\left(p^2/G\right)V_{ac}$$

$$10^{-5}\text{ft} - \text{lb} = 3/8\left[\left(1.44 \times 10^6 \text{psf}\right)^2/576 \times 10^6 \text{psf}\right]V_{ac}$$

$$10^{-5}\text{ft} - \text{lb} = \left(1{,}350 \text{ lb/ft}^2\right)\left(V_{ac}\right)$$

$$V_{ac} = 7.4 \times 10^{-9} \text{ft}^3$$

Part b:

For a rockburst, $W_s = 10^6$ ft-lb:

$$W_s = 3/8\left(p^2/G\right)V_{ac}$$

$$10^6\text{ft} - \text{lb} = 3/8\left[\left(1.44 \times 10^6 \text{psf}\right)^2/576 \times 10^6 \text{psf}\right]V_{ac}$$

$$10^6\text{ft} - \text{lb} = \left(1{,}350 \text{ lb/ft}^2\right)V_{ac}$$

$$V_{ac} = 740 \text{ ft}^3$$

Example 10.8.16

Using the chart in Figure 10.14, if a volume of broken rock of 10^6 ft^3 is observed in an opening at 6,000 ft depth after a rockburst, estimate the earthquake magnitude and the seismic energy, and classify the earthquake according to the chart.

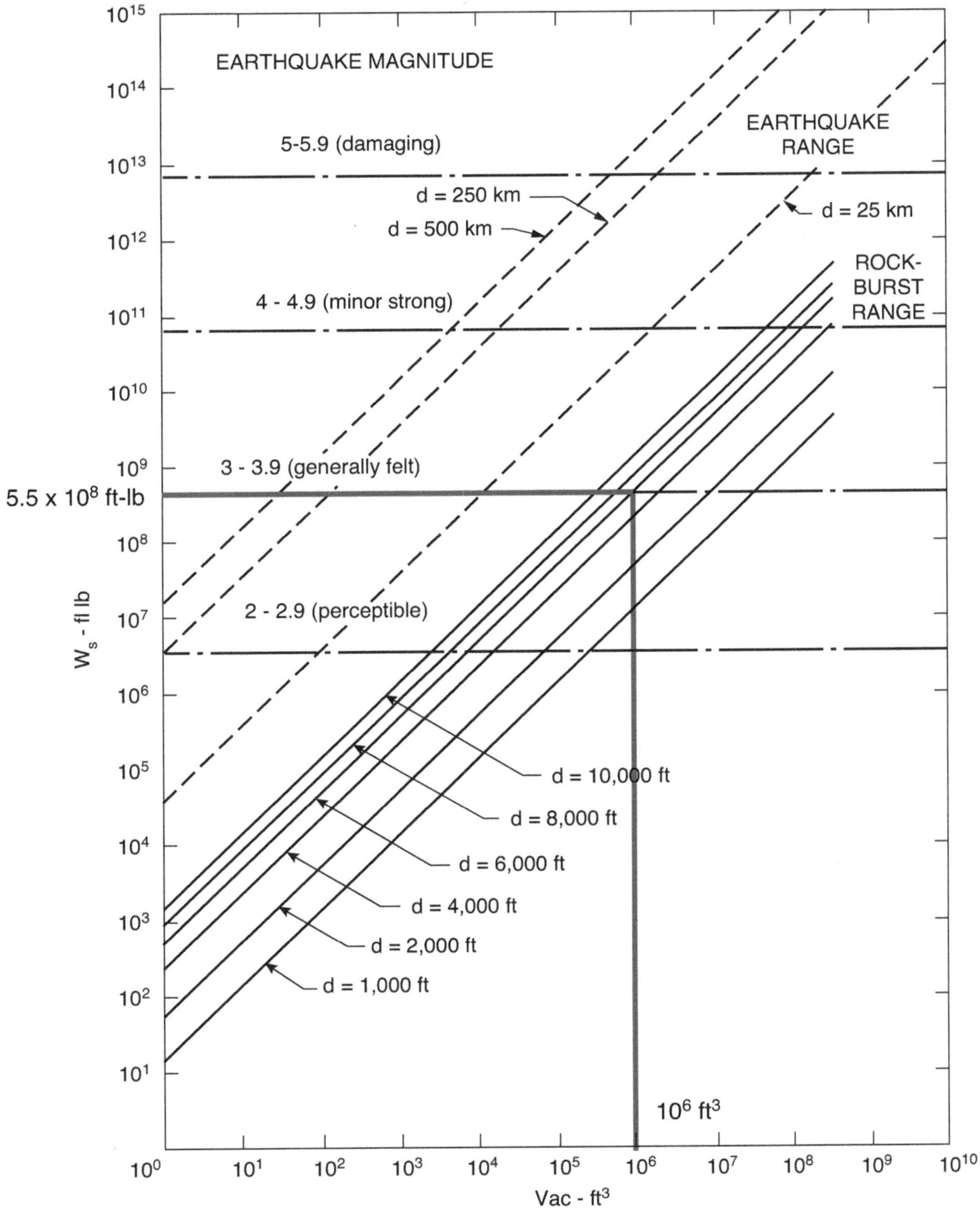

FIGURE 10.14 Plot of seismic energy as a function of depth and volume of broken rock after Duvall and Stephenson (1965).

The assumptions for constructing the chart and its complete description are given in Duvall and Stephenson (1965). The chart (using solid lines) plots the equation used in the previous example, for W_s, on log–log coordinates.

Solution:

From the plot, at a depth of 6,000 ft, the seismic energy would be 5.5×10^8 ft-lbs. The earthquake magnitude (shown by the horizontal lines) would be roughly 3.0 and classified at the border of "perceptible" to "generally felt".

Example 10.8.17

Toward detection of the location of a seismic source, Blake et al. (1974) discuss the use of a raw, iterative approach, where initial, trial guesses are made of the location and iterated until the left- and right-hand sides of the equation below equate. The approach is awkward beyond a reasonable check of other techniques and far from foolproof.

For this example, assume that the arrival time to the first geophone is 5.46 msec, and the P-wave velocity to this same geophone was 17,000 lengths/sec. For simplicity, in this example, treat a_0, b_0, and c_0 as zero. Coordinates of the first geophone are given below:

$$d_i = V_i^P t_i^P = \left[(x-a_i)^2 + (y-b_i)^2 + (z-c_i)^2\right]^{0.5} - \left[(x-a_0)^2 + (y-b_0)^2 + (z-c_0)^2\right]^{0.5}$$

where

d_i is the distance from source to *i*th geophone.
t_i^P is the relative time of first arrival of P wave to *i*th geophone.
V_i^P is the P-wave seismic velocity directed from source to geophone.
x, y, and z are unknown source coordinates.
a_i, b_i, and c_i are *i*th geophone coordinates.

Solution:

$$d_1 = V_1^P t_1^P$$

$$d_1 = (17{,}000 \text{ lengths/sec})(5.46 \text{ ms})(1 \text{ sec}/1{,}000 \text{ ms})$$

$$d_1 = 92.8 \text{ lengths}$$

Coordinates of the first geophone:
a_1: 850 lengths
b_1: 170 lengths
c_1: 200 lengths
Values of x, y, and z are desired. Take an initial guess of x = 1,000, y = –1,000, z = 0:

$$d_1 = V_1^P t_1^P = \left[(x-a_1)^2 + (y-b_1)^2 + (z-c_1)^2\right]^{0.5} - \left[(x-a_0)^2 + (y-b_0)^2 + (z-c_0)^2\right]^{0.5}$$

$$92.8 = \left[(1{,}000-850)^2 + (-1{,}000-170)^2 + (0-200)^2\right]^{0.5} - \left[(1{,}000)^2 + (-1{,}000)^2 + (0)^2\right]^{0.5}$$

$$92.8 = \left[(150)^2 + (-1{,}170)^2 + (-200)^2\right]^{0.5} - \left[(1{,}000)^2 + (-1{,}000)^2\right]^{0.5}$$

$$92.8 = -218 \quad * \text{Sides do not equate}$$

Hold one coordinate constant (i.e., y = –1,000), equate x to the original z guess, and the new z becomes the average of y and z. Therefore, x = 0, y = –1,000, z = –500.

92.8 = 488 *Sides do not equate and moving passed the solution.
Try x = 0, y = –1,000, z = 1,000
92.8 = 238 *Sides do not equate
Try x = –500, y = –1000, z = 0
92.8 = 680 *Sides do not equate and moving away from solution.

Iterating by this process in a solver, datasheet, or other numerical method reduces the effort, and it is easier to see where an appropriate guess trends toward the solution. Ultimately, the process arrives at a solution:

x = 0, y = 499, z = 1,010

$$d_1 = V_1^P t_1^P = \left[(x-a_1)^2 + (y-b_1)^2 + (z-c_1)^2\right]^{0.5} - \left[(x-a_0)^2 + (y-b_0)^2 + (z-c_0)^2\right]^{0.5}$$

$$92.8 = \left[(0-850)^2 + (499-170)^2 + (1{,}010-200)^2\right]^{0.5} - \left[(499)^2 + (1{,}010)^2\right]^{0.5}$$

$$92.8 = [722{,}500 + 108{,}241 + 656{,}100]^{0.5} - [249{,}001 + 1{,}029{,}100]^{0.5}$$

$$92.8 = 1{,}219.36 - 1{,}126.54$$

92.8 = 92.8 *Both sides roughly equate.

Example 10.8.18

In the following example, a direct solution for determining source location is shown. The approach is considered the "P-wave method". A second similar approach is the "S-P-wave method" but is not shown here. Both methods are discussed in detail by Blake et al. (1974).

Actual source location data is shown below from Blake (1971) from an actual event at an underground silver mine in Idaho. The calculation process below is not explicitly covered in Blake (1971), but the reference does provide the resolved location coordinates as x = 853, y = 236, and z = 275.

Solution:

To begin the P-wave analysis, some back calculation was used to approximate distances, d_i, because V_i^P is not given explicitly in Blake (1971). Alternatively, d_i can be approximated directly if V_i^P (P-wave velocity for the *i*th geophone) and t_i^P are known from a microseismic record as shown in the previous example:

$$d_i = V_i^P t_i^P = \left[(x-a_i)^2 + (y-b_i)^2 + (z-c_i)^2\right]^{0.5} - \left[(x-a_0)^2 + (y-b_0)^2 + (z-c_0)^2\right]^{0.5}$$

$$d_1 = \left[(853-723)^2 + (236-190)^2 + (275-160)^2\right]^{0.5} - [(853-800)^2 + (236-250)^2 + (275-305)^2]^{0.5}$$

$$d_1 = \left[(130)^2 + (46)^2 + (115)^2\right]^{0.5} - \left[(53)^2 + (-14)^2 + (-30)^2\right]^{0.5}$$

$$d_1 = 179.6 - 62.5$$

$$d_1 = 117.1$$

The remaining distances d_2, d_3, and d_4 are solved similarly as follows:

$$d_2 = 174.5$$

$$d_3 = 369.6$$

$d_4 = 607.0$ *Units for distance are not given in Blake (1971), but they should be consistent.

Three equations and three unknowns are needed. Derivation of the following matrix is shown in Blake et al. (1974):

$$H_{12} = A_{12}x + B_{12}y + C_{12}z$$

$$H_{13} = A_{13}x + B_{13}y + C_{13}z$$

$$H_{14} = A_{14}x + B_{14}y + C_{14}z$$

where H_{1i} takes the following form:

$$H_{1i} = \left(L_1/d_1 - L_i/d_i\right)$$

and

$$L_i = d_i^2 + a_0^2 + b_0^2 + c_0^2 - a_i^2 - b_i^2 - c_i^2$$

As an example, find L_1:

$$L_1 = d_1^2 + a_0^2 + b_0^2 + c_0^2 - a_1^2 - b_1^2 - c_1^2$$

$$L_1 = 117.1^2 + 800^2 + 250^2 + 305^2 - 723^2 - 190^2 - 160^2$$

$$L_1 = 224{,}800$$

Similarly,

$$L_2 = -150{,}295$$

$$L_3 = 582{,}046$$

$$L_4 = 662{,}497$$

As an example, find H_{12}:

$$H_{12} = \left(L_1/d_1 - L_2/d_2\right)$$

$$H_{12} = \left[224{,}800/117.1 - (-150{,}295)/(174.5)\right]$$

$$H_{12} = 2{,}781.5$$

Similarly,

$$H_{13} = 345.5$$

$$H_{14} = 828.9$$

The forms of A_{1i}, B_{1i}, and C_{1i} are

$$A_{1i} = 2(\alpha_1/d_1 - \alpha_i/d_i)$$

where

$$\alpha_i = a_0 - a_i$$

$$B_{1i} = 2(\beta_1/d_1 - \beta_i/d_i)$$

where

$$\beta_i = b_0 - b_i$$

$$C_{1i} = 2(\gamma_1/d_1 - \gamma_i/d_i)$$

where

$$\gamma_i = c_0 - c_i$$

As an example, find A_{12}:

$$\alpha_1 = a_0 - a_1 = 800 - 723 = 77$$

$$\alpha_2 = a_0 - a_2 = 800 - 815 = -15$$

$$A_{12} = 2(\alpha_1/d_1 - \alpha_2/d_2)$$

$$A_{12} = 2[77/117 - (-15)/174.5]$$

$$A_{12} = 1.49$$

Similarly,

$$A_{13} = -0.72$$

$$A_{14} = -0.41$$

As an example, find B_{12}:

$$\beta_1 = 250 - 190 = 60$$

$$\beta_2 = 250 - 468 = -218$$

$$B_{12} = 2(\beta_1/d_1 - \beta_2/d_2)$$

$$B_{12} = 2[60/117 - (-218)/174.5]$$

$$B_{12} = 2[0.51 + 1.25]$$

$$B_{12} = 3.52$$

Similarly,

$$B_{13} = 1.18$$

$$B_{14} = 2.09$$

As an example, find C_{12}:

$$\gamma_1 = c_0 - c_1 = 305 - 160 = 145$$

$$\gamma_2 = c_0 - c_2 = 305 - 305 = 0$$

$$C_{12} = 2(\gamma_1/d_1 - \gamma_2/d_2)$$

$$C_{12} = 2(145/117)$$

$$C_{12} = 2.48$$

Similarly,

$$C_{13} = 2.48$$

$$C_{14} = 2.48$$

The three linear equations are therefore

$$2{,}781.5 = 1.49x + 3.52y + 2.48z$$

$$345.5 = -0.72x + 1.18y + 2.48z$$

$$828.9 = -0.41x + 2.09y + 2.48z$$

In matrix form:

*Note it is important to carry significant digits beyond what is shown throughout.

A			B		C
2,781.5		1.49	3.52	2.48	x
345.5	=	−0.72	1.18	2.48	y
828.9		−0.41	2.09	2.48	z

In short form, Determinate[B] = −3.23

Substituting matrix [A] into each column of matrix [B], resolving the determinate three times (once per column), and then dividing it by determinate of [B] provide the coordinates of x, y, and z:

x		853
y	=	236
z		275

Example 10.8.19

An excavation is made in a single mining step as shown in Figure 10.15. Determine the seismic energy released due to mining (ERR) given the stresses and strains in Table 10.3. Assume an in situ state of stress of $\sigma_0 = 1{,}200$ psi. Convert the final answer to kJ/m^3.

The following equations are from Mitri and Simon (1998):

$$e_r = 1/2(\sigma_2 - \sigma_0)\varepsilon_2$$

Seismic energy released in one mining step which is also the darker area in Figure 10.15a:

$$e_r = 1/2(\sigma_1 - \sigma_0)\varepsilon_1 + 1/2(\sigma_2 - \sigma_1)(\varepsilon_2 - \varepsilon_1)$$

Solution:

Excavation in a single mining step:

$$e_r = 1/2(\sigma_2 - \sigma_0)\varepsilon_2$$

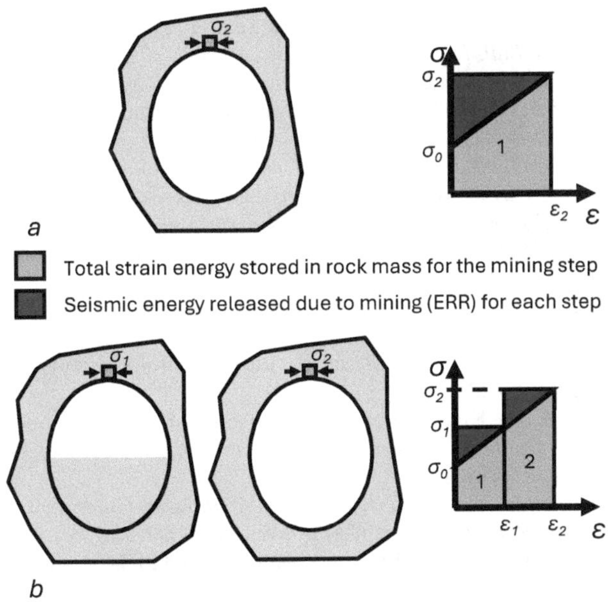

FIGURE 10.15 ERR with the mining sequence for one mining step (a) and two mining steps (b). (Created after content from Mitri and Simon (1998).)

TABLE 10.3
The Stress and Strain Data

Step	Step Description	σ_i	ε_i
1	First Excavation	2,400 psi	500 με
2	Second Excavation	3,600 psi	1,000 με

$$e_r = 1/2(3{,}600 - 1{,}200)(1{,}000\ \mu\varepsilon)\left(10^{-6}\varepsilon/\mu\varepsilon\right)$$

$$e_r = 1.2\ \text{lb} - \text{in/in}^2$$

Conversion to kJ/m³:

$$\left(1.2\ \text{lb} - \text{in/in}^3\right)\left(1.13\times10^{-4}\text{kJ/lb} - \text{in}\right)\left(1\ \text{in}^3/1.6387\times10^{-5}\text{m}^3\right) = 8.27\ \text{kJ/m}^3$$

Excavation in two mining steps:

$e_r = 1/2(\sigma_1 - \sigma_0)\varepsilon_1 + 1/2(\sigma_2 - \sigma_1)(\varepsilon_2 - \varepsilon_1)$

$e_r = 1/2(2{,}400 - 1{,}200)(500)(10^{-6}\ \varepsilon/\mu\varepsilon) + 1/2(3{,}600 - 2{,}400)(1{,}000 - 500)\ (10^{-6}\ \varepsilon/\mu\varepsilon)$

$e_r = 1/2(1{,}200)(500)(10^{-6}\ \varepsilon/\mu\varepsilon) + 1/2(1{,}200)(500)(10^{-6}\ \varepsilon/\mu\varepsilon)$

$e_r = 0.3 + 0.3$

$e_r = 0.6$ lb-in/in^2

Conversion to kJ/m^3

(0.6 lb-in/in^3)(1.13 × 10^{-4} kJ/lb-in)(1 in^3/1.6387 × 10^{-5} m^3) = 4.14 kJ/m^3

REFERENCES

Aki, K. (1966) Earthquake generating stress in Japan for the years 1961 to 1963 obtained by smoothing the first motion radiation patterns. *Bull. Earthquake Res. Inst.*, 44: 447–471.

Ansfield, V.J. (1992). A graphic representation of the Richter Scale. *J. Geol. Educ.* 40(5), 381–384.

Birch, F., Schairer, J. and Spicer, H. (1942) Handbook of physical constants. *Geol. Soc. Amer., Special Papers.* 36, 66.

Blake, W. (1971) Rock burst mechanics. Ph.D. Thesis. Colorado School of Mines, Golden, CO.

Blake, W., Leighton, F.W. and Duvall, W.I. (1974) *Microseismic Techniques for Monitoring the Behavior of Rock Structures.* US Department of the Interior, Bureau of Mines. Vol. 665, Denver, CO.

Bolt, B.A. (1976) *Nuclear Explosions and Earthquakes.* W.H. Freeman and Company, San Francisco, CA.

Bolt, B.A. (1982) *Inside the Earth.* Freeman, San Francisco, CA.

Brady, B.H. and Brown, E.T. (2004) *Rock Mechanics: For Underground Mining* (3rd Edition). Springer Science & Business Media, New York.

British Geological Survey (2008) Earthquakes. Issue 10. Retrieved from: www.earthquakes.bgs.ac.uk/education/eq_guide/earthquake_booklet.pdf

Brummer, R., Gay, N., Jager, A., McKinnon, S., Piper, P., Roberts, M., Ryder, J., Spottiswoode, S. and Wojno, L. (1988). *An Industry Guide to Methods of Ameliorating the Hazards of Rockfalls and Rockbursts.* Chamber of Mines of South Africa, Johannesburg.

Chamber of Mines Research Organization (COMRO) (1988) *An Industry Guide to Methods of Ameliorating the Hazards of Rockfalls and Rockbursts.* Chamber of Mines of South Africa, Johannesburg, South Africa.

Cook, N.G.W. and Salamon, M.D.G. (1966). Rock mechanics applied to the study of rockbursts. *J. S. Afr. Inst. Min. Metall.* 66(3), 97–108.

Cook, N.G.W., Hoek, E., Pretorius, J.P.G., Ortlepp, W.D. and Salamon, M.D.G. (1966) Rock mechanics applied to rockbursts. *J. S. Afr. Inst. Min. Metall.* 1996, 435–714.

Denkhaus, H.G. (1964) Critical review of strata movement theories and their application to practical problems. *J. S. Afr. Inst. Min. Metall.* 63(8), 310–332.

Department of the Army (1974). *Engineering Design Handbook, Explosions in Air, Part One.* Army Material Command, Alexandria, VA, No. 706-181.

Duvall, W.I. and Stephenson, D.E. (1965) Seismic energy available from rockbursts and underground explosions. *Trans. Soc. Min. Eng.* 232(3), 235–240.

Farmer, I.W. (1968) *Engineering Properties of Rocks.* E. & F. N. Spon Limited, London, England.

Hanks, T.C. and Kanamori, H. (1979) A moment magnitude scale. *J. Geophy. Res.* 84, 2348–2350.

Isaacson, E. (1962) *Rock Pressure in Mines* (2nd Edition). Mining Publications Ltd, Salisbury House, London.

Jaeger, J.C. and Cook, N.G.W. (1969) *Fundamentals of Rock Mechanics.* Methuen & Co. Ltd, London.

Kearey, P., Brooks, M. and Hill, I. (2002) *An Introduction to Geophysical Exploration* (3rd Edition). Blackwell Publishing, Hoboken, NJ.

Mendecki, A.J., van Aswegen, G. and Mountfort, P. (1999) A guide to routine seismic monitoring in mines. In: Jager, A.J. and Ryder, J.A. (eds) *A Handbook on Rock Engineering Practice for Tabular Hard Rock Mines*. Creda Communication, Cape Town, South Africa.

Mitri, H.S. and Simon, R. (1998) Modeling of energy parameters ERR and TESR around burst-prone mine openings. *Trans. Soc. Min. Metall. Exploration*. 304, 30–36.

Nkosi, N.Z. (2016) Seismological and mineralogical studies of the world's deepest gold-bearing horizon, the Carbon Leader Reef, West Wits Line goldfields (South Africa): Implications for its poor seismic reflective character. A dissertation submitted to the Faculty of Science, University of the Witwatersrand in fulfilment of the requirements for the degree of Master of Science, School of Geosciences.

Richter, C.F. (1935) An instrumental earthquake magnitude scale. *Bullet. Seismol. Soc.America*, 25, 1–32.

Richter, C.F. (1958) *Elementary Seismology*. W.H.Freeman & Co Ltd, New York.

United States Geological Survey (n.d.). Earthquake magnitude, energy release and shaking intensity. www.usgs.gov/programs/earthquake-hazards/earthquake-magnitude-energy-release-and-shaking-intensity

Villaverde, R. (2009) *Fundamental Concepts of Earthquake Engineering. Engineering Geology*. CRC Press, Boca Raton, FL.

Wagner H. (1984) Support requirements of rockburst conditions. In: Gay, N.C., Wainwright, E.H. (eds) *Rockbursts and Seismicity in Mines*. Journal of the Southern African Institute of Mining and Metallurgy, Johannesburg, pp. 209–218

Yeats, R.S. (2020) *Living with Earthquakes in the Pacific Northwest*. Oregon State University, Oregon.

11 Mine Access and Development

Reproduced and modified with kind permission of Spearing et al. (2023) and the publisher CRC Press, London.

11.1 INTRODUCTION

Underground mining is done where mineral deposits are situated beyond the economic depth of open-pit mining. It is generally applied to steeply dipping, thin, disseminated, or massive deposits for which the cost of removing the overburden and/or maintaining a slope angle in adjacent waste rock would be prohibitive.

Construction of underground openings is considered a specialized operation, time consuming, and costly. It therefore needs an appropriately skilled and trained workforce. Hence, mine development has become increasingly mechanized and efficient to reduce both capital and development costs, and the time, to develop and start mining a deposit.

11.1.1 Mine Development

Mine development is generally divided into primary and secondary development.

- Primary development
  - Basic development of excavations from surface to the orebody.
  - Consists of shafts, adits, or declines (or a combination) plus main access (such as drifts) to the actual orebody.
- Secondary development
  - Depends on the specific mining method.
  - Consists of pump chambers, workshops, ore, and waste passes for example.

11.1.2 Mine Layout Essentials

Mine layouts consist of:

- Primary and secondary development (as above).
- Shafts, adits, ramps, rock passes, and major horizontal developments.
- Production stopes in the orebody where extraction can take place safely.
- Infrastructure excavations between the stopes and the main arteries (access development).

The goal of any layout must be that it complies with all regulations, is environmentally responsible, and ensures the safe and economic extraction of the orebody.

In general, the three methods of accessing an underground mine are a shaft, adit, and decline (ramp). The shaft often remains the mine's most critical infrastructure, and downward development is frequently via ramps to allow access for the mobile equipment. A decline or ramp from the surface can facilitate easy machine movement and the transportation of people and materials. It is constrained as the development angle is limited to facilitate efficient equipment movement (typically 6°–10°). It can also be used for ore transportation by truck or conveyor, eliminating the need for hoisting shafts, but that is an economic decision as shafts are typically more effective and cheaper for moving large tonnages to surface, especially for deeper mines. Adits can only be considered if

DOI: 10.1201/9781032704845-11

the surface topography and orebody permit this. For example, an orebody in a hill (or similar) can be accessed using an adit situated near the bottom of the hill.

11.1.3 Shafts

A shaft is a vertical excavation in which suitable conveyances are used to transport miners, equipment, and ore (sometimes waste) into and out of the mine. Shafts can be inclined too, although this option is seldom used now as it is less efficient, because hoisting speeds are much less as conveyances are generally on rails. Shafts are always the option used where the deposit is located at depth.

In mining, shafts are generally used to:

- Access an orebody.
- Transport men and materials to and from underground workings.
- Hoist ore and waste from underground.
- Serve as intake and/or return airways as part of the mine's ventilation network.
- Provide a second egress in case of emergency, as typically required by the mining law.

Shafts can also provide access to nuclear waste storage, be used for hydropower generation, or provide access to underground civil structures such as basements, underground rail stations, and road, rail, or water tunnels.

Most shafts are divided into compartments each with a different use, by brattice walls or steel barriers fitted with conveyance guides. For example, one major compartment could be used for moving people and equipment, a second with two skips for taking rock to surface, and other compartments for service infrastructure (typically pipes and cables) and space for ventilation.

The main factors to establish the shaft size are the monthly tonnage requirement and the ventilation needed. At the mine planning stage, it is useful to include some excess rock-handling and ventilation capacity in case of future mining production increases. The orebody size, grade, and mining method will determine the rate of mining, and thus the tonnage (ore and waste) to be hoisted, the size of the workforce, and the material to be moved efficiently in any given shift. The modern tendency seems to use a separate shaft just for ventilation (typically upcast air) rather than splitting a shaft into a downcast section (with the conveyances for men, materials, and rock and a separate compartment for upcast air).

Once a shaft is excavated to its final depth, typically some distance below the economic depth of the orebody for ore handling and water handling, the shaft is equipped with all the necessary steel work to guide the shaft conveyances and hold the pipes and cables for the services needed underground. Figure 11.1 shows a cross section through a typical shaft divided into compartments.

Shafts are often essential for underground mines, and their location is determined based on detailed surface topography and infrastructure, the orebody, geology, rock mechanics, and environmental assessments. The location must be changed where adverse geotechnical conditions are identified in the originally planned site. Proximity to the orebody, strata conditions, and water-bearing structures (such as confined or unconfined aquifers) are the major parameters that govern the ultimate location of shafts.

The decision to locate the shaft is critical because the process to develop a shaft is very expensive and relatively slow. The vertical shaft must be well located with respect to the ore deposit and capable of handling production needs. The correct diameter and configuration of the shaft will provide optimum operational efficiency. The shaft can be rectangular, circular, or elliptical in profile, although almost all hard-rock underground mines have circular section shafts because this shape generates a better geometry for airflow and is naturally more stable from a stress concentration point of view.

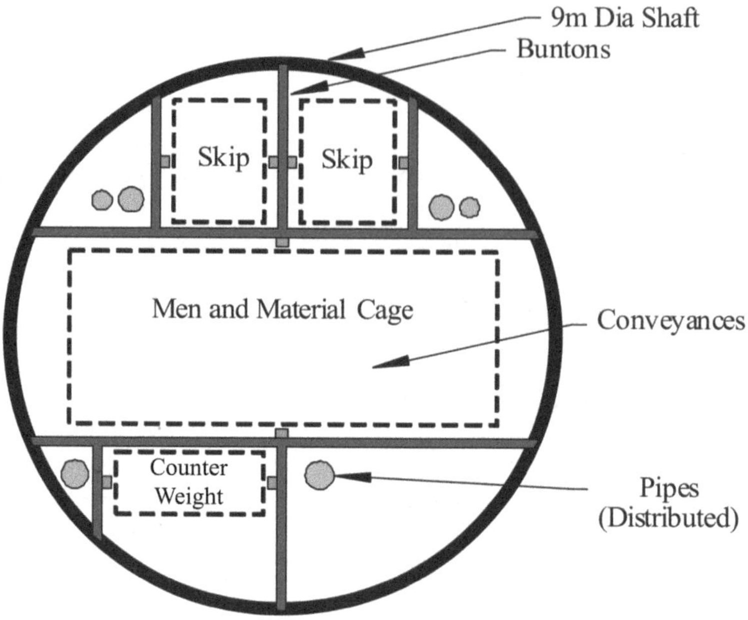

FIGURE 11.1 Cross section through a circular shaft showing all the components (Kempson et al., 2015).

11.1.4 Adits

An adit is a near-horizontal excavation that is used in mountainous areas or highwalls where the orebody is located near or above the access (valley) floor as shown in Figure 11.2. They are typically developed at a 4° incline mainly to facilitate easier water drainage out of the mine. It also helps equipment, such as trucks or rail cars, to remove rock from the mine by reducing energy when loaded due to the declination (elevation drop).

Developing an adit is the same process as developing any horizontal tunnel and is an option only where the topographic relief is considerable. Where adits can be used, ore and waste can be taken out of the mine at a low operating cost. The traditional method for developing adits is to drill and blast the face, load the material into a haulage device after the area has been cleared of dust and fumes, and then provide support and ventilation to the newly advanced face. Thus, drilling and blasting are the standard excavation method for adits and most hard-rock horizontal development. A typical adit layout is shown in Figure 11.2.

Tunnel-boring machines (TBMs) can be used to develop adits, and even long typically straight tunnels, depending typically on the length and the economics. The use of a TBM is generally safer compared to traditional blasting methods. The other main exception to the use of drilling and blasting are underground mines in relatively soft rock such as coal, salts, and limestone where the rock can be removed without the need for blasting using continuous miners or road headers.

11.1.5 Ramps (Declines)

A decline or ramp as shown in Figure 11.3 is an access tunnel usually developed at a low slope angle from the horizontal of <20° dip, or typically between 6° and 10° if mobile equipment is to be used. The design and support of declines are considered as the main challenges in underground mine development as they are generally needed for the life of the mine, which is also the same for shafts and adits. They can be straight, spiraled, or a combination of both.

Ramp access is the common selection in shallow ore bodies. A ramp from the surface can facilitate machine movements and transport of people and materials. It can also be used for ore

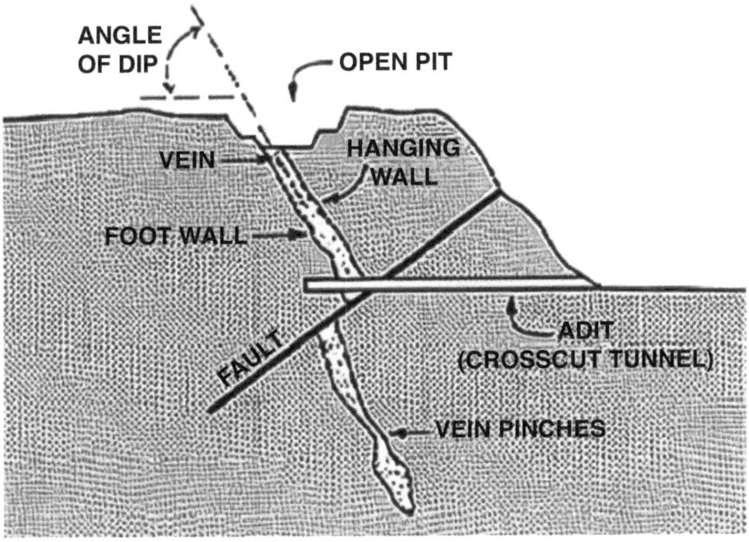

FIGURE 11.2 A mine adit to access an orebody (Purkey and Garside, 1995).

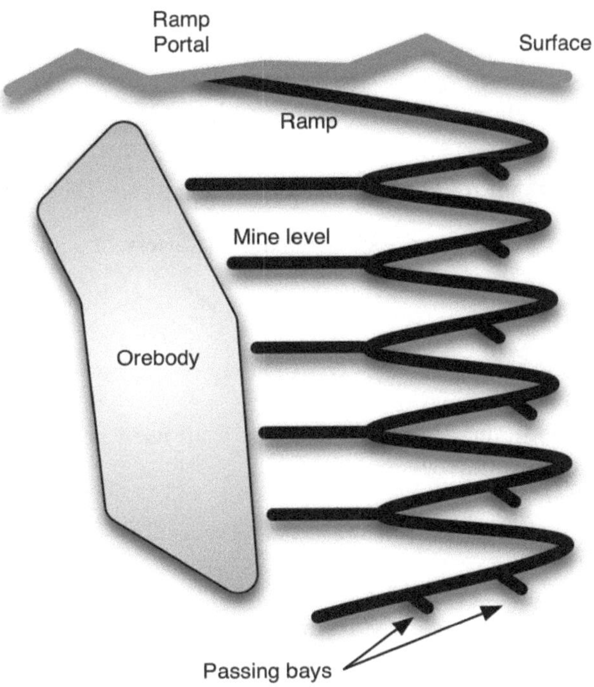

FIGURE 11.3 A ramp to access an orebody (Haviland and Marshall, 2014).

transportation by truck or conveyor, eliminating the need for hoisting shafts. Ramps are sized to include space for equipment and must incorporate a safe margin for clearance, walkways, ventilation, and other facilities and services. Cross sections typically vary from 2.2 m by 2.5 m in mines with a low degree of mechanization and low tonnage to 5.5 m by 6.0 m where large equipment is used and high ore tonnages are needed.

In many mines, the decline is used to transport ore to the surface using a conveyor belt, being associated with grade limits. For example, if utilized for conveyor belt haulage only, the maximum grade of the decline could be between 10° and 15° depending on the material to be conveyed. If trucks are used, the typical decline is between 8° and 10°.

11.1.6 Factors Influencing the Mine Access and Layout

The following general factors affect the mine access and layout:

- Physical properties of the orebody and host (surrounding) rock.
- Orebody dimensions and orientation.
- Orebody grade and distribution (value/unit volume).
- Geological structure.
- Hydrology.
- Proximity to other mining areas.
- Mine subsidence restrictions.
- Land ownership and restriction requirements.
- Waste rock and tailings disposal constraints.
- Wastewater discharge requirements.
- Surface topography and existing infrastructure.
- Climate. For example, in extreme cold conditions, all infrastructure must be covered and heated, so this can influence the mine access selection. In very high rainfall areas, if ramps are used, special considerations must be made so that huge amounts of water do not flow into the mine.

Most of the above directly relate to the mining method selected.

11.2 SHAFT SINKING AND SHAFT PROTECTION

Shaft support requirements in soil and rock depend on a variety of geotechnical and hydrological parameters.

Stability may be:

- Complete which is the condition that typically exists where:
  - Rock is strong and massive.
  - Stresses induced in the shaft walls from the shaft itself and from any nearby excavations are well below the rock mass strength.
- Dependent upon geologic structure.
- Dependent upon rock strength or in situ stress relationships or a combination of both.

Even where shafts are completely stable for the life of the mine, a cosmetic lining should be installed to reduce friction losses in ventilation or to hang shaft hardware for hoisting, support of utilities, for example. This also helps in case of unexpected seismicity, weathering, or adverse hydrology or geology.

11.2.1 Stresses around Shafts

Stress-controlled shaft wall instability (or stability) is governed by:

- Shaft size and shape.
- Rock types and properties.
- In situ stress conditions.
- Future mining-induced stresses.

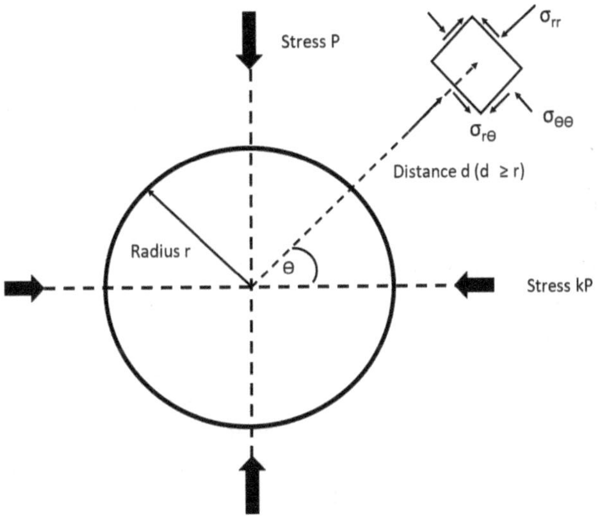

FIGURE 11.4 Stresses around a circular excavation.

The stability and hence the required lining support pressure (if any) are governed by the magnitude of the stresses induced in the shaft walls relative to the rock mass strength. The stresses at the shaft walls are dependent on the shaft shape, the magnitude of the horizontal stresses, and their relative stress uniformity.

A circular opening in a uniform stress field (horizontal stresses are equal to each other and constant) produces the least stress concentration at the shaft wall. For a non-uniform stress field (orthogonal horizontal stress components are unequal but remain constant), the stress concentration is least for an elliptical shaft opening which is oriented with its major axis parallel to the direction of the major horizontal stress direction. Oval shaft cross sections are therefore good compromises. Numerical modeling is the most useful tool to help with optimizing the design.

The Kirsch (1898) equations can be used for the shaft support design of circular and elliptical cross-sectional shafts. These equations are based on Figure 11.4 (note that in the Figure, the shear stress $\sigma_{r\theta}$ is the same as $\tau_{r\theta}$, but is just represented in a different way) which also shows the parameters used in the formula (added again for convenience):

$$\sigma_{rr} = (P/2)\left[(1+k)\left(1-r^2/d^2\right)-(1-k)\left(1-4r^2/d^2+3r^4/d^4\right)\cos 2\theta\right] \tag{11.1}$$

$$\sigma_{\theta\theta} = (P/2)\left[(1+k)\left(1+r^2/d^2\right)+(1-k)\left(1+3r^4/d^4\right)\cos 2\theta\right] \tag{11.2}$$

$$\sigma_{r\theta} = (P/2)\left[(1-k)\left(1+2r^2/d^2-3r^4/d^4\right)\sin 2\theta\right] \tag{11.3}$$

Simplifying the Kirsch equations, for a circular opening in a hydrostatic stress field (σ_h), the stresses away from the shaft wall can be estimated as follows:

$$\sigma_{rr} == \sigma_h\left[1-(r/d)^2\right] \tag{11.4}$$

$$\sigma_{\theta\theta} == \sigma_h\left[1+(r/d)^2\right] \tag{11.5}$$

where

σ_{rr} is the radial stress.
$\sigma_{\theta\theta}$ is the tangential stress.

r is the shaft radius.
d is any radius of interest beyond radius, r.

This method only produces a first estimate. Typically, some form of computer modeling is needed to finalize the design, and it should be revisited during the excavating as more field information becomes available.

Once the stress distribution and any future stress changes in the rock around a shaft have been determined, it can be compared to the rock mass strength to determine whether overstressing (failure) of the rock is likely. Triaxial strength data may be used for this purpose if available. Linear or non-linear Mohr's envelopes may be used, depending on the amount of nonlinearity the rock exhibits and the accuracy that is required. It must be emphasized that the rock mass strength rather than the rock specimen strength should be used in this type of analysis.

The depth of the failed zone is estimated by the intersection of the rock mass strength curve with the tangential stress curve. If the rock would fail without a lining, the supporting pressure of the lining would prevent rock failure from disrupting the function of the shaft by supporting the additional load. This may be determined by assuming either that the failing rock behaves plastically or that it behaves in a brittle manner (typical clastic rock behavior).

11.2.1.1 Pressure Support Lining Design in Elastic-Plastic Rock

The theoretical pressure that must be developed by the lining to support the rock assumed to be elastic-plastic (conversely the pressure that the lining must withstand without failing) is originally after Terzaghi (1943):

$$P_i = [(2/\tan\beta + 1)][\sigma_h + \{\sigma_o/(\tan\beta - 1)]\}(r/R_o)^{\tan\beta - 1} - [\sigma_o/(\tan\beta - 1)] \tag{11.6}$$

where

P_i is the required supporting pressure of the liner (psi).
tan β is the coefficient of passive pressure = (1 + Sinφ)/(1 – Sinφ).
φ is the angle of internal friction (degrees).
σ_h is the uniform horizontal ground stress (psi).
σ_o is the rock mass unconfined compressive strength (psi).
r is the shaft radius (ft).
R_o is the radius of relaxed or failed zone in the surrounding rock (ft).

It should be noted that although "R_o" is proportional to "r" in the equation, the lining pressure required is not a function of the shaft radius "r". The primary variables to be considered are tan β, σ_o, and σ_h. For $\sigma_o > 2\sigma_h$, failure will not occur. Note that as the ratio σ_o/σ_h approaches 2.0, the lining pressure approaches zero.

11.2.1.2 Pressure Support Lining Design in Brittle (Clastic) Rock

The theoretical pressure that must be developed by the lining in clastic rock is (Talobre, 1957)

$$P_i = [(C_m/\tan\phi) + \sigma_h(1 - \sin\phi)](r/R_0)^{\tan\beta - 1} - (C_m/\tan\phi) \tag{11.7}$$

where

P_i is the radial pressure between outside of lining and rock.
φ is the angle of internal friction.
C_m is the rock mass cohesion = $\sigma_o/[2(\tan\beta)^{1/2}]$.

Some points worth noting about shaft lining for rock are as follows:

- As the width of relaxed zone approaches zero, the r/R_o term approaches 1. For this case, the lining pressure is a function of σ_h and φ only.
- As σ_o approaches $2\sigma_h$, the width of the relaxed zone approaches zero, and consequently the r/R_o term again approaches 1.
- The computed lining pressure is independent of the shaft radius, r, and is directly proportional to the uniform horizontal stress σ_h. The lining pressures are functions of variables σ_o/σ_h and φ as mentioned earlier.

It is suggested that both the elastic-plastic and clastic solutions be calculated to determine which of the two cases is worse. The greater of the two pressures calculated should be used for the lining design, unless it can be reliably determined that the less conservative value is actually justified.

11.2.2 Design of Shaft Lining

The lining design phase is a structural engineering task, which is a very specialized field and beyond the scope of this book. If a concrete lining is planned, Lame's (1837) equation as given by Ostrowski (1972) details approximations for lining thickness:

$$d = a\left(\{(K/SF)/[(K/SF) - 2p]\}^{0.5} - 1\right) \tag{11.8}$$

where
K is the unidirectional compressive strength of lining.
SF is the safety factor.
p is the outside pressure.
a is the inside radius to the rockwall.
d is the lining thickness.
Some pertinent facts about lining design are given in the following (Ostrowski, 1972).

The notation K represents typically the unconfined compressive strength of the concrete. Since the lining will be exposed to a three-dimensional state of stress, the required thickness may be overestimated due to the strength increase of materials under triaxial conditions.

This conservatism is, however, offset by the fact that the horizontal stress field is never precisely uniform, and some localized bending stresses may be present in the lining. It is worth retaining this conservatism to cover any unknown factors that may increase the lining loads or create localized more adverse stresses.

If the excavation extends beyond the payline (the theoretical boundaries of the shaft), the overbreak zone will normally be filled with concrete during lining placement, increasing the effective thickness of the lining. Based on the previous shaft-lining experience (Researcher Unknown, nd), the overbreak can be empirically assumed with some confidence to be

$$\text{Overbreak} = 6.6728\ r + 0.0307 \text{ in inches} \tag{11.9}$$

where
r is the shaft radius in ft.

Due to this additional thickness needed to be filled with concrete (into any overbreak) to create a constant and smooth final finished shaft to minimize ventilation losses, the safety factor on the lining thickness can be possibly minimized. A safety factor of 1.50 is considered appropriate and is in compliance with the American Concrete Institute building code.

11.2.2.1 Shaft Support with Rock Bolts, Steel Sets, and Shotcrete

Peck (1969) developed a set of guidelines for shaft support using rock bolts, steel sets, and shotcrete. In general, the requirements for support in shafts are less than those in tunnels where the stress conditions are less uniform, and the rock is more likely to be further disturbed by mining activities.

The self-supporting capabilities of rock are therefore enhanced by the support. The final design of a rock bolt, steel set, or shotcrete support for a given shaft should be based on consideration of the possible mechanisms of failure associated with specific conditions around the shaft and the way the supports might enhance the stability.

The brief guideline below applies to shafts with diameters between 4.6 and 7.6 m. For smaller diameters, somewhat less support is required, whereas considerably more support would be required for larger diameters:

- The length of the rock bolt required is determined based on consideration of the probable size of the loosening rock blocks. A rock bolt length of 1/2–1/3 the shaft diameter, with a maximum length of 2.4–3.0 m, is recommended as a first estimate.

11.2.3 Shallow Shaft Protection Pillar Design

At shallow depths, several empirical methods can be used to determine a first estimate of the required shaft pillar dimensions. At best, they can be used as a first approximation, and as the depth increases, they tend to be too conservative (i.e., suggesting that shaft pillars be made larger than necessary). They all use a linear relationship such that the shaft pillar size increases proportionally with depth increase (reference unknown). They are based on observations and the angle of draw in the case of longwall coal mining:

$$D = H/5 \tag{11.10}$$

$$D = 150 + (H/5) \tag{11.11}$$

$$D = 0.72(H) \tag{11.12}$$

where

D is the shaft pillar diameter in m.
H is the depth below surface in m.

The final design should be determined, preferably using an appropriate and suitably calibrated computer simulation package and updated and modified where necessary.

11.3 RAISE AND ORE PASS DEVELOPMENT

Raises are steeply inclined openings linking the mine sublevels at several vertical elevations and are often developed within the orebody. They are normally placed near or in the stopes employing specialized cyclic or continuous operations. Specific applications of bored raises are transfer of material, ventilation, personnel access, and ore production. Inclination varies typically from a minimum of 55° to the vertical, which is the lowest angle of repose of blasted rock. They have variable cross sections from typically 2 to 30 m^2.

Since manual excavation of raises is a potentially more hazardous activity, raise boring is frequently utilized for developing ventilation raises, ore passes, and rock fill passes. It provides safer and more efficient mechanized excavation of circular raises up to 6 m diameter because this method eliminates the need for explosives and keeps personnel remote from the actual rock excavation.

11.3.1 Raise Boring

In mining, raise boring is the procedure for mechanically boring a vertical or inclined shaft between two or more levels (one of which could be the surface). Access must be available at the start and end of the excavation. In conventional raise boring, a downward pilot hole is drilled to the target level by the raise boring machine, where the bit is removed and replaced by a reaming head (Figure 11.5).

Typical applications of raise boring include:

- For mining:
  - Shafts (men, material, and/or ventilation).
  - Rock passes.
  - Raises.
  - Mineral production (where it has been sometimes used in the orebody to produce ore for processing).
- For civil construction:
  - Drainage.
  - Hydropower.
  - Escapeways, e.g., road tunnels.
  - Ventilation for metros, road, and rail tunnels.

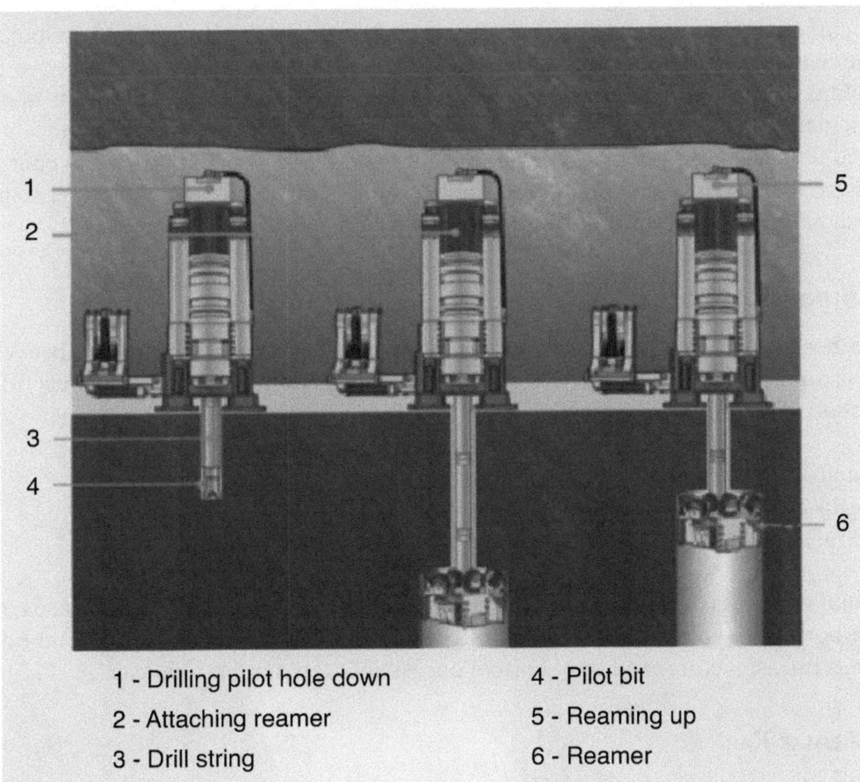

FIGURE 11.5 Raise boring process (Atlas Copco, 2008).

Typically raise boring can achieve the following:

- 0.5–7.0 m diameter.
- Lengths up to 1,200 m.
- Normally vertical or near vertical.

Mechanical boring uses two principles to remove rock:

- Cratering/crushing.
- Kerf breaking (kerf is a groove usually associated with sawing).

Important factors determining the application and success of rock boring include:

- Unconfined compressive strength of the rock mass.
- Abrasiveness of the rock types to be bored.
- Local jointing and bedding planes.
- Schistocity, which relates to the mode of foliation in certain metamorphic rocks.
- Presence of water which is a problem with any excavation development method.

The development of directional drilling has made raise boring long shafts even more feasible for longer excavations as the pilot hole can be kept correctly aligned and positioned. After the pilot hole is completed and the reamer head fitted, the machine then reams back the hole to the final diameter, rotating and pulling the reaming head upward toward the start of the pilot hole (where the raise boring machine is securely located). The cuttings fall to the lower level and are removed by any convenient method using remote control for safety considerations.

The capital cost of a raise boring machine is relatively high, but the return on investment is good particularly for longer length excavations. Advantages of raise boring are that miners are not required to enter the excavation while it is underway, the excavation process is continuous, no explosives are used, so a smooth and more stable profile is obtained, and manpower requirements are reduced. Safety is, however, the most important advantage.

11.3.2 Blind Hole Boring

Blind hole boring in mining is typically effective when developing rock passes between levels. Accuracy is a potential issue as there is no pilot hole to guide the boring, so the useful length is relatively short. The limits are typically:

- Around 1.5 m diameter.
- Lengths around a maximum of 50 m.
- Hole inclination 60°, from the horizontal to vertical.

Conventional rock pass development with drilling and blasting is hazardous and tends to be slow. Blind hole boring is typically much safer and faster. For these shorter holes, blind hole boring involves less infrastructure, and only bottom access (not top and bottom access) (Figure 11.6).

11.3.3 Alimak Raising

Alimak raising, as discussed in Moore (2014), is another excavation alternative to raise boring that is still safer than conventional raise development. For ore passes (rock passes, box holes), short raises, and short shafts, a semi-mechanized system can be used especially if only bottom access is available. The excavation must be vertical or steep dipping to use this system, which was developed

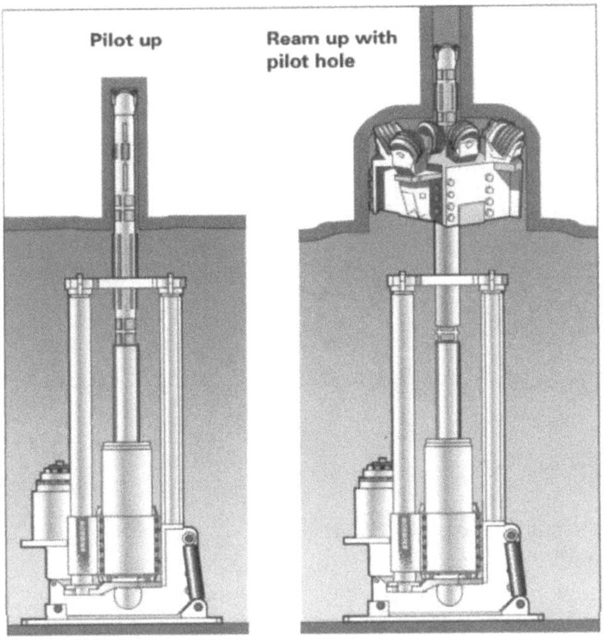

FIGURE 11.6 Blind hole borer (Atlas Copco, 2008).

in 1957 by Alimak. It was initially developed (and still used) for high-rise building construction to move light machinery and work crews to the different floors under construction. The choice of excavation method depends mainly on the following:

- Cost.
- Equipment availability.
- Access.
- Any time to completion constraints.

Typical mining applications include the following (as an alternative to conventional development or boring):

- Ventilation raises.
- Stope slot raises, for example longhole open stope first production.
- Ore and waste pass systems.
- Raise bore rehabilitation.
- Ventilation shafts.
- Manway raises.
- Emergency ladder way installations.

Alimak Raise Climbers are available with air, electric, or diesel/hydraulic drive units. One of the longest shafts driven in one step was 1,050 m long in Norway. The platform can be any shape and size. The largest work platform supplied so far measured more than 30 m^2.

The raise climber system is basically a movable working platform that runs along a monorail beam that increases productivity and improves safety when compared with other conventional blasting systems. As the excavation is developed, the monorail track is lengthened by adding sections to it and bolting them to the excavation sidewall.

With Figure 11.7 as an example, the system involves five production steps:

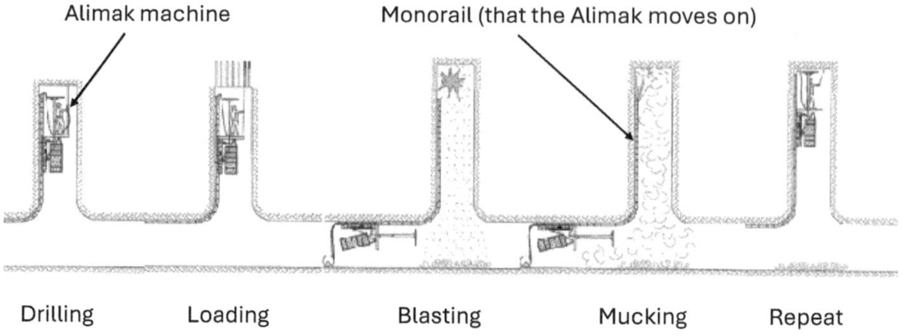

FIGURE 11.7 The Alimak raise development production steps (Tatiya, 2005).

- Scale (make safe from the personnel cage fitted with a protective canopy).
- Drill (conventionally from the cage).
- Charge the holes and then remove the cage from the excavation using the monorail which has a drive-with a toothed section that the cage engages with to safely move up and down the monorail beam.
- Blast (remotely).
- Ventilate and remove the broken ore from the bottom of the excavation.

11.4 MAIN ACCESS DEVELOPMENT AND PROTECTION

Near the shaft or decline, service excavations are typically developed for maintenance, offices, and pump chambers for example. After accessing the orebody, underground main access tunnels are developed on successive near-horizontal planes, referred to as levels a certain vertical distance apart, depending on the orebody geometry. The underground workings include production stopes and mine development infrastructure for transportation of men, materials, and broken ore (sometimes waste) to the surface for processing (or disposal). The main elements of the underground infrastructure are shown in Figure 11.8.

Between the main development levels, a mine can have sublevels that may be needed for more effective drill and blast control in the stopes. Levels and sublevels are mainly connected by inclined underground openings, called ramps, and by vertical openings, which include raises and winzes. Depending on access and other considerations, a raise is an underground opening driven upward and a winze is driven downward typically in the orebody. Raising tends to be faster as the broken ore removal is assisted by gravity.

A drive, or drift, is a horizontal or nearly horizontal underground opening developed on the underground levels along the strike of orebody. The drives are subdivided into hangingwall which is located at the upper ore-waste contact, hence hanging above the orebody, and footwall which is located at the lower ore-waste contact, hence at the foot of the orebody. A footwall drive is also commonly called an ore drive. In many hard-rock mines, they are developed in pairs: one called the haulage that is used for men, materials, and intake (fresh) ventilation and the other the Return Air Way (RAW).

A crosscut is a horizontal underground tunnel developed to intersect the orebody. The crosscuts are usually developed to connect the drives with the area in which stoping (ore production) occurs.

A stope is an underground excavation where the actual ore is produced. Development of stopes using massive mining methods often starts from blasting a slot, which is a steeply dipping excavation at one boundary of the planned orebody stope. Mining then continues by blasting rings or slices of the orebody into the slot.

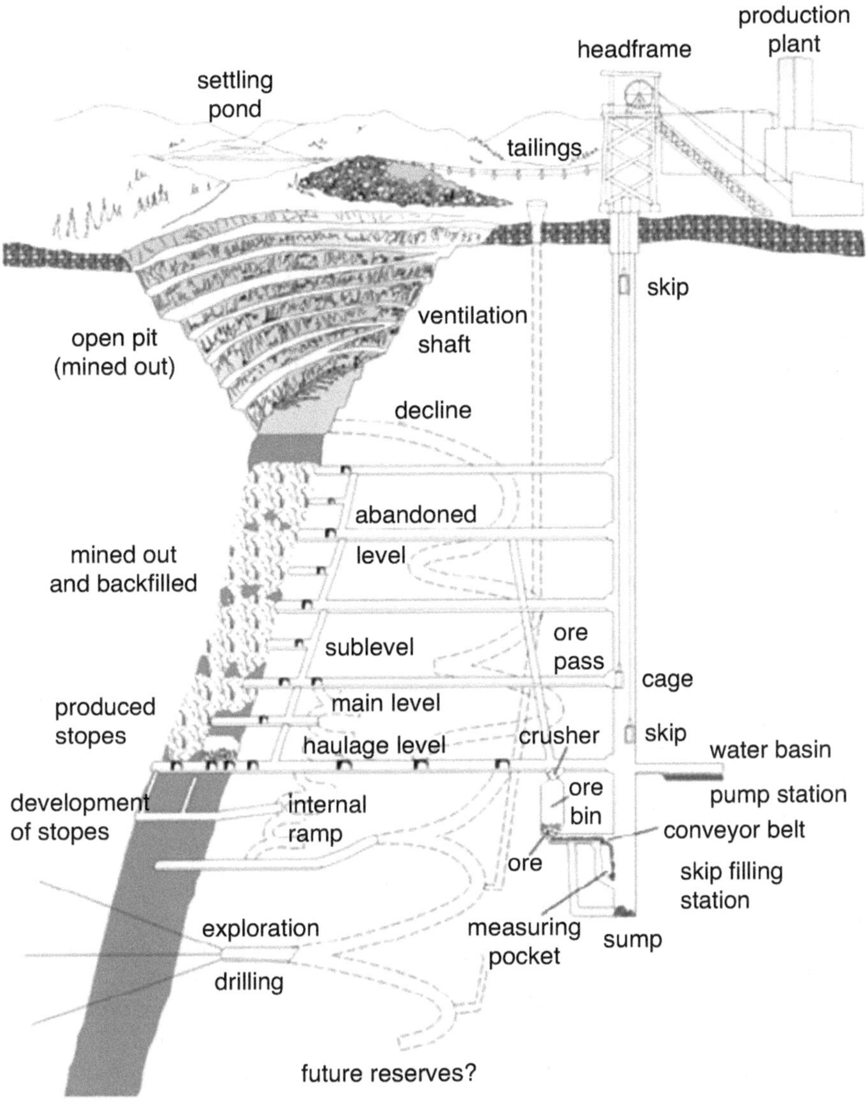

FIGURE 11.8 Basic infrastructure and terminology for an underground mine (Hamrin, 2001).

A pillar is a block of ore or barren rock left intact in the mined-out stope, between two stopes or between two drives to act as a regional support. It is required to provide structural integrity to the stoping process and prevent the stope walls from unplanned collapse. Pillars may be extracted after stopes are mined out, but some pillars may be left in place permanently to provide regional support.

A draw point is a place from which the ore is extracted from the stope and loaded onto trucks or conveyors for further transportation to surface.

An ore pass is a steeply dipping (+55°) underground opening for moving mainly ore from one level to a lower level under gravity and finally to the shaft bottom from where it is hoisted to surface. The ore is loaded through the chutes, which are the loading arrangements that utilize gravity. An important element of the ore loading and transportation system is a coarse steel grating, called a grizzly, for screening out oversize rock fragments that could block the ore passes.

In an underground mine, a significant amount of infrastructure must be built before mining begins, which requires a large capital investment. The development of a large underground mine can take as many as 5–10 years. Costs during this time will therefore be high and can comprise 30–40% of the pre-mining capital requirements before mining can start. In these cases, a dual feasibility study must be performed comparing the surface option to the best underground mining option. Generally, capital costs increase and operating costs decrease with increasing production tonnage (Atlas Copco, 2008).

An underground mine typically has a unique layout that ensures the cost-effective extraction of ore and the safety and movement of people and equipment. Therefore, each mining method requires different underground infrastructure such as access drifts to sublevels, drifts for longhole drilling, loading drawpoints, and ore passes. Together, they form an intricate network of interdependent excavations such as drifts, ramps, shafts, and raises. The mine requires three different physical installations (Atlas Copco, 2008):

- Surface process plant and infrastructure.
- Shaft and/or ramp infrastructure.
- Underground infrastructure.

The first consists of a variety of facilities to provide the mine with necessary services such as:

- Access roads and parking.
- Transportation facilities.
- Power and water supply.
- Service and maintenance buildings.
- Mineral processing plant.
- Bulk storage.
- Waste disposal facilities for air, water, and solids.

The shaft plant includes the facilities installed for material handling of ore and associated waste and the means of transport of miners and material. It generally incorporates systems for rock handling, ventilation, water handling, services and power supply, and communications.

The underground plant covers various installations to make the system work efficiently and safely, including storage bins (silos), loading pockets, rock-handling systems, power distribution equipment, underground maintenance facilities, and numerous other installations that provide auxiliary services to the underground operations.

Mine ventilation is one of the most important facilities of underground mining. Air quality in mine workings is an area of particular concern to the underground development. It must be maintained at an acceptable health standard. A continual and adequate supply of fresh air must be made available to working areas. The ventilation shafts are shown in Figure 11.9.

Underground mines use networks of fans, ventilation doors, and excavations to and from surface to move fresh air into the mine and remove the gassy and dusty exhaust air. High-pressure fans on surface extract exhaust air through the upcast shafts, and ventilation doors control the underground airflow, passing fresh air through active work areas. As most of the infrastructure is located on the footwall side of the orebody, the fresh air is normally channeled via the footwall toward the hanging wall, from where the exhaust air is routed to the surface. It is particularly important to clear the air after an underground blast, because harmful gases such as carbon monoxide or oxides of nitrogen can build up. An effective ventilation system will rapidly clear the air from a blast of dust and noxious gases (Atlas Copco, 2008).

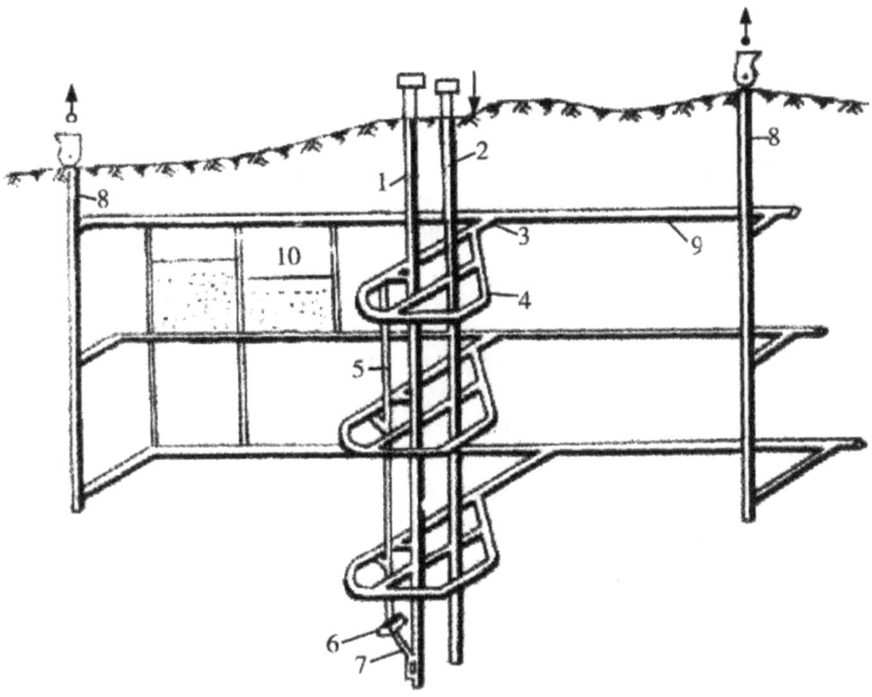

FIGURE 11.9 General layout of an underground mine development (Zhang and Wang, 2016): (1) shaft with rock-handling skips; (2) shaft with main cage for moving men, materials, and equipment; (3) main development crosscut; (4) shaft level development and crosscut; (5) waste rock or ore pass; (6) primary crusher; (7) ore bin; (8) ventilation shaft; (9) haulage level; and (10) stope (ore production area).

11.5 MAIN ACCESS PROTECTION

The inherent strength of the rock mass is generally too low for the safe and unsupported excavation of the rock mass in mines. In order to prevent deterioration and failure, the rock around the excavation must be reinforced (laterally confined) using pillars and support. The installed support is typically rockbolts and cablebolts often with liner support such as mesh (screen) or sprayed concrete (shotcrete).

The main access ramps, especially shafts, must be protected for the life of the mine. This can become a challenge as frequently the life of a mine increases significantly from the initial mine plan. This creates long-term stability issues in excavations due to corrosion and more rock deterioration over the now-extended time. The protection can require more than just installed support and may need pillars too. Even where shafts are completely stable, a cosmetic (concrete) lining may be installed to reduce friction losses in ventilation or to hang shaft hardware for hoisting, and support of utilities, for example. Shafts with hoisting need special protection from mining because movement, especially shear or tilt, cannot be easily accommodated while maintaining high and efficient hoisting speeds.

11.5.1 Shaft Protection in a Massive Orebody

If the orebody is massive, generally the shaft is sited in the footwall some distance from the deposit so that the mining-induced stresses do not cause stability issues in future. The exception to this is where the footwall strata is potentially unstable and the hangingwall is competent.

11.5.2 Shaft Protection in a Tabular Orebody

Tabular deposits need a unique approach, and there are only a few methods to protect a shaft from subsequent damaging effects induced by mining. These are as follows:

- Site the shaft away from the orebody
  - More costly and time-consuming access to the deposit and long transport distances over the mine life for men, materials, and rock.
- Leave a shaft pillar
  - Can become uneconomic especially at great depth.
  - Can become a serious stress concentrator toward the end of the mine life.
- Leave satellite pillars, which are large pillars strategically placed relatively close to the shaft (typically 4 pillars, 2 updip from the shaft, and 2 downdip).
- Mine out the orebody concurrent with the shaft sinking (early pillar removal). This involves sinking the shaft to the tabular orebody, stopping the sinking, removing the orebody symmetrically around the shaft and replacing it will stiff backfill. Sinking then continues to the final design depth. The method is briefly outlined by Spearing (1995) and an example given by Leach (1990).

For deep shaft pillar design, the main criteria to consider are the vertical strain, vertical tilt, and vertical stress. The following were postulated by Wilson and More O'Ferrall (1970) based on back analyses of shafts in hard-rock mines in South Africa:

- Shaft damage begins when the field stress exceeds a quarter of the Unconfined Compressive Strength (UCS) of the surrounding rock.
- When the field stress increases to about half the UCS, damage becomes extensive and remedial actions become very costly.
- When the field stress becomes two-thirds of the UCS, it is impractical to keep excavations open.

Salamon (1974) concluded that the maximum tolerable vertical strain was 1 millistrain, and Wagner (1980) reported a maximum tilt of 1 unit of horizontal displacement per 1,000 units of length This is useful for computer simulations to design shaft pillar protection systems, as it provides upper design limits to use for both vertical strain and tilt.

REFERENCES

Atlas Copco (2008) *Guide to Underground Methods and Applications*. Atlas Copco, Sweden.

Hamrin, H. (2001) Underground mining methods and applications. In: Hustrulid, W.A. and Bullock, R.L. (eds) *Underground Mining Methods: Engineering Fundamentals and International Case Studies*. Society for Mining Metallurgy and Exploration, Littleton, pp. 3–14.

Haviland, D. and Marshall, J. (2014) Fundamental behaviours of production traffic in underground mine haulage ramps. *Int. Jour. Min. Sci. & Tech.* 25(1), 6.

Kempson, W.J.; Webber-Youngman, R.C.W. and Meyer, J.P. (2015) Optimising shaft pressure losses through computational fluid dynamic modelling. *Appl. Therm. Eng.* 90, 1098–1108.

Kirsch, E.G. (1898) Die Theorie der Elastizität und die Bedürfnisse der Festigkeitslehre. *Zeitschrift des Vereines deutscher Ingenieure*. 42, 797–807.

Leach, A.R. (1990) *Early Shaft Reef Extraction - The Western Deep Levels South Approach to Shaft Protection*. ISRM International Symposium. Mbabane, Swaziland.

Moore, E. (2014) A cut above. CIM Magazine.

Ostrowski, W.J.S. (1972) Design considerations for modern shaft linings. *CIM Bullet.* 65(726), 58–71.

Peck, R.B. (1969) Deep excavations and tunnelling in soft ground: state of the art report. *Proceedings of the 7th International Conference on Soil Mechanics and Foundation Engineering*, Mexico, pp. 225–290.

Purkey, B.W. and Garside, L.J. (1995) *Geologic and Natural History Tours in the Reno Area, Nevada Bureau of Mines and Geology Special Publication 19*. University of Nevada, Nevada.

Salamon, M.D.G. (1974) Rock mechanics of underground excavations. Advances in rock mechanics. In *Proceedings of 3rd Congress of the International Society of Rock Mechanics*. Denver, CO, vol. 1b, pp. 91–1099.

Spearing, A.J.S. (1995) *Handbook on Hard Rock Strata Control*. SAIMM, SP-6, Johannesburg, pp. 114–115.

Spearing, A.J.S., Ma, L. and Ma, C-A. (2023) *Mine Design, Planning and Sustainable Exploitation in the Digital Age*. CRC Press, London, UK.

Talobre, J.A. (1957) *La Mecanique des Roches* (1st Edition). Dunod, Paris, pp. 228–233.

Tatiya, R.R. (2005) *Excavations in Upward Direction – Raising. Chapter 13. Surface and Underground Excavations – Methods, Techniques and Equipment*. CRC Press, Boca Raton, FL, pp. 331–349.

Terzaghi, K. (1943) *Theoretical Soil Mechanics*. John Wiley & Sons, Inc., New York.

Wagner, H. (1980) Pillar design in coal mines. *J. S. Afr. Inst. Min. Metall.* 80(1), 37–45.

Wilson, J.W. and More O'Ferrall, R.C. (1970) The application of the electrical analogue to mining operations. *J. S. Afr. Inst. Min. Metall.* 70(6), 115–148.

Zhang, Q. and Wang, X. (2016) *Underground Mining Technics of Metallic Ore Deposits*. University of Central South Press, Changsha, China.

12 Slope Stability

12.1 THE USE OF SLOPES

Slopes are common in mining. Earth slopes are associated with strip and open pit mines, highwalls, waste piles, and tailings dams, whereas rock slopes are associated with strip and open pit mines as well as access roads. Slope failure is a potential issue as shown in Figure 12.1.

Slope stability is important for obvious safety concerns and costs. To reduce costs, slopes should be generally as steep as possible to involve the least waste removal and minimize the footprint (land usage) while ensuring stability during the useful life of the slope. There must also be a long-term reclamation plan to protect the environment and the safety of the local community.

12.2 SLOPE FAILURES AND CAUSES

There are basically three main types of failure:

- Falls are characterized by a rapid movement of material. In general, they can seldom be reliably and easily analyzed.

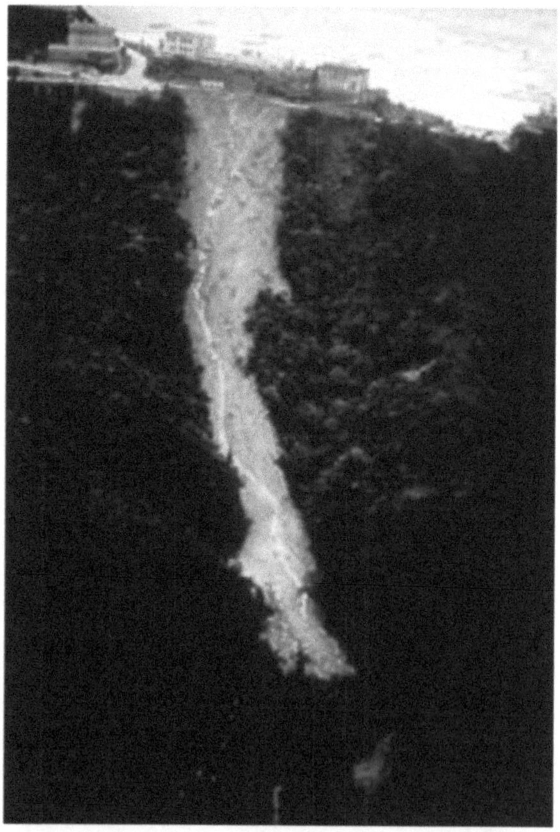

FIGURE 12.1 Massive slope failure in Australia.

Photograph reproduced with kind permission of Steven Hencher.

DOI: 10.1201/9781032704845-12

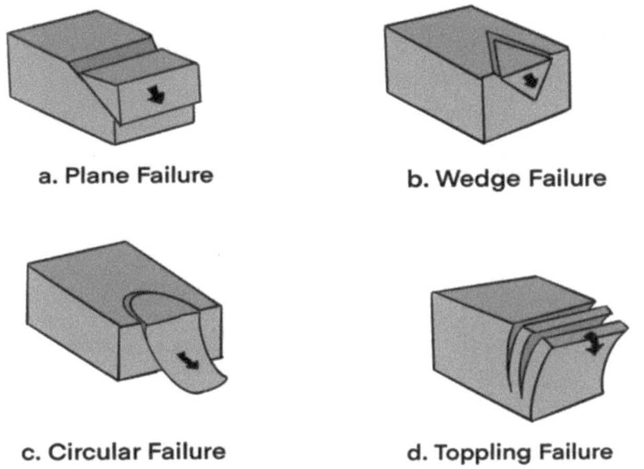

FIGURE 12.2 Simple diagrammatic representation of the common slope failure modes (Ali, 2017).

- Slides are generally induced by shear failures along one or more surfaces. Local geology plays a major role. A layer of weak stratum, e.g., clay or slickensided surfaces, often constitutes the failure surface and water can play a significant role.
- Flows are characterized by movements within the displaced mass, much like the behavior of a viscous fluid. Flows generally occur in soils with high water content but can also occur in dry conditions.

Figure 12.2 shows the common types of falls, slides, and flows.

Terzaghi (1943) divided the causes of slope failures into "internal" and "external" causes:

- Internal causes: The failure occurs without external change (including earthquakes) and is usually a result of increased pore water pressure.
- External causes: The shearing stress is increased without the shearing resistance of the material changing. This shear stress increase can typically be caused by steepening the slope angle, increasing the slope height, erosion, or some combination thereof.

Table 12.1 lists common causes for slope instability.

12.3 SLOPE DESIGN ANALYSES AND METHODOLOGIES

12.3.1 Slope Stability Analysis

Slope stability analysis is based on assumptions that were developed over a century ago by researchers including Petterson (1916, 1941), Fellenius (1936), and Terzaghi (1943) and include the following:

- Slope failure occurs along a particular sliding surface, and thus, the failure can be assumed to be a two-dimensional plane problem.
- The failure mass moves as a rigid body and any deformations, if they occur, do not influence the analysis.
- The shearing resistance of the soil mass at points along the failure surface is independent of the orientation of the failure surface (strength properties are isotropic).
- The factor of safety is defined in terms of the average shear stress along the potential failure plane and the average shear strength along the plane, rather than specific values at specific locations.

TABLE 12.1
Contributing Factors of Slope Failures

Main Factor	Specifics
Environmental factors	• Rainfall, snow, or drought (pore water) • Vegetation (density and type) or the lack of it
Geological factors	• Material types and properties • Local topography • Material grading (particle size) • Faulting in the area • Seismic risk (blasting or rockbursts) • Weathering • Hydrology
Human factors	• Poor design (e.g., poor drainage) • Mining or construction in the area

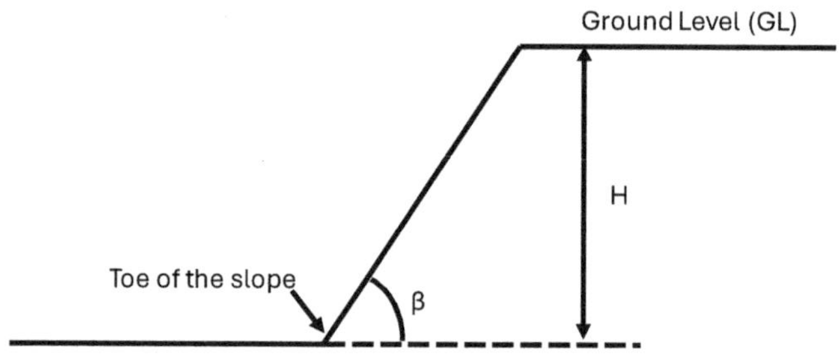

FIGURE 12.3 A typical slope with the main design parameters.

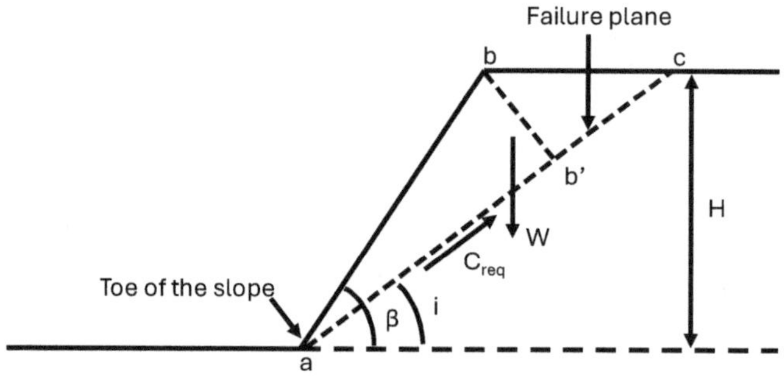

FIGURE 12.4 Slope with a potential failure plane.

A typical simplified slope stability geometry is given in Figure 12.3.

In slopes, the slope angle (β) and slope vertical height (H) are important parameters.

12.3.2 Slope Stability in Cohesive Materials

A purely cohesive material is one where the cohesive strength is non-zero and the angle of internal friction (φ) is zero. Consider Figure 12.4.

C_{req} is the cohesive (resistive) force on the potential failure surface required to maintain equilibrium (i.e., prevent sliding failure).

Based on a statics analysis,

$$C_{req} = W\sin(i) \tag{12.1}$$

The weight of the failure block W is

$$W = 1/2\gamma[H/\sin(i)][H/\sin(\beta)][\sin(\beta - i)] \tag{12.2}$$

where
- γ is the material weight density.
- $[H/\sin(i)]$ is the length ac.
- $[H/\sin(\beta)]$ is the length ab.
- $[\sin(\beta - i)]$ represents (bb′/ab).

If the average shearing stress is taken as s, along the failure surface, then

$$C_{req} = s[H/\sin(i)] \tag{12.3}$$

By substituting Equation 12.2 into Equation 12.1, the following is obtained:

$$s[H/\sin(i)] = 1/2\left\{\left(\gamma H^2\right)/[\sin(i)\sin(\beta)]\right\}[\sin(\beta - i)][\sin(i)] \tag{12.4}$$

Equation 12.4 can be simplified to

$$s/(\gamma H) = \{[\sin(i)\sin(\beta - i)]/2\}\sin(\beta) \tag{12.5}$$

$s/(\gamma H)$ is called the stability number and is dimensionless. It is the ratio of the cohesive component of shearing resistance to γH that is required to maintain stability (i.e., factor of safety of 1).

From a slope stability point of view, therefore, the most critical plane is that where $(s/\gamma H)$ is a maximum. Differentiating with respect to angle i, and equating to zero give

$$\partial/\partial i(s/\gamma H) = \partial/\partial i[\sin(i)\sin(\beta - i)] = 0 \tag{12.6}$$

from which, i = ¼ tan (β/2).

Therefore,

$$s/(\gamma H) = 1/4\tan(\beta/2) \tag{12.7}$$

where
- s is the average unit shear strength along the failure surface in lbs/ft^2.
- γ is the material weight density in lbs/ft^3.
- H is the vertical height of the slope in ft.

Equation 12.7 can therefore be used to find the maximum slope H with a given "s" and slope angle β or a maximum angle β for a given "s" and H.

12.3.3 Slope Stability in Solid Rock Masses

The stability of highwall slopes in rocks is not generally a problem unless major steeply dipping joints are present, in which case, the highwall should be created to match the angle of the major joint set. If the rock has fine intermixed material, such as soil or clay, then the slope will behave more like a soil requiring the more conservative design.

12.3.4 Slope Stability in Broken Rock

Broken rock (spoil) will naturally stand at the angle of repose of the material. An important factor in rock piles (as opposed to the rock mass) is that the stability is independent of the height of the rock pile. The typical angle of repose for broken rocks varies from 35° for fine sands, 40° for coarse sand to 45° for angular rock (Lachman and Subramanyam, 2017).

12.3.5 Slope Stability in Soils (D. Taylor, *Fundamentals of Soil Mechanics*, J Wiley, 1948)

The failure surface of a typical rotational failure can be approximated as a rotational movement (collapse) of a circular arc as shown in Figure 12.5.

Figure 12.5a shows the potential rotational sliding mass and its center of rotation (c.g. is center of gravity), Figure 12.5b shows how the sliding mass could fail only partially, and Figure 12.5c shows the total (maximum) rotational failure.

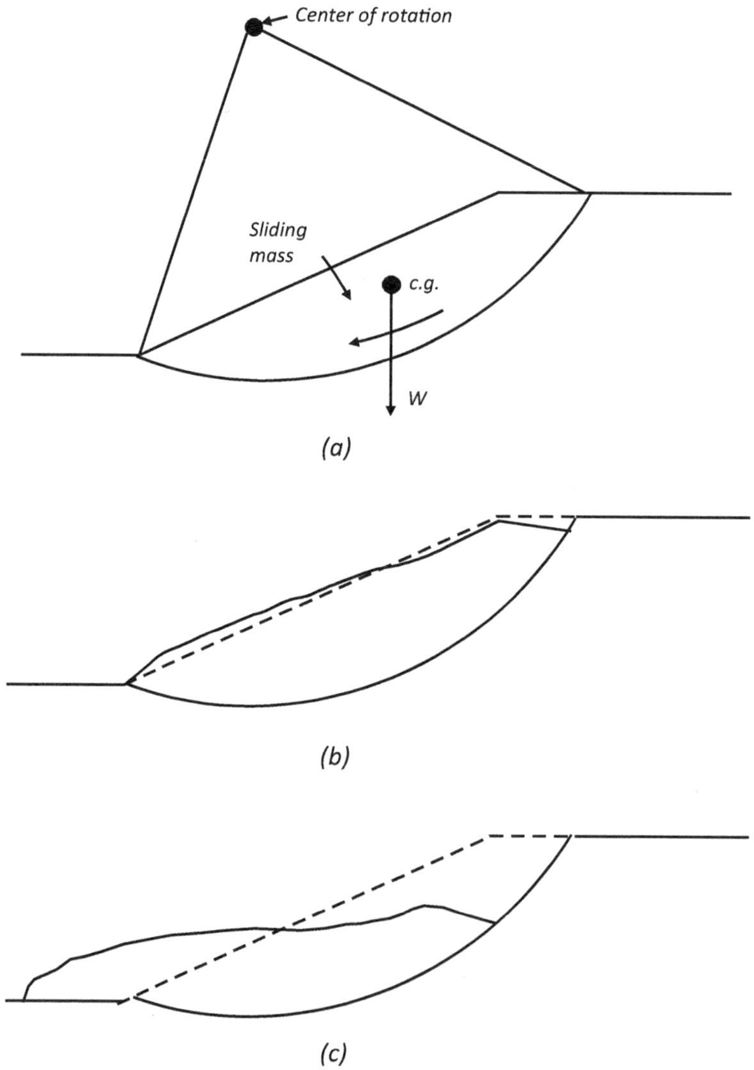

FIGURE 12.5 Rotational slope failure (Taylor, 1948).

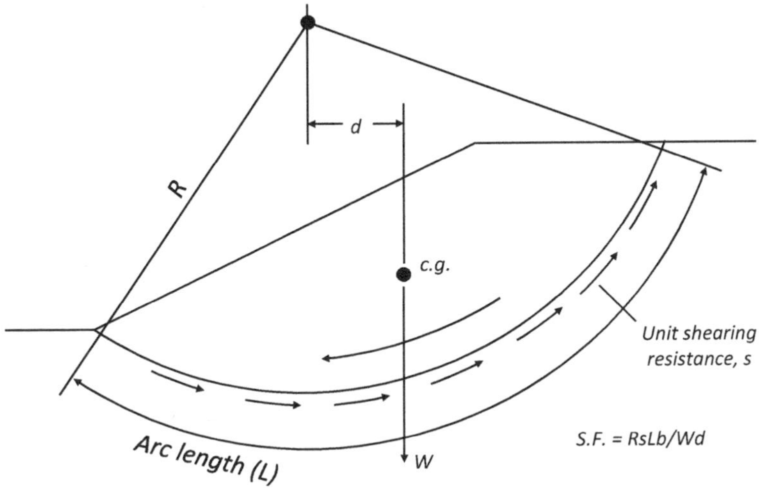

FIGURE 12.6 Worst-case rotational failure slope (Taylor, 1948).

Note: cg is the center of gravity of the potential failure volume.

To analyze a planned or existing slope before failure, a worst-case failure surface is assumed as shown in Figure 12.6.

The failure surface as shown in Figure 12.6 is also assumed to be in a uniform (consistent) soil. Generally, the failure mass is assumed to have unit thickness.

The factor of safety:

$$SF = \left(RsL'b\right)/\left(\gamma Abd\right) = \left(RsL'b\right)/\left(Wd\right) \tag{12.8}$$

where

R is the radius of the failure surface or moment arm of resisting forces.
s is the average unit shearing strength.
L′ is the arc length of the sliding/failure surface.
b is the unit thickness (generally unity).
γ is the soil unit weight.
A is the sliding mass area.
d is the moment arm of the sliding mass.
W is the weight of the sliding mass with thickness/depth of b.

This type of analysis only establishes the factor of safety for the assumed failure surfaces in the analysis, and in practice, there are many potential failure surfaces that are considered to find the least stable potential failure surface. The failure surface of interest is the one that gives the lowest factor of safety which is found usually by trial and error.

A typical example is shown in Figure 12.7. Assume that a failure surface (A) was first assumed, and the factor of safety was found to be 1.24. Another failure surface (B) to the right of (A) is now selected, and the factor of safety is found to be lower at 1.14. A third failure surface (C) is selected even further to the right, and the factor of safety is found to have increased to 1.27. The failure surface (B) is therefore assumed to be the potential failure surface, and the slope is assumed to have a factor of safety of 1.14 and therefore stable (being larger than 1.0).

The method described below can therefore be used to find the potential failure surface by this trial-and-error method. The method is obviously highly simplified, and apart from the slope material often being variable, any water presence is a major additional issue.

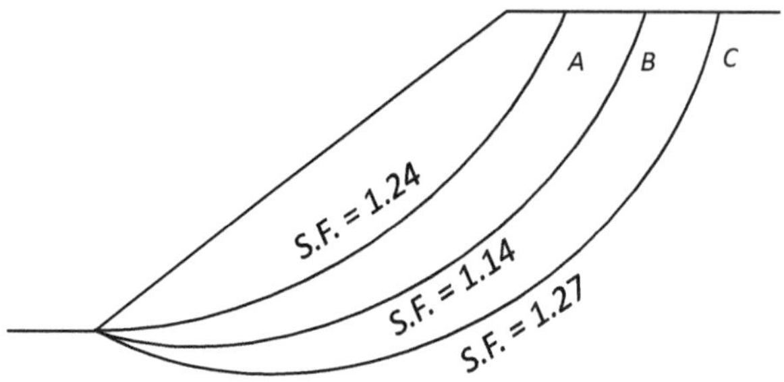

FIGURE 12.7 Different potential rotational failure surfaces (Taylor, 1948).

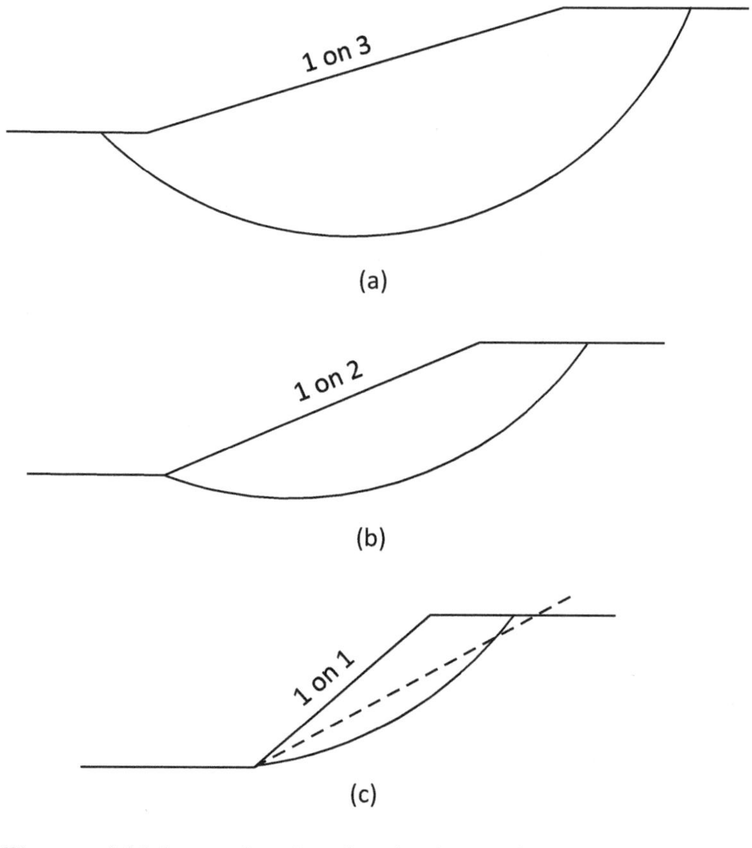

FIGURE 12.8 The potential failure surfaces based on the slope angle (Taylor, 1948).

The potential failure surface also changes with the slope dip as shown in Figure 12.8:

- Flat slopes of 1:3 vertical to horizontal (18.5°) have the worst circle/surface with high curvature dipping down into the slope base (Figure 12.8a).
- Moderate slopes of 1:2 vertical to horizontal (21.8°) to 1:1.5 ratio (34.7°) have the worst circle with less curvature passing through the slope toe (Figure 12.8b).
- Steep slopes (+34.7°) that have the worst circle is flat and can be approximated as a plane surface (as shown by a dotted straight line in Figure 12.8c).

12.3.6 Pore Water Effects

Water always exacerbates stability problems. For slope stability, water causes issues for the following reasons:

- The weight is increased by increasing water, and the effective weight affects both the driving and resisting forces.
- The cohesion is generally adversely affected (reduced) with the increasing presence of water.
- Increased pore water pressure, whether greater or less than atmospheric, significantly changes the normal stress and shearing resistance of the material due to internal friction changes.

12.3.7 Other Considerations

The condition of the slope base, sometimes referred to as the floor, is important, and the presence of weak material, such as clays, sands, and water, can cause serious foundation problems which must be considered.

As a general caution, if seismicity is a possibility near any slope, then the static solutions presented above will not be applicable and a different design approach is needed.

The stability of a potential slope failure surface can be increased by using:

- Tiebacks with screen, with the three main types of tiebacks being soil nails, grouted and mechanical tiebacks. Their purpose is to provide lateral confinement to the slope.
- Berms which can be effective in reinforcing the base at the toe (floor) of a slope with an example shown in Figure 12.9.
- Benching effectively reduces the angle of a slope but involves additional, and considerable, earth moving, especially for large pits. An example is shown in Figure 12.10.
- Dewatering can generally be effective in reducing the risk of a slope failure by reducing the pore water pressure. This is not always the case, especially if the slope is steep, weak, and saturated. Methods of dewatering include the following:
  - Sump pumping involves creating a sump typically at the toe of the slope and pumping groundwater from the slope out of the sump. It is not as effective as other methods.
  - Well points are a series of small-diameter wells strategically positioned above the slope connected to a pipe-and-pump system to lower the water table.
  - Eductor wells are similar to well points and typically consist of a series of small-diameter wells using a closed-loop pressurized system that uses a high-pressure venturi

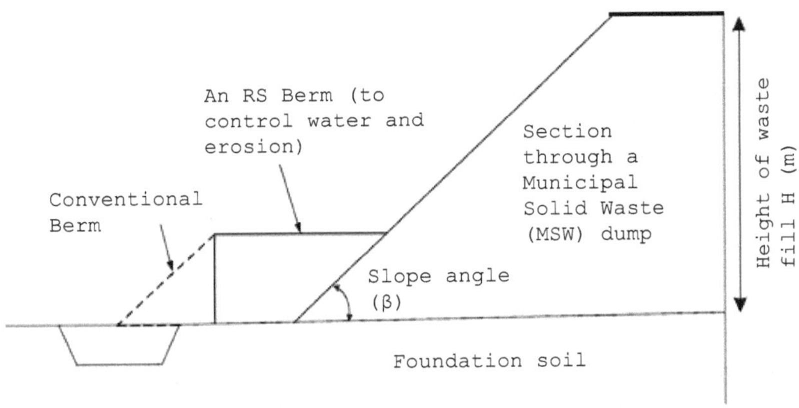

FIGURE 12.9 The use of a berm to help stabilize the toe of a slope (Agarwal et al., 2018).

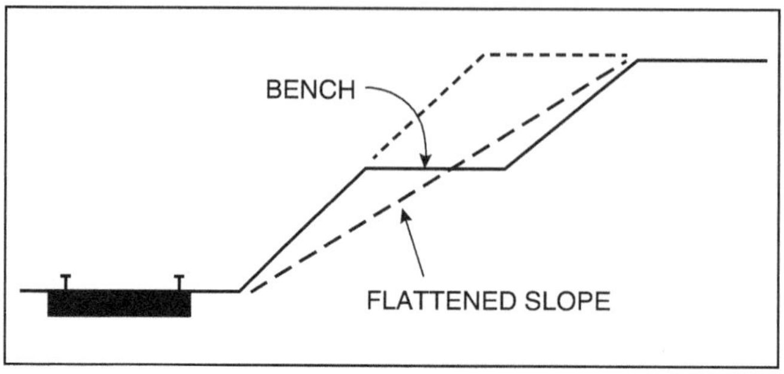

FIGURE 12.10 The use of benches to reduce the overall slope angle (Dey et al., 2010).

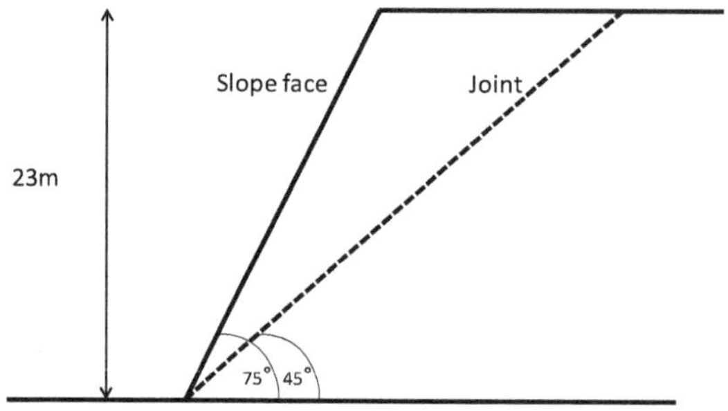

FIGURE 12.11 The bench wedge stability problem.

to create a low-pressure vacuum zone at the foot of each well. The vacuum draws in the groundwater through a valve and out of the well.

- Wick drains are used to remove pore water in loosely compacted soils and are generally long vertical drains.
- Sand drains are installed typically in predeveloped vertical holes and filled with a high-void-ratio sand.

Monitoring is a vital part of slope stability and can be achieved using the following, or a combination of the following:

- Visual inspections, or the use of drones, which can be effective.
- Suitably placed piezometers or wells to measure both the absolute pore water pressure and any changes.
- Tiltmeters, sometimes referred to as inclinometers.
- Remote sensing using radar or LiDAR, for example.
- Field tests to ensure that the correct input data were used for the initial slope design, which could include penetration tests, shear vane tests, and dilatometer tests.
- Real-time monitoring is important.

12.4 WORKED EXAMPLES

Example 12.4.1

The following example considers bench wedge stability as shown in Figure 12.11. If the cohesion and friction angle of the joint surface are estimated as 69 kPa and 35°, and the rock density is 2,560 kg/m^3, what is the weight of a 1 m section of the triangular rock resting on the joint plane?

Solution:

The block weight (W) for a simple block with no tension crack is:

$$W = H^2(\rho_r/2)\left[(\cos\psi_p/\sin\psi_p) - (\cos\psi_f/\sin\psi_f)\right]$$

where
- H is the vertical slope height.
- ρ_r is the rock density.
- ψ_p is the failure angle.
- ψ_f is the slope face angle.

$$W = H^2(\rho_r/2)\left[(\cos\psi_p/\sin\psi_p) - (\cos\psi_f/\sin\psi_f)\right]$$

$$W = 23^2(2{,}560/2)[(\cos 45/\sin 45) - (\cos 75/\sin 75)]$$

$$W = 677{,}120[1 - 0.2680]$$

$$W = 495{,}650 \text{ kg}$$

$$W = 496\text{t}$$

Example 12.4.2

A reclaimed waste rock dump consists of cohesionless material with a specific gravity of 2.0 and an internal friction angle of 37°. What is the factor of safety of the dump if it has a uniform slope angle of 2.67H:1V and it is saturated?

Solution:

For a saturated, cohesionless soil, the effective normal stress is reduced by the pore water pressure, and the resulting safety factor is:

$$SF = \left[1 - (\rho_w/\rho_r)\right](\tan\phi/\tan\psi)$$

$$SF = [1 - (1/2)][(\tan 37/\tan 20.5(1/2.67)]\tan\psi = 1/2.67 = 20.5$$

$$SF = (0.5)[(0.754)/(0.375)]$$

$$SF = 1.0$$

Example 12.4.3

For a vertical slope, where β is 90°, s is 500 psf, and γ is 110 lbs/ft³, find the maximum height H.

Solution:

$$s/(\gamma H) = 1/4\tan(\beta/2)$$

$$\left[500 \text{ lb/ft}^2\right]/\left(110 \text{ lb/ft}^3\right)H = 1/4\tan(90/2)$$

$$\left[500 \text{ lb/ft}^2\right]/\left(110 \text{ lb/ft}^3\right)H = 0.25$$

$$\left[500 \text{ lb/ft}^2\right] = \left(27.5 \text{ lb/ft}^3\right)H$$

$$H = 18.2 \text{ ft}$$

Example 12.4.4

For a slope underlain by clean, dry, cohesionless sand, the following expression for determining the factor of safety with respect to sliding is provided by Terzaghi et al. (1996), where φ' represents the angle of internal friction for sand in a loose state:

$$F = \tan(\phi')/\tan(\beta)$$

Part a:

Determine the factor of safety, F, for a sand slope possessing an angle of internal friction of 35°. The slope makes an angle of roughly 20° with the horizontal.

Solution:

$$F = \tan(\phi')/\tan(\beta)$$

$$F = \tan(35°)/\tan(20°)$$

$$F = 1.92$$

Part b:

Determine the maximum slope angle for sand with a friction angle of 30° whereby a factor of safety of 1.5 must be maintained.

Solution:

$$F = \tan(\phi')/\tan(\beta)$$

$$1.5 = \tan(30°)/\tan(\beta)$$

$$\tan(\beta) = 0.384$$

$$\beta = \tan^{-1}(0.384)$$

$$\beta = 21°$$

Part c:

How does height of the slope affect the solution in part b?

The solution is unaffected by the height of the slope. However, the slope angle, β, must always be less than the angle of internal friction to maintain a factor of safety greater than unity.

Example 12.4.5

The following example can be found in Das (2006) and applies to an infinite slope. Figure 12.12 is redrawn and modified from the example.

Soil Properties:

$$\gamma = 100 \text{ pcf}$$

$$c' = 200 \text{ psf}$$

$$\phi' = 15°$$

Slope Properties:

$$\beta = 25°$$

$$H = 8 \text{ ft}$$

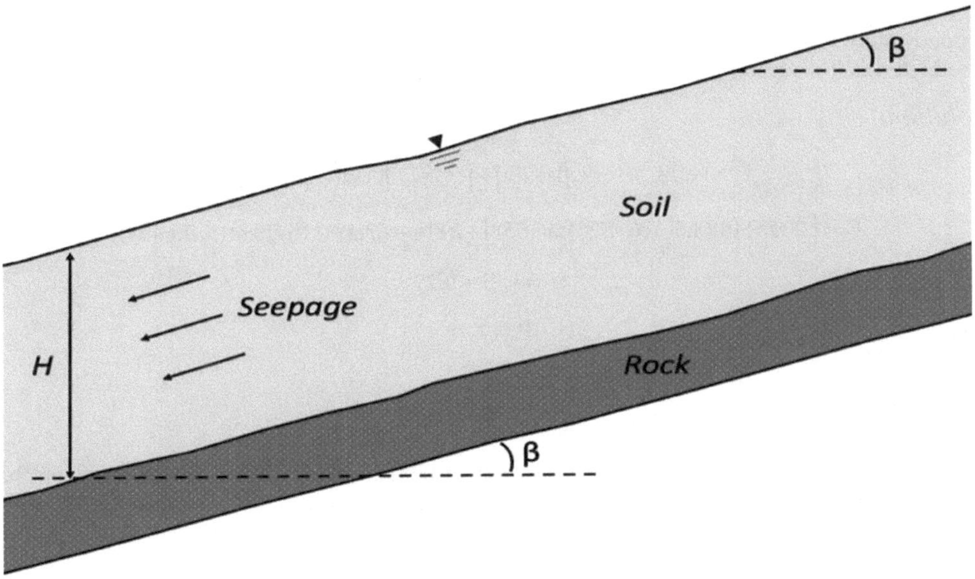

FIGURE 12.12 A section outlining the problem.

Part a

For the infinite slope shown, determine the factor of safety against sliding along the soil–rock interface.

Solution:

$$F_s = \left[c'/\gamma H \cos^2 \beta \tan \beta\right] + \tan(\phi')/\tan(\beta)$$

$$F_s = \left[(200\ \text{psf})/(100\ \text{pcf})(8\ \text{ft})\cos^2(25°)\tan(25°)\right] + \tan(15°)/\tan(25°)$$

$$F_s = [(200\ \text{psf})/(306\ \text{psf})] + 0.575$$

$$F_s = 1.23$$

Part b

For the infinite slope shown calculate the height (H) that will give a factor of safety (F_s) of 2 against sliding along the soil–rock interface.

Solution:

$$F_s = \left[c'/\gamma H \cos^2 \beta \tan \beta\right] + \tan(\phi')/\tan(\beta)$$

$$2 = [(200\ \text{psf})/(100\ \text{pcf})(H)\cos^2(25°)\tan(25°)] + \tan(15°)/\tan(25°)$$

$$2 = [(200\ \text{psf})/(38\ \text{pcf})(H)] + 0.575$$

$$2 = (5.33\ \text{ft})/(H) + 0.575$$

$$H = 3.66\ \text{ft(with rounding)}$$

Example 12.4.6

The following is a continuation of Example 12.4.5 after Das (2006).

Referring to Figure 12.12 and Example 12.4.5, if there is seepage through the soil and the groundwater table exists up to the ground surface, what is the factor of safety for H = 3.66 ft and a saturated unit weight of soil of 118 pcf?

Solution:

$$F_s = \left[c'/\gamma_{sat} H \cos^2\beta \tan\beta\right] + \left[\gamma'/\gamma_{sat}\right]\left[\tan(\phi')/\tan(\beta)\right]$$

$$F_5 = \left[200/(118)(3.66)\cos^2(25)\tan(25)\right] + [(118 - 62.4)/(118)][\tan(15)/\tan(25)]$$

$$F_s = 1.21 + 0.27$$

$$F_5 = 1.48$$

Example 12.4.7

The following example is provided but unsolved in Hough (1957). The problem statement has been summarized for succinctness. For an unretained earth slope with an inclination of 45°, determine the factor of safety for the slope given the following input parameters:

$$R = 30 \text{ ft}$$

$$s = 0.2 \text{ tons/ft}^2$$

$$L = 50 \text{ ft}$$

$$b = 1 \text{ ft}$$

$$d = 10 \text{ ft}$$

$$W = 43{,}000 \text{ lbs/ft of depth } (\text{perpendicular to the cross-section})$$

Solution:

$$SF = (Rs\acute{L}b)/(Wd)$$

$$SF = \left[(30 \text{ ft})(0.2 \text{ tons/ft}^2)(2{,}000 \text{ lbs/1 ton})(50 \text{ ft})(1 \text{ ft})\right]/[(43{,}000 \text{ lbs/ft})(1 \text{ ft})(10 \text{ ft})]$$

$$SF = (600{,}000 \text{ ft} - \text{lbs})/(430{,}000 \text{ ft} - \text{lbs})$$

$$SF = 1.4$$

Example 12.4.8

The following example is provided but also unsolved in Hough (1957). The problem statement has been summarized for succinctness. For an unretained earth slope with a factor of safety of 1.2 against a rotational mass slide, find the average value of the potential unit shearing resistance in lbs/ft² given the following:

$$SF = 1.2$$

$$b = 1 \text{ ft}$$

$$Wd = 630{,}000 \text{ ft} - \text{lbs}$$

$$L = 45 \text{ ft}$$

$$R = 27 \text{ ft}$$

$$SF = (Rs\ Lb\ )/(Wd)$$

$$1.2 = [(27 \text{ ft})(s)(45 \text{ ft})(1 \text{ ft})]/[(630{,}000 \text{ ft} - \text{lbs})]$$

$$1.2 = \left(1{,}215 \text{ ft}^3\right)(s)/(630{,}000 \text{ ft} - \text{lbs})$$

$$756{,}000 \text{ ft} - \text{lbs} = \left(1{,}215 \text{ ft}^3\right)(s)$$

$$s = 622 \text{ lbs/ft}^2$$

REFERENCES

Agarwal, A., Datta, M., Ramana, G.V. and Satyakam, R. (2018) Improving the slope stability of municipal solid waste dumps using reinforced soil berms: a case study. In *Proceedings of the 8th International Congress on Environmental Geotechnics* Volume 2, Springer, Singapore.

Ali, S. (2017) *Types of Slope Failure*. University of Azad Jammu and Kashmir Muzaffarabad, Pakistan. www.slideshare.net/slideshow/types-of-slope-failures-75401227/75401227

Das, B.M. (2006) *Principles of Geotechnical Engineering* (6th Edition). Nelson Engineering, New York.

Dey, A., Basudhar, P., Ghosh, P, Kolathayar, S. and Nainegali, L. (2010) *Reinforced Earth Design of Embankments and Cuts in Railways*. Research Designs and Standards Organization (RDSO), Lucknow.

Fellenius, W. (1936) Calculation of stability of earth dams. In: TRANS. 2nd Congresso on large dams, Washington, DC, vol. 4, p. 445.

Hough, B.K. (1957) *Basic Soils Engineering*. Ronald Press, New York.

Lachman, L. and Subramanyam, C.V.S. (2017) Practical: Determination of angle of repose, Carre's Index & Hausner Ratio of given powder. Knowledge of Pharma. Retrieved from: www.knowledgeofpharma.blogspot.com/2017/07/practical-determination-of-angle-of.html

Petterson, K.E. (1916) The quay slide in Gothenburg. *Teknisk Tidskrift*. 46, 471–478.

Petterson, K.E. (1941) A special foundation method for a structure in Gothenburg harbour. *Teknisk Tidskrift*. 71, 45–52.

Taylor, D.W. (1948) *Fundamentals of Soil Mechanics*. John Wiley & Sons Inc., New York.

Terzaghi, K. (1943) *Theoretical Soil Mechanics*. John Wiley & Sons Inc., New York, 510 p.

Terzaghi, K., Peck, R.B. and Mesri, G. (1996) *Soil Mechanics in Engineering Practice*. John Wiley & Sons, New York.

13 Tailings Disposal

13.1 THE NEED FOR TAILINGS DISPOSAL

All mines, whether surface or underground, typically produce a finely ground waste material called tailings, which is produced after the mineral, or metal, has been removed during the metallurgical process. These waste materials can be disposed of in the following ways:

- Into the sea or a lake
    - Presently unfavorable due to environmental considerations.
- Into a river behind a suitable dam wall
    - Can be a costly method.
    - Have adverse environmental considerations.
- Onto a specifically constructed tailings dams on surface, with suitable liners.
- Placing the waste back underground as a backfill material
    - Cannot be used to remove all the tailings produced, and some surface disposal is still required.

Long-term safety must always be the key and the overriding consideration.

Figure 13.1 shows a typical large hard-rock tailings disposal dam.

Tailings disposal (storage) needs are unique to a specific site and mineralogy. Considerations include the following:

- Safety during the construction and use of the disposal facility.
- Post-mine closure safety.
- Size and nature of the orebody.
- Surface topography.

FIGURE 13.1 Tailings dam. (Reproduced with kind permission of Impala Platinum Holdings Ltd.)

DOI: 10.1201/9781032704845-13

- Earthquake or rockburst potential.
- Best practices and local regulations.
- Community engagement.
- Sound operational management and monitoring.
- Mining method.
- Relative location of the tailings facility and all mining excavations.

Poor design, inadequate tailings dam management and operation, and changes in weather are the main causes of failures.

13.2 PLANNING AND ENGINEERING CONSIDERATIONS FOR TAILINGS DAM DESIGN

There are many considerations when planning a tailings facility especially with a new mining operation or in the case of major expansion. Some of the considerations for planning a tailings dam or basin are as follows:

- Life and size of the facility.
- Water reuse needs.
- Environmental issues.
- Land availability and topography.
- Cost of construction and operation.
- Reclamation.

Some of the considerations for dam design, once the location is selected, are as follows:

- Spillway location, type, and size.
- Water reclaim and reuse from the facility.
- Availability of materials for dam construction.
- Controlling seepage (dam failure and pollution).
- Foundations.
- Abutment walls (berms).
- Operations including surface and underground for the life of mine.

The considerations are all aimed at short- and long-term stability, avoiding the obvious most significant hazard of liquefaction and failure.

13.3 TAILINGS DISPOSAL METHODS

There are several common methods of tailings dam construction that are used. The selected method is usually a function of the following:

- Terrain.
- Operations.
- Availability of materials.
- Cost.

The common methods include the following:

- Conventional method.
- Downstream method.
- Upstream method.
- Centerline method.

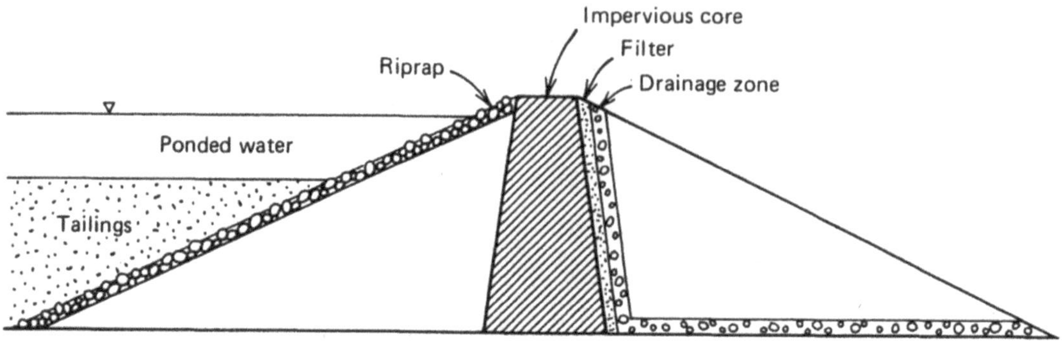

FIGURE 13.2 Example of a conventional earth dam retaining wall (Vick, 1990).

13.3.1 Conventional Method

In this method, a conventional earth dam wall is constructed in a suitable valley or ravine and the tailings stored behind the single wall. An example of a conventional earth dam wall is given in Figure 13.2.

According to Vick (1990), because construction expenditures are distributed over the life of the impoundment, initial project development costs are reduced to those necessary for construction of the starter dyke. Spacing expenditures over a longer time results in a lower discounted total cost and produces cash-flow benefits that are often important in financial considerations related to mine startup. Also, because the total volume of fill required for the ultimate embankment need not be available initially, there can be much more flexibility in the selection of materials for embankment construction.

13.3.2 Downstream Method

Over time, the volume of tailings stored increases. Therefore, the retaining wall height is increased, and the center of gravity of the wall moves away from the center of the dam, i.e., down dip.

According to Engels and Dixon-Hardy (2007), the downstream design was developed to reduce the risks associated with the upstream design (discussed next), particularly when subjected to dynamic loading as a result of earthquake shaking (ICOLD and UNEP, 2001). The installation of impervious cores and drainage zones can also allow the impoundment to hold a substantial volume of water directly against the upstream face of the embankment without jeopardizing stability.

The downstream embankment design starts with an impervious starter dyke unlike the upstream design that has a pervious starter dyke. The tailings are at first deposited behind the dyke, and as the embankment is raised, the new wall is constructed and supported on top of the downstream slope of the previous section (Figure 13.3). This shifts the centerline of the top of the dam downstream as the embankment stages are progressively raised (Vick, 1990).

An advantage of the downstream design is that the raised sections can be designed to be of variable porosity to address any problems with the phreatic surface relative to the embankment. This can be useful if a processing plant has made significant changes to increase extraction, and hence, the tailings characteristics are changed (especially the particle size distribution). This could result in either having more water in the tailings or altering the drainage characteristics of the tailings.

13.3.3 Upstream Method

This is the opposite of the downstream method as the centerline of the wall moves into the dam as the wall height increases.

According to Engels and Dixon-Hardy (2007), the upstream method has the lowest initial cost and the most popular design for a raised tailings embankment. One of the main reasons for this is

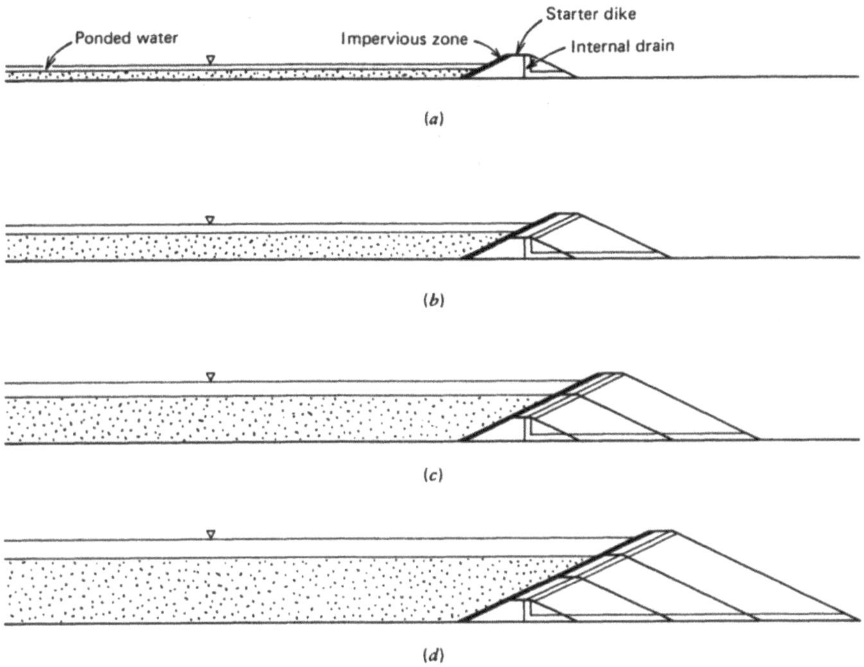

FIGURE 13.3 Downstream method (Vick, 1990).

the minimal amount of fill material required for initial construction and subsequent raising which normally consists entirely of the coarse fraction of the tailings.

The construction of an upstream-designed embankment starts with a pervious (free-draining) starter dyke foundation. The tailings are usually discharged from the top of the dam crest creating a beach that becomes the foundation for future embankment raises (Vick, 1990).

Figure 13.4 shows a simplistic diagram of the stages of construction of an upstream-raised embankment. Where the tailings properties are suitable, natural segregation of coarse material settles closest to the spigot and the fines furthest away. Cyclones can be used to accelerate this particle segregation for certain tailings characteristics to send the slime proportion to the center of the impoundment and the sand fraction to the beach behind the crest. The conventional method of upstream raises relies on no compaction of the spigotted beach that forms the embankment shell (Martin and McRoberts, 1999).

Today compaction by earthmoving equipment is common to increase the degree of safety of raised embankments. Generally, the settled coarse fraction from the spigots/discharge point is used as the raise material for the embankments. For multiple spigot discharge, a series of shallow pits are dug in front of the spigots (once the tailings have dried and consolidated) and tailings placed on the embankment crest, then they are compressed, the tailings lines lifted and reassembled, and then the normal operation commences (Vick, 1990).

The upstream method is the most common design to fail causing huge environmental consequences all over the world (International Commission on Large Dams (ICOLD) and United Nations Environmental Protection (UNEP), 2001). An example of an upstream dam construction is given in Figure 13.4.

13.3.4 Centerline Method

The centerline method is a compromise between both the upstream and downstream designs (Benckert and Eurenius, 2001). It is more stable than the upstream method but does not require as much construction material as the downstream design. Like the upstream method, the tailings are

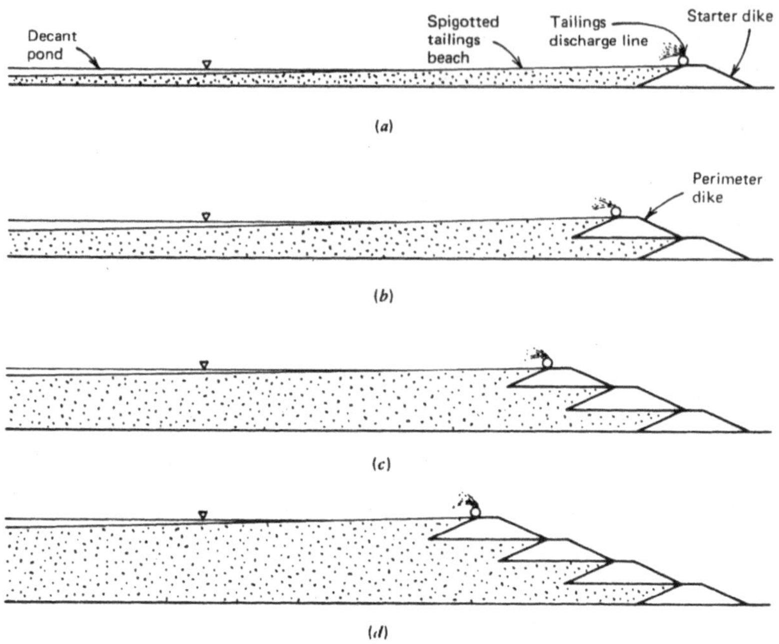

FIGURE 13.4 Upstream method (Vick, 1990).

generally discharged by spigots from the embankment crest to form a beach behind the dam wall. When subsequent raising is required, material is placed on both the tailings and the existing embankment. The embankment crest is being raised vertically and does not move in relation to the upstream and downstream directions of subsequent raises, hence the term, centerline design (Figure 13.5).

The design incorporates the internal drainage zones that are similarly found in the downstream method. Therefore, the free water can encroach closer to the dam crest than the upstream method without the worry of increasing the phreatic surface and causing a potential failure risk. However, a centerline dam cannot be used as a large water retention facility solely due to the subsequent raises being partially built on consolidated tailings. A suitable decant system needs to be installed to prevent the free water submerging the beach around the dam crest.

13.3.5 High-Density Tailings Disposal

Many mines are considering high-density (paste) disposal because of inherent historical failures, especially in the last several years. This method is called dry stacking and involves the densification of the waste metallurgical tailings prior to the final disposal. The densification is usually accomplished using a high-rate thickener producing a near-homogenous dense material with some self-supporting capability (Spearing et al., 2000).

The potential advantages of such a method are as follows:

- Placed tailings consolidate and stabilize much faster compared to conventional disposal, thus reducing the risk of liquefaction, and a settling pond is not created.
- Process reagents losses are reduced as they can be recovered during the thickening process.
- Low water in the tailings reduces seepage and losses.
- Overall water recovery and recycling are much higher.
- Rehabilitation of the dry stack can be undertaken concurrently with the tailing's disposal.

The method is not new and has been successfully used at Kidd Creek Mine in Canada since 1974. Figure 13.6 compares the conventional low-density surface tailings method with the high-density method.

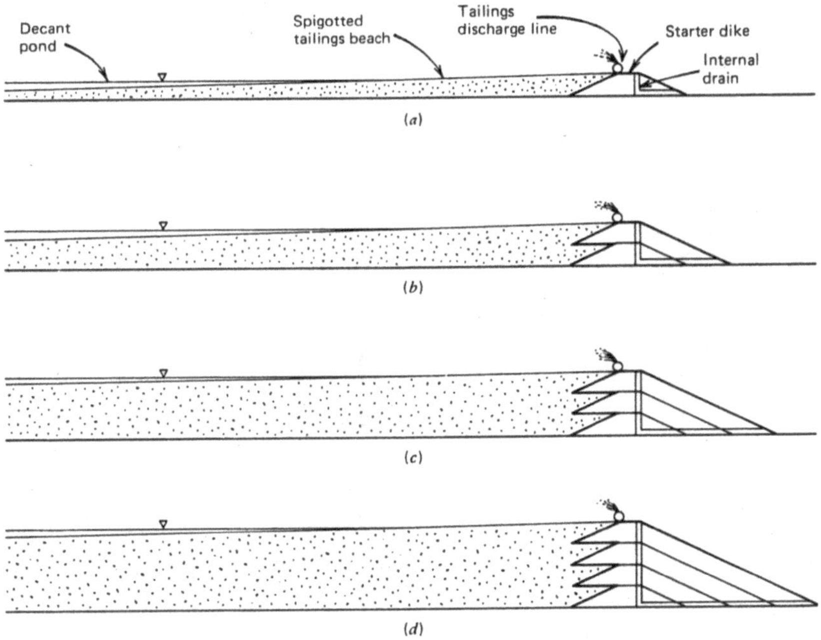

FIGURE 13.5 Centerline method (Vick, 1990).

13.3.6 Discussion on Tailings in Seismically Active Locations

Soil mechanics serves as the fundamental field for understanding liquefaction mechanisms and to date houses the most comprehensive literature on the topic.

The basis for liquefaction concepts was described by Casagrande (1936) such that when sands were subject to shear load, dense sand dilated, whereas loose sand, in similar conditions, decreased in volume (contracted). Paraphrasing by Das (1983) who also credits Casagrande, the void ratio when sand does not change its volume when subject to shear is termed the critical void ratio (also termed a critical density) and sands where the void ratio exceeds said critical value tend to decrease in volume under vibrations by a seismic effect. If the pore space is occupied by water such that the water pressure increases, and no such evacuation (drainage) measures are in place, the effect is the possibility of liquefaction.

Later, Casagrande (1976) defined sub-classes of the generalized term of liquefaction. Those terms are quoted here:

> Actual liquefaction was simply called liquefaction before the development of cyclic laboratory tests. It is the response of loose, saturated sand when subjected to strains or shocks that results in substantial loss of strength and in extreme cases leads to flow slides. Cyclic liquefaction is the response of a test specimen of dilative sand to cyclic loading in a triaxial test when the peak pore pressure rises momentarily in each cycle of the confining pressure.

In this same publication, Casagrande effectively reserved actual liquefaction for loose (contractive) sands which leads to strength loss and flow slides, whereas cyclic liquefaction pertains to dilative (dense) sands which are subject to cyclic loads. These details separated two very distinctive phenomena. Over time, these precise and fundamental definitions have become overly generalized, but now, simple terms such as static liquefaction have taken their place without much detail as to the underlying mechanisms which Casagrande showed and found important.

Overall, mine tailings liquefaction is an underemphasized topic needing studies and effective, updated policy and guidance for personnel who must manage these problems on a regular basis.

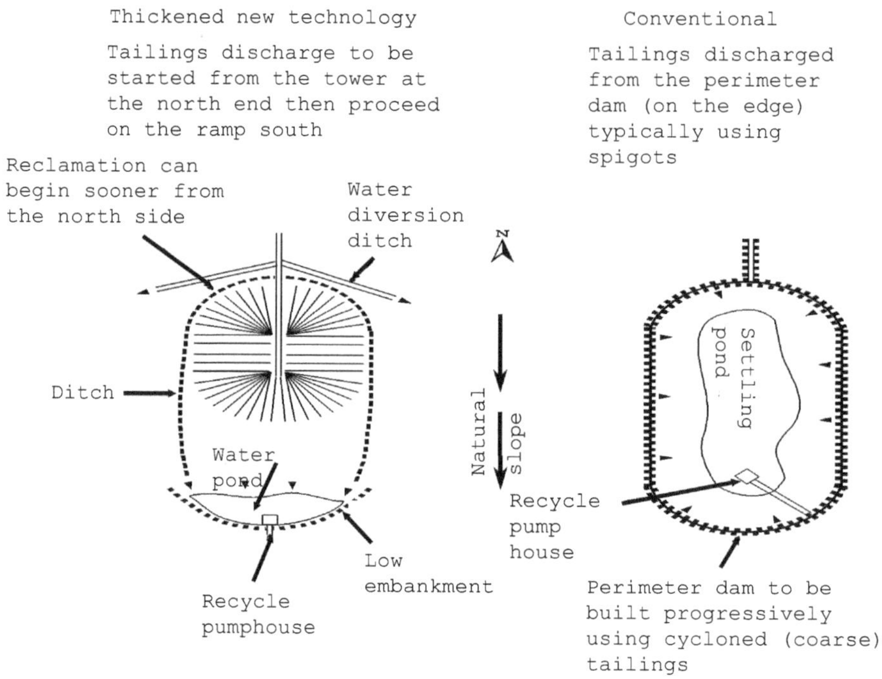

FIGURE 13.6 Aerial-view comparison of conventional tailings disposal and tailings dry stacking (Robinsky, 1999).

Defining many specific design processes for the broad range of tailing compositions will be challenging, and as a result, it should be expected that this topic will be a continual underlying concern for mining operations.

Prakash (1981) outlines several factors that have relevance toward the liquefaction character of sands and should be kept in mind:

- Grain size distribution of sand.
- Density of deposit (initial relative density D_R).
- Vibration characteristics.
- Location of drainage.
- Dimensions of deposit.
- Magnitude and nature of superimposed loads.
- Method of soil formation (soil structure).
- Period under sustained load.
- Previous strain history.
- Entrapped air.

13.4 TAILINGS FAILURE

The problems with conventional slurry tailings dams include the following:

- Inherent safety risk of liquefaction (failure).
- Serious water losses on the dams in regions where water is a scarce commodity such as in Australia and South Africa.
- Large area needed for the dam and the cost of eventual reclamation that should be done at mine closure.

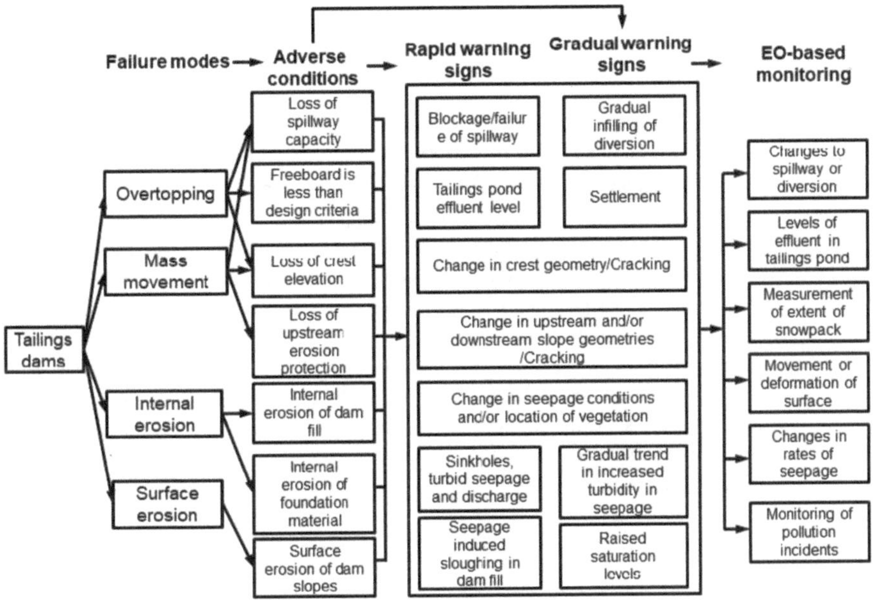

FIGURE 13.7 The main tailings dam failure modes, warning signs, and what Earth Observation (EO)-based methods can monitor (Martin and Davies, 2000).

Figure 13.7 shows the common tailings dam failure modes and potential monitoring that can be used to identify the potential failure and take action to avoid it (Martin and Davies, 2000).

Following the Vale tailings dam failure in 2019, various key players mobilized the global mining industry to take action and work to improve the stability of Tailings Storage Facilities (TSFs).

Recommendations and guidelines were prepared mainly by the International Council of Mining and Metals (ICMM) who helped create the Global Tailings Mining Institute (GTMI) and prepared the Global Industry Standard on Tailings Managements (GISTM). A copy can be obtained at https://globaltailingsreview.org/global-industry-standard/. It was created because there are over 18,000 TSFs worldwide, of which about 3,500 are active. There was also a growing need for independent oversight.

GTMI was a partnership between the UN Environment Programme, Principles for Responsible Investment, the Church of England Board, and the Swedish National Pension Fund.

This partnership resulted in the development of the following principles in Table 13.1:

In the future, TSFs will be handled as shown in Figure 13.8, which is not necessarily new, but if implemented, it should reduce massive failures in the future.

13.5 TAILINGS MANAGEMENT, MONITORING AND INSTRUMENTATION

The tailings dam operators often control the tailings dam design, with respect to the method of dyke construction and availability of materials. The volume of material to be safely placed and the hazards associated with failure make safe tailings operation critical.

Some of the considerations for tailings dam operations are as follows:

- Water pond location and elevation (within the dam).
- Spillway.
- Tailings deposition (placement).

ICMM (2021) laid out the latest best practices internationally.

TABLE 13.1
The TSF Principles (GISTM)

Principle 1	Respect the rights of project-affected people and meaningfully engage them at all phases of the tailing's facility lifecycle, including closure.
Principle 2	Develop and maintain an interdisciplinary knowledge base to support safe tailings management throughout the tailing's facility lifecycle, including closure.
Principle 3	Use all elements of the knowledge base – social environmental, local economic and technical – to inform decisions throughout the tailing's facility lifecycle, including closure.
Principle 4	Develop plans and design criteria for the tailings facility to minimize risk for all phases of its lifecycle, including closure and post-failure.
Principle 5	Develop a robust design that integrates the knowledge base and minimizes the risk of failure to people and the environment for all phases of the tailing's lifecycle, including closure and post-closure.
Principle 6	Plan, build, and operate the tailings facility to manage risk at all phases of the tailing's facility lifecycle, including closure and post-cycle.
Principle 7	Design, implement, and operate monitoring systems to manage risk at all phases of the facility lifecycle, including closure.
Principle 8	Establish policies, systems, and accountabilities to support the safety and integrity of the tailings facility.
Principle 9	Appoint and empower an engineer of record.
Principle 10	Establish and implement levels of review as part of a strong quality and risk management system for all phases of the tailings facility lifecycle, including closure.
Principle 11	Develop an organizational culture that promotes learning, communication, and early problem recognition.
Principle 12	Establish a process for reporting and addressing concerns and implement whistleblower protections.
Principle 13	Prepare for emergency response to tailings facility failures.
Principle 14	Prepare for long-term recovery in the event of a catastrophic failure.
Principle 15	Publicly disclose and provide access to information about the tailings facility to support public accountability.

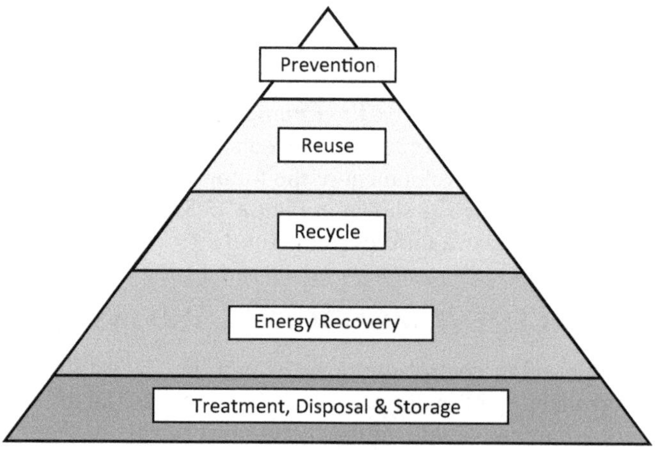

FIGURE 13.8 TSF considerations moving forward (Global Tailings Review, 2020).

Monitoring via inspections (in person or using drones) is essential, and trained professionals need to be employed to operate and monitor tailings facilities. The use of radar, LIDAR, or similar is also very useful as it gives real-time monitoring. Piezometers appropriately positioned in the retaining structure are also strongly recommended.

13.6 RECLAMATION AND CLOSURE

The planning for the design and operation of a tailing's basin should consider the reclamation to the deposition of tailings, location of spillways, and final drainage patterns.

In recent years, the costs associated with dam rehabilitation have increased dramatically due mainly to the environmental lobby in first-world countries. A single large tailings dam at the now-defunct Elliot Lake Uranium Mines, in Canada, cost over US$100 M to rehabilitate after the end of the life of the mine (Berthelot et al., 2019).

According to McKenna and van Zyl (2020), tailings facilities typically occupy 10%–40% of the area of a reclaimed mining landscape, with pits and waste rock dumps responsible for most of the rest. Typically, regulators now require reclaimed facilities to meet agreed-upon land uses and performance standards that sustain landscapes for the benefit of local communities. After mining and suitable rehabilitation, the sites are commonly used as natural areas or wildlife habitat (especially for remote mines). Near cities, they may be used for agricultural, recreational, or industrial activities.

McKenna and van Zyl (2020) further state that a good closure practice includes the following important steps, which can be reviewed from the above reference:

- Develop site-specific closure criteria.
- Identify and analyze alternative closure technologies that can meet the closure criteria.
- Meaningful and transparent community engagement from the start to the final rehabilitation.

The quality of any water drainage needs to be monitored and conditioned if needed before and after closure.

13.7 WORKED EXAMPLES

Example 13.7.1

A basic theory for understanding liquefaction is presented by Prakash (1981) toward the strength of sand (a cohesionless soil) in a saturated state as follows:

$$s = (\gamma_b Z - \Delta u)\tan\phi$$

where

s is the strength.

Z is the depth according to Figure 13.9.

Δu is the increase in pore water pressure (+) due to ground shaking = $\gamma_w h_w'$.

φ is the angle of internal friction.

γ_b is the submerged unit weight of soil.

h_w' is the water head.

If in this example γ_b is taken as 50 lb/ft^3 and the depth, Z, is 1 ft, what must be the height h_w' of the water table for the sand to have complete loss of strength?

Solution:

For complete loss of strength ($s = 0$),

$$s = (\gamma_b Z - \Delta u)\tan\phi$$

$$s = (\gamma_b Z - \gamma_w h_w')\tan\phi$$

We now see that the following is the only necessary relationship for complete loss of strength, and therefore,

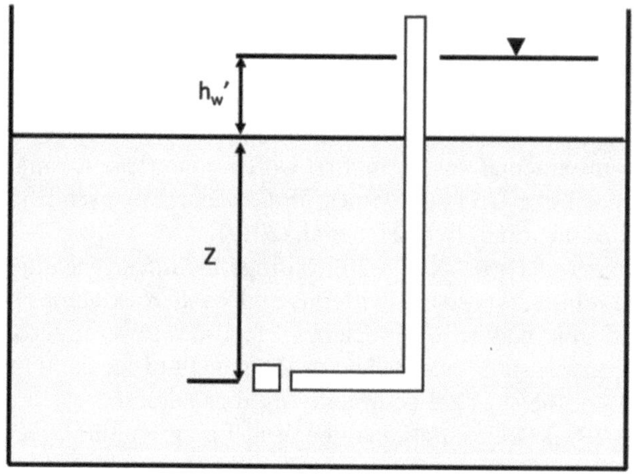

FIGURE 13.9 A saturated and cohesionless soil (Prakash, 1981).

$$\gamma_b Z = \gamma_w h_w'$$

$$(50\ \text{lb/ft}^3)(1\ \text{ft}) = (62.4\ \text{lb/ft}^3)(h_w')$$

$h_w' = 0.8$ ft or 9.6 inches of water above surface with cyclic shaking produces a complete loss of strength in the sand given.

Foundationally the concept just shown (loss of strength due to introduction of water in pore space with cyclic loading) is very important to understanding the basics of liquefaction.

Example 13.7.2

Determine the stress increase under the embankment shown, which is part of an example completed by Das (2007).

Solution:
Part a

$$\gamma H = (17.5\ \text{kN/m}^3)(7\ \text{m}) = 122.5\ \text{kN/m}^2$$

The embankment is divided as shown in Figure 13.10.

This allows for the determination of dimensional quantities of the embankment, termed,

$$B_1 = 2.5\ \text{m and } B_2 = 14\ \text{m}$$

$$B_1/z = 2.5/5 = 0.5$$

$$B_2/z = 14/5 = 2.8$$

Based on Figure 13.10 in the reference, these two quantities are used to determine and influence value, I′ for embankment loading. For this case, I′ = 0.445 (Figure 13.10):

$$\Delta\sigma = \Delta\sigma_1 + \Delta\sigma_2 = q_o\left[I'_{(\text{left side of embankment})} + I'_{(\text{right side of embankment})}\right]$$

$$\Delta\sigma = (122.5\ \text{kN/m}^2)[0.445 + 0.445]$$

$$\Delta\sigma = 109\ \text{kN/m}^3$$

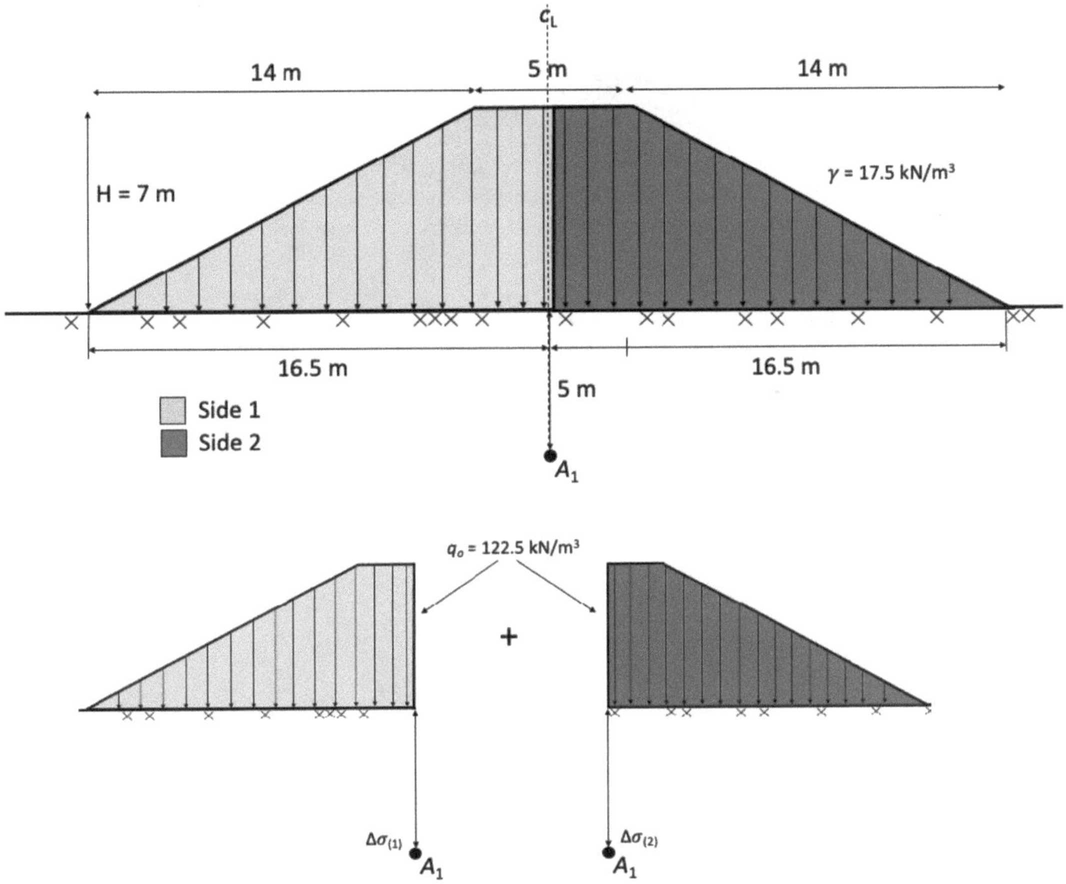

FIGURE 13.10 Section through the embankment (Das, 2007).

Example 13.7.3

Determine the earth pressure on a buried, positive projecting conduit if the conduit is flexible and then again if the conduit is rigid (refer to Figure 13.13). The equations are found in Society of Mining, Metallurgy and Exploration's (SME) Mining Engineering Handbook (Lowrie, 2002).

Given:

Unit weight, w = 115 lbs/ft^3

Depth from surface to top of buried pipe (below frost depth) = 60 inches

Diameter of pipe, D = 8 inches

Load coefficient for positive projecting conduits, C_p = 1.0 for flexible and 1.5 for rigid pipe when no site-specific data available.

Solution:

Flexible Pipe:

$$W = C_p whD$$

$$W = (1.0)(115 \text{ lbs/ft}^3)(5 \text{ ft})(8/12 \text{ ft})$$

$$W = 383 \text{ lbs/ft} \qquad \text{* per linear foot}$$

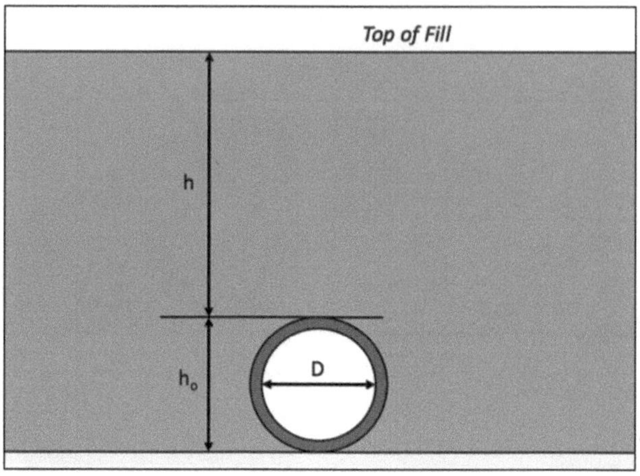

FIGURE 13.11 The buried conduit (Lowrie, 2002).

Rigid Pipe:

$$W = C_p whD$$

$$W = (1.5)(115 \text{ lbs/ft}^3)(5 \text{ ft})(8/12 \text{ ft})$$

$$W = 575 \text{ lbs/ft * per linear foot}$$

Example 13.7.4

A filter liner is recommended at the toe of an earth dam as shown in Figure 13.12 constructed after Das (2008). According to Das (2008) and Terzaghi and Peck (1948), the following two conditions should be considered:

- The size of void in the filter should be small enough to hold larger particles of the protected material in place.
- The filter material should have high enough permeability so that the buildup of large seepage forces and hydrostatic pressures in the filters doesn't occur.

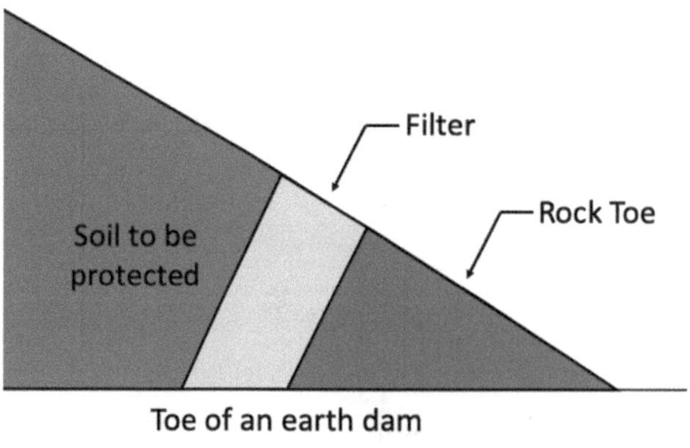

FIGURE 13.12 Section through the earth dam (Das, 2008).

According to Terzaghi and Peck (1948), the above conditions are satisfied when

$D_{15(F)}/D_{85(B)} \leq 4\text{–}5$ to satisfy condition 1

$D_{15(F)}/D_{15(B)} \geq 4\text{–}5$ to satisfy condition 2

and

$D_{15(F)}$ diameter through which 15% filter material will pass

$D_{15(B)}$ diameter through which 15% of soil to be protected will pass

$D_{85(B)}$ diameter through which 85% of soil to be protected will pass.

If the grain size distribution of the soil to be protected has the following values:

$$D_{85(B)} = 0.11\text{ mm}$$

$$D_{15(B)} = 0.009\text{ mm}$$

Determine the $D_{15(F)}$ to satisfy conditions 1 and 2.

Solution:

For condition 1,

$$D_{15(F)}/D_{85(B)} \leq 5 \text{ (The example in Das (2008) uses 5)}$$

$$D_{15(F)}/\ 0.11\text{ mm} \leq 5$$

$$D_{15(F)} = 0.55\text{ mm}$$

For condition 2,

$$D_{15(F)}/D_{15(B)} \geq 5 \text{ (The example in Das (2008) uses 5)}$$

$$D_{15(F)}/0.009 \geq 5$$

$$D_{15(F)} = 0.045\text{ mm}$$

A particle size distribution plot of the filter should accompany this result and is shown in Das (2008).

Example 13.7.5

The following figures illustrate a graphical method for approximating the phreatic surface through a dam on an impermeable surface. The method is referred to as Schaffernak's Graphical Solution (1917). The following step-wise procedure is from Das (2008) (Figures 13.3–13.20).

The solution is a rough approximation of the phreatic surface as Casagrande (1937) later showed the parabola for the phreatic line will be offset from a, some distance along the waterline and away from the earth dam. This modification is discussed in more detail in Das (2008).

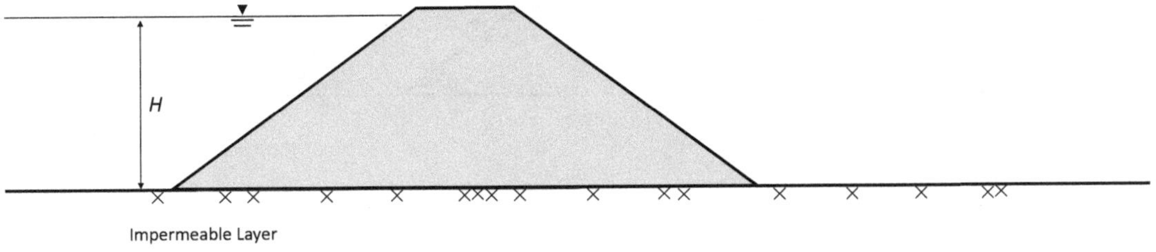

FIGURE 13.13 Section through the dam (Das, 2008).

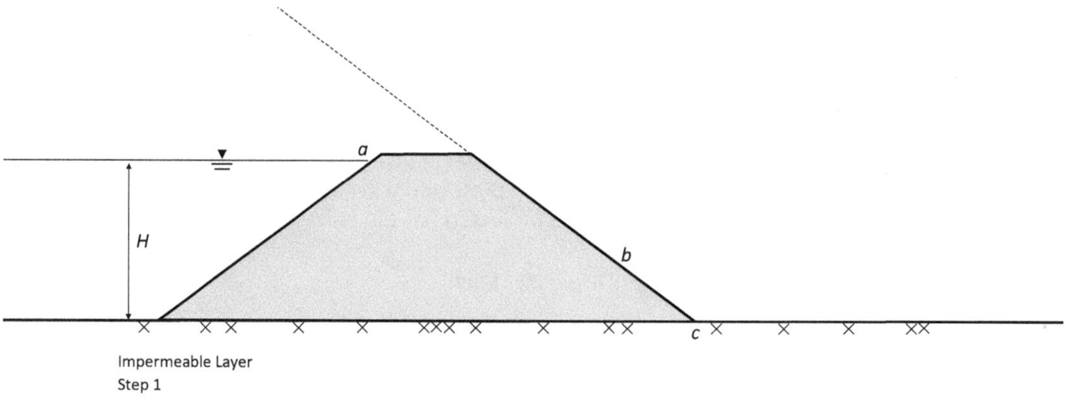

FIGURE 13.14 Step 1: Extend the downstream slope line bc upward (Das, 2008).

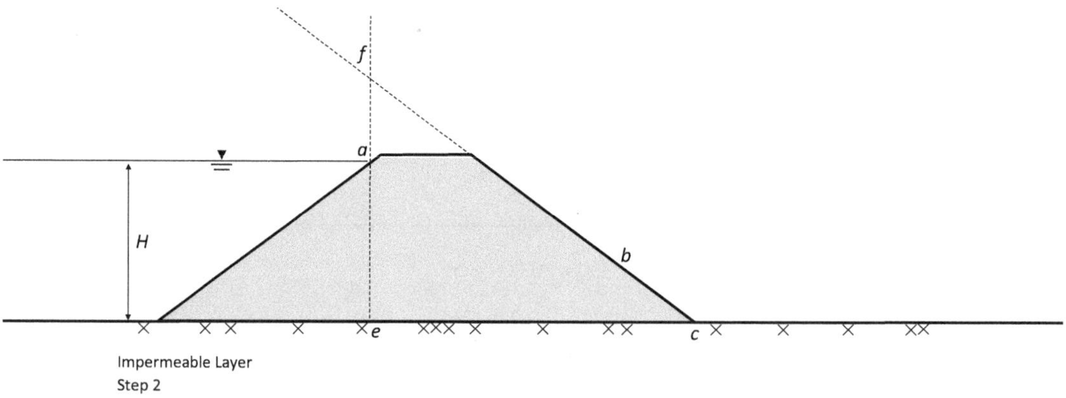

FIGURE 13.15 Step 2: Draw a vertical line (ae) through the point a which should intersect the projected line bc constructed in the previous step, to intersection point f (Das, 2008).

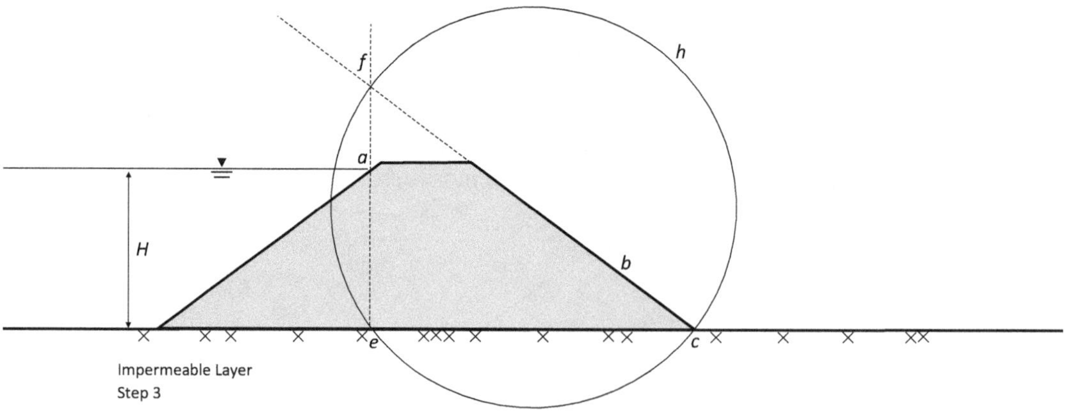

FIGURE 13.16 Step 3: The line from f to c serves as the diameter of a circle. A semicircle fhc is now constructed (Das, 2008).

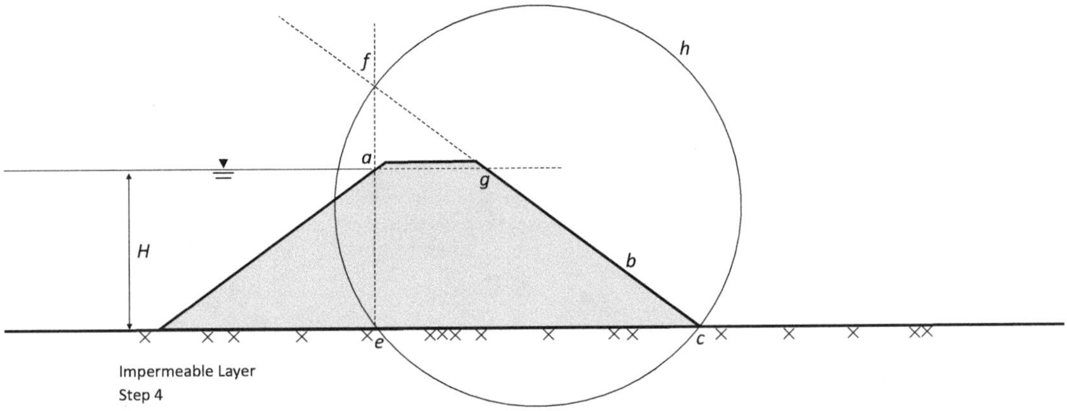

FIGURE 13.17 Step 4: Construct a horizontal line from a to g (ag) (Das, 2008).

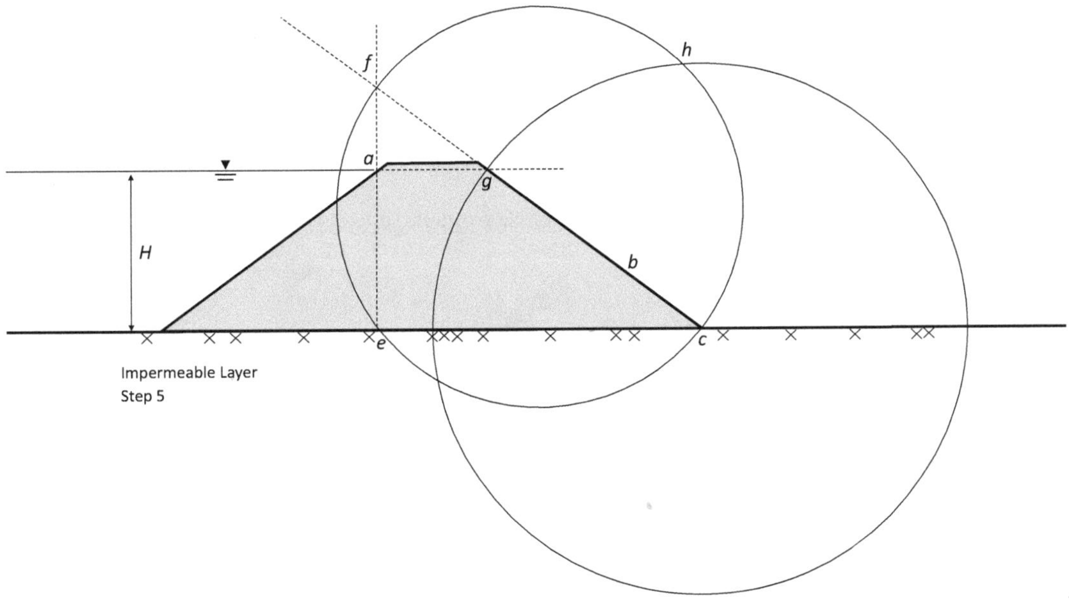

FIGURE 13.18 Step 5: Now with c serving as the center of a circle, and cg representing the radius, construct an arc, from g to h (Das, 2008).

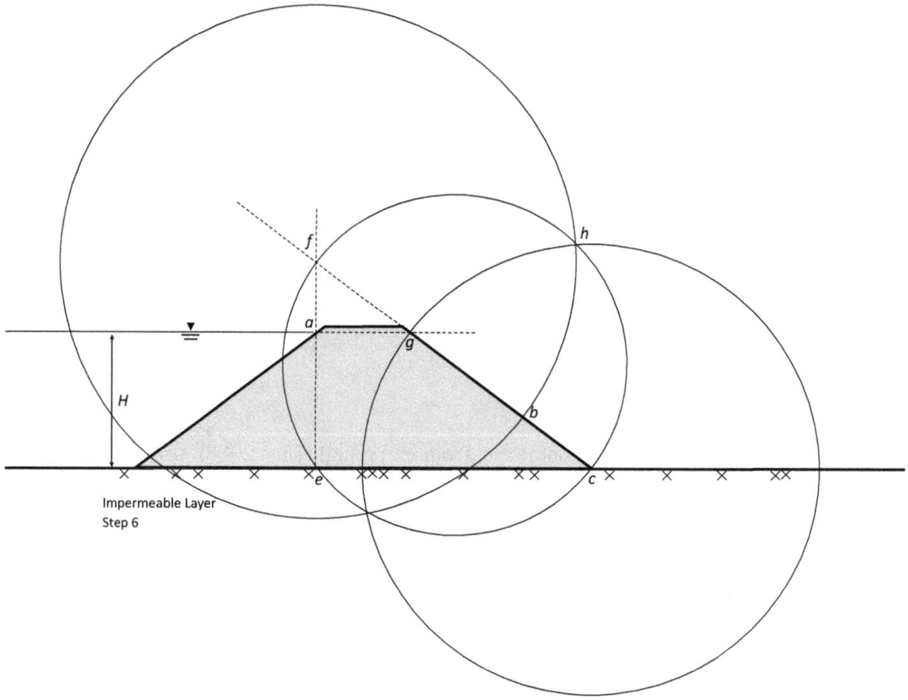

FIGURE 13.19 Step 6: With f serving as another circle's center, fh as a radius, draw another arc from h toward b (Das, 2008).

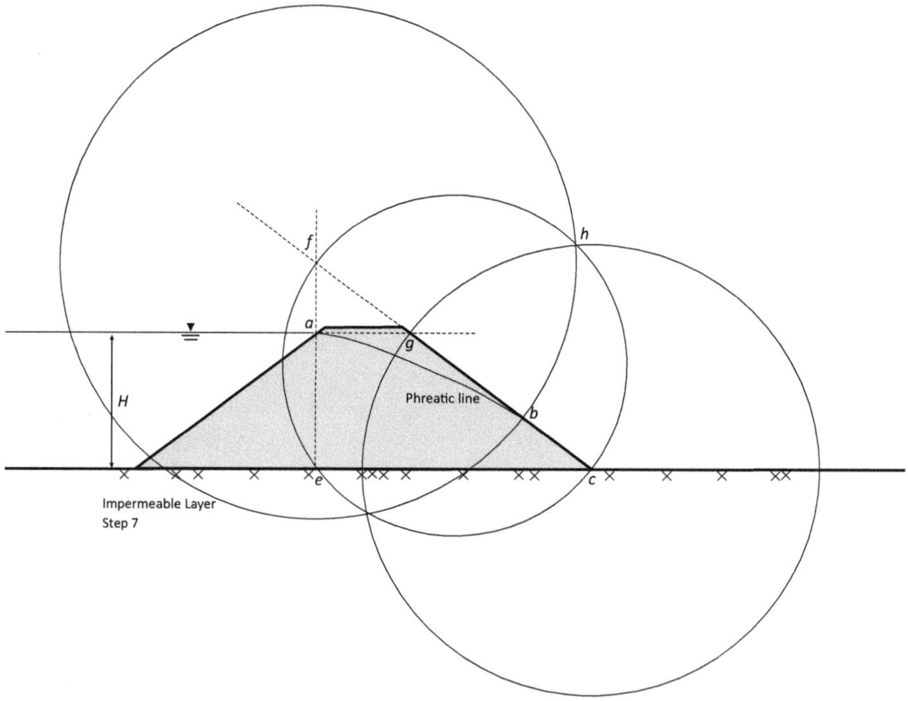

FIGURE 13.20 Step 7: The distance from b to c (bc) represents a value l or the distance from the impervious base where the phreatic surface crosses.

REFERENCES

Benckert, A. and Eurenius, J. (2001) *Tailings Dam Constructions, Seminar on Safe Tailings Dam Constructions*. Swedish Mining Association, Natur Vards Verket, European Commission, Gallivare, Sweden, pp. 30–36.

Berthelot, D.S., Place-Hoskie, D., D Willems, D. and Black, K. (2019) *Elliot Lake, Ontario Uranium Mines a Legacy Perpetual Care Case Study. Mine Closure*. Australia Centre for Geomechanics, Perth.

Casagrande, A. (1936) Characteristics of cohesionless soils affecting the stability of slopes and earthfills. *J. Boston Soc. Civil Eng*. 23, 257–276.

Casagrande, A. (1937) Seepage through dams. In *Contributions to Soil Mechanics 1925-1940*. Boston Soc. Civil Eng. Boston. p 295.

Casagrande, A. (1976) Liquefaction and cyclic deformation of sands: a critical review. In *5th Panamerican Conference on Soil Mechanics and Foundation Engineering*, Buenos Aires, Argentina.

Das, B (1983) *Fundamentals of Soil Dynamics*. Elsevier Science Publishing Co., Inc, New York.

Das, B. (2007) *Principles of Foundations Engineering* (6th Edition). Cengage Learning, Stamford CT.

Das, B. (2008) *Advanced Soil Mechanics* (3rd Edition). Taylor and Francis, New York NY.

Engels, D. and Dixon-Hardy, J. (2007) Methods for the disposal and storage of mine tailings. *Environ. Sci. Eng*. 15(3), 301–317.

Global Tailings Review (2020) Global industry standard on tailings management. Convened by: ICCM, UN Environment Programme and Principles for Responsible Investments. Retrieved from: www.globaltailingsreview.org

ICMM (2021) Tailings Management –Good practice guide by ICMM –Mining with Principles. Retrieved from: https://www.icmm.com/en-gb/guidance/environmental-stewardship/tailings-management-good-practice

ICOLD and UNEP (2001) *Bulletin 121: Tailings Dams – Risk of Dangerous Occurrences, Lessons Learnt from Practical Experiences*. UNEP, DTIE and ICOLD, Paris.

Lowrie, R. L. (Ed.) (2002) SME Mining Reference Handbook. SME, New York.

Martin, T.E. and Davies, M.P. (2000) *Development and Review of Surveillance Programs for Tailings Dams. AGRA Earth and Environmental Limited*. Burnaby, British Columbia, Canada.

Martin, T.E. and McRoberts, E.C. (1999) Some considerations in the stability analysis of upstream tailings dams. In *Proceedings of Sixth International Conference on Tailings and Mine Waste* '99, 24–27 January, Fort Collins, Colorado, USA

McKenna, G. and van Zyl, D. (2020) *Towards Zero Harm-A Compendium of Papers Prepared for the Global Tailings Review*. GlobalTailingsReview, Switzerland. Retrieved from: www.GlobalTailingsReview.org.

Prakash, S. (1981) *Soil Dynamics*. McGraw-Hil, New York.

Robinsky, E.I. (1999) *Thickened Tailings Disposal in the Mining indusTry*. E.I Robinsky Assoc. Ltd., Toronto, Canada.

Schaffernak, F. (1917) Über die Standicherheit durchläessiger geschüetteter. Dämme, Allgem. Bauzeitung.

Spearing, A.J.S., Millette, D. and Gay, F. (2000) *The Potential Use of Foam Technology in Underground Backfilling and Surface Tailings Disposal*. MassMin, Brisbane, Australia.

Terzaghi, K. and Peck, R. (1948) *Soil Mechanics in Engineering Practices*. Wiley, New York NY.

Vick, S.G. (1990) *Planning, Design and Analysis of Tailings Dams* (2nd Edition). BiTech Publishers Ltd, Vancouver.

14 Rock Mechanics Modeling and Monitoring

14.1 ROCK MECHANICS NUMERICAL MODELING

14.1.1 The Need for Modeling

Modified after the Mining Department at Queens University, Canada:

Rock is a natural, highly variable geologically disturbed material, and therefore, the range of physical and engineering properties of it must be established via testing, classification observation, and back analyses. Unlike materials like steel and concrete, it is not engineered and does not have fixed properties. The rock mass is a discontinuous, heterogenous, anisotropic, and non-elastic material. Due to dynamic (tectonic) movements of the upper crust of the earth, the rock mass is still under continually changing stresses, which result from tectonic plate movements for example. The rock mass around mining is also typically fractured and often a porous material.

Challenges in rock mechanics are getting more complex, requiring more demanding and challenging modeling requirements. Therefore, designing a rock engineering model requires comprehensive knowledge of the geomaterial, engineering properties, and parameters of the rock mass. However, the quality and quantity of the supporting data for rock engineering projects can never be complete due to the inherent variability of the rock mass properties. In most cases, the model does not need to be totally precise, but it needs to be adequate for the purpose of the study. Models are part of a "toolbox" approach and should be revisited and updated (if needed) regularly as more real in situ information becomes available.

14.1.2 Modeling Requirements and Constraints

In mining, numerical models are computer-based representations of a physical problem. Any model must adequately represent the reality that is of concern. Mining layouts are complex, and many different layouts can be adopted in practice. Computer simulations are a sensible option because they can capture large, multiple, complex layouts and their interaction with the surrounding rock. This leads to the ability to compare different layouts to maximize extraction and minimize rock-related risk.

Computer modeling is useful for simple problems even if it only helps to reinforce assumptions. The most important rule, however, is that it must be set up, used, and interpreted correctly and logically. Computers and numerical models are powerful systems which tend to exceed our basic comprehension.

With all modeling, calibration, and recalibration, of the results is essential to assure they represent the in situ situation. The results must be checked against reality, when possible, with other design approaches, and especially real measurements. The important point is that all the results should have an explanation tied to them that makes sense with the actual physical system.

In greenfield mining areas, calibration may be initially difficult so as development begins and progresses, monitoring and observations are needed to compare the model to reality and adjust the model inputs as more data become available.

Limitations with models have and will continue to be input parameters, failure criteria, boundary conditions, and meshing. Some computational limitations occur as well with very complex and large numerical modeling packages. The skills and understanding of the user are also critical.

DOI: 10.1201/9781032704845-14

In addition, since most input parameters are estimated to one or at best a few significant figures (Poisson's ratio typically only to one – e.g. 0.3), reporting stresses, strains, and other results to many significant figures just because the computer output does this is inappropriate. For example, if an average pillar stress is calculated as 109.54 MPa, rather report it as 110 MPa. Results and reports need to be simple, clear, and justified. It is important to always refer to the base assumptions made.

Most modeling can be effectively used qualitatively to compare different layouts or scenarios as large-scale modeling results are seldom quantitatively accurate and can be easily criticized. In situ, relevant measurements are the best tool that can provide justification to quantitative results. A blind quantitative single model, meaning a model with no other engineering solution to justify the results, should always be challenged.

Basically "garbage in, garbage out" is a serious potential issue for many applications (Mellin, 1957). In mining, numerical models can be of considerable assistance to mining engineers in designing underground excavations, provided that they are correctly set up and used. Still some users don't calibrate their models and don't logically review the output to see if it makes intuitive sense and can be justified.

14.1.3 Types of Models

Three main modeling approaches are as follows:

- Physical Modeling

  Involves a scale model (size and property wise) of the problem with representative stresses being applied. The advantages are that the result is generally reliable and can be used to better calibrate a numerical model. The disadvantages are that the set-up is very costly, time consuming, and typically only be used for limited excavations that can be approximated by 2-D representations. Physical modeling was used before the modern computer age and is still used to a lesser extent to help calibrate computer models. The scale-up effect can be a serious issue too. Due to the cost and time involved with setting up such models, not many are done today.
- Empirical (a form of Experiential Modeling) Modeling

  This uses simplified mathematical design equations based primarily on measured or observed behavior. The advantages of this are that it is simple and quick to calculate with few input parameters and it is generally reliable as it should be based on a large database. The disadvantages are that it is only reliable within the constraints of the database, it must be calibrated afterward (as all models should), and it does not consider the mechanisms involved.
- Mathematical and Numerical Modeling

  This uses a simplified mathematical approximation of the physical and mechanical behaviors of the orebody and the surrounding rock mass according to physical laws. The advantages are that it can be used for more complex layouts, the results can be extrapolated, and it serves as a powerful comparative tool. The disadvantages are that the input parameters are difficult to obtain, and in reality, the situation is non-elastic or plastic (far more complex) and the results need to be checked against reality.

14.1.4 Mathematical Modeling

Mathematical models can be of some use in rock mechanics, although due to the normal complexity of mining geometries, numerical modeling tends to be more effective. Mathematical modeling is more commonly used for plain strain applications such as in long tunnels with limited mining in the vicinity.

Typical applications include the following:

- Tributary area theory for pillar design (mainly an empirical and closed form method).
- Stress and closure calculations for deep narrow and relatively flat tabular orebodies.
- Kirsch solutions for mainly circular and elliptical tunnels.

14.1.5 Numerical Modeling

With complex 3-D problems, it is often helpful, if time permits, to first model the situation in 2-D to understand the possible mechanisms involved and the parameters that are of more relevance. Two-dimensional problems are quicker to set up and easier to interpret, although some problems are clearly only 3-D in nature and cannot be reliably simplified into 2-D.

Most modeling requires the following key elements:

- Correct model selection based on experience and judgment.
- Determination of representative input parameters with mean values and deviations (or a sensitivity analysis to determine the main design parameters).
- Setting of the model boundary conditions.
- Verification of the results by qualitative or quantitative means, where possible.
- Adjustment of the model if necessary or the selection of a more applicable model.

Even if the input parameters and boundary conditions can be represented, there are still certain problems such as:

- The constitutive model still may not reflect the physical processes since we are dealing with nature and not "engineered" products in mining.
- The transition from continuum to discontinuum (i.e., failing rock) may not be adequately modeled.

Mining falls typically into Region 3 in Figure 14.1 (after Holling, 1978). This is because little reliable data are available, and the problem is often "ill posed" because slight changes in say the joint properties or orientation have a significant effect on the output.

According to Feng and Hudson (2011), there are four main analysis methods as shown in Figure 14.2.

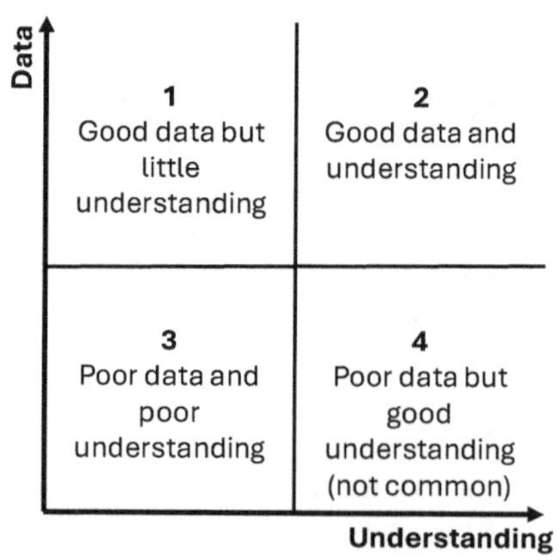

FIGURE 14.1 The relationship between data and the level of understanding (Holling, 1978).

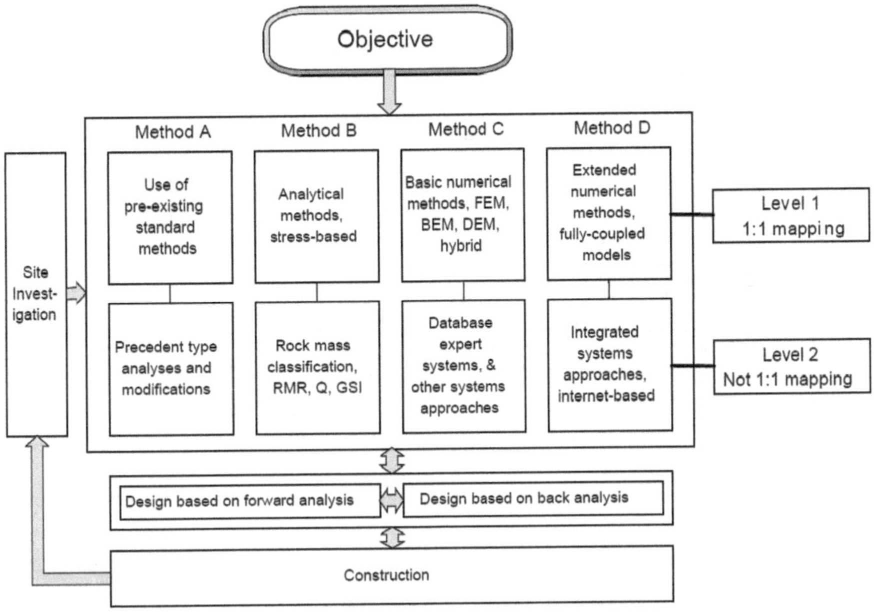

FIGURE 14.2 The four main analysis methods (Feng and Hudson, 2011).

The four columns represent the four main modeling methods as follows (Feng and Hudson, 2011):

- Method A: Engineering design based on previous empirical design and experiences.
- Method B: Engineering design based on simplified analytic solutions, e.g., Kirsch equations for stress analysis around circle excavations.
- Method C: Basic numerical modeling which captures most relevant rock mass mechanisms.
- Method D: Extensive numerical methods which involve all rock mass mechanisms.

Rock mass characterization can be obtained as part of a site investigation. One important point in rock engineering is the lack of data, and this is the reason for also applying empirical approaches, for example, classification systems.

Hudson and Feng (2015) investigated the issues further and determined risk factors within the governing framework for identification, assessment, and management of rock engineering risk. Risk is dealt with in more detail in Chapter 15 (Figure 14.3).

The way that various computer methods simulate the rock mass is shown in Figure 14.4 (Jing, 2003). The methods are discussed after the figure.

14.1.6 Boundary Element Methods

Boundary Element Method (BEM) is a computational method for solving linear partial differential equations that have been formulated as integral equations (i.e., in boundary integral form). BEM can be applied in many areas of engineering and science including rock mechanics, fluid mechanics, acoustics, electromagnetics, and fracture mechanics.

According to Eberhardt (2007), in performing an analysis, the boundary of the excavation is divided into elements and the interior of the rock mass is represented mathematically as an infinite continuum. The solution works to find a set of approximate stresses which satisfy prescribed boundary conditions, and then uses these to calculate the stresses and displacements in the rock mass.

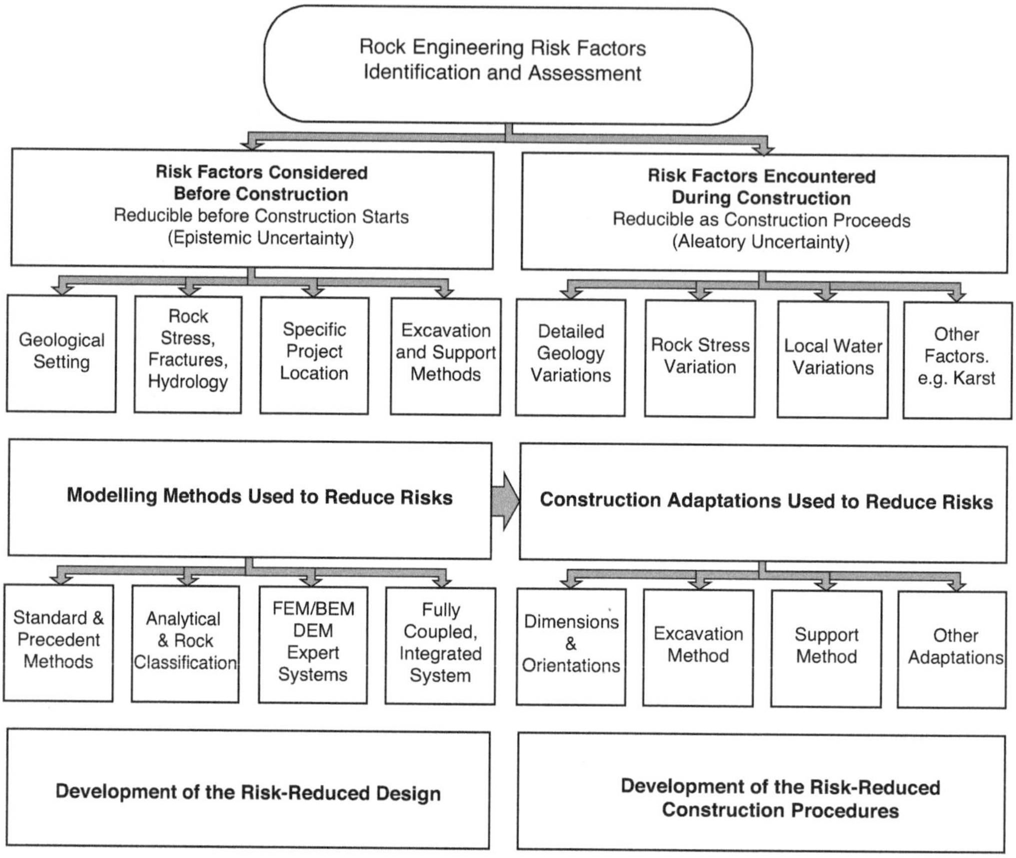

FIGURE 14.3 Flowchart of rock engineering risk factors enabling the development of risk-reduced design and risk-reduced construction (Hudson and Feng, 2015).

Figure 14.5 gives an output example of a tunnel using a BEM program (Rocscience, 2024). Common applications include the following:

- Stress and strain analyses of underground excavations.
- Brittle fracture investigations.
- Some dynamic (seismic) modeling capability.
- Water and foundation problems.

14.1.7 Displacement Discontinuity (DD) Methods

One form of the BEM used commonly for underground structural analysis is the Displacement Discontinuity (DD) method. The mining area is treated mathematically as a discontinuity or slot. Displacements are generally generated by the method.

With the DD method, only the planar area of the mining is discretized to obtain displacements and stress solutions within the area of interest. This makes the programs simple and quick to run.

Figure 14.6 shows a complex geometry using Displacement Discontinuities.

Common applications include the following:

- Modeling tabular deposits such as gold mines (in South Africa), coal, and salt deposits.
- Representing excavations and geological features.

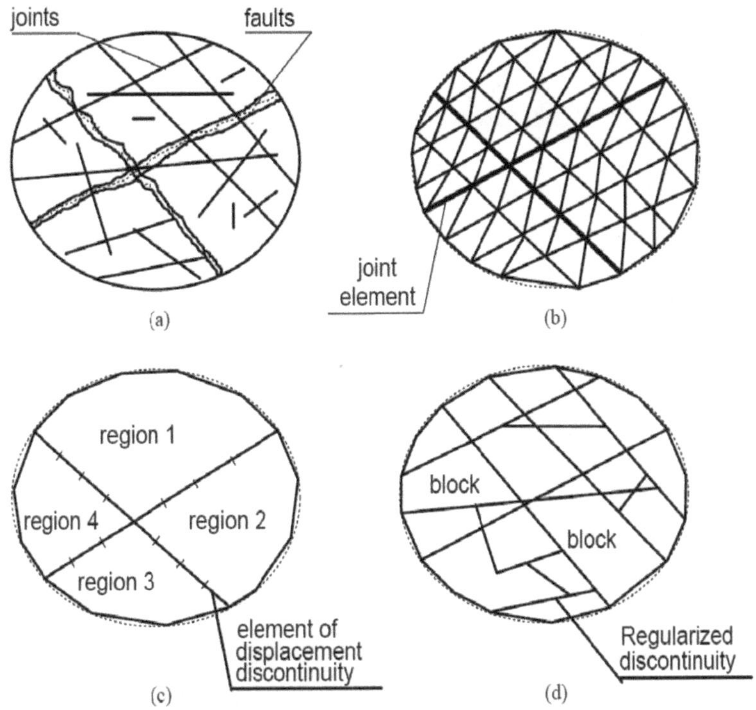

FIGURE 14.4 How different computer models represent (a) the actual rock mass, (b) finite difference model (FDM) or finite element model (FEM), (c) boundary element method (BEM), and (d) discrete element method (DEM) (Jing, 2003).

Differential methods (solving sets of differential equations) are more difficult and time consuming than BEM methods but are much more versatile and useful. They are, however, far more challenging to set up, take more computing time, and the results need very careful analyses.

Differential methods include the following (Eberhardt, 2007):

- Continuums
  - Finite Element Methods (FEMs)
    - They approximate a solution for differential equations by solving the interiors of the grid cells (elements) for the grid points. They can be used in complex geometries. An example is Phase2 from Rocscience.
  - Finite Difference Methods (FDMs)
    - They also approximate a solution for differential equations but by solving a set of points on a grid. They are generally easier to implement, but the approximation between grid points needs careful consideration or it can cause problems. An example is the Fast Lagrangian Analysis of Continua (FLAC) from Itasca Consulting Group.
- Discontinuums
  - Distinct Element Methods
    - These typically use an explicit time-marching system to solve the equations of motion directly for a system of rigid or deformable bodies and the contacts are deformable.
  - Discrete Element Methods
    - With these systems, the dynamic sets of equations for the equilibrium of each block in the model are formulated and iteratively solved until the boundary conditions

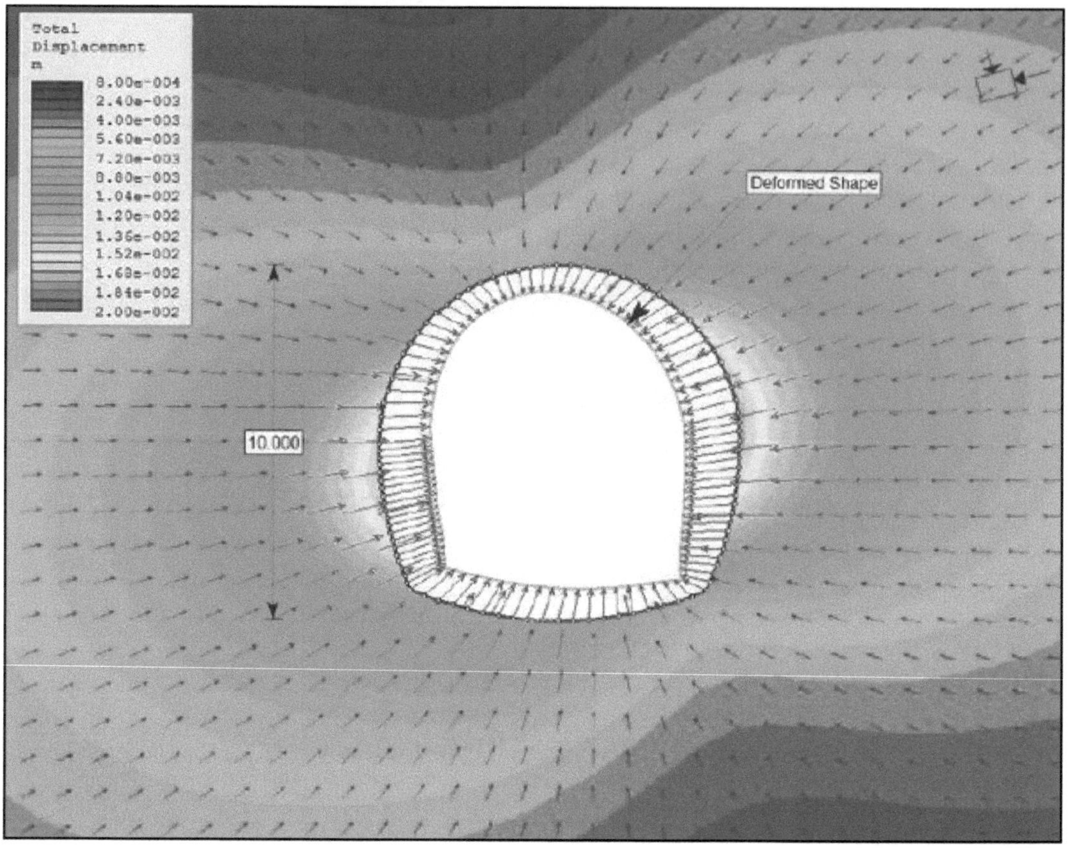

FIGURE 14.5 Example of a boundary element model using Examine 2D (Rocscience, 2024).

and laws of contact and motion are met. This allows the method to handle complex nonlinear block interactions.

For rock mechanics modeling, Figure 14.7 shows the main differences between the methods (Eberhardt, 2007).

14.1.8 Discrete and Distinct Element Methods

These methods generally use the same approach and discretize the continuum into discrete particles. The Discrete Element Method is generally used to represent granular material whilst the Distinct Element Method is used to represent blocky material and rock fractures. They are fast and computationally efficient.

They use iterative (explicit time) steps to solve the motion equations in both rigid and deformable bodies. Figure 14.8 gives an example of a model using Discrete Modeling.

14.1.9 The Finite Element Method (FEM)

FEM, also sometimes referred to as Finite Element Analysis (FEA), is a numerical technique for calculating approximate solutions to Partial Differential Equations (PDE) and Integral Equations (IE). FEM allows representation where structures deform and gives the stress tensor and the displacements. It involves discretizing the entire continuum and is computationally intensive to solve.

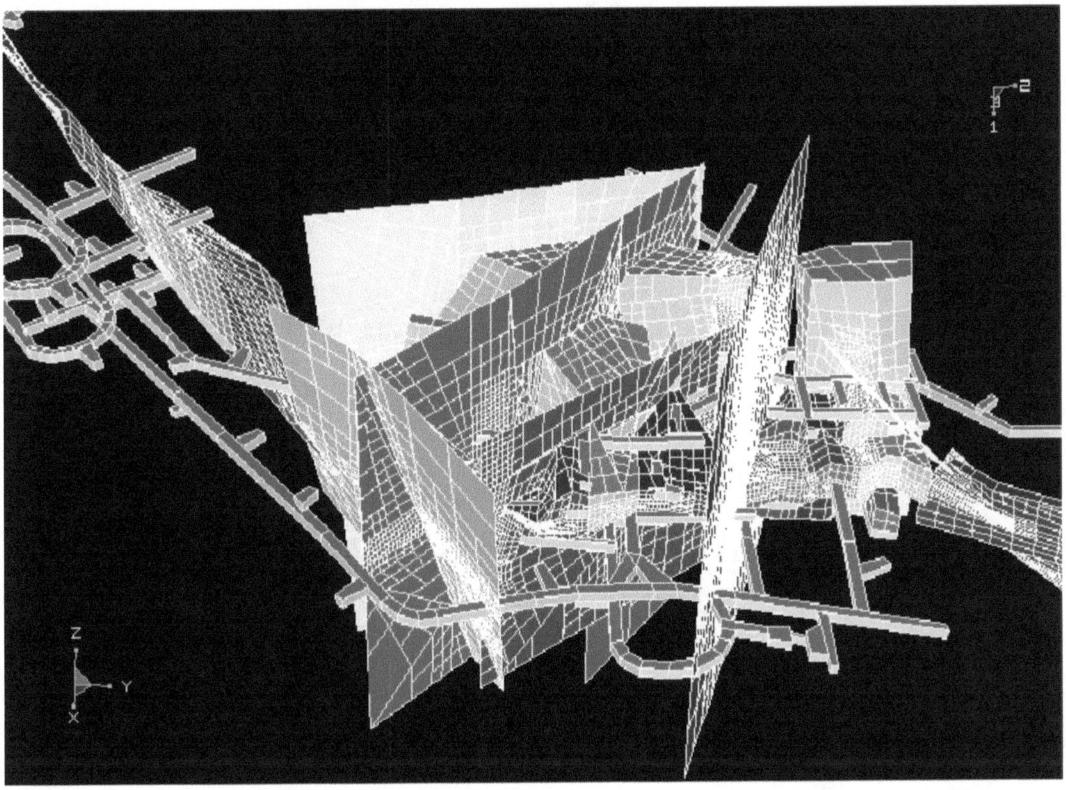

FIGURE 14.6 Using Displacement Discontinuity elements to represent multiple fault planes (Map 3D, 2024).

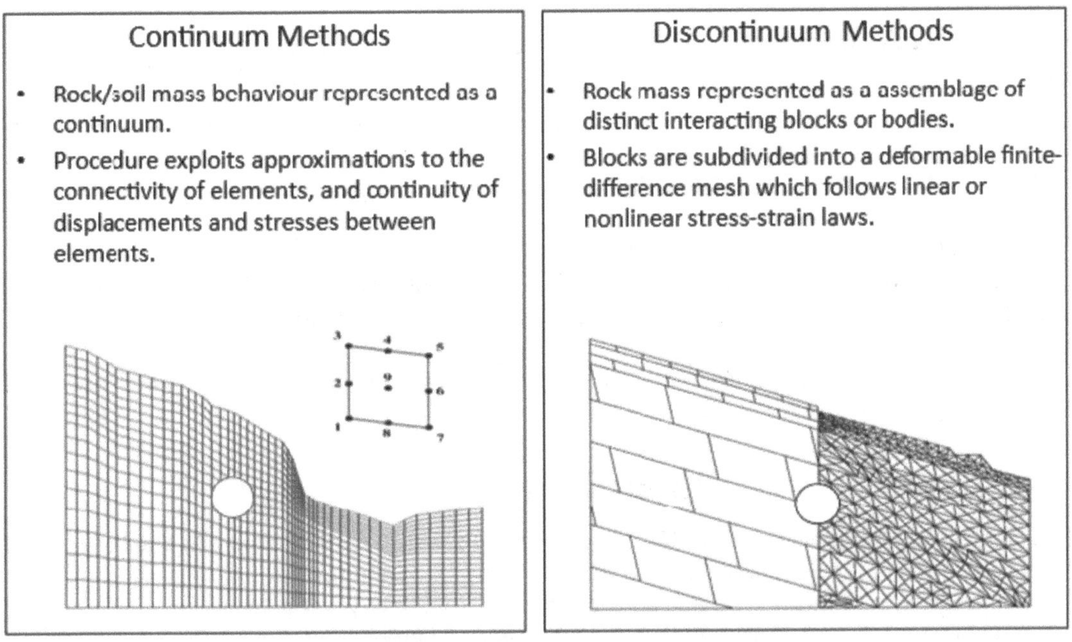

FIGURE 14.7 Continuum and discontinuum model representation (Eberhardt, 2007).

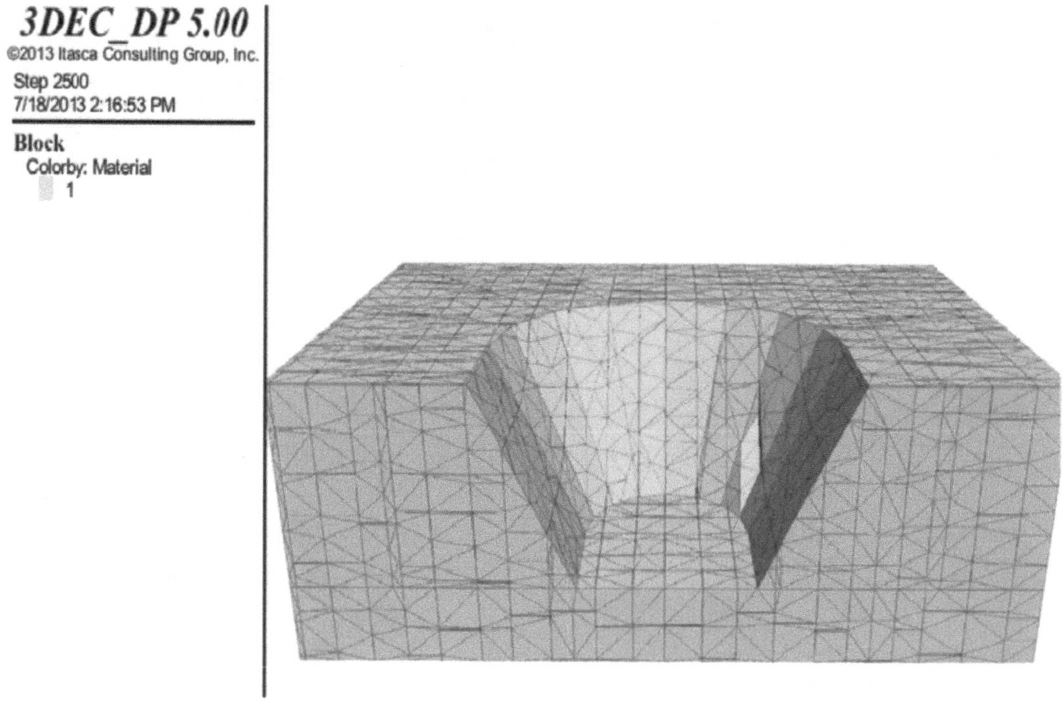

FIGURE 14.8 Discrete (distinct) element modeling using 3DEC (Itasca Consulting Group, 2024).

It is commonly used to model complex geometries and material interactions. Figure 14.9 shows a relatively simple FEM analysis.

14.1.10 The Finite Difference Method (FDM)

FDM is a basic numerical method based on the approximation of a derivative as a difference quotient. The idea is to replace the derivatives appearing in the differential equation by finite differences that approximate them.

Table 14.1 (after Numerical Modeling, Queens University) gives a comparison between the FEM and the FDM.

Figure 14.10 shows an example of an FDM.

The method discretizes the entire continuum, is computationally intensive, and is particularly useful for complex material behavior such as creep and plastic flow. It does not require a mathematical equilibrium.

Table 14.2 gives a comparison of some of the different modeling methods (after Numerical Modeling, Queens University).

14.2 ROCK MECHANICS MONITORING

This is modified and expanded from *Handbook on Hard-Rock Strata Control* by A. Spearing (1995) with permission of the South African Institute of Mining and Metallurgy.

The needs for instrumentation and testing are to improve safety, validate designs, provide early warning of a potential problem, and improve efficiency and productivity.

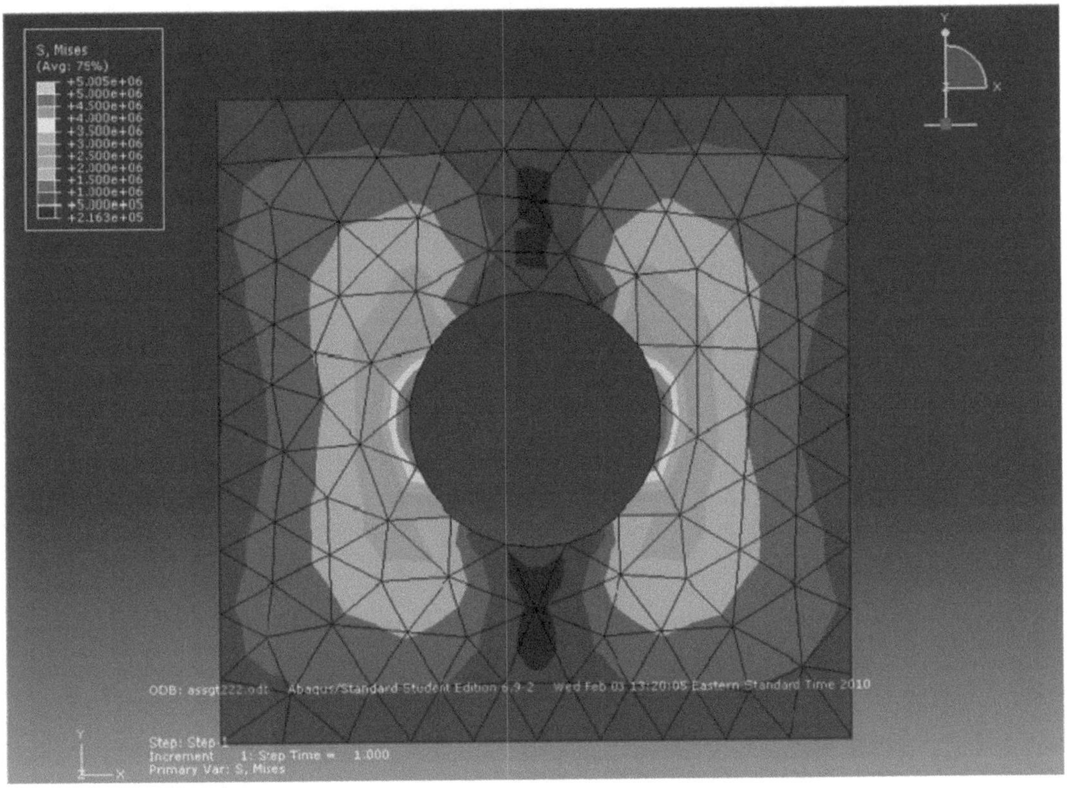

FIGURE 14.9 Stresses around a circular excavation using a finite element model (Abaqus).

TABLE 14.1
Comparison between FEM and FDM Models

Finite Element Method	Finite Difference Methods
Can use an irregular grid	Uses a more regular grid
Time steps can be large with unconditionally stable schemes	Incremental time steps need to be smaller than the critical value for stability
Large amount of calculations in each time step	Small amount of calculations in each time step
To follow a nonlinear constitutive law an iterative procedure is required	No iterations necessary to follow nonlinear constitutive law

The use of underground instrumentation, especially in stopes, is generally difficult for the following reasons:

- Constraints to physical access.
- Constraints to the manpower required to install and monitor the instrumentation.
- The rock mass is discontinuous, making the siting of instrumentation critical if representative results are to be obtained.

Owing to the heat, humidity, and space limitations, underground conditions do not generally favor sophisticated instrumentation.

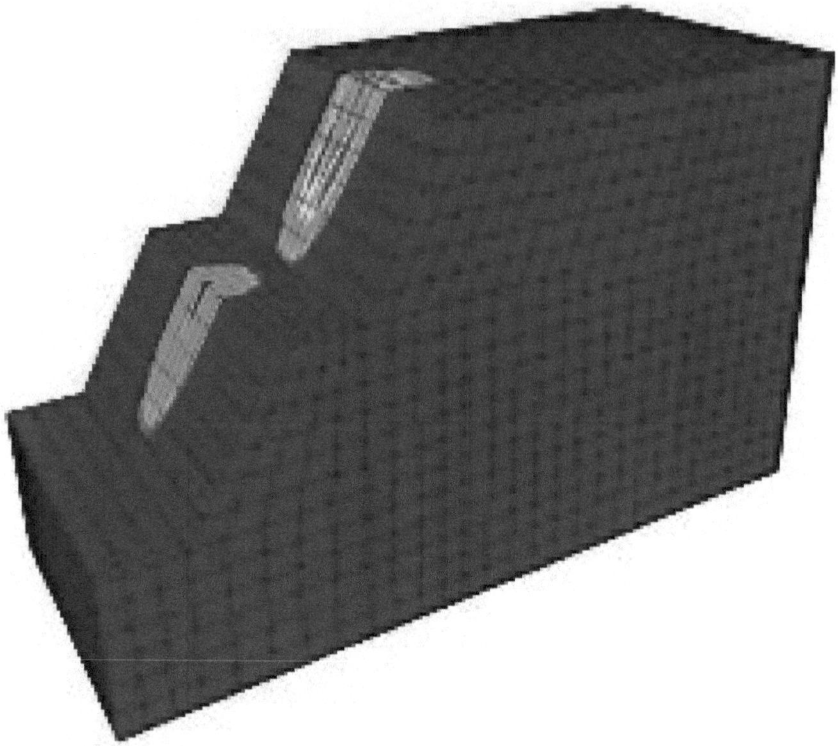

FIGURE 14.10 Finite difference model result using FLAC^{3D} (Itasca Consulting Group, 2024).

TABLE 14.2
Comparison between the Modeling Systems

Model system	Advantages	Disadvantages
BEM	Far-field conditions are inherently represented. Only boundaries need to be discretized. Smaller number of solution variables than FEM.	Limited capacity to handle heterogenous material. The solution time increases exponentially when the number of elements increases.
DEM	Excellent capability of modeling systems with high degree of nonlinearity. General constitutive relations can be used. Solution time increases linearly with the number of elements.	Results can be sensitive to assumed values of modeling parameters.
FEM and FDM	Heterogeneity of material can be easily handled.	The entire volume must be discretized which results in a larger number of solution variables in comparison with BEM. Far-field conditions are approximated. The time for solution increases exponentially when the number of elements increases.

It is difficult to achieve accuracy and reproducibility with the available equipment. The instrumentation is required mainly:

- To measure the total stresses (or stress changes) in the rock mass or in a support.
- To measure the response of the rock mass to the stresses.
- To back-analyze (or calibrate) computer analysis models.

Relatively simple, durable, cost-effective, reliable instruments should always be used where possible. The following is a list of some of the more commonly used and easiest to operate and install rock mechanics instruments for effective monitoring.

14.2.1 Extensometers

Various types of extensometers are used to measure differential movement (e.g., bed separation) from an excavation, including the spring or wire extensometer, and the solid-rod extensometer. The results from such units can be determined using a tape-measure readout, an electrical readout, or a continuously recorded readout.

14.2.2 Closure and Ride Instruments

Many sophisticated methods exist for the measurement of closure and ride (particularly in stopes), but the simplest, cheapest, and most reliable is the four-peg method, which is manpower intensive as measurements are manually taken. Ride is a differential movement between the hangingwall (roof) and footwall (floor) of typically a deep, hard-rock, tabular, and angled stope.

14.2.3 Tiltmeters

A tiltmeter is basically a more sensitive inclinometer (an instrument used mainly for measuring angles of slopes and the changes) designed to measure very small changes from the vertical or other suitable datum, either on the ground or on structures. Tiltmeters are used extensively for monitoring dams when they are being filled, the small movements of potential landslides, and the response of rock mass structures to various influences such as loading and foundation settlement. Tiltmeters may be mechanical or use vibrating wire or other sensors for electronic measurement.

Figure 14.11 shows a layout of a slope stability monitoring system.

14.2.4 Measuring Small Movements

Sometimes it is essential to measure even small movements. An example is the stability of a bench in an open-pit mine as this can warn of a potential future major failure.

The development of radar, or laser-based technology (e.g., LIDAR), for open-pit stability is now being widely used to provide real-time monitoring and advance warning signals before any minor or major slope failure occurs. Slope stability monitoring is a vital component for any surface mine and is an essential component for planning deeper and steeper open pits, in terms of safety and economics.

Using drones or similar remote-controlled equipment, underground strata movements can be measured and potential hazardous situations identified so positive actions can be taken.

Interferometric synthetic aperture radar (InSAR) is a radar technique that is very useful in geodesy and remote sensing. The method uses radar images to produce very precise maps of surface elevation changes. These are typically achieved using aircraft or satellites.

This method has proved to be very useful in predicting highwall failures in open-pit mines.

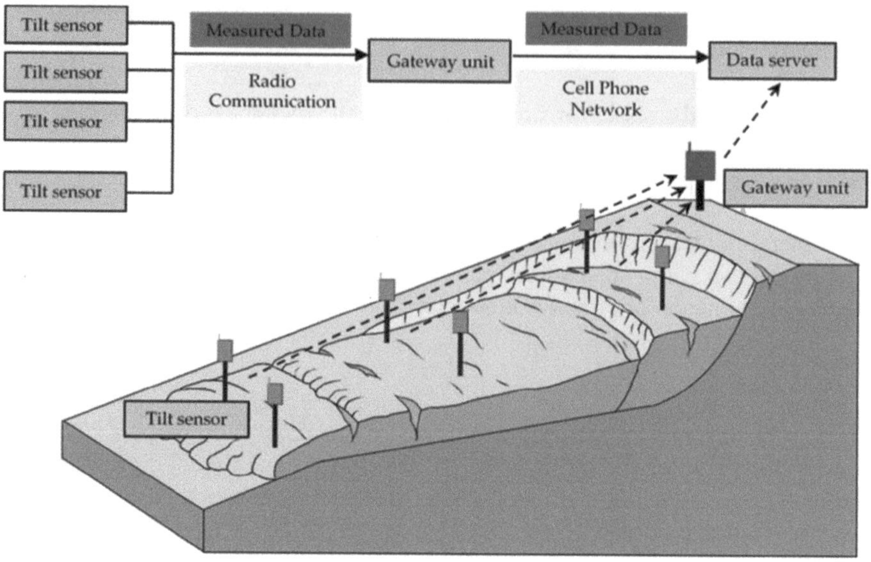

FIGURE 14.11 The use of tiltmeters to monitor slope stability (Qiao et al., 2020).

14.2.5 Stress Measurement Instruments

These are commonly used to measure absolute stress and the change of stress.

Absolute (total) stress measurements underground are very tedious, and it is often difficult to obtain reliable results. It is generally preferable to use outside contractors or consultants for this purpose. Three methods are commonly used to measure absolute stresses. Details of the methods can be obtained from Spearing (1995) and Haimson (1978).

- Door Stopper Method.
- Borehole Slotter.
- Hydrofracturing.

Hydraulic load cells and potentiometer closure meters are available that measure the stresses and strains developed in backfill.

At least three hydraulic load cells are normally used. These are secured to a wire cage before backfilling starts so that the stresses in the dip and strike directions can be measured, as well the stress perpendicular to the stope. The hydraulic load cells are fitted with flexible armored tubes to ensure that they are not broken because of high stresses or ride in the stope or longwall goaf (gob). The load cells are usually fitted with two gauges: one to measure the early build-up of stress (to about 5 MPa), and the other to measure the ultimate stress (to about 50 MPa usually). The pressure cells can also be incased in concrete for protection when high stress measurements are made.

Prior to backfilling, a closure meter should be installed near each set of hydraulic load cells. A back-up closure-measuring station can be provided at the edge of the backfill (near the gully or gate), but this does not accurately reflect the closure near the load cells (and tends to show more closure since it is not in the backfill).

Figure 14.12 shows a typical hydraulic pressure cell (also called flatjacks) and an application.

It is important that they are orientated in perpendicular to the direction of the desired stress measurement. In backfill, the vertical stress generated in the backfill is generally the most critical, so the load cells would be placed with the cell area in the horizontal plane.

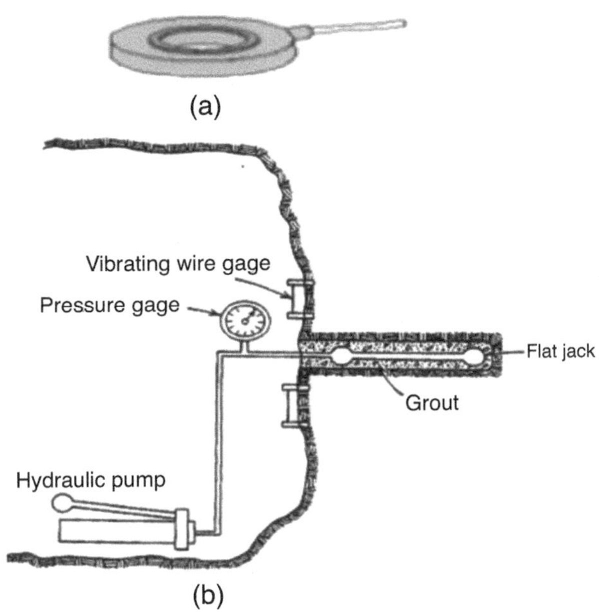

FIGURE 14.12 (a) Typical pressure (load) cell and (b) an application to measure the total stress or stress change in a sidewall abutment (The Constructor, 2021).

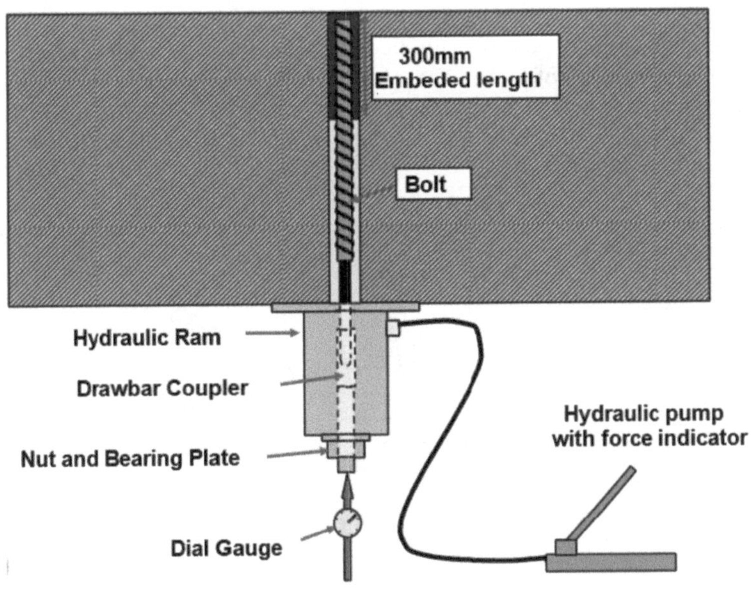

FIGURE 14.13 Testing a short encapsulated rockbolt with a hydraulic cylinder and a load cell (ASTM F432, 2020).

14.2.6 Load Cells

These are typically used on mines to measure the load generated by supports (free standing or internal) as shown in Figure 14.13. Free-standing supports are typically installed between the roof and floor of an excavation. They are commonly used in longwall mining, room-and-pillar, and narrow hard-rock tabular mining.

Load measurements on conventional support normally utilize hydraulic load cells, that either are placed on top of (or underneath) an elongate, stick or prop type of support, or are inserted within a conventional timber-based pack. This type of load cell is a relatively crude hydraulic device that has a diameter slightly larger than that of the support unit to be monitored. For ease of installation and measurement, the load cell is usually placed on the footwall (which should be level), and the support is installed in the normal manner on top of the load cell. Large errors can occur in the readings if the support unit is not placed symmetrically on the load cell.

Readings can be made with a fitted calibrated gauge or with a portable readout unit, the latter being generally more accurate but also more costly.

14.2.7 Fracture Mapping

Fracture mapping can be undertaken using different instruments including:

- A petroscope (basically similar to a periscope but having its own light source). It is a useful and cheap instrument, particularly if fitted with a vernier scale to measure the width of fractures. After the petroscope has been inserted in a hole, the insertion rods are used to locate the viewing position in the hole. Measurements can be taken with the naked eye, or with a survey level as shown in Figure 14.14. The typical depth range of such an instrument is 10–15 m.
- A borehole camera, which is similar in operation to a petroscope, is also a very useful cheap and effective instrument to use. All it needs is a regular drill hole, and it can be used to determine:
    - An idea of the geology (visual remote inspection only).
    - Location of geological features (joints, beds, and faults) or voids.
    - Presence and location of minor water ingress.
    - Orientation and growth of all fractures in the hole.

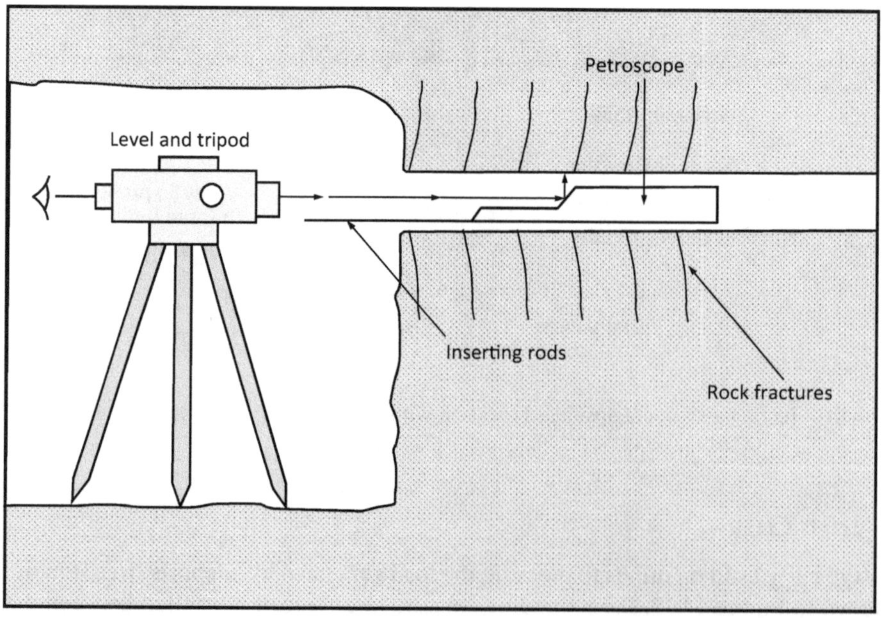

FIGURE 14.14 Using a typical petroscope (Spearing, 1995).

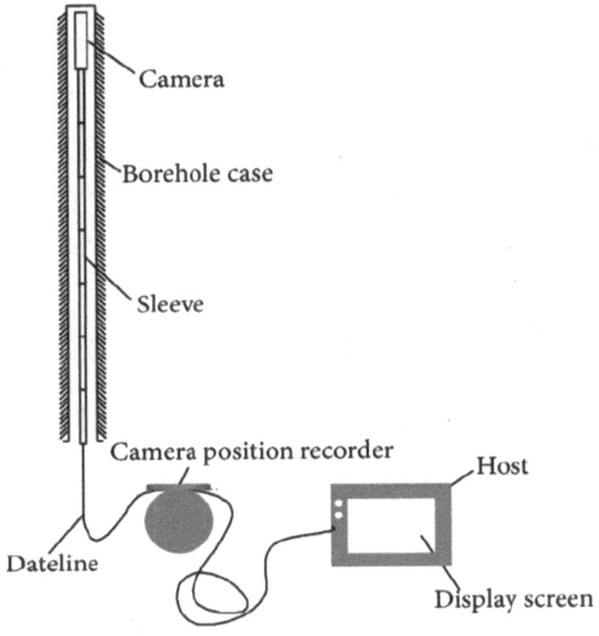

FIGURE 14.15 A borehole camera (borescope) (Zhang et al., 2016).

This is useful as it gives an easy method to record the effects and changes from mining in the immediate rock around an excavation. A borehole camera as shown in Figure 14.15 is more useful than a petroscope as it can record thus giving a permanent visual record with time.

- Ground Penetrating Radar (GPR) instruments suitable for underground use are commonplace for fracture mapping. Although more costly than a petroscope, GPS instruments have the advantage that they can fracture map fairly rapidly over a large area using antenna to a depth of about 30 m, as opposed to a single borehole.

Figure 14.16 shows a typical result in a borehole that can be obtained.

14.2.8 Personnel Monitoring

This is not directly related to rock mechanics monitoring but is indirectly related and very important. Knowing where personnel and equipment are in a mine, in real time, is important for safety and productivity. In the extreme case, after a rock-related accident for example, it is important that rescue activities can be focused in the right area as quickly as possible.

14.2.9 Important Measuring and Monitoring Terms and Definitions

Measurement work uses several terms which are defined below (source unknown):

- Instrument: a device for determining the value or magnitude of a quantity or variable.
- Accuracy: closeness with which an instrument reading approaches the true value of the variable being measured.
- Precision: a measure of the reproducibility of the measurements. Precision refers to the degree of agreement within a group of measurements or instruments (having limitations

FIGURE 14.16 View of a coal seam showing pyrite inclusions. (Photograph from the National Institute of Occupational Safety and Health, USA.)

set by the scale reading), i.e., given a fixed value of a variable, precision is a measure of the degree to which successive measurements differ from one another.

- Sensitivity: the ratio of output response of the instrument to the measured value.
- Resolution: the smallest change in measured value to which the instrument will respond.
- Error: deviation from the true value of the measure variable.
- Tolerance: the ± terms given to a measured variable, figures that are the range of doubt.

14.2.10 The Future of Monitoring

Real-time monitoring, along with automation and robotics, holds huge potential for rock mechanics, mine safety and productivity. This is because these technologies will remove workers from the advancing excavations which are the most hazardous places.

14.3 WORKED EXAMPLES

Example 14.3.1

An extensometer is fitted with eight gauges and installed in a roof to measure displacement. The measurements of each gauge are recorded by a data logger at two points in time (Table 14.3). Determine the displacement of each gauge.

TABLE 14.3
The Extensometer Measurements

Date	Gauge 1	Gauge 2	Gauge 3	Gauge 4	Gauge 5	Gauge 6	Gauge 7	Gauge 8
Date 1 (Start)	20.83	27.04	31.73	32.88	36.39	35.79	34.78	34.31
Date 2 (End)	20.7	26.87	31.38	32.63	36.2	35.66	34.64	34.24

Solution:

Find the change in displacement for each gauge relative to the starting date:

$$\Delta_1 \text{in mm} = 20.70 - 20.83 = -0.13$$

$$\Delta_2 \text{in mm} = 26.87 - 27.04 = -0.17$$

$$\Delta_3 \text{in mm} = 31.38 - 31.73 = -0.35$$

$$\Delta_4 \text{in mm} = 32.63 - 32.88 = -0.25$$

$$\Delta_5 \text{in mm} = 36.20 - 36.39 = -0.19$$

$$\Delta_6 \text{in mm} = 35.66 - 35.79 = -0.13$$

$$\Delta_7 \text{in mm} = 34.64 - 34.78 = -0.14$$

$$\Delta_8 \text{in mm} = 34.24 - 34.31 = -0.07$$

Readings must be referenced to the roof line. For Date 1, all values of displacement are 0 since it is the starting date. For Date 2, the reference is −0.07 (at Gauge 8).

Date 2:

$$\Delta_1 \text{in mm} = -0.07 - (-0.13) = 0.06 \text{ mm}$$

$$\Delta_2 \text{in mm} = -0.07 - (-0.17) = 0.10 \text{ mm}$$

$$\Delta_3 \text{in mm} = -0.07 - (-0.35) = 0.28 \text{ mm}$$

$$\Delta_4 \text{in mm} = -0.07 - (-0.25) = 0.18 \text{ mm}$$

$$\Delta_5 \text{in mm} = -0.07 - (-0.19) = 0.12 \text{ mm}$$

$$\Delta_6 \text{in mm} = -0.07 - (-0.13) = 0.06 \text{ mm}$$

$$\Delta_7 \text{in mm} = -0.07 - (-0.14) = 0.07 \text{ mm}$$

$$\Delta_8 \text{in mm} = -0.07 - (-0.07) = 0.00 \text{ mm}$$

Example 14.3.2

The following is a worked example found in lecture notes by Eberhardt (2017) but originally in Harrison and Hudson (2000).

Three flatjacks are placed in the wall of a tunnel at different orientations as shown in Figure 14.17. The measurements are 250 m underground. The axis of the tunnel dips at 7°. In a flatjack test, the cancelation pressure is the pressure applied to cancel out the in situ stresses. Cancelation pressures of each flatjack are as follows:

$$\sigma_A = 7.56 \text{ MPa}$$

$$\sigma_B = 6.72 \text{ MPa}$$

$$\sigma_C = 7.50 \text{ MPa}$$

Solution:

The example can be solved with stress transformation equations using the following relationship given in detail in the Harrison and Hudson (2000) reference:

$$\sigma_{\text{jack}} = R\sigma_{\text{global}}$$

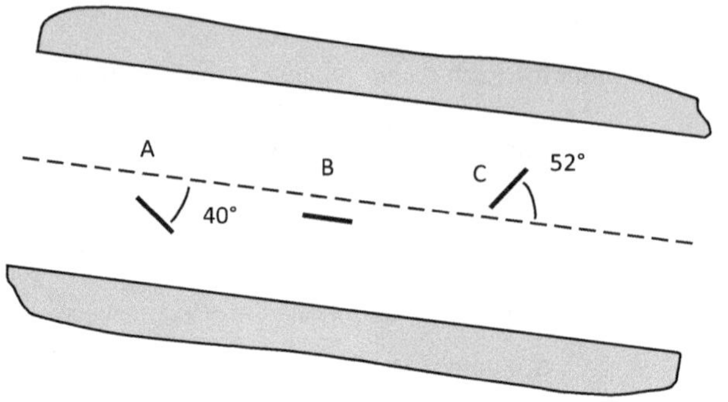

FIGURE 14.17 The tunnel and the load cells (flatjacks) (Eberhardt, 2017).

TABLE 14.4
The Solution Matrix

		Column 1	Column 2	Column 3	
σ_A		$\cos^2\theta_A$	$\sin^2\theta_A$	$2\sin\theta_A\cos\theta_A$	σ_x
σ_B	=	$\cos^2\theta_B$	$\sin^2\theta_B$	$2\sin\theta_B\cos\theta_B$	σ_y
σ_C		$\cos^2\theta_C$	$\sin^2\theta_C$	$2\sin\theta_C\cos\theta_C$	τ_{xy}

where

R represents a 3×3 matrix comprised of the individual stress transformation equation for this specific example.

σ_{global}, for this example, represents a 3×1 matrix of the global principal stresses needing to be calculated. For this case, because they are the principal stresses, theta-xy is therefore zero.

Or it can be solved as follows using the formula below from which the matrix is derived (in Table 14.4).

Corrections for the dip of the tunnel assuming:

- x-axis is orientated horizontal.
- y-axis is the vertical upward.
- measurements are counterclockwise positive from the positive x-axis:

$$\beta_{tunnel} = -7^\circ$$
$$\beta_A = -40^\circ$$
$$\beta_B = 0^\circ$$
$$\beta_C = -52^\circ$$

Corrections applied:

$$\beta_A = -40^\circ + (-7^\circ) = -47^\circ$$
$$\beta_B = 0^\circ + (-7^\circ) = -7^\circ$$
$$\beta_C = 52^\circ + (-7^\circ) = 45^\circ$$

Flatjacks measure stress normal to the direction of the slot, so another correction is applied:

$$\theta_A = \beta_A + 90^\circ = -47^\circ + 90^\circ = 43^\circ$$

$$\theta_B = \beta_B + 90^\circ = -7^\circ + 90^\circ = 83^\circ$$

$$\theta_C = \beta_C + 90^\circ = 45^\circ + 90^\circ = 135^\circ$$

Column 1 of **R** Matrix:

$$\cos^2\theta_A = \cos^2 43^\circ = 0.535$$

$$\cos^2\theta_B = \cos^2 83^\circ = 0.015$$

$$\cos^2\theta_C = \cos^2 135^\circ = 0.5$$

Column 2 of **R** Matrix:

$$\sin^2\theta_A = \sin^2 43^\circ = 0.465$$

$$\sin^2\theta_B = \sin^2 83^\circ = 0.985$$

$$\sin^2\theta_C = \sin^2 135^\circ = 0.5$$

Column 3 of **R** Matrix:

$$2\sin\theta_A\cos\theta_A = 2\sin 43^\circ\cos 43^\circ = 0.998$$

$$2\sin\theta_B\cos\theta_B = 2\sin 83^\circ\cos 83^\circ = 0.242$$

$$2\sin\theta_C\cos\theta_C = 2\sin 135^\circ\cos 135^\circ = -1$$

Filling in the above values into the matrix (Figure 14.7) creates:

		Column 1	Column 2	Column 3	
7.56		0.535	0.465	0.998	σ_x
6.72	=	0.015	0.985	0.242	σ_y
7.50		0.5	0.5	−1	τ_{xy}

The inverse of **R** must be resolved to create the Cofactor and Adjoint Matrix (Tables 14.5 and 14.6):

The cofactor matrix:

$$c_{11} = (-1)^{1+1}\left[(0.985)(-1) - (0.5)(0.242)\right] = -1.11$$

$$c_{12} = (-1)^{1+2}\left[(0.015)(-1) - (0.5)(0.242)\right] = 0.136$$

$$c_{13} = (-1)^{1+3}\left[(0.015)(0.5) - (0.5)(0.985)\right] = -0.485$$

$$c_{21} = (-1)^{2+1}\left[(0.465)(-1) - (0.5)(0.998)\right] = 0.964$$

TABLE 14.5
The Cofactor Matrix

Cofactor Matrix =	−1.11	0.136	−0.485
	0.964	−1.034	−0.035
	−0.871	−0.115	0.52

$$c_{22} = (-1)^{2+2}\left[(0.535)(-1)-(0.5)(0.998)\right] = -1.034$$

$$c_{23} = (-1)^{2+3}\left[(0.535)(0.5)-(0.5)(0.465)\right] = -0.035$$

$$c_{31} = (-1)^{3+1}\left[(0.465)(0.242)-(0.985)(0.998)\right] = -0.871$$

$$c_{32} = (-1)^{3+2}\left[(0.535)(0.242)-(0.015)(0.998)\right] = -0.115$$

$$c_{33} = (-1)^{3+3}\left[(0.535)(0.985)-(0.015)(0.465)\right] = 0.52$$

$$\text{Det}\mathbf{R} = (0.535)(-1.11)+(0.465)(0.136)+(0.998)(-0.485) = -1.015$$

$$\sigma_x = (1.094)(7.56)+(-0.951)(6.72)+(0.858)(7.50) = 8.31 \text{ MPa}$$

$$\sigma_y = (-0.134)(7.56)+(1.019)(6.72)+(0.113)(7.50) = 6.68 \text{ MPa}$$

$$\tau_{xy} = (0.478)(7.56)+(0.034)(6.72)+(-0.512)(7.50) = 0.002 \text{ MPa}$$

Tables 14.7 and 14.8 provide the inverse matrix and the final answer. The example continues stating that σ_x σ_y are the principal stresses because τ_{xy} is near zero.

Estimating the vertical stress σ_v based on depth and overburden unit weight of 27 kN/m^3,

$$\sigma_v = \left(27 \text{ kN/m}^3\right)\left(250 \text{ m}\right) = 6.75 \text{ MPa compares with } \sigma_y$$

TABLE 14.6
The Adjoint Matrix

	−1.11	0.964	−0.871
	0.136	−1.034	−0.115
Adj R =	−0.485	−0.035	0.52

TABLE 14.7
The Inverse Matrix

$\mathbf{R}^{-1}$ = 1/−1.015	−1.11	0.964	−0.871		1.094	−0.951	0.858
	0.136	−1.034	−0.115	=	−0.134	1.019	0.113
	−0.485	−0.035	0.52		0.478	0.034	−0.512

TABLE 14.8
The Resolved Stresses

σ_x		1.094	−0.951	0.858	7.56
σ_y	=	−0.134	1.019	0.113	6.72
τ_{xy}		0.478	0.034	−0.512	7.50

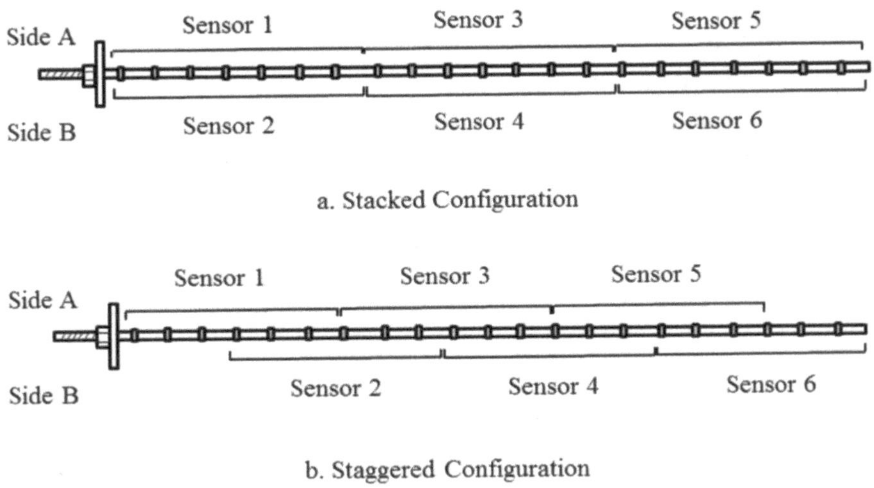

FIGURE 14.18 The rock bolt instrumentation configurations. (Spearing, Hyett, Kostecki, and Gadde, 2013)

Note that no emphasis seems to be made toward the placement of the flatjacks at the tunnel periphery and what role it may have, specifically stress concentrations, on measurement. Depth of the flatjack placement is not given in the problem statement so it is assumed that they reside outside a stress concentrated zone.

Example 14.3.3

Rock bolt instruments are attached with gauges, to measure strain along the bolt length with time. Two arrangements are to be compared, a stacked configuration, and a staggered configuration as shown in Figure 14.8. Determine the axial and bending strain for both configurations given the data below. Process for this example and the figure are found in Hyett and Spearing (2012) (Figure 14.18).

Solution:

Axial Strain at the *i*th nodal point (Tables 14.9 and 14.10)

For a stacked configuration of gauges,

$$\varepsilon^{i}_{\text{axial}} = (\varepsilon^{2i-1} + \varepsilon^{2i})/2$$

$$\varepsilon^{1}_{\text{axial}} = (\varepsilon^{1} + \varepsilon^{2})/2 \;=\; (320 + 303)/2 = 312\ \mu\varepsilon$$

TABLE 14.9
The Strain Gauge Results

Measurements (Bold) for a Rock Bolt of Stacked Gauge Configuration

Gauge i	Location (mm)	Total Strain (με)	Node	Node Location	Axial Strain (με)
1	468	**320**	1	468	312
3	998	**479**	2	998	350
5	1,528	**293**	3	1,528	157
2	468	**303**	End	1,820	0
4	998	**221**			
6	1,528	**21**			

TABLE 14.10
The Calculated Strains

Measurements (Bold) for a Rock Bolt of Staggered Gauge Configuration

Gauge i	Location (mm)	**Total Strain (με)**	Node	Node Location	Axial Strain (με)
1	400	**320**	1	400	314
3	850	**479**	2	625	351
5	1,300	**293**	3	850	371
2	625	**303**	4	1,075	303
4	1,075	**221**	5	1,300	207
6	1,525	**21**	6	1,525	84
			End	1,820	0

$$\varepsilon^2_{\text{axial}} = (\varepsilon^3 + \varepsilon^4)/2 = (479 + 221)/2 = 350\ \mu\varepsilon$$

$$\varepsilon^3_{\text{axial}} = (\varepsilon^5 + \varepsilon^6)/2 = (293 + 21)/2 = 157\ \mu\varepsilon$$

For a staggered configuration of gauges the following approach is given.

At node 1 (near the head of the bolt), the axial strain is approximated as follows:

$$\varepsilon^1_{\text{axial}} = \left(2\varepsilon^1 + \varepsilon^2\right)/3$$

At node 6 (near the toe of the bolt) and since the axial strain reaches zero at the end,

$$\varepsilon^6_{\text{axial}} = \left(2\varepsilon^6 + \varepsilon^5\right)/4$$

All other nodal points

$$\varepsilon^i_{\text{axial}} = \left(\varepsilon^{i-1} + 2\varepsilon^i + \varepsilon^{i+1}\right)/4$$

$$\varepsilon^1_{\text{axial}} = (2\varepsilon^1 + \varepsilon^2)/3 = [2(320) + 303]/3 = 314\ \mu\varepsilon$$

$$\varepsilon^2_{\text{axial}} = \left(\varepsilon^1 + 2\varepsilon^2 + \varepsilon^3\right)/4 = \left[320 + 2(303) + 479\right]/4 = 351\ \mu\varepsilon$$

$$\varepsilon^3_{\text{axial}} = \left(\varepsilon^2 + 2\varepsilon^3 + \varepsilon^4\right)/4 = \left[303 + 2(479) + 221\right]/4 = 371\ \mu\varepsilon$$

$$\varepsilon^4_{\text{axial}} = \left(\varepsilon^3 + 2\varepsilon^4 + \varepsilon^5\right)/4 = \left[479 + 2(221) + 292\right]/4 = 303\ \mu\varepsilon$$

$$\varepsilon^5_{\text{axial}} = \left(\varepsilon^4 + 2\varepsilon^5 + \varepsilon^6\right)/4 = \left[221 + 2(293) + 21\right]/4 = 207\ \mu\varepsilon$$

$$\varepsilon^6_{\text{axial}} = \left(2\varepsilon^6 + \varepsilon^5\right)/4 = \left[2(21) + 293\right]/4 = 84\ \mu\varepsilon$$

Example 14.3.4

The data in Table 14.11 represents field measurements from a 7 ft long tiltmeter. Determine the change in tilt in degrees between the two dates. Plot the tiltmeter data alongside the instrumented rock bolt data of the previous example to reveal where a plane of separation begins initiating in the mine roof.

Solution:

$$\text{Gauge } 1 = \text{Date}^{i+1} - \text{Date}^{i} = 30.5 - 30.0 = 0.5$$

$$\text{Gauge } 2 = \text{Date}^{i+1} - \text{Date}^{i} = 32.0 - 30.0 = 2.0$$

TABLE 14.11
The Tiltmeter Data

Location (mm)	153	458	763	1,068	1,373	1,678	1,983
Gauge	1	2	3	4	5	6	7
Date i	30.0	30.0	−3.3	−11.7	6.5	0.3	30.0
Date i+1	30.5	32.0	−5.5	−13.5	8.8	0.3	30.0

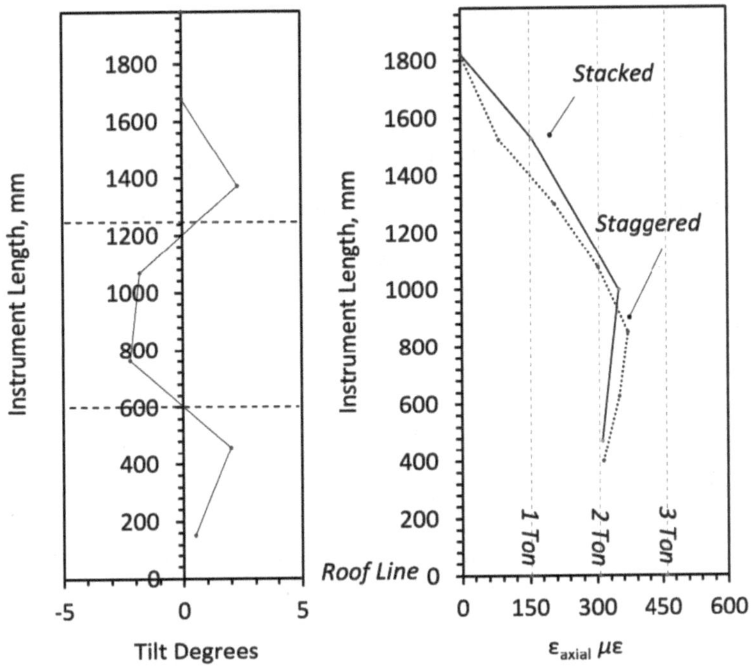

FIGURE 14.19 The tilt and axial strain.

$$\text{Gauge 3} = \text{Date}^{i+1} - \text{Date}^{i} = -5.5 - (-3.3) = -2.2$$

$$\text{Gauge 4} = \text{Date}^{i+1} - \text{Date}^{i} = -13.5 - (-11.7) = -1.8$$

$$\text{Gauge 5} = \text{Date}^{i+1} - \text{Date}^{i} = 8.8 - 6.5 = 2.3$$

$$\text{Gauge 6} = \text{Date}^{i+1} - \text{Date}^{i} = 0.3 - 0.3 = 0.0$$

$$\text{Gauge 7} = \text{Date}^{i+1} - \text{Date}^{i} = 30.0 - 30.0 = 0.0$$

The data show that there could be a plane of movement as demarcated by the dashed lines on the tiltmeter plot (Figure 14.19); however, the axial strains of the instrumented rock bolt plot do not indicate unfavorable performance of the support system, at least in terms of axial performance. Additionally, analysis or measurement of bolt bending performance, or some equivalent, would need to be carried out at this stage.

Example 14.3.5

The following worked example was found in Eberhardt (2017) and illustrates a method for estimating principal stress using the hydraulic fracturing method in granite. It is also found in greater detail and worked in Harrison and Hudson (2000). If tensile strength of the rock is 10 MPa, what are the principal stresses assuming that one is vertical and pressures already account for formation pressure, such that P_o=0. It is also assumed that the rock is in an elastic state.

Given:

Depth is 500 m.
Breakdown pressure P_b is 14 MPa.
Shut-in pressure P_s is 8 MPa.

Solution:

Find the minimum horizontal stress:

$$\sigma_h = P_s = 8 \text{ MPa}$$

Find the maximum horizontal stress:

$$\sigma_H = 3\sigma_h - P_B + \sigma_t = (3)(8 \text{ MPa}) - 14 \text{ MPa} + 10 \text{ MPa} = 20 \text{ MPa}$$

Estimate the vertical stress assuming an overburden unit weight of 27 kN/m³:

$$\sigma_v = (500 \text{ m})(27 \text{ kN/m}^3) = 13{,}500 \text{ kN/m}^2 = 13.5 \text{ MPa}$$

Therefore,

$$\sigma_1 = 20 \text{ MPa}$$

$$\sigma_2 = 13.5 \text{ MPa}$$

$$\sigma_3 = 8 \text{ MPa}$$

Example 14.3.6

Resolve the principal strains given the three measured strains from a delta rosette. The example below is solved in the Appendix of Hudson and Harrison (1997).

Given:

$$\varepsilon_P = 8 \times 10^{-4}$$

$$\varepsilon_Q = -8 \times 10^{-4}$$

$$\varepsilon_R = 2 \times 10^{-4}$$

Solution:

For a delta rosette,

$$c = 1/3(\varepsilon_P + \varepsilon_Q + \varepsilon_R)$$

$$\tan 2\theta = \left[\sqrt{3}(\varepsilon_R - \varepsilon_Q)\right] / \left[\left(2\varepsilon_P - (\varepsilon_Q + \varepsilon_R)\right)\right]$$

$$r = (\varepsilon_R - \varepsilon_Q) / (\sqrt{3}\sin 2\theta)$$

$$\tan 2\theta = \left[\sqrt{3}(\varepsilon_R - \varepsilon_Q)\right] / \left[\left(2\varepsilon_P - (\varepsilon_Q + \varepsilon_R)\right)\right]$$

$$\tan 2\theta = \left[\sqrt{3}\left(2 \times 10^{-4} - (-8 \times 10^{-4})\right)\right] / \left[2\left(8 \times 10^{-4}\right) - \left(-8 \times 10^{-4} + 2 \times 10^{-4}\right)\right]$$

$$\tan 2\theta = \left[\sqrt{3}\left(1 \times 10^{-3}\right)\right] / \left[1.6 \times 10^{-3} + 6 \times 10^{-4}\right]$$

$$\tan 2\theta = \left[1.73 \times 10^{-3}\right] / \left[2.2 \times 10^{-3}\right]$$

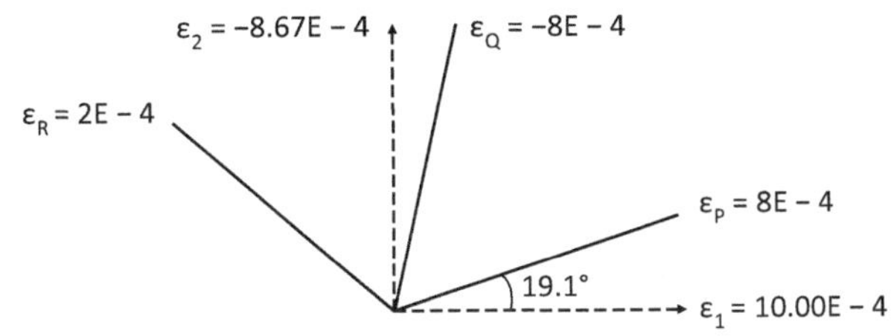

FIGURE 14.20 The solution (after Hudson and Harrison, 1997).

$$\tan 2\theta = 0.786$$

$$2\theta = \left[\tan^{-1}(0.786)\right]$$

$$2\theta = 38.2° \text{ which can be alternatively expressed as } 38.2 - 180 = -141.8$$

$$\text{As } \theta \text{ is to be between } -180° \text{ and } +180°$$

$$c = 1/3\left(\varepsilon_P + \varepsilon_Q + \varepsilon_R\right)$$

$$c = 1/3\left(8\times10^{-4} + (-8\times10^{-4}) + 2\times10^{-4}\right)$$

$$c = 6.667\times10^{-5}$$

$$r = (\varepsilon_R - \varepsilon_Q)/(\sqrt{3}\sin 2\theta)$$

$$r = \left(2\times10^{-4} - (-8\times10^{-4})\right)/\left(\sqrt{3}\sin 38.2°\right)$$

$$r = \left(1\times10^{-3}\right)/(1.071)$$

$$r = 9.333\times10^{-4}$$

Using the value of 2θ which provides a positive value of r,

$$\varepsilon_1 = c + r = 6.667\times10^{-5} + 9.333\times10^{-4} = 9.99\times10^{-4}$$

$$\varepsilon_2 = c - r = 6.667\times10^{-5} - 9.34\times10^{-4} = -8.67\times10^{-4}$$

The value of θ must now be chosen which is compatible with the calculated values of ε_1 and ε_2. Either θ = 19.1° or −70.9°. This step requires a comparison of the value of ε_1 relative to ε_P, ε_Q, and ε_R. Taking θ = 19.1° and plotting all values of ε, it is shown this is the correct choice. Note that clockwise corresponds to a positive rotation (Figure 14.20).

REFERENCES

ASTM International (2020) *ASTM-F432-Standard Specification for Roof and Rock Bolts and Accessories.* ASTM International, West Conshohocken, PE.

Constructor (2021) In situ stresses in rock and measurement methods. Retrieved from: www.theconstructor.org/building/measurement-methods-insitu-stresses-rocks/11649/

Eberhardt, E. (2007) *Topic 6-Numerical Methods. Tunneling & Underground Design Graduate Class.* UBC, Vancouver, Canada.

Eberhardt, E. (2017) Lecture on: In situ stresses and stress measurement. Geotechnical Engineering Practice & Design, EOSC433. Geological Engineering. University of British Columbia, Canada.

Feng, X.T. and Hudson, J.A. (2011) *Rock Engineering Design.* CRC Press, Leiden.

Haimson, B.C. (1978) The hydrofracturing stress measuring method and recent field results. *International Journal of Rock Mechanics and Mining Sciences.* 15, 167–178.

Harrison, J.P. and Hudson, J.A. (2000) *Engineering Rock Mechanics – Part 2: Illustrative Worked Examples.* Elsevier Science, Oxford.

Holling, C.S. (ed) (1978) *Adaptive Environmental Assessment and Management*. Wiley, Chichester, p. 377.
Hudson, J.A. and Feng, X.T. (2015) *Rock Engineering Risk*. CRC Press, Boca Raton, FL.
Hudson, J.A. and Harrison, J.P. (1997) *Engineering Rock Mechanics*. Elsevier Science, Oxford, p. 458.
Hyett, A. and Spearing, A.J.S. (2012) New technology for measuring the in situ performance of rock bolts. In *ARMA US Rock Mechanics/Geomechanics Symposium, ARMA*, Chicago, IL.
Itasca Consulting Group (2024) Retrieved from: www.itascacg.com
Jing, L. (2003) A review of techniques, advances and outstanding issues in numerical modelling for rock mechanics and rock engineering. *Inter. J. Rock Mech. Mining Sci.* 40(3), 283–353.
Map 3D (2024) Retrieved from: www.map3d.com/example_grid_one.php
Mellin, W. (1957) Work with new electronic 'brains' opens field for army math experts. *The Hammond Times*. 10, 66.
Qiao, S., Feng, C., Yu, P., Tan, J., Uchimura, T., Wang, L., Tang, J, Shen, Q. and Xie, J. (2020) Investigation on surface tilting in the failure process of shallow landslides. *Sensors*. 20, 2662.
Queens University, Mining Engineering Department, Canada (2011) Retrieved from: https://minewiki.engineering.queensu.ca/mediawiki/index.php/Numerical_modelling
Rocscience (2024) Retrieved from: www.rocscience.com/learning/
Spearing, A.J.S. (1995) *Handbook on Hard Rock Strata Control*. SAIMM Monograph series SP6, Johannesburg.
Spearing, A. J. S., Hyett, A. J., Kostecki, T., & Gadde, M. (2013) New technology for measuring the in situ performance of rock bolts. *International Journal of Rock Mechanics and Mining Sciences*, *57*, 153–166.
Zhang, G., He, F. and Jiang, L. (2016) Analytical analysis and field observation of break line in the main roof over the goaf edge of longwall coal mines. *Math. Probl. Eng*. 9, 1–11.

15 Rock Mechanics Risk Assessment

15.1 INTRODUCTION TO RISK AND RISK MANAGEMENT

15.1.1 Risk

Risk when associated with safety can be defined as a situation involving exposure to danger. It is related to the expected losses which could occur because of the unsuccessful outcome of an event and to the probability of this event occurring. In general:

- The greater the loss and the higher the probability of the event, the worse and higher the risk.
- Measuring and quantifying risk is often difficult as rare failures are hard to estimate, and loss of human life should be considered irreplaceable.

An engineering definition of risk is calculated as follows (Safety Section, 2024):

$$\text{Risk} = (\text{Probability of a Risk Event})(\text{Consequences of the Risk Event}) \tag{15.1}$$

Risk can therefore be considered as the combination of probability and consequence.

A hazard (risk) is the potential of something to cause harm or damage. People often expose themselves or others (indirect risk) to unnecessary risk because they:

- Want to save themselves or the company time and/or money.
- Are behind or pressed into completing an assignment, so they cut corners.
- Have done it in an unsafe manner before and nothing untoward happened.
- Feel that performing the task safely is frustrating, time consuming and inconvenient.
- Have not been properly tasked trained or have not been properly trained.
- Were unaware of the inherent hazard.
- Have a different view on hazards due perhaps to inexperience, different experiences, or difference social norms.

Tolerable risk is defined differently by each industry and company. As an example, the level of tolerable risk for a soldier, fire fighter, or policeman is much higher than that of an office worker. The potential outcome, however, may be the same, so the risks must still be considered, and appropriate actions taken to mitigate the risk.

Tolerable risk and risk exposure depend on the following considerations:

- Government and local regulations and guidelines.
- Current industry standards and best practices.
- Practices from other industries.
- Qualitative assessment of what is socially fair, acceptable, and reasonable.
- Insurance coverage and any related mandates.
- Comfort level for the company directors.
- Attitude of the individual.

DOI: 10.1201/9781032704845-15

15.1.2 Measuring Risk

Risk is measured using the two parameters already mentioned and further classified as follows:

The probability of a hazard occurring which can be:

- Frequent.
- Probable.
- Occasional.
- Remote.
- Improbable.

The severity (impact) if the hazard occurs which could be:

- Catastrophic (e.g., death).
- Critical (e.g., severe injury).
- Marginal.
- Negligible.

These two parameters can be combined into a generally qualitative matrix as shown in Figure 15.1. Time is also a consideration. As an example, an active volcano may erupt, but it could be in many years so the immediate risk is less and say seismic monitoring could be used to try to quantify the risk in time.

15.1.3 Identifying Risk and Methods to Handle It

Risk and methods to deal with it can be done using numerous methods (experiential, qualitative, and quantitative) related to operational risk management. One of the simplest and versatile methods for this is the Bow Tie Risk Analysis method as shown in Figure 15.2, but the method does not effectively consider the costs involved.

The Bow Tie analysis considers five components as follows (moving from left to right):

- The center is the key or the specific hazard being considered.
- The potential causes of the hazard are then listed on the left.

Likelihood ↑ / Impact →	Negligible	Minor	Moderate	Significant	Severe
Very Likely	Low Med	Medium	Med Hi	High	High
Likely	Low	Low Med	Medium	Med Hi	High
Possible	Low	Low Med	Medium	Med Hi	Med Hi
Unlikely	Low	Low Med	Low Med	Medium	Med Hi
Very Unlikely	Low	Low	Low Med	Medium	Medium

FIGURE 15.1 The risk matrix (Garvey and Lansdowne, 1998).

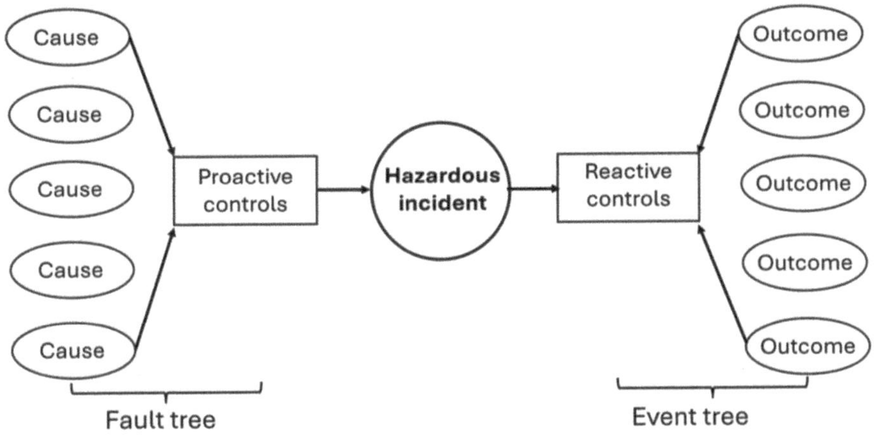

FIGURE 15.2 The basics of a Bow Tie risk analysis (Imperial Chemical Industry, 1979).

- Proactive controls are then considered that can totally or partially remove (mitigate) the potential causes of the hazard.
- Reactive controls are considered if the hazard in fact occurs and can be used to reduce the effects of the hazard.
- Outcomes are then finally considered.

Figure 15.3 gives a simple Bow Tie analysis example associated with a pillar collapse.

15.1.4 Risk Management

According to Gibson (2023), "Risk management is the systematic process of identifying, assessing, and mitigating threats or uncertainties that can affect your organization. It involves analyzing risks' probability and impact, developing strategies to minimize harm, and monitoring measures' effectiveness."

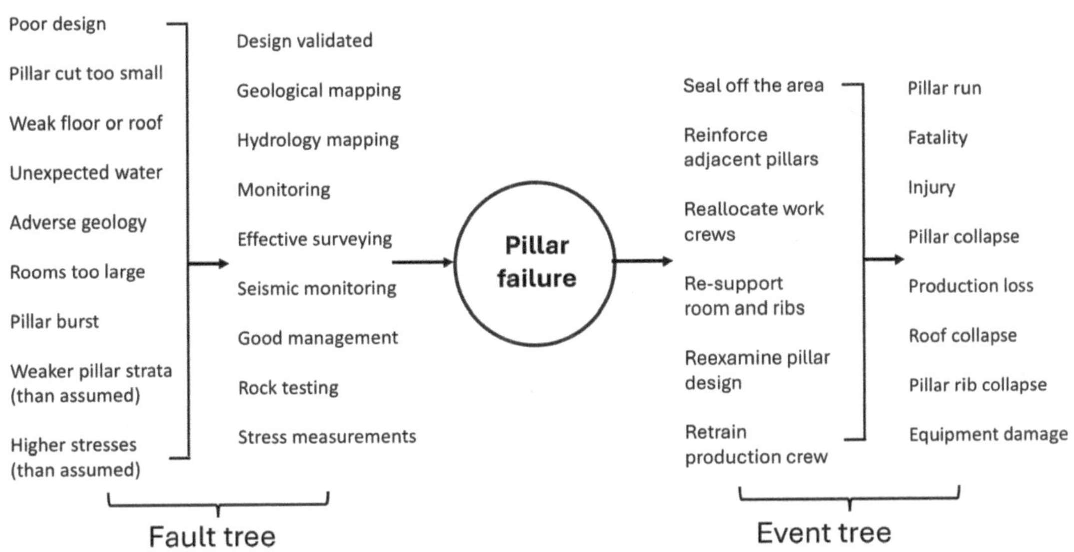

FIGURE 15.3 Bow Tie pillar failure example.

The concept of risk management was basically developed after World War II and quickly adopted by the insurance and financial institutions as they needed to quantify and understand the risks so they could make a profit from life insurance, personal loans, and other similar financial services (Snider, 1956).

Gibson (2023) states that risk management is important because it:

- Protects an organization's reputation.
- Minimizes losses.
- Encourages innovation and growth.
- Enhances decision making.

It should be noted that a "risk-free" environment does not exist, so no action is free of risk and even inaction has risk. The objective therefore is to understand the risk, and then try to eliminate or mitigate it.

Risk management typically involves the following:

- Measuring and assessing risk.
- Developing plans and strategies to manage risk.
- Implementing the plans and strategies.
- Monitoring the effectiveness of the plans and strategies and modifying them if necessary.
- Record all the components above and keep a detailed record.

The following is needed for a successful risk management mine program (McMullen, 2015):

- Management must make a firm commitment to the program.
- Management needs to give the workforce time to get the program going.
- Every person on the mine site needs to be trained on risk assessment management and the benefits of it.
- Management must actively involve their employees in risk assessment management.
- Encourage all employees not to fear change.

A representative committee with engaged members is required with the necessary authority and accountability.

15.2 ROCK-RELATED HAZARDS

In mining, like other industries, accidents in the workplace could result in the following:

- Death or injury.
- Damage and economic loss to equipment, materials, or property.
- Environmental damage.
- Unplanned delays and therefore economic loss.
- Increased costs.
- Loss of public confidence and goodwill that lasts for many years and is difficult to overcome.

Common surface mining hazards include the following:

- Damage from explosive use, such as ground vibrations and fly rock.
- Dust, fumes and gases.
- Slope/highwall instability.
- Heat.

- Noise.
- Workplace activities (mainly associated with materials loading and handling).
- Uncontrolled water.
- Surface effects such as highwall failures and surface-water interference.

Common underground mining hazards include the following:

- Explosives (if applicable).
- Bumps and rockbursts.
- Fires.
- Roof and rib instability.
- Dust, fumes, and gases (potentially radiation).
- Heat.
- Noise.
- Workplace activities (especially material loading and handling).
- Water inrush.
- Ignitions (e.g., spontaneous combustion).
- Stope, pillar, or tunnel collapse.
- Effects on the surface such as subsidence.

15.3 ROCK MECHANICS RISK ASSESSMENT AND MANAGEMENT

15.3.1 Mine Safety

The following is critical (source unknown): "Nothing in mining is more important than safety, Nothing!"

Personal safety equipment is used by mining-related personnel to reduce the risk of injuries and examples of commonly used mine personal safety equipment include the following:

- Hard hat.
- Gloves.
- Safety glasses or goggles.
- Steel-toed boots.
- Gas monitor.
- Self-rescuer (mainly for coal mines).
- Hearing protection.
- Lamp.
- Possibly a safety harness.
- Proximity devices are an important development too.

Mining is a relatively risky industry because:

- It is inherently dangerous as mines are developed in natural materials, and mining can cause the rock surrounding an excavation to fail.
- Multiple activities take place at the same time and often in confined spaces.
- Probability of a loss is relatively high.
- Workers frequently interact with heavy machinery and moving equipment.
- Infrastructure, equipment, and processes are expensive.
- Unplanned events occur routinely.
- Impacts of unplanned events can be significant.
- Cost of an accident or loss can be high.

15.3.2 Risk Management

The key to effective risk management is to minimize exposure/liability by "Always being prepared for any hazardous situation!"

An effective health, safety, and environmental program is essential and must be reviewed regularly (in writing) and updated when needed.

Effective general and tailored risk management systems are therefore essential because it can:

- Prevent accidents.
- Reduce medical costs.
- Reduce workman's compensation premiums.
- Reduce insurance premiums.
- Keep employee morale high and hence reduce worker turnover.
- Maintain a positive image with the community, investors, and the media.
- Improve profitability.

The main elements of an effective risk reduction program are therefore:

- Risk identification.
- Risk analysis.
- Risk reduction, mitigation, or elimination.
- Risk management structure with adequate financing and support.
- Data, analysis, and record keeping.
- Effective involvement of everybody and senior leadership.
- Taking action and making continuous improvements.

Elements of effective risk management on mines are given in Figure 15.4.

15.3.3 Risk Identification on Mines

Risk areas and activities in mines can be identified using:

- Federal accident records, such as from the Mining Safety & Health Administration (MSHA) in the United States.
- Citation records.
- Workman's Compensation Records or equivalents.
- Medical reportable and lost time records.
- First-aid cases.
- Records from similar mines.
- Reports of unplanned incidents or events.
- Insurance claims.

Figure 15.5 shows the components to measure health and safety.

15.3.4 Elements of a Ground Control Plan

The specific name of the rock mechanics plan depends on the country concerned and can be referred to as a Roof Control Plan, but the goal is always zero harm and a cost-effective mining operation.

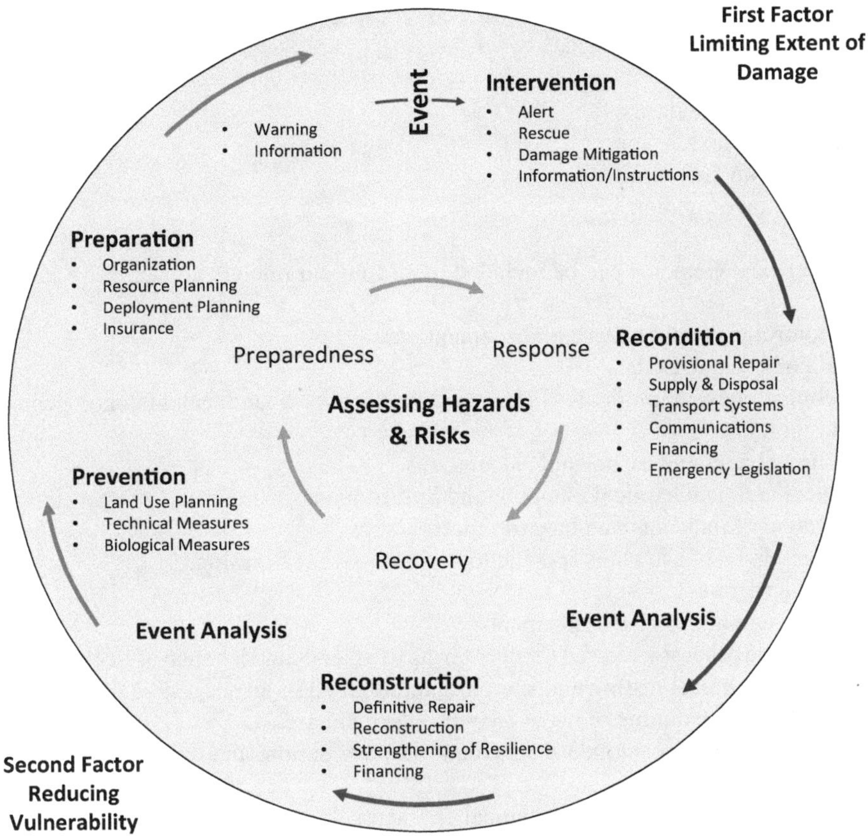

FIGURE 15.4 Risk management system outline for a mine (McMullen, 2015).

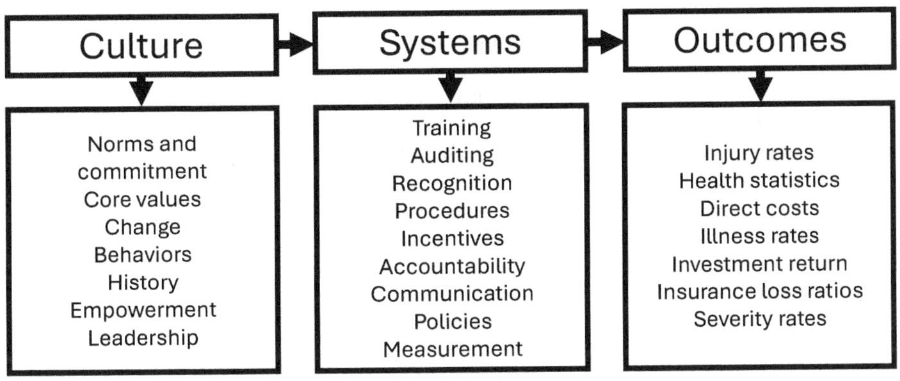

FIGURE 15.5 Components for effective health and safety measurement (Karmis, 2001).

The elements of a typical ground control plan could consist of four main parameters (Emery et al., 2020):

- Procedures and standards.
- Data collection and design.
- Implementation and monitoring.
- Review and investigation.

The following critical elements can be included in all four parameters above:

- Strata control principal hazard management plan.
- Ground control standards.
- Geotechnical design manuals and programs (software and calculators), geotechnical design reports.
- Geotechnical data and associated calculations.
- Geological and geotechnical mapping and hazard plans.
- Investigation of incidents and hazard reports.
- Trigger action response plans specific to various stages of mining.
- Monitoring regime.
- Risk management and risk assessments.
- Permit to mine process (referred to other terms in other countries such as license to operate).
- Periodic geotechnical testing and sampling maintained in an accessible database.
- Periodic third-party audits of high-risk zones (critical areas).
- Ground support product standard specifications, evaluation, and testing procedures
- Weekly reporting system.
- Training for employees and geotechnical engineers.
- Definition and appointment of suitably qualified geotechnical engineers.
- Audits and reviews of ground control management process.

REFERENCES

Emery, J., Canbulat, I. and Zhang, C. (2020) Fundamentals of modern ground control management in Australian underground coal mines. *IJMST.* 30, 573–582.

Garvey, P.R. and Lansdowne, Z.F. (1998) Risk matrix: an approach for identifying, assessing, and ranking program risks. *Air Force J. Logist.* 22, 18–21.

Gibson, K. (2023) What is risk management and why is it important? Harvard Business School Online. Retrieved from: https://online.hbs.edu/blog/post/risk-management. October 24.

Imperial Chemical Industry (1979) *Hazard Analysis (Hazan) Lecture.* University of Queensland, Australia.

Karmis, M. ed (2001) *Mine Health and Safety Management.* SME, Denver, CO.

McMullen, L.G. (2015) Risk assessment management. MSHA. *Presentation to mining staff and students at Southern Illinois University Carbondale.*

Safety Section (2024) Risk assessment calculation formula. Article dated on February 18. https://safetysection.com/risk-assessment-calculation-formula/

Snider, H.W. (1956) *Reaching Professional Status: A Program for Risk Management, in Corporate Risk Management: Current Problems and Perspectives.* American Management Association, New York.

16 Strata Control

16.1 INTRODUCTION TO STRATA CONTROL

Strata control is a subset of rock mechanics that deals with the safety and stability of all mining excavations. Support is needed depending on the:

- Stress state around the excavation periphery.
- Change in the stress state during the useful life of the excavation.
- Shape and cross-sectional area of the excavation.
- Local and regional geological features.
- In situ rock properties.

Ortlepp (1989) defines the fundamental requirement of a support as follows: "to keep the blocks or fragments in position, without itself developing such high internal stresses as to be destroyed in the process".

The design of tunnel support must take account of the "human factor". This means that the support system must make allowance for the following possibilities:

- The tunnel may not always be adequately "made safe", i.e., not all the loose rock may be removed after each blast.
- The development crews may not always be able to identify potentially hazardous conditions.
- Unplanned overbreak that may require additional or different support for example.
- Not installing additional support over and above the "support plan" due to adverse conditions. This can result from a lack of relevant training or being unaware of the potential consequences of not installing enough support.

The installed support must serve the following functions:

- Prevent local rockfalls.
- Reduce the long-term tunnel wall deterioration with time.
- Limit unplanned overbreak during the tunnel development process.
- Maintain the overall design cross-sectional area of the tunnel.
- Provide protection against violent and sudden seismic events and the potentially large induced movement.
- Be cost effective over the planned life of the excavation.

16.2 TEMPORARY SUPPORT AND THE MAJOR INFLUENCES ON THE NEED FOR SUPPORT

16.2.1 Temporary Support

Always consider when temporary support is needed during the excavation process before any final/permanent support is installed. This support needs to be removed after the permanent support has been installed and should be from a position of safety (possibly under the protection of the newly installed permanent support).

DOI: 10.1201/9781032704845-16

FIGURE 16.1 Temporary hydraulic support fitted onto a coal roof bolter.

Reproduced with kind permission from Komatsu Mining Corp.

FIGURE 16.2 Temporary (reusable) props in a tunnel before it is permanently supported.

Reproduced with kind permission of Impala Plats.

For example, when installing permanent roof bolt support in coal mines, the roof bolter is fitted with hydraulic roof jacks that provide effective temporary support to protect the equipment operators from injury and the equipment from costly damage (Figure 16.1).

Temporary support in non-mechanized hard-rock mines using explosives is typically manually installed and removed. It usually consists of mechanically set props possibly with mesh or screening. Timber can be used but is typically not temporary support as it is generally left in place (Figure 16.2).

16.2.2 The Need for Support

Support is not needed unless the rock has failed. Failure could be caused by:

- Excavation size and shape.
- Local geology.
- Weak rock and/or high stresses.
- Blasting.
- Weathering/water.
- Stress changes (most commonly from nearby mining).

Generally the installation of some localized support is considered for safety reasons, after the roof and ribs are examined. Figure 16.3 shows an example in a limestone mine where no support is needed as it is strong and geologically undisturbed.

The rock conditions in locations where the workforce travels and works should still be regularly monitored and conditions recorded in case of an unexpected change in stability.

The support of excavations is not trivial because of the risks associated with errors such as:

- Under-design is the worst-case scenario as this can result in injury or loss of life.
- Over-design (being too conservative) has a mitigated safety risk but can be an economic risk as costs were more than they needed to be.
    - Yield pillars are a special case where an "over-design" (making the pillars too large) could be an issue as the pillar may not yield as designed and could even fail violently.

FIGURE 16.3 No regular support is required in a large and competent room and pillar limestone mine.

Reproduced with kind permission of East Fairfield Coal Company.

The overall stress state is a factor as are the rock properties. For shallow excavations for example, typically the installed support (between pillars and abutments) is designed:

- To hold dead weight.
- Limit failures due to local geology (such as joints).
- To be rigid.
- To limit and control closure.

Pillars are typically essential and often required by the approved mine design and relevant regulations.

The following compares mining in weak and strong rocks under similar conditions.

The challenges in weak rock tunnels and excavations are significant, and the following should be considered:

- These excavations are difficult to analyze as failure is typically progressive and weathering can also be a factor.
- Even relatively low stresses can cause failures.
- Typical rock types include mudstones, siltstones, tuffs, clays, and some especially laminated shales.
- Deformation monitoring is essential and shotcrete (not often used in coal mines), cast concrete, steel sets, and/or arches are commonly used.

For deep excavations in hard rock, where the stresses (and blasting) often cause failure in the rock around an excavation, the following is typically needed:

- Support the excavation (failed) skin with area (liner) support between the rockbolts.
- A yielding rock anchor support system may be needed to handle the relatively high rock deformation.
- If seismicity is possible, then the support must be able to absorb energy.
- Try to avoid leaving unnecessary pillars as they could act as stress concentrators, which is generally undesirable.

The excavations must be sized to also ensure adequate ventilation.

Support in coal mines is generally classified as either primary support, secondary support, or supplemental support. Primary support is roof bolts that are installed in the development cycle. Secondary support typically consists of cable bolts, trusses, and/or standing support, and it is installed sometime after the initial development, especially in intersections. Supplemental support is support installed by exception over and above the primary and secondary support when conditions are worse than those designed for or expected, due to weathering for example or for rehabilitation of an excavation. Skin (area) support, such as screen, is often very important on coal mines too for the roof and ribs.

16.3 OVERALL SUPPORT CATEGORIES AND TYPES

16.3.1 Support Categories Based on the Load on Installation

All types of support fall into one of the two main categories:

- Passive support is by definition unstressed/unloaded during installation and only develops a reactive force when rock deformation takes place. The magnitude of the reactive force depends mainly on the:
  - Amount of rock movement.

- Stiffness of the support.
- Locking or contact mechanism between the support and the rock, for example, in the case of rock anchors.

- Active support by definition is stressed/tensioned by action of the installation process. The support therefore exerts and maintains some restraining force on the rock, the magnitude of which depends on the:
  - Stiffness of the support.
  - Tension (or compression in the case of standing support) imparted during the installation.

16.3.2 Main Rock Anchor Support Methods

There are three main simplified support conditions that account for how rock anchors work in supporting the rock around excavations, especially the roof. These are as follows:

- Key block occurs where a block of ground bounded by fractures caused by geology, blasting or stress, can fall under gravity unless supported.
- Beam building occurs where the roof is bedded and individually the beds are not self-supporting, so installed support attempts to create a "laminated" strong beam.
- Suspension occurs where the immediate roof is very weak, but a layer above is strong and stable. The support is used to basically suspend the weak roof from the overlying strong roof. This method would be considered a simple dead weight design.

They are important and simple concepts but are still useful (Figures 16.4–16.6).

Figure 16.7 shows a different example of keyblock support in jointed hard rock with discrete blocks.

Figure 16.8 shows an example of beam building in generally hard-rock sedimentary rocks with horizontal and steeply dipping bedding planes.

Clearly, the excavation shape relative to the geological features such as beds, joints, and faults and the associated stresses are key to the excavation stability. The angle that a rockbolt or a standing support intersects a failure plane is important as shown in Figure 16.9.

Hoek and Wood (1987) introduced the concept of rockbolts creating a reinforced beam of thickness "t" as shown in Figure 16.10. To create this and hence have a support system, as opposed to just random rock anchors, the spacing "s" must be less than the rockbolt length "L", and the closer the

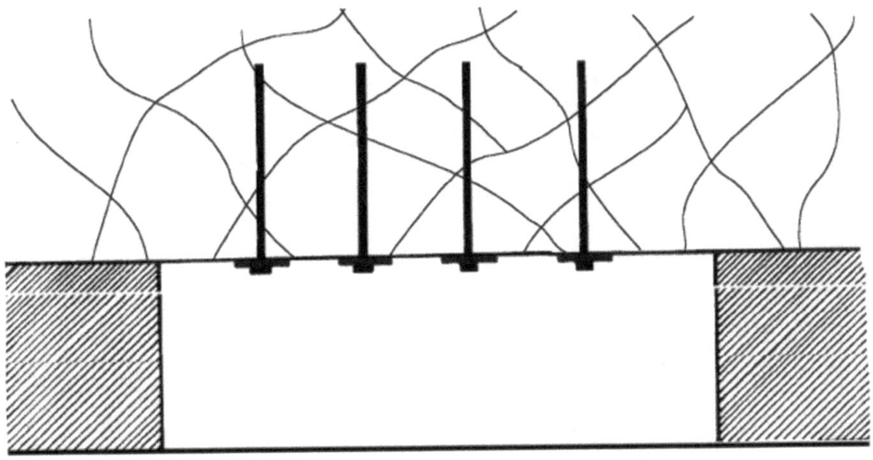

FIGURE 16.4 Key block support (conceptualized by Panek, 1956).

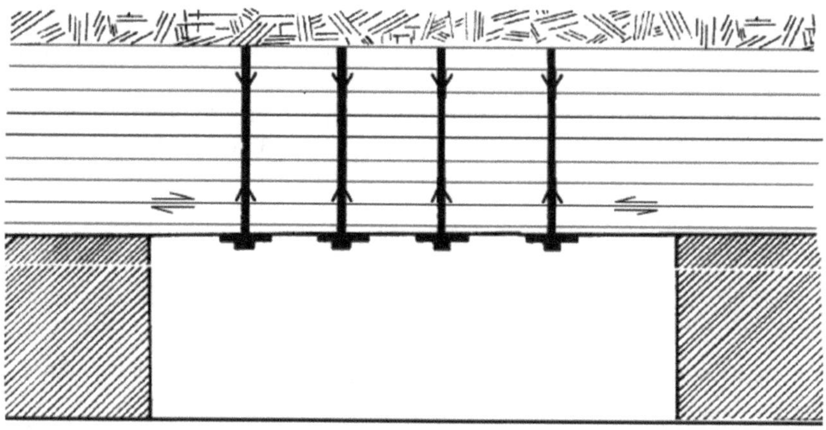

FIGURE 16.5 Beam building (conceptualized by Panek, 1956).

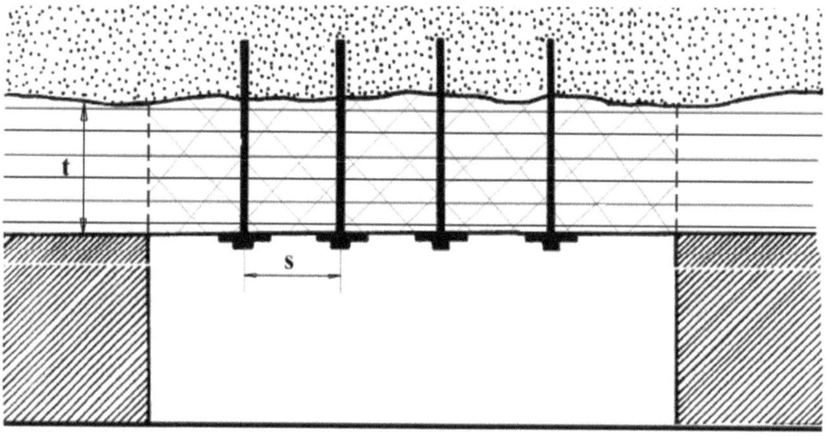

FIGURE 16.6 Suspension (conceptualized by Panek, 1956).

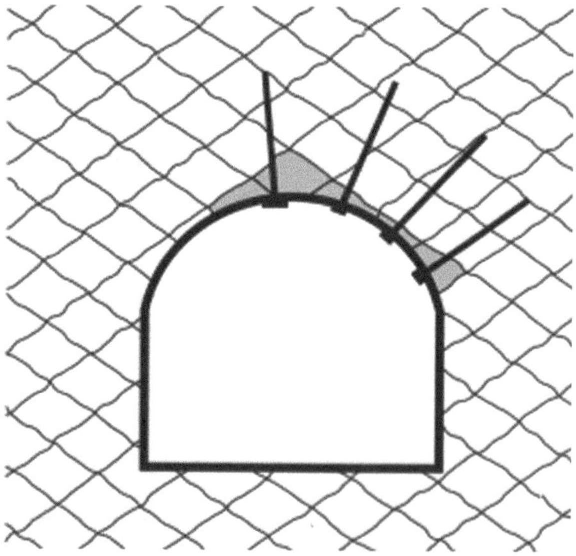

FIGURE 16.7 An example of jointed hard-rock keyblock support.

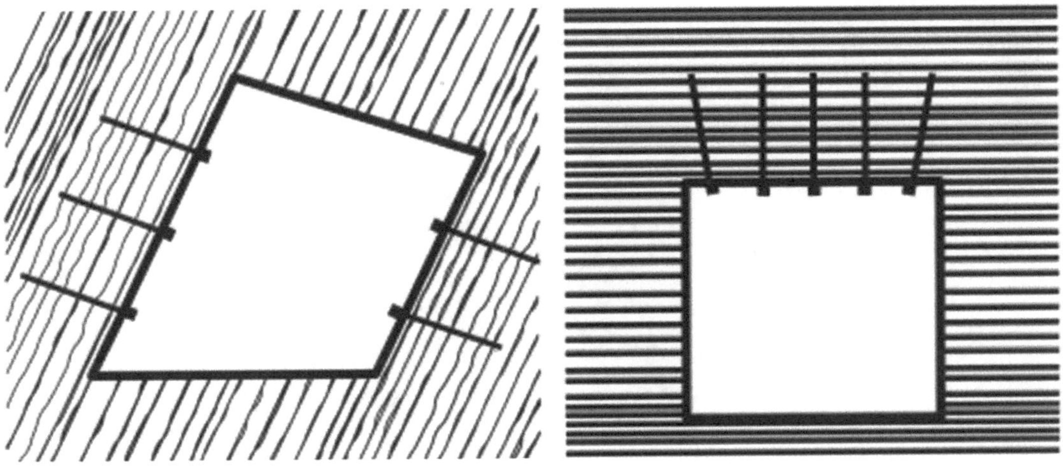

FIGURE 16.8 Other examples of beam building.

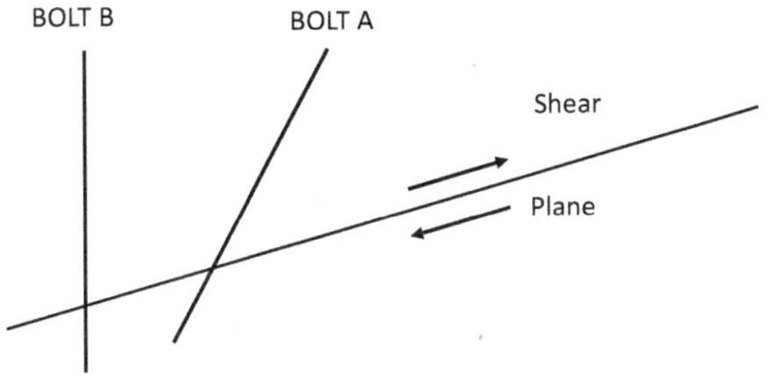

FIGURE 16.9 The importance of bolt inclination to a shear plane: Bolt B is more effective (Stillborg, 1986).

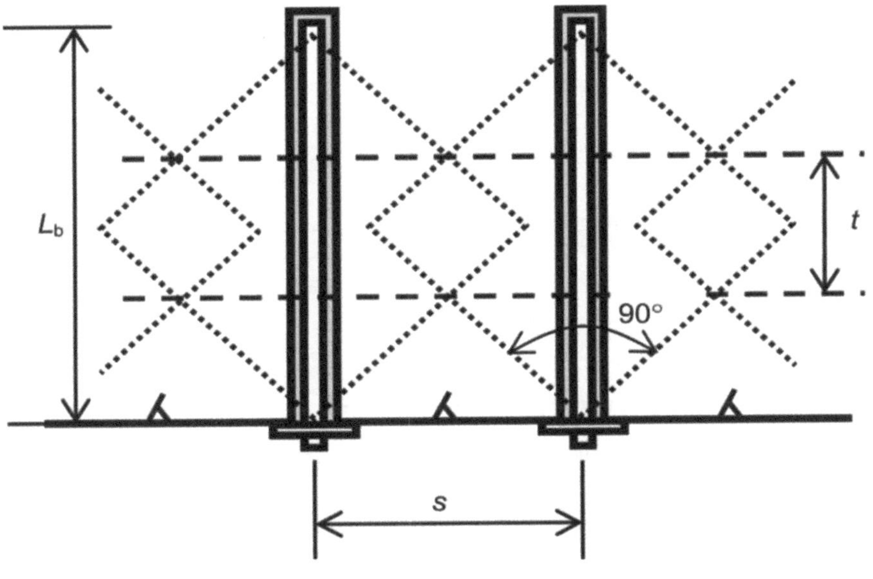

FIGURE 16.10 An artificially created rock-reinforced beam using rock anchors (Hoek and Wood, 1987).

spacing for a given rockbolt length, the thicker the reinforced arch produced. Figure 16.11 illustrates a reinforced pressure arch.

Figure 16.12 also illustrates the different rock conditions typically found underground and needing support.

FIGURE 16.11 A reinforced pressure (Voissoir) arch example (Hoek and Wood, 1987; Hoek and Brown, 1980).

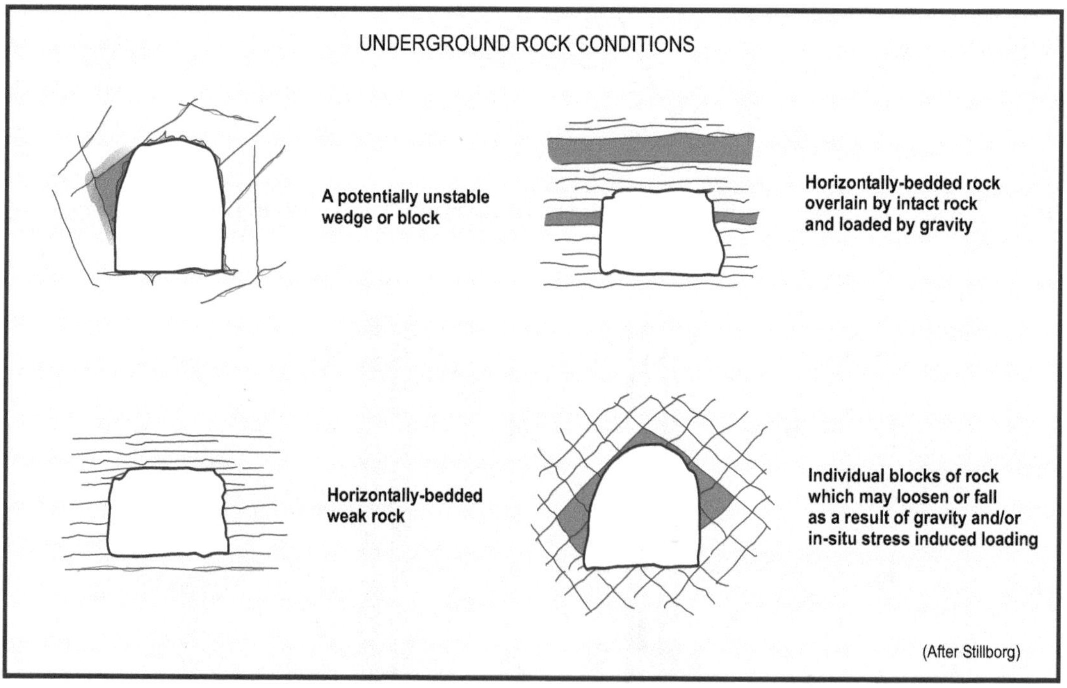

FIGURE 16.12 Underground rock conditions around excavations (Stillborg, 1986).

16.4 SUPPORT DESIGN METHODOLOGIES

For an effective rock bolting system design, rather than individual components, the following parameters must be determined correctly:

- Bolt type.
- Bolt length.

- Pattern and spacing of bolts.
- Bolt diameter.
- Anchor capacity.
- Whether pretension should be applied or not
  - If pretension is desirable, then the appropriate magnitude of pretension should be chosen, which is also controlled by the torque that the rockbolter can apply.
- Skin support.
- Installation sequence within the mining cycle.
- Excavation useful life
  - For example, weathering and corrosion can be an issue over time for steel rockbolts, plates, and mesh.

16.4.1 Basic Considerations

No support is generally required if the stresses do not exceed the strength of the rock mass and there is no adverse local geology, weathering or blast damage.

In addition assuming the same rock properties, the support needs to be increased with increased depth and more significantly with increased excavation span.

16.4.2 Design Principles

There are numerous design systems and principles used in underground construction and mining, and the following five are commonly used:

Approach 1: Civil engineering approach (Choi, 2017)

The Rational Design Procedure is a common civil engineering approach and works well for underground civil structures and includes:

- Determining the mode of failure of the structure.
- Determining the relationship between the applied loads and the most significant mode of failure factor, e.g., stress or deflection.
- Simple material tests to determine the "maximum load value" obtained by the most significant factor (tensile, shear, or a combination) before the failure occurs.
- By experiment and experience within economic and legal constraints, determine the safe, permissible, or safe value (this is the factor of uncertainty or safety).

Approach 2: Another civil engineering approach (Choi, 2017)

- Stress field determination by finite element analysis or plane strain math solutions for example.
- Selection of representative rock mass properties.
- Adoption of a simplified model of the rock response to the stress to estimate the rock failure extent and hence the support needs.
- The observation and monitoring of the rock response and the effectiveness of the support.
- Analysis of the above to determine the permanent support needs.
- Best for large excavations where the stress will not change significantly during the useful/design life of the excavation.

Approach 3: The empirical approach

- This relies on the premise that methods that worked in the past under similar conditions of failure will work again. This approach can provide a good first estimate.

- The main problem is in the assessment of similar conditions where the relationship is relevant, and whether this approach is still the most useful or if other numerical methods may be more representative.
- Still useful for many mine support applications.

Approach 4: The analytical approach

- Stress redistribution around an excavation has been often found to be closely predicted by the Laws of Elasticity, but this excludes plastic rock behavior.
- Assumes two-dimensional plane strain.
- Solved by rigorous math, photo-elastic methods, finite element or finite difference analysis, or displacement discontinuity.
- The main problem is the rock mass failure criterion used. Another is that the models cannot accommodate failed rock or significantly dilated rock correctly, and this limits further failure within the model.

Approach 5: Combination empirical and analytical approach

- Basically this is a back analysis technique.
- Stresses and stress changes can be adequately estimated.
- The support needs can be assessed for the new excavation and if more or less support is needed.
- Geological and strata changes must be identified, and if needed, the design amended based on the changes.
- It is a useful approach in mining.

Bieniawski (1984) developed a system of activities for the design, siting, developing, and supporting excavations. The following is slightly modified from Rock Mechanics Design in Mining and Tunneling after Bieniawski (1984).

Bieniawski postulated that the design principles consist of five main activities with the most critical feedback loop as shown in Figure 16.13. This is most relevant for underground construction

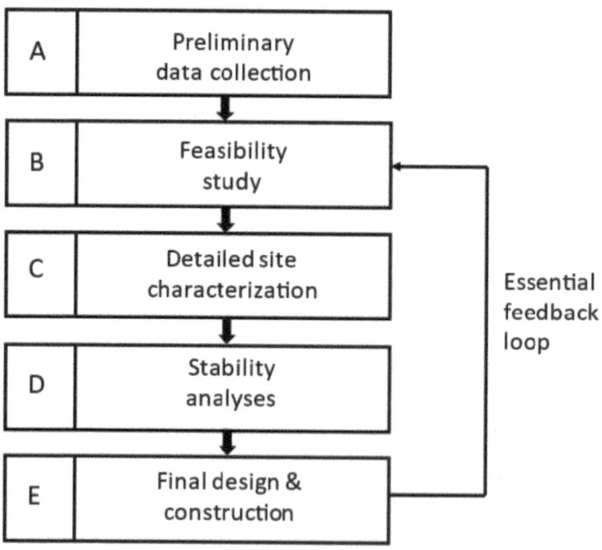

FIGURE 16.13 The overall design methodology (Bieniawski, 1984).

excavations and tunneling but is useful in mining too, especially for the orebody main access development.

The feedback loop is critical, and any design must be rechecked regularly to ensure that it remains safe and cost effective and meets the actual ground conditions underground.

Important aspects to check include the input parameters, the assumptions made, and the rock mass response.

- A – Preliminary Data Collection
  This is the vital first step, and the reliability of the other activities and the final design depends on this step. As the project is developed, additional representative field data are, or should be, made available, so the feedback loop continues to be vital (Figure 16.14).
- B – Feasibility Study
  The feasibility study is the next step that must be undertaken as shown in Figure 16.15.
- C – Detailed Site Characterization
  Figure 16.16 shows the typical steps to conduct an effective detailed site evaluation.
- D – Stability Analyses
  The stability analyses are the next stage that must be undertaken as shown in Figure 16.17.
- E – Final Design and Construction
  The final stage is the final design and construction as outlined in Figure 16.18.

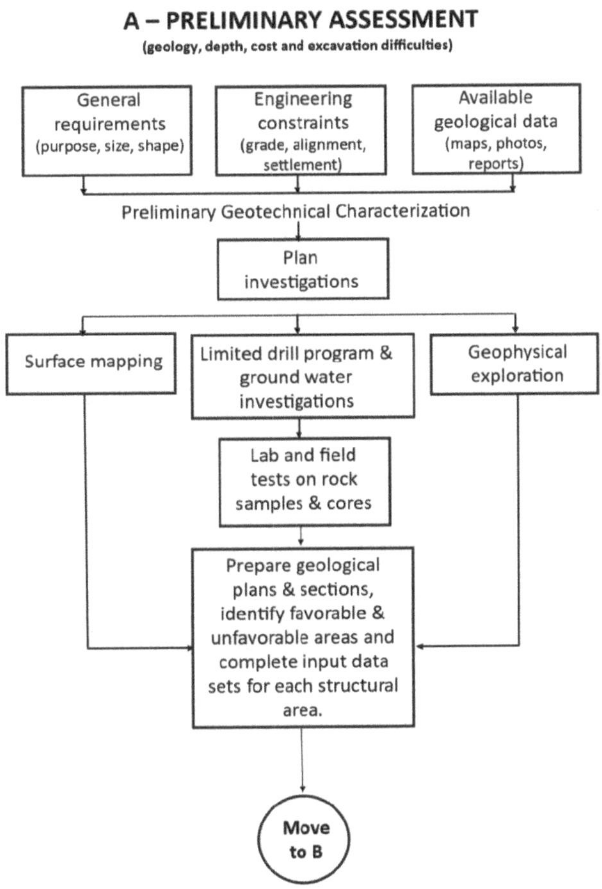

FIGURE 16.14 A – Preliminary data collection (Bieniawski, 1984).

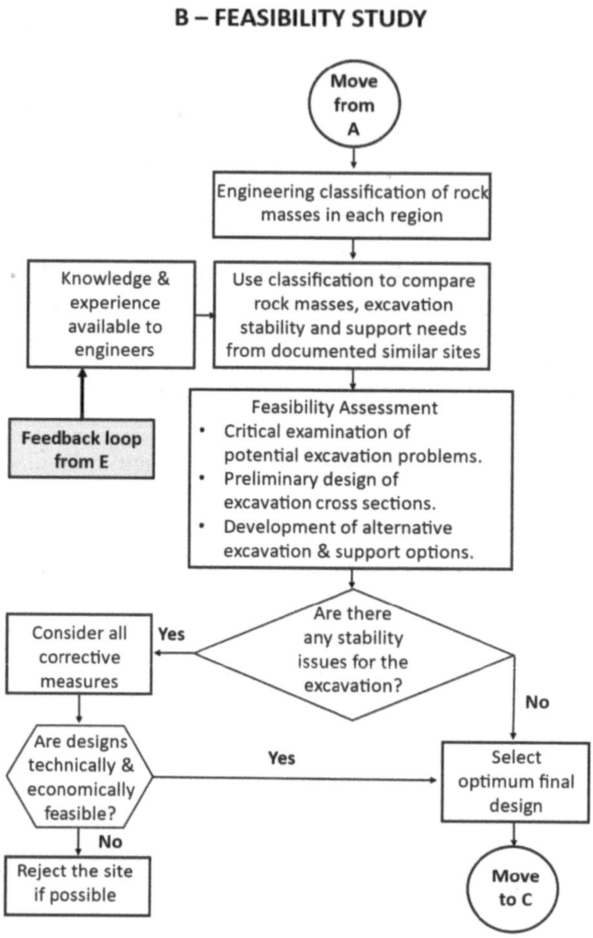

FIGURE 16.15 B – Feasibility study (Bieniawski, 1984).

16.5 SPECIFICS FOR TUNNEL SUPPORT

A typical tunnel design process would be as follows (Figure 16.19):

The choice of tunnel support can be based on various methods and principles:

- It is reasonable to assume, if a certain support system has been successful under similar conditions, that it will work again.
- If there is a chance of major seismicity in the vicinity of the tunnel, a yielding support system must be installed with a suitably designed fabric between the anchors such as meshing and lacing.
- Economics must always be a major consideration in any tunnel-support design.
- Computer analysis packages (elastic and inelastic) can be utilized, especially if some form of calibration exercise has been conducted, based on underground measurements, observations, and monitoring.
- Use can be made of the keyblock theory or any applicable system of rock mass characterization. The Rockwall Condition Factor (RCF) can also be useful (Wiseman, 1979):

$$\text{Rock Condition Factor RCF} = (3\sigma_1 - \sigma_3)/F\sigma_c \tag{16.1}$$

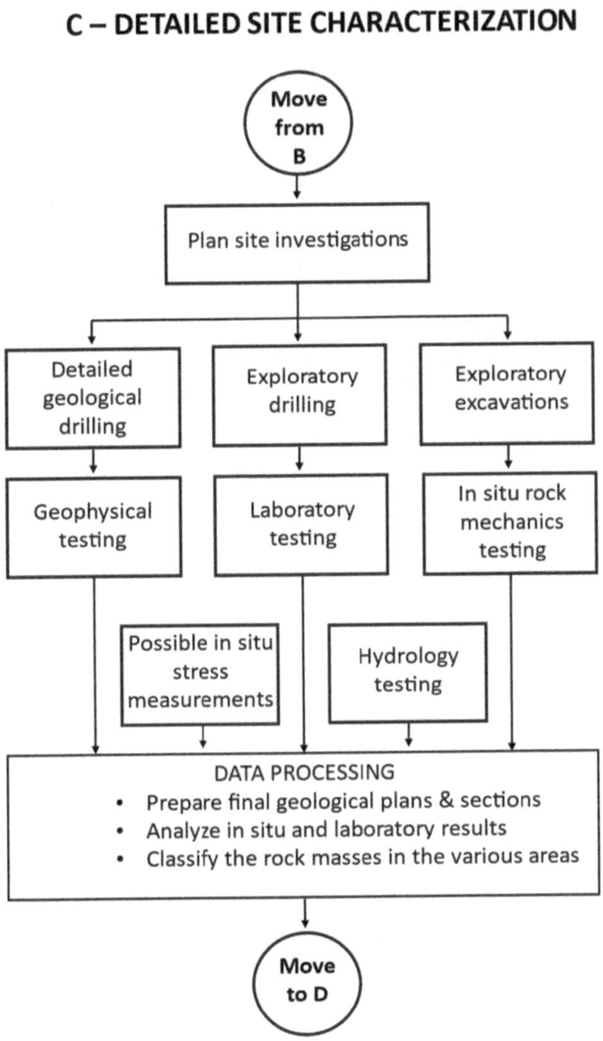

FIGURE 16.16 C – Detailed site characterization (Bieniawski, 1984).

where σ_1 and σ_3 are the major and minor principal stress components acting normal to the tunnel, F is an empirical rock mass condition factor, and σ_c is the uniaxial compressive strength of the country rock.

In highly competent rock F = 1, and this value should be decreased for incompetent rock to a value of about 0.5. The size of the tunnel also influences the value of F, and if a large tunnel cross-sectional area is planed (say more than 36 m²), F should be reduced by at least a further 10%.

If the RCF is less than 0.7, the underground stability will be good, but if the RCF is greater than 1, the underground stability will be very poor.

16.6 A SUCCESSFUL SUPPORT DESIGN

A successful design considers:

- The measure of success is not only a safe outcome, although this is the key, but the solution adopted must also meet the functional requirements of the excavation and also be cost-effective.

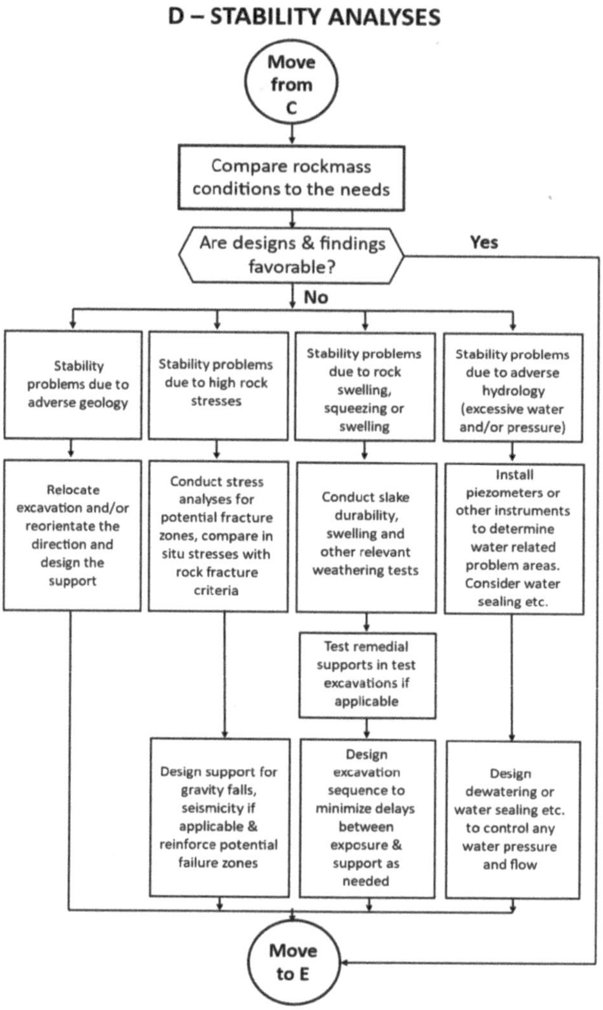

FIGURE 16.17 D – Stability analyses (Bieniawski, 1984).

- Engineering judgment based on available information is critical, and the design must be revisited and modified, if needed, as more information becomes available.

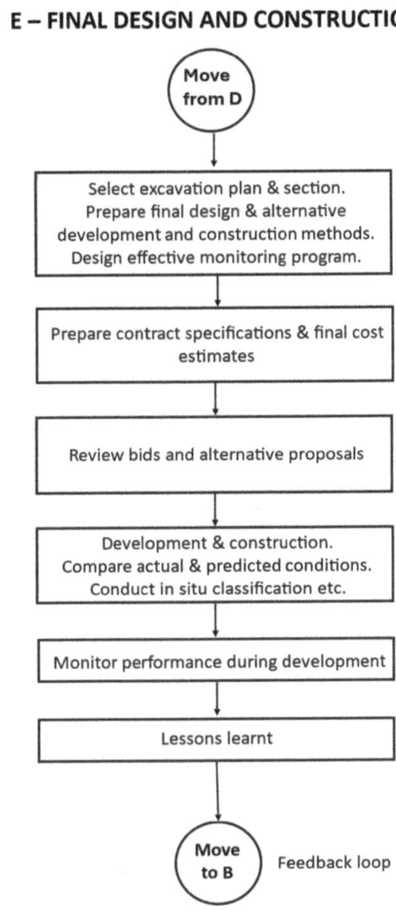

FIGURE 16.18 E – Final design and construction (Bieniawski, 1984).

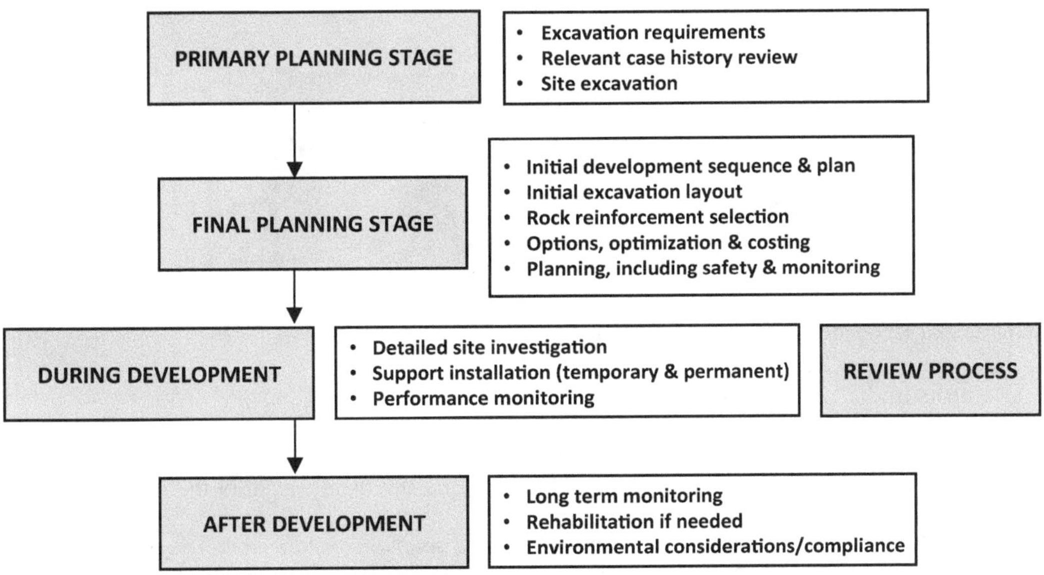

FIGURE 16.19 Tunnel design outline (Bieniawski, 1984).

16.7 WORKED EXAMPLES

Example 16.7.1

The roof rock above an entry is 10 m long, 5 m wide, and 1.5 m thick. This rock is laminated and weak, but the rock above is massive and competent. The rock relative density is 3.0. Resin-grouted rebars are to be used with a 15 t tensile strength. Assume that resin grout typically holds 4 t with every 10 cm of grout length and a factor of safety of 1.5 is needed (for the block weight and bolt length). For part a, what are the minimum number of bolts to support the block? For part b, what minimum length should they be? The example is part of course examples by Spearing (2015) (Figure 16.20).

Solution:

Part a:

$$\text{Density of water, } \rho_w = 1{,}000 \text{ kg/m}^3$$

$$1 \text{ tonne} = 1{,}000 \text{ kg}$$

$$\text{Block mass} = \left(10 \text{ m}\right)\left(5 \text{ m}\right)\left(5 \text{ m}\right)(3.0)\left(1{,}000 \text{ kg/m}^3\right)\left(1 \text{ t/1,000 kg}\right) = 750 \text{ t}$$

$$\text{Minimum number of rock bolts} = \left(750 \text{ t/15 t}\right)(1.5) = 75 \text{ bolts}$$

Part b:

Need to anchor into the massive strong rock and with the Factor of Safety (FOS) of 1.5, this means

$$\left(15 \text{ t}\right)(1.5) = 22.5 \text{ t}$$

$$x = \left(22.5 \text{ t}\right)\left(10 \text{ cm/4 t}\right) = 56.25 \text{ cm}$$

The minimum bolt length is 1.5 m + 0.5625 m = 2.06 m (for practical considerations, assume 2.1 m or even 2.2 m).

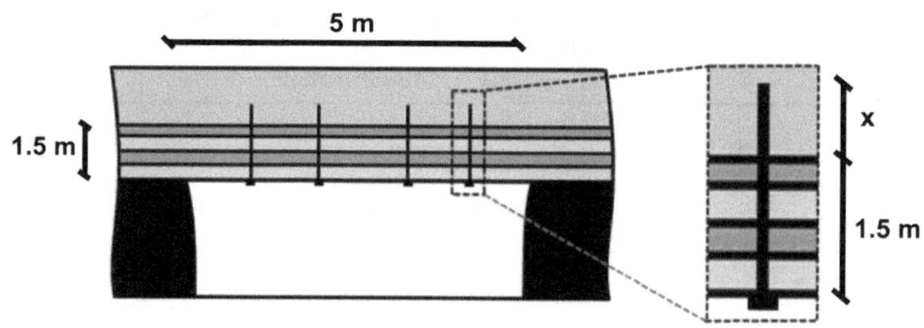

FIGURE 16.20 Section of the room and pillar entry.

Example 16.7.2

Assume that a set of bolts are installed in a square grid, spaced 4 ft apart at a depth of 8 ft into the mine roof. Find the tension in the bolt needed to support the dead weight of the rock. The unit weight of the rock is 150 lbs/ft³.

This simple example is based on stabilization of a simple Voussoir arch. The rockbolts are assumed to be ungrouted and end anchored, i.e., conventional bolt type, into a competent overlying rock (Figure 16.21). The procedure for this example is found in Hudson and Harrison (1997).

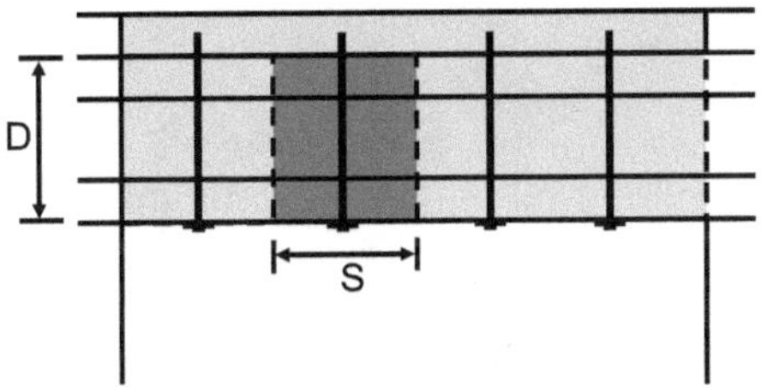

FIGURE 16.21 Section through the room showing the bolted roof.

Modified after Hudson and Harrison (1997).

Given:

Spacing (s) = 4 ft
Depth (D) = 8 ft
Unit weight of rock (γ) = 150 lbs/ft^3

Solution:

$$\text{Bolt tension, } T = \gamma D s^2$$

$$T = \left(150\text{ lbs/ft}^3\right)\left(8\text{ ft}\right)\left(4\text{ ft}\right)^2$$

$$T = \left(150\text{ lbs/ft}^3\right)\left(128\text{ ft}^3\right)$$

$$T = 19{,}200\text{ lbs}\left(1\text{ ton/2,000 lbs}\right)$$

$$T = 9.6\text{ tons}$$

This solution does not include a factor of safety based on the equation.

Example 16.7.3

Find the factor of safety of wedge reinforced against sliding on a single plane (F) for the tetrahedral block, as shown in Figure 16.22. For part a, treat $\varphi = 40°$ and the tension in the rockbolts (T) as 20 kN. The area of the sliding surface (A) is 4 m^2, and the weight of the block (W) is 50 kN. Strength properties for the problem are defined and available below. For part b, find the factor of safety (F) if the rock bolts are orientated to $\theta = 40°$.

The approach for this example is found in Hoek and Brown (1980), Hudson and Harrison (1997), and Brady and Brown (2004).

Solution:

After Hoek and Brown (1980), the input parameters are as follows:

Factor of safety of a block or wedge reinforced against sliding on a single plane (F).
Weight of the wedge (W).
Total load in the bolts or cables (T).
Base area of the sliding surface (A).
Dip of the sliding surface (Ψ) = 66°.
Angle between the plunge of the bolt or cable and the normal to the sliding surface, $\theta = 30°$.
Cohesive strength of the sliding surface, assume c = 0 MPa.
Friction angle of the sliding surface, $\varphi = 40°$.

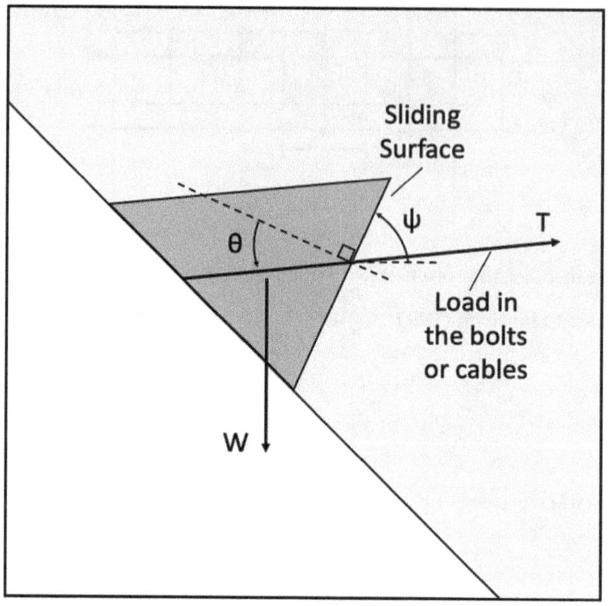

FIGURE 16.22 The sliding keyblock problem on one face.

Modified after Hudson and Harrison (1997).

Part a:

$$F = \left[cA + (W\cos\psi + T\cos\theta)\tan\varphi\right] / (W\sin\psi - T\sin\theta)$$

$$F = \left\{\left[(0\ \text{N/m}^2)(4\ \text{m}^2) + \left[(50{,}000\ \text{N})\cos 66° + (20{,}000\ \text{N})\cos 30°\right]\tan 40°\right]\right\} /$$

$$\left[(50{,}000\ \text{N})\sin 66° - (20{,}000\ \text{N})\sin 30°\right]$$

$$F = \left\{\left[20{,}337\ \text{N} + 17{,}321\ \text{N}\right]\tan 40°\right\} / \left[(45{,}677\ \text{N}) - (10{,}000\ \text{N})\right]$$

$$F = 31{,}599\ \text{N} / 35{,}677\ \text{N}$$

$$F = 0.88$$

Part b:

$$F = \left[cA + (W\cos\psi + T\cos\theta)\tan\varphi\right] / (W\sin\psi - T\sin\theta)$$

$$F = \left\{\left[(0\ \text{N/m}^2)(4\ \text{m}^2) + \left[(50{,}000\ \text{N})\cos 66° + (20{,}000\ \text{N})\cos 40°\right]\tan 40°\right]\right\} /$$

$$\left[(50{,}000\ \text{N})\sin 66° - (20{,}000\ \text{N})\sin 40°\right]$$

$$F = \left\{\left[20{,}337\ \text{N} + 15{,}321\ \text{N}\right]\tan 40°\right\} / \left[(45{,}677\ \text{N}) - (12{,}856\ \text{N})\right]$$

$$F = \left\{\left[35{,}658\ \text{N}\right]\tan 40°\right\} / \left[(45{,}677\ \text{N}) - (12{,}856\ \text{N})\right]$$

$$F = 29{,}921\ \text{N} / 32{,}821\ \text{N}$$

$$F = 0.91$$

Both factors of safety (F) would be well below the recommended range of 1.5–2.0 in Hoek and Brown (1980).

Example 16.7.4

A prism of rock requires support around an excavation as shown in Figure 16.23. What is the required force per bolt if three rockbolts are planned to be used? The background for this example is presented in Brady and Brown (2004) and originally from Hoek and Brown (1980).

Given:

Again, after Hoek and Brown (1980), the input parameters are as follows:

Factor of safety of a block or wedge reinforced against sliding on a single plane (F) For this example, use 2.0.

Weight of the wedge (W) = 50 kN.

Load in the bolts or cables (T).

Base area of the sliding surface (A).

Dip of the sliding surface (Ψ) = 45°.

Angle between the plunge of the bolt or cable and the normal to the sliding surface. Note that T required to maintain F is minimized when θ = Fcotφ.

Cohesive strength of the sliding surface, assume c = 0 MPa.

Friction angle of the sliding surface (φ) = 10°.

Solution:

$$\theta = F\cot\varphi$$

$$\theta = 2(1/\tan 10)$$

$$\theta = 11°$$

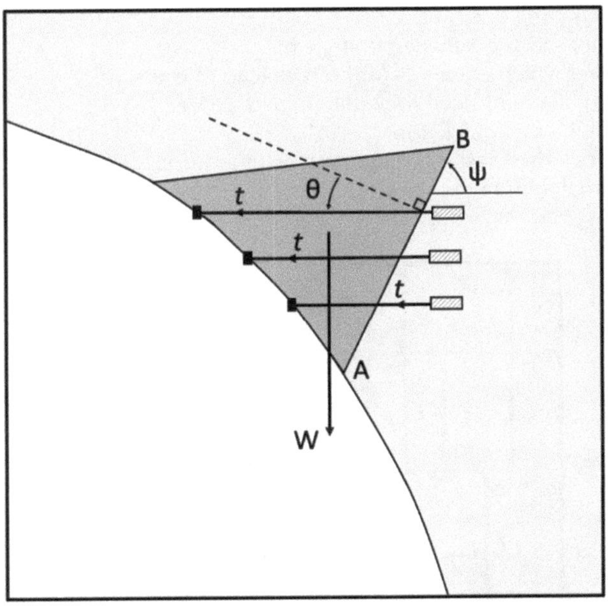

FIGURE 16.23 The sliding keyblock support problem on one face.

Modified from Hoek and Brown (1980) and Brady and Brown (2004).

$$T = [W(F\sin\Psi - \cos\Psi\tan\varphi) - cA]/[\cos\theta\tan\varphi + F\sin\theta]$$

$$T = [(50\ \text{kN})(2\sin45° - \cos45°\tan10°)]/[\cos11°\tan10° + 2\sin11°]$$

$$T = [(50\ \text{kN})(1.41 - 0.125)]/[0.173 + 0.382]$$

$$T = [64.3\ \text{kN}]/[0.555]$$

$$T = 115\ \text{kN}(0.102\ \text{ton} - \text{force}/1\ \text{kN}) = 11.8\ \text{ton} - \text{force}$$

After Hoek and Brown (1980), for three bolts, each bolt is then

$$\Sigma t = T$$

$$3t = 12\ \text{ton} - \text{force}$$

$$t = 4\ \text{ton} - \text{force}$$

Example 16.7.5

Find the tension, T, for an active bolt given the information described in Figure 16.24 and the data given. This example is produced from the equations in Brady and Brown (2005) and originally in Lang and Bischoff (1982) (Figure 16.24).

Given:

After Brady and Brown (2004) and Lang and Bischoff (1982):
L = bolt length, 2 m
D = height of destressed rock above the opening, 4 m. Note that L is usually less than D.
Z = a portion of the total destressed rock zone, D.
B = span of the opening.
A = area of roof carrying one bolt, s × s.
s = center to center spacing between bolts, 1 m.
P = shear perimeter which equals 4s for bolts spaced as (s × s) m^2.
R = shear radius of the reinforced rock determined as A/P.
γ = unit weight of the rock, 25 kN/m^3.
c = cohesion of rock mass (as a starting point assume zero).
φ = angle of internal friction of rock mass, 10°.

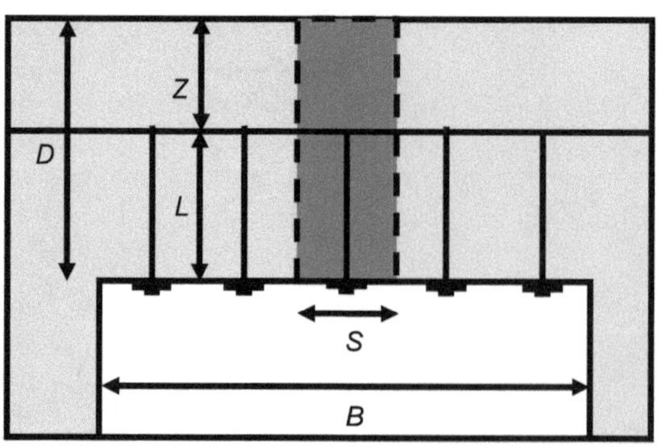

FIGURE 16.24 The room with the rock bolts.

Modified after Lang and Bischcoff (1982).

μ = coefficient of friction of the rock mass determined from Tanφ.

T = tensile load in the rockbolt.

α = factor for time of installation of rock bolts where 0.5 can be used for active support and 1.0 can be used for passive support.

k = ratio of horizontal to vertical stress, treat as 0.5.

Solution:

$$\mu = \tan\varphi$$

$$\mu = \tan 10° = 0.176$$

$$\mu = 0.176$$

$$P = 4s$$

$$P = 4(1\ \text{m})$$

$$P = 4\ \text{m}$$

$$A = s^2$$

$$A = (1\ \text{m})^2$$

$$A = 1\ \text{m}^2$$

$$R = A/P$$

$$R = 1\ \text{m}^2/4\ \text{m}$$

$$R = 0.25\ \text{m}$$

Resolve for the tensile load in the bolt (T). Note that the use of γ in the denominator of the below equation is not in Brady and Brown (2004) but is found within Lang and Bischcoff (1982). If γ is not used, the units for T will not resolve to load:

$$T/(\gamma AR) = \alpha/\mu k(1 - c/\gamma R)\{[1 - \exp(-\mu kD/R)]/[1 - \exp(-\mu kL/R)]\}$$

$$T/[(25\ \text{kN/m}^3(1\ \text{m}^2)(0.25\ \text{m}))] = [(0.5)/(0.176)(0.5)](1)$$

$$\{[1 - \exp(-(0.176)(0.5)(4\ \text{m})/(0.25\ \text{m}))]/[1 - \exp(-(0.176)(0.5)(2\ \text{m})/(0.25\ \text{m}))]\}$$

$$T/[6.25\ \text{kN}] = [5.68](1)\{[1 - 0.245]/[1 - 0.495]\}$$

$$T/6.25\ \text{kN} = 8.46$$

$$T = 53\ \text{kN} = 5.4\ \text{ton} - \text{force}$$

Example 16.7.6

A tunnel is proposed at depth in the Witwatersrand Basin with a dimension of 3 m by 3 m in quartzite. Find the Rockwall Condition Factor (RCF), and primary and secondary support requirements if the uniaxial compressive strength for the represented rock mass condition and excavation size is 190 MPa and the major and minor principal stresses are estimated at 75 and 25 MPa, respectively. Determine RCF for two cases. In the first case, F = 1.0, and in the second case, F = 0.5. Use the empirical approach and charts below found in Jager and Ryder (1999) and assume the case where seismic activity is anticipated.

The following is notable in Jager and Ryder (1999) for the empirical approach given:

"The support recommendations given in the tables [and within Jager and Ryder (1999)] are intended to be indicative and not necessarily *prescriptive*. Actual support requirements in specific geotechnical environments vary widely, and may justifiably fall short of, or exceed the values suggested here."

Solution:

Part a:

Case 1 where F = 1.0 for good-quality quartzite

$$\text{RCF} = (3\sigma_1 - \sigma_3)/F\sigma_c$$

$$\text{RCF} = \left[(3)(75\text{ MPa}) - 25\text{ MPa}\right]/\left[(1)(190\text{ MPa})\right]$$

$$\text{RCF} = \left(225 - 25\text{ MPa}\right)/\left(190\text{ MPa}\right)$$

$$\text{RCF} = 200/190\text{ MPa}$$

$$\text{RCF} = 1.05$$

Based on the RCF and Table 16.1:

TABLE 16.1
Support Recommendations for Average Ground Conditions Where 0.7 < RCF < 1.4

Case	Primary Support (Typical Requirements)	Secondary Support (Typical Requirements)
Static stress conditions	Rock bolts or tendons >1.5 m in length, installed as close to face as possible on basic 2 m pattern Support resistance: 30–50 kN/m^2.	Steel strapping or rope lacing integrated with primary support tendons; or shotcrete.
Stress changes anticipated	Fully grouted tendons or steel ropes > 1.8 m in length, installed as close to face as possible, on basic 2 or 1.5 m pattern. Support resistance: 40–60 kN/m^2	Medium strength rope lacing and moderate strength wire mesh integrated with primary support tendons.
Seismic activity anticipated	Fully grouted (pref. yield) tendons or steel ropes >1.8 m in length, installed as close to face as possible, on 1.5 m or double 2 m pattern. Support resistance: 80–110 kN/m^2	Strong rope lacing and strong, stiff wire mesh integrated with primary support tendons, plus optional shotcrete.

Source: Modified after Jager and Ryder (1999).

Case 1 primary support: fully grouted (preferably yield) tendons or steep ropes greater than 1.8 m in length. They should be installed as close to the face as possible. The pattern is 1.5 m or double 2 m pattern (see Jager and Ryder, 1999) with a support resistance of 80–100 kN/m².

Case 1 secondary support: strong rope lacing and strong, stiff wire mesh integrated with primary support tendons plus optional shotcrete.

Part b:

Case 2 where F = 0.5 for highly discontinuous rock mass,

$$RCF = (3\sigma_1 - \sigma_3)/F\sigma_c$$

$$RCF = \left[(3)(75\text{ MPa}) - 25\text{ MPa}\right]/\left[(0.5)(190\text{ MPa})\right]$$

$$RCF = (225 - 25\text{ MPa})/(95\text{ MPa})$$

$$RCF = 200/95\text{ MPa}$$

$$RCF = 2.10$$

Based on the RCF and Table 16.2:

Case 2 primary support: reinforced shotcrete to the face if necessary, fully grouted steel ropes or yielding tendons with a length greater than 2.3 m in length. They should be installed as close to the face as possible. The pattern is 1.0 m with a support resistance of 220–290 kN/m².

TABLE 16.2
Support Recommendations for Poor Ground Conditions Where RCF > 1.4

Case	Primary Support (Typical Requirements)	Secondary Support Installed by Specialist Crews Close to Face (Typical Requirements)
Static stress conditions	(If necessary) shotcrete to face, then fully grouted tendons or steel ropes, length >1.8 m on basic 1.5 m pattern, as close to face as possible. Support resistance: 80–110 kN/m²	Moderate strength steel wire mesh usually with additional tendons but may be integrated with primary support, optional shotcrete.
Stress changes anticipated	(If necessary) reinforcement shotcrete to face, then fully grouted steel ropes or yielding tendons, length >1.8 m on basic 1 m or double 2 m pattern, as close to face as possible. Support resistance: 120–230 kN/m².	Medium strength rope lacing and moderate strength wire mesh with additional tendons but may be integrated with primary support. Add integral shotcrete in long life tunnels. If necessary, additional hangingwall support comprising grouted steel ropes.
Seismic activity anticipated	(If necessary) reinforcement shotcrete to face, then fully grouted steel ropes or yielding tendons, length >2.3 m on basic 1 m pattern, as close to face as possible. Support resistance: 220–290 kN/m².	Strong rope lacing and strong, stiff wire mesh with additional tendons but may be integrated with primary support. Add integral shotcrete in long life tunnels. If necessary, additional hangingwall support comprising grouted steel ropes.

Source: Modified after Jager and Ryder (1999).

Case 2 secondary support: strong rope lacing and strong, stiff wire mesh with additional tendons which may be integrated with primary support. If the tunnel is long life, then add integral shotcrete. If needed, additional hangingwall support comprising grouted steel ropes.

REFERENCES

Bieniawski, Z.T. (1984) *The Design Process in Rock Engineering. Rock Mechanics Design in Mining and Tunnelling*. AA Balkema Publishers, Kenya.

Brady, B.H. and Brown, E.T. (2004) *Rock Mechanics: For Underground Mining*. Springer Science & Business Media, New York.

Choi, Y-K. (2017) *Principles of Applied Civil Engineering Design*. American Society of Civil Engineers, New York. https://doi.org/10.1061/9780784414736.

Hoek, E. and Brown, E.T. (1980) *Underground Excavations in Rock. 1st Edition*. E & FN SPON and imprint of Chapman & Hall. London, UK.

Hoek, E. and Wood. D.F. (1987) *Support in Underground Hard Rock Mines. Underground Support Systems*. Canadian Institute of Mining and Metallurgy, Montreal.

Hudson, J. and Harrison, J. (1997) *Engineering Rock Mechanics: An Introduction to the Principles, Part One*. Elsevier Science Ltd, Kidlington, Oxford.

Jager, A. J., and Ryder, J.A. (1999). *A Handbook on Rock Engineering Practice for Tabular Hard Rock Mines*. Creda Communications, Cape Town.

Lang, T.A. and Bischoff, J.A. (1982) Stabilization of rock excavations using rock reinforcement. *Issues in Rock Mechanics, Proceedings of the 23rd U.S. Symposium on Rock Mechanics*. Berkley, CA. Editors: R.E. Goodman and F.E. Heuze). 935–943. AIME: New York.

Ortlepp, W.D.O (1989) Guidelines for the Early Removal of the Shaft Reef Area. Anglo American Corp of SA Ltd, Johannesburg.

Panek, L. (1956) Principles of Reinforcing Bedded Mine Roof with Bolts. U.S. Bureau of Mines RI5156.

Spearing, A.J.S. (2015) Course notes, introduction to mining engineering, MNGE 270. Mining and Mineral Resources Engineering Department, Southern Illinois University Carbondale, Carbondale, IL.

Stillborg, B. (1986) *Professional User's Handbook for Rockbolting*. Clausthal-Zellerfeld. Trans Tech Publications

Wiseman, N. (1979) Factors affecting the design and conditions of mine tunnels. Research Report No. G01G10. Chamber of Mines Research Organization, Johannesburg, South Africa.

17 Rock Anchor and Standing Support Main Types and Applications

There are many different supports that are commercially available. On-going training and technical support from the supplier are essential for any support type selected. Specifically, it is the appropriate support for the application, and it is essential that it is safely and correctly installed.

17.1 COMMON ROCK ANCHOR SUPPORTS

The following are some of the more commonly used rock anchors.

17.1.1 Mechanical Expansion Shell Rockbolt

This type of rockbolt has the following features:

- This is an active point anchored bolt.
- This can be post-grouted to become an effective permanent support. In this case, ripple bar or rebar, instead of smooth bar, could be used.
- This can lose tension mainly due to adjacent blasting, and bolts should be re-tensioned, after a nearby blast.
- This is frequently used for spot bolting or as a temporary support.
- This should not be used in soft or friable ground.

An example is given in Figure 17.1.

A mechanical rockbolt installed at 30° off the perpendicular may provide less than 50% of the tension produced by a bolt equally torqued that is perpendicular to the rock face, unless a spherical washer is employed which can improve the applied effective applied tension, if the bolt is not installed perpendicular to the rock surface. Figure 17.2 shows the two main types of mechanical shells used.

17.1.2 Deformed Bar/Rebar/Ripple Bar

This is one of the most common types of rockbolts used and shown in Figure 17.3.

The following are its features:

- It can be most commonly a passive support (anchored with a single-speed cementitious or resin grout). In this case, it develops a reactive load only with rock movement.
- It can be an active support (anchored with a mechanical shell with cement or resin grout or by two-speed resin). The rockbolt is then tensioned during installation.

The cement grout can be pumped in, or capsules/cartridges can be inserted, or the resin is usually inserted in the form of capsules/cartridges. The grout quality and correctly filling the hole with

DOI: 10.1201/9781032704845-17

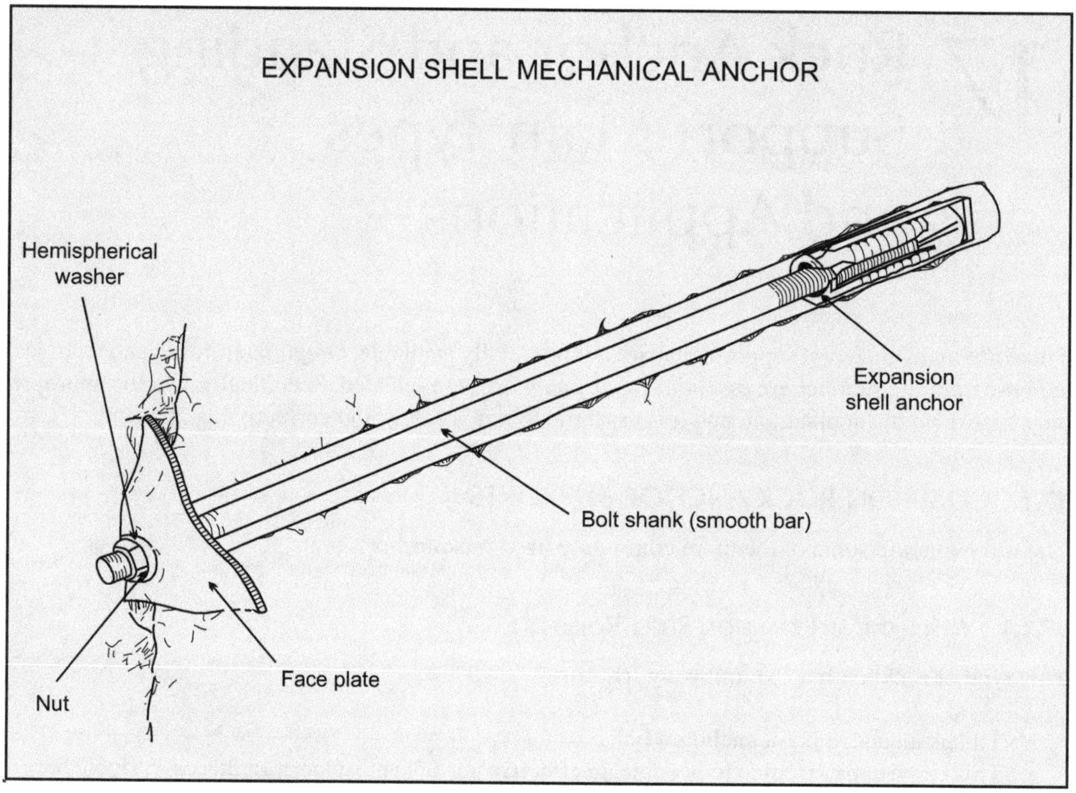

FIGURE 17.1 A mechanically anchored rockbolt. (Modified after Stillborg (1994).)

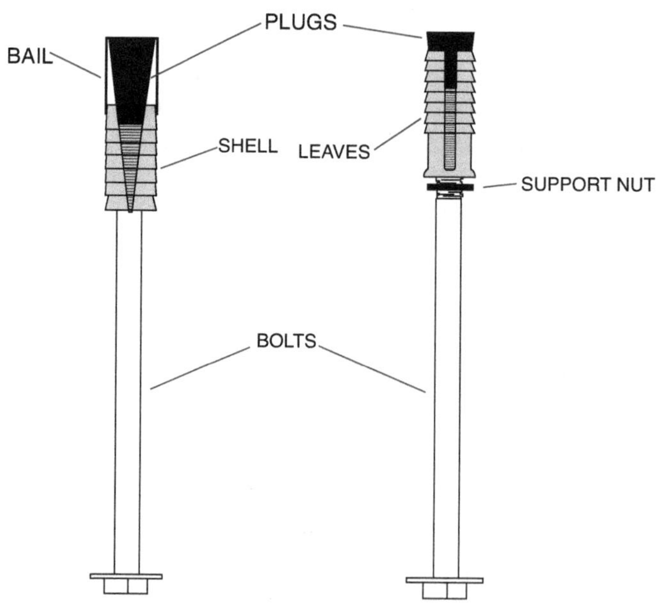

FIGURE 17.2 The two main mechanical shell types (bail and sleeve types) (Mazzoni et al., 1996).

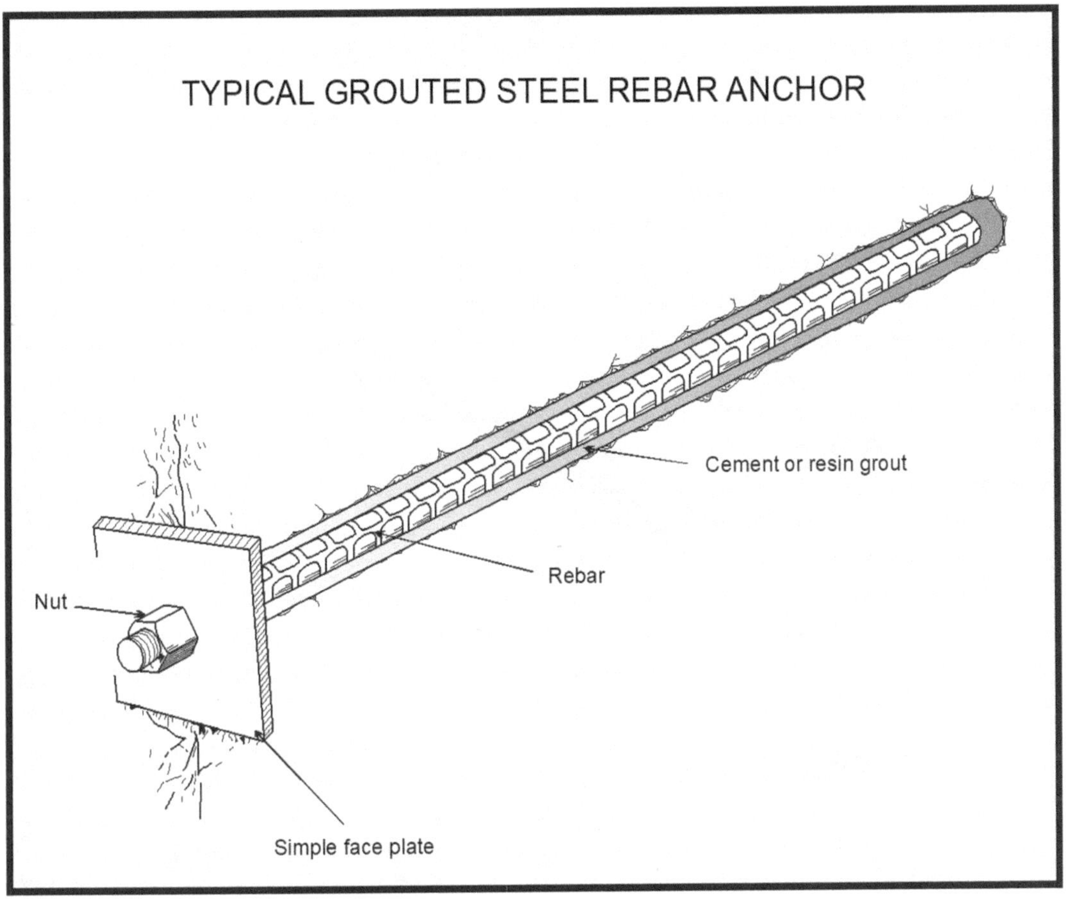

FIGURE 17.3 Typical grouted rebar. (Modified after Stillborg (1994).)

grout are the most important parameters. It is used for permanent installations but should not generally be used in the presence of running water.

Holes drilled for cement-grouted bolts should be 12–25 mm (½ to 1 inch) larger in diameter than the bolt being used. The larger gap (annular space) is especially desired in weak ground to increase the bonding area.

Note that the end of the rebar can have a thread with a nut or a forged head.

17.1.3 Friction Bolts (Split Sets)

The following are the features of friction bolts:

- These are a passive support as the reactive load generated depends on the friction of the spring steel against the rock periphery (friction rock stabilizer).
- These are rapid and easy to install.
- These are prone to corrosion, and if such conditions exist, stainless steel units or post-grouting can be used.

The hole diameter is important, and a simple "go, no-go" device (or equivalent device) for the drill steel should be used. This is important because of the following factors:

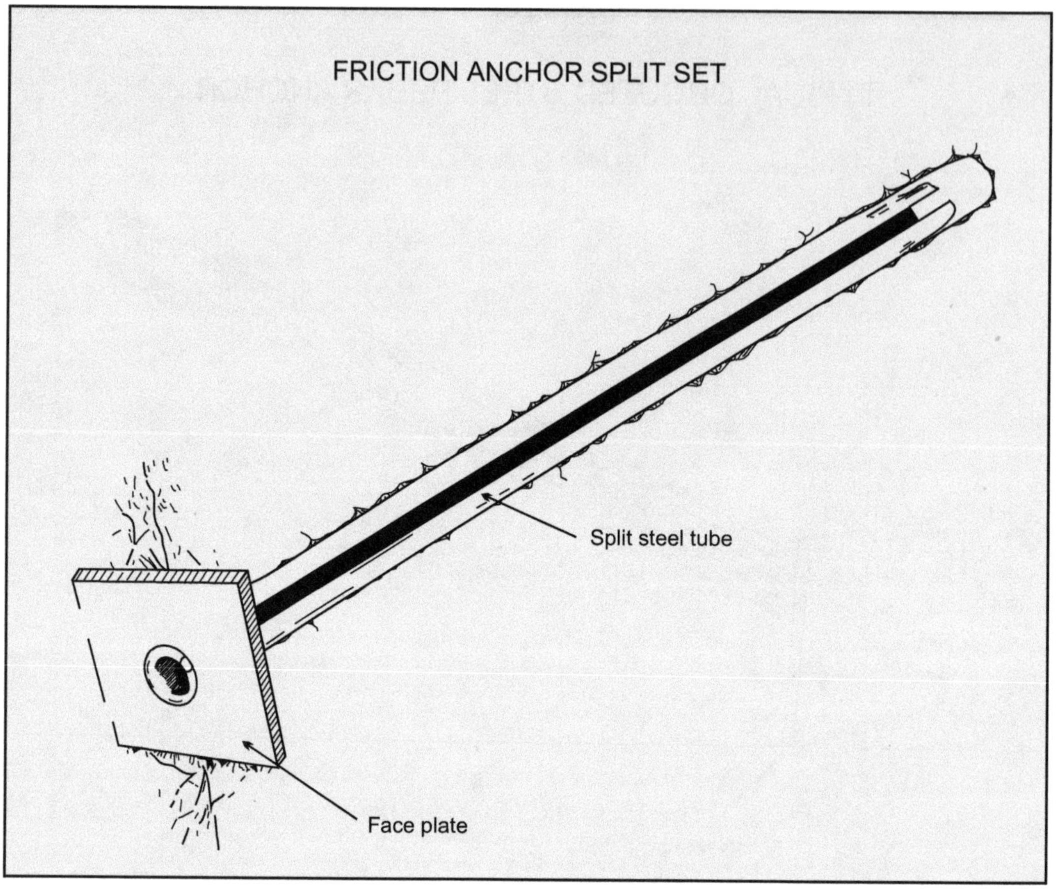

FIGURE 17.4 A friction rockbolt. (Modified after Stillborg (1994).)

- If the hole diameter is too large, the split set will not be compressed adequately on installation, and hence, the friction generated per unit length will be far too low.
- If the hole diameter is too small, the split set may not be able to be installed readily into the hole and rather collapse.

For each 30 cm (about 1 ft) of friction bolt (split set) installed, there is typically 1 ton of anchorage support. They are generally used as a primary support or for localized spot bolting on hard-rock mines. They are also often used to pin mesh (screen) onto the rock surface to reduce local falls underground, on highwalls, or before shotcreting for example. Figure 17.4 shows a typical friction bolt.

17.1.4 Swellex Bolts

The following is relevant for this type of rockbolt:

- It is quick to install.
- It requires a high-pressure installation pump to inflate/expand the bolt in the drilled support hole.
- It serves as a semi-active support.

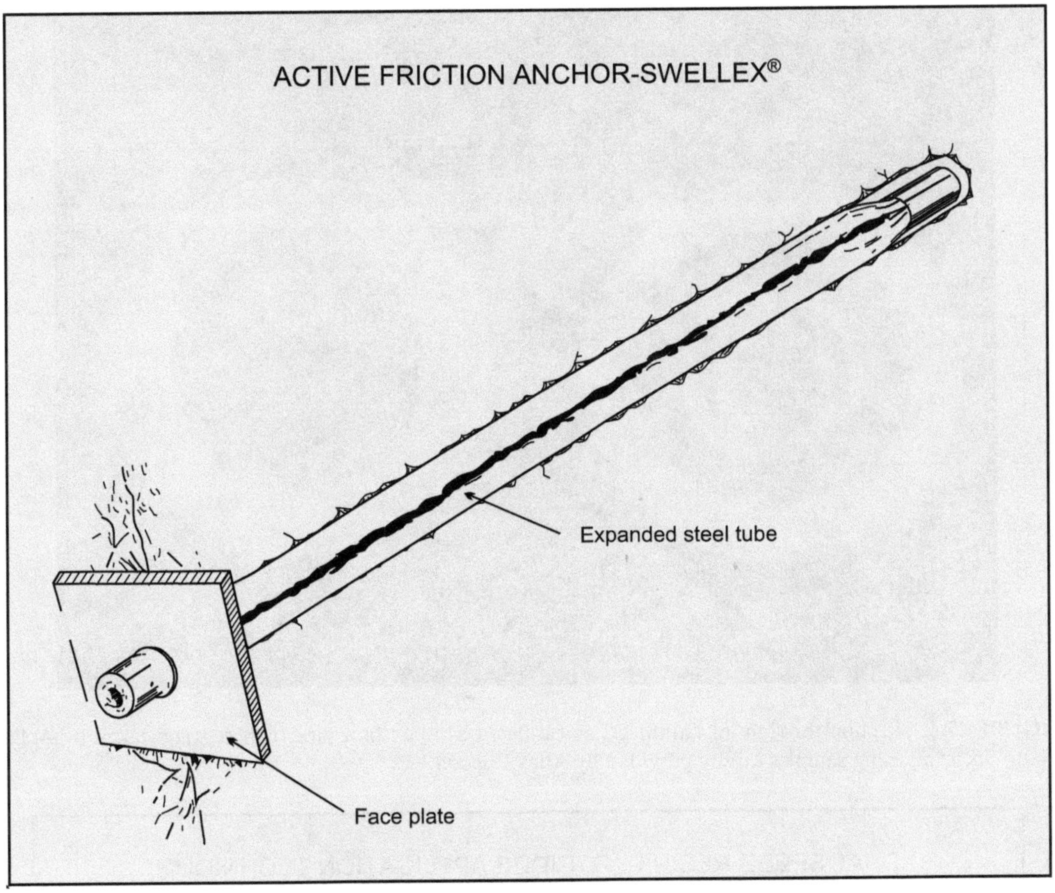

FIGURE 17.5 A Swellex Bolt from Atlas Copco. (Modified after Stillborg (1994).)

- It can corrode (stainless steel units are available at a cost premium).
- It is mainly used for high-speed mechanized development.
- Its load reaction depends mainly on the pump pressure (that is used to "inflate" the deformed tube after it has been placed in the hole).
- It can be made to yield/deform.

Figure 17.5 shows a typical Swellex Bolt, and Figure 17.6 shows an uninflated and a fully inflated bolt in a drill hole.

17.1.5 Resin-Grouted Bolts

These are a common and effective support that typically uses rebar that is anchored in place using resin (usually polyester) (Figure 17.7). They are a specific application of 17.1.2.

The use of resin cartridges is very quick and effective, but the applications must be correctly done, and operator training, and retraining, is essential. The resin cartridge has two compartments where one has the filled polyester resin (mastic) and the other has the filled catalyst/activator. The components are mixed by spinning the rockbolt through them, mixing the components which then set hard.. The resin typically shrinks marginally and creates an effective mechanical key with the rock around the hole.

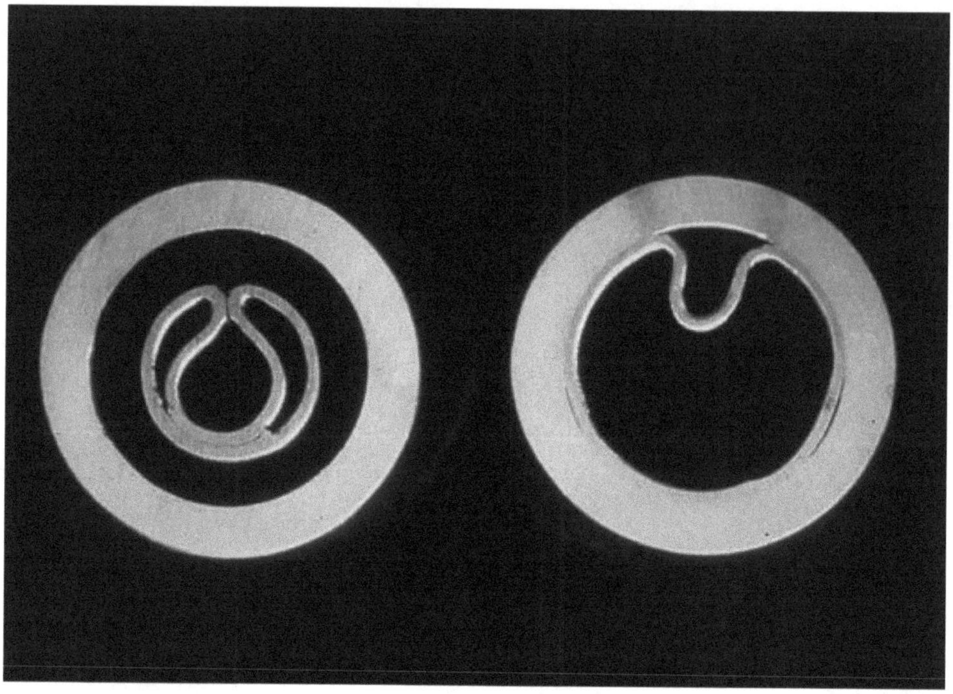

FIGURE 17.6 Section through an uninflated and inflated Swellex in a pipe. (Photograph taken by A.J.S. (Sam) Spearing using samples kindly provided by Atlas Copco.)

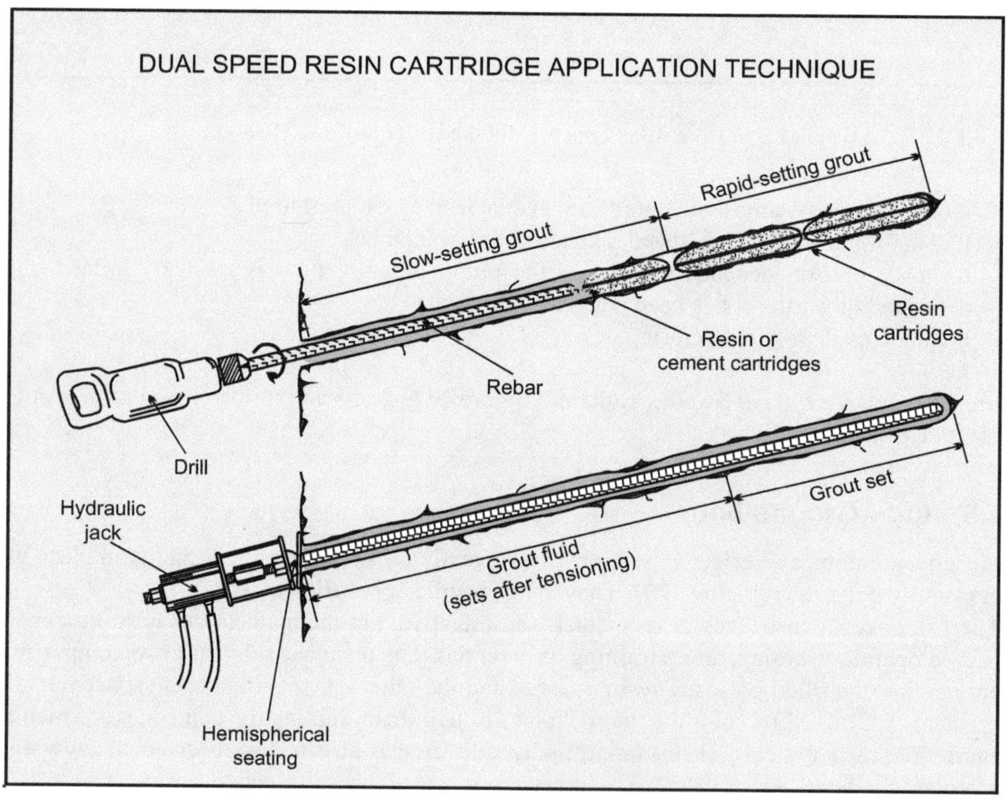

FIGURE 17.7 Using two-speed resin to make an active (tensioned) rockbolt. (Modified after Stillborg (1994).)

Installation practices for passive and active bolts mainly in coal mines are summarized below:

- The installation of a traditional fully grouted passive rebar rockbolt is relatively straightforward and simple. On installation, a single resin cartridge is typically inserted into the drill hole. As the rockbolt is inserted into the drill hole, the resin cartridge bursts, is thereafter spun for an allotted number of seconds, as recommended by the manufacturer, and is held via the upthrust of the bolter unto the bolt. During this time, the resin sets, and loading of the bolt is attributed to subsequent rock movement overtime.
- For two-speed active torque tension rockbolts, two cartridges are inserted into the drill hole. One is a small and fast-setting resin cartridge followed by a larger and slow setting resin cartridge. During the installation, the fast resin gels first and the bolt is tensioned using a threaded bolt and nut. The tension is kept in the bolt as the slower resin sets. A relatively new development is a single cartridge containing two resin gel times (fast and slow). This also reduces the exposure of the bolter operators to the unsupported top, as only one longer resin cartridge is needed instead of two for a tensioned rebar bolt. This also improves bolting productivity significantly as the time to insert the cartridges is obviously reduced.
- For use of a partially grouted resin-assisted mechanical anchor rockbolt, a mechanical shell is attached to the end of a bolt with the use of a partially grouted resin column. Generally, a partial column offsets the high insertion back pressures caused by the shell. The bolt is tensioned when the expansion shell is fixed on rotation, and once the resin sets, the tension is locked into place.

Important operational parameters for the successful and cost-effective use of resin-grouted rockbolts are as follows:

- Standard practice is to drill the bolt hole an inch longer than needed.
- The annulus (annular gap) between the rockbolt and the hole is defined as half the diametric difference between the drill hole diameter and the rockbolts diameter. Keeping this low can facilitate the mixing of the resin components during installation. The optimum annulus is 3–4 mm, and more than 8 mm should be avoided.
- Manufacturers typically recommend between 30 and 40 rotations during bolt rotation, which, with typical bolter speeds, equates to 3–5 seconds for fast-setting cartridges. Fast cartridges are used to minimize installation time per unit as in room-and-pillar mining, where support installation is often a production bottleneck.
- After spinning the bolt, if it is passive, it is held for several seconds and then the bolter boom is lowered. If tensioned with the mechanical shell, it is tensioned (torqued) immediately. If two-speed resin is used, it is held for typically 3–5 seconds and then torqued by the rockbolter.

As mentioned earlier, holes drilled for resin bolts should be about 6–8 mm larger in diameter than the bolt. If it is increased to around 10 mm, the pullout load is not significantly affected (the spin time may need to be increased though), but the stiffness of the bolt/resin assembly is lowered by more than 80%, besides wasting money on unnecessary resin. A typical resin cartridge is shown in Figure 17.8.

Pumped resin systems have been improved recently and are an alternative to using resin cartridges. This system is especially effective when using self-drilling rockbolts in very poor ground conditions.

17.1.6 Cable Bolts

Where long rock anchors are needed, cable bolts can be more easily used than solid bars, and this is highly beneficial where space is limited, because they are flexible/bendable. They have a high

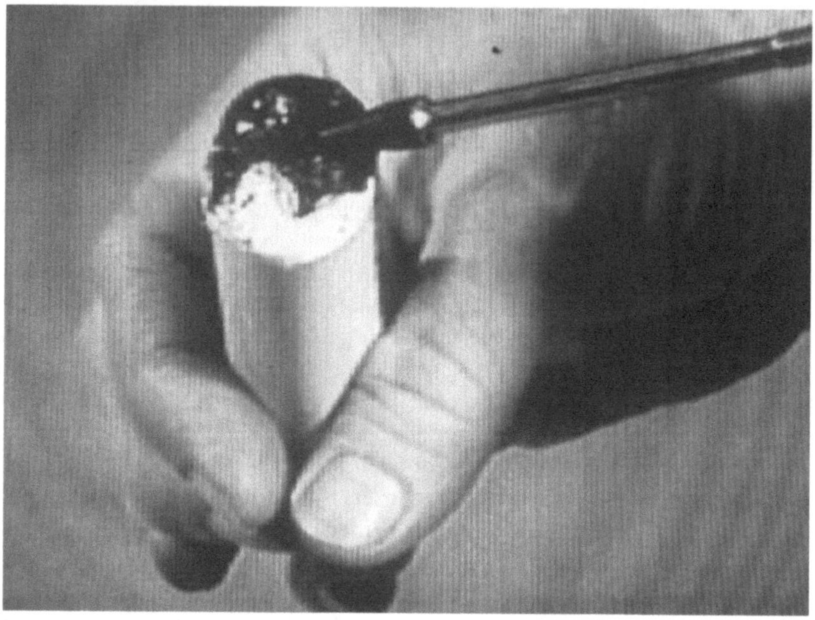

FIGURE 17.8 Typical resin cartridge components where the black is the resin mastic and the white is the catalyst mix. (Photograph taken by A.J.S. (Sam) Spearing.)

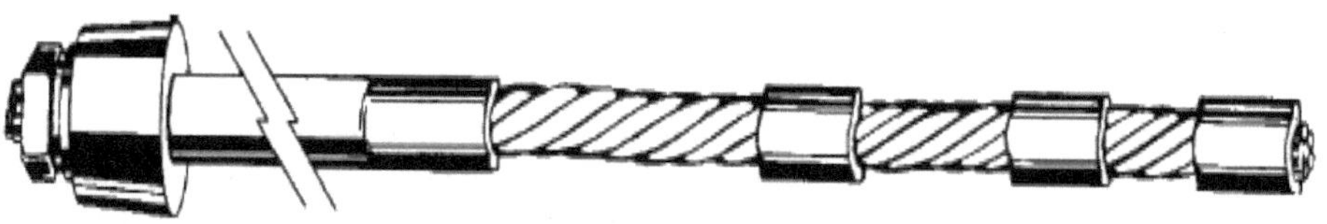

FIGURE 17.9 Typical passive cable bolt often used in coal mines (not to scale) showing the swages used to stiffen the cable bolt for easier installation and better keying with the resin.

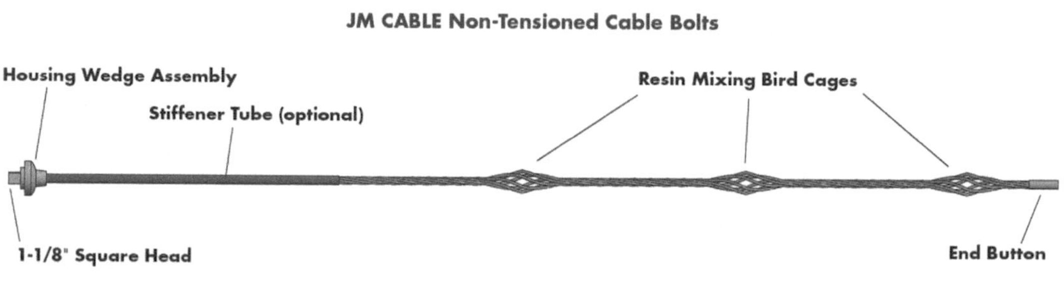

FIGURE 17.10 Example of a commercially available cablebolt. (Reproduced with kind permission of Jennmar Holdings LLC.)

tensile load-bearing capacity and can be made to virtually any length. They can also sustain more deformation that standard rebar rockbolts. They can be active or passive and can be manufactured to withstand very high loads (up to 100 tons each).

There are many different types of cable bolts, but only a few will be discussed. Figure 17.9 shows a simple passive cablebolt with swaged rings to make it more rigid for installation. These types of cable bolts are commonly used as the secondary support on coal mine intersections.

Figure 17.10 shows a specific cable bolt configuration as an example.

The cable bulb or nest where the individual strands are separated help to mix the resin components during installation (rotation) and helps provide a better key with the set resin mix.

17.1.7 Truss Bolts

Cable trusses are useful as they tend to reduce the "tensile zone" induced in the roof. They are therefore useful in coal mines, especially intersections, where the tensile strength of the rock is typically very low. Cable trusses consist of two cable bolts installed near the room corners at about 45° with a tie rod, as shown in Figure 17.11.

In truss bolt systems, the applied tension generated typically by the tie rod creates areas of compression in the tunnel roof. Figure 17.12 shows trusses being used as a support system in very weak strata.

17.1.8 Yielding Rockbolts

In areas where seismicity or large quasi-static deformation is expected, yielding rockbolts can and should be used.

There are many examples of yielding support, and only a few will be shown here.

17.1.8.1 The Cone Bolt (Chamber of Mines Research Organization)

The Cone Bolt was one of the first effective yielding rockbolts developed. It consists of a smooth bar, with a cone produced on the end (Figure 17.13). The Cone Bolt is grouted into a hole, and when

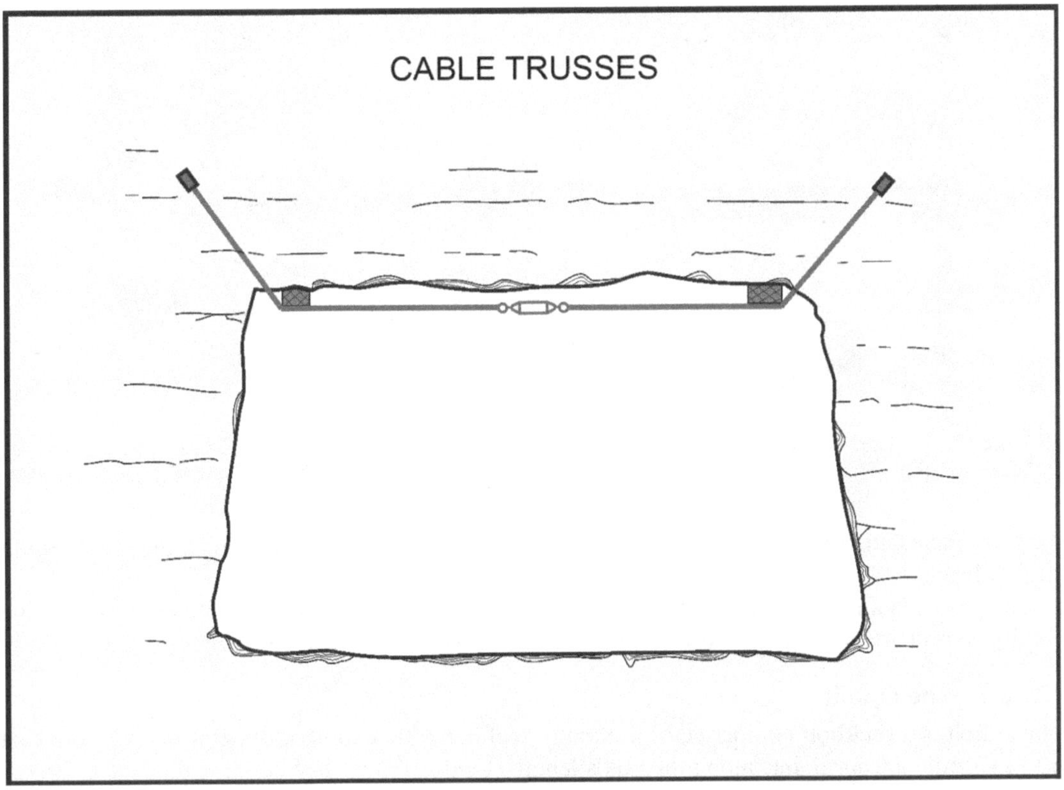

FIGURE 17.11 A simple cable truss system. (Modified after Stillborg (1994).)

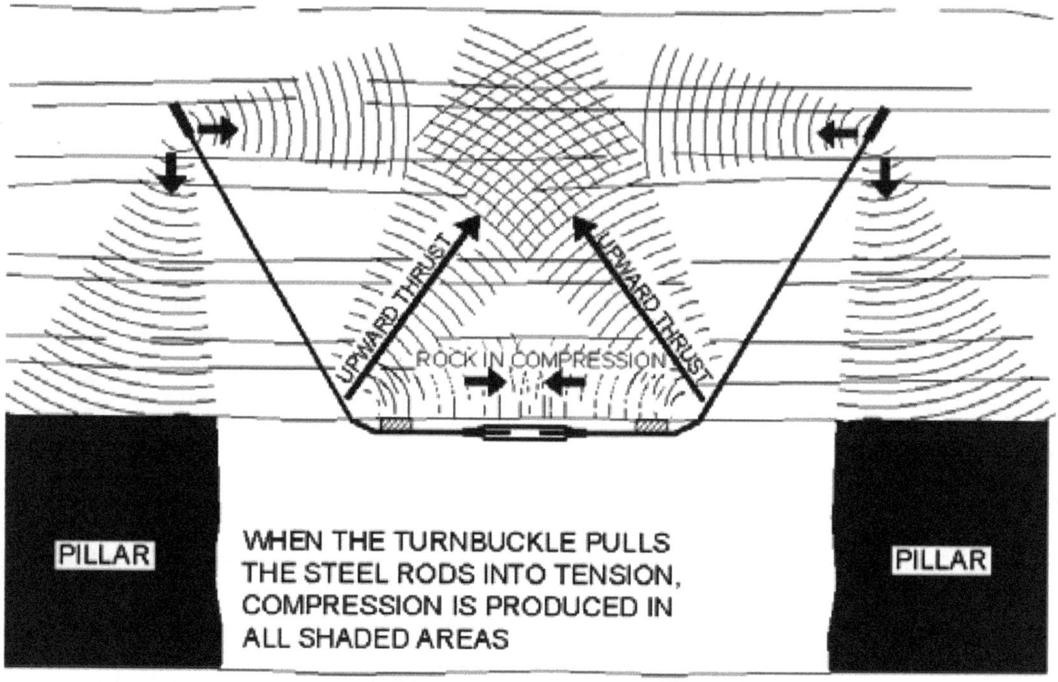

FIGURE 17.12 The induced compression zone formed by cable bolt trusses (Ghabraie et al., 2013).

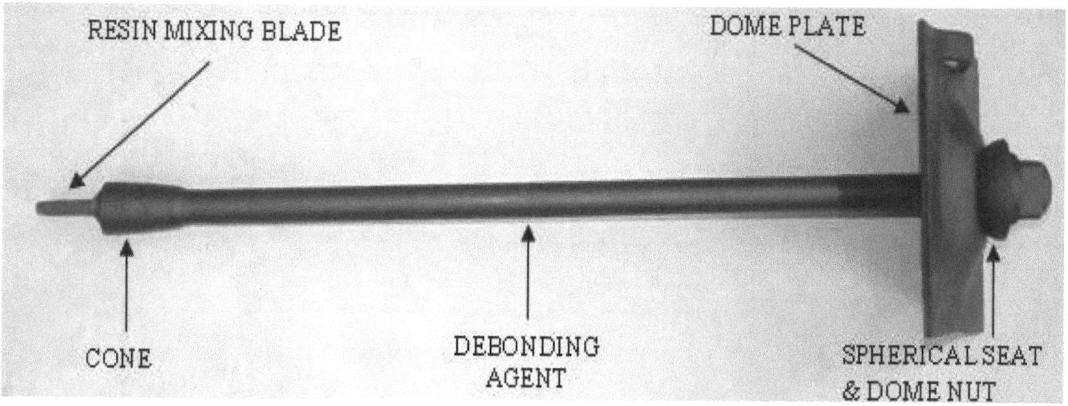

FIGURE 17.13 The smooth de-bonded Cone Bolt (Jager, 1992). (Photo from COMRO in South Africa which no longer exists.)

the rock dilates, the reactive load is generated because the cone is pulled through the grout (South Africa, Jager, 1992).

The grout installation is critical to the yielding performance of the Cone Bolt, so extra care and quality control are needed to ensure that the yield is satisfactory as shown in Figure 17.14.

17.1.8.2 The D Bolt

The D Bolt is a **rockbolt** comprised of a smooth steel bar with a number of deformed sections that act as specific anchor points along the bolt's length (Figure 17.15). The collar end of the D Bolt is threaded and is designed to be used with standard fittings (Normet, 2016).

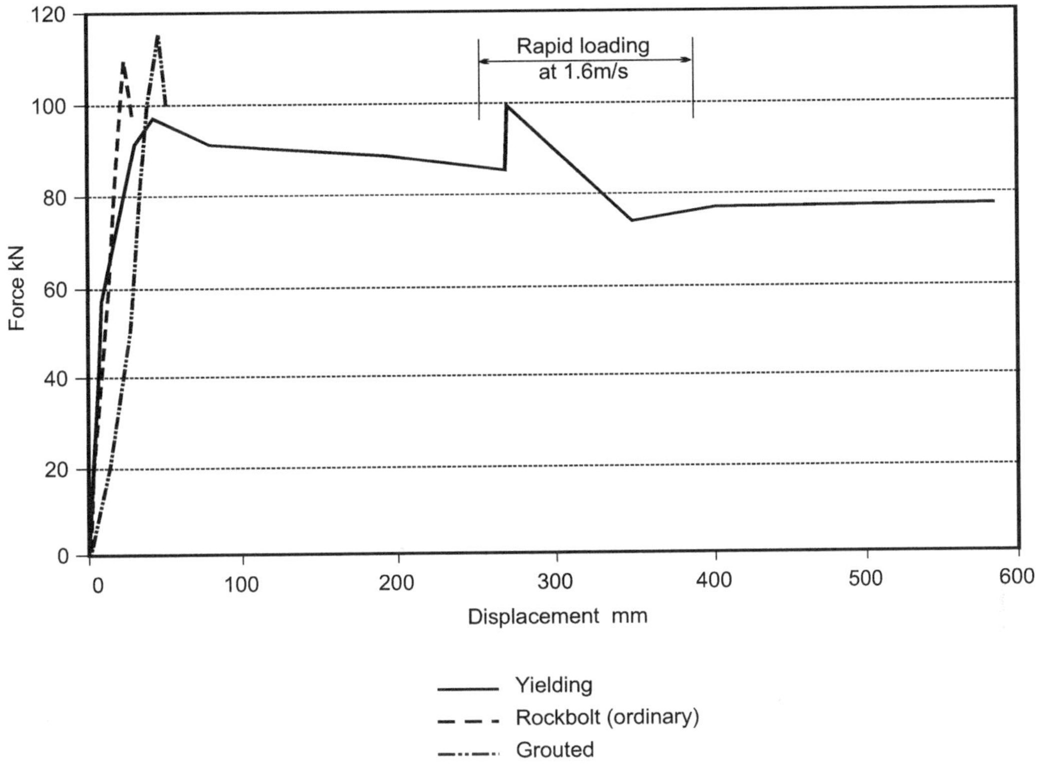

FIGURE 17.14 The Cone Bolt under large slow and rapid deformation in lab tests (Jager, 1992).

FIGURE 17.15 The D bolt. (Reproduced with kind permission of Normet.)

The yield comes from the smooth bar offering little anchorage and the main anchorage coming from the deformations in the bar that crush and fail the grout as the rock around the bolt deforms.

17.1.8.3 Garford Dynamic Solid Bolt

It is a solid bolt that has a mechanism that allows it to yield in a stable manner and thus absorb seismic movement (Hedrick, 2010). The solid bolt can be installed in both resin and cement grouts. The polyethylene sleeve acts as a debonding agent that allows the solid bolt to slip through the sleeve which acts as the dynamic device (Figure 17.16). It is mechanical, therefore enabling repeatability regarding the energy absorption process. The dynamic device is factory installed, so good-quality control is assured.

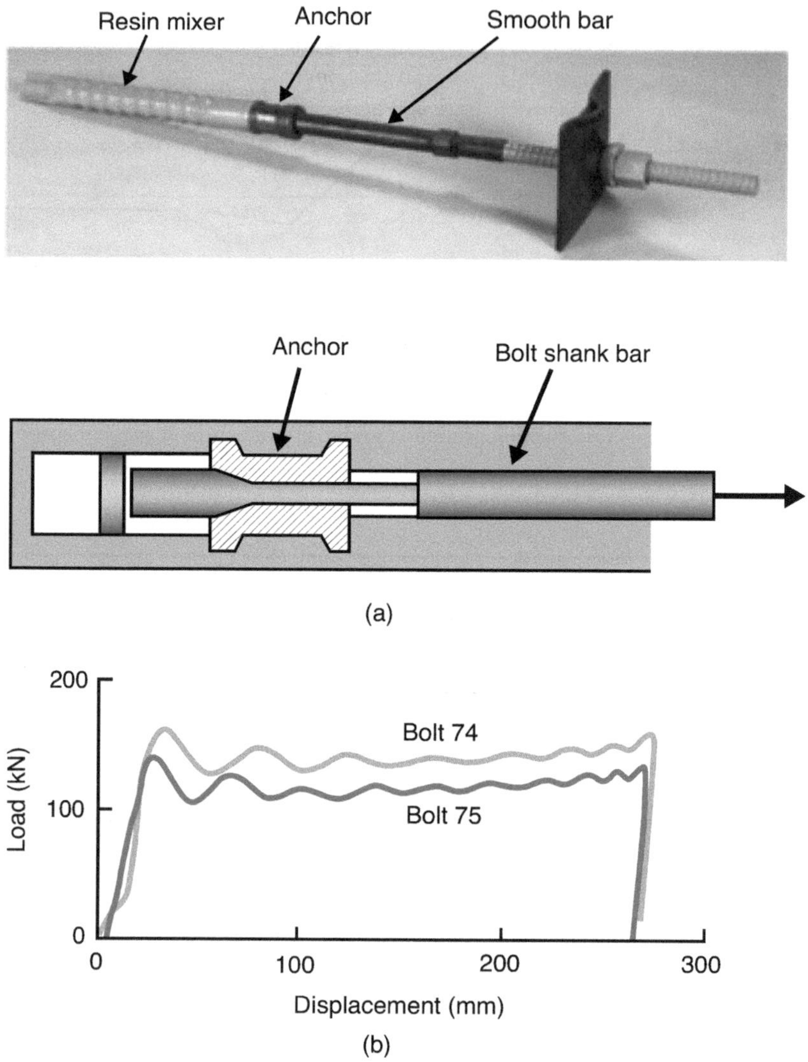

FIGURE 17.16 The garford dynamic solid bolt, yield mechanism, and performance. (Reproduced with kind permission of Garock, Australia.)

17.1.8.4 Yield-Lok Bolt

The bar is upset (stepped to a larger diameter) to specified dimensions at one end and partially, or fully, encapsulated in an engineered polymer coating to achieve designed yielding performance under dynamic loading (Jennmar, 2025). The end profile of polymer coating is angled to aid insertion of the bolt and, along with mixing/centering paddles longitudinally spaced over the length of the coating, provide shredding of resin cartridge packaging. The other end of the bar is threaded for tensioning with a nut. A dome plate and spherical washer are used for angle compensation and to load the bolt axially (Wu and Oldsen, 2010) (Figure 17.17).

17.1.8.5 Rockbolt Plates

There are many plate strengths and sizes, and the selections depend on the following:

- The tensile strength of the rockbolt as the plate should withstand the tension of the bolt until failure.

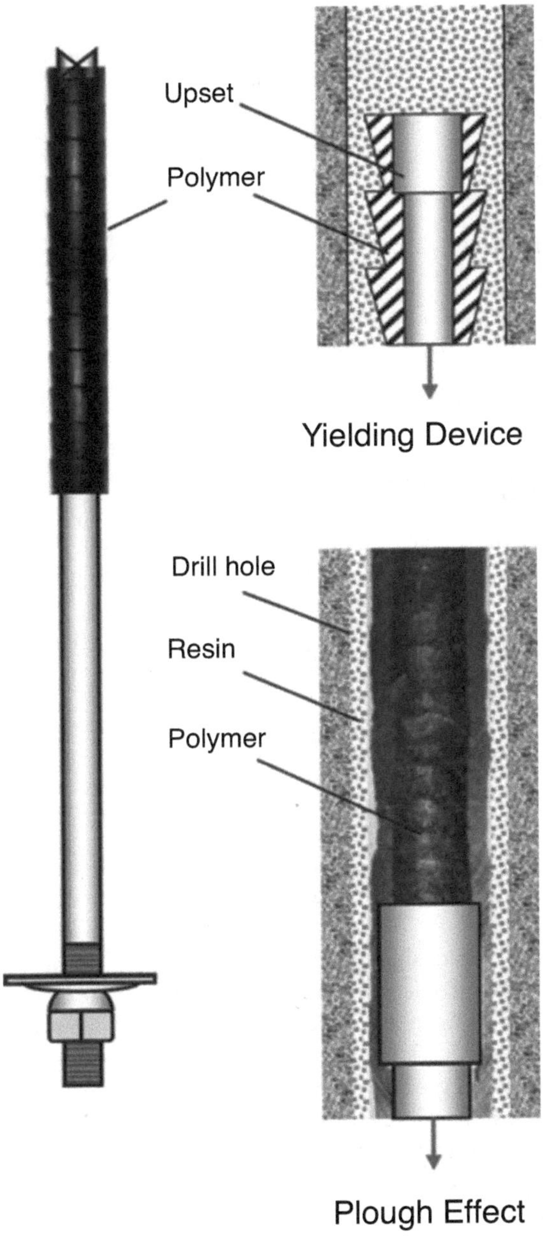

FIGURE 17.17 The Yield-Lok yielding bolt (Wu and Oldsen, 2010).

- The condition of the rock mass skin.
- If the exposed rock is prone to weathering over time.
- The useful life of the specific excavation.
- The excavation dimensions.
- The stress state on the excavation and any changes expected due to adjacent subsequent mining operations.

Figure 17.18 shows some common examples. In some coal mines, very large thin pans (>30 cm) can be used which are not shown in the figure.

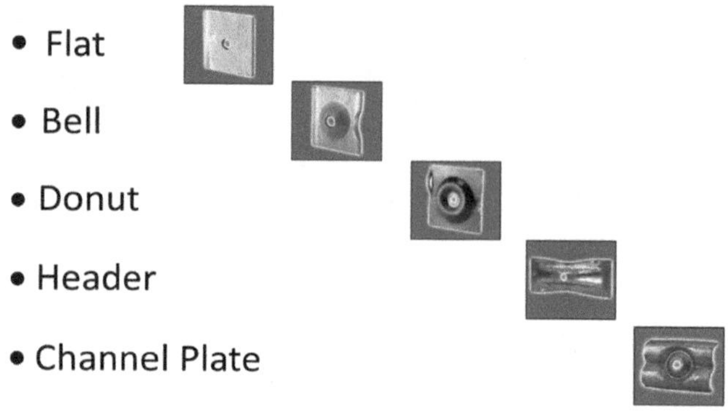

FIGURE 17.18 Examples of rockbolt plates.

17.2 FREE-STANDING SUPPORT

Free-standing supports are those that are set on the mine excavation floor/footwall and extend to the roof/hangingwall. They can be active or passive supports and are mainly used in soft rock tabular mines (coal). They are used mainly for "dead weight" applications.

Free-standing supports are secondary support, i.e., installed after the development cycle, have been used in coal mines for many years, and include the following main types:

- Hydraulic props (re-usable if safely removed).
- Pumped cementitious cribs.
- Cast cementitious cribs.
- Timber cribs.
- Timber posts or props (elongates).
- Mechanical props (some can be re-used).

Both non-yielding and yielding designs are necessary to satisfy the variety of load and closure conditions that occur in underground coal mines. Non-yielding or minimal-yielding applications (of say <75 mm) typically will include bleeder support in longwall applications and supplemental support in room-and-pillar applications. Longwall tailgate support will typically require yielding support systems, although the degree of yielding required can vary considerably based on the specific coal seam application. Other applications of standing support include longwall recovery and retreat room-and-pillar mining. Here, controlled yielding is necessary, although stiff support with preloading capability is also beneficial.

Prop supports have many advantages over bulky, crib-type, standing supports, particularly in material handling and installation. The aspect ratio of the support is also a design consideration. The installation, and removal if applicable, of all support must always be from a position of safety, i.e., a supported area.

17.2.1 Design and Performance Requirements With Emphasis on Free-Standing Support

The design requirements for any roof support system can be divided into the following fundamental requirements:

- Performance requirements
  - Safety obviously paramount as well as stiffness, load capacity, and yield.
- Installation requirements
  - Handling, installation, and height adjustment.

Syd Peng (2023) has stated: "The vast variety of roof bolting patterns adopted in coal mines of various countries including Australia, Canada, China, Germany, Mexico, and South Africa presented at ICGCM in the past 40 years was shown and discussed – there is a complete lack of a commonly acceptable roof bolting theory or theories." This needs to be addressed moving forward, to improve mine safety as it is a global issue.

Other relevant design issues include:

- Application including general accessibility.
- Ventilation issues.
- Cost.
- Durability.
- Floor and roof contact area.
- Weight.
- Flammability.

An understanding of the ground response is necessary for ground control. The ground reaction concept, as shown in Figure 17.19, has been used to provide insight into the performance characteristics required for roof support systems. Conceptually, the ground reaction curve shows that there is a component of vertical convergence that cannot be prevented by standing roof support systems. Therefore, both non-yielding (or more accurately low-yielding) and (high) yielding support designs are required depending on the application (Spearing, 2008).

Although not always required in coal mines unless bursting or bumping occurs, many hard-rock applications require an ability to withstand rapid loading cycles that are typical during a seismic event or rockburst. If such conditions exist, the capability of the support to sustain loading at convergence rates as high as 3 m/s should be required (COMRO, 1988).

A ground reaction curve (GRC) graphs the relationship between the displacement of an excavation against the load transferred and that required to maintain its stability.

The GRC can be used for a first support design by understanding how much pressure the rock will exert at different levels of deformation, so the installed support capacity is adequate to prevent too much deformation.

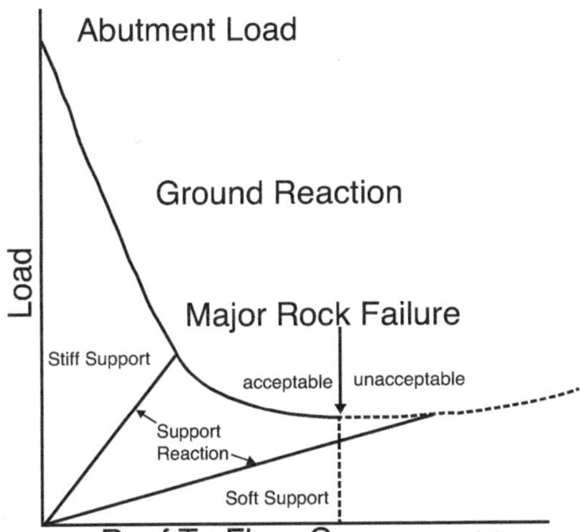

FIGURE 17.19 Ground reaction curve relates convergence to the support capacity to establish design requirements for roof support systems (Barczak, 2005).

Other design issues that must be considered or achieved are as follows:

- The shape and aspect ratio of the excavation.
- The confidence that the workforce has in the support to be installed.
- Adequate contact area to avoid damaging/punching into the roof or floor strata.
- Cost effectiveness of design including transportation and installation costs.
- Durability for long-term applications considering the effects of corrosion for example.
- Low weight.
- Ease of transportation.
- No ignition sources, such as sparking, when installed in gassy conditions.
- Minimum interference to men, material, and ventilation.
- Minimal or zero fire potential.
- The support installation should be done without exposing the miner to unsupported ground or injury during the installation.

Examples of the load versus deformation performance of some free-standing supports are shown in Figure 17.20.

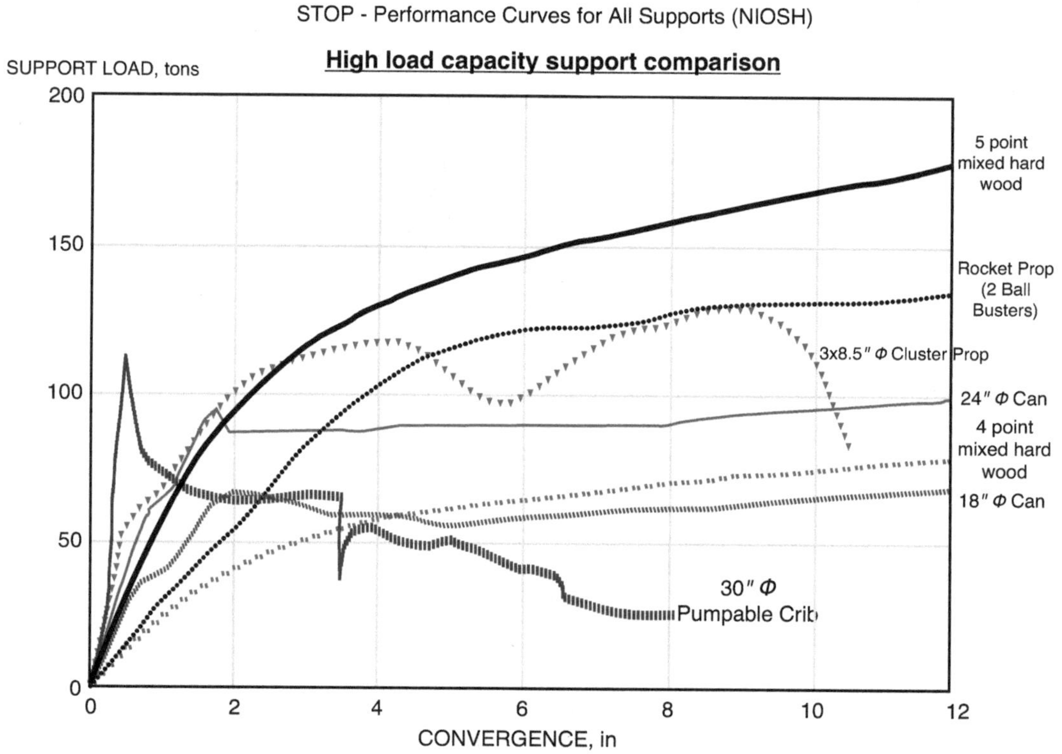

FIGURE 17.20 Performance examples of some free-standing supports (Barczak, 2000)

17.2.2 Free-Standing Support Testing

The performance of any new support being considered for mine use must first be tested under representative and safe conditions in a laboratory. Figure 17.21 shows the testing equipment used by the National Institute of Occupational Safety and Health (NIOSH), Pittsburgh, Pennsylvania, US.

17.2.3 Examples of Commonly Used Free-Standing Supports

Free-standing supports are installed between the floor and the roof.

The following sections discuss some of the commonly used free-standing supports.

17.2.3.1 Hydraulic Props

These are re-usable; they give a high pre-load if needed; with a valve, they can withstand slow and rapid (rockburts or coal bursts); and they yield when a predetermined reactive load is reached. The main disadvantage is that they need a hydraulic (portable) power pack to reset them in place each time they are moved. They are also rather heavy. They are mainly for longwall gate/entry support. Any removable support must be safely removed, which is ideally with a remote release, so the miner is never under unsupported ground (Figure 17.22).

17.2.3.2 Alternative Crib Support (ACS)

These typically consist of a box-threaded section that can raise and lower the supports as shown in Figure 17.23. They can only be used for low-closure applications and only exert very low preloads.

FIGURE 17.21 Shield testing at the NIOSH Pittsburgh Research Lab. (Photo reproduced with kind permission of the NIOSH Mining Program.)

FIGURE 17.22 Hydraulic props for coal mines. (Photographs reproduced with kind permission of Prof. Liqiang Ma.)

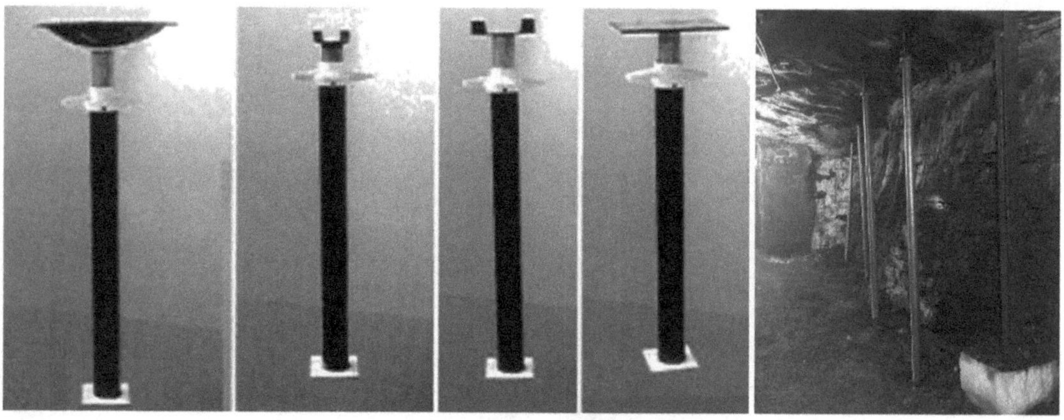

FIGURE 17.23 Various ACS props. (Reproduced with kind permission of Jennmar Holdings LLC.)

17.2.3.3 Quik Stik

This is a low-deformation (yield) prop that cannot be re-used and is very quick to install, saving considerable time (Jingle and Spearing, 2005). It uses ball bearings for height adjustment between the floor and roof and for the load-bearing capacity as shown in Figure 17.24.

The load versus displacement is shown in Figure 17.25, based on testing using the NIOSH test equipment shown earlier in Figure 17.21.

Each ball bearing sits in a reinforced and tapered slot as shown in Figure 17.24c. The taper allows the inner tube, Figure 17.24b, to be raised, but it cannot be lowered as the bearings lock the inner and outer tubes together. The Quik Stik does not yield as the reinforcing ring shown in

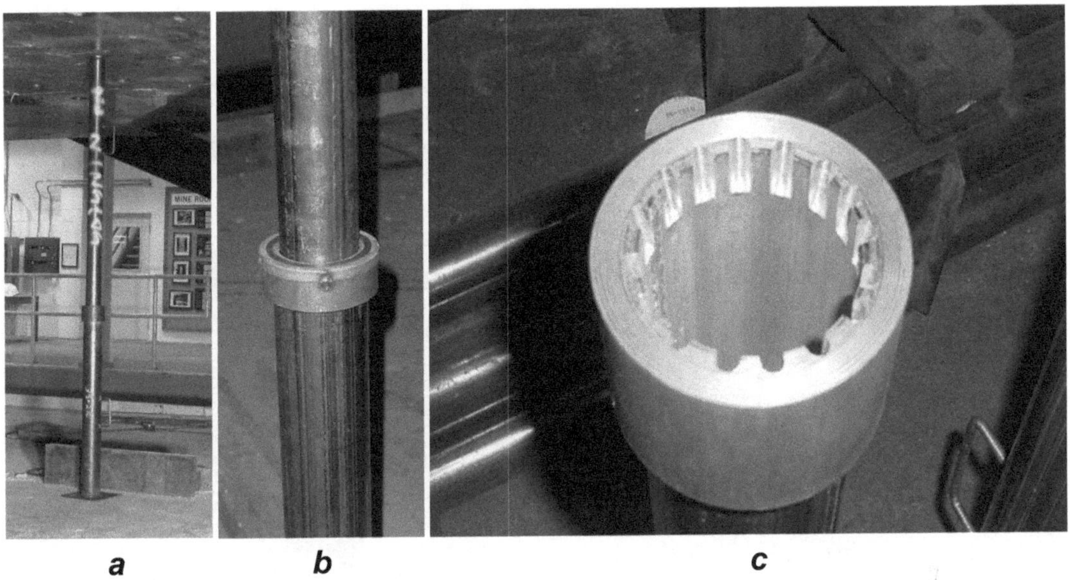

FIGURE 17.24 The Quik Stik. (Photos taken by A.J.S. (Sam) Spearing at NIOSH Mining Lab.)

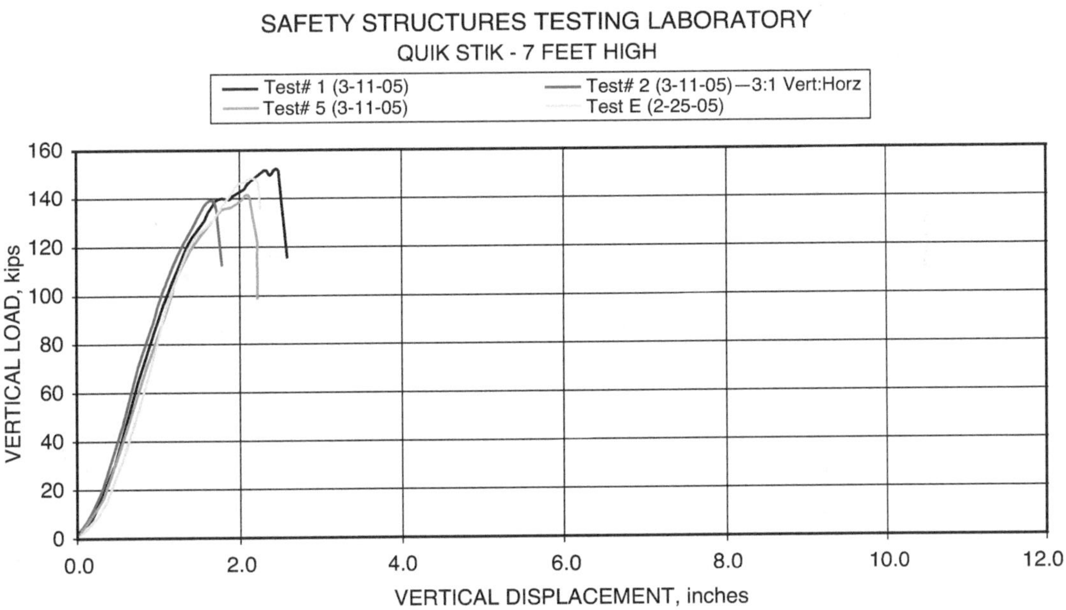

FIGURE 17.25 Quik Stik operational performance (Jennmar, 2025).

Figure 17.24b and c on the outer tube stops the outer tube deforming, and hence, it does not yield significantly.

17.2.3.4 Ball Buster

With the Ball Buster which is based on the same mechanism as the Quik Stik, the yield is achieved because the ball bearing scours (gouges) the inner and outer tubes in a controlled manner, as shown in Figure 17.25. The ball bearings are not in individual slots as in the Quik Stik, and they are placed around a wedged section (taper) around the entire pipe perimeter.

This product also uses ball bearings in the tapered slot for height adjustment, load and yielding (both quasi-static or slow closure and dynamic or rapid closure (Jingle and Spearing, 2005). Examples of the prop construction and its performance are shown in Figures 17.26 and 17.27.

a *b*

FIGURE 17.26 The Ball Buster and its yield mechanism. (Photos taken by A.J.S. (Sam) Spearing.)

FIGURE 17.27 The Ball Buster yielding props being used to prepare for a longwall move. (Reproduced with kind permission of Jennmar Holdings LLC.)

Figure 17.26a shows the prop yielding by expanding the outer pipe in a stable manner. Figure 17.26b shows the inner and outer tubes having been cut apart to illustrate the gouging of the ball bearings as the prop yields in a stable manner.

Figure 17.27 shows Ball Buster props being used for a longwall coal panel move.

The Ball Buster performance is shown in Figure 17.28, where the stable yield to 18″ is clearly shown.

17.2.3.5 The Pencil Prop

Timber poles can be modified to make them yield a certain amount, and the pencil prop is an example, where the diameter of the prop is reduced to locally reduce the diameter as a taper, to produce a lower peak load but with a relatively stable yield as shown in Figure 17.29.

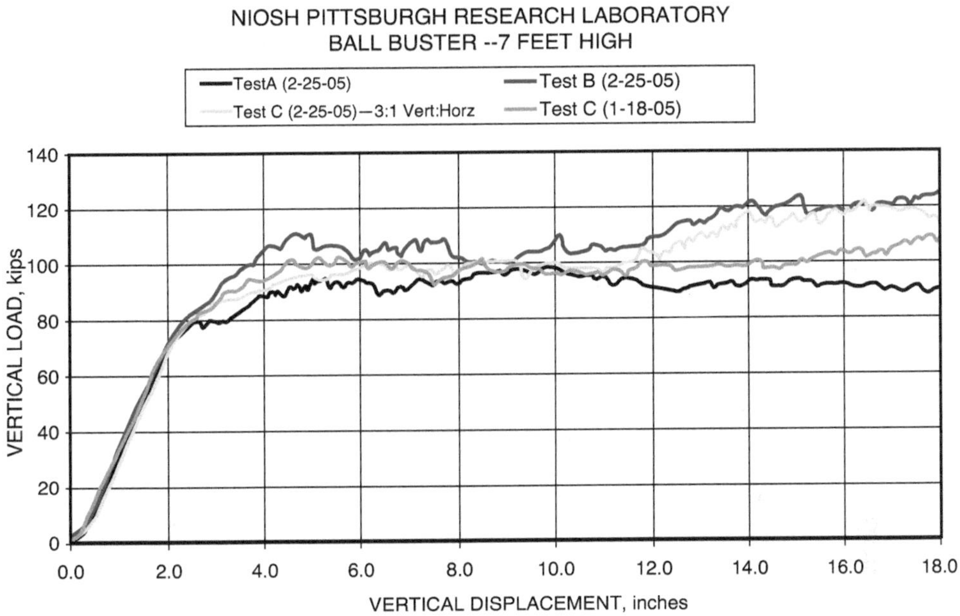

FIGURE 17.28 The Ball Buster performance (Jennmar, 2025).

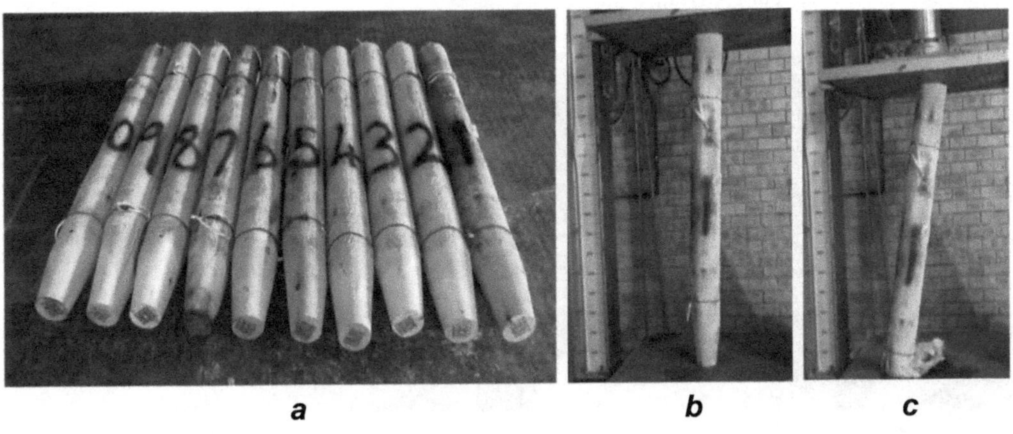

FIGURE 17.29 Pencil props: pencil props (a), a prop installed for lab testing (b), and a prop after stable yielding (c). (Reproduced with kind permission of Timrite Underground Mine Support Solutions, South Africa.)

Figure 17.30 shows results of lab tests, and it can be seen that the prop can yield in a stable manner for over 200 mm.

17.2.3.6 Pumped Cribs (Grout/Cement Filled)

Laterally reinforced and contained, grout-based cribs are frequently used especially in longwall tailgates. The mix which can be aerated (foamed) is generally pumped from a centralized mixing plant, that can even be on surface. The outer bag contains the mix as it is setting (gaining strength), and the structure offers confinement and reinforcement during loading as shown in Figure 17.31.

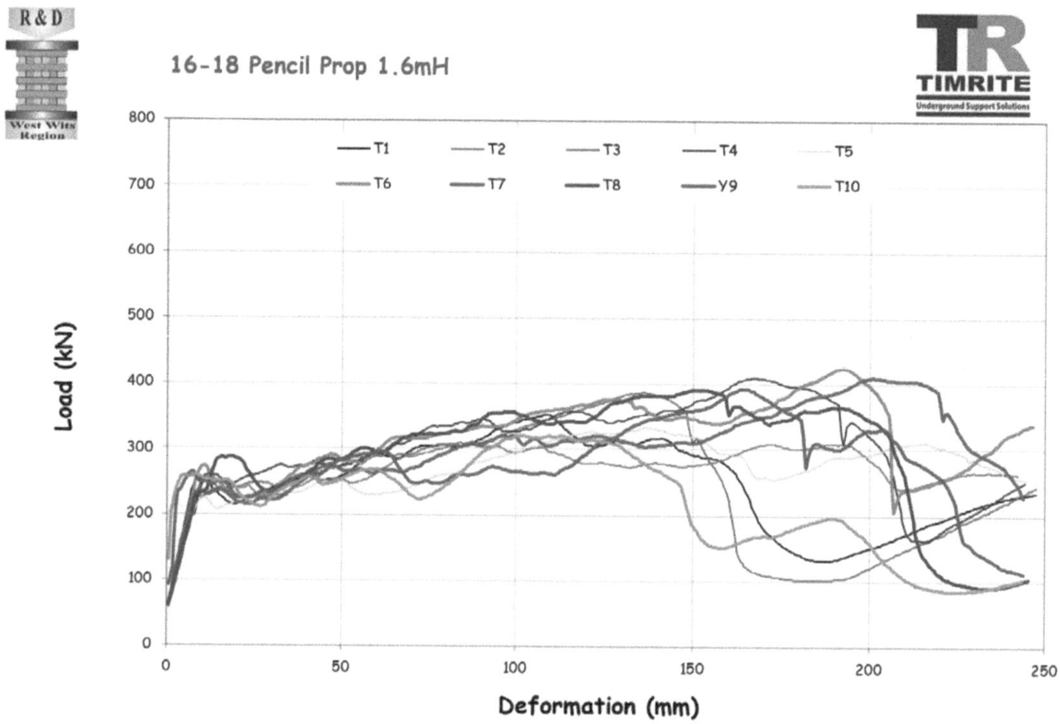

FIGURE 17.30 Pencil prop lab test results. (Reproduced with kind permission of Timrite Underground Mine Support Solutions, South Africa.)

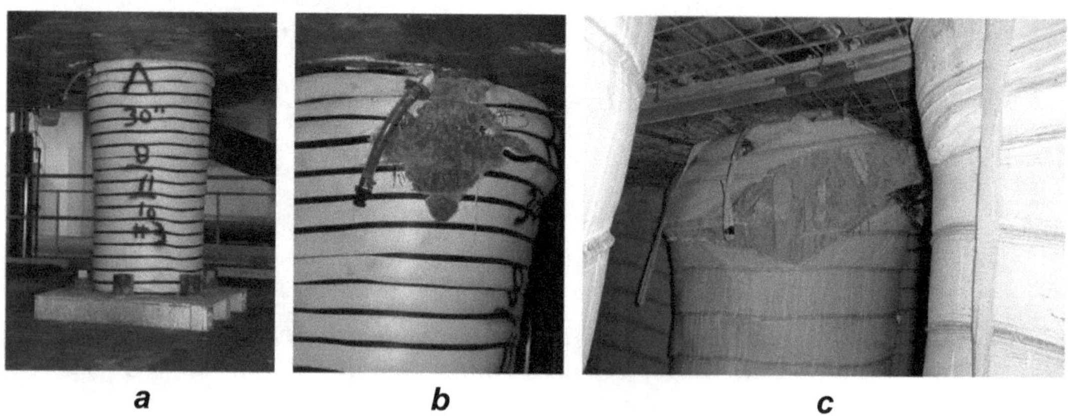

FIGURE 17.31 Pumped concrete cribs. (Reproduced with kind permission of Steve Tadolini, Minova.)

Figure 17.31a shows a filled concrete crib before lab testing, Figure 17.31b shows the crib yielding with the outer-reinforced bag (in black) giving lateral confinement, and Figure 17.31c shows a crib in a tailgate showing controlled yield.

Typical crib performance is shown in Figure 17.32, and it can be seen that after a post-peak load loss, the crib yields in a relatively consistent manner over large deformations.

17.2.3.7 Lock-N-Load (Non-Yielding Timber Mechanically Locked Prop)

This type of interlocking square posts was initially used in building construction to hold up formwork before a concrete floor is formed in multi-story buildings. It is now also used in a similar mining application to offer temporary support in mainly tabular mines as shown in Figure 17.33.

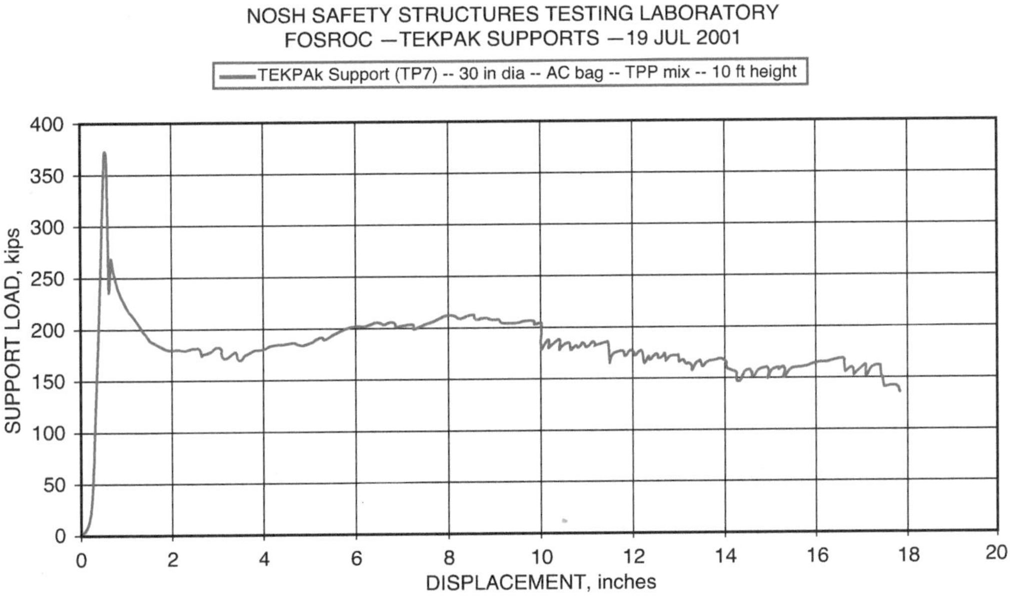

FIGURE 17.32 Typical concrete crib performance (Tadolini, Minova).

FIGURE 17.33 The lock-N-load prop. (Reproduced with kind permission of Strata Worldwide.)

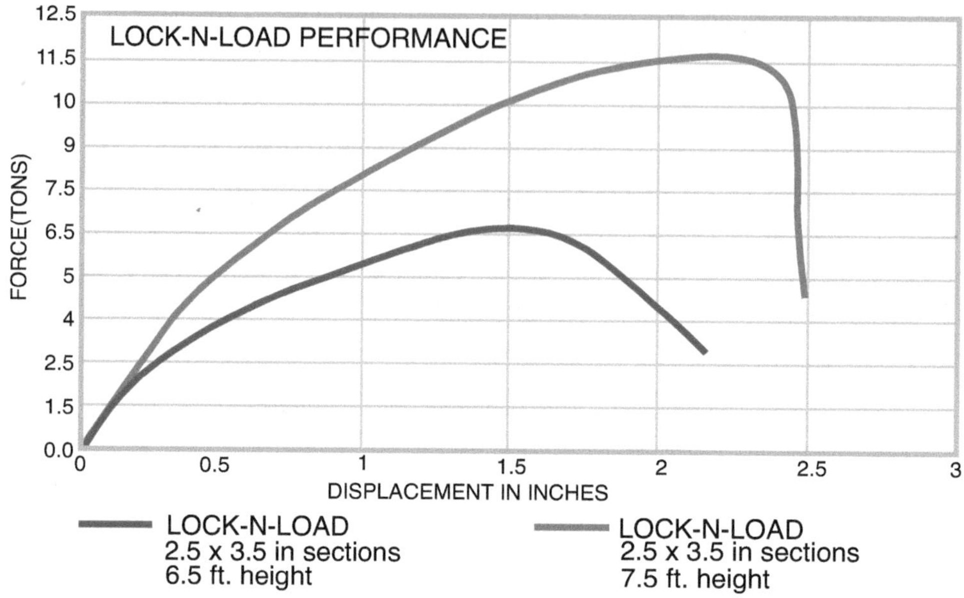

FIGURE 17.34 Lock-N-load performance (Strata Worldwide).

Figure 17.34 shows the typical performance that can be useful in low-deformation applications.

17.2.3.8 Sand Prop Non-Yielding Unit (Provisionally Patented by Spearing and Jingle, 2008, but not Completed)

The height adjustment is done via dry sand flow, and the load is generated using the same sand column as shown in Figure 17.35. The prop is very rapid and easy to install.

The performance of the Sand Prop is shown in Figure 17.36. The stable yield is obviously limited so its application is used in low closure areas such as in room-and-pillar coal mining.

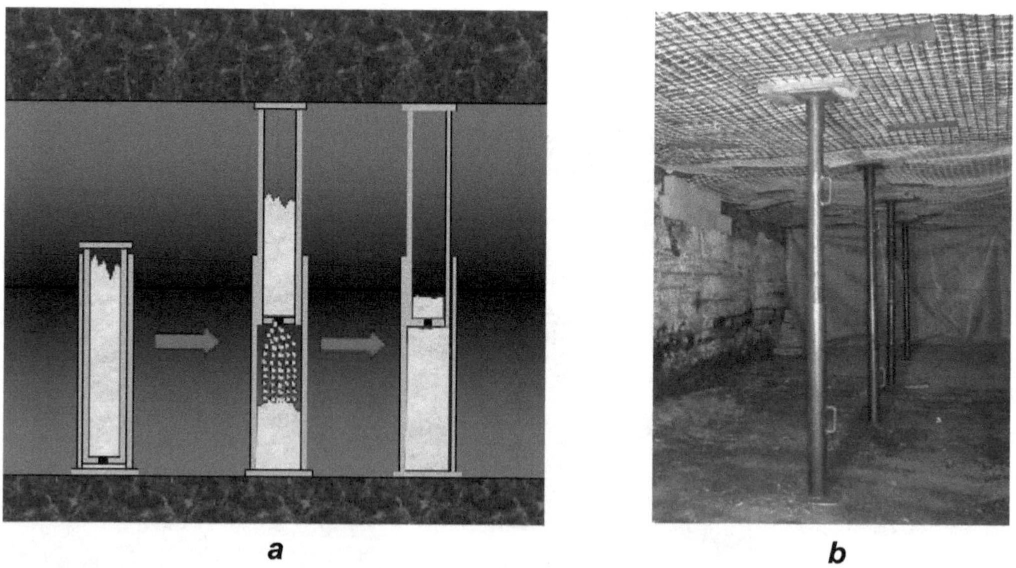

FIGURE 17.35 The sand prop. (Reproduced with kind permission of Strata Worldwide.)

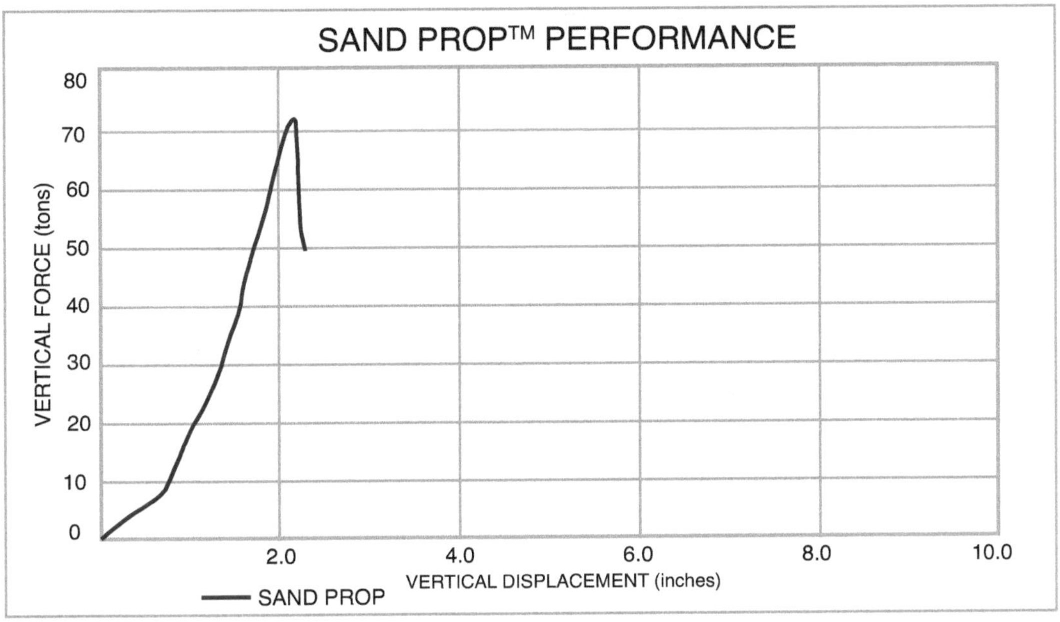

FIGURE 17.36 Sand prop performance (Strata Worldwide).

17.2.3.9 RocProp

This is a static and rapid yielding prop using the ductility of steel tubes and a steel wedge with teeth on the inside to grip the inner tube (Spearing et al., 1996). Yield is obtained by the wedge deforming the outer pipe as shown in Figure 17.37.

The split wedge between the inner and outer tubes is used to set the prop against the roof and the internal teeth on the wedge firmly grip the inner tube. The yield is obtained because the wedges cause the outer tube to expand in a stable manner, thus maintaining a load while it is deformed as shown in Figure 17.37b and c. Figure 17.37d shows a row of Rocprops installed in a longwall coal mine.

The RocProp performance is shown in Figure 17.38. For the test, the inner tube was raised about 560 mm from the outer tube, and then the wedges were set in place (Figure 17.37a). The rise in load

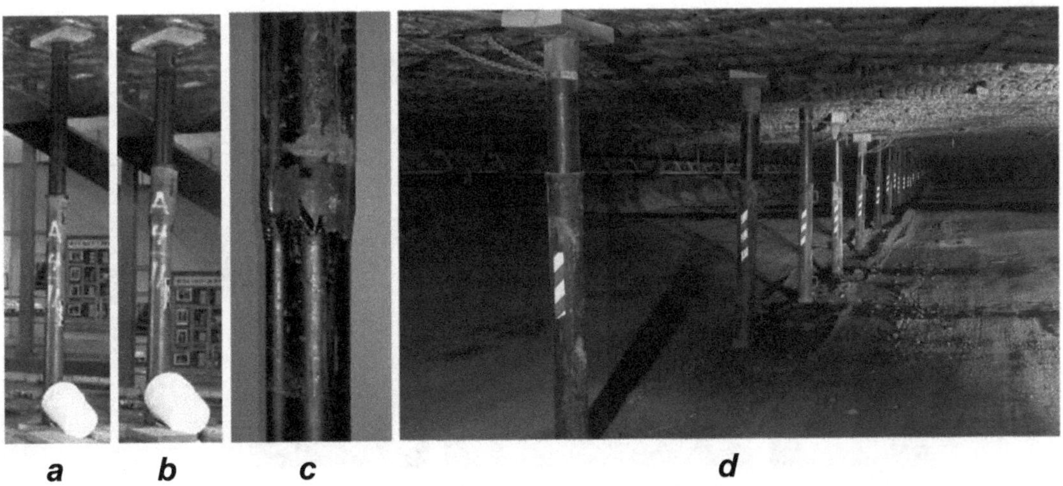

FIGURE 17.37 The RocProp. (Prop set for testing (a), yielding prop (b), close-up of the outer pipe yielding (c), and props installed underground (d). Reproduced with kind permission of Strata Worldwide.)

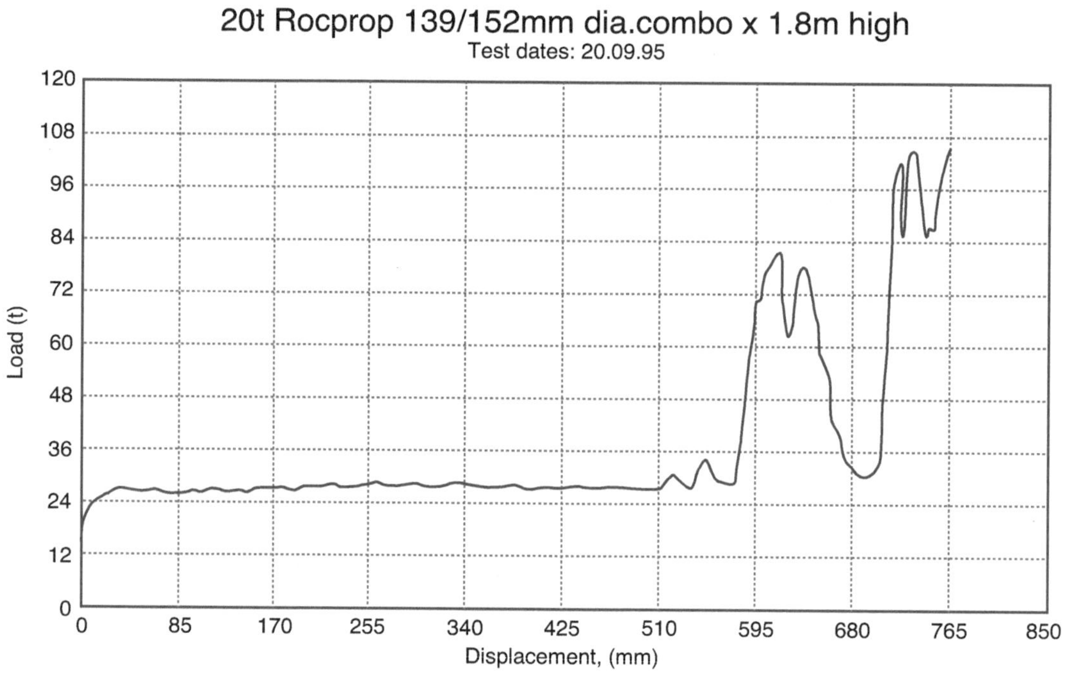

FIGURE 17.38 The RocProp performance (Strata Worldwide).

after around 560 mm of stable displacement is because the outer tube has bottomed out into the inner tube, so it is effectively now a double tube so much stronger. Its performance after this happens becomes erratic.

17.2.3.10 Propsetter (Yielding Timber Prop)

The Propsetter is an engineered timber prop with a wedge cut partially into one end and wire rings attached to keep the wedge assembly from failing as it yields; see Figure 17.39.

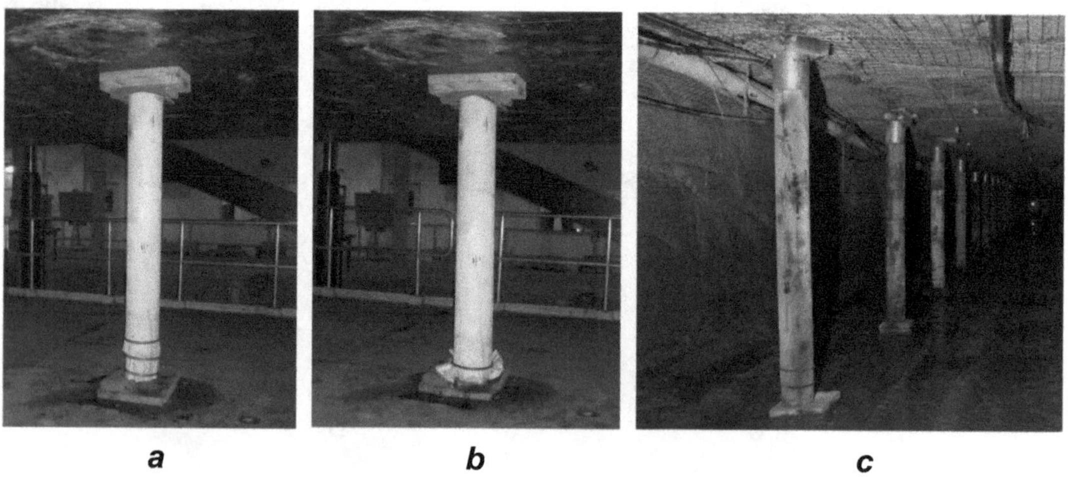

FIGURE 17.39 The Propsetter. (Reproduced with kind permission of Strata Worldwide.)

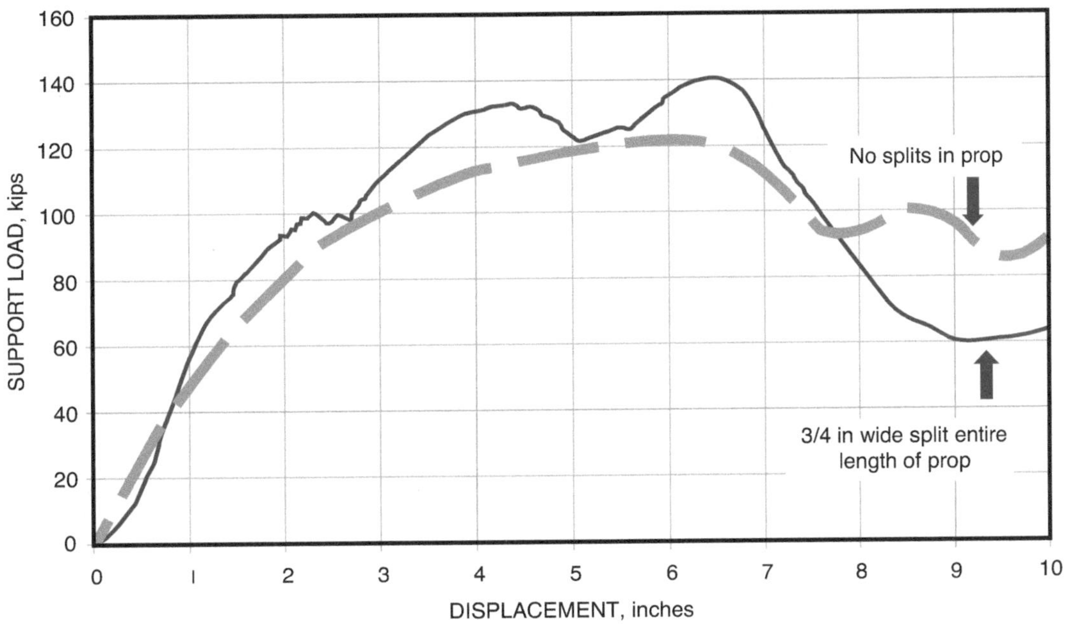

FIGURE 17.40 The Propsetter performance (Strata Worldwide).

The performance of the Propsetter is shown in Figure 17.40, which clearly shows its stable high-deformation application.

17.2.3.11 Timber-Based Cribs/Packs

Timber cribs can be made in many different variations and assemblages, with the common objectives of reducing the installed cost while maintaining safety and performance. An example of a timber crib is shown in Figure 17.41.

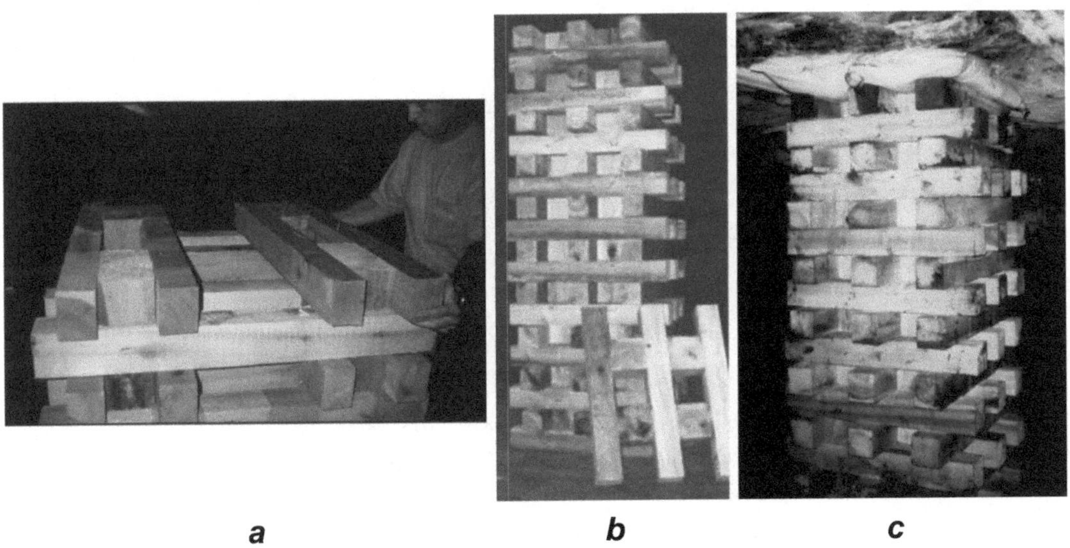

FIGURE 17.41 The Hercules timber crib. Crib constructed for lab testing (a), a crib ready for testing (b), and a crib underground (c). (Reproduced with kind permission of Strata Worldwide.)

FIGURE 17.42 Cans installed underground. (Reproduced with kind permission of Jennmar Holdings LLC.)

17.2.3.12 The Can

These are thin-walled metal outer cylinders (cans) that are prefilled with a low-density concrete foam (or similar) and used in longwall tailgates. They are ordered at a preset height and installed with timber or an inflatable device to give good contact between the floor and roof. Examples of cans underground are shown in Figure 17.42.

They are effective supports but relatively difficult to transport and install. They require special equipment like a "can setter" which can be attached onto a mobile carrier like a bobcat.

17.3 SUPPORT INSTALLATION AND COSTS

Training and retraining of support installation crews is essential, as is quality control. Support must always be installed from a position of safety that could be a location already supported or where temporary support is installed. The costs of a support system are made up of direct and indirect costs.

The direct costs include the following and are relatively easy to determine:

- Total material costs.
- Material transportation.
- Drilling costs (if applicable).
- Installation labor costs.

The indirect costs include the following and are difficult to obtain representative values for:

- Material wastage.
- Lost production due to an inappropriate support installation cycle.
- Health, safety, and rehabilitation costs if, for example, the system fails and needs reinstallation.

An optimum support system is one that minimizes the total costs (direct and indirect) while ensuring safety. It is important that support is correctly designed with the first goal being safety and the next being cost effectiveness.

17.4 SUPPORT FAILURES

Rockbolt support systems can fail for the following reasons or their combinations:

- Incorrect hole diameter.
- Inappropriate steel member diameter or strength.
- Inappropriate grout material.
- Incorrect grout filling of the hole.
- Incorrect length of hole.
- Poor installation procedure.
- Corroded or greasy steel member.
- Over-straining and/or over-stressing.
- Long-term corrosion.

17.5 WORKED EXAMPLES

Example 17.5.1

Determine the tension at yield (in tons) for a #6 grade 60 steel bar.

A grade 60 (or 60 gr) steel bar indicates a yield stress (in tension) of 60,000 psi. The #6 indicates a (6/8″) bar diameter (nominal).

Solution:

$$\text{Area } = \left[\pi\left(d^2\right)\right]/4 = \left[\pi\left(0.75^2\right)\right]/4 = 0.44\text{in}^2 \text{ (nominal area)}$$

$$\sigma_{\text{yield}} = 60,000\text{lbs/in}^2$$

$$\text{Tensile Yield Force } = \left(60,000 \text{ lbs/in}^2\right)\left(0.44 \text{ in}^2\right) = 26,400 \text{ lbs}/2,000 \text{ lbs} = 13.2\text{t}$$

Example 17.5.2

If the steel in the previous example has an elastic modulus (E) of 29 × 10^6 psi, what is the tensile yield strain in micro-strain?

Solution:

$$E = \sigma_{\text{yield}}/\varepsilon_{\text{yield}}$$

$$29,000,000 \text{ psi } = (60,000 \text{ psi})/\varepsilon_{\text{yield}}$$

$$\varepsilon_{\text{yield}} = (0.00206 \text{ in/in})\left(10^{-6}\right)$$

$$\varepsilon_{\text{yield}} = 2,060\ \mu\varepsilon$$

Example 17.5.3

A post is fitted with a cap and footer as shown in Figure 17.43. Determine the location of the center of gravity.

Solution:

$$\text{Area}_1 = \text{Area}_3 = (6 \text{ in})(1.5 \text{ in}) = 9 \text{ in}^2$$

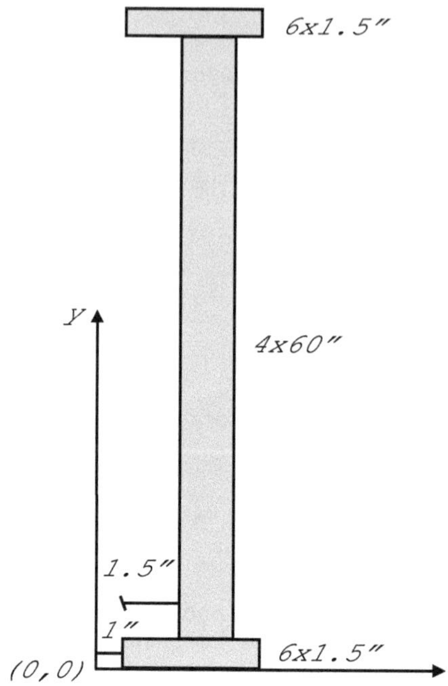

FIGURE 17.43 The post with cap and footer.

$$\text{Area}_2 = (60\text{ in})(4\text{ in}) = 240\text{ in}^2$$

Center Points:

$$x_1 = 4\text{ in},\quad y_1 = 0.75\text{ in}$$

$$x_2 = 4.5\text{ in},\quad y_2 = 31.5\text{ in}$$

$$x_3 = 4\text{ in}',\ y_3 = 62.25\text{ in}$$

Centroid of the Whole Shape:

$$x_c = \sum x_i A_i / \sum A_i$$

$$x_c = \left[(4\text{ in})\left(9\text{ in}^2\right) + (4.5\text{ in})\left(240\text{ in}^2\right) + (4\text{ in})\left(9\text{ in}^2\right)\right] / \left(9\text{ in}^2 + 240\text{ in}^2 + 9\text{ in}^2\right)$$

$$x_c = \left(1152\text{ in}^3\right) / \left(258\text{ in}^2\right)$$

$$x_c = 4.47\text{ in}$$

$$y_c = \sum y_i A_i / \sum A$$

$$y_c = \left[(0.75\text{ in})\left(9\text{ in}^2\right) + (31.5\text{ in})\left(240\text{ in}^2\right) + (62.25\text{ in})\left(9\text{ in}^2\right)\right] / \left(9\text{ in}^2 + 240\text{ in}^2 + 9\text{in}^2\right)$$

$$y_c = \left(8{,}127\text{ in}^3\right) / \left(258\text{ in}^2\right)$$

$$y_c = 31.5\text{ in}$$

Example 17.5.4

Determine the annulus of a fully grouted rockbolt if the diameter of the bolt hole is 30.5 mm drill bit plus 5% reaming and the diameter of the bolt is 24.2 mm (Maepa and Zvarivadza, 2017).

Solution:

Diameter of bolt hole before reaming: 30.5 mm
Diameter of bolt hole after remaining: (30.5 mm)(0.05) + (30.5 mm) = 32 mm
Design annulus size = (30.5 mm – 24.2 mm)/2 = 3.15 mm
Practical annulus size = (32 mm – 24.2 mm)/2 = 3.90 mm

Example 17.5.5

A strain gauge is glued to the center of a longitudinal supporting member that will be placed under tensile load. For part a, determine the strain in the 6-inch gauge if the end points change by 0.012 inches. For part b, assume now that two gauges are used and affixed in a diametrically opposed fashion (i.e., one on each side of the section) to measure strain. If gauge one measures a change of 0.012 inches, but the other gauge measures 0.010 inches, determine the axial tension and bending of the member. Assume the support has a circular cross section and is elastic.

Solution:

Part a:

$$\varepsilon = \Delta L / L_0$$

$$\varepsilon = 0.012 \text{ in}/6 \text{ in}$$

$$\varepsilon = 0.002 \text{ in}/\text{in}$$

$$\varepsilon = 2{,}000\ \mu\varepsilon$$

Part b:

From part a, the micro-strain in gauge one is ε_1 = 2,000 $\mu\varepsilon$, for gauge two:

$$\varepsilon_2 = \Delta L / L_0$$

$$\varepsilon_2 = 0.010 \text{ in}/6 \text{ in}$$

$$\varepsilon_2 = 0.0017 \text{ in}/\text{in}$$

$$\varepsilon_2 = 1{,}700\ \mu\varepsilon$$

For diametrically opposed gauges across a simple circular solid section, the axial strain is approximated as follows:

$$\varepsilon_{\text{axial}} = (\varepsilon_1 + \varepsilon_2)/2$$

$$\varepsilon_{\text{axial}} = (2{,}000\ \mu\varepsilon + 1{,}700\ \mu\varepsilon)/2$$

$$\varepsilon_{\text{axial}} = 1{,}850\ \mu\varepsilon$$

For diametrically opposed gauges across a simple circular solid section, the bending strain is approximated as follows:

$$\varepsilon_{bend} = (\varepsilon_1 - \varepsilon_2)/2$$

$$\varepsilon_{bend} = (2{,}000\ \mu\varepsilon - 1{,}700\ \mu\varepsilon)/2$$

$$\varepsilon_{bend} = 150\ \mu\varepsilon$$

Bending strain on the other side of the circular section will be -150 $\mu\varepsilon$.

Example 17.5.6

A pull test is conducted on a rockbolt. Given the following information, determine the deformation of the rockbolt during pull testing. Equation is provided from Sinha (1989). The bolt diameter is 25 mm, and the modulus of elasticity of the bolt is 160 GPa. Use an effective length of 325 mm and a pullout force of 250 kN. Values are from Thenevin et al. (2017).

Solution:

$$\Delta_B = \left[T_B/A_B\right]\left[L_B/E_B\right]$$

where

- Δ_B is the deformation of rockbolt during pullout test.
- T_B is the pulling out force.
- A_B is the cross-sectional area of the bolt.
- L_B is the effective bolt length (total length less anchoring length).
- E_B is the rockbolt modulus of elasticity.

$$\Delta_B = \left[T_B/A_B\right]\left[L_B/E_B\right]$$

$$\Delta_B = \left\{250\ \text{kN}/\left[\pi(25\ \text{mm})^2/4\right]\right\}[325\ \text{mm}/160{,}000\ \text{MPa}]$$

$$\Delta_B = \left\{[0.509\ \text{kN/mm}^2]\right\}[0.00203\ \text{mm/MPa}]$$

$$\Delta_B = \{509\ \text{MPa}\}\}[0.00203\ \text{mm/MPa}]$$

$$\Delta_B = 1\ \text{mm}.$$

Example 17.5.7

The following is an example of the joint friction approach by the U.S. Army Corps of Engineers (1980). Several scenarios are available from this reference. The following is an example for a single joint with bolt normal to the end load. Figure 17.44 is from this source.

where

- B is the force exerted by bolt.
- P is the direct force on joint.
- tan φ is the coefficient of joint friction.
- W is the weight of block.

Solution:

Determine if stability is satisfied for the following:

For the example, treat P as 50% of B which is 15,000 lbf.

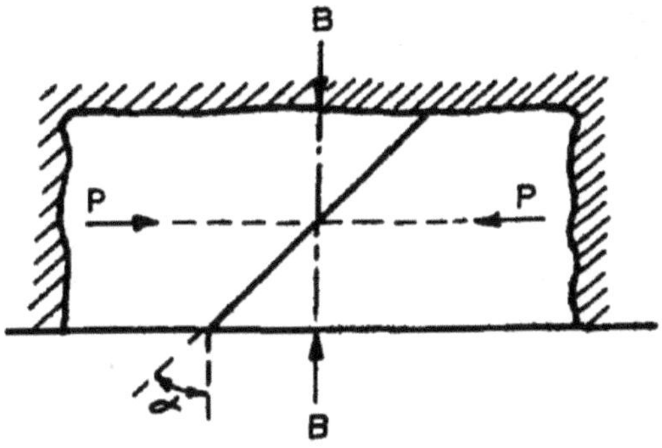

For stability:

$$\tan(\alpha - \phi) < \frac{B}{P} < \tan(\alpha + \phi)$$

FIGURE 17.44 The single joint with a bolt (U.S. Army Corps of Engineers, 1980).

Assume that φ is equal to 30°.
The joint makes an angle of 45°:

$$B/P = 15{,}000/7{,}500 = 2$$

$$\tan(45 - 30) = 0.267$$

$$\tan(45 + 30) = 3.73$$

Stability criteria is satisfied. See Figure 17.44 and U.S. Army Corps of Engineers (1980) for additional information.

Example 17.5.8

The following example shows the basic deterministic approach to the ground reaction curve. Figure 17.45 has been redrawn from Rocscience's RocSupport Introduction Manual, and the process below is developed from the manual as well (Rocscience, n.d.).

Two support systems are tested in an underground tunnel. The measured response of these systems with the ground reaction are shown in the Figure 17.45 below. The first support system is represented by the curve with associated annotations p_{sm} and p_{eq}. The second support system is represented by the curve with associated annotations p_{sm} and p'_{eq}. Based on the following quantities, determine the Factor of Safety of each support system.

p_{sm} is the maximum support pressure.

p_{eq} is the equilibrium pressure of the respective support system.

p'_{eq} is the projected equilibrium pressure of the respective support system.

Assume the following values:

$$p_{sm} = 8 \text{ MPa}$$

$$p_{sm}(\text{Second support system}) = 3\ MPa$$

$$p_{eq} = 5\ MPa$$

$$p'_{eq} = 4\ MPa$$

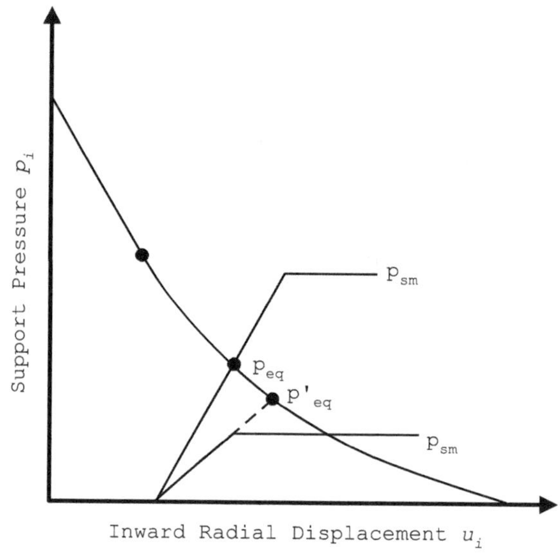

FIGURE 17.45 The ground reaction curve (Rocscience, n.d.).

Solution:

First Support System:

$$FS = p_{sm}/p_{eq}$$

$$FS = 8\ MPa/5\ MPa$$

$$FS = 1.6$$

Second Support System:

$$FS = p_{sm}/p'_{eq}$$

$$FS = 3\ MPa/4\ MPa$$

$$FS = 0.75$$

REFERENCES

Barczak, T.M. (2000) Optimizing secondary roof support with the NIOSH support technology optimization program (STOP). *Proceedings of 19th International Conference on Ground Control in Mining*, Morgantown, WV, pp. 74–84.

Barczak, T.M. (2005) *An Overview of Standing Roof Support Practices and Developments in the United States*. SARES 2005, Johannesburg, South Africa.

Chamber of Mines Research Organization (COMRO) (1988) *An Industry Guide to Methods of Ameliorating the Hazards of Rockfalls and Rockbursts*. COMRO, Johannesburg, South Africa.

Ghabraie, B., Ren, G., Ghabraie, K. and Min Xie, Y.M. (2013) A study on truss bolt mechanism in controlling stability of underground excavation and cutter roof failure. *Geotech. Geo. Eng*. 31, 667–682.

Hedrick, N. (2010) Yielding rock bolt. US patent 07645096. Garford Pty. Ltd.

Jager, A.J. (1992) Two new support units for the control of rockburst damage. *Proceedings of the International Symposium on Rock Mechanics*, Sudbury, Canada, pp. 621–631.

Jennmar (2025) Bolt systems. Retrieved from: https://www.jennmar.com/product-category/bolt-systems

Jingle, J. and Spearing, A. (2005) Quik Stik and Ball Buster. US patent 532969.

Maepa, T.G. and Zvarivadza, T. (2017) Clarification summary in response to a reader's query: Installation of resin-grouted rock bolts in hard rock mining: Challenges and solutions for improved safety. *J. S. Afr. Inst. Min. Metall.* 117(11), 1095–1096.

Mazzoni, R.A., Karabin, G.J. and Cybulski, J.A. (1996) *A Trouble-Shooting Guide for Roof Support Systems.* Mine Safety and Health Administration, US Department of Labor, Washington, DC.

Normat, D.B. (2016) Retrieved from: www.normet.com/wp-content/uploads/2016/09/d-bolt-tds-global-20200511.pdf.

Peng, S. (2023) Keynote address. *23rd International Conference on Ground Control in Mining*, Morgantown, WV. https://groundcontrolmining.org/speakers/

Rocscience. (1989). RocSupport Introduction Manual, p. 13.

Sinha, R.S. (1989) *Underground Structures: Design and Instrumentation.* Elsevier, Amsterdam, The Netherlands.

Spearing, A. (2008) Designing free-standing steel supports for coal mine roof support. *27th International Conference on Ground Control in Mining,* Morgantown, WV.

Spearing, A. et al. (1996) RocProp. US Patent 549824.

Stillborg, B. (1994) *Professional Users Handbook for Rock Bolting* (2nd Edition). Trans Tech Publications, Bäch, Switzerland.

Thenevin, I., Blanco-Martín, L., Hadj-Hassen, F., Schleifer, J., Lubosik, Z. and Wrana, A. (2017) Laboratory pull-out tests on fully grouted rock bolts and cable bolts: Results and lessons learned. *J. Rock Mech. Geotech. Eng.* 9(5), 843–855.

U.S. Army Corps of Engineers (1980) *Rock Reinforcement. EM 1110-1-2907.* U.S. Army Corps of Engineers, CECW-EG, Washington, DC.

Wu, R.Y. and Oldsen, J. (2010) *Development of a New Yielding Rock Bolt.* ARMA, Washington, DC.

18 Liner (or Area) Supports

18.1 THE NEEDS, FUNCTIONS, AND DESIGN OF LINER SUPPORTS

Liner, area, or skin support is needed to provide support between the main rock anchors. The main skin supports are as follows:

- Shotcrete which is common in hard-rock mines.
- Screen or mesh which is the most common skin support.
- Thin-skin liners which are mainly spray on polymers.

Skin failures of roof and ribs, especially in underground coal mines, continue to be a significant rock-related safety hazard. These failures are typically small and localized, and do not often involve failure of the main support elements (usually rockbolts) but result from localized collapses or spalling between them.

According to Bauer and Dolinar (2000), in many soft-rock mines, the roof is composed of draw rock (soft slate, shale, or rock), coal, bony material, and other highly stratified, thinly laminated rock. These strata are susceptible to failing in thin layers because of bedding plane weaknesses. Some of the causes of bedding plane failures that result in skin failures include sag of the strata from gravity, overburden pressure as depth increases, high horizontal stress, and moisture or temperature sensitivity. Mining-induced stresses and damage are also important in the development of skin failure in both the roof and rib. If geologic discontinuities are present, the likelihood of skin failure increases because the discontinuities weaken and compromise the structural integrity of the rock. In their study, they also reported that the following were the most common activities of workers injured by roof skin failures in coal mines for 1995–1998:

- Drilling or bolting of the roof: 39% of injuries from skin failure.
- Operating the continuous mining machine: 11%.
- General inside labor: 9%.

For injuries resulting from rib skin failures, the following were the most common worker activities:

- Operating the continuous mining machine: 18%.
- Drilling or bolting the roof: 16%.
- General inside labor: 12%.
- Walking: 9%.
- Maintenance and repair: 8%.

18.2 SHOTCRETE

Shotcrete is one of the most effective liners for most underground conditions, with the exception of coal. This is because coal is typically very weak, and if shotcrete is applied, it creates a shotcrete layer onto the immediate coal surface and the coal fails just behind the shotcrete layer, as it has little or no shear and bond strength. On some underground civil projects, particularly in Europe, shotcrete is referred to as sprayed concrete (Spearing, 2002).

DOI: 10.1201/9781032704845-18

18.2.1 The Historical Development of Shotcrete

The timeline in Table 18.1 outlines many of the developments of this effective liner support.

Figure 18.1 shows the first shotcrete machine developed by Carl Akeley to fill his wire-formed dinosaurs for his park.

TABLE 18.1
The Development of Shotcrete (Spearing, 2002)

1907	The first machine to spray concrete was developed in Pennsylvania, USA by naturalist Carl Akeley for use in construction.
1910	The machine and process were introduced to the public at large at the Cement Show in Madison Square Gardens in New York, USA.
1915	This concept was later improved by the Cement Gun Company, which later became the Allentown Gun Company. Carl Akeley also registered the term gunite for his sprayed mortar mix, and this term is still sometimes used.
1920s	Gunite was used in a mine for the first time in the United States to fireproof mine drifts which were also supported by timber sets.
1930s	The term *shotcrete* was introduced by the American Railway Engineering Association.
1940s	Coarse aggregate was introduced into the sprayed concrete mixes.
	Gunite was used to reinforce pillars at St. Joseph Lead Mine in Missouri, USA.
1954	Anton Brunner, an engineer from Salzburg in Austria, replaced heavy steel and timber support with shotcrete in a new diversion tunnel, in squeezing ground, at the Runserau Power Plant.
1955	The wet shotcrete process was introduced (discussed in detail in 18.2.2).
1957	Little major equipment development took place until 1957 when the rotary drum was perfected by Meynadier (now MEYCO Equipment) and Aliva. These machines (still in use today for dry shotcreting) have a feed hopper above a rotor with chambers that intersect feed openings allowing some material to enter under gravity. A compressed air stream forces this material into the discharge pipeline and keeps it in suspension until the shotcrete nozzle. This development was probably a result of the use of shotcrete in Austria as a civil tunnel support in the 1950s.
Ca. 1965	Gunite was also used in some Boliden Mines in Sweden. This involved the dry process, and in some applications, the machine was on surface, and the material applied some 300 m below surface (at Garpenberg Mine).
1968	Hecla Mining Company in the United States, started using shotcrete routinely as an underground support.
1975	The Norwegians were the first to identify the benefits of silica fume (microsilica) in shotcrete.
1977	The Norwegians introduce steel fibers for the first time as a replacement for mesh/screen.
1994	Master Builders Technology (MBT) is the first to introduce non-caustic alkali-free accelerators.
1999	MEYCO (then a subsidiary of MBT) starts field testing the first shotcrete equipment capable of being operated in a fully robotic mode (i.e., with no operator necessary).

FIGURE 18.1 The first shotcrete machine in 1910 – photo taken by the late K. Akeley (Sauer and Gold, 1989).

18.2.2 Shotcrete Application Processes

There are two different application shotcrete processes:

- Dry process
  - A technique in which cement and aggregate are batched, mixed, and fed into a purpose-made machine wherein the mixture is pressurized, metered into a compressed air stream, and conveyed through hoses or pipes to a nozzle where water is introduced as a spray to wet the mixture which is then projected continuously into place. The mixture may also incorporate admixtures and fibers.
- Wet process
  - A technique in which cement, aggregate, and water are batched and mixed together prior to being fed into a purpose-made machine and conveyed through a pipeline to a nozzle where the mixture is pneumatically and continuously projected into place. The mixture normally incorporates admixtures and may also include additions or fibers or a combination of these.

The two processes are discussed in more detail in the following.

18.2.2.1 The Dry Shotcrete Machine and Process

The American Concrete Institute (ACI), 2016 defines dry mix shotcrete as "the shotcrete in which most of the mixing water is added at the nozzle".

The dry mix process involves the mixing of cement and aggregates in the designed proportions before supplying it to the shotcreting equipment. The final mixed ingredients are placed in the feed hopper. During the shotcreting operation, the mix is taken from the hopper to the nozzle through the delivery hose using compressed air.

Once the dry mix reaches the application nozzle, water under high pressure is sprayed to the mix through a perforated ring attached to the nozzle assembly. During spraying, the water thoroughly "wets" the mix making it a concrete paste that is ejected at high velocity against the surface to be shotcreted. A section of a dry mix machine is shown in Figure 18.2.

The amount of water added for the dry mix shotcrete process is controlled by the nozzleman which is key to the operation and relies on experience and training. The water is controlled by means of a valve situated in the nozzle assembly. The dry mix shotcrete is typically applied in areas where the total shotcrete volume needed is relatively low, and access is difficult.

The maintenance of the pump and ancillaries is critical, and the following are the most important equipment that need to be checked:

- Feeding hopper should be kept free of concrete after use.
- Check that the re-mixer (located between the hopper and rotor) is working and rotating in the right direction.
- Wear plate and ring must be tight with no gap.
- No residual concrete build-up in lines and hoses.
- For dry equipment, keep the seals and wear components in good order.

A dry shotcrete machine is shown in Figure 18.3.

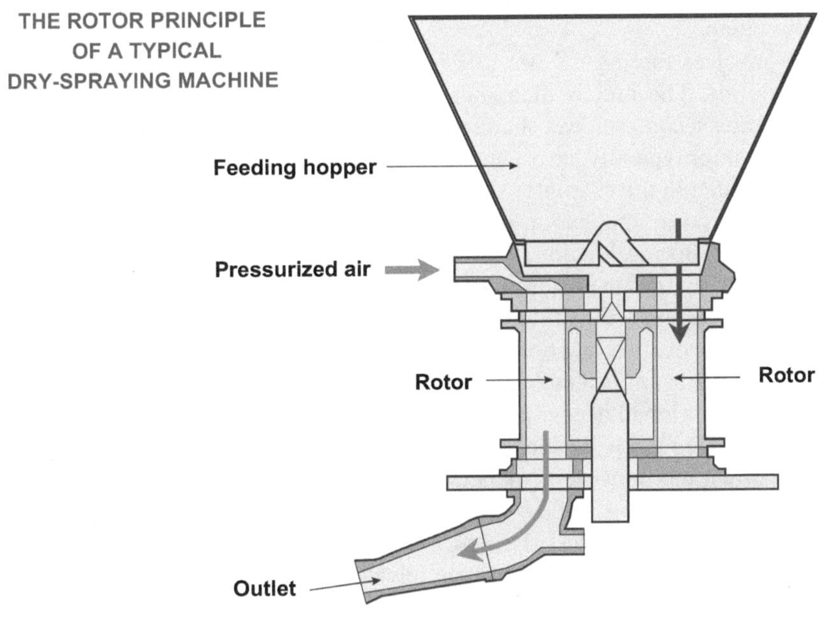

FIGURE 18.2 Section through a typical dry shotcrete machine (Vandewalle, 1997).

FIGURE 18.3 A typical dry shotcrete machine.

Reproduced with kind permission of Normet.

18.2.2.2 The Wet Shotcrete Process and Pump

The wet mix process involves mixing all ingredients to form a complete concrete mix, including the correct water, before it enters the wet shotcrete machine. The shotcrete mix is therefore prepared before placing it in the shotcrete hopper. The delivery pump is typically either a positive displacement or a worm pump.

The process involves forcing the wet mix to the nozzle through a high-pressure delivery hose using compressed air. The mix is then shot at high velocity onto the surface to be shotcreted. Figure 18.4 illustrates a common wet shotcrete pump.

Wet shotcrete pumps typically have a much higher capacity, and many different equipment variations exist as illustrated in the examples shown in Figures 18.5 and 18.6.

Special "fit-for-purpose" application arms can be made for wet shotcreting, especially when excavating using a Tunnel Boring Machine (TBM) or shaft sinking. Figure 18.7 shows the application arm to line a circular shaft with shotcrete as it is being sunk (developed).

On a TBM, the shotcrete arm is fitted to a circular beam near the cutting head and applies typically a temporary layer of shotcrete before the final lining is applied.

In shaft sinking, the beam is attached to the platform that is suspended some distance from the shaft bottom as it is developed deeper using drilling and blasting.

Accelerators are frequently added at the discharge nozzle of the shotcreter, and the dosage must be accurate and consistent. A peristaltic hose pump is often used for this application (Figure 18.8).

The pump has a rotor with rollers that squeezes the flexible hose progressively as it rotates and moves the fluid between two rollers at an accurately controlled rate.

18.2.2.3 Selecting the Best Process for a Shotcrete Application

The selection of the most appropriate process depends on local conditions, required placement rates, and logistical considerations. Table 18.2 gives a comparison between the methods.

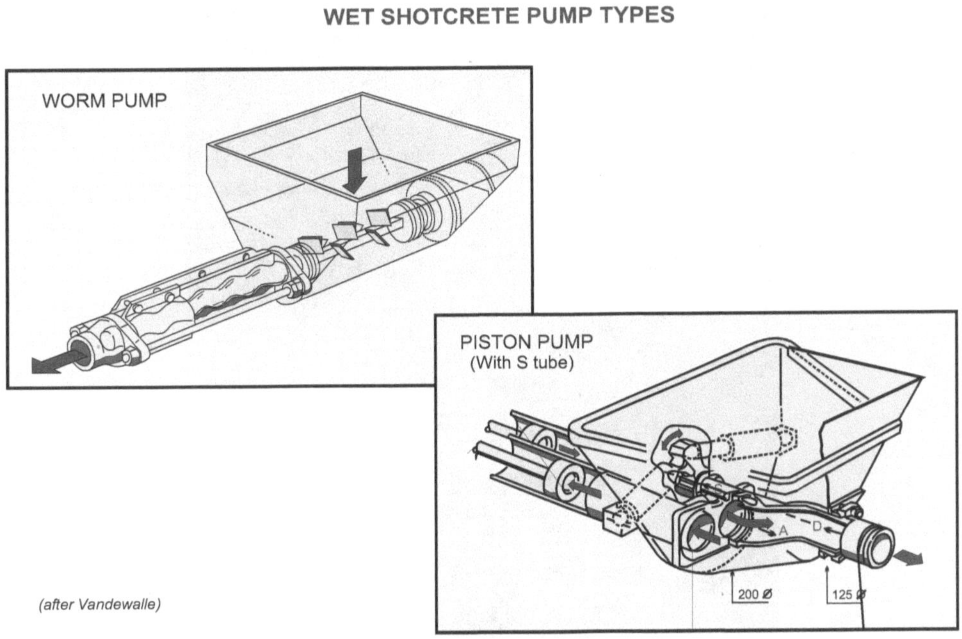

FIGURE 18.4 The common wet shotcrete pumps (Vandewalle, 1997).

FIGURE 18.5 A trailer mounted small wet shotcrete unit.

Reproduced with kind permission of Normet.

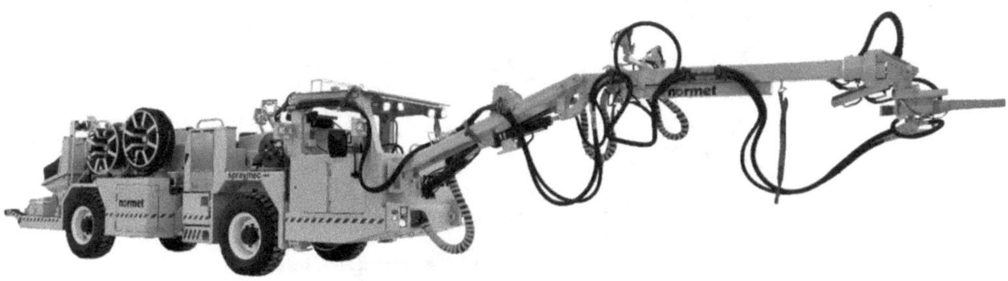

FIGURE 18.6 A wet shotcrete machine with a boom on a carrier.

Reproduced with kind permission of Normet.

FIGURE 18.7 Wet shotcrete application arm and nozzle assembly for circular shaft sinking.

Reproduced with kind permission of Tom Melbye.

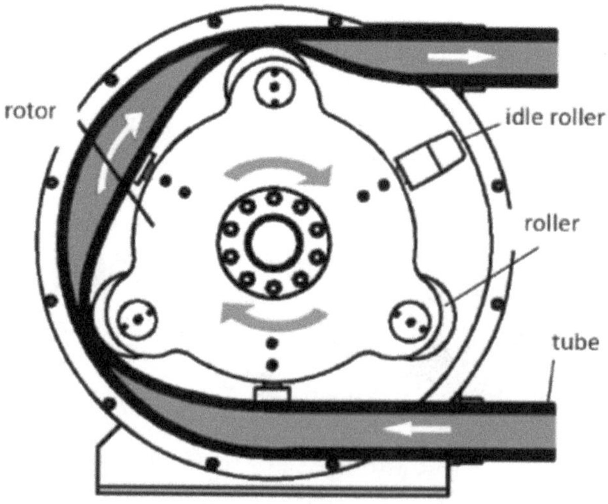

FIGURE 18.8 Section through a peristaltic pump (MechanicalBoost, 2024).

TABLE 18.2
Comparison between the Shotcrete Processes (Melbye, 1994)

Wet Process	Dry Process
Little dust	Considerable dust
Low maintenance	High maintenance
High capital	Low capital
Low rebound (≅10%)	High rebound (usually >25%–40%)
Moderate to high placement rate (3–20 m^3/h)	Low to moderate placement rate (≤ 5.0 m^3/h)
Low transport distance (≤ 200 m)	High transport distance
Moderate to high placed quality	Moderate to low placed quality

It should be noted that if fibers are used in the shotcrete mix, then the dry process is to be generally avoided, if possible, due to the high losses of fibers in the rebound.

18.2.3 Shotcrete Nozzles

Shotcrete nozzles are important to ensure thorough water mixing in dry shotcreting, accelerator addition, and good compaction of the shotcrete being sprayed. Figure 18.9 shows an examples of dry shotcrete nozzles. From Figure 18.9, it can be seen that the water enters via the hose fitted with a valve to regulate the total water addition and then is introduced via a water ring uniformly around the pipe containing the dry mix being propelled using compressed air, to ensure uniform water into the dry mix.

The nozzle assembly must have the following:

- Adequate air hose size.
- Accelerator hose (if needed).
- Proper mixing apparatus especially for dry shotcrete to ensure thorough "wetting" of the dry concrete before placement.
- Well-designed nozzle assembly.

For this to work, the compressor must be adequately sized, meaning typically a minimum of 10–12 m^3/minute of compressed air. The pressure is also important too, but mainly for post-application compaction.

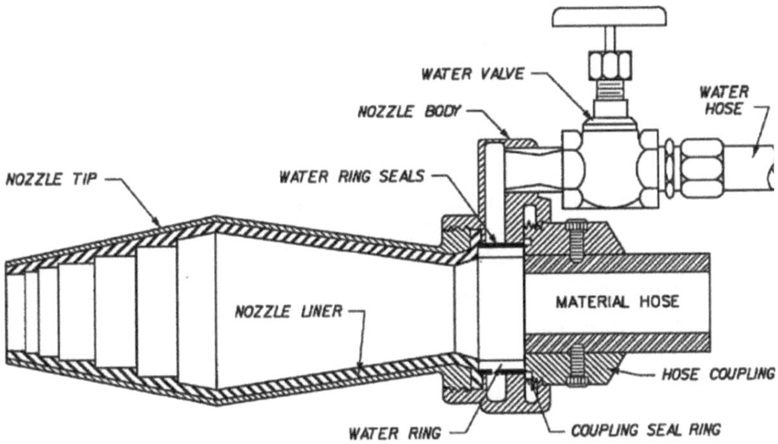

FIGURE 18.9 Dry shotcrete nozzle assembly (US Army Corps of Engineers, 1993).

18.2.4 Shotcrete Raw Materials and Preparation

Much is available concerning the design of shotcrete, but often simple aspects are frequently overlooked, at least initially, and these can create major losses, costly delays, and final sprayed linings that do not meet the desired performance.

Shotcrete design must include more than creating a laboratory mix that meets the strength gain requirements, with locally available raw materials (such as cement, sand, stone, and water) in adequate supply. While this is important, three other equally vital aspects must not be ignored:

- Strength must be achieved on the rock (not in the lab)
  - The rock could be dry, wet, or covered in mud for example, so adhesion (bonding) to the substrate (rock) is a concern.
- The time available for spraying a given volume of shotcrete (in most production cycles, the time to install the new support is the main bottleneck).
- Cost of the mix.

Shotcrete is basically concrete, so it consists of the same components, and when produced, the same quality control procedures must be applied. As a minimum, this would include the following:

- Testing the aggregates to ensure that they fit with the aggregate specifications.
- Ensure the quality of the water being used
  - Generally, it should be potable (drinkable) and at least free of oils and other impurities.
- Testing that the concrete is made with the correct percentage of each component.
- Checking that the cementitious component is correctly added and meets the predetermined specification.
- Ensuring the correct dosages of all admixtures and additives.
- Strength gain tests (over time at typically 1, 3, 7, 14, and 28 days) to ensure that the placed shotcrete meets or exceeds the minimum strength design specification.

It consists of the following materials:

- Cement/binders.
- Aggregates including sand.
- Water.
- Admixtures.
- Additives such as microsilica and fibers.

To reiterate the water quality is important, and the water should be free from oils and not heavily acidic (as cement is alkaline). If the water is potable (safe to drink), it is good for concrete and hence for shotcrete.

18.2.4.1 Binders

Cement is the bonding material (glue) that holds a cementitious material together. For most shotcreting applications, Portland cement is used (ASTM C150). This cement was invented by an English bricklayer, Joseph Aspdin, in 1824. The name was derived from the set materials color and texture that resembled a local limestone called Portland stone. The cement is produced from mainly a mixture of klinker and gypsum. The klinker is typically produced in a rotary kiln from lime, silica, alumina, and ferric oxide.

Other types of cement that are also used in shotcrete include the following:

- Sulfate-resisting cement, which typically has a lower tricalcium aluminate content than Portland cement, so is less chemically affected by potential sulfate attack.

- High-alumina cement (HAC) that is produced by fusing (melting) a mix of bauxite and limestone together. HAC is therefore not a Portland cement derivative. It is not used for standard structural shotcreted, but because of its thermal stability at high temperatures, it is often used in refractory applications where it can be troweled in place or shotcreted.

For basic shotcrete applications, a cement content by mass of 20% is typical.

Cement extenders (supplementary materials) are commonly used in shotcrete mainly for a cost reduction. These include the following:

- Fly ash
  - The most commonly used and is obtained as a waste product from coal-fired power plants. The fly ash for use in shotcrete is mainly supplied as a blend with the Portland cement already in bags or bulk.
- Ground-granulated blast-furnace slag (GGBS)
  - Another cement extender that can be used. GGBS is a glassy and granular material that is usually produced as a by-product of iron production. The molten slag is rapidly quenched and then finely ground.
- Fumed silica, silica fume, or microsilica
  - A cement extender that has unique and desirable properties to meet the design strength faster and/or more cost effectively.

Cement extenders can be pozzolans which are siliceous and aluminous materials that in themselves, possess little or no cementitious properties but if very fine and when mixed with water, will react chemically with calcium hydroxide, $Ca(OH)_2$, in cement to form cementitious compounds.

Cement extenders (pozzolans) are commonly used to replace about 30% of the cement used and up to as high as 50%. Generally, the rate of strength development is reduced, i.e., the early strength, and this is frequently undesirable in mining due mainly to safety considerations. The final strength, usually assumed to be at 28 days, is unaffected or could even be higher. The same applies to sulfate-resistant cements, and a better alternative is the use of Portland cement and microsilica that can gain strength rapidly (without an accelerator overdose).

18.2.4.2 Aggregates

As in concrete, aggregates are used to provide dimensional stability by providing a rigid skeletal structure and reducing the void space to be filled with the cementitious paste and hence reduce the cost. The main parameters to consider are as follows:

- Grading and the maximum particle size.
- Particle shape and density.
- Aggregate type especially the presence of reactive chemicals and minerals (for possible alkali-aggregate reaction).
- Compressive strength.
- Moisture content.

When applying the shotcrete by the dry process, the aggregates can be selected such that the voids are minimized, as is the common approach when producing and using concrete. As a general guide, the use of +16 mm material is to be avoided and the modern trend is to use a maximum size of 10 mm, due to rebound and wear considerations (EFNARC, 2023).

Figure 18.10 gives the aggregate envelope for shotcrete recommended by the European Federation of National Associations Representing Concrete (EFNARC).

Table 18.3 gives the recommended grading curve for the combined aggregate for use in shotcrete based on ASTM C1436-13 (2021), which is similar to that shown in Figure 18.10.

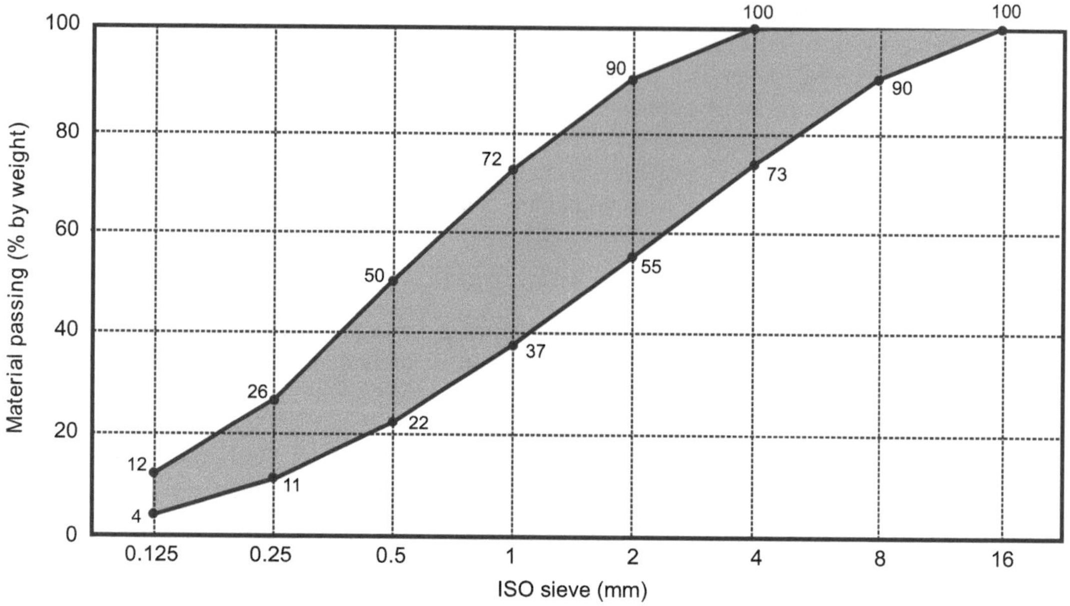

FIGURE 18.10 Recommended aggregate envelope (EFNARC).

TABLE 18.3
Recommended Aggregate Grading Curve

Sieve Size (US Standard)	Grading No. 1	Grading No. 2
12.50 mm (1/2 inch)	–	100
9.50 mm (3/8 inch)	100	90–100
4.75 mm (No. 4)	95–100	70–85
2.36 mm (No. 8)	80–98	50–70
1.18 mm (No. 16)	50–85	35–55
600 μm (No. 30)	25–60	20–35
300 μm (No. 50)	10–30	8–20
150 μm (No. 100)	2–10	2–10

It is obviously possible to use gradings outside the recommended envelopes, but the performance will be reduced, or the material cost will increase as additional binder and/or admixtures would need to be used.

A typical dry process base mix by weight could consist of the following:

- 20%–25% of cementitious binder.
- 15%–20% of coarse aggregate.
- 55%–65% of sand where natural or wash sand is preferred.

Where the dry process is used, the moisture content of the aggregates should be a maximum of 6%.

The mix design needs to be modified when using the wet process because pumpability is a major issue, and generally more fines are needed to reduce the chance of line blockages and make pumping easier, i.e., at a lower pressure.

18.2.4.3 Water

Already covered, but it should be potable.

18.2.4.4 Admixtures

The American Concrete Institute in the ACI Manual of Concrete Practice (1999 edition) defines an admixture as follows:

> A material other than water, aggregates, hydraulic cement, and fiber reinforcement, used as an ingredient of concrete or mortar, and added to the batch immediately before or during its mixing.

Many different types of admixtures can be beneficial in shotcrete, depending on the specific application and requirements. Most admixtures can be used in the wet process only, but generally only a set accelerator, usually in powdered form, is used when dry shotcreting. To achieve the quantifiable benefits from admixture usage, it is critical that the dosing equipment is reliable and appropriate and fully tested and justified before use.

The main types of admixtures are as follows:

- Accelerators

  The use of an accelerator is essential in most mining applications to ensure that the shotcrete:
  - Develops a bond with the rock as soon as possible.
  - Generates internal strength quickly.
  - Can be placed in thick single passes (if needed).

  Shotcrete accelerators generally fall into the following categories:
  - Silicates, e.g., water glass or sodium silicate.
  - Sodium or potassium aluminates.
  - Alkali-free accelerators.

  Silicates are not technically true accelerators as they only create a gelling effect rather than a rapid early strength gain. Accelerators also tend to reduce the final strength of the shotcrete. This is partly because the slower rate of strength gain in an un-accelerated mix results in a finer and a denser crystalline growth that creates a stronger final product.

  In a typical dry mix, addition of powdered accelerator between 2% and 5% based on the weight of the total cementitious addition is reasonable. Accelerator dosage rates are normally expressed as a percentage of the total cementitious content.

  There tends to be some confusion in the mining industry over the definition of an alkali-free accelerator. To understand this, the difference between alkalinity and alkali content must be noted. Alkalinity refers to a "basic" liquid with a pH between 7 and 14, whereas alkali content refers to the presence of alkali cations, and a liquid may contain alkali cations but have a neutral pH.

  To be a true alkali-free accelerator according to CEN definitions (Comite Europeen de Normalisation, 2007), an accelerator must contain <1% by weight of alkali (Na_2O) equivalents. Certain accelerators on the market claim to be alkali free but do not meet this specification and are only low alkali and hence generally not safe to use or effective.

  In wet shotcrete, the dose range is about 3%–10% based on the weight of the total cementitious content. The latest trend is toward the alkali-free (non-caustic) accelerators because they are more environmentally safe, induce more rapid strength gain, and tend to cause significantly less long-term strength loss. With such an accelerator, a dose rate of around 6%–8% is normally adequate.

  Before selecting an accelerator, compatibility tests must be carried out first. Certain accelerators react differently with different cements.
- Water reducers (plasticizers and superplasticizers)

  There are various types of water reducers available, and they tend to fall into three broad categories:

- Low range, e.g., lignosulfonates give about 10%–15% water reduction, but some can retard the strength gain.
- Medium range, e.g., melamines give about 25% water reduction.
- High range, e.g., polycarboxylates give about 45% water reduction.

These admixtures work by charging each cement particle ionically causing them to separate, thereby effectively lubricating each particle in the entire mix and thus being able to reduce the water and still obtain the same consistency (slump).

- Microsilica (silica fume)

  Microsilica is a very fine and spherical material with a high pozzolanic reactivity. The use of the product in shotcreting has the following significant benefits:
  - Improved durability and therefore more resistant to freeze/thaw cycles and improved sulfate attack resistance.
  - Improved bonding to substrates.
  - Higher strengths (compressive and flexural).
  - Reduced rebound.
  - Improved flow in the delivery hose (in the wet process).
  - Reduced wear in the pump and nozzle (in the wet process).
  - Improved mix cohesiveness.
  - Thicker single-pass applications.

  A typical dosage would be 5%–10% by weight of the cementitious binder.
- Curing agents/concrete improvers

  There is an incorrect perception that an underground environment provides good curing conditions for shotcrete. This is incorrect because the ventilation tends to cause premature drying of the shotcrete surface resulting in:
  - Poor hydration causing a weaker final product.
  - Reduced substrate bonding if the shotcrete layer is relatively thin (less than 75 mm/3 inches).
  - Significant shrinkage cracking.
- Solutions to this include:
  - Regular wetting of the placed shotcrete.
  - Application of an external curing agent such as a spray-applied wax.
  - Inclusion in the mix (wet process only) of a concrete improver that has the potential to improve curing and increase the bond with the substrate.
- Consistency controllers

  Under certain wet shotcrete applications, the use of a consistency control system can help. The first component is added into the mix (before pumping) to keep the open time and improve pumpability. The second component is added at the nozzle to stiffen the mix and improve the strength development rate.
- Hydration controllers

  The useful life of a wet shotcrete mix can be a limiting factor in underground applications due to logistic considerations. A typical batch of untreated shotcrete should be discarded as waste after about 1–2 hours (depending on the ambient temperature). Conventional retarders can extend this to about 4 hours, but hydration controllers can effectively stop the cement hydration process for up to 72 hours. This technology is of great advantage in many underground wet shotcreting applications because it helps resolve logistic problems.

  Hydration controllers' function in two ways:
  - Act as an effective dispersant, thus keeping hydrating particles apart.
  - Forming a barrier around all the cementitious particles, thus stopping the hydration process altogether (unlike conventional retarders).

  These effects are over-ridden when shotcreting, by adding an effective accelerator at the nozzle, and the hydration controller has no adverse effect on the rate of strength gain and the ultimate strength, provided that adequate accelerator is added. If a hydration controller is used, a minimum flow (to BS 1881, Part 105) of 50 cm should be used.

- Pumping aids

  Pumping aids usually also act as water reducers in shotcrete. They improve the pumpability of harsh mixes, frequently caused by poorly graded aggregates. Air-entraining agents are a common pumping aid.

18.2.4.5 Fibers

The application of fibers in construction dates back centuries to the use of horsehair, jute, sisal, and cotton. Concrete is by nature a brittle product and is weak in tension. In shotcreting, reinforcement is provided using screen/mesh or fibers. Fibers have obvious advantages over screen (mesh) including:

- Fibers are more evenly distributed throughout the shotcrete.
- Mesh is difficult and labor intensive to effectively apply, and it can also represent a safety hazard if manually installed.
- Fiber-reinforced shotcrete effectively follows the contours of the tunnel periphery, and additional shotcrete is needed when screen is used, to fill the depressions as the mesh is usually fixed at the high points (contours) of the tunnel periphery.
- Mesh can increase shotcrete rebound significantly due mainly to screen vibration during spraying and reduce the final compaction.

Figure 18.11 is a comparison between the support performance of unreinforced shotcrete and shotcrete reinforced using screen or fibers in shotcrete panels. The use of mesh or fibers significantly increases the amount of deformation that the shotcrete can withstand.

If on-going movement of the rock mass is expected, shotcrete is frequently applied over mesh/screen, or fibers are introduced into the base shotcrete mix. In extreme ground control situations, fiber-reinforced shotcrete can be used and mesh can be attached to the outside of the shotcrete, and this system can sustain very large deformations.

The trend seems to be away from mesh/screen-reinforced shotcrete to fiber-reinforced shotcrete in mining because:

- Durability is improved.
- Ductility (toughness) is increased.
- Impact resistance is increased.
- Surface cracking is reduced (not a major concern on mines).
- Rebound is reduced.
- Compaction is improved (see Figure 18.12).
- Application productivity and safety are improved, which is a major benefit.
- Final placed costs are reduced.
- Logistics are simplified.

Typical fiber reinforcement materials include the following:

- Polyolefin fiber of which polypropylene is common (mono-filament or fibrillated).
- Carbon fiber.
- Glass fiber.
- Drawn steel wire.
- Slit sheet.
- Milled steel pieces.
- Melt extract pieces.

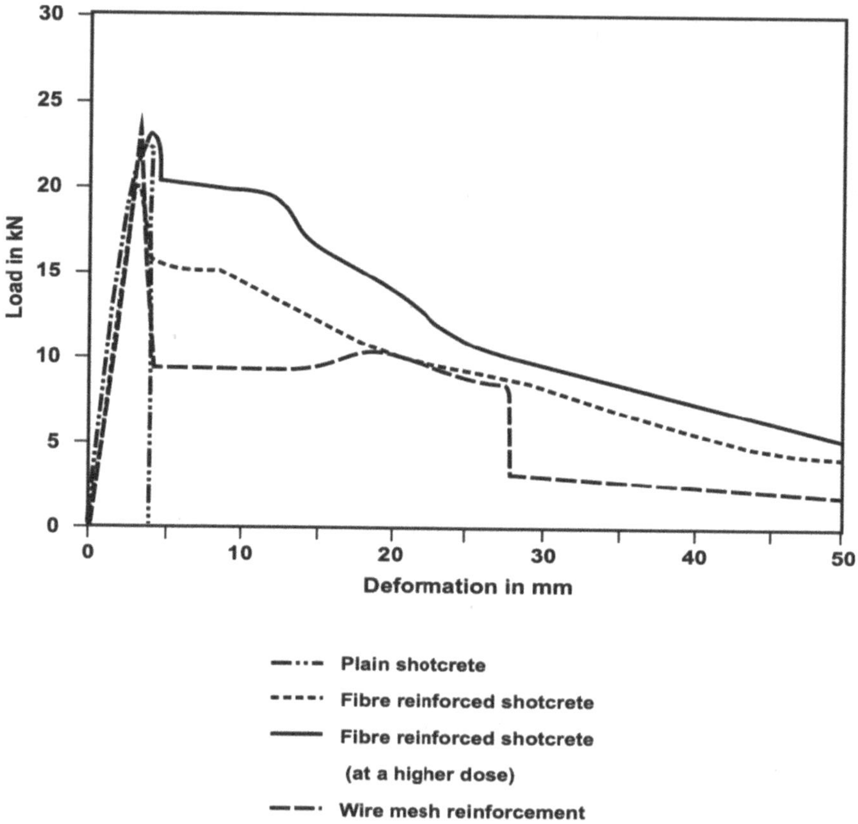

FIGURE 18.11 Comparison between mesh- and fiber-reinforced shotcrete panels (Vandewalle, 1997).

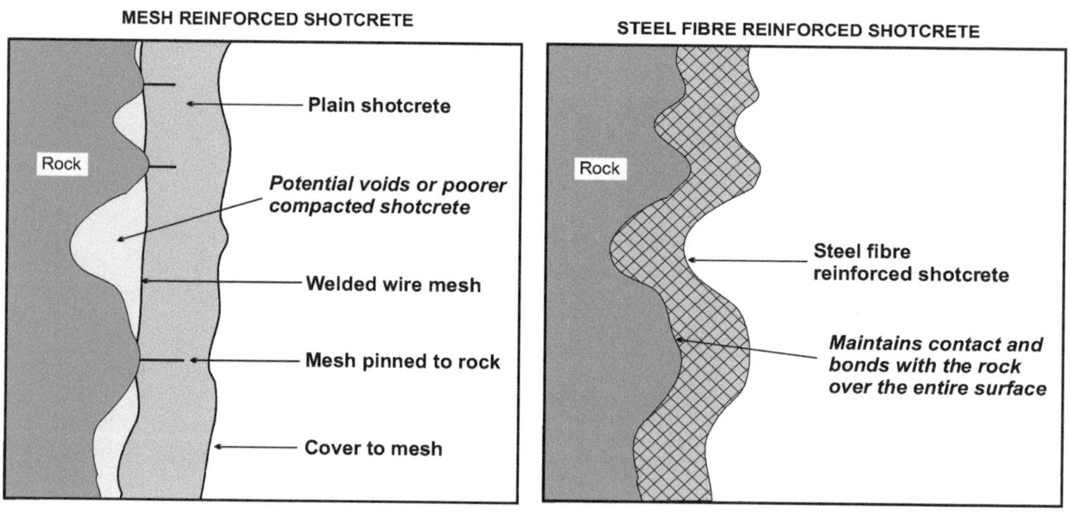

FIGURE 18.12 The compaction advantage of fiber-reinforced shotcrete over mesh-reinforced shotcrete (Vandewalle, 1997).

Two of the more commonly used types in shotcrete are the drawn wire and polypropylene fibers.

The most important parameters for fibers are as follows:

- Aspect ratio (overall ratio of the fiber length to diameter).
- Tensile strength.
- Shape.
- Number of fibers per unit mass.

An ideal fiber should have the following:

- A length such that it can overlap and bridge at least two of the largest aggregate particles used in the mix which is typically a length between 25 and 40 mm.
- A high aspect ratio, i.e., thin.
- A high tensile strength.
- A shape that results in a good anchor particularly at the fiber ends in the shotcrete.

The use of steel fibers and particularly synthetic fibers in dry shotcrete is not generally recommended due to the high fiber loss found in the rebound. The fiber loss can be a significantly larger percentage than the overall rebound, which is typically between 35% and 50%.

Where steel fiber is used, a dosage rate between 30 and 60 kg/m^3 (0.40% and 0.75% by volume) is generally used.

The potential use of polypropylene fibers in shotcrete for mining applications has increased dramatically recently with the development of high-performance polymer fibers. The high tensile strength and in some cases the crimped shape have resulted in performances similar to that obtained with steel. Dosage rates between about 7.0 and 13.5 kg/m^3 (0.75% and 1.5% by volume) are typical (Melbye, 1994).

A further advantage of synthetic fibers over steel fibers is that in case of fire, the synthetic fibers help significantly reduce the "explosive spalling" of shotcrete caused by the super-heated steam resulting from the extreme heat on the shotcrete (or concrete). In fire, the synthetic fibers melt, creating additional "free space" to help dissipate the pressures caused in the shotcrete by the super-heated steam.

Figure 18.13 shows a variety of fibers and their respective load and deflection performance that can be obtained in shotcrete panels. Typically, high-performance steel fibers have higher residual loads immediately after failure but less load at high deformations than high-performance synthetic fibers. The key to the cost-effective use of fibers is to consider the cost per joule of energy.

In future, it is likely that more polymer fibers will be used, possibly blended with steel fibers, where high-performance mining shotcrete is needed.

Figure 18.14 shows some examples of steel and synthetic fibers used with shotcrete.

For the most extreme ground conditions from very high stresses, squeezing ground and/or seismicity, fiber-reinforced shotcrete with screen installed afterward would be a viable and safe solution that could sustain high, violent deformation.

18.2.5 Shotcrete Transportation to Site

The shotcrete must be transported from where it is produced (batched) to where it will be sprayed (applied). The two main options for wet shotcrete are as follows:

- A concrete transporter (vehicle).
- Through a pipeline.

The transportation selection is mainly dependent on the mine site, the main mine access, and the volumes of shotcrete needed and when it is needed.

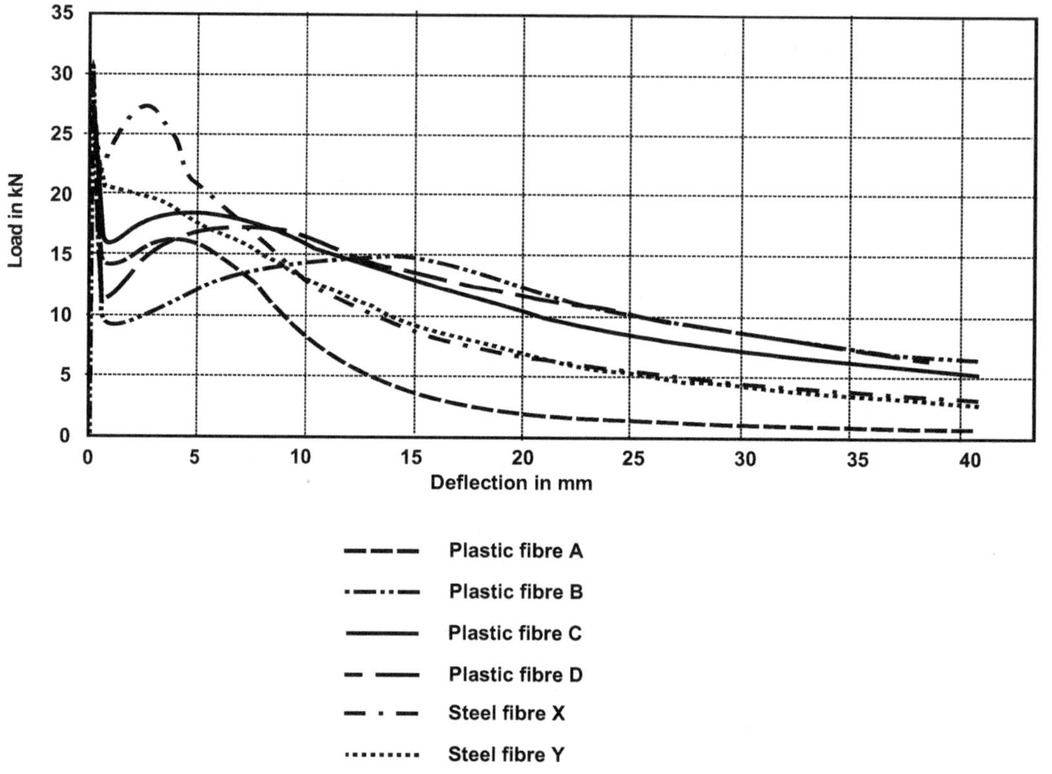

FIGURE 18.13 Examples of shotcrete performance with several different fiber types (Spearing, 2002).

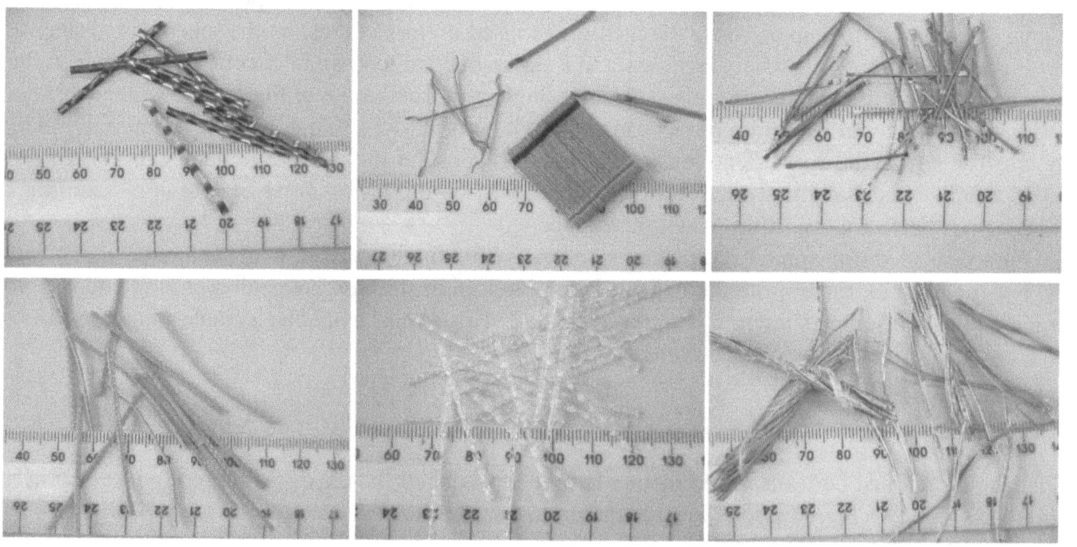

FIGURE 18.14 Examples of some shotcrete fibers.

Figure 18.15 shows a shotcrete transporter that is effective if the mine has ramp or adit access. It can also be used to transport from an underground batch plant or from a shaft delivery system (such as a pipeline).

FIGURE 18.15 A low-profile vehicle designed to carry shotcrete (or concrete) underground.

Reproduced with kind permission of Normet.

A pressure reducer (similar action to a kettle) is needed when "dropping" shotcrete vertically down a pipeline in a shaft or a borehole. Examples are shown in Figure 18.16. From the shaft, it can be transported to the site via a custom-designed, low-profile concrete carrier.

The pressure reducers work as follows:

- Referencing Figure 18.16 (i), the shotcrete fills the dead end and then the shotcrete flows out into the pipe on the side for use. The shotcrete material in the dead end acts as a pressure reducer by absorbing the kinetic energy of the shotcrete free falling down the pipe.
- Figure 18.16 (ii) is basically the same as Figure 18.16 (i), but the blank flange at the end of the pipe is protected by the sacrificial wear block that can typically be made of steel, polypropylene, or ceramic, and hence will last longer.

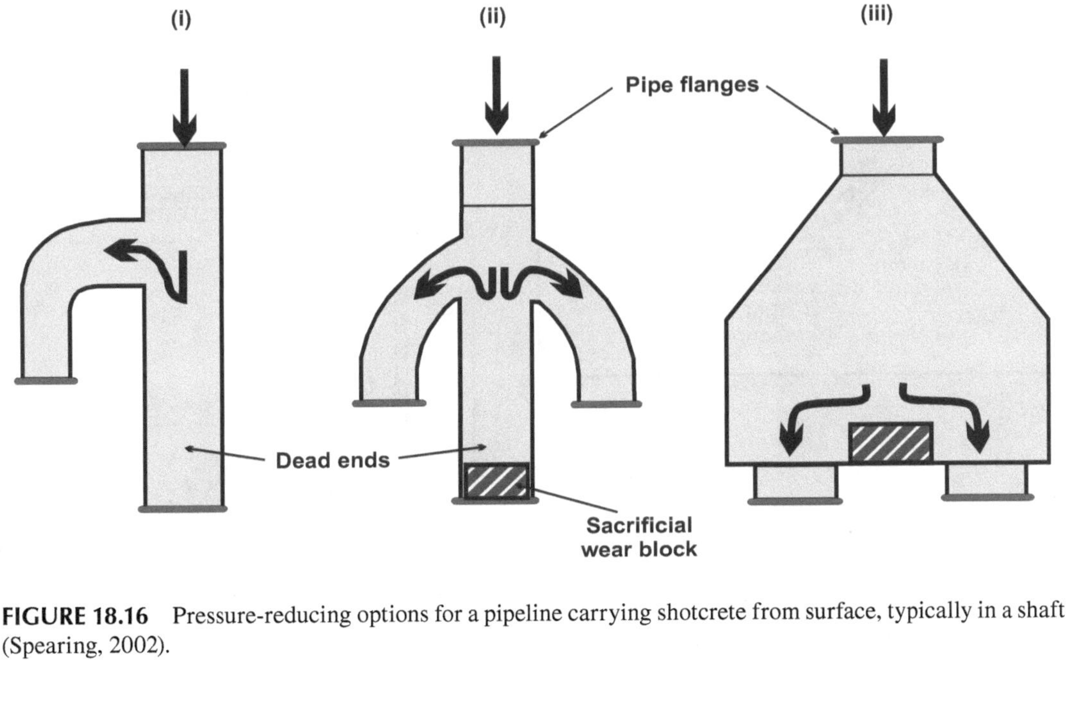

FIGURE 18.16 Pressure-reducing options for a pipeline carrying shotcrete from surface, typically in a shaft (Spearing, 2002).

- The design in Figure 18.16 (iii) relies more on the wear block than the triangular volume of shotcrete that sits on it either side of the two feeder pipes.

18.2.6 Shotcrete Application (Placement)

The angle of the nozzle to the rock surface being sprayed needs to be close to 90° and rotated in a circular manner to help compact the placed shotcrete and build a desirable thickness. This is shown in Figure 18.17.

18.2.7 Shotcrete Costs

Shotcrete total placed costs are seldom, if ever, accurately estimated. The same is evident for most underground support system applications.

The following are some of the more significant cost elements:

- Material Costs
  - Relatively easy to establish and often higher for wet mixes.
- Equipment Costs
  - The capital cost for the shotcrete and ancillary equipment is also easy to determine, and it varies significantly depending on the capacity and on the application process. Wet machines are always significantly costlier than dry machines.
- Maintenance Costs
  - The operational cost of the equipment is generally overlooked but can be substantial. Dry equipment maintenance per m^3 sprayed is between 2 and 4 times more than with wet equipment. In Canada and the United States, maintenance (and replacement) costs for the dry machine, hoses, and nozzle are typically around \$20.00/m^3 sprayed.

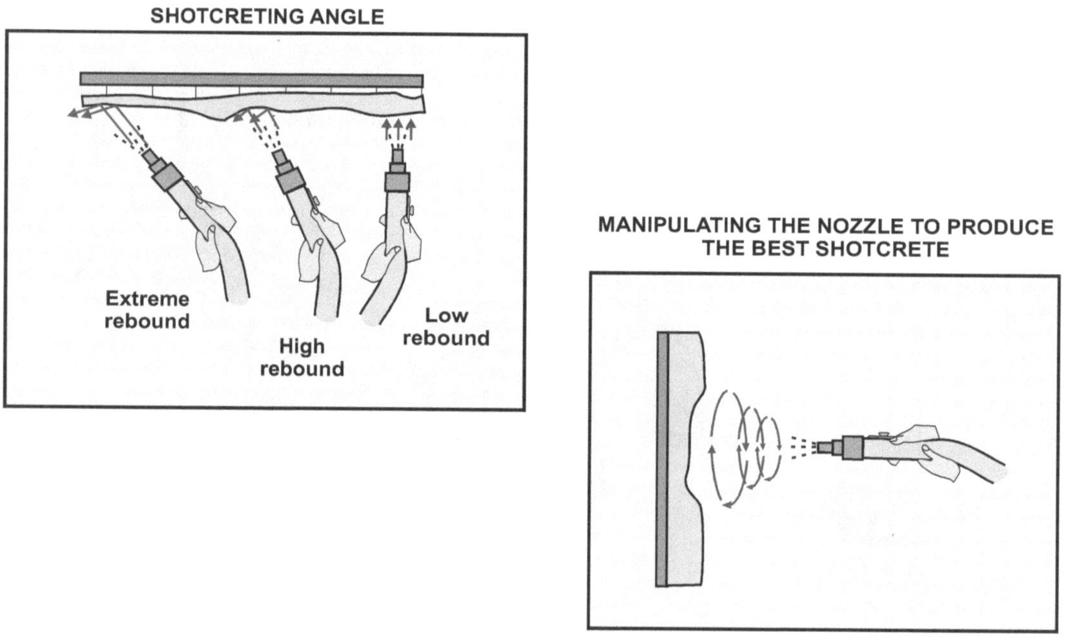

FIGURE 18.17 The correct shotcrete application (Vandewalle, 1997).

- Material and Equipment Transportation Costs
  - This important cost is generally ignored or avoided but can be significant and should be considered. Mines that bother to investigate this aspect are generally able to justify the installation of a shaft pipeline with pumps or "agitator-cars" for the horizontal transportation. This assumes that the monthly volumes needed are regular and relatively large.
- Labor Costs
  - The true labor cost involved with the entire process needs to be considered. When comparing shotcrete against other supports, any rehabilitation costs should also be estimated.
- Application Efficiency Costs
  - The rebound must be considered when shotcreting and is not too difficult to estimate. Quality wet shotcreting should be 10% or less, and quality dry shotcreting should be 25% or more. This must be accounted for in any shotcrete costing exercise.
- Time-Related Costs
  - These need to be considered if the mining cycle is critical for a specific excavation development. The effects of lost blasts, due to the support installation taking too long, is significant in many cases.

18.2.8 Shotcrete Quality Control

With shotcrete (even more important than with concrete), test everything associated with the mix and the components, and remember to set up an effective quality assurance system. Cement is a weak alkali, so check that the aggregate is not acid producing. Quality control tests should be:

- Reliable.
- Meaningful.
- Timeous (quick for most tests).
- Simple (so the operators can easily do them as part of their jobs).
- Relatively inexpensive.

The main parameters checked in any shotcrete quality program should be associated with design compliance, including mix design, water content, bond and strength, and sprayed design thickness. It is unacceptable to have a quality control system in place but fail to take effective and immediate action if issues are identified.

18.3 MESH (SCREEN)

Mesh/screen is routinely installed to prevent rockfalls and to stabilize the rock mass between rock anchors or similar. The performance of mesh/screen is different from that of shotcrete because large deformations are typically required before screen will start generating a reactive load as shown in Figure 18.18.

Figure 18.18 shows how much deformation must take place before a reactive load is generated in the mesh.

18.3.1 Mesh Installation

Mesh is versatile and can be used in many applications. It is commonly held in place by rockbolts with plates. The main problem is the labor and time needed to install the mesh and often personnel can be at some risk due to the typical semi-manual installation method. Significant progress has been made to make the process safer, faster, and more effective on both soft- and hard-rock applications, with the introduction of more mechanization.

FIGURE 18.18 An example of the inefficiency of mesh in generating an early reactive load.

Photo by A.J.S. (Sam) Spearing.

FIGURE 18.19 Mesh installation on a coal mine.

Photograph reproduced with the kind permission of Komatsu Mining Corp.

The development of the walk-through rockbolter for coal mines has made bolting and mesh installation much safer as the exposure of the operators has been significantly reduced. The sheets of mesh can now be placed atop the protective canopy and slid forward and fixed into place safely (Figure 18.19). The installation of synthetic mesh has also improved significantly.

Synthetic mesh is useful on coal mines as it is lightweight and easy to install, and can be cut relatively easily and effectively with continuous miners or shearers. The synthetic mesh is also not subject to corrosion. A common application is for longwall panel moves, where it is effective if correctly installed. Care must be taken to re-support areas where the synthetic mesh has been significantly torn (for whatever reason). An example is given in Figure 18.20.

Figures 18.21 and 18.22 give examples of mesh installation on hard-rock mines.

FIGURE 18.20 Plastic mesh on the roof and rib (held in place with the rockbolts) Tensar Earth Technologies.

FIGURE 18.21 Installing mesh panels using a rockbolter.

Reproduced with kind permission of Geobrugg.

18.3.2 Mesh Testing

Mesh performs very differently depending on parameters such as thickness, stiffness, and tensile strength. Like all safety-related products, mesh must be tested before it is considered for use. An effective testing method is shown in Figure 18.23. A test typically consists of four rockbolts in a pattern, similar to that used on the mine site. The bolts hold the mesh being tested at the corners, and the center is loaded using a large and generally curved platen.

Lacing is a steel cable that is used between shepherd crooks grouted rebar as shown in Figure 18.24. The lacing is fitted through the shepherd crooks. It can be incorporated with meshing and helps improve the skin (liner) support effectiveness.

FIGURE 18.22 Flexible mesh installation using a carousel.

Reproduced with kind permission of Geobrugg.

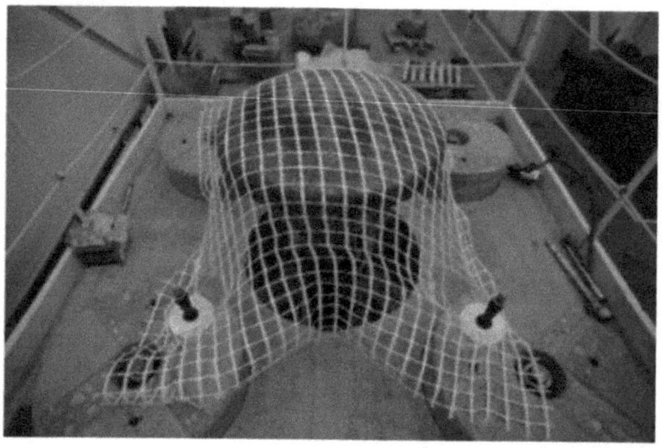

FIGURE 18.23 Mesh testing.

Photo by NIOSH.

FIGURE 18.24 Meshing and lacing skin support with grouted rebar.

Photograph taken by W.D. Ortlepp and reproduced with kind permission of the South African Institute of Mining and Metallurgy.

18.4 THIN SUPPORT LINERS

Considerable product development has been done by many mining product supply companies to develop thin liners as an alternative to mesh or shotcrete for less-demanding load applications or to seal, prevent, or limit weathering. Typical application thicknesses are between 4 and 10 mm. The advantages are the much smaller volume of material to transport per unit area of excavation supported and the very rapid strength gain compared with even highly accelerated shotcrete.

18.4.1 Possible Applications

Typical uses underground include the following:

- Temporary face support during tunnel development.
- Stope gully support (for narrow hard-rock tabular mining).
- Pillar reinforcement.
- Sealing rock prone to weathering.
- As a replacement for strapping or meshing, but only where geotechnically approved.
- As a support component to help possibly reduce rockburst damage.

Thin membranes are not generally a replacement for mesh or shotcrete where thicknesses of 25 mm or more are required. A combination could offer good potential in severe rock stability areas, as the membrane can give rapid support, and the shotcrete (generally fiber reinforced) provides the high long-term strength.

18.4.2 Potential Advantages of Thin-Skin Liners

The main advantages of support membranes over other similar types of support such as meshing and lacing or straps are as follows:

- High application productivity is achieved.
- Remote application is possible and therefore safer.
- They are not prone to corrosion.
- They are easier to keep concurrent with the face advance.
- They are less prone to blasting damage.
- Reactive loads are generated with far less rock movement.
- Weathering of the rock is virtually eliminated.

In addition, most thin membranes have the following advantages:

- Less costly application equipment.
- Low volumes to transport and apply, i.e., easy logistics.
- Most can be shotcreted over at a later and more convenient time, if required. This is frequently an important requirement.

18.4.3 Support Membrane Needs

The performance of any membrane type of liner support, including shotcrete, is governed by the following main physical parameters:

- Bond strength (adhesion) to the substrate (rock) which can be moist, dry, or have areas of both.
- Tensile strength.

- Elongation.
- Tear strength.
- Durability.
- Flammability.
- Resistance to chemicals, humidity, and temperatures in mines (generally site specific).
- The rate of gain of strength parameters.

Other important considerations are as follows:

- The health and safety considerations of the material, before, during, and after application.
- Ease of application; the ability to obtain the desired application thickness is also an important operational parameter.
- Cleaning of the equipment after use is also a consideration, unless water only can be used, because solvents tend to create additional potential hazards during handling and usage.

The physical properties of membranes are constrained by the chemistry of the raw materials available. High tensile strength and high elongation are usually desirable, but generally this becomes a trade-off, as higher elongation generally results in lower tensile strengths and vice versa.

Certain standard membranes do not handle the hot and humid conditions often found underground, so the chemistry for manufacturers is challenging.

18.4.4 Precautions and Considerations

Before deciding on using a particular thin membrane, extensive relevant testing needs to be undertaken and the following needs to be considered:

- No membrane can be readily applied over live (running) water.
- A decision must be made as to whether support is needed rapidly (in which case a two-component thin liner system is generally needed), or it can take some time to develop strength (in which case the generally cheaper single component system could rather be used).
- The useful life of the thin membrane must be established for the specific application. Certain thin-membrane systems are prone to significant embrittlement with time (particularly if they contain a cementitious addition), and this needs to be checked.
- As with any membrane application, the substrate must be free from dust, oil, and loose materials. This is best achieved using a pressurized water spray, but some thin-membrane systems do not adhere well in the presence of even a damp surface.
- If humidity is an issue, the long-term water stability of the product needs to be checked.

As with all similar activities underground, appropriate protective clothing and equipment are required and can vary from:

- Goggles, gloves, and a basic dust mask, to
- Goggles, gloves, and a charcoal filter, to
- Full body suits, respirators, and water curtains especially if isocyanates are used.

REFERENCES

American Concrete Institute (2016) Guide to Shotcrete. Report 506R-2.

ASTM C150-07 (2012) *Standard Specification for Portland Cement*. ASTM International, West Conshohocken, PE. www.astm.org/standards/c150

ASTM C1436-13 (2021) *Standard Specification for Materials for Shotcrete*. ASTM International, West Conshohocken, PE. www.astm.org/standards/c1436.

Bauer, E.R. and Dolinar, D.R. (2000) Skin failure of roof and rib and support techniques in underground coal mines. *Environmental Science*. 151, 99–109.

Comite Europeen de Normalisation (2007) Admixtures for Concrete, Mortar and Grout-Part 5 EN 934-5. Retrieved from: www.en-standard.eu

European Federation of National Associations Representing Concrete (EFNARC) (2023) Flums, Switzerland. www.efnarc.org

MechanicalBoost (2024) What is a peristaltic pump (Engineer Waqar)? www.mechanicalboost.com/peristaltic-pump/

Melbye, T.A. (1994) *Shotcrete for Rock Support. 3rd Ed.* MBT Europe (MAC Spa), Zurich, Switzerland.

Sauer, G & Gold, H. (1989) NATM Ground support concepts and their effect on contract practices. *Rapid Excavation and Tunnelling Conference*. Los Angeles, US.

Spearing, A.J.S. (2002) Shotcrete (sprayed shotcrete) for mining applications. Master Builders Technology.

US Army Corps of Engineers (1993). Standard Practice for Shotcrete. Engineering Manual EM 1110-2-2005, 31 January.

Vandewalle, M. (1997) *Tunnelling the World*. N.V.Bekaert S.A.

19 Backfilling

19.1 INTRODUCTION TO BACKFILLING

Backfilling can be considered a "green mining" (environmentally friendly) technology as it can not only reduce waste on surface but can also reduce surface subsidence and groundwater damage caused by mining.

Backfilling is the process whereby waste material, which is generally metallurgical tailings or prepared aggregate, is replaced underground into open mine workings, such as stopes or panels, or directed into open mine voids from surface. It is becoming an essential part of many underground mining operations and can resolve some of the environmental surface impacts, but in most metal mines, too much metallurgical waste is produced, so not all can be placed back underground. Overall backfilling is a more environmentally friendly technology compared with conventional tailings disposal, as it not just reduces waste on surface but also reduces surface subsidence and groundwater damage.

Typical sources of backfill material include the following:

- Waste metallurgical tailings that can be used as total or classified material.
- Underground waste rock.
- Sand.
- Quarried rock.
- Industrial waste (such as thermal power plant fly ash waste).

Often a blend of materials is used depending on the availability and need.

19.2 THE NEED FOR BACKFILLING

Backfill is typically used for the following in mining:

- Working platforms.
- Artificial roof or sidewall.
- Stope or other void filler.
- Pillar replacement.
- Surface subsidence control.
- Reduce rockburst damage.
- As regional support.
- Increased extraction such as in partial or total pillar extraction.
- Hangingwall (roof) support and/or dilution control.
- Conventional support replacement.
- Improve ventilation control.
- Reduce heat ingress.
- Reduce fire hazard.

Backfill can also improve the safety of the workforce by reducing rockburst damage or improving ventilation control for examples.

DOI: 10.1201/9781032704845-19

19.3 BACKFILL TYPES, REQUIREMENTS, AND PLACEMENT

Basic types of backfill used are as follows:

- Rockfill: Typically run-of-mine waste or quarried waste.
- Cemented rockfill: As rockfill above but with a cementitious addition.
- Uncemented classified (hydraulic slurry) tailings fill: Typically, the underflow from classified (cyclone) tailings. In the context of backfilling, classification means separating the total tailings into a dense and coarse fraction (used for the backfill) and a fine and dilute fraction that still needs to be disposed of, typically on a tailing's storage facility on surface.
- Cemented classified (hydraulic slurry) tailings fill: Classified tailings fill but with a cementitious addition.
- High-density/paste fill: A further densified material with no free water.
- Solid backfill: Dry or pre-damped material specifically for longwall mining above sensitive surface areas developed in China.

Figure 19.1 shows the performance of types of backfill. These tests are typically confined tests or at a high width to height ratio, as this is effectively how the backfill is placed underground.

The following are some common and important criteria and requirements:

- Fill should be placed at the lowest possible cost.
- Risk of fill failure (e.g., liquefaction, slip failure) must be minimized.
- Early strength development needs to be adequate and meet design requirements.
- Long-term strength should be sustainable.
- Delivery volumes must be reliable and adequate.
- Backfill must not be a production bottleneck.
- After placement, dimensional stability must be achieved as soon as possible, meaning that the placed backfill will be stable quickly and not prone to slump or even liquefy.
- Segregation should be minimized.

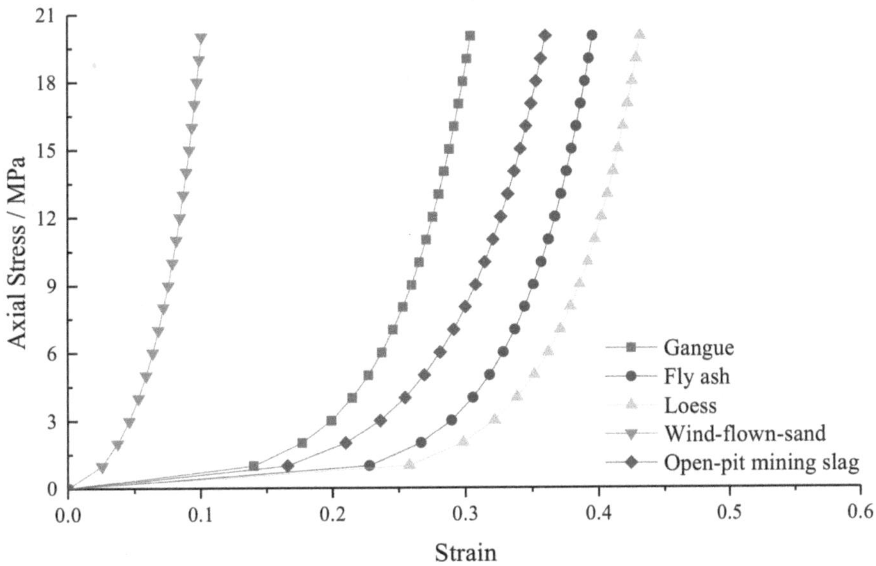

FIGURE 19.1 Performance behavior of various backfill materials (Zhang et al., 2019).

The most reliable measurement of backfill stability is its strength gain, but the placed quality also depends on the following:

- The effectiveness of the placement technique.
- Segregation after placement (partly dependent on the placement method).
- The consistency of the backfill mix (a quality control function).
- Post-filling water and solids loss if it is not a paste, especially if any binder is present.

The placed backfill strength is dependent on the following:

- Delivered density.
- Final density which occurs after water loss, solids loss, and gravity settlement.
- Particle size distribution and porosity.
- Binder type and content (if any).

Backfill is required in situ mainly under two basic placement schemes:

- Bulk placement where the fill is a working floor (platform) or roof. This placement is relatively simple, and the only major potential problems are segregation of the fill and possible binder loss in the drainage water, depending on the type of backfill mix used.
- Tight placement where the fill is typically used as an artificial sidewall and needs to have reliable contact with the roof of the stope. The potential problems are to control/eliminate post-filling shrinkage and avoid costly containment structures and/or mining at unfavorable gradients from a production standpoint.

19.4 MINING METHODS USING BACKFILL

Backfilling can be an integral part of many mining methods, such as

- Longwall mining.
- Room-and-pillar mining.
- Cut and fill and undercut and fill, where backfill must be used.
- Longhole open stoping and vertical crater retreat.
- Drift and fill, where backfill must be used.

Depending on the backfill performance required for the specific mining application, the backfill type can be selected. Figure 19.2 shows some results for different backfill trials at high strains and measuring the resulting reactive compressive stress.

19.4.1 Longwall Mining

Longwall mining is commonly used for narrow tabular mining when the orebody (seam or reef) dips at usually less than 30° from the horizontal. It can be used in hard-rock applications, such as gold mining on the Witwatersrand Basin in South Africa, or more commonly in soft-rock applications, such as coal mines. Typical classified backfilling has been used in hard-rock South African narrow deep tabular mines to try and reduce stresses, reduce the occurrence of rockbursts, and improve ventilation. In longwall coal mines in China, solid backfilling is used when mining under critical infrastructure to significantly reduce surface subsidence and groundwater disruption. More details of the method can be found in Zhang et al. (2012).

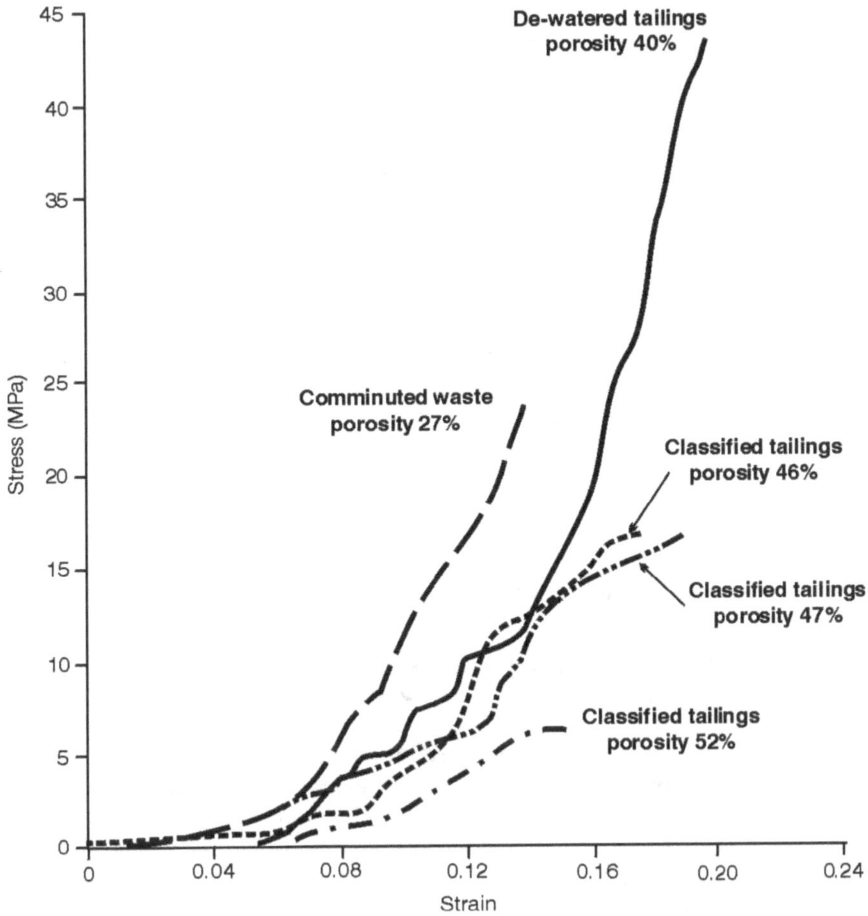

FIGURE 19.2 In situ performance of various different backfill materials at high strain (Gürtunca et al., 1989).

19.4.2 Room-and-Pillar Mining

The key to this mining method is the room width which represents the span between the pillars and the pillar dimensions. The room width is generally selected using various geotechnical evaluations, including rock characterization methods, tributary area theory, and numerical modeling, so that the room can be safely mined with the minimum of installed support and to cater for localized instability such as minor geological features. The relationship between increasing the room width and hence increasing the installed support is generally dictated by economics and relevant regulations.

This method of mining is generally applicable in orebodies where the dip is 30° or less and the orebody and the hangingwall (roof) are relatively competent. Increasing the pillar dimensions and/or reducing the room width can compensate for weak roof or floor conditions, but the overall extraction is significantly reduced.

There are three basic variations of this mining method:

- Horizontal room-and-pillar mining.
- Inclined room-and-pillar mining.
- Step room-and-pillar mining.

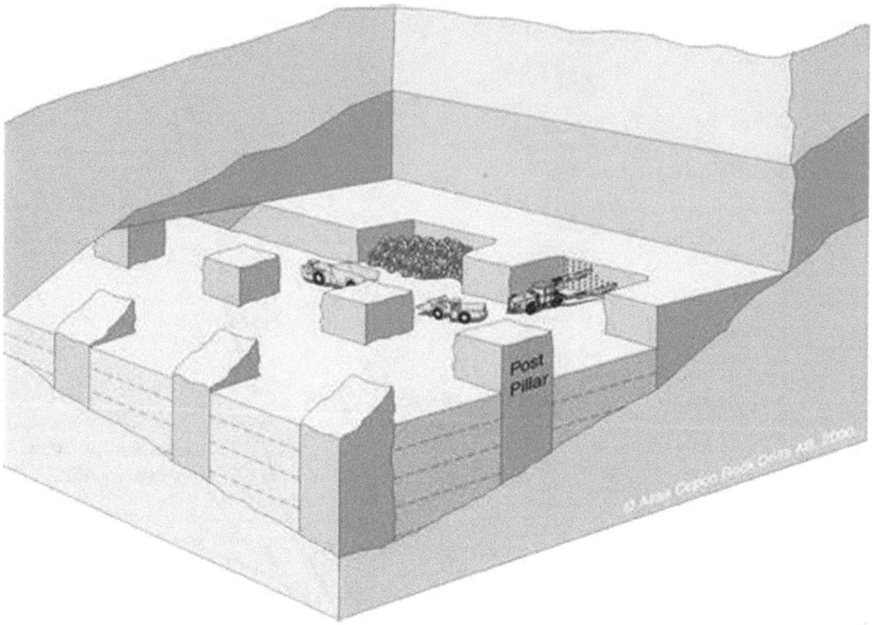

FIGURE 19.3 Room-and-pillar mining in a thick orebody with backfill (Atlas Copco, 2000).

Generally backfilling is not cost effective with this mining method, unless there are special circumstances where extra extraction is financially viable. An example is a tabular orebody that contains a high-value material such as copper or gold. It can also be used if long-term stability after mining is required. Figure 19.3 shows an example of backfilling with room-and-pillar mining to increase extraction. The backfill provides lateral confinement to the relatively slender pillars as mining progresses from the bottom of a thick orebody in slices (De Souza, 2010).

Once the primary mining has been completed by mining out the initial rooms, some form of total or partial secondary extraction can be undertaken, usually on a retreat basis. This is generally done without backfill, unless it is justified to reduce surface subsidence for example. The total extraction is seldom achieved as the secondary pillar extraction is usually difficult, more hazardous, costly, and inefficient.

19.4.3 Cut-and-Fill, Undercut-and-Fill, and Drift-and-Fill Mining

In the common form, cut-and-fill mining involves taking horizontal slices of ore, backfilling the void created, and moving progressively upward. The method can also be used to mine downward, called undercut and fill, and this requires a competent backfill that is more costly.

In this application, the backfill is used as a working platform and a support for the orebody walls. In wider orebodies, drift and fill may be used, where horizontal rooms are developed and backfilled, before the next vertical slice is removed. The backfill needs to be free draining.

Productivity is dependent on achieving the whole mining cycle of drilling and blasting, making safe and supporting, removing the ore, and backfilling. The method is relatively labor intensive and tends to be utilized in the extraction of more valuable and higher-grade minerals such as silver and gold.

It is generally used for the following conditions:

- Steep dip.
- Relatively thin orebody widths.

- Competent orebody.
- Country/host rock can be relatively incompetent.
- It can be used at almost any depth.

For undercut and fill, the conditions are similar to cut and fill except that the orebody is very weak and not self-supporting. In this application, the stoping is in slices from the top moving down, and the backfill, which must be cemented, acts as an artificial roof. This is a costly mining operation, so the applications are limited to high-grade and high-value deposits such as silver and gold.

For drift and fill, the primary drifts are backfilled using cemented fill as they need to be self-standing for when the secondary drifts are mined. After the secondary drifts are mined, depending mainly on the orebody, strata properties, geology, and hydrology, the secondary drifts can be left unfilled, filled with uncemented backfill, or filled with cemented backfill.

These mining methods are shown in Figures 19.4 and 19.5.

19.4.4 Sublevel/Open/Blasthole Stoping and Vertical Crater Retreat (VCR)

These mining methods are used in orebodies with the following characteristics:

- A steep dip, defined as more than the angle of repose of the material.
- Competent footwall (floor) and hangingwall (roof).
- Competent ore.
- Regular orebody boundaries are essential for VCR mining.

Depending on the overall conditions, the stopes can be filled or unfilled. If unfilled, then substantial rib and crown pillars are typically required.

Figure 19.6 shows VCR with backfilling after a stope is completed.

In sublevel stoping, after the ore is extracted, the stope is left empty. Stopes are typically very large with the largest dimension in the vertical. Pillars are left between stopes depending on the geometry of the orebody. In VCR mining, the stopes are typically filled after mining.

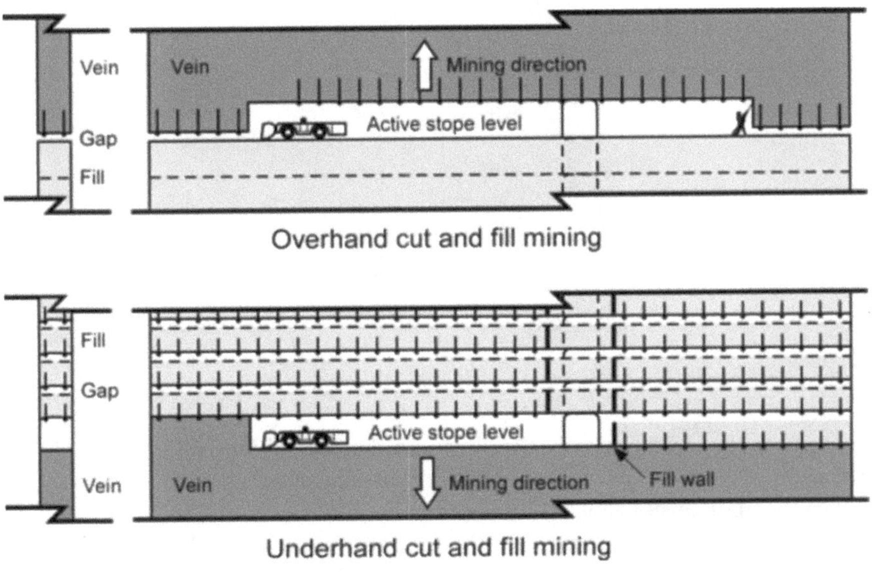

FIGURE 19.4 The differences between cut and fill and undercut and fill (Williams et al., 2007).

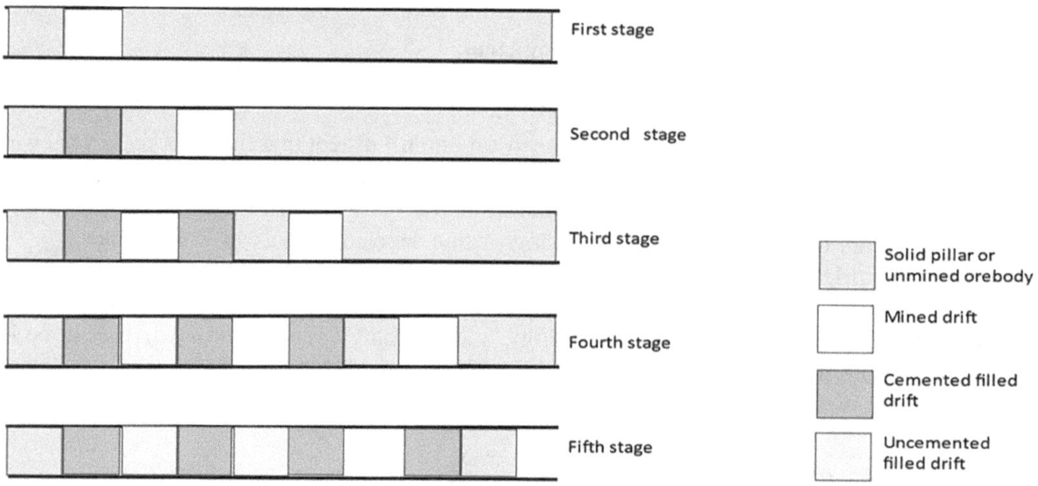

FIGURE 19.5 Drift and fill (Spearing et al., 2023).

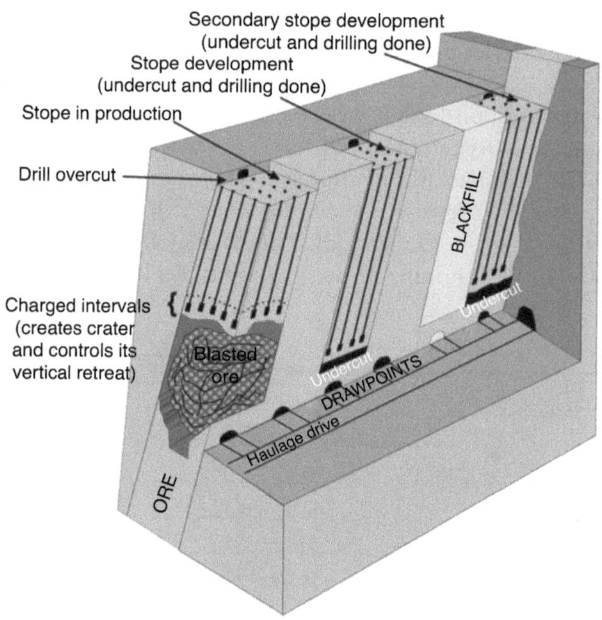

FIGURE 19.6 Vertical crater retreat (VCR) with backfill. (Modified after Hamrin, 2001.)

The mining in sublevel stoping is carried out from horizontal levels and sublevels at pre-determined vertical intervals. The ore is broken by drilling and blasting from the sublevel drifts and collection from the main level below.

The drilling in the sublevel stoping can be carried out well in advance of mining, and the blast hole length drilled is seldom more than 25 m. An outline of the method is given in Figure 19.7.

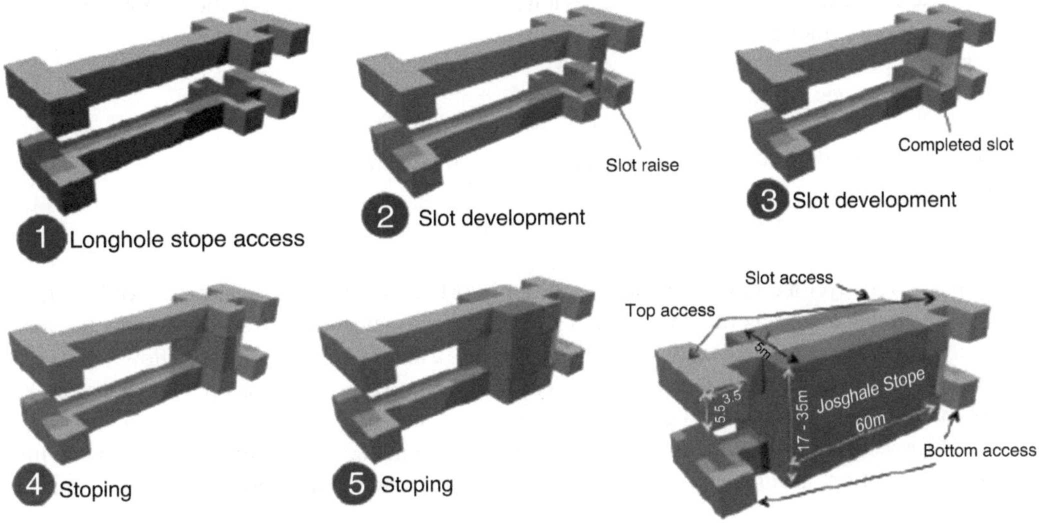

FIGURE 19.7 Longhole open stoping mining sequence between two sublevels (Goldfields, 2009).

19.5 BACKFILL PREPARATION

The main backfill types in hard-rock tabular mining are as follows:

19.5.1 Classified (Cycloned) Tailings

The components of a typical cyclone are shown in Figure 19.8. On the introduction of the dilute feed, usually directly from the metallurgical plant to the cyclone and while under pressure, the

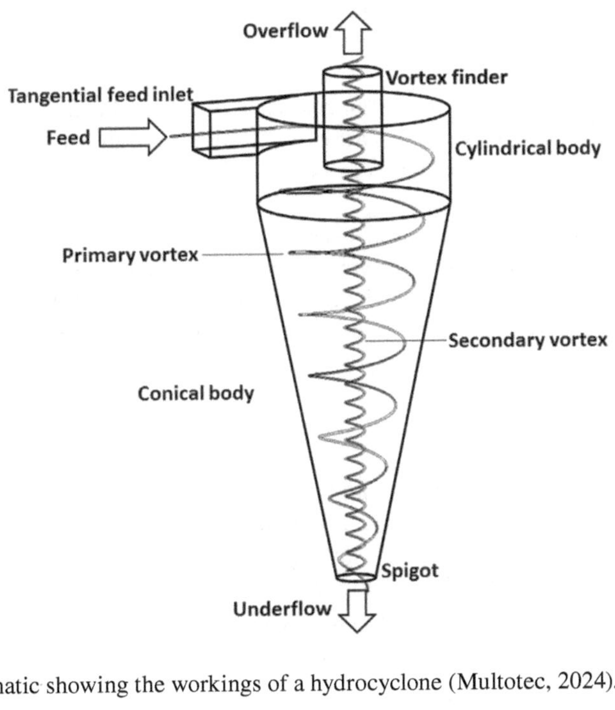

FIGURE 19.8 Schematic showing the workings of a hydrocyclone (Multotec, 2024).

coarser particles are thrown outward and gravitate down the cone and out of the spigot, due to centrifugal forces. The finer particles are forced upward through the vortex finder into the discharge chamber and report to the overflow as waste.

Therefore, cyclones produce two products:

- An underflow that is relatively coarse and densified, and it becomes the free draining backfill material.
- An overflow that is very fine and dilute, for disposal on a tailings facility/dam.

For any particular cyclone diameter, the spigot and vortex finder can be varied according to the desired performance. In general, the smaller the spigot diameter compared to that of the vortex finder, the greater the upward velocity. This results in a larger quantity of overflow solids, and hence, the underflow will contain fewer fines, less overall moisture, and less recovered mass.

The capacity and efficiency of a cyclone increase with increasing inlet pressure, because higher centrifugal forces are generated. The pressure applied in practice is a compromise between more efficient classification, at higher pressure, and reducing cyclone component wear which requires a lower pressure.

It is also essential that prior to cycloning, all foreign bodies (such as grit and timber chips) are removed by screening from the tailings to ensure a quality and consistent fill material and to avoid blockages, and damage, to the cyclone lining.

Research by Spearing (1989) has shown that the following must be achieved for successful classified (free draining) uncemented fill production:

- A solids content above 65%, preferably around 70%, should be maintained.
  - About 75% is the maximum solids content obtainable using cyclones.
- A minus 10 μm size fraction of less than 6%, with 4% being a good target but sometimes difficult to consistently achieve, to allow adequate percolation and hence drainage.
  - With typical gold plant tailings, the minus 10 μm size fraction contributes to the majority of the material surface area which adversely effects drainage of the placed fill.
- A consistent product, considering especially the density, particle size distribution, and porosity.

A cementitious component can be added if necessary, but the efficiency of the cement to function as a binding agent is relatively low because of the adverse high water to cement ratio and the fact that the cement, being very fine, tends to be a significant part of the drainage material after placement.

Even though cycloned fill produces fewer overall tailings for surface disposal, the problems and costs associated with handling the cyclone overflow can in fact be more due to the very fine material that tends not to drain, and mineralogy (typically a high clay content) of the product.

Classified tailings are a common fill type used, due mainly to cost considerations and the relative ease of operation.

19.5.2 Dewatered Full Plant Tailings Typically a Paste Backfill

This fill material is produced by densifying and using all the metallurgical plant tailings. The most common methods of densification are a vacuum belt filter (either a vertical disc or a horizontal belt) or high-rate thickeners. Generally, the tailings are densified to achieve a stiff pulp and then repulped by the appropriate water addition to the design relative density. A conventional thickener is frequently included in the backfill plant equipment, as an intermediate densifier at a relatively cheaper operational cost. A true paste has no free water, making it basically stable after placement with no post-filling shrinkage.

The method tends to be a more costly densifying process than using cyclones. Many backfill plants internationally claim to be paste producers but really should be considered as high-density fills. The difference may seem subtle, but the reality is that a high-density fill drains and therefore can produce water channels (rat holes) that cause significantly higher loads on the retaining structure and can even liquefy. A cementitious addition, if needed, tends to be more effective than in a classified tailings, as the water-to-cement ratio is more favorable and drainage should be very low (zero if a true paste is placed).

The significant cost of a paste system is a result of the paste rheology as it does not free-flow, so it is typically pumped using positive displacement pump stations.

19.5.3 Fill Containing Aggregate

Underground mines produce waste rock, so if it can rather be used as a backfilling material, it does not need to be transported to surface.

This fill material is either just aggregates or a blend of aggregate and either cycloned tailings or full plant tailings. Depending on the mining method, the waste can be co-disposed of with usually cemented hydraulic or paste fill. For this application, the waste rock must be placed such that it is fully enveloped by the other fill.

A cementitious addition can be added if needed for the specific application. Aggregate backfill is very stiff, compared with cycloned or paste backfill. Depending on the geometry of the fill distribution system, positive displacement pumps, trucks, or conveyors are used.

19.6 BACKFILLING DESIGN, TRANSPORTATION, AND PLACEMENT

19.6.1 Backfill Design

The design of a backfill plant and infrastructure must be carefully considered and implemented to ensure cost-effective filling. This includes the delivery of sufficient quality-controlled fill on demand so that mining can progress uninterrupted.

To produce a cost-effective and efficient backfill operation, a multi-disciplinary project team is required that should include at least the following skills and functions as a minimum requirement:

- Mine Planning.
- Rock Engineering.
- Metallurgy.
- Mechanical Engineering.
- Environmental Health and Safety.
- Tailings Disposal/Waste Management.

19.6.2 Backfill Storage

Backfill storage is often needed, especially if classified or paste fill, mainly on the surface, because:

- The metallurgical plant operates for 24 hours a day, while backfill is generally placed for 8–16 hours per day, and at an hourly rate of more than the fill plant processing rate.
- The storage is generally in silos above the ground near the site where the backfill is transported underground. The surface storage provides the pressure head to maintain an adequate feed to the pipelines which are usually vertical initially, down boreholes, or a shaft. If necessary, pumps can also be used to achieve this but at a significant cost increase.

Backfill storage facilities are relatively costly to build, and therefore, they must be sized correctly and include some overcapacity.

Storage tanks can have a circular or rectangular cross section, but generally circular geometry is preferred because it is easier to agitate and keep the fill homogeneous.

Agitation is required in the storage tank, particularly if slurries are stored. Agitation is achieved mechanically or pneumatically, and it is not uncommon to use mechanical agitation with a pneumatic back-up, in case of mechanical failure. Using both systems simultaneously is counter-productive because mechanical agitation mixes by pushing the material downward, and pneumatic agitation works by lifting the solids (as in an airlift pump system).

The design of the storage vessel and agitation should be professionally undertaken, but the following guidelines should be considered:

- Wall baffles are generally needed to keep the fill from vortexing and to improve mixing.
- The main mechanical impellor is generally placed about one-third of the slurry total height above the base of the tank.
- The optimal rotational speed is usually determined by tests and recommendations from the suppliers, based on the specific fill material.
- Air agitation can be used but is usually more costly; however, it is sometimes used as a back-up to prevent fill settling if the mechanical agitation system fails.

19.6.3 Backfill Transportation

There are four main transportation methods (or combinations) used to move the fill from the point of preparation which is typically on surface to the underground workings:

- Hydraulic via pipelines, with the fill as a low-density slurry or a high-density paste. This is the most common method, but as the slurry density increases toward a paste, the pressure gradient increases and pipeline tests are advisable before the diameter is specified. The pressure loss, due essentially to friction, relationship to transport velocity in a pipe is shown in Figure 19.9. It can be seen that laminar flow is preferable as it flows with less pressure loss and causes less pipe wear. The main pipeline transportation methods are shown in Figure 19.10. If paste fill is used, then pumps are typically required unless the horizontal transport distance is relatively limited.
- Mechanical via conveyor belts or trucks. Trucking and conveyors are commonly used to transport rockfill. Each has merits of its own and should be evaluated individually before a decision is made. Considerations that should be made include the number of areas requiring simultaneous placement, fill tonnage requirements, access, and ventilation considerations.
- Pneumatic via pipelines as a dry or pre-damped material. This method of transport has become unpopular mainly because of the high energy consumption and cost of using air, the severe wear rate due to impact, and the dust, even if the material is pre-damped.
- Gravity via raises. This method is commonly used to transport rockfill vertically, and the raises can be lined or unlined. Depending on the strength of the rock, lining the raise with an impact and wear-resistant liner is often cost effective if the cost of rehabilitation of an unlined raise is considered. Generally, it is advisable to run rockfill raises/passes full to reduce wear and try to limit degradation that creates more fines. This does have the potential to increase the risks of hang-ups in the raise.

Option 1: This is where the backfill is produced and transported down a vertical pipeline in a borehole or shaft and delivered to the stope requiring the backfill. If the conditions are suitable, then a pump station is not required, but if the pressure losses are too large due to the long near horizontal distance, then an intermediate pump station may be required.

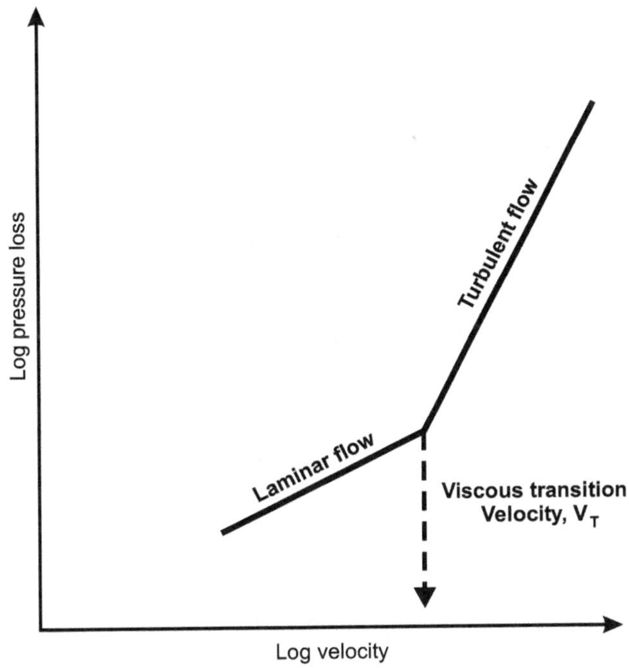

FIGURE 19.9 Pressure loss versus velocity for a homogeneous slurry (Spearing, 1989).

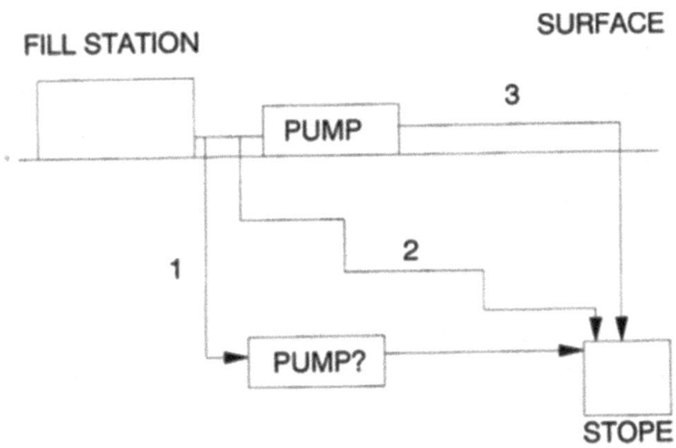

FIGURE 19.10 The three main and idealized backfill pipeline distribution systems (Thomas et al., 1979).

Option 2: This is where the backfill is delivered vertically between major haulage levels to the stope. An intermediate pump station could also be required.

Option 3: This is where the backfill is pumped on surface to above the stope to be backfilled and then dropped via a borehole into the stope.

Where piping is used as the distribution system, the correct diameter and pressure rating are key to the site specific conditions.

The following aspects must be considered when designing a backfill distribution system:

- The transport velocity must be above the depositional velocity, to stop the solids from settling out. The depositional velocity is mainly dependent on the pipe diameter, the particle size distribution, and the concentration of the solids.

- The transportation velocity must, however, be minimized to reduce the wear, since

$$\text{Wear} \propto \text{Velocity}^k \tag{19.1}$$

The exponent k lies typically between 1.5 and 3.0, about 1.8 typically (Steward and Spearing, 1992).

- Standard pipe lengths and diameters should be used to reduce the pipe and fitting costs.
- Provision must be made to flush the pipes when necessary, with water at an adequate pressure and flow rate.
- Rupture discs should be considered at locations of peak pressure, to safeguard the pipelines in the case of blockages.

For any backfill system, the following pre-design steps should be taken, as much of the performance and cost issues of backfill mix design can be accurately estimated through laboratory testing. This testing can be divided into three stages:

i. Physical and chemical properties of the aggregate(s)
   Basic physical testing of the aggregates should include determinations of particle size analyses, specific gravity, bulk density, and permeability. Chemical properties and mineral composition can be determined by some combination of the following: pH, zeta-potential measurement, optical microscopy, X-ray diffraction (XRD) which shows the crystalline structure, X-ray florescence (XRF) for specific mineral composition, and/or inductively coupled plasma (ICP) for determination of all metals simultaneously.
ii. Mechanical properties of the backfill mix
   Compressive strength testing is necessary to determine binder contents and blend ratios of binders in a mix design. Two parameters that are very important in mixing compressive strength samples are to use the same water that is intended for use in actual production, and to size the samples appropriately, if using minus 12″ (300 mm) rockfill aggregate casts a minimum sample diameter of 18″ (450 mm). Monitoring of set times of the mixes using thermocouples is very valuable. It is important to cure the samples in a similar environment to the stopes where the material is to be poured. If the stope environment is 30°C, then curing test samples at 20°C will not necessarily yield representative results.
iii. Rheological properties of the backfill mix
   For binder slurries used in rockfill applications, flow cone measurements and set time determination are normally adequate. There are many different methods of measuring rheology in paste-fill applications, but the accuracy of any of them is often questioned. Viscometers and rheometers are commonly used, as are pipe loop testing and slump cone. With standard viscometers and rheometers, the paste plug will often rotate within the container, giving false readings. Pipe loops are valuable if properly set up, but they should be adequately instrumented, be of the same diameter throughout and not a closed loop. Slump testing, although often used, is highly inaccurate when a flocculent is present in the tailings or when the slump is greater than 230 mm (9″). Slump also varies considerably with temperature changes.

The following are the design steps that should be followed:

- Calculate the monthly volume of fill required, at the planned mining rate.
- Increase this calculated amount to allow for additional mining production by at least 10%, and backfill losses due to wastage and drainage which depends on the fill system to be used.

- Calculate the flow rate required, taking account of the actual filling time per day (typically only 6 hours/day if filling on a single shift).
- Calculate the static pressure head:

$$\text{Static head}(\text{kPa}) = \text{Vertical height} \times \text{gravity} \times \text{slurry relative density} \tag{19.2}$$

- Determine the total pipeline length, from surface to the stope for a dedicated pipeline, or depending on the mining method and suitability, determine the furthest stope that will need to be backfilled.
- The piping selection is based mainly on the pressure and the velocity needed to achieve the flow rate.
- The frictional losses must be determined using lab testing and/or estimated from the previous similar experience. For a system to be balanced, the pipeline frictional losses must equal the static pressure head. If this is not the case, the following can be considered:

    - Decrease the pipe diameter selected.
    - Increase the slurry relative density.
    - Choke the system.

Figure 19.11 shows an example of a classified backfill plant, using cyclones.

As previously indicated, the cyclone underflow is used as the tailings as it is relatively coarse (free draining) and dense and is placed as a slurry underground. The cyclone overflow is a fine and high-water-content material that is placed on a tailing's facility with as much water recovery as possible. Special care must be taken as the material is not free draining and typically has a high clay content.

Figure 19.12 shows an example of a paste plant.

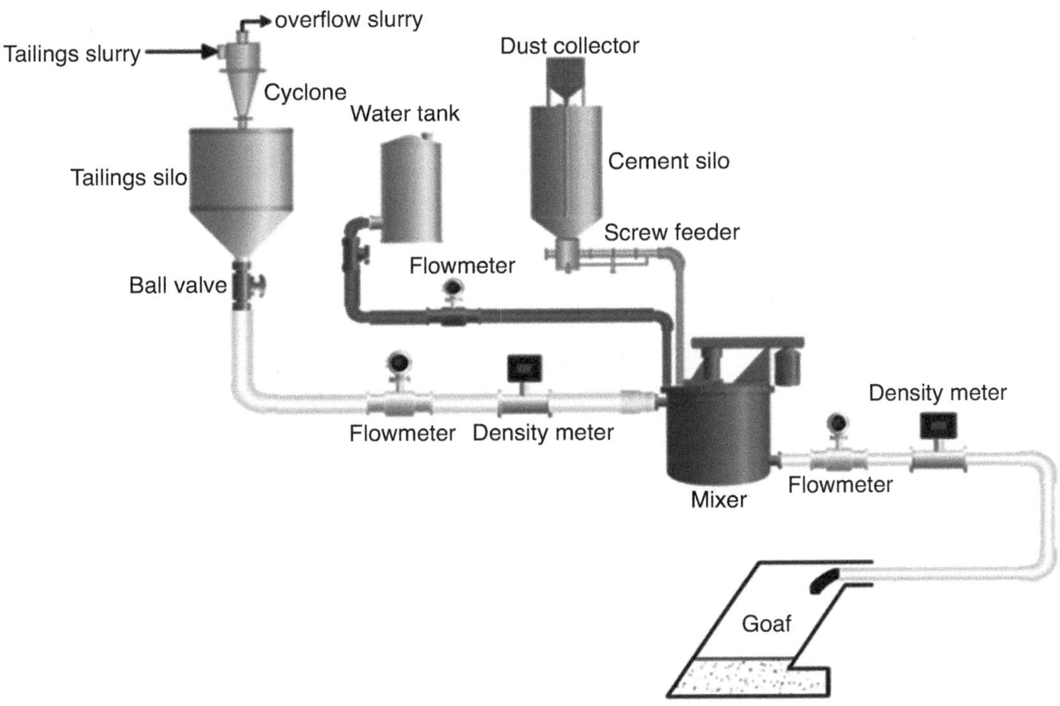

FIGURE 19.11 Classified backfill plant (Chen et al., 2023).

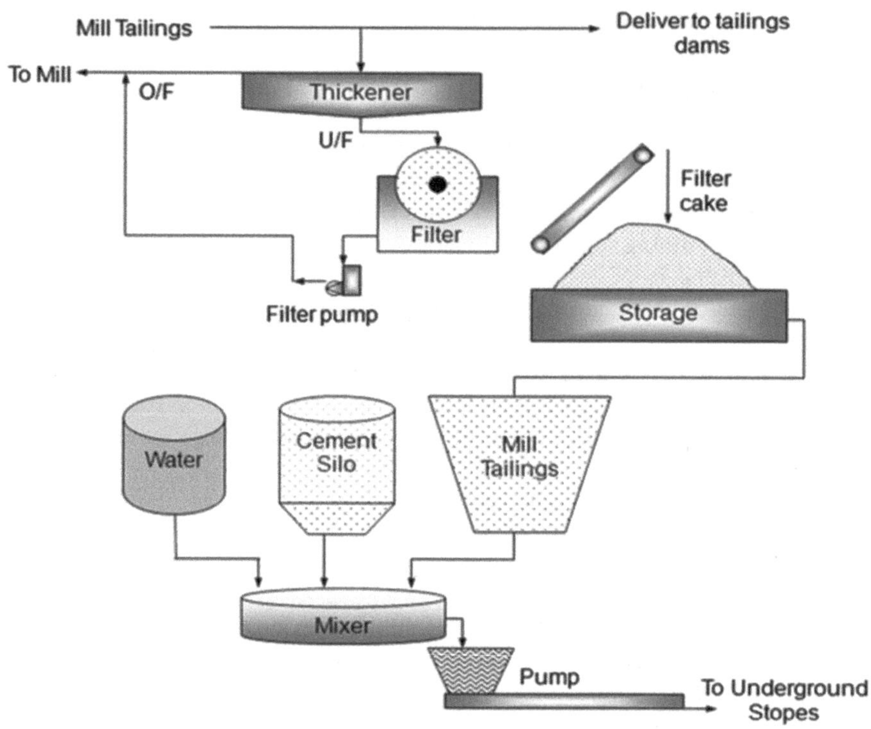

FIGURE 19.12 Paste plant flow chart example (Yilmaz and Guresci, 2017).

With a paste-fill plant, additional water is removed from the total tailings to create a paste that does not free-flow very well, so typically positive displacement pumps are needed. This type of backfill is therefore considerably more costly that classified slurry backfill but has a higher performance and is safer to use.

Figure 19.13 gives an example of a cemented rockfill plant.

With a cemented rockfill system, the cementitious addition is often added via a cement bath which is usually a pure cementitious slurry, as the fill is being placed via conveyor belt or truck.

19.6.4 Backfill Placement

The placed backfill strength is dependent on the following:

- Delivered density.
- Final density after any water and solids loss and gravity settlement.
- Particle size distribution.
- Binder type and content (if any).

Backfill is required in situ mainly under two basic placement schemes:

- Bulk placement where the fill is a working floor, platform, or roof. This placement is relatively simple, and the only major potential problems are segregation of the fill and binder loss in the drainage water.

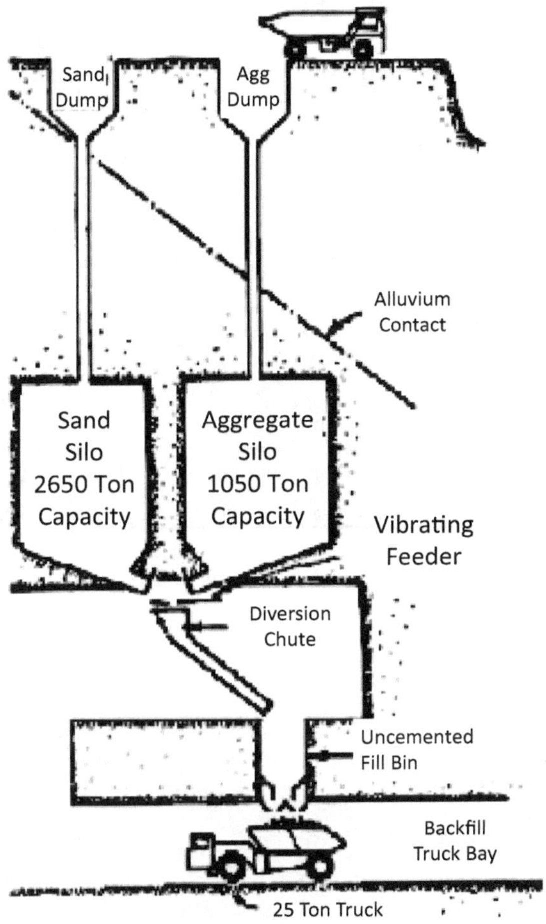

FIGURE 19.13 An example of a cemented rockfill plant (Jain, 2011).

- Tight placement where the fill is typically used as an artificial sidewall and needs to have reliable contact with the roof of the stope. The potential problems are to control/eliminate post-filling shrinkage and avoid costly containment structures and/or mining at unfavorable gradients from a production standpoint.

The backfill barricades (or retaining structures) are vital for the safe placement of backfill.

An example of a stope being filled, the barricade, and the localized stresses is shown in Figure 19.14.

Barricades can be made of numerous materials and combinations, including:

- Shotcrete.
- Concrete blocks.
- Backfill or a waste berm.
- Timber posts and planks.
- Steel shuttering.

Figures 19.15 and 19.16 show examples of two of the many different backfill barricade construction methods.

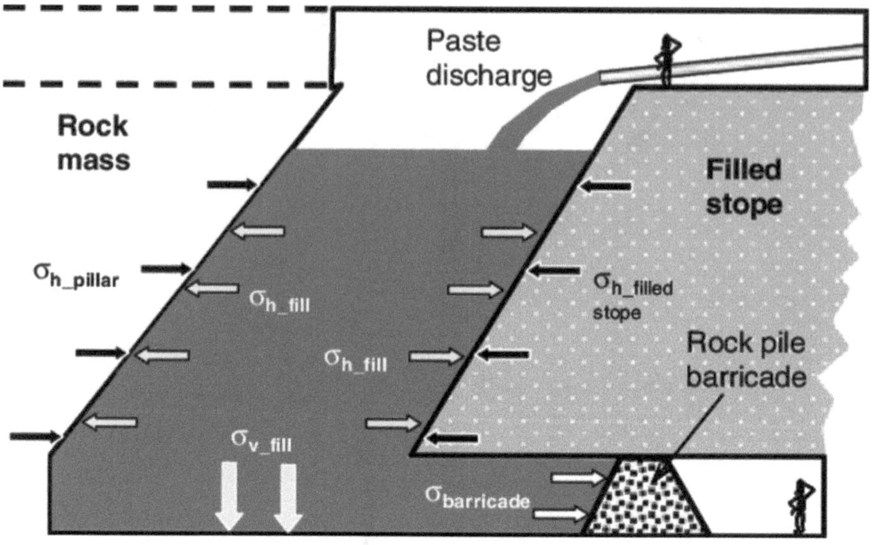

FIGURE 19.14 Schematic diagram of backfilled stope components and stress field distribution (Belem and Benzaazoua, 2007).

FIGURE 19.15 A concreted weld reinforced backfill wall, showing the inlet pipe and an overflow pipe. (Reproduced with kind permission of Paterson & Cooke.)

FIGURE 19.16 A fine weld and geofabric backfill barricade being prepared. (Reproduced with kind permission of Paterson & Cooke.)

FIGURE 19.17 Free water draining out of the backfill barricade during filling. (Reproduced with kind permission of Paterson & Cooke.)

It is important that any free water can drain out of the barricade structure, so a drainage system is essential and usually consists of a geofabric and even perforated drainage pipes. Monitoring the pressure exerted against the barricades is important to ensure that the structures are safe and don't fail. Figure 19.17 shows a void being backfilled with a slurry and the backfill barricade draining the free water.

19.7 BACKFILL QUALITY CONTROL AND MONITORING

All aspects of the backfilling process must be very carefully monitored and controlled because of the inherent potential hazards which include:

- Liquefaction.
- Barricade or retaining structure failure.
- Pipe rupture that is a particular concern in shafts.
- High pipe wear.
- Water hammer.
- Pipeline blockage.
- Fill production that does not meet design specifications.
- Handling the drainage products from the fill that are mainly water but could also include fill and cement (if used) and not placed as a true paste.

The need for effective backfill quality management is frequently overlooked, often resulting in a waste of money or even a potentially serious safety risk. Not only should a workable quality system be in place, but corrective actions must be taken if non-compliance to the pre-specified criteria is discovered.

Figure 19.18 shows a backfill barricade being constructed and fitted with instrumentation. Typically, the force exerted on the barricade and the water pressure would be measured.

As a minimum, the following should be regularly and routinely monitored:

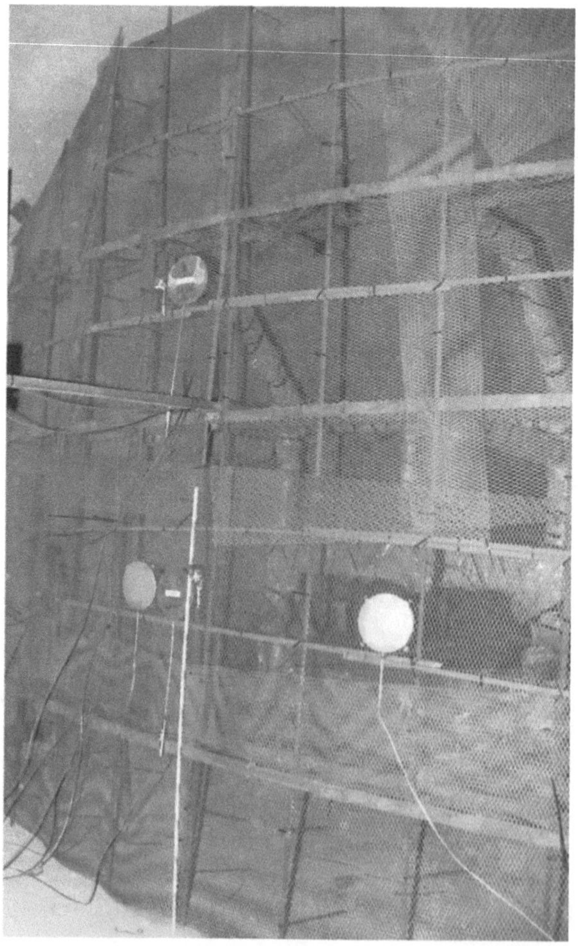

FIGURE 19.18 A backfill barricade being constructed and fitted with instrumentation. (Reproduced with kind permission of Paterson & Cooke.)

- The particle size distribution if classified tailings are used. The minus 10 μm size fraction is a useful indication of the percolation rate and can be quickly obtained if a laser diffraction instrument is available. This should ideally be measured in the surface storage, and material should be discarded if it fails the criterion.
- The relative density prior to underground discharge is very important. Care must be exercised that any recirculation system does not significantly dilute the backfill with time, due to the introduction of gland service water.
- Where a cementitious binder is used, care must be taken to ensure that an adequate quantity is actually added. Depending on the plant design, this can be routinely checked, especially if the fill mix is batched. An additional check can be readily and quickly undertaken by measuring the pH of the mix. The cementitious addition should make the final product alkali. This can be easily checked even by just using Universal Indicator Paper. Adding too much can also cause problems of wastage, extra costs, and even pipeline blockages.
- The level of any contaminants should also be monitored in the fill prior to placement and from any underground run-off.
- The performance of any barricades.

19.8 BACKFILL COSTS

Any backfill system selected must be as cost effective as possible. In most mines utilizing backfill, the fill cost is one of the highest efforts that must be made to minimize it while maintaining safety. The total (placed) backfill cost is the major consideration, and often increasing the fill preparation and/or material cost can actually reduce the overall placed cost.

Backfill operating costs typically reported by mines vary from \$3.00/t placed to \$8.00/t placed depending on the fill type used and the elements considered. For undercut and fill, the backfill cost is typically higher due mainly to the large cementitious content required to create the safe artificial roof.

Elements that need to be included in any backfill costing exercise include the following:

- Plant amortization.
- Plant operating cost, including labor, power, and consumables.
- Raw material costs (e.g., cement, sand, admixtures).
- Maintenance considering both on surface and underground.
- Fill placement preparation including bulkheads and new piping.
- The cost and frequency of breakdowns from mainly pipe blockages and pump issues.
- The influence of the fill on the mining cycle. For example, a faster setting fill material would improve overall mining efficiencies by permitting secondary extraction sooner.
- The additional cost of handling any spillage and any water and solids run-off after placement.

19.9 WORKED EXAMPLES

Example 19.9.1

A bulkhead is to be constructed to resist hydraulic tailings reaching a maximum height (H) of 5 m. Determine the full hydrostatic water pressure at this height, and the lateral pressure if the unit weight of the hydraulic fill is $\gamma_{sat} = 20$ kN/m^3. Equations are after Mitchell (1983).

Solution:

Hydrostatic Water Pressure:

$$u = \gamma_w H$$

$$u = (9.81\ \text{kN/m}^3)(5\ \text{m})$$

$$u = 49\ \text{kN/m}^2$$

Lateral Pressure (using a Saturated Unit Weight):

$$\sigma_h = \gamma H$$

$$\sigma_h = (20\ \text{kN/m}^3)(5\ \text{m})$$

$$\sigma_h = 100\ \text{kN/m}^2$$

Mitchell (1983) suggests that these approximations resulted in impractical bulkhead designs and construction. The next example illustrates a less conservative alternative approach.

Example 19.9.2

Assuming that the backfill in the previous example is not subject to liquefaction, estimate the total pressure on a bulkhead for the following two cases. The water pressure is confirmed by monitoring water pressures at piezometer tips. Assume a friction angle for the tailing sands of 30°. Equations from Mitchell (1983).

Solution:

Case 1

Stope strike length (L) > Stope height (H)

Assume L = 5 m and H = 4 m

$$u = \gamma_w H$$

$$u = (9.81\ \text{kN/m}^3)(4\ \text{m})$$

$$u = 39\ \text{kN/m}^2$$

$$\sigma_h = K_a \gamma H + u$$

$$\sigma_h = (0.33)(20\ \text{kN/m}^3)(4\ \text{m}) + (39\ \text{kN/m}^2)$$

$$\sigma_h = 26\ \text{kN/m}^2 + 39\ \text{kN/m}^2$$

$$\sigma_h = 65\ \text{kN/m}^2$$

where

K_a is the active pressure coefficient.

γ is the unit weight.

u is the water pressure.

φ' is the effective apparent angle of shearing resistance.

There are a few methods to estimate the active earth pressure coefficient, and the most suitable should be used. If we assume Rankine's earth pressure coefficient,

$$K_a = \tan^2(45 - \varphi'/2) = 0.33$$

It is not mentioned specifically in Mitchell (1983) which form of the coefficient is appropriate to use.

For example, Smith and Mitchell (1982) refer to the use of a lateral pressure coefficient:

$$K_o = 1 - \sin\varphi' = 0.5$$

Case 2

Stope height (H) > Stope strike length (L)

Assume L = 4 m and H = 5 m

$$u = \gamma_w H$$

$$u = (9.81\ \text{kN/m}^3)(5\ \text{m})$$

$$u = 49\ \text{kN/m}^2$$

$$\sigma_h = [(L\gamma)/(4\tan\varphi')] + u$$

$$\sigma_h = [(4\ \text{m})(20\ \text{kN/m}^3)/(4\tan 30)] + (49\ \text{kN/m}^2)$$

$$\sigma_h = (80\ \text{kN/m}^2)/(2.3) + (49\ \text{kN/m}^2)$$

$$\sigma_h = 83\ \text{kN/m}^2$$

Example 19.9.3

The following serves as a further contrast to the previous two examples. Equations are from Smith and Mitchell (1982). Determine the total bulkhead design pressure. Assume $\varphi' = 30°$, $\gamma = 20\ \text{kN/m}^3$, Z = 5 m. For part a, assume it is a free draining bulkhead, u = 0. For part b, use the estimate of hydrostatic pressure from Example 19.9.2.

Solution:

Total pressure, p = (solid fill pressure, $\sigma_{h'}$) + (hydrostatic water pressure, u)

For a case where height of the pour (Z) > minimum stope plan dimensions

$$\sigma_h' = K_o Z\gamma / 2\tan\varphi'$$

K_o = lateral pressure coefficient = $1 - \sin\varphi'$

Z = approximate dimension over which strong arching can take place

γ = saturated unit weight of the fill.

Note that the "saturated unit weight" is not indicated in the previous two examples, usually shown as γ_{sat}, but would seem to result in the most conservative approach.

Part a:

$$K_o = 1 - \sin\varphi' = 0.5$$

$$\sigma_h' = (0.5)(5\ \text{m})(20\ \text{kN/m}^3)/(2\tan 30)$$

$$\sigma_h' = 50\ \text{kN/m}^2/1.15$$

$$\sigma_h' = 43\ \text{kN/m}^2$$

$$\text{Total pressure, } p = \sigma_h' + u$$

$$\text{Total pressure}, p = 43\ \text{kN/m}^2 + 0\ \text{kN/m}^2$$

$$\text{Total pressure}, p = 43\ \text{kN/m}^2$$

Part b:

$$u = \gamma_w Z$$

$$u = (9.81\ \text{kN/m}^3)(5\ \text{m})$$

$$u = 49\ \text{kN/m}^2$$

$$\text{Total pressure}, p = \sigma_h{}' + u$$

$$\text{Total pressure}, p = 43\ \text{kN/m}^2 + 49\ \text{kN/m}^2$$

$$\text{Total pressure}, p = 92\ \text{kN/m}^2$$

Smith and Mitchell (1982) detail yet another method for estimating u at the bulkhead, which is detailed in the next example.

Example 19.9.4

A drift plug is used to retain backfill, and an estimate of the bulkhead design pressure to retain the plug is needed. If the solid fill pressure is already measured at 74 kPa, find the hydrostatic water pressure, after Smith and Mitchell (1982):

$$Q = (Ws/t)\Big(\big[(1-PD)/PD\big] - 0.5\big\{n/\big[(1-n)Gs\big]\big\}\Big)$$

Assumed in this example is that the fill remains 50% saturated during pour and infiltration of groundwater is negligible.

where

Ws is the mass of dry solids per day (tonnes/day) = 3,600 tonnes/day.

t is the pour time (h/day) = 18 hours.

PD is the pulp density as a decimal = 0.65.

n is the porosity of settled fill = 0.45.

Gs is the specific gravity of solids = 3.56.

Q is the tonnes/h = m^3/h = 84.7 tonnes/h.

S is the length of the drift plug = 3 m.

P is the percolation rate of drift plug material = 3.8 m/h (filled with coarse sand).

A is the total area of drainage drifts = 34.2 m^2 (two drainage bulkheads 3.8 m high by 4.5 m wide).

γ_w is the unit weight of water = 9.81 kN/m^3.

u is the water pressure acting on the drift plug, kN/m^2.

Solution:

$$Q = (Ws/t)\Big(\big[(1-PD)/PD\big] - 0.5\big\{n/\big[(1-n)Gs\big]\big\}\Big)$$

$$Q = \big[(3{,}600\ \text{tonnes/day})/18\ \text{hr}\big]\Big(\big[(1-0.65)/(0.65)\big] - 0.5\big\{0.45/\big[(1-0.45)3.56\big]\big\}\Big)$$

$$Q = (200)\Big(\big[(0.54)\big] - 0.5\{0.229\}\Big)$$

$$Q = (200)(0.425)$$

$$Q = 85\ \text{tonnes/hr}$$

$$u = QS\gamma_w/PA$$

$$u = \left[(85\ \text{tonnes/hr})(3\ \text{m})(9.81\ \text{kN/m}^3)\right]/\left[(3.8\ \text{m/hr})(34.2\ \text{m}^2)\right]$$

$$u = \left[(2{,}502\ \text{tonnes} - \text{kN/m}^2 - \text{hr})\right]/\left[(130\ \text{m}^3/hr)\right] * \text{tonnes} = \text{m}^3$$

$$u = 19.3\ \text{kPa}$$

$$\text{Total bulkhead design pressure} = u + p = 19.3\ \text{kPa} + 74\ \text{kPa} = 93.3\ \text{kPa}$$

Example 19.9.5

Assume that backfill is to be placed in a 40 ft high void with a constrained modulus of 50,000 psi. The length of the void is approximately 50 ft.

Assume a unit weight of surround rock of 140 lbs/ft^3 and a depth of 150 ft. The rock bolted horizon is 8 ft, and the bulk modulus is estimated at 0.39×10^9 lbs/ft^2. Determine if caving is likely to develop. This is after Mitchell (1983) and in part by formulations based on Hetényi (1958). Figure 19.19 has been redrawn from Mitchell (1983).

The following formula and caving approximation will be used:

$$M = (\sigma_v/\lambda^2)\left[(b-a)/c\right]$$

And caving will develop if

$$M > \sigma_h (d_b)^2/6$$

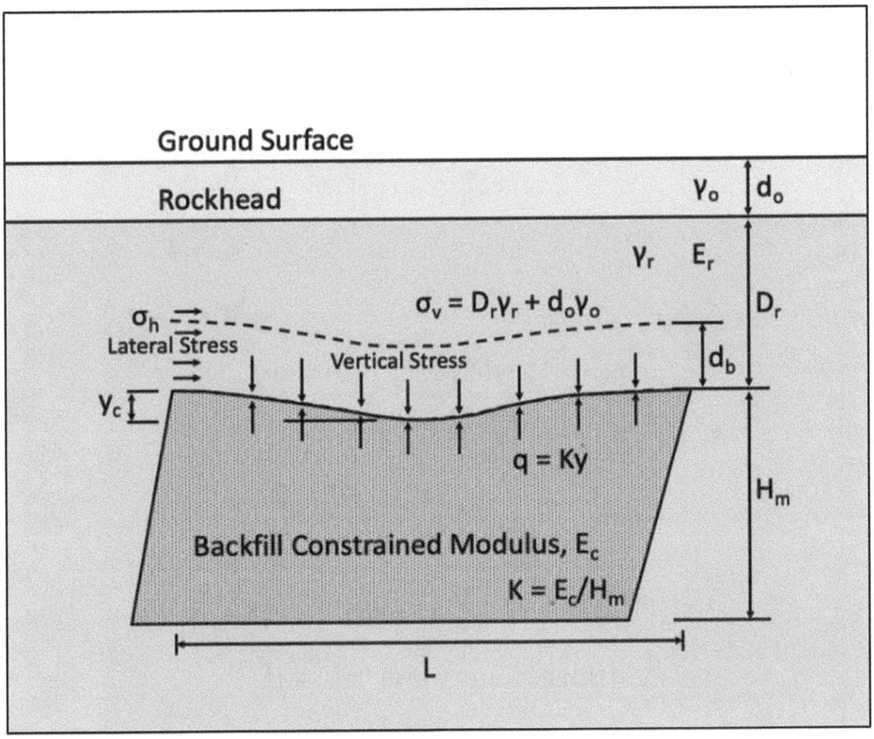

FIGURE 19.19 The backfilled area (Mitchell, 1983).

The following parameters are defined as follows:

$$I = d_b^3/12$$

K=backfill subgrade modulus
E_c=constrained modulus of fill
H_m=maximum thickness of the backfill influenced by loading (q)
E_r=assumed bulk modulus of the rock
d_b=assumed depth of intact or bolted rock strata

Solution:

Step 1:

$$K = E_c/H_m$$

$$K = (50{,}000 \text{ lbs/in}^2)/(40 \text{ ft})(12 \text{ in/1 ft})$$

$$K = 104 \text{ lb/in}^3$$

Step 2:

$$\sigma_v = D_r\gamma_r + d_o\gamma_o$$

Treating the entirety of the overburden as single rock unit, d_o=0 ft:

$$\sigma_v = D_r\gamma_r$$

$$\sigma_v = (150 \text{ ft})(140 \text{ lbs/ft}^3)$$

$$\sigma_v = (21{,}000 \text{ lbs/ft}^2)(1 \text{ ft}^2/144 \text{ in}^2)$$

$$\sigma_v = 146 \text{ lbs/in}^2$$

Step 3:

$$I = d_b^3/12$$

$$I = [(8 \text{ ft})(12 \text{ in/1 ft})]^3/12$$

$$I = 73{,}728 \text{ in}^3$$

Step 4:

$$E_r = (0.39 \times 10^9 \text{ lbs/ft}^2)(1 \text{ ft}^2/144 \text{ in}^2)$$

$$E_r = 2.7 \times 10^6 \text{ lbs/in}^2$$

Step 5:

$$\lambda = [K/(4E_rI)]^{0.25}$$

$$\lambda = [(104 \text{ lb/in}^3)]/[4(2.7 \times 10^6 \text{ lbs/in}^2)(73{,}728 \text{ in}^3)]^{0.25}$$

$$\lambda = [(104 \text{ lb/in}^3)/(7.96 \times 10^{11} \text{ lbs} - \text{in})]^{0.25}$$

$$\lambda = [1.306 \times 10^{-10}/\text{in}^4]^{0.25}$$

$$\lambda = 3.38 \times 10^{-3} \text{in}^{-1}$$

Step 6: Some of these are hyperbolic Sine and Cosines, and working in radians instead of degrees as shown in Hetényi (1958), the following is found:

$$\lambda L/2 = \left(3.38\times10^{-3}\right)\left[\left(50\text{ ft}\right)\left(12\text{ in/1 ft}\right)\right]/2 = 1.01$$

$$a = \sinh\left(\lambda L/2\right)\cos\left(\lambda L/2\right)$$

$$a = \sinh(1.01)\cos(1.01)$$

where

$$\sinh(x) = \left(e^{x} - e^{-x}\right)/2$$

$$\sinh(1.01) = \left(e^{1.01} - e^{-1.01}\right)/2$$

$$\sinh(1.01) = (2.75 - 0.36)/2$$

$$\sinh(1.01) = 1.19$$

$$\cos(1.01) = 0.532$$

$$a = (1.19)(0.532) = 0.633$$

$$b = \cosh(\lambda L/2)\sin(\lambda L/2)$$

$$b = \cosh(1.01)\sin(1.01)$$

where

$$\cosh(x) = \left(e^{x} + e^{-x}\right)/2$$

$$\cosh(1.01) = \left(e^{1.01} + e^{-1.01}\right)/2$$

$$\cosh(1.01) = (2.75 + 0.36)/2$$

$$\cosh(1.01) = 1.555$$

$$\sin(1.01) = 0.847$$

$$b = (1.555)(0.847) = 1.317$$

$$c = \sinh(\lambda L)\sin(\lambda L)$$

$$c = \sinh(2.05)\sin(2.05)$$

$$\sinh(x) = \left(e^{x} - e^{-x}\right)/2$$

$$\sinh(2.03) = \left(e^{2.03} - e^{-2.03}\right)/2$$

$$\sinh(2.03) = (7.61 - 0.131)/2$$

$$\sinh(2.03) = 3.74$$

$$c = (3.74)(0.887) = 3.3$$

Step 7:

$$M = (\sigma_v/\lambda^2)[(b-a)/c]$$

$$M = \left[(146\ \text{lbs/in}^2)/(3.38\times10^{-3}\text{in}^{-1})^2\right][(1.317-0.633)/(3.3)]$$

$$M = \left[(146\ \text{lbs/in}^2)/(1.14\times10^{-5}\text{in}^{-2})\right][0.2]$$

$$M = \left[12.78\times10^6\ \text{lbs}\right][0.2]$$

$$M = 2.56\times10^6\ \text{in}-\text{lbs/in}$$

Step 8: Check if caving will develop. Assume $\sigma_h = 3\sigma_v = (3)146\ \text{lbs/in}^2 = 438\ \text{lbs/in}^2$:

$$M > \sigma_h (d_b)^2/6$$

$$2.56\times10^6\ \text{in}-\text{lbs/in} > \left\{(438\ \text{lbs/in}^2)\left[(8\ \text{ft})(12\ \text{in/1 ft})\right]^2\right\}/6$$

$$2.56\times10^6\ \text{in}-\text{lbs/in} > (438\ \text{lbs/in}^2)\left[9{,}216\ \text{in}^2\right]/6$$

$$2.56\times10^6\ \text{in}-\text{lbs/in} > 0.67\times10^6\ \text{lbs}\quad \text{Caving will develop}$$

This example and prediction may be best for academic purposes only due to the difficulty in accurately estimating some of the material parameters needed and because the formulations represent an elastic analysis whereas caving is non-elastic.

Example 19.9.6

A confined block of backfill has the following properties and rests on a potential sliding plane, as shown in Figure 19.20. Determine the factor of safety against the plane failing. Originally from Mitchell (1983) but also featured in Brady and Brown (2004).

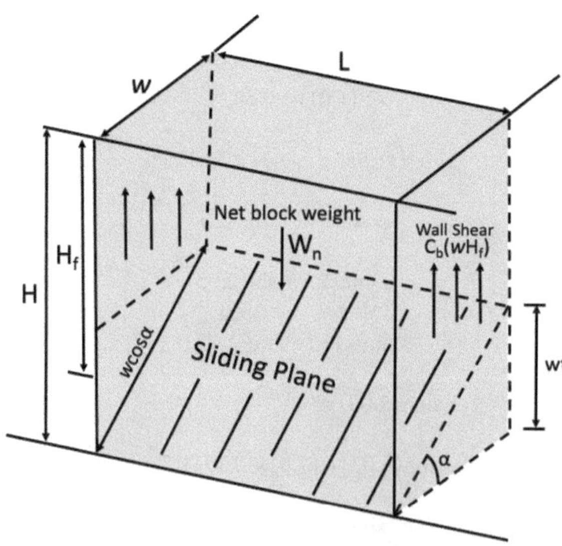

FIGURE 19.20 The backfill block with the potential failure surface (Mitchell, 1983).

Given:

$$H = \text{total height} = 20\ \text{m}$$

$$w = \text{width} = 2\ \text{m}$$

$$L = \text{length} = 10\ \text{m}$$

$$\alpha = 45°$$

$$\gamma = 20\ \text{kN/m}^3$$

$$\varphi' = 30°$$

Solution:

Step 1: Find the net weight (W_N) of the block:

$$W_N = wH_f(\gamma L - 2c')$$

where

$$H_f = H - 1/2w\tan\alpha$$

$$H_f = 20\ \text{m} - (½)(2\ \text{m})\tan 45°$$

$$H_f = 19\ \text{m}$$

$$c' = [\gamma LH_f][\sin 2\alpha/2(L + H_f\sin 2\alpha)]$$

$$c' = [(20\ \text{kN/m}^3)(10\ \text{m})(19\ \text{m})][\sin 2(45°)/2(10\ \text{m} + (19\ \text{m})\sin 2(45°))]$$

$$c' = [(3{,}800\ \text{kN/m})][1/58]$$

$$c' = 66\ \text{kN/m}^2$$

$$W_N = (2\ \text{m})(19\ \text{m})[(20\ \text{kN/m}^3)(10\ \text{m}) - 2(66\ \text{kN/m}^2)]$$

$$W_N = 2{,}584\ \text{kN}$$

Step 2: Determine the factor of safety (F):

$$F = (\tan\varphi'/\tan\alpha) + \{2c'L/[H_f(\gamma L - 2c')\sin 2\alpha]\}$$

$$F = (\tan 30°/\tan 45°) + \left([2(66\ \text{kN/m}^2)(10\ \text{m})]/\{19\ \text{m}[(20\ \text{kN/m}^3)(10\ \text{m}) - (2)(66\ \text{kN/m}^2)]\sin 2(45°)\}\right)$$

$$F = (0.577) + \left([(1{,}320\ \text{kN/m})]/\{19\ \text{m}[(200\ \text{kN/m}^2) - (132\ \text{kN/m}^2)]\}\right)$$

$$F = 0.577 + 0.359$$

$$F = 0.936 \quad \text{So too low}$$

Example 19.9.7

This is from Brady and Brown (2004) on the use of tendons of overhand cut-and-fill. From inspection via drillholes in the crown of a stope, there is fracturing at a depth of 1.5 m. If the unit weight of the rock mass in 30 kN/m³, each tendon has a yield load capacity of 260 kN, and when the grout fails against the tendon a shear surface of diameter, d, is produced that is 50 mm then determine the weight of the rock, W, as a function of spacing, a, per tendon. A factor of safety for 1.5 is needed for the load capacity of the bolts. The shear strength of the grout is 1.4 MPa.

Solution:

$$FS = 260\ \text{kN}/\text{Load}_{\text{allowable}}$$

$$1.5 = 260\ \text{kN}/\text{Load}_{\text{allowable}}$$

$$\text{Load}_{\text{allowable}} = 173.3\ \text{kN}$$

$$W = \gamma a^2 h$$

$$W = \left(30\ \text{kN/m}^3\right)\left(1.5\ \text{m}\right)a^2$$

$$173.3\ \text{kN} = \left(30\ \text{kN/m}^3\right)\left(1.5\ \text{m}\right)a^2$$

$$a = 1.96\ \text{m}$$

Check if the grout annulus passing through a potentially unstable block can support a deadweight load:

Maximum shear resistance of grout column:

$$S = \pi dh\left(1.4\ \text{MN}\right)$$

$$S = \pi\left(0.05\ \text{m}\right)\left(1.5\ \text{m}\right)\left(1.4\ \text{MN}\right)$$

$$S = 329.9\ \text{kN}$$

Factor of safety against grout column shear failure:

$$FOS = 329.9\ \text{kN}/173.3\ \text{kN} = 1.90$$

Example 19.9.8

Estimate the height of a stable free face of cemented silicate backfill in a narrow hard-rock tabular gold mine stope as shown in Figure 19.21 (Blight and Spearing, 1995).

$$h = 2c'/F\sin\alpha\left[\rho a\cos\alpha\right]\left[1 - \cot\alpha\left(\tan\varphi'/F\right)\right]$$

where

c' is the effective cohesion parameter from the Coulomb strength equation below = 10 kPa.

φ' is the effective friction angle from the Coulomb strength equation below = 30°.

F is the factor of safety of the face against failure = 1.2.

ρ is the density of the backfill in kg/m³ = 1,800 kg/m³.

a is the vertical acceleration capped at a minimum value of g. This could be larger due to blasting and seismic action = 2g = 2(9.81 m/s²) = 19.6 m/s².

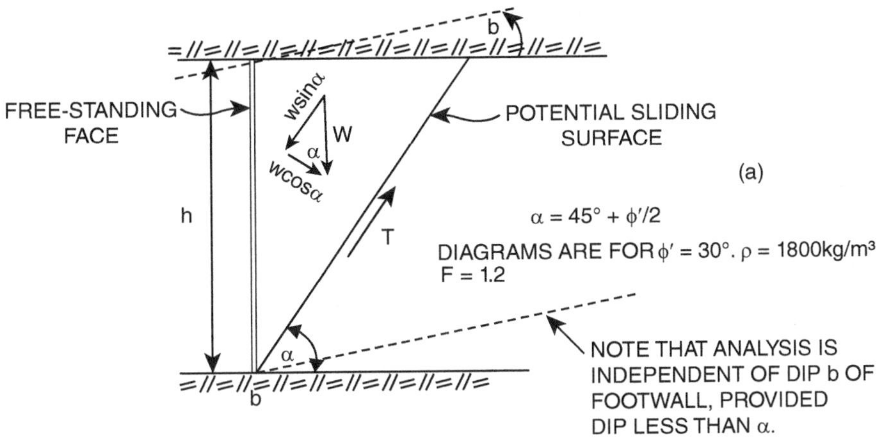

FIGURE 19.21 The silicate backfill with the free face in a narrow tabular slope (Blight and Spearing, 1995).

α is the angle of potential sliding surface with horizontal, taken at the toe of the face=60°:

$$s = c' + (\sigma - u)\tan\varphi'$$

where

s is the shear strength.
σ is the normal stress on plane of shearing.
u is the pore water pressure.

Solution:

$$h = 2c'/F\sin\alpha\left[\rho a\cos\alpha\right]\left[1 - \cot\alpha\left(\tan\varphi'/F\right)\right]$$

$$h = \left[2\left(10\ \text{kN/m}^2\right)\right]/(1.2)\sin 60°\left\{\left(1{,}800\ \text{kg/m}^3\right)\left(19.6\ \text{m/s}^2\right)\cos 60°\left[1 - \cot 60°(\tan 30°/1.2)\right]\right\}$$

$$h = \left[20\ \text{kN/m}^2\right]/(1.04)\left\{\left(1{,}7640\ \text{N/m}^3\right)\left[1 - 0.58(0.48)\right]\right\}$$

$$h = \left[20\ \text{kN/m}^2\right]/(1.04)\left\{12{,}718\ \text{N/m}^3\right\}$$

$$h = \left[20\ \text{kN/m}^2\right]/\{13{,}226\ \text{N/m}^3\}$$

$$h = \left[20{,}000\ \text{N/m}^2\right]/\{13{,}226\ \text{N/m}^3\}$$

$$h = 1.5\ \text{m}$$

REFERENCES

Atlas Copco (2000) *Guide to Underground Methods and Applications*. Atlas Copco, Sweden.

Belem, T. and Benzaazoua, M. (2007) Design and application of underground mine paste backfill technology. *Geotech. Geol. Eng*. 26–175, 147–174.

Blight, G. and Spearing, A.J.S. (1995) The properties of cemented silicated backfill for use in narrow, hard-rock tabular mines. *J. S. Afr. Inst. Mining Metallurgy*. 96(1), 17–28.

Brady, B.H. and Brown, E.T. (2004) *Rock Mechanics: For Underground Mining*. Kluwer Academic Publishers, Dordrecht, The Netherlands.

Chen, X., Guo, L., Zhou, Y., Xu, W. and Zhao, Y. (2023) Remediation of grassland subsidence and reduction of land occupation with tailings backfill technology: a case study of lead-zinc mine in Inner Mongolia, China. *Front. Environ. Sci.* 11, 1183945.

De Souza, E. (2010) *Room and Pillar*. The Robert M. Buchan Department of Mining, Queens University, Kingston, Ontario, Canada.

Goldfields (2009) Mineral resources and mineral reserves. South Deep Gold Mine-Technical Short Report. www.goldfields.com/reports/rr_2009/tech_south.php/tech_south.php

Gürtunca, R.G., Jager, A.J., Adams, D.J. and Gonlag, M. (1989) In situ measurements of backfills and surrounding rockmass behaviour in South African Gold Mines. In *4th International Symposium on Mining with Backfill*. Ontario, Canada.

Hamrin, H. (2001) Underground mining methods and applications. In: Hustrulid, W.A. and Bullock, R.L. (eds) *Underground Mining Methods: Engineering Fundamentals and International Case Studies*. Society for Mining Metallurgy and Exploration, Littleton, pp. 3–14.

Hetényi, M (1958) *Beams on Elastic Foundations*. University of Michigan Press, Chicago, IL.

Jain, S. (2011) Filling. www.slideshare.net/slideshow/filling/9844256

Mitchell, R. (1983) *Earth Structures Engineering*. Allen & Unwin Inc., Boston, MA.

Multotec (2024) Section through a hydrocyclone. https://www.multotec.com/en/news-articles/multotec-hydrocyclones-Australia

Smith, J.D. and Mitchell, R.J. (1982) Design and control of large hydraulic backfill pours. *CIM Bull.* 75(838), 102–111.

Spearing, A.J.S. (1989) The modification of Witwatersrand gold plant tailings for optimum application as an underground support medium. Thesis for MS in Civil Engineering. University of the Witwatersrand, Johannesburg, South Africa.

Spearing, A.J.S., Ma, L. and Ma, C. (2023) *Mine Design, Planning and Sustainable Exploitation in the Digital Age*. CRC Press, London, UK.

Steward, N.R. and Spearing, A.J.S. (1992) Combating pipeline wear – An advancing technology. *J. S. Afr. Inst. Min. Metal.* 92, 6.

Thomas, E.G. Nantel, J.H. and Notely, K.R. (1979) *Fill Technology in Underground Metalliferous Mines*. International Academic Services Limited, Canada, p. 293.

Williams, T.J., Brady, T.M., Bayer, D.C., Bren, M.J., Paklanis, R.T., Marjerison, J.A. and Langston, R.B. (2007) Underhand Cut and Fill Mining as Practiced in Three Deep Hard Rock Mines in the United States. NIOSH, CDC NIOSHTIC2 Number: 20033309.

Yilmaz, E. and Guresci, M. (2017) *Design and Characterization of Underground Paste fill. Paste Tailings Management*. Springer, Chem, 111–143.

Zhang, J., Li, M., Taheri, A., Zhang, W., Wu, Z. and Song, W. (2019) Properties and application of backfill materials in coal mines in China. *Minerals*. 9, 53.

Zhang, J., Zhou, Y.J. and Huang, Y.L. (2012) Integrated technology of fully mechanized solid backfill mining. *J. China Coal Soc*. 11, 10–13.

20 Specialized Support Methods and Applications

20.1 WATER SEALING

Significant water ingress is common and dangerous in mining and underground civil projects, and an example is shown in Figure 20.1.

The aim of water sealing is to:

- Stop leaks through structures.
- Control (eliminate or reduce) groundwater flows.
- Stabilize granular soils and fractured rock.
- Be cost effective.
- Seal the water when managing the water is not effective or feasible.

The safest and most cost-effective way of water sealing is to locate it before it is intersected in the excavation itself. This is achieved using probe drilling ahead of the advancing face (Figure 20.2). Handheld drilling using pneumatic jack-hammer-type drills are generally not suitable for probe drilling as the hole length tends to be too short.

Knowledge of local and regional geology and hydrology is critical to avoid potentially serious water ingress. The location of abandoned mine workings is also important.

The early location of water is essential to controlling inrushes, even if the chance of locating serious water ingress appears low based on pre-development results. As much relevant data must be obtained to help guide the water ingress management.

FIGURE 20.1 15-bar water pressure intersected by probe drilling. (Reproduced with kind permission from Knut Garshol.)

DOI: 10.1201/9781032704845-20

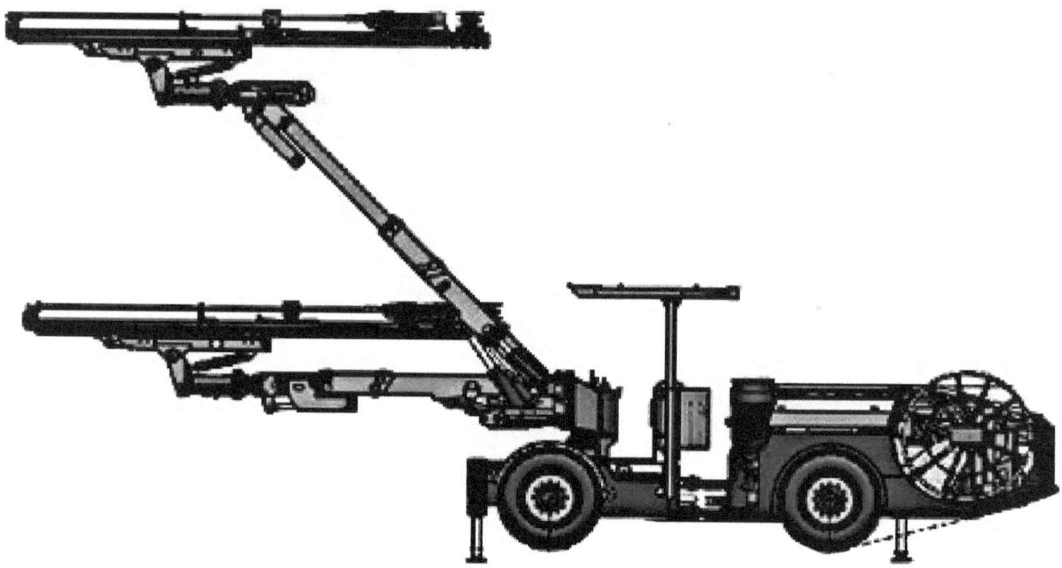

FIGURE 20.2 Example of equipment suitable for probe hole drilling equipment (Split Mountain Machine and Equipment, 2025).

20.1.1 Common Water Sealing (Injection) Materials

The following are the more commonly used water sealing materials. As part of a risk assessment, any potential environmental impact of using the selected water sealing product needs to be considered.

i. Cement

Portland cement is often selected because only the material cost of water sealing materials is considered. The issues with the use of cement only are the relatively slow setting and strength gain as well as the relatively large particle diameter compared to the other potential materials.

ii. Micro-cement

Micro-cements are ground much finer than Portland cement, so they set faster and can penetrate fractures further.

iii. Polyurethane (PU) foam of which there are two basic types: a single-component system and a two-component system:

Single component:

- Pre-polymer (methyenediphenyl diisocyanate – MDI) plus amine accelerator.
- Needs water to react.
- Pre-polymer reacts with water to form a stable polymer.
- Carbon dioxide is a by-product that forms as small, closed bubbles in the resin that set to form a rapidly expanding closed foam. The set time and the amount of expansion can be fixed and controlled.
- Foaming expansion helps seal the water.

Two components:

- Pre-polymer (isocyanate) plus polyol.
- Does not need water to react.
- Components, when mixed, react to form a stable polymer that can be rigid or flexible and be foamed if needed.

- Can be made to set faster than single-component PU.
- Tends not to break down so withstands high water flows and washout better.

iv. Acrylic grout

Acrylics are long-chain polymers and are relatively inexpensive. With all water sealants, care must be taken to ensure that they are safe and environmentally friendly.

v. Colloidal silica

This material is naturally very fine, so it has good penetration into fine cracks, and it is environmentally friendly. The good penetration is because colloidal silica exhibits Newtonian flow (it exhibits a lack of initial resistance to shearing), whereas cementitious grouts exhibit Bingham behavior (having a shear stress threshold that needs to be overcome).

20.1.2 Handling Water Intersected by a Borehole

This is relatively cheap, quick, and easy to control. In this application, the following is done:

- A packer can be placed into each strategically drilled hole.
- The packers are set in the holes (expanded in place).
- The packer valves are closed, and this stops any water ingress.
- The holes can be injected one by one until newly drilled probe boreholes show that the area is water tight. A typical packer is shown in Figure 20.3.

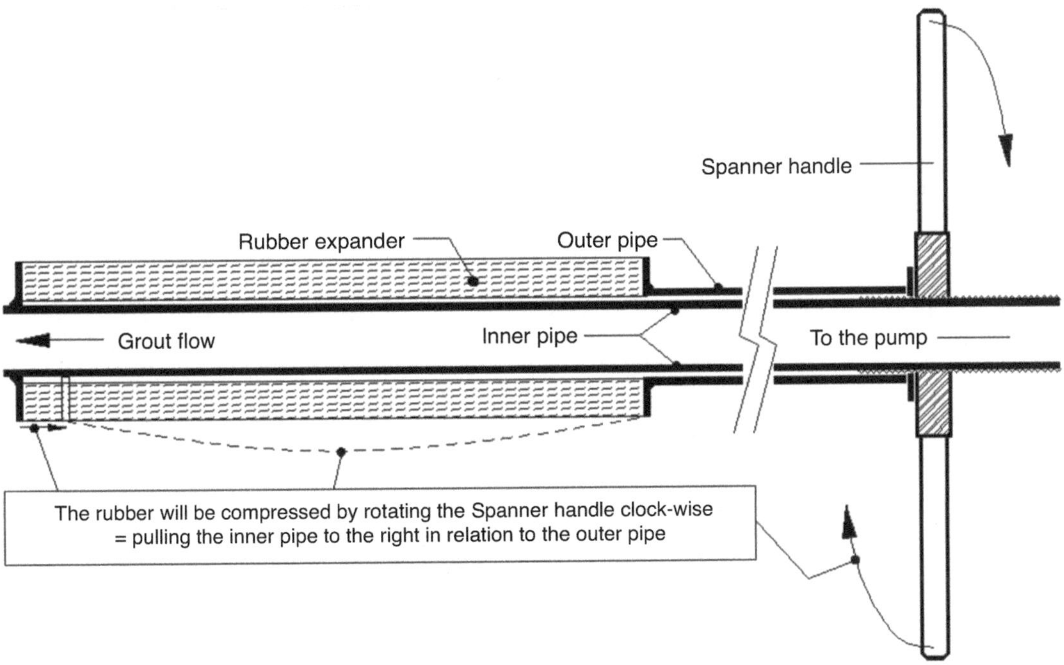

FIGURE 20.3 An expanding hand-installed rubber packer to control water in a borehole (Garshol, 2007).

20.1.3 Typical Grouting Equipment

Using the correct equipment for water sealing underground is essential. It must be able to deliver the desired volume and pressure, safely and reliably.

Where any cement-based product is used, the water content must be minimized and controlled, so a colloidal grout pump is useful. An example is shown in Figure 20.4.

Figure 20.5 shows a typical two-component water sealing pump often used for PU applications. With some two-component PU materials, the product must be heated in the pump, and hence, extra caution is required.

20.1.4 Typical Pre-Injection Methods

This is the safest method as it creates a waterproof curtain or seal, in case potential water is intersected. Examples of drilling and injection patterns are shown in Figure 20.6.

The time and cost to develop excavations, especially main access development, means that in general it is best to pre-inject so to avoid costly reactive control measures and longer-term rehabilitation.

FIGURE 20.4 Colloidal pump for cement grouting (water sealing). (Reproduced with kind permission of Chemgrout.)

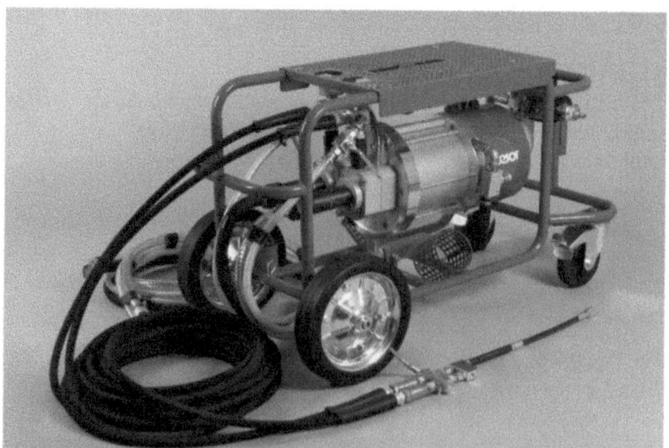

FIGURE 20.5 Two-component pump for resin grouting (water sealing). (Reproduced with kind permission of Normet.)

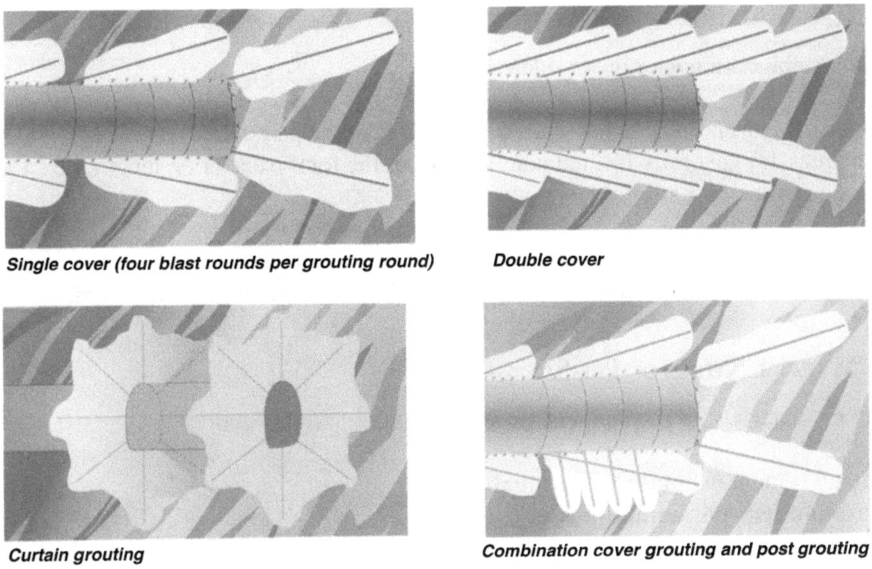

FIGURE 20.6 Typical injection patterns (Garshol, 2007).

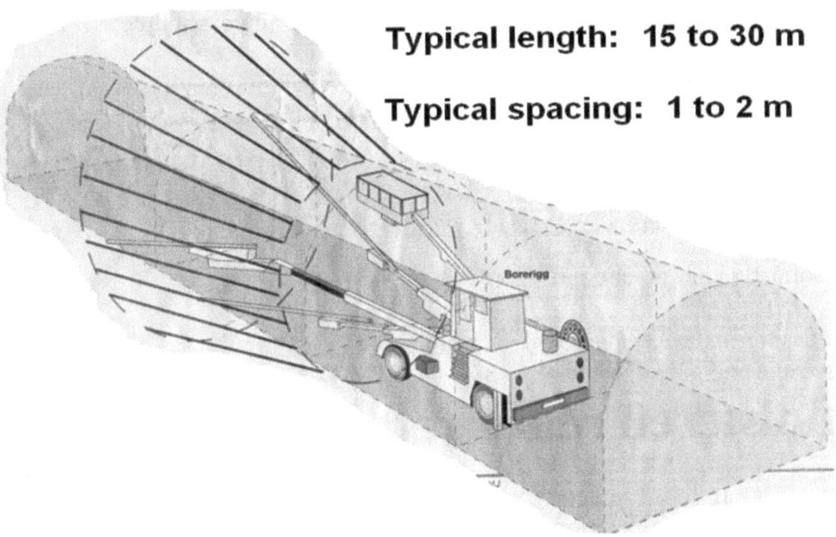

FIGURE 20.7 Probe drilling from a drilling rig (Holter and Zámečník, 2009).

Probe drilling is obviously essential in many mine development operations, and Figure 20.7 shows an example. The probe drilling is done in advance of the face, and the probe cover is kept ahead of the face (by a couple of rounds at least, if conventionally blasted). This ensures that the excavation advance is always protected from sudden water ingress because of the probe drilling in advance of the excavation development.

20.2 GROUND CONSOLIDATION AND SPILING

Water is always unwelcome in all underground operations!

20.2.1 Typical Applications

Pre-injection/stabilization is only considered when:

- Ground "stand-up time" is too short for standard support systems.
- Such areas are generally limited compared with the total excavation length, i.e., a specific geological zone or feature.
- Rock is generally weak or badly fractured with poor cohesion.
- Rock is generally wet, but not always.
- Pre-stabilization is often used to re-open/rehabilitate areas already collapsed.

20.2.2 Material Needs for Pre-Stabilization

The following materials and equipment (or a combination) are needed:

- Portland cement (admixtures essential).
- Micro-fine cements (admixtures essential).
- Chemical grouts.
- Drainage mechanism for pressure release.
- Injection bore holes.
- Application equipment and fittings.

The method of application depends on what is the main water ingress source:

- Permeation of pores, cracks, and joints.
- Fracturing and compaction.
- Void filling.

20.2.3 Selection Criteria and Possible Products to Consider

Ground consolidation has potential if the rock has:

- Clean fissures in broken hard rock (cementitious).
- Coarse sand and gravel (cementitious).
- Silty and sandy ground (no clay) using acrylates.
- Large void filling (foamed cement mix or PU foam).

It has little potential if the ground has:

- High clay and silt (hence, permeability is very low) and hydro fracturing is difficult.
- Difficulty keeping boreholes open as they tend to collapse which could be due to heavy fracturing for example.
- Large conductivity contrast and water.
- Mix of clay and silt with large open channels.

20.2.4 The Use of Spiling Bolts with Pre-Stabilization

Generally, a self-drilling anchor is used because the rock is too weak and the hole collapses, or shears closed, if it was drilled conventionally. This is because the removal of the drill rods causes the collapse or shear, so the rock anchor cannot be installed. There are many self-drilling bolts, but as an example, only one will be discussed.

The GSA anchor system and spiling rockbolt is shown in Figure 20.8 (Langer, 1997).

In operation, the following is needed (as shown on Figure 20.9):

FIGURE 20.8 The GSA bolt (Langer, 1997).

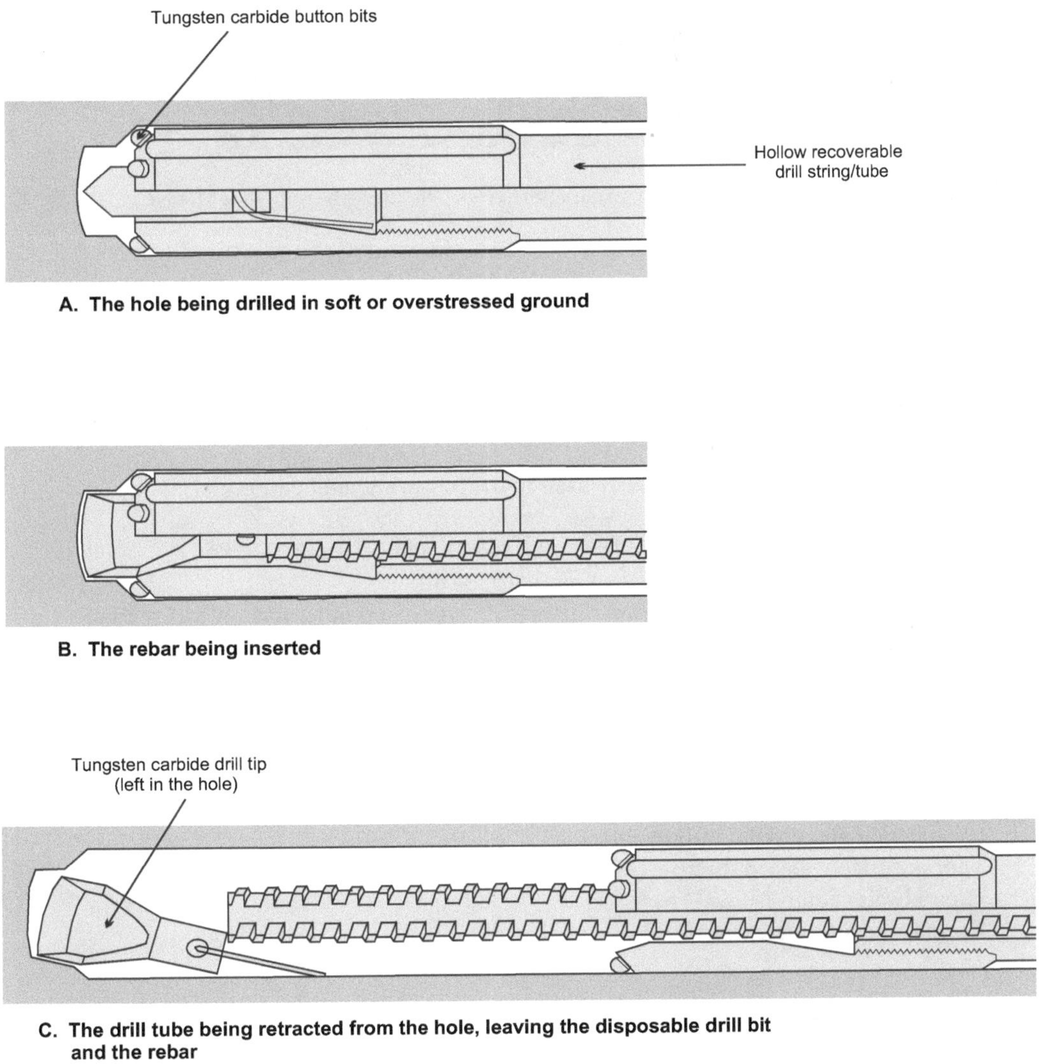

A. The hole being drilled in soft or overstressed ground

B. The rebar being inserted

C. The drill tube being retracted from the hole, leaving the disposable drill bit and the rebar

FIGURE 20.9 Using the GSA bolt (Langer, 1997).

FIGURE 20.10 The good thick concrete grouting material. (Photograph from Tom Melbye.)

- Drilling to full design depth.
- Insertion of 25 mm rebar bolt inside the drilling tool, or some other bolt material.
- Grout injection through annular spaces simultaneous to the drilling tool retraction.

A good thick cement mortar (i.e., low water content) that will not run out of the hole and will set quickly is needed for such applications. An example of a good cementitious (thixotropic) grout is shown in Figure 20.10.

Self-drilling rock anchors are typically used in the following applications:

- Where borehole stability is a problem (very short stand-up time).
- Drilling through any mixture of ground (hard to soft).
- Drilling into gravel and sand mixes, "running" dykes, and other unstable formations.
- Drilling in hard rock under very high stresses, so brittle fracturing is an issue that could block the drill hole.

Where rock anchors are needed, the following can be considered: rebars, cables, and glass fiber bolts, especially for spiling bolts under extreme conditions. They are used for:

- Compaction and ground improvement.
- Very accurately placed micro-cement or acrylic grout.
- Groundwater control.
- Drainage and/or injection.

20.2.5 Spiling Bolts and Forepoling

Spiling bolts are used to support ahead of the advancing face when the strata are very weak and have little or no stand-up time. After the blast, shotcrete is typically applied. They are typically drilled in a close fan at about 10°–20° outward from the excavation perimeter. In extreme cases of instability, spiling bolts can be drilled into the face and the face shotcreted as a temporary support to permit drilling and blasting or a suitable non-explosive method. The spiling bolts can be self-drilling or rebars conventionally grouted into the rock.

Forepoling is similar in application to spiling but is used when the strata is even weaker and the forepoles are simply forced into the strata ahead of the excavation (and not grouted). Examples of forepoles are sharped timber, fiberglass, or steel poles and steel sections, similar to steel straps frequently used on coal mines (Figure 20.11).

Figure 20.12 shows a section through an excavation using foreploes (the same process for spiling).

Figure 20.13 shows spiling bolts and a reinforcing girder that will be shotcreted over as the excavation is developed.

Figure 20.14 shows forepoling using steel sections.

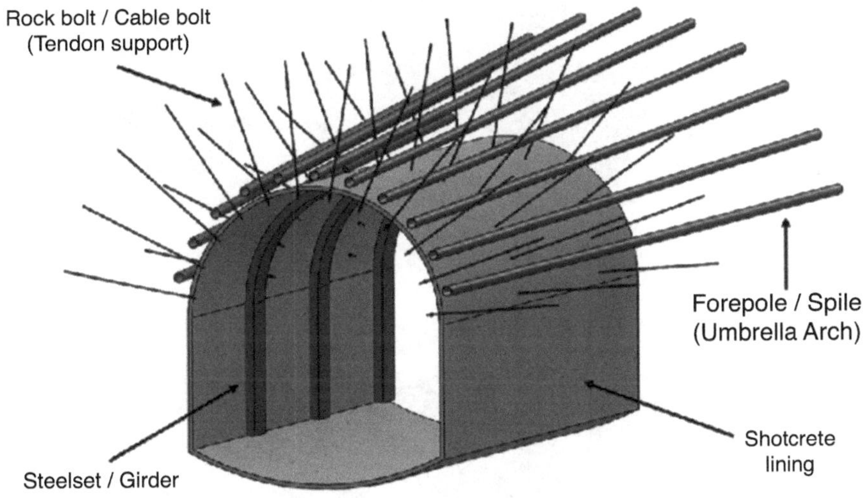

FIGURE 20.11 Forepoling example where the spiling bolts would be grouted into the holes typically (Forbes et al., 2018).

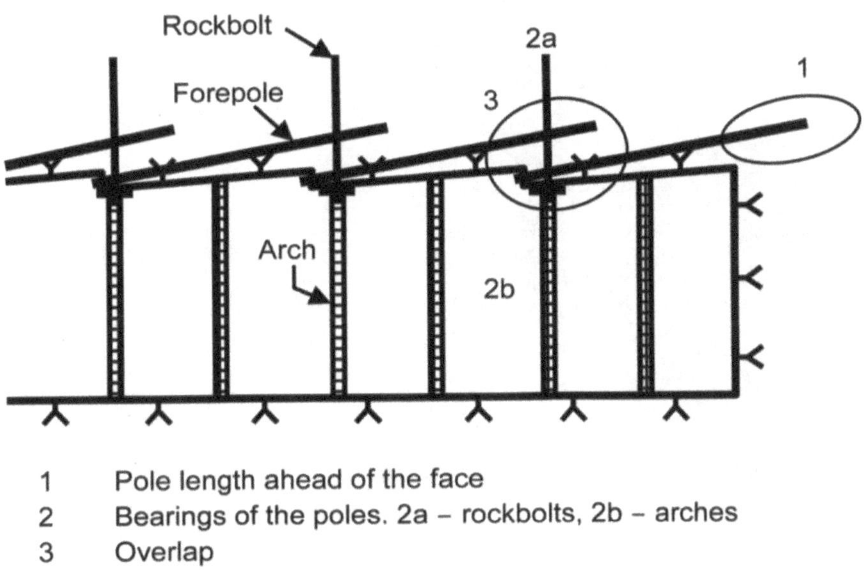

FIGURE 20.12 Section through a tunnel using forepoling (Li, 2017).

FIGURE 20.13 Underground spiling example also using steel girders (DSI Underground, 2024).

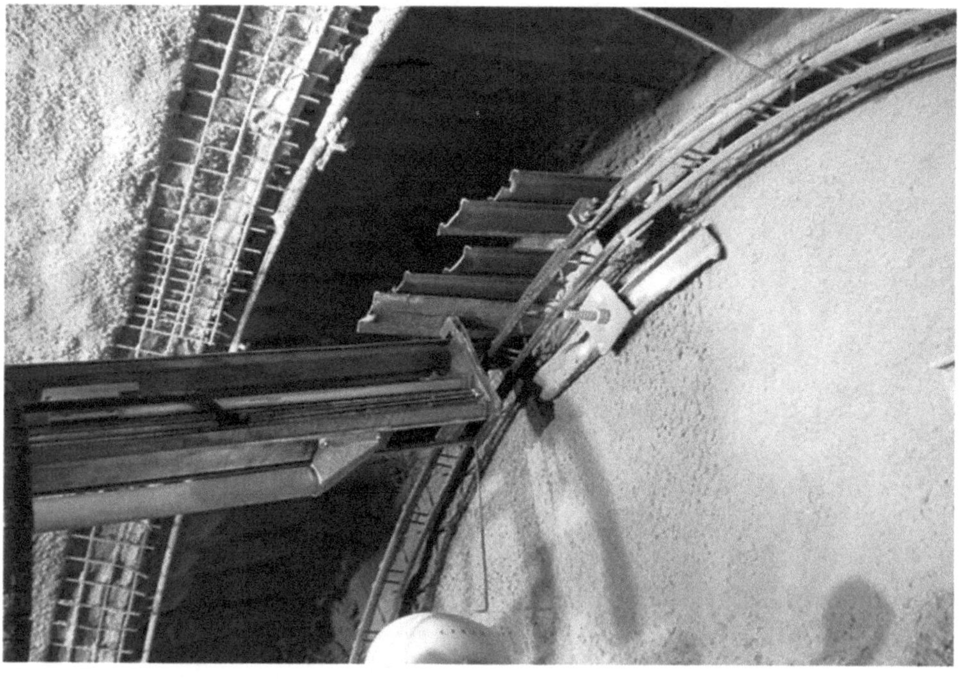

FIGURE 20.14 Installing forepoles (DSI Underground, 2024).

20.3 GROUND FREEZING

Ground freezing is the use of refrigeration to convert in situ pore water to ice. The ice then acts as an effective cement or glue, bonding together adjacent particles of soil, gravel, or blocks of rock to increase their combined strength and make them impervious. Ice is relatively strong and self-supporting. It is more common in underground civil construction but has applications in mining too. Much of the following is from Keller North America.

20.3.1 Ground Freezing Applications

- Temporary underpinning for strengthening the base or foundation of construction on very weak strata.
- Temporary support for an excavation with little or no stand-up time, before permanent support can be installed.
- Prevention of water flow into an excavation that is being developed.
- Temporary slope stabilization.

20.3.2 Feasibility Evaluation Considerations

- Excavation geometry and the planned direction.
- Soil and rock properties around the length of the planned excavation.
- Local hydrology.
- Future stress changes or external conditions.
- Sensitivity of adjacent structures and utilities.
- Time schedule for installation and subsequent operation.
- Availability of electric power and water services.
- Meaningful local community involvement.

20.3.3 Ground Freezing Design

- Refrigeration system geometry from the plant to the excavation is finalized.
- The determination of the thermal properties of the soil and rock that will be intersected during the excavation development is established using lab and field testing.
- The freezing rates in the different strata are determined.
- The plant energy requirements are finalized.
- Safety and environmental considerations and constraints are identified and mitigated where possible.

20.3.4 Structural Analyses

- Structural geometry of the excavation including the cross section and length.
- Selection of the material properties required, especially when frozen.
- Analytical model design and calibration that must be revisited as more reliable site data become available.

20.3.5 Design Considerations

- The protection from exposed frozen material is important especially for the workforce.
- The refrigerant distribution system is designed with safety being paramount.
- Instrumentation and quality management is introduced with the emphasis on safety.
- The handling, removing, or recycling of waste is implemented.

20.3.6 Plant Mobilization and Construction

- Site access and infrastructure are safely established.
- Services are laid on.
- The workforce is trained effectively.
- The holes are drilled, and the pipes for freezing pipe are installed. Drilling accuracy is important.
- Ensuring that the project can be completed on time and on budget (conditions in situ are seldom as planned).
- Project activities are scheduled and undertaken based on the project timeline.

20.3.7 Excavation Development

- Plant availability and utilization that needs to meet or exceed the design.
- Effective safety, QC, and cost control systems implemented.
- Freezing successfully completed within planned schedule.
- Both temporary and permanent support working effectively, as designed or modified as needed.

20.3.8 Demobilization

- Removal of all infrastructure and any waste created.
- Long-term quality assurance and monitoring implemented.

Figures 20.15 and 20.16 show examples of ground freezing in a tunnel and a shaft, respectively. Figure 20.17 shows an example of a ground freezing layout.

FIGURE 20.15 Ground freezing infrastructure on surface. (Reproduced with kind permission of Keller North America.)

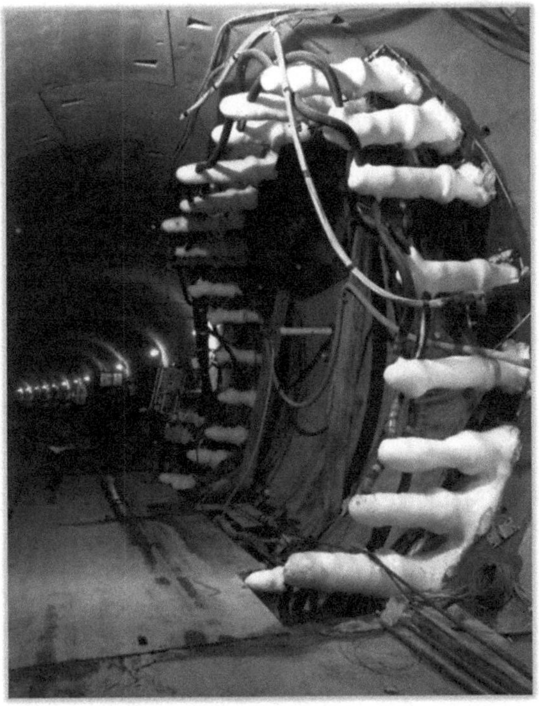

FIGURE 20.16 An excavation showing the frozen ground. (Reproduced with kind permission of Keller North America.)

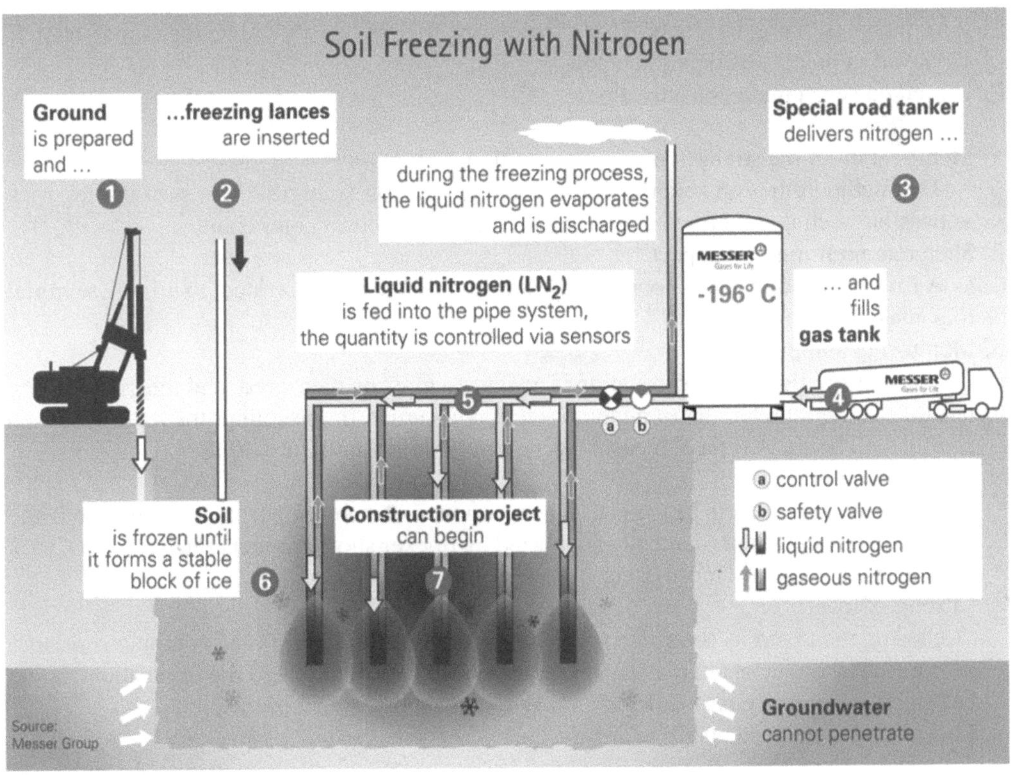

FIGURE 20.17 Example of a ground freezing layout (Hamdan, 2019).

20.4 NEW AUSTRIAN TUNNELING METHOD (NATM)

20.4.1 The Development of NATM

The background to the development of NATM is as follows (Sauer, 1990):

- 1811 – Brunel invents the first circular shield.
- 1848 – Fast setting concrete first used by Wejwanow.
- 1872 – Rziha uses steel to replace timber.
- 1908 – Akeley develops the first dry shotcrete machine.
- 1948 – Rabcewicz develops the dual-lining system.
- 1954 – Bruner uses shotcrete in squeezing ground.
- 1960 – Muller recognizes the importance of systematic deformation measurements.
- 1962 – Rabcewicz presents NATM in the XIII Geomachanics Colloquium in Salzburg, Austria.

The New Austrian Tunneling Method (NATM) was developed between 1957 and 1965 in Austria. It was given its name in Salzburg in 1962, to distinguish it from earlier Austrian tunneling approach. The main contributors to the development of NATM were Leopold Muller, Franz Pacher, and Ladislaus von Rabcewicz (1965).

Von Rabcewicz developed the dual-lining excavation system of an initial liner followed by the final and thicker liner. His concept was to permit some controlled deformation before the application of the final liner so that the stresses in the final liner were reduced (helping with long-term stability).

20.4.2 The Concept of NATM

The main concept is to use the stress of the surrounding rock mass to stabilize the tunnel itself. How this is achieved in practice is detailed below.

It is based on seven main considerations:

i. Mobilization of the strength of the surrounding rock mass

   The method relies on keeping as much of the inherent strength of the surrounding rock as possible such that it effectively becomes the main support component.

ii. Shotcrete preliminary support

   A thin layer is applied as soon as possible after the face is excavated, to minimize initial rock mass deformation.

iii. Monitoring and measuring

   The deformation as the excavation is advanced must be measured, and this requires the installation of complex instrumentation, that is installed in the lining, the rock, and even nearby drill holes. A typical measuring system is shown in Figure 20.18.

iv. Flexible support

   The primary shotcrete lining is thin as mentioned. Active support is typically used, and the excavation can be permanently supported by thicker shotcrete, concrete segments, rock bolts (cable bolts), wire mesh, steel ribs, or a combination.

v. Closing of invert

   Closing the invert as soon as possible to complete a total reinforced continuous permanent lining is important. This is even important in soft rock where the excavation should be sealed as soon as possible. Figure 20.19 shows an invert in a tunnel.

vi. Contractual arrangements (dynamic design)

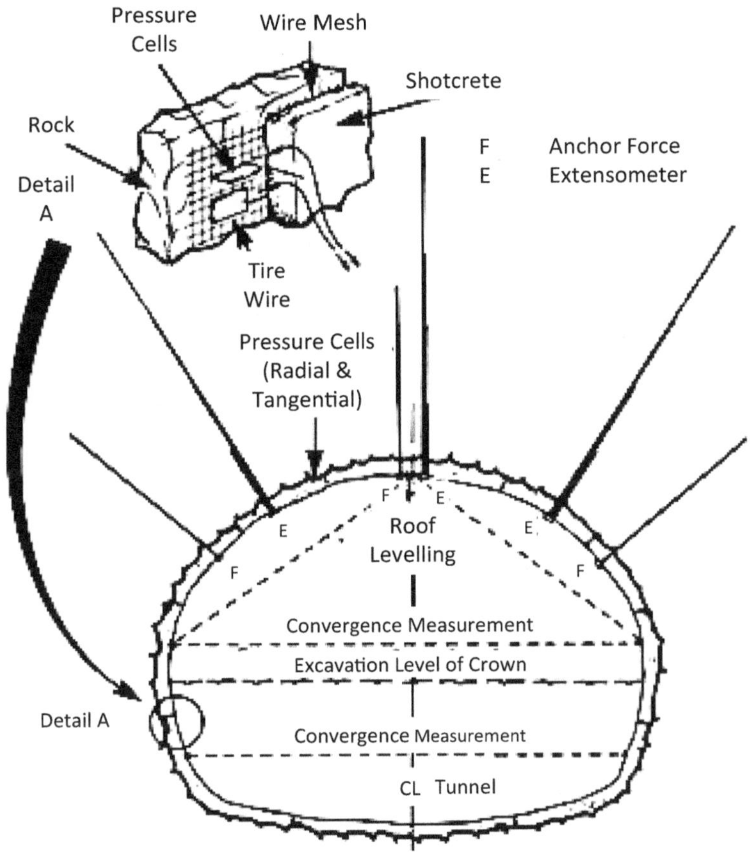

FIGURE 20.18 Typical NATM measuring layout (Sauer and Gold, 1989).

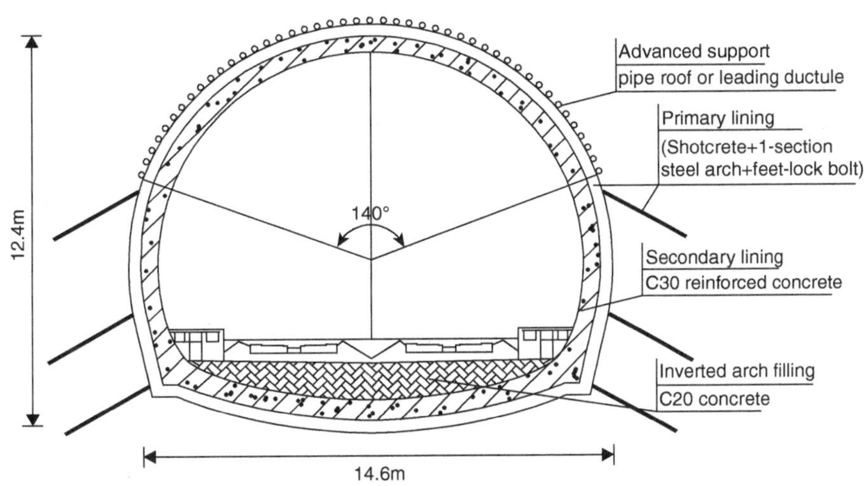

FIGURE 20.19 A tunnel section showing the invert (Liu et al., 2019).

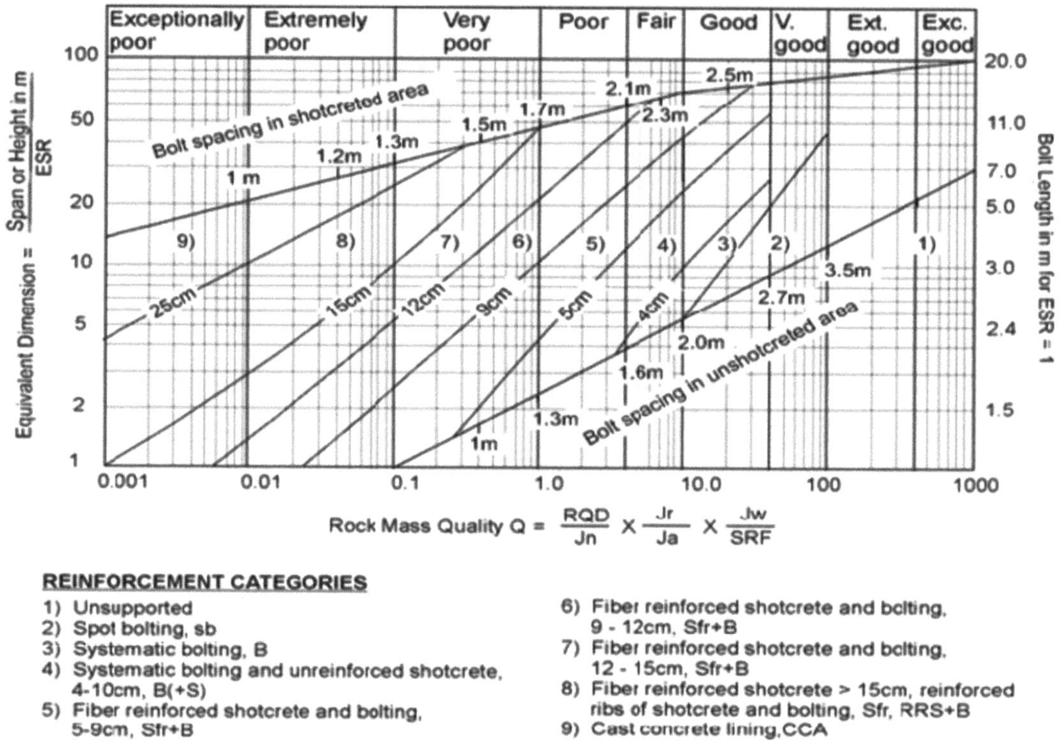

FIGURE 20.20 Support design using the Q system (NGI Handbook, 2022).

NATM is based on in situ monitoring, so changes in support and excavation method can be needed. The contract must therefore be flexible enough for such changes to be possible.

vii. Rock mass classification

This helps determine the support system needed.

These considerations serve as the initial guidelines for tunnel reinforcement. Figure 20.20 shows the support design that is often used for NATM.

The following are the main considerations for the final lining:

- Ground reaction curve.
- Groundwater must be accounted for (consider drainage or sealing).
- Additional support may be needed (such as bolts, spiling bolts, lattice girders, and screen).
- On-going monitoring of especially deformation but also stress monitoring.
- Liner design (using empirical, experiential, computational and/or physical models).

20.4.3 NATM Design

NATM design philosophy is based on technical requirements and external constraints as shown in Figure 20.21.

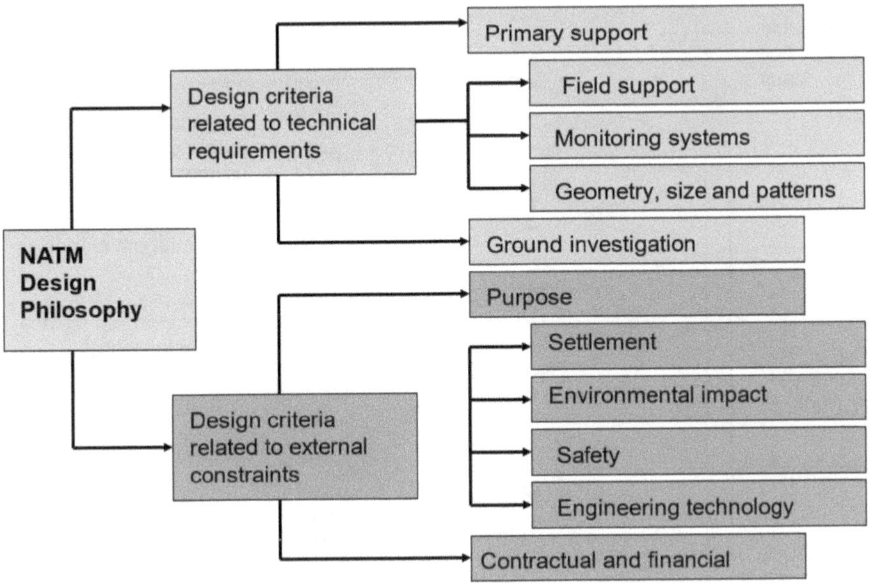

FIGURE 20.21 NATM general design aspects (Panwar, 2021).

20.5 SOIL NAILS

20.5.1 Introduction to Soil Nails

Soil nails are used to reinforce soil slopes and walls by installing usually steel nails on a tight pattern that create a friction-based internal reinforcement that helps stabilize the soil structure. The steel nails are typically rebar, tubes, or right-angled plates (modified after Keller, 2025).

They can be un-grouted, conventionally grouted, or jet grouted. Due to the steel-based support being installed in soil, corrosion is often an issue and protection is frequently required for long-term stability. A drainage system needs to be incorporated into the design to stop pore water pressure build-up.

They represent part of a support system which could also include:

- Shotcrete.
- Retaining wall of concrete or timber.
- Geotechnical structures.
- External typically angled posts.

20.5.2 Soil Nail Design and Quality Control

Soil nail design requires the following steps:

- Perform geotechnical site investigation.
- Evaluate excavation/nailed wall geometry.
- Check wall stability – internal/external.
- Incorporate drainage control system with nail wall design.
- Finalize nailing layout and work sequence.

Post-installation monitoring is often required and could include the use of inclinometers, fixed survey targets, and piezometers.

Figure 20.22 shows a soil nail retaining structure and the potential slope failure surface.

Figure 20.23 shows a highwall supported with soil nails, steel beams, and heavy screen.

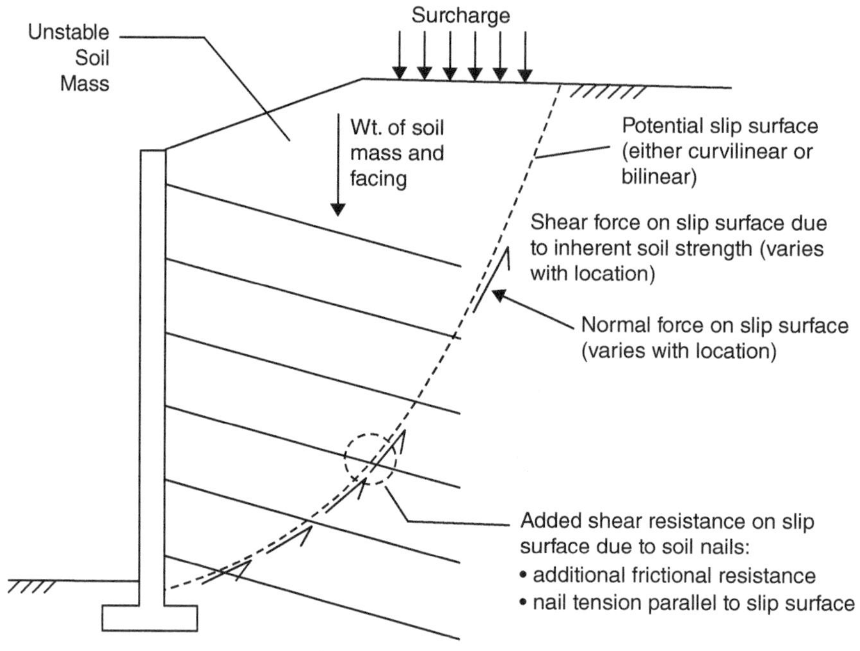

FIGURE 20.22 Slip surface approach for the design of soil nails (FHWA, 1996).

FIGURE 20.23 A highwall supported using soil nails, steel beams, and heavy screen. (Reproduced with kind permission of Keller North America.)

REFERENCES

DSI Underground (2024) www.dsiunderground.com/our-markets/tunneling

Federal Highway Administration (1996) Manual for design and construction monitoring of soil nail walls. FHWA Rep. No. FHWA-SA-96-069, Washington, DC, p. 468.

Forbes, B., Vlachopoulos, N. and Hyett, A.J. (2018) The application of distributed optical strain sensing to measure the strain distribution of ground support members. *Facets*, 3, 195–226.

Garshol, K. (2007) *Injection*. UGC International, Division of BASF Construction Chemicals Ltd, Switzerland.

Hamdan, M. (2019) Ground improvement and stabilization techniques: Ground freezing. Retrieved from: www.curiouscivileng.com/2019/07/ground-improvement-and-stabilization.html

Holter, K.G. and Zámečník, M. (2009) Modern pre-injections in underground construction for water ingress reduction and ground improvement. *Tunnel*. 2, 34–40.

Keller (2025) *Soil nailing, earth retention*. Haywood Baker, Keller, TX. Retrieved from: www.keller-na.com/hayward-baker-keller-company

Langer, F. (1997) *GSA Stability for Difficult Ground*. Dr Sauer & Partners, Herndon, VA.

Li, C.C. (2017) Principles of rockbolting design. *J. Rock Mech. Geotech. Eng*. 9, 396–414.

Liu, D., Zhang, D., Fang, Q., Sun, Z., Cao, L. and Li, A. (2019) Displacement characteristics of shallow-buried large-section loess tunnel with different types of pre-supports: a case study of new badaling tunnel. *Appl. Sci.* 10(1), 195.

NGI (2022) *Using the Q-System Rock Mass Classification and Support Design*. NGI, Oslo, Norway.

Panwar, R. (2021) New Austrian tunneling method: Design and operation. The Constructor. Retrieved from: www.theconstructor.org/geotechnical/new-austrian-tunneling-method/63500/

Rabcewicz, L.V. (1965) The new Austrian tunnelling method. *Water Power*. 453–457, 19–24.

Sauer, G. (1990) Design concept for large underground openings in soft ground using NATM. In *International Symposium on Unique Underground Structures*, Denver, Colorado.

Sauer, G and Gold, H. (1989) NATM Ground support concepts and their effect on contract practices RETC. In Rapid Excavation and Tunneling Conference, Springer, Cham.

Split Mountain Machine and Equipment Pty. Ltd (2025) Retrieved from: www.splitmountainmachine.com/mining-jumbo-abu.html

Appendix A
Terminology and Definitions

Abrasion: Abrasive wear is also known as cutting wear and involves the wearing particle moving across the surface in constant contact with the surface or impinging the surface at angles less than 45°.

Aggregate: Crushed or naturally occurring rock/stone that is typically added to concrete, shotcrete, or backfill.

Angle of Internal Friction: The angle (ϕ) between the axis of normal stress and the tangent to the Mohr envelope at a point representing a given failure to stress condition for a solid material.

Angle of Repose: The maximum angle with respect to a horizontal plane that the surface of a pile of a loose material will assume in its natural state.

Anisotropy: Having different properties in different directions.

Attenuation: Reduction in the amplitude of a wave as the distance from the source of propagation increases.

Backfill: Waste material, usually metallurgical tailings or waste rock that is used to fill underground voids formed by mining.

Bedrock: The more or less continuous body of rock that underlies the overburden soils.

Blastibility: The index value of the resistance of a rock formation to blasting.

Brittle Fracture: The sudden failure associated with the complete loss of cohesion across a plane.

Buckling: Instability of a column or plate under sufficiently high load due to sudden deflection of the structure.

Bulk (Incompressibility) Modulus: Ratio of the hydrostatic pressure to the volumetric strain that it produces.

Burden: Distance between the explosive charge and the free surface in the direction of throw.

Cavitation: The formation and subsequent collapse of vapor bubbles that occur when a slurry pressure drops and then rises above the vapor pressure of water. Cavitation can occur in pipes when the pressure is reduced and the velocity increases, or when backfill is supplied intermittently. It can cause serious damage and erosion of the pipe material.

Cement: It is usually defined as a binder, glue, or adhesive. In hardened concrete, it forms the matrix that binds the aggregates together to form the strong and rigid composite structure.

Closure: The reduction in the dimensions of an opening.

Coefficient of Friction: A constant of proportionality (μ) relating the normal stress and the corresponding critical shear stress at which sliding starts between two surfaces.

Cohesion: The shear resistance at zero normal stress (intrinsic shear strength in rocks). The shear strength is imparted to the material through the bonding of adjacent particles by physical or chemical means.

Compaction: The reduction in specific volume of the placed backfill arising from gravity or external stresses.

Compressive Stress: Normal stress tending to shorten a body in the direction in which it acts.

Consolidation: The process by which water drains from saturated backfill by percolation. This causes a reduction in the specific volume and an increase in the density.

Contraction: The linear strain associated with a decrease in length of a body.

Controlled Blasting: All forms of blasting designed to preserve the integrity of the remaining rocks (e.g., smooth blasting, pre-splitting, post-splitting).

Convergence: The shortening between two basically parallel surfaces which is usually the hangingwall and the footwall.

Creep: Time-dependent deformation.

Damping: The reduction in the amplitude of vibration of a body or system due to dissipation of energy internally or by radiation.

Decay Time: The interval of time required for a pulse to decay from its maximum value to some specified fraction of that value.

Decoupling: The ratio of the radius of the blasthole to the radius of the charge. This causes a reduction in the strain wave amplitude by increasing the space between the charge and the blasthole wall.

Deformation: A change in shape or size of a solid body.

Detonation: An extremely rapid and violent chemical reaction causing the production of a relatively large volume of gas.

Dilatancy: The property of volume increase under loading.

Dilation (Volumetric Strain): The quotient of the change in volume and the original volume of an element of the material under stress.

Dip: The angle at which a stratum or other planar feature is inclined from the horizontal.

Discontinuity Surface: Any surface across which some property for a rock mass is discontinuous (e.g., bedding planes, fractures).

Dispersion: The phenomenon of varying speed of transmission waves depending on the relevant frequency.

Displacement: A change in position of a point in a material.

Distortion: The change in shape of a solid body.

Ductility: The condition in which a material can sustain permanent deformation without losing its ability to resist load.

Elastic Limit: The point on a stress versus strain curve at which transition from elastic to inelastic behavior occurs.

Elasticity: The property of a material whereby it returns to its original form or condition after an applied force is removed.

Failure: Where the maximum strength of a material is exceeded by an applied load.

Fatigue: Decrease in the strength of a material caused by repetitive loading.

Footwall: The mass of rock beneath a discontinuity surface (in tabular mining, the rock below the reef plane).

Force: An action that tries to move an object from a stationary position or to change its rate of movement or its direction of movement.

Fracture: A mechanical discontinuity in the rock.

Friction: The force between two surfaces tending to resist relative motion. It is caused by the roughness of the surfaces of the bodies and acts tangentially between them.

Gap Graded: Material (e.g., backfill) with a solids particle size distribution that has a deficiency of particles in a specific size range.

Gravity: The force that tends to accelerate any object toward the center of the earth at a constant rate. The acceleration due to gravity is 9.81 m/s^2.

Hangingwall: The mass of rock above a discontinuity surface. In tabular mining, it is typically the rock above the reef plane.

Heterogeneity: Having different properties at different points within a material.

Homogeneity: Having the same properties at all points within a material.

Hysteresis: The incomplete recovery of strain during unloading due to energy consumption.

Inelastic Deformation: The portion of deformation under stress that is not annulled by the removal of the stress.

Isotropy: Having the same properties in all directions.

Liquefaction: The phenomenon whereby the shear resistance of a placed backfill mass decreases when subjected to loading, and as a result, the mass begins to mobilize and flow.

Normal Force: A force directed normal (perpendicular) to the surface element across which it acts.

Normal Stress: The component of stress normal to the plane on which it acts.

Overbreak: The quantity of rock that is removed beyond the planned perimeter of the final excavation.

Percolation: The movement of water, under hydrostatic pressure, through the small interstices of rock or soil.

Permanent (Residual) Strain: The strain remaining in a solid with respect to its initial condition after the application and removal of a stress greater than the yield stress.

Permeability: The capacity of a material to allow the passage of liquid or gas.

Plane Stress (or Strain): This is a state of stress (or strain) within a body in which all the stress (or strain) components normal to a certain plane are zero.

Plasticity: Property of a material to continue to deform indefinitely while sustaining a constant stress.

Poisson's Ratio: The ratio of the shortening in the transverse direction, to the elongation in the direction of an applied force in a body under tension, and while staying below the proportional limit.

Pore Water Pressure: This is the pressure produced by groundwater contained in the gaps (pores) in soils or rocks.

Porosity: The ratio of the aggregate volume of voids or interstices in a rock or soil to its total volume.

Portland Cement (PC or OPC): An inter-ground or separately ground mixture of clinker and gypsum, with generally not more than 5% by mass of an inorganic mineral addition. The particle relative density is about 3.14, and the loose bulk density is about 1.30. The typical specific surface (an indication or the fineness of grind of the product) is around 300 m^2/kg. As with all cement types, the exact specification varies from country to country.

Potential Energy: The work a body is capable of doing when moving from one position to another relative to the gravitational effect of the earth.

Pozzolan: A siliceous or siliceous and aluminous material which in itself possesses little or no cementitious value, but will in finely divided form, and in the presence of moisture, react with calcium hydroxide to form compounds possessing cementitious properties.

Primitive (Virgin) Stress: The state of stress in a geological formation before it is disturbed by man-made operations.

Principal Stress (or Strain): The stress (or strain) normal to one of three mutually perpendicular planes on which the shear stress (or shear strain) at the point in the body is zero.

Rapid-Hardening Portland Cement (RHPC): This has the same typical chemical composition of Portland cement but is ground finer, having a specific surface (area) of typically around 400–500 m^2/kg.

Rock: Any natural formed aggregate of mineral matter occurring in large masses or fragments.

Rockburst: A seismic event that causes damage to underground workings.

Rockfall (Fall of Ground): A fall of a rock fragment or a portion of the fractured rock mass without the simultaneous occurrence of a seismic event.

Rock mass: The rock as it occurs in situ, including its discontinuities.

Rock Mechanics: The theoretical and applied science of the mechanical behavior of rock.

Scalar: A scalar has only magnitude that does not change with directional changes.

Seismic Event: A transient earth motion caused by a sudden release of the strain energy stored in the rock.

Spalling: The longitudinal splitting in uniaxial compression or the breaking-off of plate-like pieces from a free rock surface.

Spitting Rock: The violent ejection of splinters of rock from the surface of an excavation.

Springline: The imaginary line where the curved portion of a tunnel roof meets the top of the wall. In a circular tunnel, the springlines are at the opposite ends of the horizontal centerline.

Stiffness: The force versus the displacement ratio.

Strain Burst: Rockbursts at the lower end of the spectrum of violent events occurring essentially at the surface of an excavation.

Stress: The force acting across a surface element divided by the area of the element.

Stress (or Strain) Field: The ensemble of stress (or strain) states defined at all points of an elastic solid.

Stress (or Strain) Rate: The rate of change of stress (or strain) with time.

Strength: The maximum stress that a material can resist without failing for any given loading regime.

Stress Relaxation: Stress release due to creep.

Strike: The direction of azimuth of a horizontal line in the plane of an inclined stratum (or other planar feature) within a rock mass.

Subsidence: The downward movement of the overburden (soil and/or rock) lying above an underground excavation or adjoining a surface excavation.

Support: A structure or structural feature built into or around an underground excavation for maintaining its stability.

Swelling: The constitutive mineralogy of the rock whereby water is absorbed causing a measurable volume increase. Swelling can exert very large time-dependent forces on rock or support systems and reduce the size of excavations.

Tangent Modulus: Slope of the tangent to the stress versus strain curve at a given stress value (generally a stress equal to half the compressive strength).

Tensile Stress: The normal stress tending to lengthen the body along the direction in which it acts.

Thickness: The perpendicular distance between bounding surfaces (e.g., bedding planes).

Transverse (Shear) Wave: A wave in which the displacement at each point of the medium is parallel to the wave front.

Triaxial Compression: Compression caused by the application of normal stress in three perpendicular (orthogonal) directions.

Uniaxial (Unconfined) Compression: Compression caused by the application of normal stress in a single direction.

Vector: A quantity having magnitude and direction but no fixed position.

Wavefront: A continuous surface over which the phase of a wave that progresses in three dimensions is constant.

Weathering: The process of disintegration and decomposition as a consequence of exposure to the atmosphere, to chemical action and to the action of frost, water, and heat.

Weight: The force with which the earth attracts a body.

Yield Stress: The stress beyond which the induced deformation is not fully annulled after complete destressing.

The definitions above are largely based on:

i. ISRM Commission on "Terminology, Symbols and Graphic Presentations."
ii. Rockburst Terminology – South African Chamber of Mines Circular 28/82.

Appendix B
General Information

B.1 STANDARD INTERNATIONAL (SI) UNITS

Basic units

Quantity	Unit	Symbol
Length	meter	m
Mass	kilogram	kg
Time	second	s
Temperature	Kelvin	K

Derived units

Quantity	Unit	Symbol	Derivation
Force	Newton	N	kg m/s^2
Pressure	Pascal	Pa	kg m/s^2m^2
Energy (work)	Joule	J	Nm
Power	Watt	W	Nm/s
Frequency	Hertz	Hz	s^{-1}

B.2 PREFIXES

Power	Prefix	Abbreviation
10^{12}	Tera	T
10^{9}	Giga	G
10^{6}	Mega	M
10^{3}	kilo	K
10^{2}	hecto	h
10^{1}	deca	da
1	Unity	
10^{-1}	deci	d
10^{-2}	centi	c
10^{-3}	milli	m
10^{-6}	micro	μ
10^{-9}	nano	n
10^{-12}	pico	p

B.3 TEMPERATURE CONVERSIONS

$$C = 5/9\ (F - 32)$$

$$F = \left(\frac{9}{5}\right) C + 32$$

$$K = C + 273$$

where
C is Celsius (centigrade).
F is Fahrenheit.
K is Kelvin.

B.4 COMMON UNIT CONVERSIONS

	To obtain	
Multiply by	To convert	To
2.54	inches	centimeters
30.48	feet	centimeters
0.914	yards	meters
1609.3	miles	meters
1853.3	nautical miles	meters
6.45	square inches	square centimeter
0.093	square feet	square meters
0.836	square yards	square meters
16.39	cubic inches	cubic centimeters
28.3	cubic feet	liters
0.765	cubic yards	cubic meters

B.5 MESH SIZES AND EQUIVALENT APERTURE SIZES

Mesh Size	Aperture Size (μm)
400	38
350	45
300	53
240	63
200	75
170	90
150	106
120	126
100	150
85	180
72	212
52	300
36	426

B.6 BASIC TRIGONOMETRY

Considering the right-angled triangle below, the following trigonometrical functions can be defined and derived:

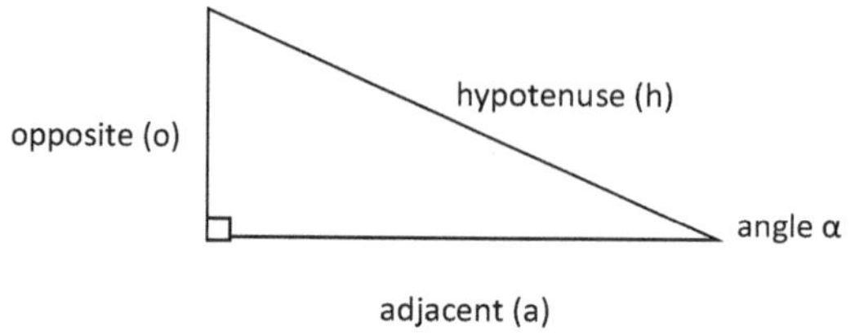

hypotenuse (h)
opposite (o)
angle α
adjacent (a)

$$\sin\alpha = o/h$$

$$\cos\alpha = a/h$$

$$\tan\alpha = o/a$$

$$\cos ec\alpha = h/o = 1/\sin\alpha$$

$$\sec\alpha = h/a = 1/\cos\alpha$$

$$\cot\alpha = a/o = 1/\tan\alpha$$

Also form Pythagoras' theorem:

$$h^2 = a^2 + o^2$$

From this theorem, the following can be derived:

$$\sin^2\alpha + \cos^2\alpha = 1$$

$$1 + \tan^2\alpha = \sec^2\alpha$$

$$1 + \cot^2\alpha = \text{cosec}^2\alpha$$

B.7 SOME INTERNATIONAL PROFESSIONAL MINING- AND ROCK MECHANICS–RELATED CONTACTS

- Australasian Institute of Mining and Metallurgy (AusIMM)
  www.ausimm.com
- Canadian Institute of Mining (CIM)
  www.cim.org
- Institution of Materials, Minerals and Mining (IOM3), UK
  www.iom3.org
- International Society of Rock Mechanics (ISRM)
  www.isrm.net

- Society of Mining Engineers (SME)
  www.smenet.org
- Society of Mining Professors (SOMP)
  www.miningprofs.org
- South African Institute of Mining and Metallurgy (SAIMM)
  www.saimm.co.za
- The Mining, Geological and Metallurgical Institute of India
  www.mgmiindia.in

References

Agarwal, A., Datta, M., Ramana, G.V. and Satyakam, R. (2018) Improving the slope stability of municipal solid waste dumps using reinforced soil berms: a case study. In: *Proceedings of the 8th International Congress on Environmental Geotechnics Volume 2: Towards a Sustainable Geoenvironment*. Springer, Singapore.

Ahlawat, A. (2018) *Lecture 9-Calculation of Capillarity Rise and Fall*. Mechanical Education Hub. Retrieved from: www.modemech.blogspot.com

Akazawa, T. (1943) New test method for evaluating internal stress due to compression of concrete: the splitting tension test. *J. Japan Soc. Civil Eng.* 29, 777–787.

Aki, K. (1966) Earthquake generating stress in Japan for the years 1961 to 1963 obtained by smoothing the first motion radiation patterns. *Bull. Earthquake Res. Inst.* 44, 447–471.

Ali, S. (2017) *Types of Slope Failure*. University of Azad Jammu and Kashmir Muzaffarabad, Pakistan. Retrieved from: www.slideshare.net/slideshow/types-of-slope-failures-75401227/75401227

American Concrete Institute (2016) *Guide to Shotcrete*. Report 506R-2.

American Insurance Services Group, Inc (1972) *Blasting Damage and Other Structural Cracking: A Guide for Adjusters and Engineers*. American Insurance Services Group, New York.

American Insurance Services Group, Inc (1990) *Blasting Damage and Other Structural Cracking: A Guide for Adjusters and Engineers* (3rd Edition). American Insurance Services Group Inc., New York.

Ansfield, V.J. (1992) A graphic representation of the Richter Scale. *J. Geol. Educ.* 40(5), 381–384.

ASTM C1436-13 (2021) *Standard Specification for Materials for Shotcrete*. ASTM International, West Conshohocken, PE. www.astm.org/standards/c1436.

ASTM C150-07 (2012) *Standard Specification for Portland Cement*. ASTM International, West Conshohocken, PE. www.astm.org/standards/c150

ASTM D 5731-95 (1995) *Standard Test Method for Determination of the Point Load Strength Index of Rock*. American Society for Testing and Materials, West Conshohocken, PE.

ASTM D2487 (2006) *Standard Practice for Classification of Soils for Engineering Purposes (Unified Soil Classification System)*. American Society for Testing and Materials (ASTM), West Conshohocken, PA.

ASTM D2488 (2023) *Description and Identification of Soils*. American Society for Testing and Materials (ASTM), West Conshohocken, PA.

ASTM D427 (2004) *Standard Test Method for Shrinkage Factors of Soils by the Mercury Method (Withdrawn 2008)*. American Society for Testing and Materials (ASTM), West Conshohocken, PA.

ASTM D4943 (2018) *Standard Test Method for Shrinkage Factors of Cohesive Soils by the Water Submersion Method. Annual Book of ASTM Standards, Vol. 04.08*. American Society for Testing and Materials (ASTM), West Conshohocken, PA.

ASTM D653 (2021) *Standard Terminology Relating to Soil, Rock and Contained Fluids*. American Society for Testing Materials (ASTM) Standards, West Conshohocken, PA.

ASTM International (2020) *ASTM-F432-Standard Specification for Roof and Rock Bolts and Accessories*. ASTM International, West Conshohocken, PE.

Atlas Copco (2008) *Guide to Underground Methods and Applications*. Atlas Copco, Sweden.

Atterberg, A. (1911) Die Plastizität der Tone. *Int. Mitt. Bodenkd.* 1, 4–37.

Bals, R. (1931) *Beitrag zur Frage der Vorausberechnung Bergbaulicher Senkugen (A Contribution to the Problem of Precalculating Mine Subsidence)*. Mitteilungen aus dem Markscheidewesen.

Bandis, S., Lumsden, A. and Barton, N. (1981) Experimental studies of scale effects on the shear behaviour of rock joints. *Int. J. Rock Mech. Min. Sci. Geomech. Abstr.* 18(1), 1–21. https://doi.org/10.1016/0148-9062(81)90262-X.

Barron, K. and Pen, K. (1992) A revised model for coal pillars. In: *Proceedings of the Workshop on Coal Pillar Mechanics and Design USBM IC 9315*, Pittsburgh, PA, pp. 144–157.

Barton, N., Lien, R. and Lunde J. (1974) Engineering classification of rock masses for the design of tunnel support. *Rock Mech.* 6, 189–236.

Bauer, E.R. and Dolinar, D.R. (2000) Skin failure of roof and rib and support techniques in underground coal mines. *Environ. Sci.* 151, 99–109.

Bauschinger, J. (1876) *Mitteilungen aus dem Mechanisch Technishen Laboratorium der Königlichen Technischen Hochschule in München, vols 1–23*. Theodor Ackermann, München.

Bear, J. (1979) *Hydraulics of Groundwater*. McGraw-Hill, New York, p. 569.

Benckert, A. and Eurenius, J. (2001) *Tailings Dam Constructions, Seminar on Safe Tailings Dam Constructions*. Swedish Mining Association, Natur Vards Verket, European Commission, Gallivare, Sweden, pp. 30–36.

Berthelot, D.S., Place-Hoskie, D., D Willems, D. and Black, K. (2019) *Elliot Lake, Ontario Uranium Mines a Legacy Perpetual Care Case Study. Mine Closure*. Australia Centre for Geomechanics, Perth.

Bhasin, R. and Grimstad, E. (1996): The use of stress – strength relationships in the assessment of tunnel stability. Tunnelling and Underground Space Technology, 11(1): 93-98.

Bieniawski, Z.T. (1968) In situ strength and deformation characteristics of coal. *Eng. Geol*. 2, 325–340.

Bieniawski, Z.T. (1968) The effect of specimen size on compressive strength of coal. *Int. J. Rock Mech. Min. Sci*. 5, 327–335.

Bieniawski, Z.T. (1973) Engineering classification of jointed rock masses. *Trans. S. Afr. Civ. Engrs*, 15.

Bieniawski, Z.T. (1984) *The Design Process in Rock Engineering. Rock Mechanics Design in Mining and Tunnelling*. AA Balkema Publishers, Kenya.

Birch, F., Schairer, J. and Spicer, H. (1942) Handbook of physical constants. *Geol. Soc. Amer., Special Papers*. 36, 66.

Blake, W. (1971) *Rock Burst Mechanics*. Ph.D. Thesis. Colorado School of Mines, Golden, CO.

Blake, W., Leighton, F.W. and Duvall, W.I. (1974) *Microseismic Techniques for Monitoring the Behavior of Rock Structures*. US Department of the Interior, Bureau of Mines, Denver, CO, Vol. 665.

Blight, G. and Spearing, A.J.S. (1995) The properties of cemented silicated backfill for use in narrow, hard-rock tabular mines. *J. S. Afr. Inst. Mining Metallurgy*. 96(1), 17–28.

Bolt, B.A. (1976) *Nuclear Explosions and Earthquakes*. W.H. Freeman and Company, San Francisco, CA.

Bolt, B.A. (1982) *Inside the Earth*. Freeman, San Francisco, CA.

Booth, C.J. (1998) Impacts of mine subsidence on groundwater. In: Hooks, C.L., Vories, K.C. and Throgmorton, D. (eds) *Proceedings of Prime, Farmland Interactive Forum*. Department of Agronomy, Illinois Agricultural Experiment Station, University of Illinois at Urbana-Champaign, University of Southern Indiana, Evansville, pp. 143–148.

Brady, B.H. and Brown, E.T. (2004a) *Rock Mechanics: For Underground Mining* (3rd Edition). Springer Science & Business Media, New York.

Brady, B.H. and Brown, E.T. (2004b) *Rock Mechanics: For Underground Mining*. Kluwer Academic Publishers, Dordrecht, Netherlands.

Brady, B.H. and Brown, E.T. (2005) *Rock Mechanics: For Underground Mining* (3rd Edition). Springer Science & Business Media, New York.

British Geological Survey (2008) *Earthquakes*. Issue 10. Retrieved from: www.earthquakes.bgs.ac.uk/education/eq_guide/earthquake_booklet.pdf

British Standards Institution (2007). *BS 5930:1999 Code of Practice for Site Investigations*. BSI, London.

Broch, E. and Franklin, J.A. (1972) The point load test. *Inst. J. Rock Mech. Mining Sci*. 1972, 669–697.

Brown, E.T. (1970) Strength of models of rock with intermittent joints. *J. Soil Mech. Found. Div. ASCE*. 96, 1935–1949.

Bruhn, R.W., Magnuson, M.O. and Gray, R.E. (1978) Subsidence over the mined-out Pittsburgh coal. ASCE Preprint 3293, American Society of Civil Engineers Spring Convention, Pittsburgh, pp. 26–55.

Brummer, R., Gay, N., Jager, A., McKinnon, S., Piper, P., Roberts, M., Ryder, J., Spottiswoode, S. and Wojno, L. (1988). *An Industry Guide to Methods of Ameliorating the Hazards of Rockfalls and Rockbursts*. Chamber of Mines of South Africa, Johannesburg.

Budavari, S. (1983) *Rock Mechanics in Mining Practice*. SAIMM monograph Series 5, Johannesberg.

Bunting, D. (1911) Chamber pillars in deep anthracite mines. *Trans AIME*. 42, 236–245.

Carneiro, F.L.L.B. (1943) A new method to determine the tensile strength of concrete. *Paper presented at the Proceedings of the 5th meeting of the Brazilian Association for Technical Rules* ("Associação Brasileire de Normas Técnicas—ABNT"), 3D Section.

Casagrande, A. (1936) Characteristics of cohesionless soils affecting the stability of slopes and earthfills. *J. Boston Soc. Civil Eng*. 23, 257–276.

Casagrande, A. (1937) Seepage through dams. In *Contributions to Soil Mechanics 1925-1940*. Boston Soc. Civil Eng. Boston. p 295.

Casagrande, A. (1976) Liquefaction and cyclic deformation of sands: a critical review. In: *5th Panamerican Conference on Soil Mechanics and Foundation Engineering*, Buenos Aires, Argentina.

Chamber of Mines Research Organization (COMRO) (1988) *An Industry Guide to Methods of Ameliorating the Hazards of Rockfalls and Rockbursts*. Chamber of Mines of South Africa, Johannesburg, South Africa.

Chem Europe (2025) *Hydrology*. Retrieved from: https://www.chemeurope.com/en/encyclopedia/Hydrology.html

Chen, X., Guo, L., Zhou, Y., Xu, W. and Zhao, Y. (2023) Remediation of grassland subsidence and reduction of land occupation with tailings backfill technology: a case study of lead-zinc mine in Inner Mongolia, China. *Front. Environ. Sci.* 11, 1183945.

Choi, Y-K. (2017) *Principles of Applied Civil Engineering Design*. American Society of Civil Engineers, New York. https://doi.org/10.1061/9780784414736.

Christiansen, E.H. and Hamblin, W.K. (2015) *Dynamic Earth: An Introduction to Physical Geology*. Jones and Bartlett Learning, LLC, Burlington.

Church, H.K. (1981) *Excavation Handbook*. McGraw-Hill Book Company, New York.

Colback, P. and Wiid, B. (1965) Influence of moisture content on the compressive strength of rock. In: *Proceedings Rock Mechanics Symposium Canadian Department of Mines and Technology Surveys*, Ottawa.

Comite Europeen de Normalisation (2007) *Admixtures for Concrete, Mortar and Grout-Part 5 EN 934-5*. Retrieved from: www.en-standard.eu.

Constructor (2021) *In Situ Stresses in Rock and Measurement Methods*. Retrieved from: www.theconstructor.org/building/measurement-methods-insitu-stresses-rocks/11649/.

Contreras, C, Elmo, D, Jakubec, J and Thomas, A (2022) Reviewing Laubscher's empirical method to estimate subsidence limits. In: Potvin, Y. (ed) *Caving 2022: Proceedings of the Fifth International Conference on Block and Sublevel Caving, Australian Centre for Geomechanics*, Perth, pp. 805–818, https://doi.org/10.36487/ACG_repo/2205_55.

Contreras, C.P. (2022) *Reviewing Laubscher's Empirical Method to Estimate Subsidence Limits (T)*. University of British Columbia. Retrieved from: https://open.library.ubc.ca/collections/ubctheses/24/items/1.0421707.

Cook, N.G.W. and Salamon, M.D.G. (1966). Rock mechanics applied to the study of rockbursts. *J. S. Afr. Inst. Min. Metall.* 66(3), 97–108.

Cook, N.G.W., Hoek, E., Pretorius, J.P.G., Ortlepp, W.D. and Salamon, M.D.G. (1966) Rock mechanics applied to rockbursts. *J. S. Afr. Inst. Min. Metall.* 1996, 435–714.

Coulomb, C.A. (1776) Essai sur une application des regies de maximis & minimis a quelques problemes de statique, relatifs a l'architecture, Memoires de Mathematique et de Physique. *Presentes a V Academie Roy ale des Sciences par divers Savans*, Vol. 1, pp. 343–382 (written, 1773).

Darcy, H. (1856) *Les fontaines publiques de la ville de Dijon*. Dalmont, Paris.

Das, B (1983) *Fundamentals of Soil Dynamics*. Elsevier Science Publishing Co., Inc, New York.

Das, B. (2007) *Principles of Foundations Engineering* (6th Edition). Cengage Learning, Stamford CT.

Das, B. (2008) *Advanced Soil Mechanics* (3rd Edition). Taylor and Francis, New York NY.

Das, B.M. (2010) *Principles of Geotechnical Engineering* (7th Edition). Nelson Engineering, New York.

Das, B.M. and Sivakugan, N. (2016) *Introduction to Geotechnical Engineering* (2nd Edition). Cengage Learning Inc., Stamford, CT.

Das, B.M. and Sivakugan, N. (2019) *Principles of Foundation Engineering* (9th Edition). Cengage Learning Inc., Stamford, CT.

Das, M.N. (1986) Influence of width to height ratio on post-failure behaviour of coal. *Int. J. Min. Geol. Eng.* 4, 79–87.

De Souza, E. (2010) *Room and Pillar*. The Robert M. Buchan Department of Mining, Queens University, Kingston, Ontario, Canada.

Deere, D.U. (1964) Technical description of rock cores for engineering purposes. *Rock Mech. Eng. Geol.* 1(1), 17–22.

Deere, D.U. (1989) *Rock Quality Designation (RQD) after Twenty Years*. U.S. Army Corps of Engineers Contract Report GL-89-1. Waterways Experiment Station, Vicksburg, MS, p. 67.

Deere, D.U., Hendron, A.J., Jr, Patton, F.D. and Cording, E.J. (1967) Design of surface and near surface construction in rock. Failure and breakage of rock. In: Fairhurst, C. (ed) *Society of Mining Engineers of AIME*. AIME, New York, pp. 237–302.

Deere, D.U., Merritt, A.H. and Coon, R.F. (1969) *Engineering Classification of In-Situ Rock*. Air Force Weapons Laboratory, Air Force Systems Command, Kirtland Air Force Base, Montgomery County, OH, Vol. 67.

Denkhaus, H.G. (1964) Critical review of strata movement theories and their application to practical problems. *J. S. Afr. Inst. Min. Metall.* 63(8), 310–332.

Department of the Army (1974). *Engineering Design Handbook, Explosions in Air, Part One*. No. 706-181, Army Material Command, Alexandria, VA.

Dey, A., Basudhar, P., Ghosh, P, Kolathayar, S. and Nainegali, L. (2010) *Reinforced Earth Design of Embankments and Cuts in Railways*. Research Designs and Standards Organization (RDSO), Lucknow.

DSI Underground (2024) www.dsiunderground.com/our-markets/tunneling

Dupuit, J. (1863) *Etudes Théoriques et Pratiques sur le Mouvement des Eaux dans les Canaux Découverts et à Travers les Terrains Perméables [Theoretical and Practical Studies on the Movement of Water in Open Channels and Permeable Ground]*. Dunod, Paris, France.

Dutto, P. (2014) *Modelling of Landslides Propagation with SPH: Effects of Rheology and Pore Water Pressure*. Thesis. Universidad Politecnica De Madrid, Spain.

Duvall, W.I. and Stephenson, D.E. (1965) Seismic energy available from rockbursts and underground explosions. *Trans. Soc. Min. Eng*. 232(3), 235–240.

Eberhardt, E. (2007) *Topic 6-Numerical Methods. Tunneling & Underground Design Graduate Class*. UBC, Vancouver, Canada.

Eberhardt, E. (2017) *Lecture On: In Situ Stresses and Stress Measurement*. Geotechnical Engineering Practice & Design, EOSC433. Geological Engineering. University of British Columbia, Canada.

Elementary Engineering (2019) *Particle size Distribution Curve-Civil Engineering*. Retrieved from: www.elementaryengineeringlibrary.com/civil-engineering/soil-mechanics/particle-size-distribution-curve.

Emery, J., Canbulat, I. and Zhang, C. (2020) Fundamentals of modern ground control management in Australian underground coal mines. *IJMST*. 30, 573–582.

Engels, D. and Dixon-Hardy, J. (2007) Methods for the disposal and storage of mine tailings. *Environ. Sci. Eng*. 15(3), 301–317.

Esterhuizen, G.S., Dolinar, D.R. and Ellenberger, J.L. (2011) Pillar strength in underground stone mines in the United States. *Int. J. Rock Mech. Min. Sci*. 48(1), 42–50.

Esterhuizen, G.S., Tyrna, P.L. and Murphy, M.M. (2019) A case study of the collapse of slender pillars affected by through-going discontinuities at a limestone mine in Pennsylvania. *Rock Mech. Rock Eng*. 52(12), 4941–4952.

European Federation of National Associations Representing Concrete (EFNARC) (2023) Flums, Switzerland. Retrieved from: www.efnarc.org.

Farmer, I.W. (1968) *Engineering Properties of Rocks*. E. & F.N. Spon Ltd, London.

Federal Highway Administration (1996) *Manual for Design and Construction Monitoring of Soil Nail Walls*. FHWA Rep. No. FHWA-SA-96-069, Washington, DC, p. 468.

Fellenius, W. (1936) Calculation of stability of earth dams. In: *TRANS. 2nd Congresso on large dams*, Washington, DC, Vol. 4, p. 445.

Feng, X.T. and Hudson, J.A. (2011) *Rock Engineering Design*. CRC Press, Leiden.

Forbes, B., Vlachopoulos, N. and Hyett, A.J. (2018) The application of distributed optical strain sensing to measure the strain distribution of ground support members. *Facets*. 3, 195–226.

Fredlund, D.G. (1993) *Soil Mechanics for Unsaturated Soils*. Wiley, New York.

Fredlund, D.G. and Rahardjo, H. (1993) *Soil Mechanics for Unsaturated Soils*. John Wiley and Sons, New York.

Gadde, M.M. (2009) *Weak Floor Stability in the Illinois Basin Underground Coal Mines*. West Virginia University, Virginia.

Gaddy, F.L. (1956) *Study of the Ultimate Strength of Coal as Related to Absolute Size of the Cubical Specimens Tested*. Series No. 112. Virginia Polytechnic Institute Bulletin, pp. 1–27.

Galperin, M., Zayetsev, V.S. and Normatov, Y.A. (1993) *Hydrogeology and Engineering Geology: Geotechnika*. CRC Press, Taylor & Francis Group, Boca Raton, FL.

Galvin, J.M., Hebblewhite, B.K. and Salamon, M.D.G. (1999) *University of New South Wales Pillar Strength Determinations for Australian and South African Conditions*. A.A. Balkema, Rotterdamn, pp. 63–71.

Garshol, K. (2007) *Injection*. UGC International, Division of BASF Construction Chemicals Ltd, Switzerland.

Garvey, P.R. and Lansdowne, Z.F. (1998) Risk matrix: an approach for identifying, assessing, and ranking program risks. *Air Force J. Logist*. 22, 18–21.

Gay, N.C. (1975) In situ stress measurements in southern Africa. *Tectonophys*. 29, 1–4.

Gibson, K. (2023) *What is Risk Management and Why Is It Important?* Harvard Business School Online. Retrieved from: www.online.hbs.edu/blog/post/risk-management.

Global Tailings Review (2020) Global industry standard on tailings management. Convened by: ICCM, UN Environment Programme and Principles for Responsible Investments. Retrieved from: www.globaltailingsreview.org

Goldfields (2009) *Mineral Resources and Mineral Reserves*. South Deep Gold Mine-Technical Short Report. Retrieved from: www.goldfields.com/reports/rr_2009/tech_south.php/tech_south.php

Goodman, R.E. (1976) *Methods of Geological Engineering*. West Publications, New York.

Gray, R.E., Gamble, J.C., McLaren, R.J. and Rogers, D.J. (1974) *State of the Art of Subsidence Control*. Appalachian Reg. Comm., ARC-73-111-2550.

Greenwald, H.P., Howarth, H.C. and Hartmann, I. (1939) *Experiments on Strength of Small Pillars of Coal in the Pittsburgh Bed*. U.S. Department of Interior, Bureau of Mines, Washington, DC.

Griffith, A.A. (1921) The phenomena of rupture and flow in solids. *Philos. Trans. R. Soc. London. A*. 221, 163–198.

Gürtunca, R.G., Jager, A.J., Adams, D.J. and Gonlag, M. (1989) In situ measurements of backfills and surrounding rockmass behaviour in South African Gold Mines. In: *4th International Symposium on Mining with Backfill*. Ontario, Canada.

Hagen, G.H.L. (1839) Uber die Bewegung des Wassers in engen cylindrischen R6hren. *Poggendorf's Annalen der Physik und Chemie*. 46, 423–442.

Haimson, B.C. (1978) The hydrofracturing stress measuring method and recent field results. *International Journal of Rock Mechanics and Mining Sciences*. 15, 167–178.

Hall, M. (2015) *What is Anisotropy?* Agile Scientific Blog-Geoscience & Technology. Retrieved from: www.agilescientific.com/blog/2015/2/9/what-is-anisotropy.

Halley, E. (1686) An historical account of the trade winds, and monsoons, observable in the seas between and near the Tropics, with an attempt to assign the physical cause of the said winds. *Philos. Trans*. 16, 153–168

Hamdan, M. (2019) *Ground Improvement and Stabilization Techniques-Ground Freezing*. Retrieved from: www.curiouscivileng.com/2019/07/ground-improvement-and-stabilization.html.

Hamrin, H. (2001) Underground mining methods and applications. In: Hustrulid, W.A. and Bullock, R.L. (eds) *Underground Mining Methods: Engineering Fundamentals and International Case Studies*. Society for Mining Metallurgy and Exploration, Littleton, pp. 3–14.

Hanks, T.C. and Kanamori, H. (1979) A moment magnitude scale. *J. Geophy. Res*. 84, 2348–2350.

Harrison, J.P. and Hudson, J.A. (2000) *Engineering Rock Mechanics – Part 2: Illustrative Worked Examples*. Elsevier Science, Oxford.

Hartman, H. L. (1992) *SME Mining Engineering Handbook* (Vol. 2). Society for Mining, Metallurgy, and Exploration, Denver, CO.

Haviland, D. and Marshall, J. (2014) Fundamental behaviours of production traffic in underground mine haulage ramps. *Int. Jour. Min. Sci. & Tech*. 25(1), 6.

Heasley, K.A. (2014) *LaModel – A Program for Coal Mine Modelling*. AusIMM, Melbourne.

Heath, R.C. (1983) *Basic Ground-Water Hydrology*, U.S. Geological Survey Water-Supply Paper 2220, p. 86.

Hedley, D.G.F. and Grant, F. (1972) Stope and pillar design for the Elliot Lake uranium mines. *Can. Inst. Min. Metall. Bull*. 65, 37–44.

Hendricks, J.R. (2019) *Geological Time. Digital Atlas of Ancient Life*. Paleontological Research Institution, New York. www.digitalatlasofancientlife.org/learn/geological-time/geological-time-scale/.

Hetényi, M (1958) *Beams on Elastic Foundations*. University of Michigan Press, Chicago, IL.

Hoek E (1968) Brittle failure of rock. In: Stagg, K.G. and Zienkiewicz, O.C. (eds) *Rock Mechanics in Engineering Practice*. Wiley, New York, pp. 99–124.

Hoek, E. (1994) Strength of rock and rock masses. *ISRM News J*. 2, 4–16.

Hoek, E. and Bray, J.W. (1981) *Rock Slope Engineering* (3rd Edition). Institution of Mining and Metallurgy, London, UK.

Hoek, E. and Brown, E.T. (1980a) *Underground Excavations in Rock*. Institution of Mining and Metallurgy, London, p. 527.

Hoek, E. and Brown, E.T. (1980b) Empirical strength criterion for rock masses. *J. Geotech. Eng. Div*. 1980, 1013–1035.

Hoek, E. and Brown, E.T. (1997) Practical estimates of rock mass strength. *J. Rock Mech. Min. Sci. Geomech*. 34, 1165–1186.

Hoek, E. and Brown, E.T. (1998) The Hoek-Brown failure criteria – A 1998 update. In: Curran, J.H. (ed) *15th Canadian Rock Mechanics Symposium*. University of Toronto, Toronto, Canada, pp. 31–38

Hoek, E. and Wood. D.F. (1987) *Support in Underground Hard Rock Mines. Underground Support Systems*. Canadian Institute of Mining and Metallurgy, Montreal, Vol. 35.

Holland, C.T. (1964) The strength of coal in mine pillars. In: *Proceedings of the Sixth Symposium on Rock Mechanics*. University of Missouri, Rolla, MO, pp. 450–456.

Holland, C.T. (1973a) Mine pillar design. In: Cummins, A.B. and Given, I.A. (eds) *SME Mining Engineering Handbook*. AIME, New York, Vol. 1, pp. 97–118.

Holland, C.T. (1973b). Pillar design for support of the over burden in coal mines. In: *Proceedings of the 9th Canadian Symposium on Rock Mechanics*. Mines Branch of the Department of Energy, Mines and Resources, Ottawa, pp. 114–139.

Holland, C.T. and Gaddy, F.L. (1957). Some aspects of permanent support of overburden on coal beds. *Proc. West Virginia Coal Min. Inst.* 5, 43–65.
Holling, C.S. (ed) (1978) *Adaptive Environmental Assessment and Management*. Wiley, Chichester, p. 377.
Holter, K.G. and Zámečník, M. (2009) Modern pre-injections in underground construction for water ingress reduction and ground improvement. *Tunnel*. 2, 34–40.
Hough, B.K. (1957) *Basic Soils Engineering*. Ronald Press, New York.
Hsuing, S.M. and Peng, S.S. (1985) Chain pillar design for U. S. longwall panels. *Min. Sci. Technol.* 2, 279–305.
Hudson, J. and Harrison, J. (1997) *Engineering Rock Mechanics: An Introduction to the Principles, Part One*. Elsevier Science Ltd, Kidlington, Oxford.
Hudson, J.A. and Feng, X.T. (2015) *Rock Engineering Risk*. CRC Press, Boca Raton, FL.
Hustrulid, W.A. (1976) A review of coal pillar strength formulas. *Rock Mech.* 8, 115–145.
Hyett, A. and Spearing, A.J.S. (2012) New technology for measuring the in situ performance of rock bolts. In: *ARMA US Rock Mechanics/Geomechanics Symposium, ARMA*, Chicago, IL.
Iannacchione, A.T., Dolinar, D.R. and Mucho, T.P. (2002) High stress mining under shallow overburden in underground U.S. stone mines. In: *1st International Seminar on Deep and High Stress Mining, Section 32*. Australian Center for Geomechanics, Nedlands, Australia.
ICMM (2021) *Tailings Management –Good Practice Guide by ICMM –Mining with Principles*. Retrieved from: https://www.icmm.com/en-gb/guidance/environmental-stewardship/tailings-management-good-practice
ICOLD and UNEP (2001) *Bulletin 121: Tailings Dams – Risk of Dangerous Occurrences, Lessons Learnt from Practical Experiences*. UNEP, DTIE and ICOLD, Paris.
Imperial Chemical Industry (1979) *Hazard Analysis (Hazan) Lecture*. University of Queensland, Australia.
International Society of Rock Mechanics (1981) *Rock Characterization, Testing & Monitoring: ISRM Suggested Methods*. Pergamon Press, Oxford.
International Society of Rock Mechanics (2007) *The Complete ISRM Suggested Methods for Rock Characterization, Testing & Monitoring: 1974–2006*. Elsevier, Amsterdam, Netherlands.
International Society of Rock Mechanics (2015) *The ISRM Suggested Methods for Rock Characterization, Testing and Monitoring: 2007–2014*. Ulusay, R. (ed). Springer, Switzerland.
Isaacson, E. (1962) *Rock Pressure in Mines* (2nd Edition). Mining Publications Ltd, Salisbury House, London.
International Society of Rock Mechanics (1981) *Suggested Methods for Rock Characterization, Testing, and Monitoring*. Pergamon Press, Oxford, UK.
International Society of Rock Mechanics (2007) The complete ISRM Suggested methods for rock characterization, testing and monitoring: 1974–2006. Elsevier, Amsterdam, Netherlands.
Itasca Consulting Group (2024) Retrieved from: www.itascacg.com.
Jaeger, J.C. (1960) Shear failure of anisotropic rock. *Geol. Magazine*. 97(1). 65–72. https://doi.org/10.1017/S0016756800061100.
Jaeger, J.C. and Cook, N.G.W. (1969) *Fundamentals of Rock Mechanics*. Methuen & Co Ltd, London.
Jager, A. J., and Ryder, J.A. (1999). *A Handbook on Rock Engineering Practice for Tabular Hard Rock Mines*. Creda Communications, Cape Town.
Jain, S. (2011) *Filling*. Retrieved from: www.slideshare.net/slideshow/filling/9844256.
Jakubec, J. and Esterhuizen, G.S. (2007) *Use of the Mining Rock Mass Rating (MRMR) Classification: Industry Experience*. CDC, Canada. Downloads/cdc_8482_DS1%20(1).pdf
Jakubec, J. and Laubscher, D.H. (2000) *The MRMR Rock Mass Classification System in Mining Practice*. MassMin, Brisbane, Australia.
Jeran, P.W. and Trevits, M.A. (1995) *Mining Publication: Timing and Duration of Subsidence Due to Longwall Mining*. US Bureau of Mines (USBM RI 9552). January.
Jessu, K.V. and Spearing, A.J.S. (2018) Effect of dip on pillar strength. *J. S. Afr. Inst. Min. Metall.* 118(7), 765–776.
Jessu, K.V., Spearing, A.J.S. and Sharifzadeh, M. (2018) A parametric study of blast damage on hard rock pillar strength. *Energies*. 11(7), 190–191.
Jessu, K.V., Spearing, A.J.S. and Sharifzadeh, M. (2022) An improved pillar design methodology. *J. S. Afr. Inst. Min. Metall.* 122(10), 587–596.
Jillavenkatesa, A., Dapkunas, S.J. and Lum, L-S.H. (2001) *Particle Size Characterization*. NIST Special Publication, Washington, DC, pp. 960–961.
Jing, L. (2003) A review of techniques, advances and outstanding issues in numerical modelling for rock mechanics and rock engineering. *Inter. J. Rock Mech. Mining Sci.* 40(3), 283–353.
Journal of Petroleum Technology. (1977) The SI Metric System of Units and SPE's Tentative Metric Standard, Vol. 29, Issue 12, 1575–1616.
Karmis, M. ed (2001) *Mine Health and Safety Management*. SME, Denver, CO.

Kearey, P., Brooks, M. and Hill, I. (2002) *An Introduction to Geophysical Exploration* (3rd Edition). Blackwell Publishing, Hoboken, NJ.

Keller (2025) *Soil Nailing, Earth Retention*. Haywood Baker, Keller, TX. Retrieved from: www.keller-na.com/hayward-baker-keller-company.

Kempson, W.J.; Webber-Youngman, R.C.W. and Meyer, J.P. (2015) Optimising shaft pressure losses through computational fluid dynamic modelling. *Appl. Therm. Eng.* 90, 1098–1108.

Kim, B.H., Cai, M., Kaiser, P.K. and Yang, H.S. (2007) Estimation of block sizes for rock masses with nonpersistent joints. *Rock Mech. Rock Eng.* 40(2), 169–192.

Kindred, G. (2022) Creative Commons by 4.0 Deed.

Kious, W.J. and Tilling, R.I. (1996) *This Dynamic Earth: The Story of Plate Tectonics*. U.S. Geological Survey, Washington, DC. https://doi.org/10.3133/7000097.

Kirsch, E.G. (1898) Die Theorie der Elastizität und die Bedürfnisse der Festigkeitslehre. *Zeitschrift des Vereines deutscher Ingenieure*. 42, 797–807.

Kratzsch, H. (1983) *Mining Subsidence Engineering*. Springer, Berlin, Germany.

Krauland, N. and Soder, P.E. (1987) Determining pillar strength from pillar failure observations. *Eng. Min. J.* 8, 34–40.

Kripakov, N.P. (1981) *Analysis of Pillar Stability on Steeply Pitching Seam Using the Finite Element Method*. USBM Report RI 8579.

Lachman, L. and Subramanyam, C.V.S. (2017) *Practical: Determination of Angle of Repose, Carre's Index & Hausner Ratio of Given Powder*. Knowledge of Pharma. Retrieved from: www.knowledgeofpharma.blogspot.com/2017/07/practical-determination-of-angle-of.html

Lambe, T.W. and Whitman, R.V. (1969) *Soil Mechanics*. John Wiley & Sons, Inc., New York, p. 553.

Lang, T.A. and Bischcoff, J.A. (1982) Stabilization of rock excavations using rock reinforcement. In: Goodman, R.E. and Heuze, F.E. (eds) *Issues in Rock Mechanics, Proceedings of the 23rd U.S. Symposium on Rock Mechanics, Berkley, CA*. AIME, New York, pp. 935–943.

Langer, F. (1997) *GSA Stability for Difficult Ground*. Dr Sauer & Partners, Herndon, VA.

Laubscher, D.H. (1975) Class distinction in rock masses. *Coal Gold Base Min. S. Afr.* 23, 119–128.

Laubscher, D.H. (1990) A geomechanics classification system for the rating of rock mass in mine design. *Trans. S Afr. Inst. Min. Metal.* 9(10), 257–272.

Laubscher, D.H. (2000) *Block Caving Manual. International Caving Study*. JKMRC and Itasca Consulting Group, Inc, Brisbane.

Laubscher, D.H. and Jakubec, J. (2001) The MRMR rock mass classification for jointed rock masses. In: Hustrulid, W.A. and Bullock, R.L. (eds) *Underground Mining Methods: Engineering Fundamentals and International Cases Studies*. Society for Mining, Metallurgy, and Exploration, Inc., Littleton, CO, pp. 475–481.

Leach, A.R. (1990) *Early Shaft Reef Extraction - The Western Deep Levels South Approach to Shaft Protection*. ISRM International Symposium, Mbabane, Swaziland.

Li, C.C. (2017) Principles of rockbolting design. *J. Rock Mech. Geotech. Eng.* 9, 396–414.

Likos, W.J. and Lu, N. (2004) *Unsaturated Soil Mechanics*. John Wiley and Sons Inc., New Jersey.

Liu, D., Zhang, D., Fang, Q., Sun, Z., Cao, L. and Li, A. (2019) Displacement characteristics of shallow-buried large-section loess tunnel with different types of pre-supports: a case study of new badaling tunnel. *Appl. Sci.* 10(1), 195.

Lowrie, R. L. (ed.) (2002) *SME Mining Reference Handbook*. SME, New York.

Lunder, P.J. (1994) *Hard Rock Pillar Strength Estimation an Applied Empirical Approach*, Master's Thesis, University of British Columbia, Department of Mining and Mineral Process Engineering.

Lunder, P.J. and Pakalnis, R. (1997) Determining the strength of hard rock mine pillars. *CIM Bulletin*. 90, 51–55.

Madden, B.J. (1991) A re-assessment of coal pillar design. *J. S. Afr. Inst. Min. Metall.* 91, 27–37.

Map 3D (2024) Retrieved from: www.map3d.com/example_grid_one.php

Marais, C.S and Spearing, A.J.S. (1990) *Internal Training Module*. Anglogold Division, Anglo American Corp, South Africa.

Marinos, P. and Hoek, E. (2000) GSI: a geologically friendly tool for rock mass strength estimation. In: *Proceedings of the Eighth International Conference on Deep and High Stress Mining*, Australian Centre for Geomechanics, Perth, pp. 1422–1442

Mariotte, E. (1679) Essais de physique, ou Mémoires pour servir à la science des choses naturelles attribué à Roberval (1679–1681).

Mark, C. and Agioutantis, Z. (2018) Analysis of coal pillar stability (ACPS): a new generation of pillar design software. *Int. J. Min. Sci. Technol.* 29, 87–91.

Mark, C. and Barczak, T.M. (2000) Fundamentals of coal mine roof support. In: Mark, C., Dolinar, D.R., Tuchman, R.J., Barczak, T.M., Signer, S.P. and Wopat, P.F. (eds) *Proceedings: New Technology for Coal Mine Roof Support.* , DHHS (NIOSH) Publication No. 2000-151, IC 9453. U.S. Department of Health and Human Services, Public Health Service, Centers for Disease Control and Prevention, National Institute for Occupational Safety and Health, Pittsburgh, PA, pp. 23–42.

Mark, C. and Bieniawski, Z.T. (1986) Field measurements of chain pillar response to longwall abutment loads. In: *Proceedings of the Fifth Conference on Ground Control in Mining*, Morgantown, WV, pp. 114–122.

Mark, C. and Chase, F. (1997). Analysis of retreat mining pillar stability. In: *Proceedings of New Technology for Ground Control in Retreat Mining*, NIOSH IC 9446. National Institute of Occupational Safety and Health, Pittsburgh, PA, pp. 17–34.

Mark, C. and Molinda, G.M. (2005) The coal mine roof rating (CMRR)—a decade of experience. *Int. J. Coal Geol.* 64, 85–103.

Martin, T.E. and Davies, M.P. (2000) *Development and Review of Surveillance Programs for Tailings Dams. AGRA Earth and Environmental Limited.* Burnaby, British Columbia, Canada.

Martin, T.E. and McRoberts, E.C. (1999) Some considerations in the stability analysis of upstream tailings dams. In: *Proceedings of Sixth International Conference on Tailings and Mine Waste '99,* 24–27 January, Fort Collins, Colorado, USA.

McCarthy, D.D. (1996) *Technical Note 21.* International Earth Rotation Services (IERS), Paris France.

McCarty, L.B., Hubbard, L.R. and Quisenberry, V.L. (2016) *Applied Soil Physical Properties, Drainage, and Irrigation Strategies.* Springer International Publishing, Switzerland.

McClintock, F.A. and Walsh, J (1962) Friction on Grifith cracks in rock under pressure. *Int. Cong. Appl. Mech.*, Berkeley, California, USA.

McKenna, G. and van Zyl, D. (2020) *Towards Zero Harm-A Compendium of Papers Prepared for the Global Tailings Review.* GlobalTailingsReview, Switzerland. www.GlobalTailingsReview.org.

McMullen, L.G. (2015) Risk assessment management. MSHA. *Presentation to mining staff and students at Southern Illinois University Carbondale*, Illinois.

MechanicalBoost (2024) *What is a Peristaltic Pump (Engineer Waqar)?* Retrieved from: www.mechanical-boost.com/peristaltic-pump/.

Melbye, T.A. (1994) *Shotcrete for Rock Support* (3rd Edition). MBT Europe (MAC Spa), Zurich, Switzerland.

Mellin, W. (1957) Work with new electronic 'brains' opens field for army math experts. *The Hammond Times.* 10, 66.

Mendecki, A.J., van Aswegen, G. and Mountfort, P. (1999) A guide to routine seismic monitoring in mines. In: Jager, A.J. and Ryder, J.A. (eds) *A Handbook on Rock Engineering Practice for Tabular Hard Rock Mines.* Creda Communication, Cape Town, South Africa.

Michaud, D. (2018) *Longwall Mining Method and Design.* 911 Metallurgist. Retrieved from: www.911metallurgist.com/longwall-mining-methods/.

Mine Subsidence Engineering Consultants. (2007) *Introduction to Longwall Mining and Subsidence.* Retrieved from: https://www.minesubsidence.com/index_files/files/Intro_Longwall_Mining_and_Subs.pdf.

Mitri, H.S. and Simon, R. (1998) Modeling of energy parameters ERR and TESR around burst-prone mine openings. *Trans. Soc. Min. Metall. Exploration.* 304, 30–36.

Mogi, K. (1966) Pressure dependence of rock strength and transition from brittle fracture to ductile flow. *Bull. Earthquake Res. Inst.* 44, 215–232.

Mohr, O.C. (1870) Beitrag zur Theorie der elastischen Bogentrager. *Zeitschrift des Architekten und Ingenieur-Vereins zu Hannover.* 16, 389.

Molinda, G.M. and Mark, C. (1994) *Coal Mine Rock Rating (CMRR): A Practical Rock Mass Classification for Coal Mines.* Bureau of Mines Information Circular (IC) 9387, Pittsburgh, PA.

Moore, E. (2014) A cut above. *CIM Magazine.*

Morelli, G.H. (2013) On joint roughness: measurements and use in rock mass characterization. *Geotech. Geol. Eng.* 32, 345–362.

Multotec (2024) *Section through a Hydrocyclone.* Retrieved from: https://www.multotec.com/en/news-articles/multotec-hydrocyclones-Australia.

Muntahar, A.S. (2009) Influence of plastic waste fibers on the strength of lime-rice husk ash stabilized clay soil. *Civil Eng. Dimen.* 11(1), 32–40.

NASA.(2022).MotionofFreeFallingObject.NASA.https://www1.grc.nasa.gov/beginners-guide-to-aeronautics/motion-of-free-falling-object/

National Institute of Occupational Safety and Health (NIOSH) (2024) *S-Pillar Windows Software for Stone Mine Pillar Design.* Retrieved from: www.cdc.gov/niosh/mining/tools/s-pillar.html

Navier, C.L. (1823) Rapport à Monsieur Becquey, Conseiller d'État, Directeur Général des Ponts et Chaussées et des Mines; et mémoire sur les ponts suspendus. Paris, France: L'Imprimerie Royale

Nelson, A. and Nelson, K.D. (1967) *Dictionary of Applied Geology, Mining & Civil Engineering*. Philosophical Library Inc., New York.

Nelson, S.A. (2024) *The Age of the Earth*. Geological Time, Earth Science Australia. Retrieved from: www.earthsci.org/education/teacher/basicgeol/geotim/geotim.html.

NGI (2022) *Using the Q-System Rock Mass Classification and Support Design*. NGI, Oslo, Norway.

Nkosi, N.Z. (2016) *Seismological and Mineralogical Studies of the World's Deepest Gold-Bearing Horizon, the Carbon Leader Reef, West Wits Line Goldfields (South Africa): Implications for Its Poor Seismic Reflective Character*. A Dissertation Submitted to the Faculty of Science, University of the Witwatersrand in fulfilment of the Requirements for the Degree of Master of Science, School of Geosciences.

Norwegian Geotechnical Institute (2022) *Using the Q-System*. Retrieved from: www.ngi.no.

Obert, L. and Duvall, W.I. (1967) *Rock Mechanics and the Design of Structures in Rock*. John Wiley & Sons, New York, pp. 542–545.

Ortlepp, W.D.O (1989) *Guidelines for the Early Removal of the Shaft Reef Area*. Anglo American Corp of SA Ltd, Johannesburg.

Ostrowski, W.J.S. (1972) Design considerations for modern shaft linings. *CIM Bullet.* 65(726), 58–71.

Ozbay, M.U. and Roberts, M.K.C (1988) Yield pillars in stope support. In: *Proceedings of the SANGORM Symposium in Africa*. South African National Group on Rock Mechanics, Johannesburg, Swaziland, pp. 317–326.

Pallister, G.F., Gay, N.C. and Cook, N.G.W. (1970) Measurement of the virgin state of stress in rock at depth. In: *Proceedings of 2nd Congress of the International Society for Rock Mechanics (ISRM)*, Belgrade, Serbia.

Palmström, A. (2005) Measurements of and correlations between block size and rock quality designation (RQD). *Tunn. Underg. Space Technol.* 20(4), 362–377.

Pan, R., Li, Y., Wang, H., Xu, Y. and Yang, H. (2020) Assessment of ground instability risk of lower seam longwall panels during crossing overlying remnant pillars. *R. Soc. Open Sci.* 7, 201249.

Panek, L. (1956) *Principles of Reinforcing Bedded Mine Roof with Bolts*. U.S. Bureau of Mines RI5156.

Panwar, R. (2021) *New Austrian Tunneling Method: Design and Operation*. The Constructor. Retrieved from: www.theconstructor.org/geotechnical/new-austrian-tunneling-method/63500/

Patchet, S.J. (1970) *Rock Mechanics Studies Associated with the Development of a Deep Potash Mine*. Ph.D Thesis, University of Newcastle-upon-Tyne, United Kingdom.

Peck, R.B. (1969) Deep excavations and tunnelling in soft ground: state of the art report. In: *Proceedings of the 7th International Conference on Soil Mechanics and Foundation Engineering*, Mexico, pp. 225–290.

Peng, S. (2020) Longwall Mining (3rd edition) CRC Press. https://doi.org/10.1201/9780429260049

Peng, S.S., Luo, Y. and Zhang, Z.M. (1995) Subsidence parameters – their definitions and determination. In: *Society for Mining, Metallurgy & Exploration Annual Meeting Transactions*, Littleton, CO, Vol. 300, pp. 60–65.

Pennsylvania Department of Environmental Protection (n.d.) Retrieved from: https://files.dep.state.pa.us/Mining/District%20Mining/DistrictMinePortalFiles/California/Underground%20Mining/MineSubsidence/

Perrault, P. (1966). *On the Origin of Springs*. Translation of "De l'origine des fontaines" (1674). Hafner. ASIN B0026MBIHE

Petterson, K.E. (1916) The quay slide in Gothenburg. *Teknisk Tidskrift*. 46, 471–478.

Petterson, K.E. (1941) A special foundation method for a structure in Gothenburg harbour. *Teknisk Tidskrift*. 71, 45–52.

Philipps, F. (1971) *The Use of Stereographic Projection in Structural Geology*. Edward Arnold, London.

Pillalamarry, M. and Spearing, A.J.S. (2009) *Internal Notes for Rock Mechanics Class*. Southern Illinois University Carbondale, Carbondale, IL.

Prakash, S. (1981) *Soil Dynamics*. McGraw-Hil, New York.

Prassetyo, S.H., Irnawan, M.A., Simangunsong, G.M., Wattimena, R.K., Arif, I. and Rai, M.A. (2019) New coal pillar strength formulae considering the effect of interface friction. *Int. J. Rock Mech. Min. Sci.* 123, 104102.

Priest, S.D. (1985) *Hemispherical Projection Methods in Rock Mechanics*. Allen and Unwin, New York.

Purkey, B.W. and Garside, L.J. (1995) *Geologic and Natural History Tours in the Reno Area*, Nevada Bureau of Mines and Geology Special Publication 19. University of Nevada, Nevada.

Qiao, S., Feng, C., Yu, P., Tan, J., Uchimura, T., Wang, L., Tang, J, Shen, Q. and Xie, J. (2020) Investigation on surface tilting in the failure process of shallow landslides. *Sensors*. 20, 2662.

Queens University, Mining Engineering Department, Canada (2011) Retrieved from: https://minewiki.engineering.queensu.ca/mediawiki/index.php/Numerical_modelling

Rabcewicz, L.V. (1965) The new Austrian tunnelling method. *Water Power*. 453–457, 19–24.

Read, H.H. and Watson, J. (1975) *Introduction to Geology*. Halsted, New York.

Richter, C.F. (1935) An instrumental earthquake magnitude scale. *Bullet. Seismol. Soc.America*, 25, 1–32.

Richter, C.F. (1958) *Elementary Seismology*. W.H.Freeman & Co Ltd, New York.

Robinsky, E.I. (1999) *Thickened Tailings Disposal in the Mining Industry*. E.I Robinsky Assoc. Ltd., Toronto, Canada.

Rockaway, J.D, Stephenson, R.W. (1979) *Investigation of the effects of weak floor conditions on the stability of coal pillars*. Report No. BUMINES-ofr-12-81.

Rocscience (2024) Retrieved from: www.rocscience.com/learning/

Rusnak, J. and Mark, C. (2000) Using the point load test to determine the uniaxial compressive strength of coal measure rock. In: Peng, S.S. and Mark, C. (eds) *Proceedings of the 19th International Conference on Ground Control in Mining*. West Virginia University, Morgantown, WV, pp. 362–371.

Safety Section (2024) *Risk Assessment Calculation Formula*. Retrieved from: https://safetysection.com/risk-assessment-calculation-formula/

Saint-Venant, A.J. (1855) Memoire sur la Torsion des Prismes. *Memoires Divers Savants*. 14, 233–560.

Salamon, M.D.G. (1974) Rock mechanics of underground excavations. Advances in rock mechanics. In: *Proceedings of 3rd Congress of the International Society of Rock Mechanics*. Denver, CO, Vol. 1b, pp. 91–1099.

Salamon, M.D.G. and Munro, A.H. (1967) A study of the strength of coal pillars. *J. S. Afr. Inst. Min. Metall.* 68, 55–67.

Salamon, M.D.G. and Wagner, H. (1985) Practical experiences in the design of coal pillars. In: *Proceedings of the 21st International Conference on Mining Software Repositories*. A.A. Balkema, Rotterdam.

Sauer, G & Gold, H. (1989) NATM Ground support concepts and their effect on contract practices. In: *Rapid Excavation and Tunnelling Conference*. Los Angeles, US.

Sauer, G. (1990) Design concept for large underground openings in soft ground using NATM. In: *International Symposium on Unique Underground Structures*, Denver, Colorado.

Schaffernak, F. (1917) Über die Standicherheit durchläessiger geschüetteter. Dämme, Allgem. Bauzeitung.

Schlitt, W. J. (1984) *The Role of Solution Management in Heap and Dump Leaching. Au and Ag Heap and Dump Leaching Practice*. SME-AIME, Denver, CO, pp. 69–83.

Sepehri, M., Apel, D.B. and Hall, R.A. (2017) Prediction of mining-induced surface subsidence and ground movements at a Canadian diamond mine using an elastoplastic finite element model. *Inter. J. Rock Mech. Mining Sci.* 100, 73–82.

Sheorey, P.R., Das, M.N., Barat, D., Prasad, R.K. and Singh, B. (1987) Coal pillar strength estimation from failed and stable cases. *Int. J. Rock. Mech. Min. Sci. Geomech. Abstr*. 24, 347–355.

Singh, B., Jethwa, J.L., Dube, A.K. and Singh, B. (1992): Correlation between observed support pressure and rock mass quality. Tunnelling and Underground Space Technology 7(1): 59-74.

Singh, M.M. (1992) *Mine Subsidence (Chapter 10.6) in SME Mining Engineering Handbook*. SME-AIME, Denver, CO, pp. 938–971

Sinha, R.S. (1989) *Underground Structures: Design and Instrumentation*. Elsevier, Amsterdam, The Netherlands.

Skinner, W.J. (1959). *Experiments on the Compressive Strength of Anhydrite*. The Engineer, London.

Smith, J.D. and Mitchell, R.J. (1982) Design and control of large hydraulic backfill pours. *CIM Bull.* 75(838), 102–111.

Snider, H.W. (1956) *Reaching Professional Status: A Program for Risk Management, in Corporate Risk Management: Current Problems and Perspectives*. American Management Association, New York.

Soliman, M. (1998) *Environmental Hydrogeology*. CRC Press LLC, Boca Raton, FL.

Spearing, A.J.S. (1987) *A Practical Rock Mechanics Manual*. Anglo American Corp. of SA. Internal Report.

Spearing, A.J.S. (1989) *The Modification of Witwatersrand Gold Plant Tailings for Optimum Application as an Underground Support Medium*. Thesis for MS in Civil Engineering. University of the Witwatersrand, Johannesburg, South Africa.

Spearing, A.J.S. (1995) *Handbook on Hard Rock Strata Control*. SAIMM, SP-6, Johannesburg, pp. 114–115.

Spearing, A.J.S. (2002) *Shotcrete (Sprayed Shotcrete) for Mining Applications*. Master Builders Technology, Mannheim, Germany.

Spearing, A.J.S. (2015) *Course Notes, Introduction to Mining Engineering*. MNGE 270. Mining and Mineral Resources Engineering Department, Southern Illinois University Carbondale, Carbondale, IL.

Spearing, A. J. S., Hyett, A. J., Kostecki, T., & Gadde, M. (2013) New technology for measuring the in situ performance of rock bolts. *International Journal of Rock Mechanics and Mining Sciences*, *57*, 153–166.

Spearing, A.J.S., Ma, L. and Ma, C. (2023) *Mine Design, Planning and Sustainable Exploitation in the Digital Age*. CRC Press, London, UK.

Spearing, A.J.S., Millette, D. and Gay, F. (2000) *The Potential Use of Foam Technology in Underground Backfilling and Surface Tailings Disposal*. MassMin, Brisbane, Australia.

Split Mountain Machine and Equipment Pty. Ltd (2025) Retrieved from: www.splitmountainmachine.com/mining-jumbo-abu.html

Steart, F.A. (1954) Strength and stability of pillars in coal mines. *J. Chem. Metall. Min. Soc. S. Afr*. 54, 309–325.

Steward, N.R. and Spearing, A.J.S. (1992) Combating pipeline wear - an advancing technology. *J. S. Afr. Inst. Min. Metal.* 92, 6.

Stillborg, B. (1986) *Professional User's Handbook for Rockbolting*. Trans Tech Publications, Clausthal-Zellerfeld.

Suryakanta, P. (2015) *How to calculate bearing capacity of soil from a plate load test?* Retrieved from: CivilBlog.org.

Talobre, J.A. (1957) *La Mecanique des Roches* (1st Edition). Dunod, Paris, pp. 228–233.

Tarbuck, E.J. and Lutgens, F.K. (2014) *Illustrated by Dennis Tasa. Earth An Introduction to Physical Geology, Custom Edition for Southern Illinois University – Carbondale*. Pearson, Upper Saddle River, NJ.

Tatiya, R.R. (2005) Excavations in upward direction – raising. In: *Surface and Underground Excavations – Methods, Techniques and Equipment*. CRC Press, Boca Raton, FL, pp. 331–349.

Taylor, D.W. (1948) *Fundamentals of Soil Mechanics*. John Wiley & Sons Inc., New York.

Terzaghi, K. (1925) *Erdbaumechanik auf Bodenphysikalischer Grundlage*. Franz Deuticke, Liepzig-Vienna, Austria.

Terzaghi, K. (1943) *Theoretical Soil Mechanics*. John Wiley & Sons Inc., New York, 510 p.

Terzaghi, K. (1946) Chemical defects of rocks. In: Proctor, R.V. and White, T.L. (eds) *Rock Tunneling with Steel Supports with an Introduction to Tunnel Geology*. The Commercial Shearing and Stamping Co., New York, pp 39–46.

Terzaghi, K. and Richart, F.E. (1952) Stresses in rock around cavities. *Geotechnique*. 3(2), 57–90.

Terzaghi, K., Peck, R.B. and Mesri, G. (1996) *Soil Mechanics in Engineering Practice*. John Wiley & Sons, New York.

Thomas, E.G. Nantel, J.H. and Notely, K.R. (1979) *Fill Technology in Underground Metalliferous Mines*. International Academic Services Limited, Canada, p. 293.

Timsoshenko, S, & Goodier, J.,N. (1951). *Theory of Elasticity*. McGraw-Hill Book Company, Inc. 2nd ed.

Tschebotarioff, G.P. (1951) *Soil Mechanics, Foundations and Earth Structures*. McGraw-Hill Book Company, New York.

United States Geological Survey (n.d.) *Earthquake Magnitude, Energy Release and Shaking Intensity*. Retrieved from: www.usgs.gov/programs/earthquake-hazards/earthquake-magnitude-energy-release-and-shaking-intensity.

US Army Corps of Engineers (1993) Standard Practice for Shotcrete. Engineering Manual EM 1110-2-2005, 31 January.

US Geological Survey (n.d.) *Water Science School*. Retrieved from: www.usgs.gov/special-topic/water-science-school.

USGS (1999) *Cutaway View Showing the Internal Structure of the Earth*. Volcano Hazards Program. Retrieved from: www.usgs.gov/media/images/cutaway-views-showing-internal-structure-earth-left.

Van As, A. (2003) *Subsidence Definitions for Block Caving Mines*. Technical Report. Rio Tinto Technical Services, Sydney, p. 59.

van der Merwe, J.N. (2003) New pillar strength formula for South African coal. *J. S. Afr. Inst. Min. Metall.* 103, 281–292.

Vandewalle, M. (1997) *Tunnelling the World*. N.V.Bekaert S.A, Belgium.

Vick, S.G. (1990) *Planning, Design and Analysis of Tailings Dams* (2nd Edition). BiTech Publishers Ltd, Vancouver.

Villaverde, R. (2009) *Fundamental Concepts of Earthquake Engineering. Engineering Geology*. CRC Press, Boca Raton, FL.

Volkholz, J. (2021) *River Flood Modelling*. ISIpedia, Germany.

Wagner H. (1984) Support requirements of rockburst conditions. *J. South. Afr. Inst. Min. Metall.* 88(3), 209–218.

Wagner, H. (1974). Determination of complete load deformation characteristics of coal pillars. In: *Proceedings of the 3rd ISRM Congress*, Denver, CO, pp. 1076–1081.

Wagner, H. (1980) Pillar design in coal mines. *J. S. Afr. Inst. Min. Metall.* 80(1), 37–45.

Wagner, H. and Madden, B.J. (1984) Fifteen years' experience with the design of coal pillars in shallow South African collieries: an evaluation of the performance of the design procedure and recent improvements. Design and Performance of Underground Excavations. In: *ISRM International Symposium*, British Geotechnical Society, Cambridge, UK, pp. 391–399.

Walker, T. (2026) The rock cycle-How do we handle it? *Creation Ministries International*. October 8. www.creation.com/en/articles/the-rock-cycle

Wang, F.D., Skelly, W.A., and Wolgamott, J. (1977). In-situ coal pillar strength study. Proceedings 18th U.S. Symposium on Rock Mechanics, Keystone, CO, pp. 2B5-1:9.

Wikiwand - Constitutive relationships (n.d.) Retrieved from: www.wikiwand.com/en/articles/Constitutive_relationship.

Williams, T.J., Brady, T.M., Bayer, D.C., Bren, M.J., Paklanis, R.T., Marjerison, J.A. and Langston, R.B. (2007) *Underhand Cut and Fill Mining as Practiced in Three Deep Hard Rock Mines in the United States*. NIOSH, CDC NIOSHTIC2 Number: 20033309.

Williamson, D.A. (1984) Unified rock classification system. *Bull. Assoc. Eng. Geol.* 21, 345–354.

Wilson, A.H. (1981) Stress and stability in coal ribsides and pillars. In: *Proceedings, 1st Conference on Ground Control in Mining*, Morgantown, WV, pp. 1–12.

Wilson, J.W. and More O'Ferrall, R.C. (1970) The application of the electrical analogue to mining operations. *J. S. Afr. Inst. Min. Metall.* 70(6), 115–148.

Wiseman, N. (1979) *Factors Affecting the Design and Conditions of Mine Tunnels*. Research Report No. G01G10. Chamber of Mines Research Organization, Johannesburg, South Africa.

Yeats, R.S. (2020) *Living with Earthquakes in the Pacific Northwest*. Oregon State University, Oregon.

Yilmaz, E. and Guresci, M. (2017) *Design and Characterization of Underground Paste fill. Paste Tailings Management*. Springer, Cham, pp. 111–143.

Yu, C., Loureiro, C., Cheng, J., Jones, L.G., Wang, Y.Y., Chia, Y.P. and Faillace, E. (1993) *Data Collection Handbook to Support Modeling Impacts of Radioactive Material in Soil*. Environmental Assessment and Information Sciences Division-Argonne National Laboratory, Argonne, IL.

Zhang, Q. and Wang, X. (2016) *Underground Mining Technics of Metallic Ore Deposits*. University of Central South Press, Changsha, China.

Zhang, G., He, F. and Jiang, L. (2016) Analytical analysis and field observation of break line in the main roof over the goaf edge of longwall coal mines. *Math. Probl. Eng.* 9, 1–11.

Zhang, J., Li, M., Taheri, A., Zhang, W., Wu, Z. and Song, W. (2019) Properties and application of backfill materials in coal mines in China. *Minerals*. 9, 53.

Zhang, J., Zhou, Y.J. and Huang, Y.L. (2012) Integrated technology of fully mechanized solid backfill mining. *J. China Coal Soc.* 11, 10–13.

Zhang, Y., Zhang, Q.-Y., Zhou, X.-Y. and Xiang, W. (2021) Direct tensile tests of red sandstone under different loading rates with self-developed centering device. *Geotech. Geol. Eng.* 39, 709–718.

Index

FSC
www.fsc.org
MIX
Papier | Fördert gute Waldnutzung
FSC® C083411

Kolibri 360 GmbH
Ferdinand-Jühlke-Straße 7
99095 Erfurt, Deutschland
produktsicherheit@kolibri360.de